Lecture Notes in Mathematics 1651

Springer
Berlin
Heidelberg
New York
Barcelona
Budapest
Hong Kong
London
Milan
Paris
Santa Clara
Singapore
Tokyo

Michael Drmota   Robert F. Tichy

Sequences, Discrepancies and Applications

Authors

Michael Drmota
Department of Discrete Mathematics
Technical University of Vienna
Wiedner Hauptstr. 8–10
A-1040 Vienna, Austria
e-mail: m.drmota@tu-wien.ac.at

Robert F. Tichy
Department of Mathematics
Technical University of Graz
Steyrergasse 30
A-8010 Graz, Austria
e-mail: tichy@weyl.math.tu-graz.ac.at

Cataloging-in-Publication Data applied for

Die Deutsche Bibliothek - CIP-Einheitsaufnahme

Drmota, Michael:
Sequences, discrepancies and applications / Michael Drmota ; Robert F. Tichy. - Berlin ; Heidelberg ; New York ; Barcelona ; Budapest ; Hong Kong ; London ; Milan ; Paris ; Santa Clara ; Singapore ; Tokyo : Springer, 1997
(Lecture notes in mathematics ; 1651)
ISBN 3-540-62606-9
NE: Tichy, Robert F.:; GT

Mathematics Subject Classification (1991): K11

ISSN 0075-8434
ISBN 3-540-62606-9 Springer-Verlag Berlin Heidelberg New York

Typesetting: Camera-ready TEX output by the author
SPIN: 10520256 46/3142-543210 - Printed on acid-free paper

Dedicated to Professor Edmund Hlawka

on the occasion of his 80th birthday

Preface

KRONECKER's approximation theorem says that the fractional parts of the multiples of an irrational lie dense in the unit interval. This result was the starting point of a long and fruitful development of the theory of uniformly distributed sequences. At the beginning of the 20th century first refinements and generalizations of KRONECKER's theorem were established by various authors such as BOHL, SIERPINSKI, BERNSTEIN, HARDY and LITTLEWOOD, and WEYL who was the first establishing a systematic treatment of uniformly distributed sequences in his famous paper *'Über die Gleichverteilung von Zahlen mod. Eins'* [1953]. A historic survey on the theory of uniform distribution until 1916 is given by HLAWKA and BINDER [837] whereas the development of this subject after 1916 is contained in several monographs and survey articles. Some chapters of the classical books *'Aufgaben und Lehrsätze aus der Analysis'* by PÒLYA and SZEGÖ [1450] and *'Diophantische Approximationen'* by KOKSMA [941] are devoted to the early stage of the theory of uniform distribution. A detailed survey of the whole subject until 1960 was given by CIGLER and HELMBERG [369]. The first exhausting monograph *'Uniform Distribution of Sequences'* is due to KUIPERS and NIEDERREITER [983]. Some years later HLAWKA published his monograph [804, 815] on the theory of uniform distribution.

The beginning of the theory was the discovery that the fractional parts of the multiples of an irrational are not only dense in the unit interval but they are uniformly distributed. This means that the empirical distribution of the sequence is asymptotically equal to the uniform distribution. Therefore the main root of this theory is diophantine approximation and number theory, however there are strong connections to various fields of mathematics such as measure and probability theory, harmonic analysis, topological groups, summability theory, discrete mathematics, and numerical analysis. In the twenties and thirties several authors, for instance BEHNKE, VAN DER CORPUT, KOKSMA, and OSTROWSKI, established quantitative results concerned with the distribution behaviour of special sequences. As a quantitative measure of the distribution behaviour of a sequence VAN DER CORPUT considered the so-called discrepancy, i.e. the maximal deviation between the empirical distribution of the sequence and the uniform distribution. One problem is to find upper bounds for the discrepancy of special sequences. The main tool for proving such bounds is to apply analytic tools for estimating exponential sums. Another important problem is to prove general lower bounds for the discrepancy of sequences. This subject is called *Theory of Irregularities of Distribution* since it turns out that the distribution of a sequence cannot be too smooth. First significant results in this direction are due to VAN AARDENNE-EHRENFEST, ROTH, and SCHMIDT. There are two monographs on this subject, namely by SCHMIDT [1634] and more recently by BECK and CHEN [143].

In pioneering papers HLAWKA [782, 783] generalized the theory of uniform distribution to the setting of compact topological spaces and groups. These abstract aspects and connections to summability theory can be found in [983, chapters 3,4]. More recently the abstract theory of uniform distribution was further extended by several authors such as NIEDERREITER, LOSERT, and RINDLER. The special case of discrete spaces was extensively studied, mainly sequences of integers modulo m were considered. Distribution problems for integer sequences are surveyed by NARKIEWICZ [1260].

A very important application of uniformly distributed sequences is numerical integration since the approximation error can be estimated in terms of the discrepancy. Hence it is important to construct low-discrepancy sequences. Constructions of such sequences are due to HLAWKA and KOROBOV. A concise treatment of this so-called *Good Lattice Point Method* can be found in the monographs by KOROBOV [958] and by HUA and WANG [852]; see also HLAWKA, FIRNEIS, and ZINTERHOF [838]. More recently low-discrepancy and related sequences are used for several other applications: for simulation of random numbers, Quasi-Monte Carlo optimization, etc. In the meantime there exists a huge literature on various applications of Quasi-Monte Carlo methods to different kinds of problems; an excellent survey is NIEDERREITER's book [1336].

The present book attempts to summarize special developements and methods of the theory of uniform distribution since 1974 when KUIPERS' and NIEDERREITER's book [983] appeared. We emphasize on such topics which are not covered by some of the above mentioned monographs. Every section of this book consists of two parts, a self-contained one where main results and methods are established and a notes part where the corresponding literature is discussed. References of papers published before 1974 are only taken into account if they are necessary for the presentation and proofs of the results. For references of other papers we explicitely refer to the extensive bibliography in [983] and to a recent manuscript by HELMBERG [773].

In chapter 1 we discuss the classical theory of uniform distribution in the unit interval and in the k-dimensional unit cube. We present an improved version of the famous ERDÖS-TURÁN-KOKSMA inequality as well as BECK's proof of ROTH's lower bound for the discrepancy. Furthermore estimates for the discrepancy of special sequences are established, e.g. for the $(n\alpha)$-sequence, higher dimensional analogues, digital sequences, and exponential sequences. Here we also survey on BECK's recent metric result on the KRONECKER sequence as well as on special results on normal numbers. In a concluding section metric bounds for the discrepancy of sequences are proved.

In chapter 2 we shortly demonstrate BECK's Fourier-Transform approach for finding general lower bounds of the discrepancy. For a detailed presentation of this method we refer to BECK and CHEN [143]. However, a new method due to ALEXANDER is discussed in more details. Furthermore a quantitative treatment of the discrepancy with respect to summation methods is given. Continuous analogues are investigated as well as some new applications of BECK' method not contained in [143]. In a final section we study distribution problems in finite sets, in particular we present concepts of discrepancy for sequences in finite sets and some statistical results. This involves combinatorial methods and generating functions. The combinatorial discrepancy theory is not discussed exhaustively since there exists an excellent survey by BECK and SÓS [150] on this subject. Then we shortly deal with uniform distribution in integers and generalizations. For a detailed presentation of this topic we refer to NARKIEWICZ [1260]. However, we emphasize on recent results on the uniform distribution of linear recurring sequences.

The final chapter 3 is devoted to various applications of uniformly distributed sequences such as numerical integration and numerical solution of differential equations, random number generation, and Quasi-Monte Carlo methods. Furthermore we include as a very recent application some aspects of Mathematical Finance. During the last years all these applications have been a rapidly growing area of research. There were several important conferences on these topics, e.g. one in Lamprecht (Germany), subsequent ones in Fairbanks (Alaska, 1990), in Las Vegas (Nevada, 1994) and in Salzburg (Austria, 1996). NIEDERREITER's book [1336] is an extended version of his letures given at the Fairbanks conference. We try to focus on some application problems which we have selected following our own taste. We also include the explicit computation of the L^2-discrepancy which is the basic quantity for the average case analysis of numerical integration. For a detailed survey on average case analysis of numerical integration we refer to the monographs by NOVAK [1371] and by TRAUB, WASILKOWSKI and WOŹNIAKOWSKY [1872].

We are indebted to M. Blümlinger, M. Goldstern, P. Grabner, C. Heuberger, B. Klinger, A. Knappe, I. Radovic, T. Siegl, M. Skałba, J. Thuswaldner, G. Turnwald, M. Unterguggenberger, R. Winkler for helpful discussions and to Mrs. H. Panzenböck for typesetting. Especially we thank the Springer-Verlag for the kindness and support during the time we have written this volume. Finally we want to thank our wives for their patience and encouragement.

Vienna and Graz, 1996. M. Drmota and R.F. Tichy

Contents

Chapter 1

Discrepancy of Sequences

1.1 Basic Concepts

1.1.1 Basic Definitions

Let us consider the k-dimensional Euclidean space $\mathbf{R}^k$. We will identify two points $\mathbf{x}, \mathbf{y} \in \mathbf{R}^k$ if their difference $\mathbf{x} - \mathbf{y}$ is an integral lattice point $\in \mathbf{Z}^k$. Equivalently we can say that we consider the space $\mathbf{R}^k$ modulo 1 or we deal with the k-dimensional torus $\mathbf{T}^k = \mathbf{R}^k/\mathbf{Z}^k$.

Obviously $\mathbf{T}^k$ can be identified with the unit cube $\mathbf{U}^k = [0,1)^k$. Formally this can be done by using the notion of fractional parts. The fractional part $\{x\}$ of a real number x is defined by $\{x\} = x - [x]$, where $[x]$ denotes the integral part of x, that is, the greatest integer $\leq x$. The fractional part $\{\mathbf{x}\}$ and the integral part $[\mathbf{x}]$ for $\mathbf{x} \in \mathbf{R}^k$ are defined componentwise.

Let $J = [a_1, b_1) \times \cdots \times [a_k, b_k) \subseteq \mathbf{R}^k$ be an interval (or a rectangle with sides parallel to the axes) in the k-dimensional space $\mathbf{R}^k$ with $0 < b_i - a_i \leq 1$, $i = 1, \ldots, k$. Then the reduction modulo 1, $I = J/\mathbf{Z}^k$ is called an interval (or a rectangle with sides parallel to the axes) of the torus $\mathbf{T}^k = \mathbf{R}^k/\mathbf{Z}^k$, e.g. $I = [0, \frac{1}{2}) \cup [\frac{2}{3}, 1) = [\frac{2}{3}, \frac{3}{2})/\mathbf{Z}$ is such an interval. The volume $\lambda_k(I)$ of an interval $I \subseteq \mathbf{R}^k/\mathbf{Z}^k$ is given by $\prod_{i=1}^k (b_i - a_i)$. (Of course, λ_k denotes the k-dimensional LEBESGUE measure.) For such an interval $I \subseteq \mathbf{R}^k/\mathbf{Z}^k$ and a sequence $(\mathbf{x}_n)_{n\geq 1}$, $\mathbf{x}_n \in \mathbf{R}^k$, let $A(I, N, \mathbf{x}_n)$ be the number of points $\mathbf{x}_n$, $1 \leq n \leq N$, for which $\{\mathbf{x}_n\} \in I$, i.e.

$$A(I, N, \mathbf{x}_n) = \sum_{n=1}^{N} \chi_I(\{\mathbf{x}_n\}), \tag{1.1}$$

where χ_I is the characteristic function of I.

Using these notations it is easy to define the notion of uniformly distributed sequences.

Definition 1.1 *A sequence $(\mathbf{x}_n)_{n\ge1}$ of points in the k-dimensional space $\mathbf{R}^k$ is said to be* uniformly distributed modulo 1 *(for short u.d. mod 1) if for every interval $I \subseteq \mathbf{R}^k/\mathbf{Z}^k$ we have*

$$\lim_{N\to\infty} \frac{A(I,N,\mathbf{x}_n)}{N} = \lambda_k(I). \tag{1.2}$$

Furthermore $(\mathbf{x}_n)_{n\ge1}$ is called well distributed modulo 1 *(for short w.d. mod 1) if for every interval $I \subseteq \mathbf{R}^k/\mathbf{Z}^k$ we have*

$$\lim_{N\to\infty} \frac{A(I,N,\mathbf{x}_{n+\nu})}{N} = \lambda_k(I) \tag{1.3}$$

uniformly for all $\nu \ge 0$.

Remark. Of course, well distribution is a stronger concept then uniform distribution. We postpone a systematic study of this notion to Section 2.2 ; some special w.d. sequences will be considered in Section 1.4.

Note that (1.2) is equivalent to

$$\lim_{N\to\infty} \frac{1}{N}\sum_{n=1}^{N} \chi_I(\{\mathbf{x}_n\}) = \int_{\mathbf{T}^k} \chi_I \, d\lambda_k \tag{1.4}$$

or for short

$$\lim_{N\to\infty} m_N(\chi_I) = m(\chi_I), \tag{1.5}$$

where

$$m_N(f) = \frac{1}{N}\sum_{n=1}^{N} f(\{\mathbf{x}_n\}) \tag{1.6}$$

and

$$m(f) = \int_{\mathbf{T}^k} f \, d\lambda_k \tag{1.7}$$

are positive linear functionals on the space of LEBESGUE integrable functions $f : \mathbf{T}^k \to \mathbf{R}$. Hence a sequence $(\mathbf{x}_n)_{n\ge1}$ of points in $\mathbf{R}^k$ is u.d. mod 1 if and only if (1.5) holds for all characteristic functions χ_I of intervals $I \subseteq \mathbf{T}^k$. By linearity,

$$\lim_{N\to\infty} m_N(f) = m(f) \tag{1.8}$$

holds for all step functions f, too. Furthermore we have the following property.

Lemma 1.2 *Let m_N $(N \in \mathbf{Z},\ N \ge 0)$ and m be positive functionals on some space $\mathcal{F}$ of real-valued functions $f : X \to \mathbf{R}$ $(X \ne \emptyset)$ and let $\mathcal{L} \subseteq \mathcal{F}$ the subspace of these functions f satisfying (1.8). Suppose that $f \in \mathcal{F}$ has the property that for every $\varepsilon > 0$ there exist functions $g_1, g_2 \in \mathcal{L}$ with $g_1 \le f \le g_2$ and $m(g_2) - m(g_1) < \varepsilon$. Then $f \in \mathcal{L}$, too.*

Proof. By $m_N(g_1) \le m_N(f) \le m_N(g_2)$ and $m(g_1) \le m(f) \le m(g_2)$ we immediately get

$$\begin{aligned} m(g_1) &= \liminf_{N\to\infty} m_N(g_1) \le \liminf_{N\to\infty} m_N(f) \\ &\le \limsup_{N\to\infty} m_N(f) \le \limsup_{N\to\infty} m_N(g_2) \\ &= m(g_2) \end{aligned}$$

which implies

$$\left|m(f) - \liminf_{N\to\infty} m_N(f)\right| < \varepsilon$$

and

$$\left|m(f) - \limsup_{N\to\infty} m_N(f)\right| < \varepsilon$$

for every $\varepsilon > 0$. Thus $\lim_{N\to\infty} m_N(f) = m(f)$. □

This Lemma can be used to prove two criteria for sequences u.d. mod 1. Recall that the k-dimensional torus $\mathbf{T}^k$ can be identified with the cube $[0,1)^k \subseteq [0,1]^k$.

Theorem 1.3 (Criterion A) *A sequence $(\mathbf{x}_n)_{n\ge1}$ of points in the k-dimensional space $\mathbf{R}^k$ is u.d. mod 1 if and only if*

$$\lim_{N\to\infty} \frac{1}{N}\sum_{n=1}^{N} f(\{\mathbf{x}_n\}) = \int_{[0,1]^k} f\, d\lambda_k \tag{1.9}$$

holds for all RIEMANN *integrable functions $f : [0,1]^k \to \mathbf{R}$.*

Theorem 1.4 (Criterion B) *A sequence $(\mathbf{x}_n)_{n\ge1}$ of points in the k-dimensional space $\mathbf{R}^k$ is u.d. mod 1 if and only if (1.9) holds for all continuous functions $f : [0,1]^k \to \mathbf{R}$.*

Proof. Let $\mathcal{L}$ be the space of all LEBESGUE integrable functions $f : [0,1]^k \to \mathbf{R}$ satisfying (1.9). Then $\mathcal{L}$ contains all stepfunctions. Now, if f is RIEMANN integrable then by definition for every $\varepsilon > 0$ there exist step functions g_1, g_2 with $g_1 \le f \le g_2$ and

$$\int_{[0,1]^k} (g_2 - g_1)\, d\lambda_k < \varepsilon. \tag{1.10}$$

Hence by Lemma 1.2 $f \in \mathcal{L}$. On the other hand all characteristic functions χ_I are RIEMANN integrable. This proves Criterion A.

Since every continuous function f is RIEMANN integrable (1.9) surely holds for continuous functions. Conversely it is easy to see that for every characteristic function χ_I and every $\varepsilon > 0$ there exist continuous functions g_1, g_2 with $g_1 \le \chi_I \le g_2$ and (1.10). Thus a second application of Lemma 1.2 proves Criterion B. □

Remark 1. It should be noted that Criterion B is a proper version to generalize the definition of u.d. sequences to compact topological spaces whereas the first definition has no direct analogue at a first glance.

Remark 2. There is no sequence satisfying (1.9) for all LEBESGUE integrable functions.

Remark 3. Property (1.9) is also the starting point to apply u.d. sequences for numerical integration.

1.1.2 Discrepancies

In order to quantify the convergence in (1.2) the discrepancy D_N of a sequence $(\mathbf{x}_n)_{n\geq 1}$ has been introduced.

Definition 1.5 *Let $\mathbf{x}_1,\ldots,\mathbf{x}_N$ be a finite sequence of points in the k-dimensional space $\mathbf{R}^k$. Then the number*

$$D_N = D_N(\mathbf{x}_1,\ldots,\mathbf{x}_N) = \sup_{I\subseteq \mathbf{T}^k} \left| \frac{A(I,N,\mathbf{x}_n)}{N} - \lambda_k(I) \right| \tag{1.11}$$

is called the discrepancy of the given sequence. For an infinite sequence $(\mathbf{x}_n)_{n\geq 1}$ $D_N(\mathbf{x}_n)$ should denote the discrepancy of $(\mathbf{x}_n)_{n=1}^N$ and is called discrepancy, too.

The essential point of the concept of discrepancy is that the notion of uniform distribution can be covered by it; i.e. the convergence in (1.2) is uniform with respect to all intervals $I\subseteq \mathbf{T}^k$.

Theorem 1.6 *A sequence $(\mathbf{x}_n)_{n\geq 1}$ is u.d. mod 1 if and only if*

$$\lim_{N\to\infty} D_N(\mathbf{x}_n) = 0. \tag{1.12}$$

Proof. (1.12) immediately implies (1.2) for all intervals $I\subseteq \mathbf{T}^k$. So we only have to show that every u.d. sequence satisfies (1.12). For this reason let M be an arbitrary positive integer and set

$$I_{m_1,\ldots,m_k} = \left[\frac{m_1}{M},\frac{m_1+1}{M}\right) \times\cdots\times \left[\frac{m_k}{M},\frac{m_k+1}{M}\right)$$

for $0\leq m_i < M$, $i=1,\ldots,k$. By (1.2) there exists a positive integer N_0 such that

$$\frac{1}{M^k}\left(1-\frac{1}{M}\right) \leq \frac{A(I_{m_1,\ldots,m_k},N,\mathbf{x}_n)}{N} \leq \frac{1}{M^k}\left(1+\frac{1}{M}\right) \tag{1.13}$$

for $N\geq N_0$ and for all cubes $I_{m_1,\ldots,m_k}$. Now consider an arbitrary interval $I\subseteq \mathbf{T}^k$. Clearly there exist intervals $\underline{I},\overline{I}$, finite unions of cubes $I_{m_1,\ldots,m_k}$, such that $\underline{I}\subseteq I\subseteq \overline{I}$ and

$$\lambda_k(\overline{I}) - \lambda_k(\underline{I}) \leq 1-\left(1-\frac{2}{M}\right)^k = \frac{2k}{M} + \mathcal{O}\left(\frac{2}{M^2}\right). \tag{1.14}$$

From (1.13) we get

$$\begin{aligned} \lambda_k(\underline{I})\left(1-\frac{1}{M}\right) &\leq \frac{A(\underline{I},N,\mathbf{x}_n)}{N} \leq \frac{A(I,N,\mathbf{x}_n)}{N} \\ &\leq \frac{A(\overline{I},N,\mathbf{x}_n)}{N} \leq \lambda_k(\overline{I})\left(1+\frac{1}{M}\right) \end{aligned} \tag{1.15}$$

for $N \geq N_0$. Thus 1.14 and 1.15 immediately imply

$$\left|\frac{A(I,N,\mathbf{x}_n)}{N} - \lambda_k(I)\right| \leq \frac{2k+1}{M} + \mathcal{O}\left(\frac{2}{M^2}\right) \tag{1.16}$$

for $N \geq N_0$. Hence (1.12) holds.□

Sometimes it is of some importance to consider only intervals of the form $I = [0,\mathbf{y}) = [0,y_1) \times \cdots \times [0,y_k)$, $\mathbf{y} = (y_1,\ldots,y_k)$, $0 < y_i \leq 1$, $i = 1,\ldots,k$, and the so-called *star discrepancy*

$$D_N^*(\mathbf{x}_n) = \sup_{\mathbf{y}\in[0,1]^k} \left|\frac{A([\mathbf{0},\mathbf{y}),N,\mathbf{x}_n)}{N} - \lambda_k([\mathbf{0},\mathbf{y}))\right|. \tag{1.17}$$

Again we have $\lim_{N\to\infty} D_N^*(\mathbf{x}_n) = 0$ if and only if $(\mathbf{x}_n)_{n\geq 1}$ is u.d. mod 1.

Lemma 1.7 *For any sequence $(\mathbf{x}_n)_{n\geq1}$, $\mathbf{x}_n \in \mathbf{R}^k$, we have*

$$D_N^*(\mathbf{x}_n) \leq D_N(\mathbf{x}_n) \leq 4^k\, D_N^*(\mathbf{x}_n). \tag{1.18}$$

Proof. The first inequality is trivial. In order to prove the second one we first note that any interval $I \subseteq \mathbf{T}^k$ can be identified with a disjoint union of at most 2^k intervals contained in $[0,1)^k$. Now consider an arbitrary interval $I = [a_1,b_1) \times \cdots \times [a_k,b_k) \subseteq [0,1)^k$. By the principle of inclusion and exclusion we have

$$A(I,N,\mathbf{x}_n) = \sum_{\mathbf{y}\in\prod_{i=1}^k\{a_i,b_i\}} (-1)^{s(\mathbf{y})} A([\mathbf{0},\mathbf{y}),N,\mathbf{x}_n),$$

where $s(\mathbf{y})$ is the number of components y_i, $i = 1,\ldots,k$ with $y_i = a_i$. Similarly we have

$$\lambda_k(I) = \sum_{\mathbf{y}\in\prod_{i=1}^k\{a_i,b_i\}} (-1)^{s(\mathbf{y})} \lambda_k([\mathbf{0},\mathbf{y})).$$

Thus a direct application of the triangle inequality proves the second inequality $D_N \leq 4^k D_N^*$. □

The star discrepancy can be interpreted as the supremum (or L^∞) norm of the function

$$g(\mathbf{y}) = \frac{A([\mathbf{0},\mathbf{y}),N,\mathbf{x}_n)}{N} - \lambda_k([\mathbf{0},\mathbf{y})). \tag{1.19}$$

Thus it is quite natural to introduce a L^p*-discrepancy* for $0 < p < \infty$ by

$$D_N^{(p)}(\mathbf{x}_n) = \left(\int_{[0,1]^k} \left|\frac{A([\mathbf{0},\mathbf{y}),N,\mathbf{x}_n)}{N} - \lambda_k([\mathbf{0},\mathbf{y}))\right|^p d\mathbf{y}\right)^{\frac{1}{p}}. \tag{1.20}$$

These concepts of discrepancy can be used to describe u.d. sequences, too. This property was quantified by PROINOV [1467, 1469] in the one dimensional case and by NIEDERREITER, TICHY, and TURNWALD [1366] in the higher dimensional case (see also NIEDERREITER [1278, 1287] and YURINSKII [1981]).

Theorem 1.8 *For any sequence* $(\mathbf{x}_n)_{n\geq 1}$, $\mathbf{x}_n \in \mathbf{R}^k$, *and for* $0 < p < \infty$ *we have*

$$D_N^{(p)}(\mathbf{x}_n) \leq D_N^*(\mathbf{x}_n) \leq k(3k+4)\, c(k,p)^{\frac{-1}{p+k}}\, D_N^{(p)}(\mathbf{x}_n)^{\frac{p}{p+k}}, \tag{1.21}$$

where

$$c(k,p) = \frac{1}{(p+1)(p+2)\cdots(p+k)} \sum_{i=1}^{k} \binom{k}{i} (-1)^{k-i} i^{p+k}. \tag{1.22}$$

In the one dimensional case ($k = 1$) *the second inequality can be sharpened to*

$$D_N^*(x_n) \leq (p+1)^{\frac{1}{p+1}}\, D_N^{(p)}(x_n)^{\frac{p}{p+1}}. \tag{1.23}$$

This Theorem can be proved in a more general setting. Set $\|\mathbf{x}\|_1 = \sum_{i=1}^k |x_i|$ for $\mathbf{x} = (x_1, \ldots, x_k) \in \mathbf{R}^k$.

Proposition 1.9 *Let* F *and* G *be probability distribution functions concentrated on* $[0,1]^k$ *and suppose that* F *satisfies the* LIPSCHITZ *condition*

$$|F(\mathbf{x}) - F(\mathbf{y})| \leq L\|\mathbf{x} - \mathbf{y}\|_1 \tag{1.24}$$

for all $\mathbf{x}, \mathbf{y} \in [0,1]^k$. *Let* $D = G - F$ *and let* φ *be a nondecreasing nonnegative function on* $[0,1]$. *Define*

$$\Phi(t) = \int_{[0,t)^k} \varphi(\|\mathbf{y}\|_1)\, d\mathbf{y} \tag{1.25}$$

for $0 \leq t \leq 1/k$. *Then*

$$\int_{[0,1]^k} \varphi(|D(\mathbf{y})|)\, d\mathbf{y} \geq \frac{1}{L^k} \Phi\left(\frac{\|D\|_\infty}{k(3k+4)}\right). \tag{1.26}$$

In the one dimensional case ($k = 1$) *this inequality can be sharpened to*

$$\int_0^1 \varphi(|D(y)|)\, dy \geq \frac{1}{L} \Phi(\|D\|_\infty). \tag{1.27}$$

Applying this Proposition for $\varphi(t) = t^p$ we get $\Phi(t) = c(k,p)t^{p+k}$, where

$$c(k,p) = \int_{[0,1]^k} (\|\mathbf{y}\|_1)^p\, d\mathbf{y} = \frac{1}{(p+1)(p+2)\cdots(p+k)} \sum_{i=1}^{k} \binom{k}{i} (-1)^{k-i} i^{p+k},$$

and thus (1.21) follows. (1.23) can be can be shown along the same lines.

For simplicity we will first prove the one dimensional case ($k = 1$) of Proposition 1.9. The higher dimensional case will require more sophisticated observations.

Proof of Proposition 1.9 ($k = 1$). We will show that for every $a \in (0, \|D\|_\infty)$ we have

$$\int_0^1 \varphi(|D(y)|)\, dy \geq \frac{1}{L}\Phi(a).$$

For every $a \in (0, \|D\|_\infty)$ there exists a $y_0 \in [0,1]$ with $|D(y_0)| = |F(y_0) - G(y_0)| > a$. This means that either $G(y_0) > F(y_0) + a$ or $G(y_0) < F(y_0) - a$. We only consider the first case. The second one can be treated similarly. From $G(y_0) > F(y_0) + a$ and from the LIPSCHITZ condition it follows that $[y_0, y_0 + (a/L)] \subseteq [0,1]$. For $y \in [y_0, y_0 + (a/L)]$ we have

$$D(y) \leq F(y) - G(y_0) < F(y) - F(y_0) - a \leq L(y - y_0) - a \leq 0$$

and therefore $|D(y)| \geq a - L(y - y_0)$. This implies

$$\int_0^1 \varphi(|D(y)|)\, dy \geq \int_{y_0}^{y_0+(a/L)} \varphi(a - L(y - y_0))\, dy = \frac{1}{L}\Phi(a).$$

This proves the one dimensional case.□

Lemma 1.10 *Let* $0 < \alpha < \frac{1}{2}$, $\underline{\alpha} = (\alpha, \ldots, \alpha) \in [0,1]^k$, *and* $\underline{\beta} = (1 - \alpha, \ldots, 1 - \alpha) \in [0,1]^k$. *If D is as in Proposition 1.9, we put*

$$m(\alpha) = \sup_{\mathbf{x} \in [\underline{\alpha}, \underline{\beta}]} |D(\mathbf{x})|. \tag{1.28}$$

Then

$$m(\alpha) \geq \frac{\|D\|_\infty}{2k+2} - \frac{k(k+2)}{2k+2} L\alpha. \tag{1.29}$$

Proof. Let P, Q be the probability measures corresponding to F, G, respectively, i.e. $P([\mathbf{0}, \mathbf{x})) = F(\mathbf{x})$ and $Q([\mathbf{0}, \mathbf{x})) = G(\mathbf{x})$. For fixed $\mathbf{x} = (x_1, \ldots, x_k) \in [0,1]^k$ we set $\mathbf{y} = (y_1, \ldots, y_k)$ with $y_i = \min\{x_i, 1 - \alpha\}$, $i = 1, \ldots, k$. Then

$$0 \leq F(\mathbf{x}) - F(\mathbf{y}) \leq P(A), \quad 0 \leq G(\mathbf{x}) - G(\mathbf{y}) \leq Q(A)$$

with $A = [0,1]^k \setminus [\mathbf{0}, \underline{\beta}]$. (Note that $[\mathbf{0}, \mathbf{x}] \setminus [\mathbf{0}, \mathbf{y}] \subseteq A$.) Hence

$$|D(\mathbf{x}) - D(\mathbf{y})| \leq \max\{P(A), Q(A)\}.$$

We have

$$P(A) = 1 - F(\underline{\beta}) \leq Lk\alpha$$

by the LIPSCHITZ condition and

$$Q(A) = 1 - G(\underline{\beta}) = 1 - F(\underline{\beta}) - D(\underline{\beta}) \leq Lk\alpha + m(\alpha).$$

Thus

$$|D(\mathbf{x})| \leq |D(\mathbf{y})| + Lk\alpha + m(\alpha). \tag{1.30}$$

Set $\mathbf{z} = (z_1, \ldots, z_k)$ with $z_i = \max\{y_i, \alpha\}$, $i = 1, \ldots, k$. As above we get

$$|D(\mathbf{z}) - D(\mathbf{y})| \leq \max\{P(B), Q(B)\}$$

with $B = [0,1]^k \setminus [\alpha,1]^k$. Since $B = \bigcup_{i=1}^k [0, \underline{\alpha}_i)$, where $\underline{\alpha}_i = (1,\dots,\alpha,\dots,1)$ with α appearing in the i-th coordinate, we have

$$\begin{aligned} P(B) &\le \sum_{i=1}^k F(\underline{\alpha}_i) \le Lk\alpha \\ Q(B) &\le \sum_{i=1}^k F(\underline{\alpha}_i) + \sum_{i=1}^k |D(\underline{\alpha}_i)| \le Lk\alpha + k(2m(\alpha) + Lk\alpha), \end{aligned}$$

where we have applied (1.30) with $\mathbf{x} = \underline{\alpha}_i$ in the last step. Hence

$$|D(\mathbf{z}) - D(\mathbf{y})| \le 2km(\alpha) + k(k+1)L\alpha.$$

Together with (1.30) we obtain

$$|D(\mathbf{x})| \le (2k+2)m(\alpha) + k(k+2)L\alpha$$

and the result follows.□

Proof of Proposition 1.9 ($k > 1$). We assume that $\Delta = \|D\|_\infty > 0$. Let $\varepsilon > 0$ with $(2k+2)\varepsilon < \Delta$ and put

$$\alpha = \frac{\Delta - (2k+2)\varepsilon}{Lk(3k+4)}.$$

Note that $\alpha < \frac{1}{2}$ since $\Delta \le 1$ and

$$1 = F(1,\dots,1) - F(0,\dots,0) \le Lk.$$

Then by Lemma 1.10 we may choose $\mathbf{x}_0 \in [\underline{\alpha}, \underline{\beta}]$ with

$$|D(\mathbf{x}_0)| > \frac{\Delta}{2k+2} - \frac{k(k+2)}{2k+2} L\alpha - \varepsilon = Lk\alpha.$$

We remark that

$$\begin{aligned} \Phi(t) &= t^k \int_{[0,1]^k} \varphi(t(y_1 + \dots + y_k))\, d\mathbf{y} \\ &= t^k \int_{[0,1]^k} \varphi(t((1-y_1) + \dots + (1-y_k)))\, d\mathbf{y} \qquad (1.31) \\ &= t^k \int_{[0,1]^k} \varphi(t(k - \|\mathbf{y}\|_1))\, d\mathbf{y}. \end{aligned}$$

If $D(\mathbf{x}_0) \ge 0$ then for $\mathbf{y} \in [0, \underline{\alpha})$ we have

$$\begin{aligned} |D(\mathbf{x}_0 + \mathbf{y})| &\ge G(\mathbf{x}_0 + \mathbf{y}) - F(\mathbf{x}_0 + \mathbf{y}) \ge G(\mathbf{x}_0) - F(\mathbf{x}_0) - L\|\mathbf{y}\|_1 \\ &= D(\mathbf{x}_0) - L\|\mathbf{y}\|_1 > Lk\alpha - L\|\mathbf{y}\|_1. \end{aligned}$$

Hence

$$\begin{aligned}\int_{[0,1]^k} \varphi(|D(\mathbf{y})|)\,d\mathbf{y} &\geq \int_{[\mathbf{x}_0,\mathbf{x}_0+\underline{\alpha})} \varphi(|D(\mathbf{y})|)\,d\mathbf{y} \geq \int_{[0,\underline{\alpha})} \varphi(Lk\alpha - L\|\mathbf{y}\|_1)\,d\mathbf{y} \\ &= \alpha^k \int_{[0,1]^k} \varphi(Lk\alpha - L\alpha\|y\|_1)\,dy = \frac{1}{L^k}\Phi(L\alpha)\end{aligned}$$

by (1.31).

If $D(\mathbf{x}_0) < 0$ then for $\mathbf{y} \in [0,\underline{\alpha})$ we have

$$\begin{aligned}|D(\mathbf{x}_0 + \mathbf{y} - \underline{\alpha})| &\geq F(\mathbf{x}_0 + \mathbf{y} - \underline{\alpha}) - G(\mathbf{x}_0 + \mathbf{y} - \underline{\alpha}) \\ &\geq F(\mathbf{x}_0) - L\|\mathbf{y} - \underline{\alpha}\|_1 - G(\mathbf{x}_0) \\ &= |D(\mathbf{x}_0)| - L\|\mathbf{y} - \underline{\alpha}\|_1 > Lk\alpha - L\|\mathbf{y} - \underline{\alpha}\|_1.\end{aligned}$$

Hence

$$\begin{aligned}\int_{[0,1]^k} \varphi(|D(\mathbf{y})|)\,d\mathbf{y} &\geq \int_{[\mathbf{x}_0,\mathbf{x}_0-\underline{\alpha})} \varphi(|D(\mathbf{y})|)\,d\mathbf{y} \\ &\geq \int_{[0,\underline{\alpha})} \varphi(Lk\alpha - L\|\mathbf{y} - \underline{\alpha}\|_1)\,d\mathbf{y} \\ &= \alpha^k \int_{[0,1]^k} \varphi(Lk\alpha - L\alpha\|y\|_1)\,dy = \frac{1}{L^k}\Phi(L\alpha)\end{aligned}$$

by (1.31). Thus in both cases we have

$$\int_{[0,1]^k} \varphi(|D(\mathbf{y})|)\,d\mathbf{y} \geq \frac{1}{L^k}\Phi\left(\frac{\Delta - (2k+2)\varepsilon}{k(3k+4)}\right).$$

Letting $\varepsilon \to 0$ and using the continuity of Φ, we get the result of Proposition 1.9.□

Another possibility to quantify the convergence in (1.2), resp. in (1.9) is to use convex subsets C instead of intervals, resp. characteristic functions $f = \chi_C$ of convex subsets $C \subseteq \mathbf{T}^k$.

Definition 1.11 *Let $(\mathbf{x}_n)_{n=1}^N$ be a finite sequence of points in the k-dimensional space $\mathbf{R}^k$. The* isotropic discrepancy $J_N = J_N(\mathbf{x}_1, \ldots, \mathbf{x}_N)$ *is defined by*

$$J_N = \sup_{C \subseteq \mathbf{T}^k} \left|\frac{A(C,N,\mathbf{x}_n)}{N} - \lambda_k(C)\right|, \tag{1.32}$$

where the supremum is taken over all convex subsets $C \subseteq \mathbf{T}^k$. As above this definition can be extended to infinite sequences $(\mathbf{x}_n)_{n\geq 1}$.

Here we have

Theorem 1.12 *For any sequence of N points in $\mathbf{R}^k$ we have*

$$D_N(\mathbf{x}_n) \leq J_N(\mathbf{x}_n) \leq (4k\sqrt{k}+1)\,D_N(\mathbf{x}_n)^{\frac{1}{k}}. \tag{1.33}$$

The *proof* is due to NIEDERREITER (see [983, p. 95 ff.]).

The above concepts of discrepancy are quantitative measures of the order of convergence in the limit relation (1.2) defining u.d. of a given sequence. Of course it is of great interest to get information on the order of convergence in 1.9, too. This yields estimates for the difference of the integral $m(f)$ and its approximation by the arithmetic mean $m_N(f)$. Such an estimate is provided by KOKSMA-HLAWKA's inequality. It shows that the order of convergence in (1.9) can be estimated in terms of the variation of f and the star discrepancy. For this purpose we must define the variation of a functions $f : [0,1]^k \to \mathbf{R}$.

By a partition P of $[0,1]^k$ we mean a set of k finite sequences $\eta_i^{(0)}, \dots, \eta_i^{(m_i)}$, $i = 1, \dots, k$, with $0 = \eta_i^{(0)} \le \eta_i^{(1)} \le \dots \le \eta_i^{(m_i)} = 1$. In connection with such a partition we define for each $i = 1, \dots, k$ an operator Δ_i by

$$\begin{aligned} \Delta_i f(x_1, \dots, x_{i-1}, \eta_i^{(j)}, x_{i+1}, \dots, x_k) &= \\ = \quad & f(x_1, \dots, x_{i-1}, \eta_i^{(j+1)}, x_{i+1}, \dots, x_k) \\ - \quad & f(x_1, \dots, x_{i-1}, \eta_i^{(j)}, x_{i+1}, \dots, x_k) \end{aligned} \tag{1.34}$$

for $0 \le j < m_i$. Operators with different subscripts obviously commute and $\Delta_{i_1,\dots,i_l}$ will stand for $\Delta_{i_1} \cdots \Delta_{i_l}$.

Definition 1.13 *For a function $f : [0,1] \to \mathbf{R}$ we set*

$$V^{(k)}(f) = \sup_P \sum_{j_1=0}^{m_1-1} \cdots \sum_{j_k=0}^{m_k-1} \left| \Delta_{1,\dots,k} f(\eta_1^{(j_1)}, \dots, \eta_k^{(j_k)}) \right|, \tag{1.35}$$

where the supremum is extended over all partitions P of $[0,1]^k$. If $V^{(k)}(f)$ is finite then f is said to be of bounded variation on $[0,1]^k$ in the sense of VITALI. *Furthermore if in addition the restriction $f^{(F)}$ of f to each face F of $[0,1]^k$ of dimensions $1, 2, \dots, k-1$ is of bounded variation on F in the sense of* VITALI *then f is said to be of bounded variation on $[0,1]^k$ in the sense of* HARDY *and* KRAUSE.

So we can state

Theorem 1.14 (KOKSMA-HLAWKA's Inequality) *Let f be of bounded variation on $[0,1]^k$ in the sense of* HARDY *and* KRAUSE. *Let $(\mathbf{x}_n^{(i_1,\dots,i_l)})_{n=1}^N = (\mathbf{x}_n^{(F)})_{n=1}^N$ denote the projection of the sequence $(\mathbf{x}_n)_{n=1}^N$, $\mathbf{x}_n \in [0,1]^k$, on the $(k-l)$-dimensional face F of $[0,1]^k$ defined by $x_{i_1} = \dots = x_{i_l} = 1$. Then we have*

$$\left| \frac{1}{N} \sum_{n=1}^N f(\mathbf{x}_n) - \int_{[0,1]^k} f(\mathbf{x})\, d\mathbf{x} \right| \le \sum_{l=0}^{k-1} \sum_{F_l} D_N^*(\mathbf{x}_n^{(F_l)}) V^{(k-l)}(f^{(F_l)}), \tag{1.36}$$

where the second sum is extended over all $(k-l)$-dimensional faces F_l of the form $x_{i_1} = \dots = x_{i_l} = 1$. The discrepancy $D_N^*(\mathbf{x}_n^{(F_l)})$ is computed in the face of $[0,1]^k$ in which $(\mathbf{x}_n^{(F_l)})_{n=1}^N$ is contained.

A *proof* can be found in KUIPERS and NIEDERREITER [983, p. 147 ff.]. The original proof is given in HLAWKA [786].

Remark 1. Trivially $D_N^*(\mathbf{x}_n^{(F_l)})$ can be bounded by $D_N^*(\mathbf{x}_n)$. Hence we get from (1.36)

$$\left|\frac{1}{N}\sum_{n=1}^{N} f(\mathbf{x}_n) - \int_{[0,1]^k} f(\mathbf{x})\,d\mathbf{x}\right| \le V(f) D_N^*(\mathbf{x}_n),$$

where

$$V(f) = \sum_{l=0}^{k-1}\sum_{F_l} V^{(k-l)}(f^{(F_l)})$$

is called the variation of HARDY and KRAUSE.

Remark 2. If the partial derivative

$$\frac{\partial^l f^{(F_l)}}{\partial x_{i_1}\cdots\partial x_{i_l}}$$

is continuous on the $(k-l)$-dimensional face F_l of $[0,1]^k$, defined by $x_{i_1} = \cdots = x_{i_l} = 1$, the variation can be computed by the integral

$$V^{(k-l)}(f^{(F_l)}) = \int_{F_l}\left|\frac{\partial^l f^{(F_l)}}{\partial x_{i_1}\cdots\partial x_{i_l}}\right| dx_{i_1}\cdots dx_{i_l}.$$

1.1.3 Dispersions

It is obvious that for every u.d. sequence $(\mathbf{x}_n)_{n\ge1}$ the fractional parts $(\{\mathbf{x}_n\})_{n=1}^{\infty}$ are dense in $\mathbf{T}^k$. Similarly to the notion of discrepancy which measures the uniform distribution of N points $\mathbf{x}_1,\ldots,\mathbf{x}_N$ it is possible to quantify the denseness of N points in the torus (or unit cube).

Definition 1.15 *Let $\delta(\mathbf{x},\mathbf{y})$ denote a distance function on $\mathbf{T}^k$. Then the* dispersion *$d_N^{(\delta)}(\mathbf{x}_n)$ of N points $\mathbf{x}_1,\ldots,\mathbf{x}_N\in\mathbf{T}^k$ is defined by*

$$d_N^{(\delta)}(\mathbf{x}_n) = \max_{\mathbf{x}\in\mathbf{T}^k}\ \min_{n=1,\ldots,N}\ \delta(\mathbf{x},\mathbf{x}_n). \tag{1.37}$$

Remark 1. Let $\delta'(\mathbf{x},\mathbf{y})$ denote a translation invariant metric on $\mathbf{T}^k$. Then a metric on $\mathbf{R}^k$ is induced by

$$\delta(\mathbf{x},\mathbf{y}) = \min_{\mathbf{m}\in\mathbf{Z}^k} \delta'(\mathbf{x},\mathbf{y}+\mathbf{m}). \tag{1.38}$$

For example, take the Euclidean distance $|\mathbf{x}-\mathbf{y}|$ or the maximum distance $\|\mathbf{x}-\mathbf{y}\|_\infty$ for $\delta'(\mathbf{x},\mathbf{y})$. The corresponding dispersions will be denoted by d_N and d_N^∞.

The central Theorem is

Theorem 1.16 *Let $(\mathbf{x}_n)_{n\geq 1}$ be a sequence in $\mathbf{R}^k$. Then the sequence of fractional parts $(\{\mathbf{x}_n\})_{n=1}^{\infty}$ is dense in $\mathbf{T}^k$ if and only if*

$$\lim_{N\to\infty} d_N(\mathbf{x}_n) = 0. \tag{1.39}$$

Proof. Trivially $\lim_{N\to\infty} d_N(\mathbf{x}_n) = 0$ implies that $(\{\mathbf{x}_n\})_{n=1}^{\infty}$ is dense in $\mathbf{T}^k$. Now let $\varepsilon > 0$ be given. Then there are finitely many discs with radius $< \varepsilon$ covering $\mathbf{T}^k$. Thus there is some positive integer N such that each of these finitely many discs contains at least one point $\{\mathbf{x}_n\}$, $n = 1\ldots,N$. Hence $d_N(\mathbf{x}_n) < \varepsilon$. So we are done since $d_M(\mathbf{x}_n) \leq d_N(\mathbf{x}_n)$ for $M \geq N$. □

Remark 2. It is apparent that the same theorem holds if we use d_N^∞ instead of d_N. $\|\mathbf{x}-\mathbf{y}\|_\infty \leq |\mathbf{x}-\mathbf{y}| \leq \sqrt{k}\,\|\mathbf{x}-\mathbf{y}\|_\infty$ immediately implies

$$d_N^\infty(\mathbf{x}_n) \leq d_N(\mathbf{x}_n) \leq \sqrt{k}\, d_N^\infty(\mathbf{x}_n). \tag{1.40}$$

Remark 3. It is also very easy to generalize the concept of dispersion to arbitrary compact metric spaces. Again dense sequences are characterized there by $\lim_{N\to\infty} d_N(\mathbf{x}_n) = 0$.

As mentioned above each u.d. sequence is dense. This qualitative statement can be quantified by means of dispersion and discrepancy.

Theorem 1.17 *Let $\mathbf{x}_1,\ldots,\mathbf{x}_N \in \mathbf{R}^k$. Then*

$$d_N^\infty(\mathbf{x}_n) \leq \frac{1}{2} D_N(\mathbf{x}_n)^{\frac{1}{k}}. \tag{1.41}$$

Proof. Let Q be the largest subcube of $\mathbf{R}^k$ such that $Q^\circ \cap \{\{\mathbf{x}_1\},\ldots,\{\mathbf{x}_N\}\} = \emptyset$. Then $\lambda_k(Q) \leq D_N(\mathbf{x}_n)$ and $2\, d_N^\infty(\mathbf{x}_n) \leq \lambda_k(Q)^{1/k}$ imply (1.41). □

Remark 4. Well dispersed sequences are quite useful for optimization problems. Let $f : \mathbf{T}^k \to \mathbf{R}$ be a continuous function with modulus of continuity

$$\omega_f^{(\delta)}(c) = \max_{\delta(\mathbf{x},\mathbf{y})\leq c} |f(\mathbf{x}) - f(\mathbf{y})|. \tag{1.42}$$

Then by definition

$$\left|\max_{\mathbf{x}\in\mathbf{T}^k} f(\mathbf{x}) - \max_{n=1,\ldots,N} f(\mathbf{x}_n)\right| \leq \omega_f^{(\delta)}(d_N(\mathbf{x}_n)). \tag{1.43}$$

For example, if it is known that f is LIPSCHITZ continuous with (known) LIPSCHITZ constant L then the maximum of f can be calculated arbitrarily exactly.

Remark 5. It should be noted that the dispersion of a sequence $(\mathbf{x}_n)_{n\geq 1}$ cannot converge to 0 arbitrarily fast. Consider N points $\mathbf{x}_1,\ldots,\mathbf{x}_N \in \mathbf{R}^k$ and set $\varepsilon = [N^{1/k}+1]^{-1}$. Now subdivide $\mathbf{T}^k$ into ε^{-k} subcubes. Since $\varepsilon^{-k} > N$ there is at least one subcube that contains no point $\mathbf{x}_1,\ldots,\mathbf{x}_N$. Thus

$$d_N^\infty \geq \frac{1}{2(N^{1/k}+1)}. \tag{1.44}$$

Notes

The general definition of uniform distribution and Criteria A and B are due to Weyl [1953]. For the early literature on u.d. sequences we refer to the historical paper by Hlawka and Binder [837]; see also Hlawka [836]. A converse of Criterion A is due to de Bruijn and Post [436]: If (1.9) holds for all u.d. sequences mod 1 then $f : [0,1] \to \mathbf{R}$ is Riemann integrable. Binder [220, 221] gives a generalization to compact metric spaces satisfying some technical conditions.

The first intensive study of discrepancy is contained in Van der Corput and Pisot [1899]. For instance they prove

$$|D_N(x_n) - D_N(y_n)| \leq \mathrm{dist}(\{x_1, \ldots, x_N\}, \{y_1, \ldots, y_N\}),$$

where $\mathrm{dist}(A,B) = \min_f \max_{a \in A} |f(a) - a|$ and the minimum is extended over all bijections $f : A \to B$. An obvious consequence is the continuity of the discrepancy function $(x_1, \ldots, x_N) \mapsto D_N(x_n)$. A converse inequality is given by Myerson [1228]:

$$\mathrm{dist}(\{x_1, \ldots, x_N\}, \{y_1, \ldots, y_N\}) \leq D_N(x_n) + D_N(y_n) - \frac{1}{N};$$

for related results see Myerson [1228] and Winkler [1966]. A first detailed investigation of the L^2-discrepancy is due to Halton [729]. More recently L^p-discrepancies were extensively studied in several papers by Faure and Proinov. Most of these results are concerned with special sequences such as the $(n\alpha)$-sequence and the Van der Corput sequence (see Faure [587, 588], Proinov [1471, 1465] and the Notes of Section 1.4). Horbowicz [844] established an asymptotic expression for the difference of the usual discrepancy and the L^p-discrepancy. Generalizing the L^p-discrepancy, Proinov [1475] considered the so-called φ-discrepancy, where φ is a non-decreasing positive function on $[0,1]$ tending to 0 (for $x \to 0+$) replacing the p-th power in the definition of the L^p-discrepancy (as given by the integral in (1.26)). For instance he proved an error bound for the approximation of integrals of smooth functions in terms of the φ-discrepancy. A generalization of Theorem 1.8 to compact metric spaces were given by Tichy [1841]. Explicit formulae for L^p-discrepancies (p positive even integers) were established by Kuipers and Shiue [994] ($p = 4$), Proinov and Andreeva [1481], Winkler [1962], Andreeva [39], Nagasaka and Shiue [1240], and Pasteka [1408] in the more general context of weighted means (see Section 2.2). Hellekalek [762] studied the convergence rate of the L^p-discrepancy to the usual discrepancy and proved an explicit formula in arbitrary dimensions (for equal weights). For an explicit formula for the usual star discrepancy we refer to Kuipers and Niederreiter [983, p. 91 f.]. Multidimensional analogues of such explicit formulae are not easy to obtain: Partial results were given in Niederreiter [1271] and de Clerck [439, 438]; a complete solution is due to Bundschuh and Zhu [315, 314], Zhu [1996] and Lin Achan [2] (see Section 3.1). The isotroptic discrepancy were investigated by Hlawka [786], Zaremba [1986], Niederreiter [1271]. Niederreiter and Wills [1367] generalized Theorem 1.12 to difference of probability measures obtaining a refinement of an earlier result of Mück and Philipp [1214].

Theorem 1.18 *Let λ and μ be two finitely additive non-negative bounded measures defined at least on the algebra of Jordan measurable sets on U^k $(k \geq 2)$ and suppose that $\lambda(S) \leq M\lambda_k(S)$ for all Jordan measurable sets and $\mu(U^k) \leq M$ for some absolute constant M. Let*

$$J = \sup_C |\lambda(C) - \mu(C)| \quad \text{and} \quad D = \sup_B |\lambda(B) - \mu(B)|,$$

where C runs through all convex sets and B through all k-dimensional intervals. Then

$$J \leq k \left(\frac{4Mk}{k-1}\right)^{(k-1)/k} D^{1/k}.$$

This result is of importance for Monte Carlo integration over Jordan sets (see Section 3.1). Recently, Laczkovich [1003] obtained a refinement of Theorem 1.18. Koksma-Hlawka's inequality gives an error bound for numerical integration by using low discrepancy sequences. Of course for this purpose it is necessary to find efficient constructions for such sequences. This is outlined in Sections 1.4 and 3.1. In Niederreiter [1309] (see also [1336, p. 20]) it is shown that Koksma-Hlawka's inequality is best possible. Zaremba [1986] gave a Koksma-Hlawka type inequality in terms for the isotropic discrepancy. Proinov [1474] established the following variant of Koksma-Hlawka's inequality:

$$\left|\frac{1}{N}\sum_{n=1}^{N} f(\mathbf{x}_n) - \int_{[0,1]^k} f(\mathbf{x})\,d\mathbf{x}\right| \leq 4\omega_f^\infty(D_N^*(\mathbf{x}_n)^{1/k}),$$

where ω_f^∞ denotes the modulus of continuity with respect to the supremum norm. This improves the constant obtained given by Hlawka [791] and by Shi [1669, 1670]. Further applications of discrepancy to approximation theory are also due to Proinov [1468, 1473] and due to Proinov and Khirov [1485]. Other Koksma-Hlawka type inequalities can be found in Niederreiter [1278] and in de Clerck [437]. In the recent literature there exist various concepts of discrepancy measuring the distribution behaviour of sequences in the context of specific applications, e.g. Hlawka's "Green function discrepancy" [822, 823, 824, 820, 821] which will be discussed in Sections 3.2 and 3.3. Sobol' and Nuzhdin [1384] introduce the concept of *range* of a given sequence $x_1, \ldots, x_N$

$$\psi = \psi(x_1, \ldots, x_N) = \sup_M \psi_M,$$

where $M = (m_1, \ldots, m_k)$ is a k-tupel of non-negative integers and

$$\psi_M = \max_B \sum_{n=1}^{N} \chi_B(x_n) - \min_B \sum_{n=1}^{N} \chi_B(x_n),$$

(the maximum – resp. the minimum – is taken over all $2^{m_1+\cdots+m_k}$ dyadic boxes B of volume $\lambda_k(B) = 2^{-(m_1+\cdots+m_k)}$ which cover U^k). A sequence is u.d. mod 1 if and only if $\lim_{N\to\infty} \psi/N = 0$. In an earlier paper Sobol' [1700] has introduced a very related notion, the so-called *non-uniformity*. For a survey on different kinds of quantities measuring u.d. we refer to Shparlinskij [1687]. In this context we also mention Myerson and Pollington [1230, 1231].

The dispersion was introduced by Hlawka [800] and later investigated in more general form by Niederreiter [1302]. The application of well dispersed sets to global optimization was developed by Niederreiter [1308] and by Niederreiter and Peart [1350]. A recent progress was obtained by Biester, Grabner, Larcher, and Tichy [219]. A very similar concept of dispersion was considered in Rote and Tichy [1566] in a much more general setting: Let (X, μ) a probability space and $\mathcal{D}$ a system of measurable sets. Then the "dispersion" of a finite subset $A \subseteq X$ is defined by

$$d_{X,\mu,\mathcal{D}}(A) = \sup_{D\in\mathcal{D}, A\cap D=\emptyset} \mu(D).$$

This is related to the Vapnik-Chervonenkis dimension [1900] and can be applied for approximating spatial curves by polygons as used in robotics; see Sections 2.4 and 3.3. Furthermore there exist several papers investigating the dispersion of special sequences (see Section 1.4). For relations between u.d. and geometry of numbers and algebraic number theory we refer to Everest [574], Everest and Gagen [575], Skriganov [1692, 1693] and Temirgaliev [1795].

1.2 Exponential Sums

1.2.1 WEYL's Criterion

In section 1.1.1 it is proved that a sequence $(\mathbf{x}_n)_{n\geq 1}$ is u.d. mod 1 if and only if

$$\lim_{N\to\infty} \frac{1}{N}\sum_{n=1}^{N} f\{\mathbf{x}_n\} = \int_{[0,1]^k} f\,d\lambda_k \tag{1.45}$$

holds for all continuous functions $f : [0,1]^k \to \mathbf{R}$ (Criterion B). Trivially this criterion can be extended to continuous functions $f : [0,1]^k \to \mathbf{C}$. WEYL's ingenious observation was that is sufficies to inspect (1.45) for exponential functions.

Set $e(x) = e^{2\pi i x}$ and denote the usual inner product in $\mathbf{R}^k$ by $\mathbf{x} \cdot \mathbf{y} = \sum_{i=1}^k x_i y_i$.

Theorem 1.19 (Criterion C, WEYL's criterion) *A sequence $(\mathbf{x}_n)_{n\geq 1}$ of points in the k-dimensional space $\mathbf{R}^k$ is u.d. mod 1 if and only if*

$$\lim_{N\to\infty} \frac{1}{N} \sum_{n=1}^{N} e(\mathbf{h} \cdot \mathbf{x}_n) = 0 \qquad (1.46)$$

holds for all non-zero integral lattice points $\mathbf{h} \in \mathbf{Z}^k \setminus \{(0,\ldots,0)\}$.

Proof. It is clear that every u.d. sequence $(\mathbf{x}_n)_{n\geq 1}$ satisfies (1.46). On the other hand let $\mathcal{L}$ be the set of all (RIEMANN integrable) functions $f : [0,1] \to \mathbf{C}$ satisfying (1.45). By (1.46) all trigonometric polynomials are contained in $\mathcal{L}$. Thus by Lemma 1.2 and by WEIERSTRASS's approximation theorem $\mathcal{L}$ contains all continuous functions with period 1. So we can proceed as in the proof of Criterion B that $\mathcal{L}$ contains all characteristic functions of intervals. This completes the proof of WEYL's criterion. □

WEYL [1953] applied his theorem to $(n\underline{\alpha})_{n=1}^{\infty}$, $\underline{\alpha} \in \mathbf{R}^k$, to give a new proof of KRONECKER's approximation theorem.

Corollary 1.20 (KRONECKER's Approximation Theorem) *Let $\underline{\alpha} = (\alpha_1,\ldots,\alpha_k) \in \mathbf{R}^k$ such that $1,\alpha_1,\ldots,\alpha_k$ are linearly independent over $\mathbf{Q}$. Then the sequence of fractional parts $(\{n\underline{\alpha}\})_{n=1}^{\infty}$ is dense in $[0,1)^k$.*

Remark. Observe that WEYL's criterion implies that a sequence of the form $(n\underline{\alpha})_{n=1}^{\infty}$ is u.d. mod 1 if and only if $1,\alpha_1,\ldots,\alpha_k$ are linearly independent over $\mathbf{Q}$. Hence it follows that $(n\underline{\alpha})_{n=1}^{\infty}$ is u.d. mod 1 if and only if $(\{n\underline{\alpha}\})_{n=1}^{\infty}$ is dense in $[0,1)^k$.

1.2.2 ERDŐS-TURÁN-KOKSMA's Inequality

Exponential sums are not only useful to check whether a sequence is u.d. mod 1 or not but it is also possible to estimate the discrepancy.

Theorem 1.21 *Let $\mathbf{x}_1,\ldots,\mathbf{x}_N$ be points in the k-dimensional space $\mathbf{R}^k$ and H an arbitrary positive integer. Then*

$$D_N(\mathbf{x}_n) \leq \left(\frac{3}{2}\right)^k \left(\frac{2}{H+1} + \sum_{0<\|\mathbf{h}\|_\infty \leq H} \frac{1}{r(\mathbf{h})} \left| \frac{1}{N} \sum_{n=1}^{N} e(\mathbf{h}\cdot\mathbf{x}_n) \right| \right), \qquad (1.47)$$

where $r(\mathbf{h}) = \prod_{i=1}^{k} \max\{1,|h_i|\}$ for $\mathbf{h} = (h_1,\ldots,h_k) \in \mathbf{Z}^k$.

The proof of this version of ERDŐS-TURÁN-KOKSMA's inequality is due to GRABNER [673], who has adapted VAALER's [1887] method to the higher dimensional case. The central theorem is the inequality

$$|f(x) - (f * j_H)(x)| \leq \frac{1}{2H+2} (dV_f * k_N)(x),$$

where $j_H(x)$ is a special trigonometric polynomial of degree H, V_f the total variation of f, and $*$ the usual convolution on $[0,1]$. We will first prove this inequality due to VAALER and will get Theorem 1.21 as a corollary.

Set

$$H(z) = \left(\frac{\sin \pi z}{\pi}\right)^2 \left(\sum_{n\in\mathbf{Z}} \frac{\operatorname{sgn}(n)}{(z-n)^2} + \frac{2}{z}\right). \tag{1.48}$$

Then we can prove

Lemma 1.22 *For $x \in \mathbf{R}$ we have*

$$|H(x)| \leq 1 \tag{1.49}$$

and

$$|\operatorname{sgn}(x) - H(x)| \leq \left(\frac{\sin \pi x}{\pi x}\right)^2. \tag{1.50}$$

Proof. Set $K(z) = \left(\frac{\sin \pi z}{\pi z}\right)^2$. Then it sufficies to prove

$$1 - K(x) \leq H(x) \leq 1$$

for real $x > 0$ since $\operatorname{sgn}(x)$ and $H(x)$ are odd functions. Using

$$\sum_{n\in\mathbf{Z}} \frac{1}{(z-n)^2} = \left(\frac{\pi}{\sin \pi z}\right)^2$$

we get

$$H(x) = 1 + \left(\frac{\sin \pi x}{\pi}\right)^2 \left(\frac{2}{x} - \frac{1}{x^2} - 2\sum_{n=1}^{\infty} \frac{1}{(x+n)^2}\right)$$

for $x > 0$. Applying the inequality between the arithmetic and the geometric mean we have

$$\begin{aligned}
\frac{1}{x^2} + 2\sum_{n=1}^{\infty} \frac{1}{(x+n)^2} &= \sum_{n=0}^{\infty} \left(\frac{1}{(x+n)^2} + \frac{1}{(x+n+1)^2}\right) \\
&\geq 2\sum_{n=0}^{\infty} \frac{1}{(x+n)(x+n+1)} = \frac{2}{x}.
\end{aligned}$$

On the other hand we can estimate

$$\sum_{n=1}^{\infty} \frac{1}{(x+n)^2} \leq \sum_{n=1}^{\infty} \frac{1}{(x+n)(x+n-1)} = \frac{1}{x}.$$

Thus we get

$$1 - \left(\frac{\sin \pi z}{\pi z}\right)^2 \leq H(x) \leq 1.$$

□

Now set $J(z) = \frac{1}{2}H'(z)$. By construction, $J(z)$ and $H(z)$ are entire functions of exponential type 2π. $J(z)$ has the following basic properties.

Lemma 1.23 *$J(x)$ is absolutely integrable in* $\mathbf{R}$ *and its* FOURIER *transform is given by*

$$\hat{J}(t) = \begin{cases} 1 & \text{for } t = 0 \\ \pi t(1 - |t|)\cos \pi t + |t| & \text{for } 0 < |t| < 1 \\ 0 & \text{otherwise.} \end{cases} \tag{1.51}$$

Furthermore, $\hat{J}(t)$ is decreasing on $[0, 1]$.

Proof. Let

$$H_N(z) = \left(\frac{\sin \pi z}{\pi}\right)^2 \left(\sum_{n=-N}^{N} \frac{\operatorname{sgn}(n)}{(z-n)^2} + \frac{2}{z}\right),$$

such that

$$\lim_{N\to\infty} H_N(z) = H(z), \quad \lim_{N\to\infty} \frac{1}{2} H_N'(z) = J(z)$$

hold uniformly on compact subsets of $\mathbf{C}$. From

$$K(z) = \int_{-1}^{1} (1 - |t|) e(zt)\, dt$$

and

$$zK(z) = \frac{1}{2\pi i} \int_{-1}^{1} \operatorname{sgn}(t) e(zt)\, dt$$

we get

$$\begin{aligned} H_N(z) &= \sum_{n=-N}^{N} \left(\frac{\sin \pi(z-n)}{\pi(z-n)}\right)^2 + \frac{2(\sin \pi z)^2}{\pi^2 z} \\ &= \int_{-1}^{1} (1 - |t|) \left(\sum_{n=-N}^{N} \operatorname{sgn}(n) e(-nt)\right) e(zt)\, dt \\ &+ \frac{1}{\pi i} \int_{-1}^{1} \operatorname{sgn}(t) e(zt)\, dt. \end{aligned}$$

Differentiating with respect to z and using the identity

$$\sum_{n=-N}^{N} \operatorname{sgn}(n) e(-nt) = -i \cot \pi t + i \frac{\cos \pi(2N+1)t}{\sin \pi t}$$

yields

$$\begin{aligned} \frac{1}{2} H_N'(z) &= \int_{-1}^{1} ((1 - |t|)\pi t \cot \pi t + |t|) e(zt)\, dt \\ &- \int_{-1}^{1} \cos \pi(2N+1)t \frac{\pi t e(zt)}{\sin \pi t}\, dt. \end{aligned}$$

By the RIEMANN-LEBESGUE lemma the second integral converges to 0 for $N \to \infty$. Thus we have

$$J(z) = \int_{-1}^{1} ((1 - |t|)\pi t \cot \pi t + |t|)e(zt)\, dt.$$

Now set $\varphi(t) = (1 - t)\pi t \cot \pi t + t$. Thus the above representation can be rewritten as

$$J(z) = 2 \int_0^1 \varphi(t) \cos(2\pi z t)\, dt$$

and after three times of partial integration we have

$$J(z) = \frac{1}{(2\pi z)^3} \left(2 \int_0^1 \varphi'''(t) \sin(2\pi z t)\, dt - \frac{4\pi^3}{3} \sin(2\pi z) \right).$$

Hence

$$|J(x)| \leq \frac{C}{(1 + |x|)^3}$$

for some proper constant C and therefore $J(x)$ is absolutely integrable in $\mathbf{R}$. Since

$$\varphi'(t) = \pi(1 - 2t) \cot \pi t - t(1 - t) \frac{\pi^2}{\sin^2 \pi t} + 1,$$

we have for $0 < t < 1$ we have $\varphi'(t) < 0$. Thus $\hat{J}(t)$ is decreasing on $[0, 1]$. □

We will use the following notation. We set $F_H(x) = HF(Hx)$. Hence the FOURIER transform satisfies $\hat{F}_H(t) = \hat{F}(\frac{t}{H})$. Furthermore we need the convolution

$$(f * g)(x) = \int_{-1/2}^{1/2} f(x - t) g(t)\, dt \tag{1.52}$$

$$(f * d\mu)(x) = \int_{-1/2}^{1/2} f(x - t)\, d\mu(t) \tag{1.53}$$

for periodic functions f, g and for periodic measures μ with period 1.

By POISSON's summation formula and Lemma 1.23 we get

$$j_H(x) = \sum_{h \in \mathbf{Z}} J_{H+1}(x + h) = \sum_{h=-H}^{H} \hat{J}_{H+1}(h) e(hx). \tag{1.54}$$

Furthermore, denote

$$k_H(x) = \sum_{h=-H}^{H} \left(1 - \frac{|h|}{H + 1}\right) e(hx) \tag{1.55}$$

the FEJÉR kernel and

$$\psi(x) = \begin{cases} \{x\} - \frac{1}{2} & \text{for } x \notin \mathbf{Z} \\ 0 & \text{for } x \in \mathbf{Z} \end{cases} \tag{1.56}$$

the periodically extended first BERNOULLI polynomial. Then we have

Lemma 1.24 *The trigonometric polynomial*

$$\psi * j_H(x) = \sum_{h=-H, h\neq 0}^{H} -\frac{1}{2\pi i h}\hat{J}_{H+1}(h)e(hx)$$

satisfies

$$\operatorname{sgn}(\psi * j_H(x)) = \operatorname{sgn}(\psi(x))$$

and

$$|\psi * j_H(x) - \psi(x)| \leq \frac{1}{2H+2} k_H(x). \tag{1.57}$$

Proof. First we prove

$$S(m,x) = \sum_{k=1}^{m} \frac{\sin 2\pi k x}{k\pi} > 0 \tag{1.58}$$

for $0 < x < \frac{1}{2}$ by induction. Trivially $S(1,x) > 0$ for $0 < x < \frac{1}{2}$. Since

$$\frac{d}{dx} S(m,x) = \frac{2\cos(m+1)\pi x \sin m\pi x}{\sin \pi x}$$

$S(m,x)$ attains its minima at rational numbers of the form $\frac{k}{m}$ with $0 < 2k < m$. But by induction

$$S(m, \frac{k}{m}) = S(m-1, \frac{k}{m}) + \frac{\sin 2k\pi}{\pi m} = S(m-1, \frac{k}{m}) > 0.$$

Thus (1.58) follows.

Now we will prove $\psi * j_H(x) < 0$ for $0 < x < \frac{1}{2}$. By Lemma 1.23 $\hat{J}_{H+1}(h) - \hat{J}_{H+1}(h+1) > 0$. Hence

$$\begin{aligned} \psi * j_H(x) &= -\sum_{h=1}^{\infty} \hat{J}_{H+1}(h) \frac{\sin 2\pi h x}{\pi h} \\ &= -\sum_{h=1}^{\infty} \left(\hat{J}_{H+1}(h) - \hat{J}_{H+1}(h+1)\right) \sum_{k=1}^{h} \frac{\sin 2\pi k x}{k\pi} \\ &< 0. \end{aligned}$$

for $0 < x < \frac{1}{2}$. This proves the first part of Lemma 1.24.

In order to prove (1.57) set $E(x) = H(x) - \operatorname{sgn}(x)$. Here we have

$$\hat{E}(t) = \begin{cases} 0 & \text{for } t = 0 \\ \frac{1}{\pi i t}(\hat{J}(t) - 1) & \text{for } t \neq 0. \end{cases}$$

Hence

$$\begin{aligned}\psi * j_H(x) - \psi(x) &= \sum_{h=-\infty, h\neq 0}^{\infty} -\frac{1}{2\pi i h}\left(\hat{J}_{H+1}(h) - 1\right) e(hx) \\ &= -\frac{1}{2H+2}\sum_{h=-\infty}^{\infty} \hat{E}_{H+1}(h) e(hx) \\ &= -\frac{1}{2H+2}\sum_{h=-\infty}^{\infty} E_{H+1}(h)\end{aligned}$$

together with Lemma 1.22 gives

$$|\psi * j_H(x) - \psi(x)| \leq \frac{1}{2H+2}\sum_{h=-\infty}^{\infty} K_{H+1}(x+h) = \frac{1}{2H+2} k_H(x).$$

This proves (1.57).□

Now denote V_f the total variation

$$V_f([a,b]) = \sup \sum_{i=0}^{n-1} |f(x_{i+1}) - f(x_i)| \tag{1.59}$$

of f on $[a,b]$, where the supremum is taken over all partitions $a = x_0 < x_1 < \cdots < x_n = b$. V_f is of course a measure on $\mathbf{R}$. If f has period 1 then V_f has period 1, too.

After these preparations we are now able to prove

Theorem 1.25 *Let f be a real function of bounded variation and period 1 satisfying $|2f(x_0) - f(x_0-) - f(x_0+)| \leq |f(x_0-) - f(x_0+)|$ for all $x_0 \in [0,1]$. Then the trigonometric polynomials $f * j_H(x)$ and $dV_f * k_H(x)$ are at most of degree H and satisfy*

$$|f(x) - f * j_H(x)| \leq \frac{1}{2H+2} dV_f * k_H(x). \tag{1.60}$$

Proof. First we assume that $f(x_0) = (f(x_0-) + f(x_0+))/2$. Differentiating

$$\psi * j_H(x) = \sum_{h=-H, h\neq 0}^{H} -\frac{1}{2\pi i h}\hat{J}_{H+1}(h) e(hx)$$

we get

$$\frac{d}{dx}\psi * j_H(x) = 1 - j_H(x).$$

Hence

$$\begin{aligned}&\int_{-1/2}^{1/2} f(x-t) d(\psi * j_H(t) - \psi(t)) \\ &= \int_{-1/2}^{1/2} f(x-t)(1 - j_H(t))\, dt - \int_{-1/2}^{1/2} f(x-t)\, d\psi(t) \\ &= f(x) - f * j_H(x)\end{aligned}$$

at continuity points x of f. After partial integration and an application of Lemma 1.24 we get

$$\begin{aligned}
|f(x) - f * j_H(x)| &= \left|\int_{-1/2}^{1/2} (\psi * j_H(x-t) - \psi(x-t))\, df(t)\right| \\
&\leq \int_{-1/2}^{1/2} |(\psi * j_H(x-t) - \psi(x-t))|\, dV_f(t) \\
&\leq \frac{1}{2H+2}\int_{-1/2}^{1/2} k_H(x-t)\, dV_f(t) \\
&= \frac{1}{2H+2} k_H * dV_f(x).
\end{aligned}$$

Due to the continuity of the right side of this inequality and due to the assumption about the points of discontinuity this inequality remains true even there.□

Corollary 1.26 *Let $\mathbf{x}_1, \ldots, \mathbf{x}_N$ be a finite sequence of points in the k-dimensional space $\mathbf{R}^k$ and H an arbitrary positive integer. Then*

$$D_N(\mathbf{x}_n) \leq \left(1 + \frac{1}{H+1}\right)^k - 1 \tag{1.61}$$
$$+ \sum_{0<\|\mathbf{h}\|_\infty \leq H} \left(\frac{1}{(H+1)^{k-\alpha(\mathbf{h})}}\left(1 + \frac{1}{H+1}\right)^{\alpha(\mathbf{h})} + \frac{1}{r(\pi\mathbf{h})}\right)\left|\frac{1}{N}\sum_{n=1}^{N} e(\mathbf{h}\cdot\mathbf{x}_n)\right|,$$

where $\alpha(\mathbf{h})$ denotes the number of zeros $\sum_{l=1}^{k}\delta_{h_l 0}$ of $\mathbf{h} = (h_1, \ldots, h_k) \in \mathbf{Z}^k$.

Proof. Applying (1.60) to $f = \chi_I$ for an interval $I \subseteq \mathbf{R}$ with $\lambda_1(I) \leq 1$ we get

$$\begin{aligned}
|\chi_I(x) - \chi_I * j_H(x)| &\leq \frac{1}{2H+2} dV_{\chi_I} * k_H(x) \\
&= \frac{1}{H+1}\sum_{h=-H}^{H} \hat{K}_{H+1}(h) C_h e(hx),
\end{aligned}$$

where $C_h = \frac{1}{2}\int_0^1 e(-hx)\, dV_{\chi_I}(x)$ satifies $|C_h| \leq 1$. Using

$$\left|\prod_{j=1}^{k} b_j - \prod_{j=1}^{k} a_j\right| \leq \sum_{\emptyset \neq J \subseteq \{1,\ldots,k\}} \prod_{j \notin J} |a_j| \prod_{j \in J} |b_j - a_j|$$

we immediately get

$$\begin{aligned}
\left|\prod_{j=1}^{k} \chi_{I_j}(x_j) - \prod_{j=1}^{k} f_j(x_j)\right| &\leq \sum_{\emptyset \neq J \subseteq \{1,\ldots,k\}} \prod_{j \in J} |\chi_{I_j}(x_j) - f_j(x_j)| \\
&= \prod_{j=1}^{k} (1 + |\chi_{I_j}(x_j) - f_j(x_j)|) - 1,
\end{aligned}$$

where $f_j(x_j) = \chi_{I_j} * j_H(x_j) = \sum_{h=-H}^{H} \hat{\chi}_{I_j}(x_j)\hat{J}_{H+1}(h)e(hx_j)$.

Now set $I = \prod_{j=1}^{k} I_j$ and $\mathbf{x}_n = (x_n^{(1)}) \ldots, x_n^{(k)}$.

Then we have

$$\left|\sum_{n=1}^{N} \chi_I(\mathbf{x}_n) - N\lambda_k(I)\right| \le \left|\sum_{n=1}^{N}\left(\prod_{j=1}^{k} f_j(x_n^{(j)}) - \lambda_k(I)\right)\right|$$
$$+\sum_{n=1}^{N}\left(\prod_{j=1}^{k}\left(1+\frac{1}{H+1}\sum_{h_j=-H}^{H}\hat{K}_{H+1}(h_j)C_{h_j j}e(h_j x_n^{(j)})\right)-1\right)$$
$$= \left|\sum_{n=1}^{N}\left(\prod_{j=1}^{k}\left(\sum_{h_j=-H}^{H}\hat{\chi}_{I_j}(h_j)\hat{J}_{H+1}(h_j)e(h_j x_n^{(j)})\right)-\lambda_k(I)\right)\right|$$
$$+\sum_{n=1}^{N}\left(\prod_{j=1}^{k}\left(1+\frac{1}{H+1}\sum_{h_j=-H}^{H}\hat{K}_{H+1}(h_j)C_{h_j j}e(h_j x_n^{(j)})\right)-1\right)$$
$$= \left|\sum_{0<\|\mathbf{h}\|_\infty\le H}\prod_{j=1}^{k}\left(\hat{\chi}_{I_j}(h_j)\hat{J}_{H+1}(h_j)\right)\sum_{n=1}^{N}e(\mathbf{h}\cdot\mathbf{x}_n)\right|$$
$$+\sum_{n=1}^{N}\left(\prod_{j=1}^{k}\left(1+\frac{1}{H+1}+\frac{1}{H+1}\sum_{h_j=-H,h_j\ne 0}^{H}\hat{K}_{H+1}(h_j)C_{h_j j}e(h_j x_n^{(j)})\right)-1\right),$$

where $C_{h_j j} = \frac{1}{2}\int_0^1 e(-h_j x)\,dV_{\chi_{I_j}}(x)$. Using the inequalities $|\hat{\chi}_I(h)| = |\frac{\sin h\pi\lambda_1(I)}{h\pi}| \le \frac{1}{|h|\pi}$ for $h \ne 0$, $|\hat{K}_{H+1}(h)| \le 1$, and $|\hat{J}_{H+1}(h)| \le 1$ we directly get (1.61).□

Proof of Theorem 1.21 We only have to apply (1.61) and to use the estimates

$$\left(1+\frac{1}{H+1}\right)^k - 1 \le \left(\frac{3}{2}\right)^k \frac{2}{H+1}$$

and

$$\frac{1}{(H+1)^{k-\alpha(\mathbf{h})}}\left(1+\frac{1}{H+1}\right)^{\alpha(\mathbf{h})} + \frac{1}{\pi^{k-\alpha(\mathbf{h})}r(\mathbf{h})}$$
$$\le \frac{1}{r(\mathbf{h})}\left(\frac{1}{\pi^{k-\alpha(\mathbf{h})}} + \left(1+\frac{1}{H+1}\right)^{\alpha(\mathbf{h})}\right)$$
$$\le \frac{1}{r(\mathbf{h})}\left(\frac{1}{\pi}+\left(\frac{3}{2}\right)^{k-1}\right) \le \frac{1}{r(\mathbf{h})}\left(\frac{3}{2}\right)^k.$$

This completes the proof of Theorem 1.21.□

Remark. It should be noted that the constant $(3/2)^k$ can be sharpened by restricting H. Let $q > 1$. Then we have for $H > \frac{q}{q-1}$

$$D_N(\mathbf{x}_n) \le \frac{q^k - 1}{q-1} \left(\frac{1}{H+1} + \sum_{0<\|h\|_\infty \le H} \frac{1}{r(\mathbf{h})} \left| \frac{1}{N} \sum_{n=1}^{N} e(\mathbf{h}\cdot\mathbf{x}_n) \right| \right). \tag{1.62}$$

1.2.3 LEVEQUE's Inequality and Diaphony

In 1965 LEVEQUE [1066] proved the following inequality that is quite similar to ERDŐS-TURÁN-KOKSMA's inequality (Theorem 1.21).

Theorem 1.27 (LEVEQUE's Inequality) *The discrepancy $D_N(x_n)$ of N real numbers $x_1, \ldots, x_N \in \mathbf{R}$ can be estimated by*

$$D_N(x_n) \le \left(\frac{6}{\pi^2} \sum_{h=1}^{\infty} \frac{1}{h^2} \left| \frac{1}{N} \sum_{n=1}^{N} e(hx_n) \right|^2 \right)^{\frac{1}{3}}. \tag{1.63}$$

Remark. It is known that the constant $6/\pi^2$ and the exponent $\frac{1}{3}$ are best possible (see KUIPERS and NIEDERREITER [983]).

A higher dimensional analogon has been proved by STEGBUCHNER [1737] using ERDŐS-TURÁN-KOKSMA's inequality.

Theorem 1.28 *The discrepancy $D_N(\mathbf{x}_n)$ of N points $\mathbf{x}_1, \ldots, \mathbf{x}_N$ in the k-dimensional space $\mathbf{R}^k$ can be estimated by*

$$D_N(\mathbf{x}_n) \le 6 \left(\frac{3}{2} \right)^k \left(\sum_{0\ne\mathbf{h}\in\mathbf{Z}^k} \frac{1}{r(\mathbf{h})^2} \left| \frac{1}{N} \sum_{n=1}^{N} e(\mathbf{h}\cdot\mathbf{x}_n) \right|^2 \right)^{\frac{1}{k+2}}. \tag{1.64}$$

Proof. Set

$$F_N = \left(\sum_{0\ne\mathbf{h}\in\mathbf{Z}^k} \frac{1}{r(\mathbf{h})^2} \left| \frac{1}{N} \sum_{n=1}^{N} e(\mathbf{h}\cdot\mathbf{x}_n) \right|^2 \right)^{\frac{1}{2}}$$

and assume that $F_N \le 3^{-k/2}$. If $F_N > 3^{-k/2}$ then (1.64) is surely satisfied because the discrepancy D_N is always bounded above by 1. Set $\hat{H} = (3^{k/2} F_N)^{-2/(k+2)} \ge 1$ and $H = [\hat{H}] \ge 1$. By CAUCHY-SCHWARZ's inequality

$$\sum_{0<\|\mathbf{h}\|_\infty \le H} 1 \cdot \frac{1}{r(\mathbf{h})} \left| \frac{1}{N} \sum_{n=1}^{N} e(\mathbf{h}\cdot\mathbf{x}_n) \right| \le (2H+1)^{k/2} F_N.$$

Thus ERDŐS-TURÁN-KOKSMA's inequality yields

$$D_N(\mathbf{x}_n) \le \left(\frac{3}{2} \right)^k \left(\frac{1}{H+1} + (2H+1)^{k/2} F_N \right)$$

$$\leq \left(\frac{3}{2}\right)^k \left(\frac{1}{\hat{H}} + 3^{k/2}\hat{H}^{k/2}F_N\right)$$
$$= 2\left(\frac{3}{2}\right)^k 3^{\frac{k}{k+2}} F_N^{\frac{2}{k+2}},$$

which proves (1.64).□

It has been become usual (see ZINTERHOF [1997]) to introduce a new notation for the infinite sum that appears in LEVEQUE's inequality.

Definition 1.29 (Version A) *Let* $\mathbf{x}_1, \ldots, \mathbf{x}_N$ *be a finite sequence of points in the* k*-dimensional space* $\mathbf{R}^k$*. The number*

$$F_N(\mathbf{x}_n) = \left(\sum_{0 \neq \mathbf{h} \in \mathbf{Z}^k} \frac{1}{r(\mathbf{h})^2} \left|\frac{1}{N}\sum_{n=1}^{N} e(\mathbf{h}\cdot\mathbf{x}_n)\right|^2\right)^{\frac{1}{2}} \tag{1.65}$$

is called diaphony *of the given sequence.*

Set $g(x) = 1 - \frac{\pi^2}{6} + \frac{\pi^2}{2}(1-2\{x\})^2$, $x \in \mathbf{R}$, and $G_k(\mathbf{x}) = G_k(x_1, \ldots, x_k) = \prod_{j=1}^{k} g(x_j) - 1$. Then it is easy to verify that $G(\mathbf{x})$ has the absolutely convergent FOURIER expansion

$$G_k(\mathbf{x}) = \sum_{0 \neq \mathbf{h} \in \mathbf{Z}^k} \frac{e(\mathbf{h}\cdot\mathbf{x})}{r(\mathbf{h})^2}. \tag{1.66}$$

Therefore we can define the diaphony of sequences of points more directly.

Definition 1.30 (Version B) *Let* $\mathbf{x}_1, \ldots, \mathbf{x}_N$ *be a finite sequence of points in the* k*-dimensional space* $\mathbf{R}^k$*. The number*

$$F_N(\mathbf{x}_n) = \left(\frac{1}{N^2}\sum_{n=1}^{N}\sum_{m=1}^{N} G_k(\mathbf{x}_n - \mathbf{x}_m)\right)^{\frac{1}{2}} \tag{1.67}$$

is called diaphony *of the given sequence.*

So we can reformulate LEVEQUE's inequality to

Theorem 1.31 *For any sequence of points* $\mathbf{x}_1, \ldots, \mathbf{x}_N \in \mathbf{R}^k$ *we have*

$$D_N(\mathbf{x}_n) \leq 6\left(\frac{3}{2}\right)^k F_N(\mathbf{x}_n)^{\frac{2}{k+2}}, \tag{1.68}$$

where the constant $6\left(\frac{3}{2}\right)^k$ *can be sharpened to* $3^{1/2}/\pi$ *in the one dimensional case* $k = 1$.

On the other hand, we can use KOKSMA-HLAWKA's inequality to prove

Theorem 1.32 *For any sequence of points* $\mathbf{x}_1, \dots, \mathbf{x}_N \in \mathbf{R}^k$ *we have*

$$F_N(\mathbf{x}_n) \le 11^{\frac{k}{2}} D_N(\mathbf{x}_n)^{\frac{1}{2}}. \tag{1.69}$$

Proof. We apply KOKSMA-HLAWKA's inequality for $f(\mathbf{x}) = G_k(\mathbf{x} - \mathbf{x}_m) + 1 = \prod_{j=1}^k g(x_j - x_{mj})$. Trivially $V_g = \pi^2 < 10$ and so we get

$$\begin{aligned} \left| \frac{1}{N} \sum_{n=1}^N G_k(\mathbf{x}_n - \mathbf{x}_m) \right| &\le \sum_{j=1}^k \binom{k}{j} 10^j D_N^*(\mathbf{x}_n - \mathbf{x}_m) \\ &\le 11^k D_N(\mathbf{x}_n). \end{aligned}$$

Thus $F_N^2 \le 11^k D_N$.□

So we have found a new condition to characterize u.d. sequences.

Theorem 1.33 *A sequence* $(\mathbf{x}_n)_{n\ge1}$ *in the k-dimensional space* $\mathbf{R}^k$ *is u.d. mod 1 if and only if*

$$\lim_{N\to\infty} F_N(\mathbf{x}_n) = 0. \tag{1.70}$$

Like the discrepancy the diaphony can be used to estimate the difference between the mean value $\frac{1}{N}\sum_{n=1}^N f(\mathbf{x}_n)$ and the integral $\int_{\mathbf{T}^k} f\, d\lambda_k$. But it is natural that such estimates will only work when something is known about the FOURIER expansion of $f(\mathbf{x})$.

Definition 1.34 *Let* $\mathbf{E}_\alpha^k(C)$ *be the set of all periodic functions* $f : \mathbf{R}^k \to \mathbf{C}$ *with* FOURIER *expansion*

$$f(\mathbf{x}) = \sum_{\mathbf{h}\in\mathbf{Z}^k} c_{\mathbf{h}} e(\mathbf{h}\cdot\mathbf{x})$$

whose FOURIER *coefficients* $c_{\mathbf{h}}$ *satisfy*

$$|c_{\mathbf{h}}| \le \frac{C}{r(\mathbf{h})^\alpha}. \tag{1.71}$$

Theorem 1.35 *Let* $f \in \mathbf{E}_\alpha^k(C)$ *with* $\alpha > 3/2$ *and* $\mathbf{x}_1, \dots, \mathbf{x}_N \in \mathbf{R}^k$. *Then we have*

$$\left| \frac{1}{N}\sum_{n=1}^N f(\mathbf{x}_n) - \int_{\mathbf{T}^k} f\, d\lambda_k \right| \le C \left(\frac{4\alpha - 4}{2\alpha - 3} \right)^{k/2} F_N(\mathbf{x}_n). \tag{1.72}$$

Proof. We directly get

$$\begin{aligned} \left| \frac{1}{N}\sum_{n=1}^N f(\mathbf{x}_n) - \int_{\mathbf{T}^k} f\, d\lambda_k \right| &\le C \sum_{0\ne\mathbf{h}\in\mathbf{Z}^k} \frac{1}{r(\mathbf{h})^\alpha} \left| \frac{1}{N}\sum_{n=1}^N e(\mathbf{h}\cdot\mathbf{x}_n) \right| \\ &\le C F_N \left(\sum_{0\ne\mathbf{h}\in\mathbf{Z}^k} \frac{1}{r(\mathbf{h})^{2(\alpha-1)}} \right)^{\frac{1}{2}} \\ &\le C(\zeta(2(\alpha-1)))^{k/2} F_N \\ &\le C \left(2\frac{2(\alpha-1)}{2(\alpha-1)-1} \right)^{\frac{k}{2}} F_N, \end{aligned}$$

where we have used CAUCHY-SCHWARZ's inequality and the trivial estimate

$$\zeta(s) = \sum_{n=1}^{\infty} \frac{1}{n^s} \le 1 + \int_1^{\infty} \frac{dx}{x^s} = \frac{s}{s-1} \quad (s > 1)$$

for RIEMANN's zeta function.□

Bearing in mind that the diaphony $F_N(\mathbf{x}_n)$ can be defined by (1.67) it is not surprising that the integral $\int_{\mathbf{T}^k} f\, d\lambda_k$ can be approximated essentially better by using N^2 points $\mathbf{x}_n - \mathbf{x}_m$, $n, m = 1, \dots, N$.

Theorem 1.36 *Let $f \in \mathbf{E}_\alpha^k(C)$ with $\alpha > 1$ and $\mathbf{x}_1, \dots, \mathbf{x}_N \in \mathbf{R}^k$. Then we have*

$$\left| \frac{1}{N} \sum_{n=1}^{N} f(\mathbf{x}_n) - \int_{\mathbf{T}^k} f\, d\lambda_k \right| \le \begin{cases} C F_N^2(\mathbf{x}_n) & \text{for } \alpha \ge 2 \\ C \left(\frac{8}{\varepsilon}\right)^{k(2-\alpha)} F_N^{2\alpha-2-\varepsilon(2-\alpha)} & \text{for } \frac{3}{2} \le \alpha \le 2 \\ C \left(\frac{4}{\varepsilon(\alpha-1)}\right)^k F_N^{2\alpha-2-\varepsilon(\alpha-1)} & \text{for } 1 < \alpha \le \frac{3}{2}. \end{cases} \tag{1.73}$$

For a *proof* see ZINTERHOF and STEGBUCHNER [2000].

Notes

A first proof of Weyl's criterion (Theorem 1.19) is given in the classical paper [1953]. However, Hardy and Littlewood have used exponential sums for proving uniform distribution of sequences of type αn^k [734] even earlier. For further historic remarks and references see Hlawka and Binder [837]. A Tauberian theorem related to Weyl's criterion is given by Vaaler [1884, 1885].

There exist similar criteria for other notions of uniform distribution (well distribution, weakly well distribution) and its generalizations to abstract spaces; see Section 2.4 and Kuipers and Niederreiter [983].

The Erdős-Turán-Koksma inequality (Theorem 1.21) can be viewed as a variant of Berry-Esseen-type bounds, which translate deviations of Fourier transforms to deviations of the corresponding distribution functions, see for instance Niederreiter and Philipp [1351, 1352], Niederreiter [1288] and the Notes of Section 1.5. The original theorem in the one-dimensional case is due to Erdős and Turán [569, 570] and in the higher dimensional case due to Koksma [942]. Vaaler's approximation kernel (1.48) was first used by Cochrane [371] to obtain discrepancy bounds for higher dimensions. Grabner [674] extended this method to establish bounds for the spherical cap discrepancy; see also Grabner and Tichy [683]. Recently, Holt [842] proved a version of the Erdős-Turán-Koksma inequality with respect to balls on the torus. For related results and important analytic tools see Holt and Vaaler [843] and Vaaler [1887, 1888].

Ruzsa [1583, 1584] investigated how sharp the discrepancy bound provided by the Erdős-Turán inequality (in one dimension) is. Let

$$B = \min_H \left(\frac{1}{H} + \sum_{h=1}^{H} \left| \frac{1}{N} \sum_{n=1}^{N} e(hx_n) \right| \right).$$

Then Ruzsa showed that $D_N \gg B^{3/2}$ and that there are sequences (x_n) with $D_N \ll B^{3/2}$. The proof of the second part is essentially based on a lower bound for the L^2-norm of Rudin-Shapiro polynomials.

Hellekalek [768, 769, 770] established an analogon of Erdős-Turán inequality replacing the orthogonal system of exponential functions by the system of Walsh and Haar functions; the corresponding discrepancy bound is important for analyzing random number generators (see Section 3.4). Similar results are due to Sloss and Blyth [1698].

Fleischer and Stegbuchner [612, 613] proved bounds for the diophany in terms of the discrepancy; see also [1737, 1738] and Zinterhof and Stegbuchner [2000] and Zinterhof [1997, 1998]. A sharp lower discrepancy bound in terms of exponential sums (with respect to weighted means) is due to Horbowicz and Niederreiter [1347, 847]. Using results of Montgomery and Niederreiter [1199] the authors supercede the bound which can be proved via Koksma's inequality.

Obviously, the Erdős-Turán-Koksma inequality can be used to estimate the discrepancy of special sequences. For some types of sequences (e.g. digital sequences, see Section 1.4) this yields satisfactory results. In general an application of the Erdős-Turán-Koksma inequality leads to non optimal discrepancy bounds (e.g. for the $(n\alpha)$-sequence). Clearly, the standard tools from analytic number theory can be applied to estimate exponential sums. The method of Van der Corput is shortly described in Kuipers and Niederreiter [983]. A detailed discussion, especially of the method of exponential pairs and its applications in analytic number theory, can be found in Graham and Kolesnik's book [687]. For an application of the Erdős-Turán-Koksma inequality to the circle problem we refer to Nowak [1378] where a short proof of Sierpiński's remainder term is provided. Laczkovich [1000] used special sequences, a version of Erdős-Turán-Koksma's inequality and Van der Corput's method for Tarski's circle squaring problem. The problem is to decide whether a disc can be decomposed into finitely many parts which can be rearranged to obtain a partition of a square, see Tarski [1781]. The problem was motivated by the well-known Banach-Tarski theorem stating that in $\mathbf{R}^3$ any two bounded sets with non-empty interior are equidecomposable. On the other hand, in the plane a disc and a square can be equidecomposable only of they have the same area; for the history on Tarski's problem see Wagon [1933]. Laczkovich's proof is essentially based on the following criterion involving the discrepancy of a special sequence in the plane.

Theorem 1.37 *Let ψ be a non-negative function defined on $\mathbf{N}$ such that*

$$\sum_{k=0}^{\infty} \psi(2^k)/2^k < \infty.$$

Let $H_i (i = 1,2)$ be measurable subsets of $[0,1]^2$ of equal positive Lebesgue measure and let $u, x, y \in \mathbf{R}^2$. Let $s_N(u, x, y, H_i)$ be the sequence of points $\{u + nx + ky\}$, $0 \le n, k \le N$ that are contained in $H_i (i = 1,2)$ and let $D_N\left(s^{(i)}(u,x,y)\right)$ denote its discrepancy. Suppose that there are points $x, y \in \mathbf{R}^2$ such that

(i) the vectors $x, y, (0,1), (1,0)$ are linearly independent over the rationals, and

(ii) $D_N(s^{(i)}(u,x,y)) \le \frac{\psi(N)}{N^2}$ holds for every $u \in \mathbf{R}^2, N \in \mathbf{N}, i = 1,2$.

Then H_1 and H_2 are equidecomposable (via translations).

In order to estimate the involved discrepancies, exponential sums (Van der Corput's method) and a variant of the Erdős-Turán inequality are applied. The existence of the linearly independent points x, y is proved via metric arguments following Schmidt; for refinements see Kruse [965]. Further applications of discrepancy theory to the decomposition of sets are obtained by Laczkovich [1001]. For results on "uniformly spread" discrete sets see Laczkovich [1002]; these results allow estimating the number of lattice cubes that are necessary to construct a finite union of unit cubes by the operations of disjoint unions and proper differences.

In [10] Akita, Goto and Kano gave an explicit construction of a trigonometric series $s(x) = \sum(a_n \cos nx + b_n \sin nx)$ that diverges almost everywhere but $s(x) \in L^{2+\epsilon}$ $(\epsilon > 0)$. For instance, one can take

$$a_n = \frac{\cos(2\pi\alpha n \log n)}{\sqrt{n}}, \quad b_n = \frac{\sin(2\pi\alpha n \log n)}{\sqrt{n}}$$

with $\alpha = \frac{1}{\log 2}$. The proof is based on Van der Corput's method, see also [11].

In a series of papers the distribution behaviour mod 1 of the sequence (p^α), as p runs trough the sequence of prime values and $\alpha > 0$ is a non-integral number, was investigated; see for instance Vaughan [1902].

More precisely, the discrepancy

$$D(N) = \sup_{0\le x\le 1} \left| \sum_{p\le N, \{p^\alpha\}\le x} 1 - \pi(N)x \right|$$

was considered, where $\pi(N)$ denotes as usual the number of primes $\le N$. In the case $\alpha > 1$ Leitmann [1058] showed that $D(N) = \mathcal{O}(N^{1-\delta})$ as $N \to \infty$, for some $\delta > 0$. Earlier, Vinogradov obtained similar results; see [1912, 1913] and as further references on Vinogradov's method [1911] and [904]. For a special exponential sum see also Ismatullaev [861]. In [47, 48] Arkhipov, Karatsuba and Chubarikov give an excellent survey on multidimensional sums and various kinds of applications to the diophantine equations and distribution problems for polynomial sequences, see also [49], Pustylnikov [1491, 1490], and Chubarikov [365]. Korobov and Mitkin [963] provided lower bounds for trigonometrical sums $\sum_{x=1}^{p} e((a_1x + \ldots + a_nx^n)/p)$, p prime, $p > n \ge 3a_i \in \mathbf{Z}$. For further results of Korobov on trigonometrical sums and their applications we refer to [959, 960, 961]. Metric properties of exponential sums are investigated by Judin [878]. An application of Vinogradov's method to estimate the Hausdorff dimension related to a certain diophantine inequality is due to Bernik and Pereverseva [192].

Vinogradov's method was refined by Baker and Kolesnik [96], who obtained (in the bound for the above discrepancy) $\delta = (15000\alpha^2)^{-1}$ for $N > N_1(\alpha)$. For $\alpha = 3/2$ the estimate can be improved to $D(N) = O(N^{157/168+\epsilon})$. This sharpens the work of Fomenko and Golubeva [614].

Quite different methods have to be used in the case $0 < \alpha < 1$; see Balog [98], Harman [738] and Kaufman [914]. Here we mention also Harman [746], Glazunov [658], Gritsenko [699], Nair and Perelli [1243], Stux [1760] and Wolke [1971]. Further results on the distribution behaviour of prime number sequences are shown in Allakov [15], Baker and Harman [95], Harman [736], Ghosh [656] and Balog [99]; see also Schoißengeier [1649], Policott [1440] Rhin [1536], Deshouillers [453], Vaughan [1902] and Liu and Tsang [1091].

In [1537, 1538, 1535, 1539] Rhin investigated sequences $f(n)$ and $f(p)$, where f is an entire function; see also Rauzy [1505, 1513]. For improvements we refer to Baker [86, 88]. Continuous analogues of such results are discussed in Section 2.3. Extensions to primes in arithmetic progressions are due to Wodzak [1969].

Harman [736] improved bounds for exponential sums over primes of the Vinogradov type and derived results for the fractional parts of $f(p)$ where f is a real polynomial with irrational leading coefficient. The author continued these investigations in [740]. Heath-Brown [758] gave an excellent survey on Weyl and Hua type inequalities. For a recent contribution in this direction we refer to Pustylnikov [1492]. Goto and Kano [668] investigated the sequence $\alpha f(p)$ modulo 1 for $\alpha \ne 0$ and f satisfying suitable analytic conditions; see further Goto and Kano [670] and Too [1867].

Balog [100] showed for any irrational α that $\|\alpha p\| \le p^{-\frac{1}{3}+\epsilon}$ holds for infinitely many primes p. This improves an earlier result of Harman [739]. A further improvement is due to Jia [875], who reduced the exponent to $-4/13$. For another contribution to diophantine approximation by prime numbers we refer to Harman [745, 750, 749, 751].

Simultaneous diophantine approximation using primes was investigated by Balog and Friedlander [101]. Tolev [1866] considered the k-dimensional sequence $p^{\alpha_1}, \ldots, p^{\alpha_k}$ where p runs over the set of prime numbers and $\alpha_1, \ldots, \alpha_k$ are distinct positive numbers < 1. For the discrepancy

$$D(N) = \sup_{\substack{0\le x_j\le 1\\ j=1,\ldots,k}} \left| \sum_{\substack{p\le N\\ \{p^{\alpha_j}\}\le x, j=1,\ldots,k}} 1 - \pi(N)x_1\ldots x_k \right|$$

Tolev obtained the bound

$$D(N) = O\left(N^{1-\delta}\log^{k+9} N\right) \quad \text{with} \quad \delta = \frac{1}{3}\min_{1\le i<j\le k}\left(\frac{1}{4}, 1-\alpha_j, \alpha_i, |\alpha_i - \alpha_j|\right).$$

The main tools of the proof are the Erdős-Turán-Koksma inequality, a partial summation technique for estimating exponential sums and an application of zero-density estimates for the Riemann zeta function. A combination of Tolev's method [1866] with the Baker-Kolesnik-method is worked out by Srinivasan and Tichy [1733]:

Theorem 1.38 *Let $\alpha_1, \ldots, \alpha_k$ be distinct positive non-integral numbers. Then the discrepancy $D(N)$ of the sequence $(p^{\alpha_1}, \ldots, p^{\alpha_k})$ satisfies an estimate $D(N) = O(N^{1-\delta})$ as $N \to \infty$, for some positive δ.*

Further results concerning the distribution of prime number sequences and estimates for related Klosterman sums are due to Karatsuba [900, 901, 903]. In [902] the distribution of inverses in residue class rings is investigated.

1.3 Lower Bounds

1.3.1 ROTH's Theorem

The aim of this section is to give lower bounds for the discrepancy. A very easy first observation gives

Proposition 1.39 *For any points $\mathbf{x}_1, \ldots, \mathbf{x}_N \in \mathbf{R}^k$ we have*

$$D_N(\mathbf{x}_n) \geq \frac{1}{N}. \tag{1.74}$$

Proof. Consider $\mathbf{x}_1 = (x_1^{(1)}, \ldots, x_1^{(k)})$ and $I = [x_1^{(1)}, x_1^{(1)} + \varepsilon) \times \cdots \times [x_1^{(k)}, x_1^{(k)} + \varepsilon)$ for any $\varepsilon > 0$. Thus

$$D_N(\mathbf{x}_n) \geq \frac{1}{N} - \varepsilon^k$$

and so (1.74) follows.□

On the other hand the finite sequence $x_n = \frac{n}{N}$, $n = 1, \ldots, N$ satisfies $D_N(\mathbf{x}_n) = \frac{1}{N}$. But sequences of this kind can only exist in the one-dimensional case. The following theorem is due to ROTH [1567].

Theorem 1.40 *Let $k \geq 2$. Then the discrepancy $D_N(\mathbf{x}_n)$ of N points $\mathbf{x}_1, \ldots, \mathbf{x}_N$ in the k-dimensional space $\mathbf{R}^k$ is bounded below by*

$$D_N(\mathbf{x}_n) \geq D_N^*(\mathbf{x}_n) \geq D_N^{(2)}(\mathbf{x}_n) \geq \frac{1}{2^{4k}} \frac{1}{((k-1)\log 2)^{\frac{k-1}{2}}} \frac{\log^{\frac{k-1}{2}} N}{N}. \tag{1.75}$$

This theorem is the best known for dimensions $k > 3$. A *proof* can be found in Kuipers and Niederreiter [983]. The method there uses the orthogonality relation of RADEMACHER functions.

It is interesting that there exists a completely different proof due to BECK [143] using FOURIER transform techniques. It provides a worse constant than ROTH's method. However, it claims more than ROTH's theorem.

Theorem 1.41 *Let $k \geq 2$ and $\mathbf{x}_1, \ldots, \mathbf{x}_N$ an arbitrary finite sequence of $N > 1$ points in the k-dimensional space $\mathbf{R}^k$. Then there is some k-dimensional cube $Q \subseteq \mathbf{R}^k/\mathbf{Z}^k$ with sides parallel to the axes satisfying*

$$\left| \frac{1}{N} \sum_{n=1}^{N} \chi_Q(\mathbf{x}_n) - \lambda_k(Q) \right| \geq \frac{c_k (\log N)^{\frac{k-1}{2}} - d_k}{N},$$

where

$$c_k = \frac{(\log 2)^{\frac{2-k}{4}} 2^{-\frac{7}{2}} 3^{\frac{k}{2}} \pi^{-\frac{k}{4}}}{((k-1)!)^{\frac{1}{2}} k^{\frac{k}{4}} (2k+1)^{\frac{k-1}{2}} (6k+1)^{\frac{k}{2}}}$$

and

$$d_k = k^{\frac{1}{2}} 2^{k+2} 3^{\frac{k-1}{2}}.$$

In order to prove Theorem 1.41 we will need the following notations and some preparatory Lemmata.

Let $\mathbf{y}_1, \ldots, \mathbf{y}_N$ be $N > 1$ points in the unit cube $\mathbf{U}_k = [0,1)^k$ and

$$\nu = \frac{1}{N} \sum_{n=1}^{N} \delta_{\mathbf{y}_n} - \mu,$$

where $\delta_{\mathbf{y}}$ denotes the DIRAC measure concentrated on $\mathbf{y}$ and μ the restriction of the k-dimensional LEBESGUE measure to $\mathbf{U}_k$.

For any real number $\rho \in (0,1]$ and $g \in \mathrm{L}^2(\mathbf{R}^k)$ set

$$g_\rho(\mathbf{x}) = g(\rho^{-1}\mathbf{x})$$

and consider the function

$$F_g = g * (d\nu),$$

where * denotes the usual convolution. More explicitly we have

$$\begin{aligned} F_g(\mathbf{x}) &= \int_{\mathbf{R}^k} g(\mathbf{x}-\mathbf{y}) d\nu(\mathbf{y}) \\ &= \frac{1}{N} \sum_{n=1}^{N} g(\mathbf{x}-\mathbf{y}_n) - \int_{\mathbf{R}^k} g(\mathbf{x}-\mathbf{y})\, d\mu(\mathbf{y}). \end{aligned}$$

Next set

$$\Delta_1(g) = \int_{\mathbf{R}^k} |F_g(\mathbf{x})|^2 d\mathbf{x} \quad \text{and} \quad \Delta_2(g) = \int_0^1 \int_{\mathbf{R}^k} |F_{g_\rho}(\mathbf{x})|^2 d\mathbf{x}\, d\rho.$$

By PARSEVAL's theorem for FOURIER transforms we have

$$\Delta_1(g) = \int_{\mathbf{R}^k} |\hat{g}(\mathbf{t})|^2 |\hat{\nu}(\mathbf{t})|^2\, d\mathbf{t} \tag{1.76}$$

and

$$\Delta_2(g) = \int_{\mathbf{R}^k} \left(\int_0^1 |\hat{g}_\rho(\mathbf{t})|^2 d\rho \right) |\hat{\nu}(\mathbf{t})|^2\, d\mathbf{t}. \tag{1.77}$$

As usual $\hat{f}$ denotes the FOURIER transform

$$\hat{f}(\mathbf{t}) = (2\pi)^{-\frac{k}{2}} \int_{\mathbf{R}^k} f(\mathbf{x}) e^{-i\mathbf{x}\cdot\mathbf{t}}\, d\mathbf{x}$$

of f.

Given $\mathbf{l} = (l_1, \ldots, l_k) \in \mathbf{Z}^k$, let

$$h_{\mathbf{l}}(\mathbf{x}) = \prod_{j=1}^{k} e^{-\frac{1}{2}l_j^2 x_j^2}.$$

Its FOURIER transform is given by

$$\hat{h}_{\mathbf{l}}(\mathbf{t}) = \prod_{j=1}^{k} \frac{1}{l_j} e^{-t_j^2/(2l_j^2)}$$

$(\mathbf{t} = (t_1, \ldots, t_k))$.

Finally, let L be the integer power of 2 satisfying $4(2\pi)^{\frac{k}{2}} N \le L < 8(2\pi)^{\frac{k}{2}} N$ and the set

$$Z(L,m) = \left\{ \mathbf{l} = (l_1, \ldots, l_k) : l_j = 2^{s_j} \ge m, s_j \in \mathbf{Z}, \prod_{j=1}^{k} l_j = L \right\},$$

where $m = N^{2/(2k+1)} \ge 1$.

Lemma 1.42 *Let L, m be defined as above. Then we have*

$$|Z(L,m)| > \frac{(\log N)^{k-1}}{(\log 2)^{k-1}(k-1)!(2k+1)^{k-1}}. \tag{1.78}$$

Proof. Set $L' = \frac{\log L}{\log 2}$ and $m' = -\left[-\frac{\log m}{\log 2}\right]$. Then $|Z(L,m)|$ is the number of integral lattice points $(s_1, \ldots, s_k)$ such that $\sum_{j=1}^{k} s_j = L'$ and $s_j \ge m'$ $(1 \le j \le k)$. Hence

$$Z(L,m) = \binom{L' - (m'-1)k - 1}{k-1} \ge \frac{(L' - m'k - 1)^{k-1}}{(k-1)!}.$$

Since

$$L' > \frac{\log N}{\log 2} + k\frac{\log(2\pi)}{2\log 2} + 1 > \frac{\log N}{\log 2} + k + 1$$

and

$$m' < \frac{2\log N}{(2k+1)\log 2} + 1,$$

we immediately get (1.78).□

Lemma 1.43 *For every $\mathbf{l} \in Z(L,m)$,*

$$\Delta_1(h_{\mathbf{l}}) \ge 2^{-7} \left(\frac{\pi \log 2}{k}\right)^{\frac{k}{2}} N^{-2}. \tag{1.79}$$

Proof. Set

$$\begin{aligned} Y &= \left\{ \mathbf{x} \in \mathbf{R}^k : \prod_{j=1}^{k} e^{-\frac{1}{2} l_j^2 x_j^2} \geq \frac{1}{2} \right\} \\ &= \left\{ \mathbf{x} \in \mathbf{R}^k : \sum_{j=1}^{k} l_j^2 x_j^2 \leq \log 4 \right\}. \end{aligned}$$

Then we can estimate $F_{h_l}(\mathbf{x})$ by

$$\begin{aligned} F_{h_l}(\mathbf{x}) &= \frac{1}{N} \sum_{n=1}^{N} h_l(\mathbf{x} - \mathbf{y}_n) - \int_{\mathbf{U}_k} h_l(\mathbf{x} - \mathbf{y})\, d\mathbf{y} \\ &\geq \frac{1}{2N} \sum_{n=1}^{N} \chi_{Y+\mathbf{x}}(\mathbf{y}_n) - \int_{\mathbf{R}^k} h_l(\mathbf{x} - \mathbf{y})\, d\mathbf{y} \\ &= \frac{1}{2N} \sum_{n=1}^{N} \chi_{Y+\mathbf{x}}(\mathbf{y}_n) - \prod_{j=1}^{k} \frac{(2\pi)^{\frac{1}{2}}}{l_j} \\ &\geq \frac{1}{2N} \sum_{n=1}^{N} \chi_{Y+\mathbf{x}}(\mathbf{y}_n) - \frac{1}{4N} \\ &\geq \frac{1}{4N} \sum_{n=1}^{N} \chi_{Y+\mathbf{x}}(\mathbf{y}_n). \end{aligned}$$

Hence

$$\begin{aligned} \Delta_1(h_l) &\geq \frac{1}{16N^2} \int_{\mathbf{R}^k} \left(\sum_{n=1}^{N} \chi_{Y+\mathbf{x}}(\mathbf{y}_n) \right)^2 d\mathbf{x} \\ &\geq \frac{1}{16N^2} \int_{\mathbf{R}^k} \left(\sum_{n=1}^{N} \chi_{Y+\mathbf{x}}(\mathbf{y}_n) \right) d\mathbf{x} \\ &= \frac{\lambda_k(Y)}{16N}. \end{aligned}$$

Since

$$\begin{aligned} \lambda_k(Y) &= (\log 4)^{\frac{k}{2}} \frac{V_k}{L} \\ &\geq (\log 4)^{\frac{k}{2}} \frac{2^k k^{-k/2}}{8(2\pi)^{k/2} N}, \end{aligned}$$

where V_k denotes the volume of the unit ball $B(1) \subseteq \mathbf{R}^k$ with radius 1, we have proved (1.79).□

Lemma 1.44 *We have*

$$\sum_{\mathbf{l}\in Z(L,m)} \Delta_1(h_\mathbf{l}) \le 3^k\pi^{-k}(6k+1)^{2k}\Delta_2(\chi_{1/m}), \tag{1.80}$$

where χ_β denotes the characteristic function of the cube $[-\beta,\beta]^k$.

Proof. By (1.76) and (1.77) it suffices to show

$$\sum_{\mathbf{l}\in Z(L,m)} |\hat{h}_\mathbf{l}(\mathbf{t})|^2 \le 3^k\pi^{-k}(6k+1)^{2k}\int_0^1 |\hat{\chi}_{\rho/m}(\mathbf{t})|^2\, d\rho.$$

First, we have

$$\begin{aligned}\sum_{\mathbf{l}\in Z(L,m)} |\hat{h}_\mathbf{l}(\mathbf{t})|^2 &= \sum_{\mathbf{l}\in Z(L,m)} \prod_{j=1}^k \frac{e^{-t_j^2/l_j^2}}{l_j^2} \\ &\le \sum_{2^{s_1}\ge m}\cdots\sum_{2^{s_k}\ge m}\prod_{j=1}^k 4^{-s_j}e^{-t_j^2 4^{-s_j}} \\ &= \prod_{j=1}^k\left(\sum_{2^s\ge m} 4^{-s}e^{-t_j^2 4^{-s}}\right).\end{aligned}$$

For $t\le m$ we will use the trivially estimate

$$\sum_{2^s\ge m} 4^{-s}e^{-t^2 4^{-s}} \le \sum_{2^s\ge m} 4^{-s} \le \frac{4}{3m^2}.$$

Now suppose that $t\ge m$ and set $f(x)=e^{-x}$. Here we have

$$\sum_{2^s\ge m} 4^{-s}e^{-t^2 4^{-s}} \le \frac{1}{t^2}\sum_{j\ge 0} f(4^{-j}) < \frac{4}{3t^2}.$$

Hence

$$\sum_{2^s\ge m} 4^{-s}e^{-t^2 4^{-s}} \le \frac{8}{3}\frac{1}{m^2+t^2},$$

and consequently

$$\sum_{\mathbf{l}\in Z(L,m)} \Delta_1(h_\mathbf{l}) \le \left(\frac{8}{3}\right)^k \prod_{j=1}^k \frac{1}{m^2+t_j^2}. \tag{1.81}$$

Next we will give an upper bound for

$$\int_0^1 |\hat{\chi}_{\rho/m}(\mathbf{t})|^2\, d\rho = \left(\frac{2}{\pi}\right)^k \int_0^1 \left(\prod_{j=1}^k \frac{\sin^2(\frac{\rho}{m}t_j)}{t_j^2}\right) d\rho.$$

If $|t| \leq \frac{\pi}{2}m$ then $|\sin(\frac{\rho}{m}t)| \geq \frac{2}{\pi}\frac{\rho}{m}|t|$. Therefore, if $J = \{j : |t_j| \leq \frac{\pi}{2}m\}$ then we have

$$\int_0^1 |\hat{\chi}_{\rho/m}(\mathbf{t})|^2\, d\rho \geq \left(\frac{2}{\pi}\right)^k \left(\frac{2}{m\pi}\right)^{2|J|} \prod_{j\notin J} \frac{1}{t_j^2} \int_0^1 \rho^{2|J|} \prod_{j \notin J} \sin^2(\rho m t_j)\, d\rho.$$

By elementary means we get for any $\alpha \geq \frac{\pi}{2}$ that

$$\lambda_1\left(\left\{\rho \in [0,1] : \sin^2(\alpha\rho) \leq \left(\frac{2}{6k+1}\right)^2\right\}\right) \leq \frac{1}{2k}$$

which implies

$$\lambda_1\left(\left\{\rho \in \left[\frac{1}{4},1\right] : \sin^2\left(\frac{\rho}{m}t_j\right)^2 > \left(\frac{2}{6k+1}\right)^2 \text{ for all } j \notin J\right\}\right) > \frac{1}{4}.$$

Hence we get

$$\int_0^1 \rho^{2|J|} \prod_{j\notin J} \sin^2(\rho m t_j)\, d\rho \geq 4^{-2|J|}\frac{3}{4}\left(\frac{2}{6k+1}\right)^{2(k-|J|)},$$

and finally

$$\begin{aligned} \int_0^1 |\hat{\chi}_{\rho/m}(\mathbf{t})|^2\, d\rho &\geq \frac{3}{4}\left(\frac{2}{\pi}\right)^k \left(\frac{1}{2\pi}\right)^{2|J|} \left(\frac{2}{6k+1}\right)^{2(k-|J|)} \frac{1}{m^{2|J|}} \prod_{j\notin J}\frac{1}{t_j^2} \\ &\geq \left(\frac{2}{\pi}\right)^k \left(\frac{2}{6k+1}\right)^{2k} \prod_{j=1}^{k} \frac{1}{m^2+t_j^2}. \end{aligned} \tag{1.82}$$

Now (1.81) and (1.82) imply (1.80).□

Now we are able to finish the proof of Theorem 1.41.

Proof of Theorem 1.41. By Lemmata 1.42, 1.43, and 1.44 we immediately get

$$\Delta_2(\chi_{1/m}) \geq c_k^2 \frac{(\log N)^{k-1}}{N^2}.$$

We will distinguish two cases. First assume that for some $\rho \in [0,1]$ and $\mathbf{x} \in \mathbf{R}^k$,

$$\begin{aligned} |F_{\chi_{\rho/m}}(\mathbf{x})| &= \left|\frac{1}{N}\sum_{n=1}^{N} \chi_{\rho/m}(\mathbf{x}-\mathbf{y}_n) - \int_{\mathbf{U}_k} \chi_{\rho/m}(\mathbf{x}-\mathbf{y})\, d\mathbf{y}\right| \\ &> 2\left(\frac{2}{m}\right)^k. \end{aligned}$$

This means that there is a translated image Q of the cube $[-\frac{1}{m}, \frac{1}{m}]$ such that Q contains $\left([-\frac{\rho}{m}, \frac{\rho}{m}]^k + \mathbf{x}\right) \cap \mathbf{U}_k$ and at least $2\left(\frac{2}{m}\right)^k$ points. Hence

$$\frac{1}{N}\sum_{n=1}^{N} \chi_{Q \bmod 1}(\mathbf{y}_n) - \lambda_k(Q) > \left(\frac{2}{m}\right)^k > N^{-\frac{2k}{2k+1}}.$$

It is easily verified that $N^{-\frac{2k}{2k+1}} > c_k(\log N)^{\frac{k-1}{2}}/N$. So we can concentrate on the second case, where $|F_{\chi_{\rho/m}}(\mathbf{x})| \leq 2\left(\frac{2}{m}\right)^k$. Thus

$$\begin{aligned} \int_{[-1-\frac{2}{m},21+\frac{2}{m}]^k \setminus [-1+\frac{2}{m},21-\frac{2}{m}]^k} \int_0^1 |F_{\chi_{\rho/m}}(\mathbf{x})|^2 \, d\rho \, d\mathbf{x} &\leq 4\left(\frac{2}{m}\right)^{2k} \frac{4m}{k}\left(1+\frac{2}{m}\right)^{k-1} \\ &\leq k2^{2k+4}3^{k-1}N^{-2} \\ &= d_k^2 N^{-2} \end{aligned}$$

implies that there exists a cube $Q = \rho\left[-\frac{1}{m}, \frac{1}{m}\right] + \mathbf{x} \subseteq \mathbf{U}_k$ such that

$$\left|\frac{1}{N}\sum_{n=1}^N \chi_Q(\mathbf{y}_n) - \lambda_k(Q)\right|^2 \geq \frac{c_k^2(\log N)^{k-1} - d_k^2}{N^2}.$$

Since $A \geq B \geq 0$ implies $A^2 - B^2 \geq (\sqrt{A} - \sqrt{B})^2$ we have proved Theorem 1.41. □

As mentioned above ROTH's theorem follows from BECK's one (but with a worse constant). It is an interesting question to decide whether the converse statement is true or not. Only in the 2-dimensional case there exists an astonishing proof by RUZSA [1581].

Theorem 1.45 *Let $\mathbf{x}_1, \ldots, \mathbf{x}_N$ be a finite sequence of points in the 2-dimensional space $\mathbf{R}^2$. Then there exists a cube $Q \subseteq \mathbf{R}^2/\mathbf{Z}^2$ with sides parallel to the axes and*

$$\left|\frac{1}{N}\sum_{n=1}^N \chi_Q(\mathbf{x}_n) - \lambda_2(Q)\right| \geq \frac{1}{11} D_N(\mathbf{x}_n). \tag{1.83}$$

Proof. Assume that

$$\left|\frac{1}{N}\sum_{n=1}^N \chi_Q(\mathbf{x}_n) - \lambda_2(Q)\right| < \frac{D_N(\mathbf{x}_n)}{11} \tag{1.84}$$

holds for all cubes $Q \subseteq \mathbf{R}^2/\mathbf{Z}^2$. Now choose a rectangle $I = [a_1, b_1) \times [a_2, b_2) \subseteq \mathbf{R}^2/\mathbf{Z}^2$ with

$$\left|\left|\frac{1}{N}\sum_{n=1}^N \chi_I(\mathbf{x}_n) - \lambda_2(I)\right| - D_N(\mathbf{x}_n)\right| < \frac{1}{22} D_N(\mathbf{x}_n).$$

W.l.o.g. we can assume that $b_1 - a_1 > b_2 - a_2$ and that

$$\left|\frac{1}{N}\sum_{n=1}^N \chi_I(\mathbf{x}_n) - \lambda_2(I) - D_N(\mathbf{x}_n)\right| < \frac{1}{22} D_N(\mathbf{x}_n).$$

If $(b_1 - a_1) + (b_2 - a_2) > 1$ then set $I' = [a_1, b_1 - (b_2 - a_2)) \times [a_2, b_2)$ and $Q' = [b_1 - (b_2 - a_2), b_1) \times [a_2, b_2)$. Then we have

$$\frac{1}{N}\sum_{n=1}^N \chi_I(\mathbf{x}_n) - \lambda_2(I) =$$

$$= \frac{1}{N}\sum_{n=1}^{N}\chi_{I'}(\mathbf{x}_n) - \lambda_2(I') + \frac{1}{N}\sum_{n=1}^{N}\chi_{Q'}(\mathbf{x}_n) - \lambda_2(Q')$$

and therefore

$$\left|\frac{1}{N}\sum_{n=1}^{N}\chi_{I'}(\mathbf{x}_n) - \lambda_2(I') - D_N\right| < \left(\frac{1}{22}+\frac{1}{11}\right)D_N = \frac{3}{22}D_N.$$

So we can assume w.l.o.g. that there exists a rectangle $I_1 = [a_1, b_1) \times [a_2, b_2) \subseteq \mathbf{R}^2/\mathbf{Z}^2$ with $(b_1 - a_1) + (b_2 - a_2) \le 1$, $b_1 - a_1 > b_2 - a_2$, and

$$\left|\frac{1}{N}\sum_{n=1}^{N}\chi_{I_1}(\mathbf{x}_n) - \lambda_2(I_1) - D_N\right| < \frac{3}{22}D_N.$$

Now set

$$\begin{aligned} I_2 &= [a_1 - (b_2 - a_2), a_1) \times [b_2, b_2 + (b_1 - a_1)) \\ Q_1 &= [a_1 - (b_2 - a_2), a_1) \times [a_2, b_2) \\ Q_2 &= [a_1, b_1) \times [b_2, b_2 + (b_1 - a_1)) \\ Q_3 &= I_1 \cup I_2 \cup Q_1 \cup Q_3. \end{aligned}$$

As above we get

$$\left|\frac{1}{N}\sum_{n=1}^{N}\chi_{I_1}(\mathbf{x}_n) - \lambda_2(I_1) + \frac{1}{N}\sum_{n=1}^{N}\chi_{I_2}(\mathbf{x}_n) - \lambda_2(I_2)\right| < \frac{3}{11}D_N,$$

and so

$$\left|\lambda_2(I_2) - \frac{1}{N}\sum_{n=1}^{N}\chi_{I_2}(\mathbf{x}_n) - D_N\right| < \frac{9}{22}D_N. \tag{1.85}$$

Consider

$$\begin{aligned} I_3 &= [a_1 - (b_2 - a_2), b_1 - (b_2 - a_2)) \times [a_2, b_2) \\ Q_4 &= [b_1 - (b_2 - a_2), b_1) \times [a_2, b_2). \end{aligned}$$

Since $I_1 \cup Q_1 = I_3 \cup Q_4$ we get

$$\left|\frac{1}{N}\sum_{n=1}^{N}\chi_{I_3}(\mathbf{x}_n) - \lambda_2(I_3) - D_N\right| < \frac{7}{22}D_N.$$

As above use

$$\begin{aligned} I_4 &= [a_1 - 2(b_2 - a_2), a_1 - (b_2 - a_2)) \times [b_2, b_2 + (b_1 - a_1)) \\ Q_5 &= [a_1 - 2(b_2 - a_2), a_1 - (b_2 - a_2)) \times [a_2, b_2) \\ Q_6 &= [a_1 - (b_2 - a_2), b_1 - (b_2 - a_2)) \times [b_2, b_2 + (b_1 - a_1)) \\ Q_7 &= I_3 \cup I_4 \cup Q_5 \cup Q_6 \end{aligned}$$

to obtain

$$\left|\lambda_2(I_4) - \frac{1}{N}\sum_{n=1}^{N}\chi_{I_4}(\mathbf{x}_n) - D_N\right| < \frac{13}{22}D_N. \tag{1.86}$$

Set $I_5 = [a_1 - 2(b_2 - a_2), a_1) \times [b_2, b_2 + (b_1 - a_1)) = I_2 \cup I_4$. Now (1.85) and (1.86) imply

$$\lambda_2(I_5) - \frac{1}{N}\sum_{n=1}^{N} \chi_{I_5}(\mathbf{x}_n) > \frac{13}{22} D_N + \frac{9}{22} D_N = D_N,$$

which is of course a contradiction to the definition of the discrepancy. So (1.83) holds. □

Unfortunately the preceding proof cannot be generalized to the higher dimensional case $k > 2$. We have to use FOURIER transform techniques to obtain a similar but weaker theorem for arbitrary dimensions.

Theorem 1.46 *Let $\mathbf{x}_1, \ldots, \mathbf{x}_N$ be a finite sequence of points in the k-dimensional space $\mathbf{R}^k$. Then there exists a cube $Q \subseteq \mathbf{R}^k/\mathbf{Z}^k$ with sides parallel to the axes and*

$$\left| \frac{1}{N}\sum_{n=1}^{N} \chi_Q(\mathbf{x}_n) - \lambda_k(Q) \right| \gg D_N(\mathbf{x}_n)^{1+\frac{k}{2}}, \tag{1.87}$$

where the constant implied by $\gg$ only depends on k.

It is an immediate consequence of Theorem 1.8 (for $p = 2$) and of the following observation that the L^2-discrepancy $D_N^{(2)}(\mathbf{x}_n)$ and a proper L^2-discrepancy with respect to cubes are comparable.

Proposition 1.47 *Let $\mathbf{x}_1, \ldots, \mathbf{x}_N$ be a finite sequence of points in the k-dimensional space $\mathbf{R}^k$ and set*

$$D_N^{(2)}(\mathbf{x}_n, \mathcal{Q}) = \left(\int_0^1 \int_{[0,1]^k} \left| \frac{1}{N}\sum_{n=1}^{N} \chi_{Q(\rho)+\mathbf{y}}(\mathbf{x}_n) - \lambda_k(Q) \right|^2 d\mathbf{y}\, d\rho \right)^{\frac{1}{2}},$$

where $Q(\rho) = [-\frac{1}{2}\rho, \frac{1}{2}\rho]^k$ denotes a cube of side length ρ. Then

$$D_N^{(2)}(\mathbf{x}_n, \mathcal{Q}) \gg\ll D_N^{(2)}(\mathbf{x}_n),$$

where the constants implied by $\gg\ll$ only depend on k.

Proof. First, let us note that for technical reasons we will work with

$$\tilde{D}_N^{(2)}(\mathbf{x}_n) = \left(\int_{[0,1)^k} \int_{\mathbf{R}^k/\mathbf{Z}^k} \left| \frac{1}{N}\sum_{n=1}^{N} \chi_{R(\mathbf{a})+\mathbf{y}}(\mathbf{x}_n) \right|^2 d\mathbf{y}\, d\mathbf{a} \right)^{\frac{1}{2}} \tag{1.88}$$

instead of $D_N^{(2)}(\mathbf{x}_n)$, where $R(\mathbf{a})$ denotes the rectangle $R(\mathbf{a}) = [-\frac{a_1}{2}, \frac{a_1}{2}] \times \cdots \times [-\frac{a_k}{2}, \frac{a_k}{2}]$, $\mathbf{a} = (a_1, \ldots, a_k) \in [0,1)^k$. By obvious reasoning they are related by the inequalities

$$D_N^{(2)}(\mathbf{x}_n) \le 4^k \tilde{D}_N^{(2)}(\mathbf{x}_n) \quad \text{and} \quad \tilde{D}_N^{(2)}(\mathbf{x}_n) \le 4^k D_N^{(2)}(\mathbf{x}_n).$$

We will identify $\mathbf{R}^k/\mathbf{Z}^k$ with the unit cube $[0,1)^k \subseteq \mathbf{R}^k$ and consider the periodically expanded discrete set

$$\mathbf{X} = \{\mathbf{x}_1, \ldots, \mathbf{x}_N\} + \mathbf{Z}^k \subseteq \mathbf{R}^k$$

instead of $\mathbf{x}_1, \ldots, \mathbf{x}_N \in [0,1)^k$. The distribution properties of $\mathbf{x}_1, \ldots, \mathbf{x}_N \in \mathbf{R}^k/\mathbf{Z}^k$ with respect to rectangles and cubes in $\mathbf{R}^k/\mathbf{Z}^k$ are exactly the same as the distribution properties of $\mathbf{X}$ with respect to rectangles and cubes in $\mathbf{R}^k$.

For any positive integer $M \geq 1$ set

$$\nu_M = \frac{1}{N} \sum_{\mathbf{x} \in \mathbf{X} \cap Q(2M)} \delta_{\mathbf{x}} - \mu_M,$$

where $\delta_{\mathbf{x}}$ denotes the DIRAC measure concentrated on $\mathbf{x}$ and μ_M is the restriction of the LEBESGUE measure λ_k to $Q(2M)$. Furthermore define

$$\Delta^{(2)}(M, \mathcal{Q}) = (2M)^{-k} \int_0^1 \int_{\mathbf{R}^k} |\nu_M(Q(\rho) + \mathbf{y})|^2 \, d\mathbf{y} \, d\rho$$

and

$$\Delta^{(2)}(M, \mathcal{R}) = (2M)^{-k} \int_{[0,1)^k} \int_{\mathbf{R}^k} |\nu_M(R(\mathbf{a}) + \mathbf{y})|^2 \, d\mathbf{y} \, d\mathbf{a}.$$

Using rather crude estimates around the boundary of $Q(2M)$ we immediately get

$$\lim_{M \to \infty} \Delta^{(2)}(M, \mathcal{Q}) = \int_0^1 \int_{\mathbf{R}^k/\mathbf{Z}^k} |\nu_M(Q(\rho) + \mathbf{y})|^2 \, d\mathbf{y} \, d\rho = D_N^{(2)}(\mathbf{x}_n, \mathcal{Q})^2 \tag{1.89}$$

and

$$\lim_{M \to \infty} \Delta^{(2)}(M, \mathcal{R}) = \int_{[0,1)^k} \int_{\mathbf{R}^k/\mathbf{Z}^k} |\nu_M(R(\mathbf{a}) + \mathbf{y})|^2 \, d\mathbf{y} \, d\mathbf{a} = \tilde{D}_N^{(2)}(\mathbf{x}_n)^2. \tag{1.90}$$

Since $\nu_M(S + \mathbf{y})$ can be rewritten as a convolution

$$\nu_M(S + \mathbf{y}) = \chi_{(-S)} * (d\nu_M(\mathbf{y})),$$

we obtain by using PARSELVAL's identity (and by observing that $-Q(\rho) = Q(\rho)$ and $-R(\mathbf{a}) = R(\mathbf{a})$)

$$\begin{aligned}
\Delta^{(2)}(M, \mathcal{Q}) &= (2M)^{-k} \int_0^1 \int_{\mathbf{R}^k} \left|(\chi_{Q(\rho)} * (d\nu_M(\mathbf{y}))\right|^2 d\mathbf{y} \, d\rho \\
&= (2M)^{-k} \int_0^1 \int_{\mathbf{R}^k} |\hat{\chi}_{Q(\rho)}(\mathbf{t})|^2 \cdot |\hat{\nu}_M(\mathbf{t})|^2 \, d\mathbf{t} \, d\rho \\
&= (2M)^{-k} \int_{\mathbf{R}^k} \left(\int_0^1 |\hat{\chi}_{Q(\rho)}(\mathbf{t})|^2 \, d\rho \right) |\hat{\nu}_M(\mathbf{t})|^2 \, d\mathbf{t}
\end{aligned}$$

and

$$\Delta^{(2)}(M,\mathcal{R}) = (2M)^{-k} \int_{\mathbf{R}^k} \left(\int_{[0,1)^k} \left|\hat{\chi}_{R(\mathbf{a})}(\mathbf{t})\right|^2 d\mathbf{a} \right) \left|\hat{\nu}_M(\mathbf{t})\right|^2 d\mathbf{t},$$

The final step of the proof is to compare the quadratic mean values of

$$\hat{\chi}_{Q(\rho)}(\mathbf{t}) = \prod_{i=1}^{k} \frac{2\sin(\rho t_i)}{\sqrt{2\pi} t_i} \quad \text{and} \quad \hat{\chi}_{R(\mathbf{a})}(\mathbf{t}) = \prod_{i=1}^{k} \frac{2\sin(a_i t_i)}{\sqrt{2\pi} t_i}.$$

By (1.82)

$$\int_0^1 \left|\hat{\chi}_{Q(\rho)}(\mathbf{t})\right|^2 d\rho \gg \prod_{i=1}^{k} \frac{1}{(1+|t_i|)^2}$$

uniformly for $\mathbf{t} \in \mathbf{R}^k/\mathbf{Z}^k$. On the other hand, by using the inequality $|\sin x| \le \min\{1,|x|\}$, we get

$$\int_0^1 \left|\hat{\chi}_{Q(\rho)}(\mathbf{t})\right|^2 d\rho \ll \prod_{i=1}^{k} \frac{1}{(1+|t_i|)^2}.$$

In the same fashion (even more easily) we obtain

$$\int_{[0,1)^k} \left|\hat{\chi}_{R(\mathbf{a})}(\mathbf{t})\right|^2 d\mathbf{a} \gg\ll \prod_{i=1}^{k} \frac{1}{(1+|t_i|)^2}$$

which implies

$$\int_0^1 \left|\hat{\chi}_{Q(\rho)}(\mathbf{t})\right|^2 d\rho \gg\ll \int_{[0,1)^k} \left|\hat{\chi}_{R(\mathbf{a})}(\mathbf{t})\right|^2 d\mathbf{a}$$

and consequently

$$\Delta^{(2)}(M,\mathcal{Q}) \gg\ll \Delta^{(2)}(M,\mathcal{R}). \tag{1.91}$$

Hence (1.89), (1.90), and (1.91) prove the proposition. □

It is remarkable that ROTH's theorem is best possible for the L^2-discrepancy $D_N^{(2)}(\mathbf{x}_n)$ (see [1570]). This means that there are constants c_k such that for every $N \ge 1$ there exists a finite sequence $\mathbf{x}_1, \ldots, \mathbf{x}_N \in \mathbf{R}^k$ with

$$D_N^{(2)}(\mathbf{x}_n) \le c_k \frac{\log^{\frac{k-1}{2}} N}{N}.$$

This assertion seems not to be true for the usual discrepancy. It is conjectured that there are constants c_k such that for every sequence $\mathbf{x}_1, \ldots, \mathbf{x}_N \in \mathbf{R}^k$

$$D_N(\mathbf{x}_n) \ge c_k \frac{\log^{k-1} N}{N}. \tag{1.92}$$

This lower bound would be best possible as examples show. But (1.92) is only proved in the two-dimensional case by SCHMIDT [1630] (see section 1.3.2) and ROTH's bound has been sharpened a little bit by BECK [137] for $k = 3$ (see section 1.3.3).

It should be noted that a generalization of ROTH's method by SCHMIDT [1633] provides optimal lower bounds for $D_N^{(p)}(\mathbf{x}_n)$ $(p > 1)$. (The optimality is shown by CHEN [342].)

Theorem 1.48 *Let $k \geq 2$ and $p > 1$. Then the L^p-discrepancy $D_N^{(p)}(\mathbf{x}_n)$ of N points $\mathbf{x}_1, \ldots, \mathbf{x}_N$ in the k-dimensional space $\mathbf{R}^k$ is bounded below by*

$$D_N^{(p)}(\mathbf{x}_n) \geq c_k \frac{\log^{\frac{k-1}{2}} N}{N}. \tag{1.93}$$

1.3.2 SCHMIDT's Bound

In section 1.3.1 it is proved that the discrepancy of any sequence $\mathbf{x}_1, \ldots, \mathbf{x}_N \in \mathbf{R}^k$ is bounded below by $D_N(\mathbf{x}_n) \geq \frac{1}{N}$ and that the one-dimensional sequence $x_n = \frac{n}{N}$, $n = 1, \ldots, N$, satisfies $D_N(x_n) = \frac{1}{N}$. (In the higher dimensional case such examples cannot exist by ROTH's theorem.) Note that in this example for every N a new sequence $x_1, \ldots, x_N$ is constructed. This observation leads to the question whether there exists an infinite sequence $(x_n)_{n\geq 1}$, $x_n \in \mathbf{R}$, such that the discrepancy $D_N(x_n)$ is small for all N, e.g. $D_N(x_n) = \mathcal{O}\left(\frac{1}{N}\right)$, $N \geq 1$. It turns out that such sequences do not exist. (This was first observed by VAN AARDENNE-EHRENFEST [1889, 1890].)

It is amazing that these two problems, the problem to find lower bounds for the discrepancy which are true for any finite sequence of N points when N is fixed and to find lower bounds for the discrepancy which are true for infinitely many N, where an infinite sequence is fixed, are essentially the same, in the following sense.

Theorem 1.49 *Suppose that for any finite sequence of points $\mathbf{x}_1, \ldots, \mathbf{x}_N \in \mathbf{R}^{k+1}$, $k \geq 1$, we have $ND_N^*(\mathbf{x}_n) \geq f(N)$. Then for any sequence of points $\mathbf{y}_1, \ldots, \mathbf{y}_N \in \mathbf{R}^k$ there exist m with $1 \leq m \leq N$ such that $mD_m^*(\mathbf{y}_n) \geq f(N) - 1$.*

Conversely, assume that for any finite sequence of points $\mathbf{y}_1, \ldots, \mathbf{y}_N \in \mathbf{R}^k$ there exist m with $1 \leq m \leq N$ such that $mD_m^*(\mathbf{y}_n) \geq g(N)$. Then for any sequence $\mathbf{x}_1, \ldots, \mathbf{x}_N \in \mathbf{R}^{k+1}$ we have $ND_N^*(\mathbf{x}_n) \geq \frac{1}{2}g(N)$.

For a *proof* see KUIPERS and NIEDERREITER [983, p. 105 f].

Thus VAN AARDENNE-EHRENFEST's theorem follows from ROTH's theorem. In general we have

Theorem 1.50 *For any sequence $(\mathbf{x}_n)_{n\geq 1}$, $\mathbf{x}_n \in \mathbf{R}^k$, $k \geq 1$, we have*

$$D_N(\mathbf{x}_n) \geq \frac{1}{2^5} \frac{1}{2^{4k}} \frac{1}{(k \log 2)^{\frac{k}{2}}} \frac{\log^{\frac{k}{2}} N}{N} \tag{1.94}$$

for infinitely many N.

We are able to reformulate conjecture (1.92) to:
There exist constants c_k such that for any infinite sequence $(\mathbf{x}_n)_{n\geq 1}$, $\mathbf{x}_n \in \mathbf{R}^k$, $k \geq 1$, we have

$$D_N(\mathbf{x}_n) \geq c_k \frac{\log^k N}{N} \tag{1.95}$$

for infinitely many N.

This conjecture has been proved by SCHMIDT [1630] for $k = 1$. (Of course this result sharpens ROTH's theorem for $k = 2$.)

Theorem 1.51 *For any infinite real sequence $(x_n)_{n\ge1}$, $x_n \in \mathbf{R}$, we have*

$$D_N(x_n) \ge C\,\frac{\log N}{N} \tag{1.96}$$

for infinitely many N, where

$$C = \max_{a\ge3} \frac{1}{8}\frac{a-2}{a\log a} = 0.04667\ldots.$$

The constant $C = 0.04667\ldots$ is better than SCHMIDT's [1630] and better than that given in KUIPERS and NIEDERREITER [983]. We follow LIARDET [1083].

We also want to mention that Theorem 1.51 is best possible. There are several examples of (one dimensinal) sequences $(x_n)_{n\ge1}$ with $N\,D_N(x_n) = \mathcal{O}(\log N)$. For example, let $b \ge 2$ be a fixed integer and let

$$n = \sum_{k\ge0} d_k(n) b^k$$

the digital representation of (the non-negative integer) n, i.e. $0 \le d_k(n) < b$. If we set

$$\gamma_b(n) = \sum_{k\ge0} d_k(n) b^{-k-1}$$

then it is an easy example to show that the sequence $(\gamma_b(n))_{n\ge1}$ satisfies $N\,D_N(\gamma_b(n)) = \mathcal{O}(\log N)$ (see KUIPERS and NIEDERREITER [983]). The sequence $(\gamma_b(n))_{n\ge1}$ is called VAN DER CORPUT sequence in base b. (For higher dimensional generalizations such as the HALTON and HAMMERSLEY sequence we refer to Section 3.1.)

Lemma 1.52 *Suppose that $u \le v$. Then we have*

$$\int_u^v |ax+b|\,dx \ge |a| \left(\frac{v-u}{2}\right)^2. \tag{1.97}$$

The *proof* is trivial.

Now let $J \subseteq \{1,2,3,\ldots\}$ be a set of positive integers and set

$$F_J(x) = \sup_{j\in J} \left(\sum_{n=1}^{j} \chi_{[0,x)}(x_n) - jx \right) \tag{1.98}$$

$$G_J(x) = \inf_{j\in J} \left(\sum_{n=1}^{j} \chi_{[0,x)}(x_n) - jx \right). \tag{1.99}$$

Lemma 1.53 *Let $K \subseteq \{1,2,3,\dots\}$, $L \subseteq K$, $L' \subseteq K$, then we have*

$$F_K - G_K \geq \frac{1}{2}(F_L - G_L) + \frac{1}{2}(F_{L'} - G_{L'}) \tag{1.100}$$

$$+ \frac{1}{2}(|F_L - F_{L'}| + |G_L - G_{L'}|) \quad . \tag{1.101}$$

Proof. From $F_K = F_K - F_L + F_L \geq (F_{L'} - F_L) + F_L$ it follows

$$2F_K \geq F_L + F_{L'} + (F_{L'} - F_L)$$

and by interchanging L and L'

$$2F_K \geq F_L + F_{L'} + |F_{L'} - F_L|.$$

Similarly

$$2G_K \leq G_L + G_{L'} - |G_{L'} - G_L|.$$

This proves the Lemma.□

Now set $K = \{p+1, p+2, \dots, p+[a^{t+1}]\}$, $L = \{p+1, \dots, p+[a^t]\}$, $L' = \{p+[a^{t+1}]-[a^t]+1, \dots, p+[a^{t+1}]\}$, where $a \geq 3$ and p, t are some non-negative integers.

Lemma 1.54 *With the notation above we have*

$$\int_0^1 |F_L(x) - F_{L'}(x)| \, dx \geq \frac{1}{4}\frac{a-2}{a} \tag{1.102}$$

and

$$\int_0^1 |G_L(x) - G_{L'}(x)| \, dx \geq \frac{1}{4}\frac{a-2}{a}. \tag{1.103}$$

Proof. Set

$$A_j(x) = \sum_{n=p+2}^{j} \chi_{[0,x)}(x_n) - (j-p+1)x.$$

Then

$$|F_L - F_{L'}| = \left|\sup_{j \in L} A_j - \sup_{j' \in L'} A_{j'}\right|.$$

Let $(y_0 =)0 < y_1 < \cdots < y_{M-1} < 1(= y_M)$ be the ordered set of points $\{\{x_j\} : p+2 \leq j \leq p+[a^{t+1}]\}$. In each interval (y_i, y_{i+1}), $0 \leq i < M$, the function $x \mapsto |F_L(x) - F_{L'}(x)|$ is of the kind $x \mapsto |ax+b|$, where $|a|$ is uniformly bounded below by

$$\inf_{j \in L, j' \in L'} |(j-p+1) - (j'-p+1)| = [a^{t+1}] - 2[a^t] + 1.$$

Hence by Lemma 1.52

$$\int_{y_i}^{y_{i+1}} |F_L(x) - F_{L'}(x)|\,dx \geq \frac{[a^{t+1}] - 2[a^t] + 1}{4}(y_{i+1} - y_i)^2.$$

Since

$$\min \sum_{i=0}^{M-1} (y_{i+1} - y_i)^2 = \frac{1}{M} \geq \frac{1}{[a^{t+1}]}$$

we immediately get

$$\begin{aligned}\int_0^1 |F_L(x) - F_{L'}(x)|\,dx &\geq \frac{1}{4}\frac{[a^{t+1}] - 2[a^t] + 1}{[a^{t+1}]} \\ &\geq \frac{1}{4}\frac{a-2}{a}.\end{aligned}$$

The second inequality can be shown along the same lines. □

Lemma 1.55 *If* $K = \{p+1, p+1, \ldots, p+[a^t]\}$ *then*

$$\int_0^1 (F_K(x) - G_K(x))\,dx \geq \frac{1}{4}\frac{a-2}{a}t. \tag{1.104}$$

Proof. First consider the case $t = 1$. Here $K = \{p+1, \ldots, p+[a]\}$, $L = \{p+1\}$, $L' = \{p+[a]\}$, furthermore $F_L = G_L$ and $F_{L'} = G_{L'}$. Hence we get

$$\int_0^1 (F_K(x) - G_K(x))\,dx \geq \frac{1}{4}\frac{a-2}{a}.$$

Now suppose that (1.104) is true for some positive integer t. Then by Lemma 1.53 and 1.54 we can show inductively

$$\begin{aligned}\int_0^1 (F_K(x) - G_K(x))\,dx &\geq \frac{1}{4}\frac{a-2}{a}t + \frac{1}{2}\left(\int_0^1 |F_L - F_{L'}|\,dx + \int_0^1 |G_L - G_{L'}|\,dx\right) \\ &\geq \frac{1}{4}\frac{a-2}{a}t + \frac{1}{4}\frac{a-2}{a} = \frac{1}{4}\frac{a-2}{a}(t+1).\end{aligned}$$

□

Proof of Theorem 1.51. By Lemma 1.55 there exists $x \in [0,1]$ with

$$F_K(x) - G_K(x) \geq \frac{1}{4}\frac{a-2}{a}t,$$

where $K = \{1, 2, \ldots, [a^t]\}$. Hence there exist $j_0, i_0 \in K$ with

$$\left(\sum_{n=1}^{j_0} \chi_{[0,x)}(x_n) - j_0 x\right) - \left(\sum_{n=1}^{i_0} \chi_{[0,x)}(x_n) - i_0 x\right) \geq \frac{1}{4}\frac{a-2}{a}t,$$

which implies that there is some $j \in K$ satisfying

$$\left| \sum_{n=1}^{j} \chi_{[0,x)}(x_n) - jx \right| \geq \frac{1}{8} \frac{a-2}{a} t.$$

Since $j \leq a^t$ we get

$$jD_j^*(x_n) \geq \frac{1}{8} \frac{a-2}{a} \frac{\log j}{\log a}$$

for all $a \geq 3$. The maximal value is obtained for $a = 5.356694\ldots$.□

As a corollary we get

Theorem 1.56 *For any finite sequence* $\mathbf{x}_1, \ldots, \mathbf{x}_N \in \mathbf{R}^2$ *we have*

$$D_N^*(\mathbf{x}_n) \geq C \frac{\log N}{N},$$

where

$$C = \max_{a \geq 3} \frac{1}{16} \frac{a-2}{a \log a} = 0.02333\ldots.$$

It should be noted that by a variation of ROTH's method due to HALÁSZ it is possible to prove

$$D_N^*(\mathbf{x}_n) \geq \frac{1}{2^{16} \log 2} \frac{\log N}{N}.$$

The constant $1/(2^{16} \log 2) = 2.013\ldots \cdot 10^{-5}$ is of course worse than LIARDET's constant, but HALÁSZ [718] was able to prove by his method

Theorem 1.57 *Let* $k \geq 2$. *Then for every finite sequence* $\mathbf{x}_1, \ldots, \mathbf{x}_N \in \mathbf{R}^k$ *we have*

$$D_N^{(1)}(\mathbf{x}_n) \gg \frac{\log^{\frac{1}{2}} N}{N}.$$

This result is best possible for $k = 2$. But it is the only one known for L^1-discrepancies. It is conjectured that $D_N^{(1)}(\mathbf{x}_n)$ behaves like any L^p-discrepancy $D_N^{(p)}(\mathbf{x}_n)$ $(0 < p < \infty)$.

1.3.3 BECK's Theorem

The first and only improvement of ROTH's theorem in the higher dimensional case $k \geq 3$ is BECK's theorem [137] for $k = 3$.

Theorem 1.58 *For any finite sequence* $\mathbf{x}_1, \ldots, \mathbf{x}_N \in \mathbf{R}^3$ *and for any* $\varepsilon > 0$ *we have*

$$D_N^*(\mathbf{x}_n) \geq \frac{\log N (\log \log N)^{\frac{1}{8}-\varepsilon}}{N}$$

for sufficiently large $N \geq N_1(\varepsilon)$.

We will give an outline of BECK's proof. Set

$$D(\mathbf{x}) = \frac{1}{N}\sum_{n=1}^{N}\chi_{[\mathbf{0},\mathbf{x})}(\mathbf{x}_n) - \lambda_3([\mathbf{0},\mathbf{x})).$$

The underlying idea is to find an auxiliary function $F(\mathbf{x})$ in $U^3 = [0,1)^3$ satisfying

$$\int_{U^3} F(\mathbf{x})D(\mathbf{x})\,d\mathbf{x} > \frac{\log N(\log\log N)^{\frac{1}{8}-\varepsilon}}{N} \tag{1.105}$$

and

$$\int_{U^3} |F(\mathbf{x})|\,d\mathbf{x} < 2+\varepsilon \tag{1.106}$$

provided that $N \geq N_1(\varepsilon)$. Obviously (1.105) and (1.106) imply

$$D_N^*(\mathbf{x}_n) = \sup_{\mathbf{x}\in U^3} D(\mathbf{x}) > \frac{1}{2+\varepsilon}\frac{\log N(\log\log N)^{\frac{1}{8}-\varepsilon}}{N}.$$

In particular, let

$$x = \sum_{j=0}^{\infty}\beta_j(x)2^{-j-1}$$

be the dyadic expansion of $x \in [0,1)$ such that the sequence $(\beta_j(x))_{j\geq 0}$ does not end with $1,1,1,\ldots$. Now RADEMACHER functions $R_r(x)$, $r \geq 0$, $x \in [0,1)$ are defined by

$$R_r(x) = (-1)^{\beta_r(x)}.$$

By an r-interval we mean a dyadic interval of the form $I = [m2^{-r},(m+1)2^{-r})$, where $0 \leq m < 2^r$. If $\mathbf{r} = (r_1,r_2,r_3)$ then set

$$R_{\mathbf{r}}(\mathbf{x}) = R_{r_1}(x_1)R_{r_2}(x_2)R_{r_3}(x_3)$$

for any $\mathbf{x} = (x_1,x_2,x_3) \in U^3$. By an $\mathbf{r}$-box we mean the cartesian product $I_1\times I_2\times I_3 \subseteq U^3$ of r_i-intervals I_i $(i=1,2,3)$ and by an $\mathbf{r}$-function we mean a function $f(\mathbf{x})$ defined in U^3 such that for every $\mathbf{r}$-box either $f(\mathbf{x}) = R_{\mathbf{r}}(\mathbf{x})$ or $f(\mathbf{x}) = -R_{\mathbf{r}}(\mathbf{x})$.

Now by Lemma 2.5 of BECK and CHEN [143] there always exists an $\mathbf{r}$-function $g_{\mathbf{r}}(\mathbf{x})$ satisfying

$$\int_{U^3} g_{\mathbf{r}}(\mathbf{x})D(\mathbf{x}) \geq 2^{-n-7},$$

where $r_1+r_2+r_3 = n$ and $2^n \geq 2N$. In what follows we will assume that n is uniquely determined by $2N \leq 2^n < 4N$.

Let $\mathcal{X}$ be the set

$$\mathcal{X} = \{\mathbf{r} = (r_1,r_2,r_3) : r_i \geq 0\ (i=1,2,3),\ r_1+r_2+r_3 = n\}.$$

Furthermore let $q = \left[(\log n)^{\frac{1}{2}-\varepsilon}\right]$ and $\mathcal{A}_1, \ldots, \mathcal{A}_q$ disjoint subsets of $\mathcal{X}$ satisfying

$$\sum_{l=1}^{q} |\mathcal{A}_l| \geq \frac{|\mathcal{X}|}{2} \tag{1.107}$$

and

$$\lambda_3\left(\left\{\mathbf{x} \in U^3 : \sum_{\mathbf{r}\in\mathcal{A}_i} g_{\mathbf{r}}(\mathbf{x}) - \frac{1}{q}\sum_{\mathbf{r}\in\mathcal{X}_{i-1}} g_{\mathbf{r}}(\mathbf{x}) \leq \frac{n+1}{2(q^{\frac{1}{4}-\varepsilon})}\right\}\right) \leq e^{-\rho^2/4} \tag{1.108}$$

for every $1 \leq i \leq q$, where $\log q \leq \rho \leq 2(n+1)q^{-\frac{1}{2}}$ and $\mathcal{X}_{i-1} = \mathcal{X} \setminus (\mathcal{A}_1 \cup \mathcal{A}_2 \cup \ldots \cup \mathcal{A}_{i-1})$. The existence of such $\mathcal{A}_i$ can be checked by probabilistic arguments (see BECK [137]).

Using the above notations it is possible to define a proper function

$$F(\mathbf{x}) = \sum_{l=1}^{q} \rho^l \sum_{1\leq\nu_1<\cdots<\nu_l\leq q} \sum{}' g_{\mathbf{r}_1}(\mathbf{x}) \cdots g_{\mathbf{r}_l}(\mathbf{x}), \tag{1.109}$$

where

$$\rho = \frac{q^{\frac{1}{4}-\varepsilon}}{n+1}$$

and the summation $\sum'$ extends over all l-tuples $(\mathbf{r}_1, \ldots, \mathbf{r}_l)$ such that

$$\mathbf{r}_j = (r_{j1}, r_{j2}, r_{j3}) \in \mathcal{A}_{\nu_j} \qquad \text{for all } 1 \leq j \leq l$$

and

$$r_{j1} \neq r_{k1}, r_{j2} \neq r_{k2}, r_{j3} \neq r_{k3} \qquad \text{for all } 1 \leq j < k \leq l.$$

It is quite easy to prove that (1.107) implies

$$\int_{U^3} F(\mathbf{x}) D(\mathbf{x})\, d\mathbf{x} > 2^{-12} q^{\frac{1}{4}-\varepsilon} \frac{n+1}{N}$$

and hence (1.105) but it rather difficult to verify (1.106). For this purpose you have to use probablistic estimations related to (1.108) and the orthogonality relation for r-functions. (For details see BECK [137].)

Notes

Discrepancy bounds from below are essentially quantitative measures for the (unavoidable) irregularity of point distributions. Originally the study of irregularities of distributions began with the work of Van der Corput on distribution functions in 1935/36 [1891, 1892, 1893, 1894, 1895, 1896, 1897, 1898]. He made the conjecture that there are no sequences on an interval with a just distribution, i.e. there is no real sequence $(x_n)_{n\geq 1}$ satisfying $D_N(x_n) = \mathcal{O}(1/N)$ as $N \to \infty$. This problem was solved by van

Aardenne-Ehrenfest [1889, 1890]. She even obtained a quantitative lower bound for the discrepancy: there are infinitely many N with

$$D_N(x_n) \gg \frac{\log\log N}{N \log\log\log N}.$$

In 1954 Roth [1567] sharpended this bound to $\sqrt{\log N/N}$ and provided a generalization to higher dimensional sequences (see Theorem 1.40). In 1956 Davenport [429] showed that Roth's theorem for the L^2-discrepancy is optimal in the two dimensional case. Later Roth himself [1569, 1570] observed that his lower bound for the L^2-discrepancy is optimal in any dimension (compare also with Dobrovolskij [470]).

In 1968 Schmidt [1624, 1625, 1626, 1627, 1628, 1629, 1630, 1631, 1632, 1633] started a series of papers on irregularities of distribution (see also [1634]). The most prominent result (for dimension 1) is Theorem 1.51 [1630]. (The proof presented here is due to Liardet [1083].) His lower bound $\log N/N$ is best possible. Hence Van der Corput's original question is completely solved, even from a quantitative point of view. Tijdeman and Wagner [1860] employed a variant of Schmidt's method (used by Wagner [1919, 1921, 1923] for solving a problem of Erdős) to obtain a refinement; see also Beck [119] and Wagner [1921, 1927]. Another generalization is due to Wagner [1920]. By a variation of Roth's method Schmidt also provided lower bounds for the L^p-discrepancy (Theorem 1.48) for arbitrary $p > 1$, which are of the same order of magnitute as Roth's estimate Theorem 1.40. (A more general version is due to Chen [345, 346].) Halász obtained the same bound for the L^1-discrepancy in the two dimensional case. (All these lower bounds for the L^p-discrepancy, $1 \le p < \infty$, are best possible, see Chen [342, 343, 344].) Schmidt [1625, 1626, 1627, 1628] was the first who considered more general concepts of discrepancies (arbitrary rectangles, balls, spherical caps) and developed an integral equation method in order to obtain non-trivial lower bounds. (see Notes of Section 2.1).

The most prominent open problem in the theory of irregularities of distribution is to determine the optimal lower bound for the usual discrepancy $D_N(x_n)$. As mentioned above Roth's bound $N\,D_N(x_n) \ge c_k(\log N)^{(k-1)/2}$ (Theorem 1.40, [1567]) is only known for $k > 3$. Conversely there are various examples of k-dimensional (finite) sequences $x_1, \ldots, x_n$ satisfying $N\,D_N(x_n) \le c_k'(\log N)^{k-1}$ (e.g. provided by Halton [729] or Faure [582]). Thus Schmidt's result [1630] is not only better than Roth's bound but optimal. It solves the case $k = 2$. Another proof is due to Halász [718]. The only bound which is better than Roth's bound in dimensions $k > 2$ is due to Beck [137] (Theorem 1.58).

A far reaching improvement of Wagner's original solution of the above mentioned Erdős problem was given by Beck [139], who applied the Roth-Halász approach to obtain the following result: There are absolute constants $c, c_0 > 0$ such that

$$\max_{1\le n\le N} \max_{|z|=1} |P_n(z)| > c_0 N^c$$

for all N and all complex numbers $\xi_1, \xi_2, \ldots$ of modulus 1, where $P_n(z) = \prod_{j=1}^{n}(z - \xi_j)$. For further results concerning polynomials on the unit circle see Beck [140].

The first results concerning irregularities of distribution with respect to cubes with sides parallel to the axes are due to Beck and Halász (see [143]). Halász generalized his proof of Schmidt's bound [718] and showed that $\log N/N$ is also a lower bound for the discrepancy with respect to squares with sides parallel to axes. However, Ruzsa [1581] showed that the usual discrepancy and the discrepancy with respect to squares are in fact comparable (Theorem 1.45). Therefore Halász' result can also be obtained by applying Ruzsa's theorem and Schmidt's bound. Theorem 1.41 is a version of a theorem by Beck [143] where all constants are made explicit. Another way of obtaining an explicit bound of this kind is to use a refined version of Proposition 1.47. Theorem 1.46 is due to Drmota [476]. For a special contribution on irregularities of distribution we refer to Chung and Graham [367, 368].

The lower bounds of the usual discrepancy $D_N(\mathbf{x}_n)$ gave rise to various general concepts of discrepancies and irregularities of distribution. Some of them are discussed in Section 2.1. As a general reference to the theory of irregularities of distribution we want to mention the monograph by Beck and Chen [146]. Survey articles are due to Tijdeman [1856, 1857] and Beck [122, 123, 126].

1.4 Special Sequences

1.4.1 $(n\alpha)$-Sequence

The most prominent example of u.d. sequences is the linear sequence $(n\alpha)$ for irrational α. It is clear that for rational $\alpha = p/q$ the sequence $(n\alpha)$ is periodic with period q and therefore not u.d.

Theorem 1.59 *The sequence $(n\alpha)$ is u.d. mod 1 if and only if α is irrational.*

Proof. If $\alpha \notin \mathbf{Q}$ then $h\alpha \notin \mathbf{Z}$ for any integer $h \neq 0$. Hence

$$\left|\sum_{n=1}^{N} e(hn\alpha)\right| = \left|\frac{1 - e(hN\alpha)}{1 - e(h\alpha)}\right| \le \frac{2}{|1 - e(h\alpha)|}$$

and by WEYL's criterion (Theorem 1.19) the result follows. □

The main goal of this section is to provide the necessary methods to prove an explicit formula for the discrepancy $D_N^*(n\alpha)$ for irrational α (Theorem 1.60). For the sake of shortness we will only prove a weak version (Corollary 1.63) rigorously, where the asymptotic leading term is calculated. Since $\alpha \mapsto D_N^*(n\alpha)$ is an even and periodic functions we may assume that $0 < \alpha < \frac{1}{2}$. (We mainly follow SCHOISSENGEIER [1655].)

Let $\alpha = [a_0; a_1, a_2, \ldots]$ denote the continued fraction expansion of α with convergents $r_n = p_n/q_n$. Note that $0 < \alpha < \frac{1}{2}$ implies $a_0 = 0$ and $a_1 > 1$. Let $m \ge 0$ be chosen that $q_m \le N < q_{m+1}$ and let N be (uniquely) represented by

$$N = \sum_{j=0}^{m} b_j q_j$$

such that the integers (= digits) $b_0, b_1, \ldots, b_m$ satisfy $0 \le b_j \le a_{j+1}$, $0 \le b_0 < a_1$, and $b_{j-1} = 0$ if $b_j = a_{j+1}$. (This representation is also called OSTROWSKI representation of N with respect to basis $\alpha = [a_0; a_1, a_2, \ldots]$.) It is clear that the digits b_j may be determined by the following algorithm:

$$\begin{aligned} N &= b_m q_m + N_{m-1}; \qquad 0 \le N_{m-1} < q_m, \\ N_{m-1} &= b_{m-1} q_{m-1} + N_{m-2}; \qquad 0 \le N_{m-2} < q_{m-1}, \\ &\ldots \\ N_0 &= b_0 q_0. \end{aligned}$$

(Observe that $N_j = \sum_{i=0}^{j} b_i q_i$.) Furthermore set

$$A_j = N_{j-1}(\alpha - r_j) + \sum_{t=j}^{m} b_t (q_t \alpha - p_t)$$

for $0 \le j \le m+2$ ($N_{-1} = 0$, $N_m = N_{m+1} = N$). Let $i_N = \min\{j \ge 0 : b_j \ne 0\}$ and set

$$
\begin{aligned}
s &= \min\{j : 2 \not| j \text{ such that } b_{j+1} < a_{j+2} \text{ or } A_j > 0 \text{ or } A_{j+2} > 0, 1 \le j \le m\}, \\
t &= \min\{j : 2 \not| j \text{ s.th. } b_{j+1} < a_{j+2} - 1 \text{ or } A_{j-1} < 0 < A_{j+1} \text{ or } A_{j+2} > 0, 1 \le j \le m\}
\end{aligned}
$$

where $\min \emptyset = \infty$. Finally set

$$
u = \begin{cases} 0 & \text{if } 2|i_N \text{ and } (b_0 < a_1 - 1 \text{ or } A_1 < 0) \\ \min\{s,t\} & \text{otherwise.} \end{cases}
$$

Then the following theorem holds (which will not be proved in detail).

Theorem 1.60 *Suppose that $0 < \alpha < \frac{1}{2}$ is irrational. Then*

$$
\begin{aligned}
N\, D_N^*(n\alpha) &= \sum_{j=0}^{[m/2]} b_{2j}(1 - q_{2j}A_{2j}) + \sum_{j \in S_1} q_j A_j \\
&\quad - \sum_{j \in S_2} q_j A_j - \sum_{j \in S_3} a_{j+1} q_j A_j \qquad (1.110) \\
&\quad + (\delta_{u,0} - 1) q_u A_u + \max\left(0, A_0 - \sum_{j=0}^{m} b_j((-1)^j - q_j A_j)\right),
\end{aligned}
$$

where

$$
\begin{aligned}
S_1 &= \{j : 2|j, 0 \le j \le m, A_{j+1} < 0 < A_{j-1}\}, \\
S_2 &= \{j : 2|j, 0 \le j \le m, A_{j-1} \le 0 < A_{j+1}\}, \\
S_3 &= \{j : 2|j, 0 \le j \le m, A_j < 0\}.
\end{aligned}
$$

Corollary 1.61 *Suppose that $0 < \alpha < \frac{1}{2}$ is irrational and set $P = \{j : 0 \le j \le m, A_j > 0\}$. Then*

$$
\begin{aligned}
N\, D_N^*(n\alpha) &= \max\left(\sum_{j \in P} a_{j+1} q_j A_j - \sum_{j=0}^{[m/2]} b_{2j} q_{2j} A_{2j}, \right. \qquad (1.111) \\
&\qquad \left. \sum_{j=0}^{[(m-1)/2]} b_{2j+1} q_{2j+1} A_{2j+1} - \sum_{j \notin P} a_{j+1} q_j A_j \right) + \theta_N,
\end{aligned}
$$

where $|\theta_N| \le 4$.

In order to derive Corollary 1.61 from Theorem 1.60 we need some information about the behaviour of A_j.

Lemma 1.62 *Suppose that $0 \le j \le m$. If $b_j \ne 0$ then $(-1)^j A_j > 0$. Furthermore the estimates*

$$-\frac{1}{q_{j+1}} < (-1)^j A_j < \frac{1}{q_j} \tag{1.112}$$

and

$$q_j A_j = b_j \frac{(-1)^j}{a_{j+1}} + \mathcal{O}\left(\frac{1}{a_{j+1}}\right) \tag{1.113}$$

hold, especially $a_{j+1} q_j A_j = \mathcal{O}(1)$ if $b_j = 0$.

Proof. We assume that j is even. (The proof is similar if j is odd.) From

$$\begin{aligned}
-(q_j\alpha - p_j) &= \sum_{i=1}^{\infty}((q_{j+2i} - q_{j+2i-2})\alpha - (p_{j+2i} - p_{j+2i-2})) \\
&= \sum_{i=1}^{\infty} a_{j+2i}(q_{j+2i-1}\alpha - p_{j+2i-1}) \\
&< \sum_{1 \le i \le (m+1)/2} b_{j+2i-1}(q_{j+2i-1}\alpha - p_{j+2i-1}) \\
&\le \sum_{t=j}^{m} b_t(q_t\alpha - p_t) < \sum_{i=0}^{\infty} a_{j+2i+1}(q_{j+2i}\alpha - p_{j+2i}) \qquad (1.114)\\
&= \sum_{i=0}^{\infty}((q_{j+2i+1} - q_{j+2i-1})\alpha - (p_{j+2i+1} - p_{j+2i-1})) \\
&= -(q_{j+1}\alpha - p_{j-1}) \qquad (1.115)
\end{aligned}$$

it follows that

$$A_j = \sum_{t=j}^{m} b_t(q_t\alpha - p_t) + N_{j-1}(\alpha - r_j) > -(q_j\alpha - p_j) + b_j(q_j\alpha - p_j) + N_{j-1}(\alpha - r_j) > 0$$

if $b_j \ne 0$ and that

$$A_j < b_j(q_j\alpha - p_j) + p_{j+1} - q_{j+1}\alpha + N_{j-1}(\alpha - r_j).$$

If $b_j < a_{j+1}$ then

$$A_j < (a_{j+1} - 1)(q_j\alpha - p_j) + p_{j+1} - q_{j+1}\alpha + (q_j\alpha - p_j) = -(q_{j-1}\alpha - p_{j-1}) < 1/q_j.$$

If $b_j = a_{j+1}$ then $b_{j-1} = 0$ and therefore $N_{j-1} < q_{j-1}$. Thus

$$A_j < (q_{j+1} - q_{j-1})\alpha - (p_{j+1} - p_{j-1}) + p_{j+1} - q_{j+1}\alpha + q_{j-1}(\alpha - r_j) = 1/q_j,$$

too. On the other hand

$$A_j > \sum_{t=j}^{m} b_t(q_t\alpha - p_t) > -(q_j\alpha - p_j) > -1/q_{j+1}$$

which completes the proof of (1.112).

In order to prove (1.113) set $\alpha_0 = \alpha$ and $\alpha_{j+1} = 1/(\alpha_j - a_j)$, $j \geq 0$. Then

$$\begin{aligned}
q_j A_j &= N_j(q_j\alpha - p_j) + q_j \sum_{t=j+1}^{m} b_t(q_t\alpha - p_t) \\
&= b_j q_j(q_j\alpha - p_j) + \mathcal{O}(N_{j-1}|q_j\alpha - p_j| + q_j|q_j\alpha - p_j|) \\
&= b_j q_j(-1)^j|q_j\alpha - p_j| + \mathcal{O}(q_j|q_j\alpha - p_j|) \\
&= b_j q_j(-1)^j \frac{1}{q_{j+1}} + \mathcal{O}\left(b_j q_j\left(\frac{1}{q_{j+1}} - \frac{1}{q_j\alpha_{j+1} + q_{j-1}}\right) + \frac{q_j}{q_{j+1}}\right) \\
&= b_j \frac{(-1)^j}{a_{j+1}} + \mathcal{O}\left(\frac{1}{a_{j+1}} - \frac{q_j}{q_{j+1}} + \frac{b_j q_j^2}{\alpha_{j+2} q_{j+1}^2} + \frac{1}{a_{j+1}}\right) \\
&= b_j \frac{(-1)^j}{a_{j+1}} + \mathcal{O}\left(\frac{b_j}{a_{j+1}^2} + \frac{1}{a_{j+1}}\right) = b_j \frac{(-1)^j}{a_{j+1}} + \mathcal{O}\left(\frac{1}{a_{j+1}}\right).
\end{aligned}$$

□

Proof of Corollary 1.61. First note that for any integer $\varepsilon \geq -N$ we have $|(N+\varepsilon)D^*_{N+\varepsilon}(n\alpha) - N D^*_N(n\alpha)| \leq |\varepsilon|$. Therefore it does not really matter if we replace N by $N+\varepsilon$ for small ε. Especially we will use the fact that if $u = u_N \neq 0$ then there is an $|\varepsilon| \leq 3$ such that $u_{N+\varepsilon} = 0$.

If $u = u_N \neq 0$ and i_N is even then $b_0 = a_1 - 1$ and $A_1 > 0$. If $a_1 \geq 3$ then we replace N by $N-1$. Clearly $0 \neq b_0' = b_0 - 1 < a_1 - 1$ gives $i_{N-1} = 0$ and $u_{N-1} = 0$. If $a_1 = 2$ we have to use the fact that $A_1 > 0$ which implies $b_1 = 0$ (see Lemma 1.62). Hence, if we replace N by $N+2$ we obtain $b_0'' = b_1'' = 1$ and consequently $i_{N+2} = 0$ and $A_1 < 0$. Thus $u_{N+2} = 0$. If i_N is odd then $b_0 = 0$. If $a_1 \geq 3$ then we can use $N+1$ instead of N; we have $i_{N+1} = 0$ and $b_0' = 1 < a_1 - 1$ which gives $u_{N+1} = 0$. If $a_1 = 2$ and $A_1 < 0$ then we can choose $N+1$, too. Finally, if $a_1 = 2$ and $A_1 > 0$ then $b_1 = 0$. Hence, for $N+3$ we have $b_0''' = b_0''' = 1$ and $A_1 < 0$ which leads to $u_{N+3} = 0$.

Therefore we may assume that $u = 0$. Within this proof let $[a,b]$ denote an interval in $\mathbf{Z}$. If j is even and $j \notin P$, then $j-1 \notin P$ and $j+1 \notin P$. If j is odd and $j \in P$ then $j-1 \in P$ and $j+1 \in P$. Therefore P may be written in the form $P = \cup_{i=1}^{k}[r_i, s_i]$, where $2|r_i$, $2|s_i$ for $1 \leq i \leq k$. We get

$$\begin{aligned}
\sum_{\substack{2|j \\ A_{j+1}<0<A_{j-1}}} q_j A_j - \sum_{\substack{2|j \\ A_{j-1}\leq 0<A_{j+1}}} q_j A_j &= \sum_{\substack{i=1 \\ r_i<s_i}}^{k} q_{s_i}A_{s_i} - \sum_{\substack{i=1 \\ r_i<s_i}}^{k} q_{r_i}A_{r_i} \\
&= \sum_{i=1}^{k} \sum_{\substack{j=r_i+1 \\ 2\nmid j}}^{s_i-1} (q_{j+1}A_{j+1} - q_{j-1}A_{j-1}) \\
&= \sum_{2\nmid j\in P} (q_{j+1}A_{j+1} - q_{j-1}A_{j-1}) \\
&= \sum_{2\nmid j\in P} a_{j+1}q_j A_j.
\end{aligned}$$

Therefore

$$\begin{aligned} N\,D_N^*(\alpha) &= \sum_{2|j} b_j(1-q_jA_j) + \sum_{2|j\notin P} a_{j+1}q_jA_j \\ &\quad + \max\left(0, A_0 - \sum_{j=0}^{m} b_j((-1)^j - q_jA_j)\right) + \theta', \end{aligned} \tag{1.116}$$

where $|\theta'| \le 1$. The last sum of

$$\sum_{2\nmid j\in P} a_{j+1}q_jA_j - \sum_{2|j\in P} a_{j+1}q_jA_j = \sum_{j\in P} a_{j+1}q_jA_j - \sum_{2|j} a_{j+1}q_jA_j$$

is equal to

$$\sum_{2|j}(b_j + q_{j+1}A_{j+1} - q_{j-1}A_{j-1}) = \sum_{2|j} b_j + q_{2[m/2]+1}A_{2[m/2]+1}.$$

Similarly

$$\begin{aligned} &\sum_{2\nmid j\in P} a_{j+1}q_jA_j - \sum_{2|j\notin P} a_{j+1}q_jA_j \\ &\qquad = -\sum_{2\nmid j} b_j - \sum_{j\notin P} a_{j+1}q_jA_j + q_{2[m/2]}A_{2[m/2]} - q_0A_0. \end{aligned}$$

This proves (1.111). □

Corollary 1.63 *Suppose that* $0 < \alpha < \frac{1}{2}$ *is irrational. Then*

$$N\,D_N^*(n\alpha) = \max\left(\sum_{j=0}^{[m/2]} b_{2j}\left(1 - \frac{b_{2j}}{a_{2j+1}}\right), \sum_{j=0}^{[(m-1)/2]} b_{2j+1}\left(1 - \frac{b_{2j+1}}{a_{2j+2}}\right)\right) + \mathcal{O}(m), \tag{1.117}$$

where the $\mathcal{O}$*-constant is an absolute one.*

Proof. Note that $j \in S_1\cup S_2\cup S_3$ implies $b_j = 0$ (see Lemma 1.62) and consequently by (1.112) $|q_jA_j| \le 1$. Hence (1.117) follows from (1.110) and (1.113). □

Note that for almost all α we have $m = \mathcal{O}(\log N)$, while $N\,D_N^*(n\alpha) \ne \mathcal{O}(\log N)$ (see Theorem 1.72 and Proposition 1.74).

Another consequence is a rather crude but (despite of the constants) optimal bound for $D_N^*(n\alpha)$.

Corollary 1.64 *Let* α *be irrational and* $N \ge 1$. *Then*

$$C_1\left(\sum_{j=1}^{m} a_j + b_m\right) \le \max_{N'\le N} N'\,D_{N'}^*(n\alpha) \le C_2\left(\sum_{j=1}^{m} a_j + b_m\right), \tag{1.118}$$

where $C_1, C_2 > 0$ *are absolute constants.*

Proof. For the proof of the upper bound note that for $0 \le x \le a$ we have $x(1 - x/a) \le a/4$. Hence, by Corollary 1.63

$$N\,D_N^*(n\alpha) = \mathcal{O}\left(\sum_{j=1}^{m} \frac{a_j}{4} + b_m + m\right) = \mathcal{O}\left(\sum_{j=1}^{m} a_j + b_m\right).$$

In order to prove the lower bound set $N' = \sum_{j=0}^{m} b_j' q_j \le N$, where $b_j' = [a_{j+1}/2] + 1$ for $j < m$ and $b_m' = b_m - 1$. Suppose for a moment that m is even. (The case of odd m may be treated similiarly.) Then

$$\begin{aligned} N'\,D_{N'}^*(n\alpha) &= \frac{1}{4}\max\left(\sum_{j=0}^{[m/2]-1} a_{2j+1} + b_m,\ \sum_{j=0}^{[(m-1)/2]} a_{2j+2}\right) + \mathcal{O}(m) \\ &\ge \frac{1}{8}\left(\sum_{j=1}^{m} a_j + b_m\right) - Cm, \end{aligned}$$

in which $C > 0$ is an absolute constant. If $\sum_{j=1}^{m} a_j + b_m \ge 16Cm$ then we are done. On the other hand, if $\sum_{j=1}^{m} a_j + b_m < 16Cm$ we can use SCHMIDT's theorem (see Theorem 1.51) to obtain

$$\max_{N < q_{m+1}} N\,D_N^*(n\alpha) \gg \log q_{m+1} \gg m \gg \sum_{j=1}^{m} a_j + b_m,$$

which completes the proof of Corollary 1.64. □

As a special case we obtain the following criterion.

Corollary 1.65 *Let α be irrational. Then $N\,D_N^*(n\alpha) = \mathcal{O}(\log N)$ if and only if the sequence*

$$a_m^{(1)} = \frac{1}{m}\sum_{j=1}^{m} a_j$$

is bounded.

As mentioned above, we will not prove Theorem 1.60 but a little bit weaker version, namely Corollary 1.63. We present a complete proof of Corollary 1.63 without using Theorem 1.60. The proof is organized in the following way. Proposition 1.66 reduces the problem to determining the maximal value of $|N\{\alpha k\} - \sigma^{-1}(k) + 1/2|$, where σ is a permuation on $\{1, \ldots, N\}$ such that $\{\alpha\sigma(k)\} < \{\alpha\sigma(k+1)\}$, $1 \le k < N$. Proposition 1.67 and its Corollary 1.68 provide a further (elementary) reduction of the problem. It remains to evaluate the maximal value of $(\sigma^{-1}(k) - N\{k\alpha\})$ and the sum $\sum_{n=1}^{N}\{n\alpha\}$. The main part of the proof is the derivation of an explicit formula for

$\sigma^{-1}(k)$ in terms of the OSTROWSKI expansion of N and k (see Proposition 1.69). With help of this representation of $\sigma^{-1}(k)$ it is possible to approximate the maximal value of $(\sigma^{-1}(k) - N\{k\alpha\})$ (see Proposition 1.70) and $\sum_{n=1}^{N}\{n\alpha\}$ (see Proposition 1.71).

Proposition 1.66 *Let $0 \le x_n < 1$, $1 \le n \le N$, and let σ be a permutation of the set $\{1,\ldots,N\}$ for which $x_{\sigma(k)} \le x_{\sigma(k+1)}$, $1 \le k < N$. Then*

$$N\,D_N^*(x_n) = \frac{1}{2} + \max_{1\le k\le N}\left|Nx_k - \sigma^{-1}(k) + \frac{1}{2}\right|. \tag{1.119}$$

Proof. W.l.o.g. we may assume that $x_i \le x_{i+1}$, $1 \le i < N$. Furthermore, for notational convenience, we set $x_0 = 0$ and $x_{N+1} = 1$. The distinct values of the numbers x_i, $0 \le i \le N+1$, define a subdivision of $[0,1]$. Therefore,

$$\begin{aligned} N\,D_N^*(x_n) &= \max_{\substack{i=0,\ldots,N\\ x_i<x_{i+1}}} \sup_{x_i<\alpha\le x_{i+1}} \Big|A([0,\alpha);N) - N\alpha\Big| \\ &= \max_{\substack{i=0,\ldots,N\\ x_i<x_{i+1}}} \sup_{x_i<\alpha\le x_{i+1}} \Big|i - N\alpha\Big| . \end{aligned} \tag{1.120}$$

Whenever $x_i < x_{i+1}$, the function $g_i(\alpha) = |i - N\alpha|$ attains its maximum in $[x_i, x_{i+1}]$ at one of the endpoints of the interval. Consequently, we have

$$N\,D_N^*(x_n) = \max_{\substack{i=0,\ldots,N\\ x_i<x_{i+1}}} \max\left(\Big|i - Nx_i\Big|, \Big|i - Nx_{i+1}\Big|\right) \tag{1.121}$$

We show now that we may drop the restriction $x_i < x_{i+1}$ in the first maximum. So suppose we have $x_i < x_{i+1} = x_{i+2} = \cdots = x_{i+r} < x_{i+r+1}$ with some $r \ge 2$. The indices not admitted in the first maximum in 1.121 are the integers $i+j$ with $1 \le j \le r-1$. We shall prove that the numbers

$$|(i+j) - Nx_{i+j}| \quad \text{and} \quad |(i+j) - Nx_{i+j+1}|$$

with $1 \le j \le r-1$, which are excluded in 1.121, are in fact dominated by numbers already occuring in 1.121. For $1 \le j \le r-1$, we get by the same reasoning as above (consider the function $h_{i+1}(y) = |y - Nx_{i+1}|$):

$$\begin{aligned} |(i+j) - Nx_{i+j}| &= |(i+j) - Nx_{i+1}| < \max\left(|i - Nx_{i+1}|, |(i+r) - Nx_{i+1}|\right) \\ &= \max\left(|i - Nx_{i+1}|, |(i+r) - Nx_{i+r}|\right), \end{aligned}$$

and both numbers in the last maximum occur in 1.121. Exactly the same argument holds for $|(i+j) - Nx_{i+j+1}|$, $1 \le j \le r-1$. Thus, we arrive at

$$\begin{aligned} N\,D_N^*(x_n) &= \max_{i=0,\ldots,N} \max\left(|i - Nx_i|, |i - Nx_{i+1}|\right) \\ &= \max_{i=0,\ldots,N} \max\left(|i - Nx_i|, |(i-1) - Nx_i|\right). \end{aligned}$$

The last step is valid because the only terms we dropped are $|0 - Nx_0|$ and $|N - Nx_{N+1}|$, both of which are zero. Since $\max(|x|, |x-1|) = \frac{1}{2} + |x - \frac{1}{2}|$ we finally obtain (1.119). □

The next proposition shows that if $\max_{1\le n\le N}(\sigma^{-1}(n) - N\{n\alpha\})$ is attained at k then the minival value is attained at $k' = N - k + 1$.

Proposition 1.67 *Let $1 \le k \le N$. Then*

$$\sigma^{-1}(k) - N\{k\alpha\} + \sigma^{-1}(N-k+1) - N\{(N-k-1)\alpha\} = N + 1 - 2\sum_{n=1}^{N}\{n\alpha\}.$$

Proof. Obviously the following formula is valid for $x, y \in \mathbf{R}$:

$$\chi_{[0,\{x-y\})}(\{x\}) = \{x-y\} - \{x\} + \{y\}.$$

Taking $x = n\alpha$ and $y = (n-k)\alpha$ we get

$$\chi_{[0,\{k\alpha\})}(\{n\alpha\}) = \{k\alpha\} - \{n\alpha\} + \{(n-k)\alpha\}$$

and therefore by (1.125)

$$\begin{aligned}&\sigma^{-1}(k) - N\{k\alpha\} + \sigma^{-1}(N-k+1) - N\{(N-k+1)\alpha\}\\ &\quad = 2 - 2\sum_{n=1}^{N}\{n\alpha\} + \sum_{n=1}^{N}\{(n-k)\alpha\} + \sum_{n=1}^{N}\{(n-N+k-1)\alpha\}.\end{aligned}$$

The last sum is equal to

$$\begin{aligned}\sum_{n=-N+k}^{k-1}\{n\alpha\} &= \sum_{\substack{n=-N+k\\ n\ne 0}}^{k-1}(1 - \{-n\alpha\})\\ &= N - 1 - \sum_{n=1-k}^{N-k}\{n\alpha\} = N - 1 - \sum_{n=1}^{N}\{(n-k)\alpha\}\end{aligned}$$

which completes the proof of Proposition 1.67. □

Corollary 1.68

$$N\,D_N^*(n\alpha) = \max_{1\le k\le N}(\sigma^{-1}(k) - N\{k\alpha\}) + \max\left\{0, 2\sum_{n=1}^{N}\{n\alpha\} - N\right\}. \qquad (1.122)$$

Proof. By Propositions 1.66 and 1.67 we get

$$\begin{aligned}ND_N^*(\alpha) &= \frac{1}{2} + \max_{1\le k\le N}\left|\sigma^{-1}(k) - N\{k\alpha\} - \frac{1}{2}\right|\\ &= \max\big(\max_{1\le k\le N}(\sigma^{-1}(k) - N\{k\alpha\}), 1 + \max_{1\le k\le N}(N\{k\alpha\} - \sigma^{-1}(k))\big)\end{aligned}$$

$$= \max\Big(\max_{1\le k\le N}(\sigma^{-1}(k) - N\{k\alpha\}),$$

$$1 + \max_{1\le k\le N}(\sigma^{-1}(k) - N\{k\alpha\}) - N - 1 + 2\sum_{n=1}^{N}\{n\alpha\}\Big)$$

$$= \max_{1\le k\le N}(\sigma^{-1}(k) - N\{k\alpha\}) + \max\left(0, 2\sum_{n=1}^{N}\{n\alpha\} - N\right). \qquad (1.123)$$

□

As mentioned above, the essential step of the proof of Corollary 1.63 is to find a formula for $\sigma^{-1}(k)$, where the permutation σ is (uniquely) determined by $\{\alpha\sigma(k)\} < \{\alpha\sigma(k+1)\}$, $1 \le k < N$. In what follows we will frequently use the (digit) expansion

$$k = \sum_{j=0}^{m} c_j q_j,$$

where $0 \le c_0 < a_1$, $0 \le c_j \le a_{j+1}$, and $c_{j-1} = 0$ if $c_j = a_{j+1}$, and the notations

$$k_j = \sum_{i=0}^{j} c_i q_i, \qquad 0 \le j \le m.$$

Proposition 1.69 *Let $1 \le k \le N$. Then*

$$\begin{aligned}\sigma^{-1}(k) &= N\{k\alpha\} + \sum_{j=0}^{m}\left((-1)^j \min(b_j, c_j) - c_j q_j A_j\right)\\ &+ |\mathcal{A}_k| - |\mathcal{B}_k| + \frac{1}{2}(1 - (-1)^{i_k}),\end{aligned}$$

where

$$\begin{aligned}\mathcal{A}_k &= \{j : 0 \le j \le m, 2|j, N_{j-1} < k_{j-1}, k_j \le N_j, k_{j+1} \le N_{j+1}\},\\ \mathcal{B}_k &= \{j : 0 \le j \le m, 2|j, k_{j-1} \le N_{j-1}, k_j \le N_j, N_{j+1} < k_{j+1}\}.\end{aligned}$$

Proof. For $a, k, l \in \mathbf{Z}$, $k, l \ge 0$, set

$$S(a, k, l) = \sum_{a < n \le a+k} \chi_{[0,\{l\alpha\})}(\{n\alpha\}). \qquad (1.124)$$

It is easy to verify that for $1 \le k \le N$

$$\sigma^{-1}(k) = 1 + S(0, N, k) = 1 + \sum_{j=0}^{m} S(N_{j-1}, b_j q_j, k). \qquad (1.125)$$

Since

$$\{1, 2, \ldots, k-1\} = \sigma^{-1}\left(\{j : 1 \le j \le N, \{\alpha j\} < \{\alpha\sigma(k)\}\}\right)$$

we have $k-1 = |\{j : 1 \le j \le N, \{\alpha j\} < \{\alpha\sigma(k)\}\}|$ and hence $\sigma^{-1}(k) = 1+S(0,N,k)$. The second representation of $\sigma^{-1}(k)$ in (1.125) is obvious by definition.

In order to evaluate $S(0,N,k)$ we need some properties of $S(a,k,l)$. The first one is a reciprocity law: (We use the abbreviation $\delta(Expr) = 1$ if $Expr$ is true and $\delta(Expr) = 0$ otherwise.)

$$S(a,k,l) = S(-a-1,l,k)+k\{l\alpha\}-l\{k\alpha\}+\delta(a<0\le a+k<l)-\delta(0\le a<l\le a+k). \tag{1.126}$$

By using the FOURIER series ($0 \le x < 1$, $y \in \mathbf{R}$)

$$\chi_{[0,x)}(\{y\}) = x + \sum_{0\ne p\in\mathbf{Z}} \frac{1}{2\pi i p}(1-e^{2\pi i p x})e^{2\pi i p y} + \frac{1}{2}\chi_{\mathbf{Z}}(y) - \frac{1}{2}\chi_{\mathbf{Z}+x}(y)$$

we obtain

$$\begin{aligned} S(a,k,l) &= k\{l\alpha\} + \sum_{0\ne p\in\mathbf{Z}} \frac{1}{2\pi i p}(1-e^{-2\pi i p\alpha l}) \sum_{n=a+1}^{a+k} e^{2\pi i p\alpha n} \\ &\quad + \frac{1}{2}\sum_{n=a+1}^{a+k} (\chi_{\mathbf{Z}}(n\alpha) - \chi_{\mathbf{Z}}((n-l)\alpha)) \\ &= k\{l\alpha\} + \sum_{0\ne p\in\mathbf{Z}} \frac{1}{2\pi i p}(e^{2\pi i p k\alpha}-1)e^{2\pi i p\alpha(a+1-l)} \frac{e^{2\pi i l p\alpha}-1}{e^{2\pi i p\alpha}-1} \\ &\quad + \frac{1}{2}\delta(a+1\le 0\le a+k) - \frac{1}{2}\delta(a<l\le a+l) \\ &= k\{l\alpha\} + \sum_{n=0}^{l-1}\sum_{0\ne p\in\mathbf{Z}} \frac{1}{2\pi i p}(e^{2\pi i p k\alpha}-1)e^{2\pi i p(a-n)\alpha} \\ &\quad + \frac{1}{2}\delta(a<0\le a+k) - \frac{1}{2}\delta(a<l\le a+l). \end{aligned}$$

Thus, by using the above FOURIER expansion again, we directly get (1.126).

Next, let $j \ge 0$, $1 \le b \le a_{j+1}$ resp. $b < a_1$ if $j=0$, and suppose that $0 \le k < q_{j+1}$ and $bq_j - q_{j+1} \le a < q_{j+1} - k$. We will show that if $a+k<0$ then

$$S(a,k,bq_j) = \frac{1}{2}(1-(-1)^j)k, \tag{1.127}$$

if $a+1 \le 0 \le a+k$ then

$$\begin{aligned} S(a,k,bq_j) &= \frac{1}{2}(1-(-1)^j)k + (-1)^j \min\left(b, \left[\frac{a+k}{q_j}\right]\right) \\ &\quad + \frac{1}{2}(1-(-1)^j)\delta(a+k<bq_j), \end{aligned} \tag{1.128}$$

and if $a+1>0$ then

$$S(a,k,bq_j) = \frac{1}{2}(1-(-1)^j)k + (-1)^j \min\left(b, \left[\frac{a+k}{q_j}\right]\right) \tag{1.129}$$

$$-(-1)^j \min\left(b, \left[\frac{a}{q_j}\right]\right) + \frac{1}{2}(1-(-1)^j)\delta(a+k<bq_j).$$

The basic idea for the proof of (1.127)-(1.129) is that the ordering of $\{j\alpha\}$ is essentially the same as that of $\{jr_{m+1}\}$, more precisely for $\max(|j|,|k|,|j-k|) < q_{m+1}$ we have

$$\{j\alpha\} < \{k\alpha\} \quad \text{if and only if} \quad \{jr_{m+1}\} < \{kr_{m+1}\}. \tag{1.130}$$

Let $0 \le j < q_{m+1}$ and suppose that m is even. Here

$$r_{m+2}j - r_{m+1}j = -\frac{j}{q_{m+1}q_{m+2}} > -\frac{1}{q_{m+2}}$$

which gives $[jr_{m+1}] = [jr_{m+2}]$. Since $r_{m+1} < \alpha < r_{m+2}$ we also obtain $[j\alpha] = [jr_{m+1}]$. The same holds is m is odd. Furthermore, if $-q_{m+1} < j \le -1$ we can use $[-x] = -[x]-1$ and obtain $[j\alpha] = [jr_{m+1}]$, too. Now suppose that $\{j\alpha\} < \{k\alpha\}$. Then $\{j\alpha\} - \{jr_{m+1}\} = j(\alpha - r_m)$ and $\{k\alpha\} - \{kr_{m+1}\} = k(\alpha - r_{m+1})$. Therefore,

$$\begin{aligned}\{jr_{m+1}\} - \{kr_{m+1}\} &= \{j\alpha\} - j(\alpha - r_{m+1}) + k(\alpha - r_{m+1}) + \{k\alpha\} \\ &< (k-j)(\alpha - r_{m+1}).\end{aligned}$$

Since $|\{jr_{m+1}\}-\{kr_{m+1}\}| \ge 1/(q_{m+1})$ and $|(k-j)(\alpha-r_{m+1})| < 1/(q_{m+2})$ we directly obtain $\{jr_{m+1}\} - \{kr_{m+1}\} \le 0$, where equality is impossible. Hence $\{jr_{m+1}\} < \{kr_{m+1}\}$. Similarly $\{jr_{m+1}\} < \{kr_{m+1}\}$ implies $\{j\alpha\} < \{k\alpha\}$. This completes the proof of (1.130).

We will now prove (1.127)-(1.129). If $a+1 \le n \le a+k$ then $-q_{j+1} < a+1 \le n < q_{j+1}$ and $-q_{j+1} < a+1-bq_j \le n-bq_j < q_{j+1}$. (1.130) implies

$$S(a,k,bq_j) = \sum_{a<n\le a+k} c_{[0,\{bq_jr_{j+1}\}]}(\{nr_{j+1}\}).$$

Let us assume that j is even (the argument is similar if j is odd). Then $\{bq_jr_{j+1}\} = \{b/q_{j+1}\}$. We have to find the number of n's with $a < n \le a+k$ and $np_{j+1} \equiv 0,1,\ldots,b-1 (\operatorname{mod} q_{j+1})$, that is $n \equiv 0, q_j, \ldots, (b-1)q_j (\operatorname{mod} q_{j+1})$.

If $a+k<0$ then $n+q_{j+1} = sq_j$ for some s, $0 \le s < b$. This implies $q_j|(n+q_{j-1})$. Therefore for some $u \in \mathbf{Z}$ $n+q_{j-1} = uq_j$. We get $a+1 \le uq_j - q_{j-1}$, that is $(a+q_{j-1})/q_j < u$ and $a_{j+1}+u = s$, that is $u < b - a_{j+1}$. Therefore $a + q_{j-1} < bq_j - a_{j+1}q_j$, a contradiction. This proves (1.127).

If $a+1 \le 0 \le a+k$ then (1.127) implies

$$S(a,k,bq_j) = \sum_{n=0}^{a+k} \chi_{\{[0,b/q_{j+1}\}]}(\{nr_{j+1}\}).$$

The number of n's with $0 \le n \le a+k$ and $0 \le n \le (b-1)q_j$, $q_j|n$, equals $1+\min(b-1, [(a+k)/q_j])$. Hence (1.128) follows.

If $0 < a+1$ we use

$$S(a,k,bq_j) = \sum_{n=0}^{a+k} \chi_{[0,\{bq_{j+1}\}]}(\{nr_{j+1}\}) - S(a,k,bq_j) = \sum_{n=0}^{a} \chi_{[0,b/q_{j+1}\}]}(\{nr_{j+1}\})$$

and (1.128) to get (1.129).

Now we can start to evaluate $\sigma^{-1}(k)$. By (1.125) and (1.126) we obtain

$$\begin{aligned}\sigma^{-1}(k) &= 1+\sum_{j=0}^{m} S(N_{j-1}, b_j q_j, k)\\ &= 1+\sum_{j=0}^{m} (S(-N_{j-1}-1, k, b_j q_j) + b_j q_j\{k\alpha\} - k\{b_j q_j \alpha\})\\ &\quad -\sum_{j=0}^{m} \delta(N_{j-1} < k \le N_j)\\ &= N\{k\alpha\} - k\sum_{j=0}^{m}\{b_j q_j \alpha\}\\ &\quad +\sum_{j=0}^{m} (S(-N_{j-1}-1, k_j, b_j q_j) + S(-N_{j-1}-1+k_j, k-k_j, b_j q_j)).\end{aligned}$$

In order to evaluate $\sum_{j=0}^{m} S(-N_{j-1}-1, k_j, b_j q_j)$ observe that $0 \le k_j < q_{j+1}$ and $b_j q_j - q_{j+1} \le -N_{j-1} - 1 < q_j + 1 - k_j$ for $0 \le j \le m$. Together with (1.127) and (1.128) we get

$$\begin{aligned}\sum_{j=0}^{m} S(-N_{j-1}-1, k_j, b_j q_j) = &\sum_{\substack{j=0\\ k_j \le N_{j-1}}}^{m} \frac{1}{2}(1-(-1)^j) k_j \operatorname{sgn} b_j\\ + &\sum_{\substack{j=0\\ N_{j-1} < k_j}}^{m} \left(\frac{1}{2}(1-(-1)^j) k_j \operatorname{sgn} b_j + (-1)^j \min\left(b_j, \left[\frac{k_j - N_{j-1} - 1}{q_j}\right]\right)\right.\\ &\left. +\frac{1}{2}(1+(-1)^j)\delta(k_j \le N_j)\right).\end{aligned}$$

Now

$$\left[\frac{k_j - N_{j-1} - 1}{q_j}\right] = \left[c_j - \delta(k_{j-1} \le N_{j-1})\right]$$

and therefore

$$\min\left(b_j, \left[\frac{k_j - N_{j-1} - 1}{q_j}\right]\right) = \min(b_j, c_j) - \delta(k_{j-1} \le N_{j-1}, c_j \le b_j).$$

Next observe that "$k_{j-1} \le N_{j-1} < k_j$ and $c_j \le b_j$" is equivalent to "$k_{j-1} \le N_{j-1} < k_j \le N_j$" and that $k_j \le N_{j-1}$ implies $c_j = 0 = \min(b_j, c_j)$. Thus

$$\begin{aligned}&\sum_{\substack{j=0\\ N_{j-1} < k_j \le N_j}}^{m} \frac{1}{2}(1+(-1)^j) - \sum_{k_{j-1} \le N_{j-1} < k_j \le N_j}^{m} (-1)^j\\ &= \sum_{k_{j-1} < N_{j-1} \le k_j \le N_j}^{m} \frac{1}{2}(1+(-1)^j) + \sum_{k_{j-1} \le N_{j-1} < k_j \le N_j}^{m} \frac{1}{2}(1-(-1)^j)\end{aligned}$$

proves

$$\sum_{j=0}^{m} S(-N_{j-1}-1,k_j,b_jq_j)$$
$$=\sum_{j=0}^{m}\left(\frac{1}{2}(1-(-1)^j)k_j\operatorname{sgn} b_j+(-1)^j\min(b_j,c_j)\right)+|\mathcal{A}_k|+|\mathcal{D}_k|$$
$$+\sum_{\substack{j=0\\ k_{j-1}\le N_{j-1}<k_j\le N_j}}^{m}\frac{1}{2}(1-(-1)^j),$$

where

$$|\mathcal{D}_k|=\{j\,:\,0\le j\le m, 2|j, N_{j-1}<k_{j-1}, k_j\le N_j, N_{j+1}<k_{j+1}\}.$$

Similarly, by using (1.126) we obtain

$$S(-N_{j-1}+k_j-1,k-k_j,b_jq_j)$$
$$=\sum_{t=j+1}^{m} S(k_{t-1}-N_{j-1}-1,c_tq_t,b_jq_j)=\sum_{t=j+1}^{m}(c_tq_t\{b_jq_j\alpha\}-b_jq_j\{c_tq_t\alpha\})$$
$$+\sum_{t=j+1}^{m}\delta(k_{t-1}\le N_{j-1}<k_t\le N_j)-\sum_{t=j+1}^{m}\delta(N_{j-1}<k_{t-1}\le N_j<k_t)$$
$$+\sum_{r=j+1}^{m} S(N_{j-1}-k_{t-1},b_jq_j,c_tq_t).$$

Next (1.127)–(1.129) yield

$$\sum_{t=j+1}^{m} S(N_{j-1}-k_{t-1},b_jq_j,c_tq_t)=\sum_{t=j+1}^{m}\frac{1}{2}(1-(-1)^t)b_jq_j\operatorname{sgn} c_t$$
$$+\sum_{\substack{t=j+1\\ N_{j-1}<k_{t-1}\le N_j<k_t}}^{m}\frac{1}{2}(1+(-1)^t)-\sum_{\substack{t=j+1\\ k_{t-1}\le N_{j-1}<k_t\le N_j}}^{m}\frac{1}{2}(1+(-1)^t)$$
$$-\sum_{\substack{t=j+1\\ k_{t-1}\le N_{j-1}}}^{m}(-1)^t\min\left(c_t,\left[\frac{N_{j-1}-k_{t-1}}{q_t}\right]\right)+\sum_{\substack{t=j+1\\ k_{t-1}\le N_j}}^{m}(-1)^t\min\left(c_t,\left[\frac{N_j-k_{t-1}}{q_t}\right]\right).$$

Now $t>j$ and $k_{t-1}\le N$ imply $0\le N_j-k_{t-1}<q_t$; therefore the two last sums vanish. Furthermore, $t>j$ and $k_{t-1}\le N_{j-1}<k_t\le N_j$ imply $k_t<q_{j+1}$ and therefore $c_t=0$. This contradiction proves that the third sum is empty and that

$$\sum_{t=j+1}^{m}\delta(k_{t-1}\le N_{j-1}<k_t\le N_j)=0.$$

Thus

$$\sum_{j=0}^{m} S(-N_{j-1}+k_j-1, k-k_j, b_jq_j)$$

$$=\sum_{j-0}^{m}\sum_{t=j+1}^{m}(c_tq_t\{b_jq_j\alpha\}-b_jq_j\{c_tq_t\alpha\})+\sum_{j=0}^{m}\sum_{t=j+1}^{m}\frac{1}{2}(1-(-1)^t)b_jq_j\operatorname{sgn}c_t$$

$$+\sum_{j=0}^{m}\sum_{\substack{t=j+1\\ N_{j-1}<k_{t-1}\leq N_j<k_t}}^{m}\frac{1}{2}((-1)^t-1).$$

Now use that $(\frac{1}{2}(1-(-1)^t)-\{c_tq_t\alpha\})\operatorname{sgn}c_t=-c_t(q_t\alpha-p_t)$ and that the last sum is equal

$$\sum_{t=0}^{m}\sum_{\substack{j=0\\ N_{j-1}<k_{t-1}\leq N_j<k_t}}^{t-1}\frac{1}{2}((-1)^t-1).$$

The existence of a j with $0\leq j<t$ and $N_{j-1}<k_{t-1}\leq N_j<k_t$ is equivalent to $0<k_{t-1}\leq N_{t-1}<k_t$. Therefore this sum is equal to

$$\sum_{\substack{j=0\\ 0<k_{j-1}\leq N_{j-1}<k_j}}^{m}\frac{1}{2}((-1)^j-1)$$

$$=\sum_{\substack{j=0\\ k_{j-1}\leq N_{j-1}<k_j}}^{m}\frac{1}{2}((-1)^j-1)-\sum_{\substack{j=0\\ k_{j-1}=0<k_j}}^{m}\frac{1}{2}((-1)^j-1)$$

$$=\sum_{\substack{j=0\\ k_{j-1}\leq N_{j-1}<k_j\leq N_j}}^{m}\frac{1}{2}((-1)^j-1)+\sum_{\substack{j=0\\ k_{j-1}\leq N_{j-1}\leq N_j<k_j}}^{m}\frac{1}{2}((-1)^j-1)+\frac{1}{2}(1-(-1)^{i_k}).$$

The second sum is equal to

$$-\sum_{\substack{j=0\\ k_j\leq N_j\leq N_{j+1}<k_{j+1},2|j}}^{m-1}1=-|\mathcal{B}_k|-|\mathcal{D}_k|.$$

This finally gives

$$\sum_{j=0}^{k} S(-N_{j-1}+k_j-1, k-k_j, b_jq_j)$$

$$=\sum_{j=0}^{m}\left((k-k_j)\{b_jq_j\alpha\}-b_jq_j\sum_{t=j+1}^{m}c_t(q_t\alpha-p_t)\right)-|\mathcal{B}_k|-|\mathcal{D}_k|$$

$$+\frac{1}{2}(1-(-1)^{i_k})-\sum_{\substack{j=0\\ k_{j-1}\leq N_{j-1}<k_j\leq N_j}}^{k}\frac{1}{2}(1-(-1)^j).$$

Hence

$$\sigma^{-1}(k) = N\{k\alpha\} + \sum_{j=0}^{m}\left(\frac{1}{2}(1-(-1)^j)k_j \operatorname{sgn} b_j + (-1)^j \min(b_j,c_j) - k_j\{b_jq_j\alpha\} \right.$$
$$\left. - b_jq_j \sum_{t=j+1}^{m} c_t(q_t\alpha - p_t)\right) + \frac{1}{2}(1-(-1)^{i_k}) + |\mathcal{A}_k| - |\mathcal{B}_k|.$$

Finally, if we use that $\frac{1}{2}(1-(-1)^j)\operatorname{sgn} b_j - \{b_jq_j\alpha\} = -b_j(q_j\alpha - p_j)$ and that

$$\sum_{j=0}^{m}(k_jb_j(q_j\alpha - p_j) + b_jq_j \sum_{t=j+1}^{m} c_t(q_t\alpha - p_t)$$
$$= \sum_{t=0}^{m} c_tq_t \sum_{j=t}^{m} b_j(q_j\alpha - p_j) + \sum_{t=1}^{m} c_t(q_t\alpha - p_t)N_{t-1}$$

we have completed the proof of Proposition 1.69. □

With help of this representation of $\sigma^{-1}(k)$ we can detemine the first part of (1.122) approximately.

Proposition 1.70

$$\max_{1\le k\le N}(\sigma^{-1}(k) - N\{k\alpha\}) = \sum_{j=0}^{[m/2]} b_{2j}(1 - q_{2j}A_{2j}) + \mathcal{O}(m), \tag{1.131}$$

where the $\mathcal{O}$-constant is an absolute one.

Proof. By Proposition 1.69 we just have to approximate the maximum of

$$\sum_{j=0}^{m}\left((-1)^j \min(b_j,c_j) - c_jq_jA_j\right)$$

since the remaining terms are $\mathcal{O}(m)$.

If $b_j = 0$ then by Lemma 1.62 $|c_jq_jA_j| \le |a_{j+1}q_jA_j| = \mathcal{O}(1)$. Hence the choice $c_j = b_j = 0$ gives a proper approximation.

If $b_j > 0$ and j is even then by (1.112) $0 < q_jA_j < 1$. Thus

$$\max_{0\le c_j\le a_{j+1}}(\min(b_j,c_j) - c_jq_jA_j) = b_j(1 - q_jA_j)$$

and consequently $c_j = b_j$ is a proper choice.

Finally suppose that j is odd and $b_j > 0$. In this case we have by Lemma 1.62 $q_jA_j < 0$. Hence

$$\max_{0\le c_j\le b_j}(-\min(b_j,c_j) - c_jq_jA_j) = 0$$

and using (1.113)

$$\max_{b_j < c_j \le a_{j+1}} (-\min(b_j, c_j) - c_j q_j A_j) = b_j - a_{j+1} q_j A_j = \mathcal{O}(1).$$

Note that there might be situations where we are forced to assume that $c_j \le b_j$, e.g. if m is even then one should choose $c_m = b_m > 0$ and therefore $m \le N$ implies $c_{m-1} \le b_{m-1}$. Hence we have covered all possible cases and (1.131) follows. □

Another application of Proposition 1.69 combined with Proposition 1.67 enables us to determine the second term in (1.122).

Proposition 1.71

$$2\sum_{n=1}^{N}\{n\alpha\} - N = A_0 - \sum_{j=0}^{m} b_j((-1)^j - q_j A_j).$$

Proof. We take $k = 1$ in Proposition 1.67. Note that $\mathcal{A}_1 = \{j : j \text{ even}, 0 < j < m, N_{j-1} = 0 < N_j\}$ and therefore $|\mathcal{A}_1| = 0$ if $i_N = 0$. Obviously $\mathcal{B}_1 = \mathcal{A}_N = \mathcal{B}_N = \emptyset$. We get

$$\begin{aligned}\sigma^{-1}(1) - N\alpha &= \min(b_0, 1) - q_0 A_0 + |\mathcal{A}_1| \\ &= \frac{1}{2}(1 + (-1)^{i_N})\operatorname{sgn} i_N + 1 - \operatorname{sgn} i_N - q_0 A_0 = 1 - q_0 A_0 - \frac{1}{2}(1 - (-1)^{i_N})\end{aligned}$$

and

$$\sigma^{-1}(N) - N\{N\alpha\} = \sum_{j=0}^{m} b_j((-1)^j - q_j A_j) + \frac{1}{2}(1 - (-1)^{i_N})$$

which proves the proposition. □

Now a combination of Corollary 1.68, Propositions 1.70 and 1.71, and the relation $A_0 = \mathcal{O}(1)$ completes the proof of Corollary 1.63.

We already mentioned that the optimal case in which $N D_N^*(n\alpha) = \mathcal{O}(\log N)$ is not typical. A theorem by KHINTCHINE [921, 922] gives a complete answer.

Theorem 1.72 *Suppose that $\psi(n)$ is a positive increasing function. Then*

$$\max_{N' \le N} N' D_{N'}^*(n\alpha) = \mathcal{O}(\log N \cdot \psi(\log N))$$

for almost all $\alpha \in \mathbf{R}$ if and only if

$$\sum_{n=1}^{\infty} \frac{1}{n\psi(n)} < \infty.$$

The proof of KHINTCHINE's theorem relies on two metrical properties of continued fraction expansions.

Proposition 1.73 *Suppose that* $\psi(n)$ *is a positive increasing function. Then*

$$\sum_{n=1}^{\infty} \frac{1}{n\psi(n)} < \infty$$

if and only if

$$a_1 + a_2 + \ldots a_m = \mathcal{O}(m\psi(m))$$

for almost all $\alpha = [a_0; a_1, a_2, \ldots] \in \mathbf{R}$.

Proposition 1.74 *For almost all* $\alpha \in \mathbf{R}$ *we have*

$$\lim_{m\to\infty} \frac{\log q_m}{m} = \frac{\pi^2}{12\log 2},$$

in which q_m *denotes the denominator of the m-the convergent* p_m/q_m *of* α.

Now Theorem 1.72 follows directly from Propsitions 1.73 and 1.74 and from Corollary 1.64.

Remark. We finally want to mention that KHINTCHINE's theorem may be reformulated in the following way: *Let* $\varphi(n)$ *be a positive increasing function. Then*

$$\max_{N'\le N} N'\, D^*_{N'}(n\alpha) = \mathcal{O}(\log N \cdot \varphi(\log\log N))$$

for almost all $\alpha \in \mathbf{R}$ *if and only if*

$$\sum_{n=1}^{\infty} \frac{1}{\varphi(n)} < \infty$$

You only have to set $\varphi(\log x) = \psi(x)$ and use that fact that $\sum 1/(n\psi(n)) < \infty$ if and only if $\sum 1/\varphi(n) < \infty$.

We complete this section with an interesting property of the symmetrisized $(n\alpha)$-sequence $(y_n) = (\alpha, -\alpha, 2\alpha, -2\alpha, \ldots)$. If the partial quotients a_n of α are bounded then the discrepancy $D_N(y_n)$ essentially has the same upper bound as $D_N(n\alpha)$, i.e.

$$N\, D_N(y_n) = \mathcal{O}(\log N)$$

which is best possible by SCHMIDT's theorem [1630] (Theorem 1.51). However, the situation changes if we consider the L^2-discrepancy. The asymptotic behaviour of $D_N^{(2)}(y_n)$ is different from that of $D_N^{(2)}(n\alpha)$, where we have $N\, D_N^{(2)}(n\alpha) = \Omega(\log N)$.

Theorem 1.75 *Suppose that the irrational number* α *has bounded continued fractions* $a_n \le C$. *Then the* L^2*-discrepancy of the symmetrisized* $(n\alpha)$*-sequence* $(y_n) = (\alpha, -\alpha, 2\alpha, -2\alpha, \ldots)$ *is bounded by*

$$N\, D_N^{(2)}(y_n) \le \sqrt{2}C(\log N)^{1/2}$$

for $N \ge N_0(C)$.

By ROTH's theorem 1.40 this estimate is best possible.

Proof. By PARSEVAL's theorem the L^2-discrepancy of a sequence (x_n) satisfies

$$\left(D_N^{(2)}(x_n)\right)^2 = \sum_{h\in\mathbb{Z}} |c_h|^2,$$

where c_h are the FOURIER coefficients

$$c_h = \int_0^1 \left(x - \frac{1}{N}\sum_{n=1}^N \chi_{[0,x)}(x_n) \right) e(hx)\, dx,$$

i.e.

$$c_0 = \frac{1}{N}\sum_{n=1}^N \left(\frac{1}{2} - \{x_n\}\right)$$

and

$$c_h = \frac{1}{2\pi i h N}\sum_{n=1}^N e(hx_n)$$

for $h \neq 0$. If $(x_n) = (y_n) = (\alpha, -\alpha, 2\alpha, -2\alpha, \ldots)$ denote the symmetrisized $(n\alpha)$-sequence we obviously obtain (for even N) $c_0 = 0$ and

$$|c_h| \le \frac{1}{\pi |h| N}\, \frac{1}{\max(1, \|h\alpha\|)}.$$

Since

$$\sum_{|h|>N} |c_h|^2 \le \frac{2}{\pi^2 N} \sum_{|h|>N} \frac{1}{h^2} \le \frac{2}{\pi^2 N^2}$$

it suffices to estimate

$$S = \sum_{h=1}^N \frac{1}{h^2 \|h\alpha\|^2}.$$

Partial summation provides

$$S = \sum_{h=1}^{N-1} \left(\frac{1}{h^2} - \frac{1}{(h+1)^2}\right) S_h + \frac{S_N}{N^2},$$

where

$$S_h = \sum_{j=1}^h \frac{1}{\|j\alpha\|^2}.$$

For $0 \le p < q \le h$ we have

$$\Big| \|q\alpha\| - \|p\alpha\| \Big| \ge \|q\alpha \pm p\alpha\| = \|(q \pm p)\alpha\| \ge \frac{1}{2hC}$$

which implies that in each of the intervals

$$\left[0, \frac{1}{2hC}\right), \left[\frac{1}{2hC}, \frac{2}{2hC}\right), \ldots, \left[\frac{h}{2hC}, \frac{h+1}{2hC}\right)$$

there is at most one number of the form $\|j\alpha\|$, $1 \le j \le h$, with no such number lying in the first interval. Therefore

$$S_h = \sum_{j=1}^{h} \frac{1}{\|j\alpha\|^2} \le \sum_{j=1}^{h} \left(\frac{2hC}{j}\right)^2 \le \frac{2\pi^2 C^2}{3} h^2$$

which gives

$$|S| \le \sum_{h=1}^{N-1} \frac{1}{h^3} \frac{2\pi^2 C^2}{3} h^2 + \frac{2\pi^2 C^2}{3} \le \pi^2 C^2 \log N$$

for $N \ge N_0$. Hence

$$\left(D_N^{(2)}(y_n)\right)^2 \le \frac{2C^2}{N^2} \log N$$

for sufficiently large N. Since $|ND_N^{(2)} - (N+1)D_{N+1}^{(2)}| \le 1$ we can cover the case of odd N, too. □

1.4.2 KRONECKER-Sequence

Let $\alpha_1, \ldots, \alpha_k \in \mathbf{R}$. Then the k-dimensional sequence $\mathbf{x}_n = (\alpha_1 n, \ldots, \alpha_k n)$ is called KRONECKER sequence. The direct generalization of Theorem 1.59 is the following one, due to WEYL [1953].

Theorem 1.76 *The k-dimensional sequence $(\alpha_1 n, \ldots, \alpha_k n)$ is u.d. mod 1 if and only if $1, \alpha_1, \ldots, \alpha_k$ are linearly independent over* $\mathbf{Z}$.

Its *Proof* is exactly the same as that of Theorem 1.59, it is a direct application of WEYL's criterion (Theorem 1.19).

However, the essential difference between the one-dimensional case and the higher dimensional one is that there is no analogue to the continued fraction expansion which solves the problem of diophantine approximation as explicit as in the one-dimensional case. Nevertheless there are methods to obtain bounds for the discrepancy. We want to present two of them, one for the usual discrepancy D_N and one for the isotropic discrepancy J_N. It is interesting that those bounds obtained for J_N are optimal for badly approximable $(\alpha_1, \ldots, \alpha_k)$ but the corresponding bounds for D_N seem to be far away from optimality. We present here results of LARCHER (see Theorem 1.81), which mainly use methods from the geometry of numbers. The proofs in this section are based on methods from the geometry of numbers; as a general reference we mention here GRUBER and LEKKERKERKER [707]. The main part of this section will be a generalization of KHINTCHINE's Theorem 1.72 due to BECK [142].

Suppose that $1, \alpha_1, \ldots, \alpha_k$ are linearly independent over $\mathbf{Z}$. Let $h_1, \ldots, h_k$ be integers (not all of them zero). Then (one) main problem in the theory of diophantine approximation is to find lower bounds for

$$\|h_1\alpha_1 + \cdots + h_k\alpha_k\|.$$

Definition 1.77 *For $\underline{\alpha} = (\alpha_1, \ldots, \alpha_k) \in \mathbf{R}^k$ and $q \geq 1$ set*

$$\delta_q(\underline{\alpha}) = \max_{1 \leq j \leq k} \|q\alpha_j\|.$$

(As usual $\|x\| = \min(\{x\}, 1 - \{x\})$ denotes the distance to the nearest integers.) The best simultaneous approximation denominators $q_m = q_m(\underline{\alpha})$, $m \geq 0$, to $\underline{\alpha}$ with respect to the maximum norm are inductively defined by $q_0 = 1$ and q_{m+1} is the least positive integer such that $\delta_{q_{m+1}}(\underline{\alpha}) < \delta_{q_m}(\underline{\alpha})$. A rational vector $\mathbf{r}_m = (p_{1m}/q_m, \ldots, p_{km}/q_m)$ is called convergent of $\underline{\alpha}$ if $\|q_m(\underline{\alpha} - \mathbf{r})\|_\infty = \delta_{q_m}(\underline{\alpha})$.

The successive minima of the k-dimensional lattice Γ_m generated by $\mathbf{Z}^k$ and $\mathbf{r}_m$ are denoted by $\mu_{1m}, \ldots, \mu_{km}$, i.e.

$$\mu_{jm} = \min\{\mu > 0 : \text{there are } j \text{ linear independent lattice points } |\mathbf{g}| \leq \mu\}.$$

The determinant of Γ_m equals $1/q_m$.

Furthermore set

$$a_m = \left[\frac{q_{m+1}}{q_m}\right]$$

$\underline{\alpha} = (\alpha_1, \ldots, \alpha_k) \in \mathbf{R}^k$ is called badly approximable if there is a constant $C > 0$ such that for all $q > 0$

$$\delta_q(\underline{\alpha}) \geq \frac{C}{q^{1/k}}.$$

Remark 1. The lattice Γ_m is exactly the set $\{\{n\mathbf{r}_m\} : n \in \mathbf{N}\} + \mathbf{Z}^k$. Furthermore, since $\mathbf{r}_m$ is a convergent we have $\gcd(p_{1m}, \ldots, p_{km}, q_m) = 1$ which implies that the sequence of fractional parts $(\{n\mathbf{r}_m\})$ has period q_m. Therefore, the determinant of Γ_m equals $1/q_m$.

Remark 2. Note that in the one dimensional case $k = 1$ the best approximation denominators are the denominators q_m of the convergents p_m/q_m of the continued fraction expansion of $\alpha = \alpha_1$. Furthermore

$$\frac{1}{q_{m+2}} < \delta_{q_m}(\alpha_1) < \frac{1}{q_{m+1}}.$$

This means that α is badly approximable if and only if the sequence a_m is bounded. In general we only have that badly approximable $\underline{\alpha}$ have bounded a_m but the converse need not be true.

Lemma 1.78 **(DIRICHLET's Approximation Theorem)** *Suppose that $\alpha_1, \ldots, \alpha_k$ are real numbers. Then for any $Q > 0$ there exists a positive integer $q \le Q^k$ and integers $q_1, \ldots, q_k$ such that*

$$|q\alpha_1 - q_1| < \frac{1}{Q}, \ldots, |q\alpha_k - q_k| < \frac{1}{Q} \tag{1.132}$$

resp. there exist integers $h, h_1, h_2, \ldots, h_k$ satisfying $|h_1| \le Q^{1/k}, \ldots, |h_k| \le Q^{1/k}$ and

$$\|h_1\alpha_1 + \cdots + h_k\alpha_k\| < \frac{1}{Q}. \tag{1.133}$$

Proof. We will only prove (1.132). The proof of (1.133) is quite similar. For $0 \le q \le Q^k$ set $\mathbf{x}_q = (\{q\alpha_1\}, \ldots, \{q\alpha_k\})$. By pigeonhole principle there exist $0 \le q' < q'' \le Q^k$ such that $\|\mathbf{x}_{q''} - \mathbf{x}_{q'}\|_\infty < 1/Q$. Hence we have proved (1.132) with $q = q'' - q'$. □

Observe that the definition of the best simultaneous approximation denominators q_m and Lemma 1.78 directly imply that for every m

$$\alpha_j = \frac{p_{jm}}{q_m} + \frac{\theta_{mj}}{q_m(q_{m+1}-1)^{1/k}}, \qquad 1 \le j \le k,$$

with $|\theta_{mj}| \le 1$, i.e. $\delta_{q_m}(\underline{\alpha})q_m^{1/k} \le \delta_{q_m}(\underline{\alpha})q_{m+1}^{1/k} = \mathcal{O}(1)$. Furthermore, note that Lemma 1.78 is best possible for badly approximable $\underline{\alpha} = (\alpha_1, \ldots, \alpha_k)$.

It is easy to verify that badly approximable $\underline{\alpha}$ actually exist, e.g. if $1, \alpha_1, \ldots, \alpha_k$ are linearly independent over $\mathbf{Z}$ and if the degree of the extension $[\mathbf{Q}(\alpha_1, \ldots, \alpha_k) : \mathbf{Q}] = k+1$ then $\underline{\alpha}$ is badly approximable. Let $\alpha_j^{(i)}$, $1 \le i \le k+1$, denote the conjugates of $\alpha_j = \alpha_j^{(1)}$. Then

$$\begin{aligned} 1 &\le |N(h_1\alpha_1 + \cdots + h_k\alpha_k)| \\ &= |h_1\alpha_1 + \cdots + h_k\alpha_k| \cdot \prod_{i=2}^{k+1} |h_1\alpha_1^{(i)} + \cdots + h_k\alpha_k^{(i)}| \\ &\le |h_1\alpha_1 + \cdots + h_k\alpha_k| \cdot k \max_{1\le i,j\le k} |\alpha_j^{(i+1)}|^k \cdot \max_{1\le j\le k} |h_j|^k. \end{aligned}$$

By Lemma 1.79 this implies that $\underline{\alpha}$ is badly approximable.

Lemma 1.79 *Let $\underline{\alpha} = (\alpha_1, \ldots, \alpha_k) \in \mathbf{R}^k$ and suppose that $1, \alpha_1, \ldots, \alpha_k$ are linearly independent over $\mathbf{Z}$. Then the following properties are equivalent.*

- *For every $c_1 > 0$ there are infinitely many integers $q \ge 1$ with*

$$\delta_q(\underline{\alpha})q^{1/k} \le c_1. \tag{1.134}$$

- *For every $c_2 > 0$ there are infinitely many integral points $(h_1, h_2, \ldots, h_k, h) \in \mathbf{Z}^{k+1}$ with*

$$|h_1\alpha_1 + h_2\alpha_2 + \cdots h_k\alpha_k - h| \cdot \max_{1\le j\le k} |h_j|^k \le c_2. \tag{1.135}$$

Proof. Suppose that (1.134) is satisfied, i.e. there exist $p_1, \ldots, p_k \in \mathbf{Z}$ such that

$$|q\alpha_j - p_j| \le c_1 q^{-1/k}, \qquad 1 \le j \le k.$$

Consider the linear forms

$$\begin{aligned} L_1(x_1, \ldots, x_k, x) &= x_1, \\ &\vdots \\ L_k(x_1, \ldots, x_k, x) &= x_k, \\ L_{k+1}(x_1, \ldots, x_k, x) &= p_1 x_1 + \cdots + p_k x_k - qx \end{aligned}$$

with q as absolute value of the determinant. By MINKOWSKI's theorem on linear forms there exists a non-zero integral point $(h_1, \ldots, h_k, h) \in \mathbf{Z}^{k+1}$ such that

$$|L_j(h_1, \ldots, h_k), h| = |h_j| \le q^{+1/k}, \qquad 1 \le j \le k,$$

and

$$|L_{k+1}(h_1, \ldots, h_k, h)| = |p_1 h_1 + \cdots + p_k h_k - qh| < 1.$$

Since $p_1 h_1 + \cdots + p_k h_k - qh \in \mathbf{Z}$ we also have $p_1 h_1 + \cdots + p_k h_k - qh = 0$. Hence

$$\begin{aligned} |h_1\alpha_1 + \cdots + h_k\alpha_k - h| \cdot \max_{1 \le j \le k} |h_j|^k &\le |h_1\alpha_1 + + \cdots h_k\alpha_k - h|q \\ &= |h_1(q\alpha_1 - p_1) + \cdots + h_k(q\alpha_k - p_k)| \\ &\le kq^{1/k} c_1 q^{-1/k} = kc_1 \end{aligned}$$

which proves (1.135) for $c_2 = kc_1$. Furthermore it is impossible that there are only finitely many integral points $(h_1, \ldots, h_k, h) \in \mathbf{Z}^{k+1}$ satisfying (1.135) since

$$0 < |h_1\alpha_1 + \cdots + h_k\alpha_k - h|q \le kc_1$$

would lead to a contradiction for sufficiently large q.

Now suppose that (1.135) is satisfied. W.l.o.g. we may assume that $Q = |h_k| = \max_{1 \le j \le k} |h_j|$. Now consider the linear forms

$$\begin{aligned} L_1(x_1, \ldots, x_k, x) &= -x_1 + \alpha_1 x, \\ &\vdots \\ L_{k-1}(x_1, \ldots, x_k, x) &= -x_{k-1} + \alpha_{k-1} x, \\ L_k(x_1, \ldots, x_k, x) &= x_k, \\ L_{k+1}(x_1, \ldots, x_k, x) &= h_1 x_1 + \cdots + h_k x_k - hx \end{aligned}$$

with determinant $|h_k| = Q$. Again by MINKOWSKI's theorem on linear forms there exists a non-zero integral point $(p_1, \ldots, p_k, q) \in \mathbf{Z}^{k+1}$ with

$$|L_j(p_1, \ldots, p_k, q)| = |q\alpha_j - p_j| \le cQ^{-1}, \qquad 1 \le j \le k-1,$$

$$|L_k(p_1,\dots,p_k,q)| = |q| \le c^{-(k-1)}Q^k,$$

and

$$|L_{k+1}(p_1,\dots,p_k,q)| = |h_1p_1+\cdots+h_kp_k-hq| < 1,$$

where $c = c_2^{1/k}$. As above, we also have $h_1p_1+\cdots+h_kp_k-hq=0$. Hence

$$|q\alpha_j - p_j|\cdot|q|^{1/k} \le cQ^{-1}c^{-(k-1)/k}Q = c_2^{1/k^2}, \qquad 1\le j\le k-1.$$

Furthermore from

$$\sum_{j=1}^{k} h_j(q\alpha_j - p_j) = q\left(\sum_{j=1}^{k} h_j\alpha_j - h\right)$$

we obtain

$$\begin{aligned}|h_k(q\alpha_k - p_k)| &= \left|q\left(\sum_{j=1}^{k} h_j\alpha_j - h\right) - \sum_{j=1}^{k-1} h_j(q\alpha_j-p_j)\right| \\ &\le c_2c^{-(k-1)} + (k-1)c = kc_2^{1/k}\end{aligned}$$

which implies

$$|q\alpha_k - p_k|\cdot|q|^{1/k} \le kc_2^{1/k^2}.$$

This proves (1.134) with $c_1 = kc_2^{1/k^2}$. □

The first theorem provides an upper bound for the usual discrepancy D_N by the use of the inequality of ERDŐS-TURAN-KOKSMA (Theorem 1.21).

Theorem 1.80 *Suppose that* $1,\alpha_1,\dots,\alpha_k$ *are linearly independent over* $\mathbf{Z}$ *and that there exists a function* $\varphi:\mathbf{R}^+\to\mathbf{R}^+$ *such that* $\varphi(t)/t$ *is monotonically increasing and that*

$$\|h_1\alpha_1+\cdots+h_k\alpha_k\| \ge \frac{1}{\varphi(\max\{|h_1|,\dots,|h_k|\})}$$

for all $(h_1,\dots,h_k)\in\mathbf{Z}^k\setminus\mathbf{0}$. *Then*

$$D_N(n(\alpha_1,\dots,\alpha_k)) = \mathcal{O}\left(\frac{\log N\,\log\gamma(N)}{\gamma(N)}\right), \tag{1.136}$$

where $\gamma(x)$ *denotes the inverse function of* $\varphi(x)$.

Proof. By the inequality of ERDŐS-TURÁN-KOKSMA (Theorem 1.21) we have

$$D_N(n\underline{\alpha}) \le 3^k\left(\frac{1}{H} + \sum_{0<\|\mathbf{h}\|_\infty\le H}\frac{1}{r(\mathbf{h})}\frac{2}{N\|\mathbf{h}\cdot\underline{\alpha}\|}\right),$$

where $\underline{\alpha}$ denotes $(\alpha_1,\dots,\alpha_k)$, $\mathbf{h}$ denotes $(h_1,\dots,h_k)$. For any $\mathbf{r}=(r_1,\dots,r_k)$ $1\le r_j\le[(\log H)/(\log 2)]$, $1\le j\le k$, let $P_\mathbf{r}$ denote the set of $\mathbf{h}=(h_1,\dots,h_k)$, $0<\|\mathbf{h}\|_\infty\le H$, with

$$2^{r_j-1}\le|h_j|<2^{r_j}, \qquad 1\le j\le k,$$

resp. with $h_j = 0$ if $r_j = 1$. Hence

$$S = \sum_{0<\|\mathbf{h}\|_\infty \le H} \frac{1}{r(\mathbf{h})} \frac{2}{N\|\mathbf{h}\cdot\underline{\alpha}\|} \le \frac{2^{1-k}}{N} \sum_{\mathbf{r}} 2^{-(r_1+\cdots+r_k)} \sum_{\mathbf{h}\in P_\mathbf{r}} \frac{1}{\|\mathbf{h}\cdot\underline{\alpha}\|}.$$

Now consider a fixed set $P_\mathbf{r}$ and assume that $r_1 = \max_{1\le j\le k} r_j$. For every $\mathbf{h} \in P_\mathbf{r}$ we have

$$\frac{1}{2} \ge \|\mathbf{h}\cdot\underline{\alpha}\| \ge \frac{1}{\varphi(\|\mathbf{h}\|_\infty)} \ge \frac{1}{\varphi(2^{r_1})}$$

which implies that there exists a positive integer $l \le [\varphi(2^{r_1})]$ with

$$l/\varphi(2^{r_1}) \le \|\mathbf{h}\cdot\underline{\alpha}\| < (l+1)/\varphi(2^{r_1}). \tag{1.137}$$

The next step is to show that for any l there are at most two integral points $\mathbf{h}$ satisfying (1.137). If there were three of them then there are two points $\mathbf{h}, \mathbf{h}'$ such that $\|\mathbf{h}\cdot\underline{\alpha}\| = \{\mathbf{h}\cdot\underline{\alpha}\}$ and $\|\mathbf{h}'\cdot\underline{\alpha}\| = \{\mathbf{h}'\cdot\underline{\alpha}\}$ or $\|\mathbf{h}\cdot\underline{\alpha}\| = 1-\{\mathbf{h}\cdot\underline{\alpha}\}$ and $\|\mathbf{h}'\cdot\underline{\alpha}\| = 1-\{\mathbf{h}'\cdot\underline{\alpha}\}$. Hence it would follow that $\|(\mathbf{h}-\mathbf{h}')\cdot\underline{\alpha}\| < 1/\varphi(2^{r_1})$ and $0 < \|\mathbf{h}-\mathbf{h}'\|_\infty < 2^{r_1-1}$ and thus

$$\frac{1}{\varphi(2^{r_1})} > \|(\mathbf{h}-\mathbf{h}')\cdot\underline{\alpha}\| \ge \frac{1}{\|\mathbf{h}-\mathbf{h}'\|_\infty} > \frac{1}{\varphi(2^{r_1-1})}$$

which contradicts the assumption on φ. Therefore we obtain

$$\sum_{\mathbf{h}\in P_\mathbf{r}} \frac{1}{\|\mathbf{h}\cdot\underline{\alpha}\|} \le 2\varphi(2^{r_1}) \sum_{1\le l\le \varphi(2^{r_1})} \frac{1}{l} = \mathcal{O}(\varphi(2^{r_1})\log(\varphi(2^{r_1}))),$$

and consequently

$$\begin{aligned} S &\le \frac{c_1}{N} \sum_{r_1\ge\cdots\ge r_k} 2^{-(r_1+\cdots+r_k)}\varphi(2^{r_1})\log(\varphi(2^{r_1})) \\ &\le \frac{c_2}{N} \sum_{r_1} \frac{\varphi(2^{r_1})}{2^{r_1}} \log(\varphi(2^{r_1})) \\ &\le \frac{c_3}{N}\frac{\varphi(H)}{H}\log(\varphi(H))\log H \end{aligned}$$

for some constants $c_1, c_2, c_3 > 0$. This implies

$$D_N(n\underline{\alpha}) \le 3^k \left(\frac{1}{H} + \frac{c_3}{N}\frac{\varphi(H)}{H}\log(\varphi(H))\log H\right).$$

Now by choosing $H = [\gamma(N)]$ we directly obtain (1.136). □

Especially if $\underline{\alpha} = (\alpha_1, \ldots, \alpha_k)$ is badly approximable (i.e. $\varphi(x) = cx^k$) then we obtain $D_N = \mathcal{O}(N^{-1/k}(\log N)^2)$. However, this bound is surely not best possible. We now present a sharp bound for the isotropic discrepancy.

Theorem 1.81 *Suppose that $k \ge 2$ and that $1, \alpha_1, \ldots, \alpha_k$ are linearly independent over* **Z**. *Then $\underline{\alpha} = (\alpha_1, \ldots, \alpha_k)$ is badly approximable if and only if*

$$J_N(n\alpha_1, \ldots, n\alpha_k) = \mathcal{O}\left(N^{-1/k}\right).$$

In order to prove Theorem 1.81 we need a series of lemmata.

Lemma 1.82 *Let* $\mathbf{x}_n = (x_n^{(1)}, \ldots, x_n^{(k)})$, $\mathbf{y}_n = (y_n^{(1)}, \ldots, y_n^{(k)})$, $1 \le n \le N$. *be two finite sequences in* $\mathbf{R}^k$ *and set* $d = \max_{1\le n\le N} \max_{1\le j\le k} \|x_n^{(j)} - y_n^{(j)}\|$. *Then*

$$J_N(\mathbf{x}_n) \le 3^k J_N(\mathbf{y}_n) + 2kd(1+2d)^{k-1}.$$

Proof. Let P be any convex subset or $\mathbf{R}^k/\mathbf{Z}^k$. W.l.o.g. we may assume that $P \subseteq [0,1)^k \equiv \mathbf{R}^k/\mathbf{Z}^k$. Let P_0 and P_1 be outer and inner parallel regions to P in distance d in $\mathbf{R}^k$. (P_1 of course can be empty.) We have $\lambda_k(P \setminus P_1) \le 2kd$ and $\lambda_k(P_0 \setminus P) \le \sigma(P_0)d$, where $\sigma(P_0)$ denotes the $(k-1)$-dimensional surface measure of P_0. Since $P_0 \subseteq [-d, 1+d]^k$ we have $\sigma(P_0) \le 2k(1+2d)^{k-1}$ and $\lambda_k(P_0 \setminus P) \le 2kd(1+2d)^{k-1}$.

Let $\mathbf{Y} = \{\mathbf{y}_1, \ldots, \mathbf{y}_N\} + \mathbf{Z}^k$. Then

$$\begin{aligned}\sum_{n=1}^{N} \chi_{P_1}(\mathbf{y}_n) - N\lambda_k(P_1) - N\lambda_k(P\setminus P_1) &\le \sum_{n=1}^{N} \chi_P(\mathbf{x}_n) - N\lambda_k(P)\\ &\le \sum_{\mathbf{y}\in\mathbf{Y}} \chi_{P_0}(\mathbf{y}) - N\lambda_k(P_0) + N\lambda_k(P_0\setminus P)\end{aligned}$$

and so

$$\begin{aligned}&\left|\sum_{n=1}^{N} \chi_P(\mathbf{x}_n) - N\lambda_k(P)\right|\\ &\qquad \le \max\left(N\,J_N(\mathbf{y}_n) + 2kdN, \left|\sum_{\mathbf{y}\in\mathbf{Y}} \chi_{P_0}(\mathbf{y}) - N\lambda_k(P_0)\right| + 2kd(1+2d)^{k-1}\right).\end{aligned}$$

Since $d < 1$ we have $P_0 \subseteq [-1,2)^k$. We subdivide $[-1,2)^k$ into 3^k (translated) unit cubes W_i, $1 \le i \le 3^k$. Then $P_0 \cap W_i$ are convex and therefore

$$\left|\sum_{\mathbf{y}\in\mathbf{Y}} \chi_{P_0}(\mathbf{y}) - N\lambda_k(P_0)\right| \le \sum_{i=1}^{3^k} \left|\sum_{\mathbf{y}\in\mathbf{Y}} \chi_{P_0\cap W_i}(\mathbf{y}) - N\lambda_k(P_0\cap W_i)\right| \le 3^k N\,J_N(\mathbf{y}_n).$$

This completes the proof of Lemma 1.82. □

Lemma 1.83 *Let* $\mathbf{r} = (\frac{p_1}{q}, \ldots, \frac{p_k}{q})$ *be a (non-zero) vector of rational numbers such that* $\gcd(p_1, \ldots, p_k, q) = 1$. *Then*

$$\frac{c_1}{\mu_1\mu_2\cdots\mu_{k-1}} \le q\,J_q(n\mathbf{r}) \le \frac{c_2}{\mu_1\mu_2\cdots\mu_{k-1}},$$

where c_1, c_2 *are constants only depending on the dimension* k *and* $\mu_1 \le \mu_2 \le \cdots \le \mu_k$ *are the successive minima of the lattice generated by* $\mathbf{Z}^k$ *and* $\mathbf{r}$.

Proof. From $\gcd(p_1, \ldots, p_k, q) = 1$ it follows that the lattice Γ generated by $\mathbf{Z}^k$ and $\mathbf{r}$ has determinant $1/q$. Let $\mathcal{F}$ be a covering of $\mathbf{R}^k$ by fundamental regions F of Γ then every $F \in \mathcal{F}$ contains exactly one point of the periodically expanded set $\{\{n\mathbf{r}\} : n \in \mathbf{N}\} + \mathbf{Z}^k$. Consider a convex set $P \subseteq [0,1)^k$. Since the diameter $\operatorname{diam}(F)$ of $F \in \mathcal{F}$ obviously satisfies $\operatorname{diam}(F) \le k\mu_k$ the area of the set of all $F \in \mathcal{F}$ for which the intersection with the boundary of P is not empty is at most $c_2\mu_k$ with an absolute constant c_2 (only depending on the dimension k). Hence, since $\lambda_k(F) = 1/q$ we have

$$\left|\sum_{n=1}^{q} \chi_P(n\mathbf{r}) - q\lambda_k(P)\right| \le c_2 q\mu_k$$

which implies

$$q\, J_q(n\mathbf{r}) \le c_2 q\mu_k \le \frac{c_2}{\mu_1\mu_2\cdots\mu_{k-1}}.$$

Now, let F_0 denote the fundamental region of Γ generated by those lattice points $\mathbf{m}_1, \ldots, \mathbf{m}_k$ corresponding to the successive minima $\mu_1, \ldots, \mu_k$. Let H denote the hyperplane generated by $\mathbf{m}_1, \ldots, \mathbf{m}_{k-1}$. Then the distance d_H between H and $H + \mathbf{m}_k$ is bounded by

$$d_H \ge \frac{1}{q\mu_1\mu_2\cdots\mu_{k-1}}$$

and the space L between H and $H + \mathbf{m}_k$ contains no lattice points of Γ. Furthermore, by using a simple averaging argument it follows that there exists a lattice point $\mathbf{g} \in \Gamma$ such that the convex set $C = (\mathbf{g} + L) \cap [0,1)^k$ has volume $\lambda_k(C) \ge c_1 d_H$, where $c_1 > 0$ is a constant depending on the dimension k. Since C contains no lattice points of Γ we finally obtain $q\, J_q(n\mathbf{r}) \ge c_1 q d_H$ which completes the proof of Lemma 1.83. □

Corollary 1.84 *Let $q = q_m$ be a best approximation denominator of $\underline{\alpha} = (\alpha_1, \ldots, \alpha_k)$ and $\mathbf{r} = \mathbf{r}_m = \left(\frac{p_1}{q}, \ldots, \frac{p_k}{q}\right)$ its corresponding approximation of $\underline{\alpha}$. Furthermore, let q' be the largest best approximation denominator of $\mathbf{r}$ less than q. Then, with certain constants $c_1, c_2 > 0$,*

$$\frac{c_1}{q(\delta_{q'}(\mathbf{r}))^{k-1}} \prod_{j=2}^{k-1} \frac{\mu_1}{\mu_j} \le J_q(n\mathbf{r}) \le \frac{c_2}{q(\delta_{q'}(\mathbf{r}))^{k-1}} \prod_{j=2}^{k-1} \frac{\mu_1}{\mu_j}.$$

Proof. Since q' is a best approximation denominator of $\mathbf{r}$, i.e. $\delta_{q'}(\mathbf{r}) = \min_{n \le q'} \delta_n(\mathbf{r})$, and for any $\mathbf{x} \in \mathbf{R}^k$ we have $\|\mathbf{x}\|_\infty \le |x| \le \sqrt{k}\|\mathbf{x}\|_\infty$ we immediately obtain $\delta_{q'}(\mathbf{r}) \le \mu_1 \le \sqrt{k}\delta_{q'}(\mathbf{r})$ which proves the Corollary. □

The next step is to approximate $\delta_{q'}(\mathbf{r})$.

Lemma 1.85 *In addition to the assumptions of Corollary 1.84 assume that $q = q_m > 2^k$ and set $\overline{q} = q_{m-1}$. Then*

$$\frac{1}{2}\delta_{\overline{q}}(\underline{\alpha}) \le \delta_{q'}(\mathbf{r}) \le 2\delta_q(\underline{\alpha}).$$

Proof. We have $\alpha_j = p_j/q + r_j$, in which $|qr_j| \le q^{-1/k} < 1/2$. Hence

$$\begin{aligned}\delta_{q'}(\mathbf{r}) &= \max_{1\le j\le k}\left\|q'\frac{p_j}{q}\right\| \le \max_{1\le j\le k}\left\|\overline{q}\frac{p_j}{q}\right\| \\ &= \max_{1\le j\le k}\|\overline{q}\alpha_j - \overline{q}r_j\| \le \max_{1\le j\le k}\|\overline{q}\alpha_j\| + \max_{1\le j\le k}|\overline{q}r_j| \\ &\le \delta_q(\underline{\alpha}) + \max_{1\le j\le k}|qr_j| = \delta_q(\underline{\alpha}) + \max_{1\le j\le k}\|qr_j\| \\ &= \delta_q(\underline{\alpha}) + \max_{1\le j\le k}\|q\alpha_j\| \le 2\delta_q(\underline{\alpha}).\end{aligned}$$

On the other hand we have

$$\begin{aligned}\delta_{\overline{q}}(\underline{\alpha}) &\le \delta_{q'}(\underline{\alpha}) = \max_{1\le j\le k}\left\|q'\frac{p_j}{q} + q'r_j\right\| \\ &\le \max_{1\le j\le k}\left\|q'\frac{p_j}{q}\right\| + \max_{1\le j\le k}|q'r_j| = \delta_{q'}(\mathbf{r}) + \frac{q'}{q}\max_{1\le j\le k}\|qr_j\| \\ &= \delta_{q'}(\mathbf{r}) + \frac{q'}{q}\delta_q(\underline{\alpha}) \le \delta_{q'}(\mathbf{r}) + \frac{q'}{q}\delta_{\overline{q}}(\underline{\alpha}).\end{aligned}$$

Thus, if $q' \le q/2$ then $\delta_{\overline{q}}(\underline{\alpha}) \le 2\delta_{q'}(\mathbf{r})$. Conversely, if $q' > q/2$ then $0 < q - q' < q/2$. In the same manner we obtain $\delta_{\overline{q}}(\underline{\alpha}) \le 2\delta_{q-q'}(\mathbf{r}) = 2\delta_{q'}(\mathbf{r})$. □

Proposition 1.86 *Let q_m, $m \ge 1$, be the best approximation denominators of $\underline{\alpha} = (\alpha_1, \ldots, \alpha_k)$. If $N = b_r q_r + b_{r-1}q_{r-1} + \cdots + b_1 q_1 + b_0$ with $b_m \le a_m$, $0 \le m \le r$, then we have*

$$N\,J_N(n\underline{\alpha}) \le c_3 \sum_{m=1}^{r} \frac{b_m}{(\delta_{q_{m-1}}(\underline{\alpha}))^{k-1}} \prod_{j=2}^{k} \frac{\mu_{1m}}{\mu_{jm}} + c_4 \sum_{m=1}^{r} \frac{b_m q_m^{1-1/k}}{a_m^{1/k}}$$

with constants $c_3, c_4 > 0$ just depending on the dimension k.

Proof. Obviously we have

$$N\,J_N(n\underline{\alpha}) \le \sum_{m=0}^{r} b_m q_m J_{q_m}(n\underline{\alpha}).$$

Therefore it suffices to estimate $J_{q_m}(n\underline{\alpha})$. By Lemma 1.82, Corollary 1.84, and Lemma 1.85 we obtain

$$\begin{aligned}J_{q_m}(n\underline{\alpha}) &\le 3^k J_{q_m}(n\mathbf{r}_m) + \frac{k2^k}{(q_{m+1}-1)^{1/k}} \\ &\le \frac{c_3'}{q_m(\delta_{q'}(\mathbf{r}_m))^{k-1}} \prod_{j=2}^{k} \frac{\mu_{1m}}{\mu_{jm}} + \frac{c_4}{q_m^{1/k} a_m^{1/k}} \\ &\le \frac{c_3}{q_m(\delta_{q_{m-1}}(\underline{\alpha}))^{k-1}} \prod_{j=2}^{k} \frac{\mu_{1m}}{\mu_{jm}} + \frac{c_4}{q_m^{1/k} a_m^{1/k}},\end{aligned}$$

in which $\delta_{q_{-1}}$ stands for 1. □

Proposition 1.87 *Let q_m be a sufficiently large best simultaneous approximation denominator of $\underline{\alpha} \in \mathbf{R}^k$ such that*

$$8\sqrt{k}\mu_{1m}\cdots\mu_{k-1,m}q_m\delta_{q_m}(\underline{\alpha}) \le 1. \tag{1.138}$$

Then $N = q_m[1/(8\sqrt{k}\mu_{k-1,m}q_m\delta_{q_m}(\underline{\alpha}))]$ satisfies

$$N\, J_N(n\underline{\alpha}) \ge \frac{c_5}{q_m\delta_{q_m}(\underline{\alpha})(\mu_{1m}\cdots\mu_{k-1,m})^2}$$

with an absolute constant $c_5 > 0$.

Proof. Set $N = Bq_m$, where

$$B = \left[\frac{1}{8\sqrt{k}\mu_{1m}\cdots\mu_{k-1,m}q_m\delta_{q_m}(\underline{\alpha})}\right]$$

First we consider the sequence $(n\mathbf{r}_m)$, $1 \le n \le q_m$. As in the proof of Lemma 1.83 let H denote the hyperplane generated by $\mathbf{m}_1, \ldots, \mathbf{m}_{k-1}$, where $\mathbf{m}_j \in \Gamma_m$ is the lattice point corresponding to μ_{jm}, i.e. $|\mathbf{m}_j| = \mu_{jm}$. The distance d_H between H and $H+\mathbf{m}_k$ is bounded by

$$d_H \ge \frac{1}{q\mu_1\mu_2\cdots\mu_{k-1}}$$

and the space L between H and $H+\mathbf{m}_k$ contains no lattice points of Γ_m. Furthermore there exists a lattice point $\mathbf{g} \in \Gamma_m$ such that for every hyperplane H'' parallel to H and lying between $\mathbf{g}+H$ and $\mathbf{g}+\mathbf{m}_k+H$ the $(k-1)$-dimensional volume of $H''\cap[0,1)^k$ is larger than a constant $c > 0$. Now for every $b \le B$ we have for sufficiently large q_m

$$\begin{aligned}
\|n\mathbf{r}_m - (n+bq_m)\underline{\alpha}\| &\le \sqrt{k}\cdot\max_{1\le j\le k}\left\| n\frac{p_{jm}}{q_m} - (n+bq_m)\left(\frac{p_{jm}}{q_m} + \frac{\theta_{jm}}{q_m q_{m+1}^{1/k}}\right)\right\| \\
&= \sqrt{k}\cdot\max_{1\le j\le k}\left\|(n+bq_m)\frac{\theta_{jm}}{q_m q_{m+1}^{1/k}}\right\| \\
&\le \sqrt{k}\,(n+bq_m)\max_{1\le j\le k}\left\|\frac{\theta_{jm}}{q_m q_{m+1}^{1/k}}\right\| \\
&= \sqrt{k}\,\frac{n+bq_m}{q_m}\delta_{q_m}(\underline{\alpha}) \le \sqrt{k}\,2B\delta_{q_m}(\underline{\alpha}).
\end{aligned}$$

Thus if L'' denotes the inner parallel region of L with distance $\sqrt{k}\,2B\delta_{q_m}(\underline{\alpha})$ then $C = L''\cap[0,1)^k$ contains no points $\{n\underline{\alpha}\}$, $1 \le n \le N = Bq_m$ and has volume

$$\lambda_k(C) \ge c(d_H - 4\sqrt{k}B\delta_{q_m}(\underline{\alpha})) \ge \frac{c}{4q_m\mu_{1m}\cdots\mu_{k-1,m}}$$

and so

$$N\,J_N(n\underline{\alpha}) \ge N\lambda_k(C) \ge \frac{cB}{4\mu_{1m}\cdots\mu_{k-1,m}}$$

which proves Proposition 1.87. □

Lemma 1.88 *Let q_m be the best simultaneous approximation denominators of $\underline{\alpha} = (\alpha_1, \ldots, \alpha_k) \in \mathbf{R}^k \setminus \mathbf{Q}^k$. Then*

$$q_{m+2^{k+1}} \geq 2q_{m+1} + q_m.$$

Proof. Suppose that

$$q_{m+2^{k+1}} < 2q_{m+1} + q_m. \tag{1.139}$$

By the pigeonhole principle there exist $0 \leq i_1 < i_2 \leq 2^{k+1}$ such that

$$p_{j,m+i_1} \equiv p_{j,m+i_2} \bmod 2, \quad 1 \leq j \leq k \quad \text{and} \quad q_{m+i_1} \equiv q_{m+i_2} \bmod 2.$$

Set $q = (q_{m+i_2} - q_{m+i_1})/2$ and $p_j = (p_{j,m+i_2} - p_{j,m+i_1})/2$, $1 \leq j \leq k$. By (1.139) we have

$$0 < q \leq \frac{1}{2}(q_{m+2^{k+1}} - q_m) < q_{m+1},$$

but

$$\begin{aligned}
\delta_q(\underline{\alpha}) &\leq \max_{1\leq j\leq k} |q\underline{\alpha} - p_j| \\
&\leq \frac{1}{2}\left(\max_{1\leq j\leq k} |q_{m+i_2}\underline{\alpha} - p_{j,m+i_2}| + \max_{1\leq j\leq k} |q_{m+i_1}\underline{\alpha} - p_{j,m+i_1}|\right) \\
&= \frac{1}{2}\left(\delta_{q_{m+i_1}}(\underline{\alpha}) + \delta_{q_{m+i_2}}(\underline{\alpha})\right) \leq \frac{1}{2}\left(\delta_{q_m}(\underline{\alpha}) + \delta_{q_{m+1}}(\underline{\alpha})\right) \\
&< \delta_{q_m}(\underline{\alpha})
\end{aligned}$$

contradicts the definition of q_{m+1}. Hence $q_{m+2^{k+1}} \geq 2q_{m+1} + q_m$ which proves Lemma 1.88. □

We will also have to use the following criterion for badly approximable $\underline{\alpha}$.

Proposition 1.89 $\underline{\alpha} = (\alpha_1, \ldots, \alpha_k) \in \mathbf{R}^k$, $k \geq 2$, *is badly approximable if and only if one of the following three sequences is bounded.*

$$\begin{aligned}
A_m &= \frac{a_m^{1/k}}{\mu_{1m}\mu_{2m}\cdots\mu_{k-1,m}q_m^{1-1/k}}, \\
B_m &= \frac{\min\left(a_m^{1/k}, (q_m\delta_{q_m}(\underline{\alpha})\mu_{1m}\mu_{2m}\cdots\mu_{k-1,m})^{-1/k}\right)}{\mu_{1m}\mu_{2m}\cdots\mu_{k-1,m}q_m^{1-1/k}}, \\
C_m &= \frac{a_m^{1/k}}{q_m^{1-1/k}(\delta_{q_{m-1}}(\underline{\alpha}))^{k-1}}.
\end{aligned}$$

Proof. $\underline{\alpha}$ is badly approximable if and only if $q_m^{1/k}\delta_{q_m}(\underline{\alpha}) \geq C$ for some $C > 0$. Since $q_{m+1}^{1/k}\delta_{q_m}(\underline{\alpha}) = \mathcal{O}(1)$ it follows that $a_m = [q_{m+1}/q_m]$ is bounded. Hence C_m is bounded. Conversely, if C_m is bounded then a_m and $1/(q_m^{1/k}\delta_{q_{m-1}}(\underline{\alpha}))$ are bounded. Thus $q_m^{1/k}\delta_{q_m}(\underline{\alpha}) \geq C$ for some $C > 0$.

Now suppose that C_m is bounded. By Lemma 1.85 $\frac{1}{2}\delta_{q_{m-1}}(\underline{\alpha}) \le 2\delta_{q'}(\mathbf{r}_m) \le \mu_{1m} \le \mu_{2m} \le \cdots$ and therefore the sequence

$$A_m \le \frac{a_m^{1/k}}{\mu_{1m}^{k-1} q_m^{1-1/k}} \le 2^{k-1} C_m$$

is bounded, too. Let now A_m be bounded. From MINKOWSKI's theorem on successive minima it follows that there is a constant $c_1 > 0$ such that $1 \le q_m \mu_{1m} \cdots \mu_{km} \le c_1$. Hence $q_m^{1/k} \mu_{km} \ge 1$ and therefore $q_m^{1-1/k} \mu_{1m} \cdots \mu_{k-1,m}$ is bounded. Thus a_m and $1/(q_m^{1-1/k} \mu_{1m} \cdots \mu_{k-1,m})$ are bounded, too. By Lemma 1.85 $\mu_{1m} \le \sqrt{k}\delta_{q'}(\mathbf{r}_m) \le 2\sqrt{k}\delta_{q_m}(\underline{\alpha})$ and therefore it remains to show that $1/(q_m^{1/k} \mu_{1m})$ is bounded. Since $1/(q_m^{1-1/k} \mu_{1m} \cdots \mu_{k-1,m})$ is bounded it follows from MINKOWSKI's theorem that $q_m^{1/k} \mu_{km}$ is bounded. Finally, since $q_m \mu_{1m} \cdots \mu_{km} \ge 1$ it follows that $q_m^{1/k} \mu_{1m}$ is bounded below.

By definition $B_m \le A_m$. Furthermore, since $q_m^{1-1/k} \mu_{1m} \cdots \mu_{k-1,m} \le c_2$ for some constant $c_2 > 0$ and $(q_m a_m)^{1/k} \delta_{q_m}(\underline{\alpha}) = \mathcal{O}(1)$ it follows that

$$A_m \le \frac{c_3}{\mu_{1m} \cdots \mu_{k-1,m} q_m \delta_{q_m}(\underline{\alpha})} \tag{1.140}$$

for some $c_3 > 0$. Hence

$$B_m \ge \min\left(A_m, \frac{(A_m/c_3)^{1/k}}{c_2}\right)$$

which finally implies that A_m is bounded if and only if B_m is bounded. □

Proof of Theorem 1.81. If $\underline{\alpha}$ is badly approximable then a_m and C_m (cf. Proposition 1.89) are bounded. Hence by Proposition 1.86 and by using $b_r q_r \le N$

$$\begin{aligned}
N^{1/k} J_N(n\underline{\alpha}) \le\ & c_3 \sum_{m=1}^{r} \frac{b_m}{(b_r q_r)^{1-1/k} (\delta_{q_{m-1}}(\underline{\alpha}))^{k-1}} + c_4 \sum_{m=1}^{r} \frac{b_m q_m^{1-1/k}}{(b_r q_r)^{1-1/k} a_m^{1/k}} \\
=\ & c_3 \sum_{m=1}^{r} \frac{b_m^{1/k}}{q_m^{1-1/k} (\delta_{q_{m-1}}(\underline{\alpha}))^{k-1}} \left(\frac{b_m}{b_r}\right)^{1-1/k} \left(\frac{q_m}{q_r}\right)^{1-1/k} \\
& + c_4 \sum_{m=1}^{r} \left(\frac{b_m}{a_m}\right)^{1/k} \left(\frac{b_m}{b_r}\right)^{1-1/k} \left(\frac{q_m}{q_r}\right)^{1-1/k} \\
\le\ & \sum_{m=0}^{r} \left(c_3 C_m a_m^{1-1/k} + c_4 a_m^{1-1/k}\right) \left(\frac{q_m}{q_r}\right)^{1-1/k}.
\end{aligned}$$

By Lemma 1.88 we have $q_{m+2^{k+1}} \ge 3q_m$ for all $m \ge 0$ and thus

$$\sum_{m=0}^{r} \left(\frac{q_m}{q_r}\right)^{1-1/k} \le 2^{k+1} \sum_{m=0}^{\infty} 3^{-m(k-1)/k} < \infty.$$

Thus $N^{1/k}J_N(n\underline{\alpha})$ is bounded.

If $\underline{\alpha}$ is not badly approximable then for every $\varepsilon > 0$ there is a m with $q_m^{1/k}\delta_{q_m}(\underline{\alpha}) < \varepsilon$. Furthermore, by MINKOWSKI's theorem on succesive minima we have $\mu_{1m}\mu_{2m}\cdots\mu_{k-1,m}q_m^{1-1/k} < c'$ for every q_m (with some constant c' only depending on the dimension k; compare with the proof of Proposition 1.89). Let $\varepsilon < (8\sqrt{k}c')^{-1}$. Then for q_m as above we have $8\sqrt{k}\mu_{1m}\mu_{2m}\cdots\mu_{k-1,m}q_m\delta_q(\underline{\alpha}) \le 1$. Therefore Lemma 1.87 holds with $N = Bq_m$, where $B = [1/(8\sqrt{k}\mu_{1m}\mu_{2m}\cdots\mu_{k-1,m}q_m\delta_q(\underline{\alpha}))]$, and we obtain

$$N^{1/k}J_N(n\underline{\alpha}) \ge \frac{1}{(Bq_m)^{1-1/k}} \cdot \frac{c'}{q_m\delta_q(\underline{\alpha})(\mu_{1m}\mu_{2m}\cdots\mu_{k-1,m})^2} \ge \frac{c''}{\varepsilon^{1/k}}.$$

Thus $N^{1/k}J_N(n\underline{\alpha})$ is unbounded. □

In some sense the notion of isotropic discrepancy seems to be the proper generalization to the higher dimensional case if one is interested in bounds like Corollary 1.64. By using exactly the same methods as in the proof of the first part of Theorem 1.81 we obtain the following estimate.

Theorem 1.90 *Suppose that there exists $\sigma \ge 1/2$ such that*

$$q_{m+1}^{\sigma}\delta_{q_m}(\underline{\alpha}) \ge c_1, \qquad m \ge 0,$$

for some constant $c_1 > 0$ (and q_m denote the best simultaneous approximation denominators of $\underline{\alpha} \in \mathbf{R}^k$). Then there exists a constant $c_2 > 0$ such that for every N with $q_m \le N \le q_{m+1}$

$$N^{1-\sigma(k-1)}J_N(n\underline{\alpha}) \le c_2 \max_{0\le j\le m} a_j^{1-\sigma(k-1)}.$$

We now come back to the usual discrepancy D_N. It seems to be a very difficult problem to describe the behaviour of $D_N(n(\alpha_1,\ldots,\alpha_k))$ in terms of $(\alpha_1,\ldots,\alpha_k) \in \mathbf{R}^k$. The only known upper bounds are surely not optimal. However, there is a remarkable analogue to KHINTCHINE's theorem 1.72 to the more dimensional case which is due to BECK [142].

Theorem 1.91 *Let $\varphi(n)$ be a positive increasing function and $k \ge 1$. Then*

$$\max_{N'\le N} N' D_{N'}(n\alpha_1,\ldots,n\alpha_k) = \mathcal{O}((\log N)^k \cdot \varphi(\log\log N))$$

for almost all $\underline{\alpha} \in \mathbf{R}^k$ if and only if

$$\sum_{n=1}^{\infty} \frac{1}{\varphi(n)} < \infty.$$

We will give a sketch of the proof. The first step is to find a proper representation of the discrepancy function

$$\Delta_N(n\underline{\alpha},\mathbf{x}) = \sum_{n=1}^{N} \chi_{[\mathbf{0},\mathbf{x})}(n\underline{\alpha}) - Nx_1\cdots x_k,$$

where $\mathbf{x} = (x_1,\ldots,x_k) \in [0,1)^k$. The only disadvantage is that the function $\mathbf{x} \mapsto \Delta_N(n\underline{\alpha},\mathbf{x})$ is not continuous. In order to avoid this disadvantage we replace $\Delta_N(n\underline{\alpha},\mathbf{x})$ by $\overline{\Delta}_N(n\underline{\alpha},\mathbf{x})$ which may be interpreted as a "roof-like average" of translated boxes. Set

$$\begin{aligned}\overline{\Delta}_N(n\underline{\alpha},\mathbf{x}) &= \left(\frac{N^2}{2}\right)^k \frac{1}{2}\int_{-2}^{2} \int_{[-\frac{2}{N^2},\frac{2}{N^2}]^k} \prod_{j=1}^{k}\left(1-\frac{N^2}{2}|u_j|\right)\cdot\left(1-\frac{|u_{k+1}|}{2}\right)\\ &\quad\times \Delta_{u_{k+1},N+u_{k+1}}(n\underline{\alpha},\mathbf{u},\mathbf{u}+\mathbf{x})\,d\mathbf{u}\,du_{k+1},\end{aligned}$$

in which

$$\Delta_{a,b}(n\underline{\alpha},\mathbf{u},\mathbf{y}) = \sum_{a<n\le b}\prod_{j=1}^{k}\chi_{(u_j,y_j)}(\{n\alpha_j\}) - (b-a)\prod_{j=1}^{k}(y_j-u_j).$$

By using the FEJÉR kernel identity

$$\frac{N^2}{2}\int_{-\frac{2}{N^2}}^{\frac{2}{N^2}}\left(1-\frac{N^2}{2}|y|\right)e(ky)dy = \left(\frac{\sin(2\pi\frac{k}{N^2})}{2\pi\frac{k}{N^2}}\right)^2$$

and POISSON's summation formula

$$\sum_{\mathbf{m}\in\mathbf{Z}^{k+1}} f(\mathbf{m}) = \sum_{\mathbf{h}\in\mathbf{Z}^{k+1}}\int_{\mathbf{R}^{k+1}} f(\mathbf{y})e(-\mathbf{h}\cdot\mathbf{y})\,d\mathbf{y}$$

the following representation of $\overline{\Delta}_N(n\underline{\alpha},\mathbf{x})$ can be derived.

Proposition 1.92 *For almost all $\underline{\alpha} = (\alpha_1,\ldots,\alpha_k) \in \mathbf{R}^k$ we have*

$$\begin{aligned}\overline{\Delta}_N(n\underline{\alpha},\mathbf{x}) &= i^{k-1}\sum_{\mathbf{h}\in\mathbf{Z}^{k+1}\setminus\{\mathbf{0}\}}\prod_{j=1}^{k}\left(\frac{1-e(h_jx_j)}{2\pi h_j}\left(\frac{\sin(2\pi\frac{h_j}{N^2})}{2\pi\frac{h_j}{N^2}}\right)^2\right)\\ &\quad\times\frac{1-e((h_1\alpha_1+\cdots+h_k\alpha_k-h_{k+1})N)}{2\pi(h_1\alpha_1+\cdots+h_k\alpha_k-h_{k+1})}\\ &\quad\times\left(\frac{\sin(2\pi(h_1\alpha_1+\cdots+h_k\alpha_k-h_{k+1}))}{2\pi(h_1\alpha_1+\cdots+h_k\alpha_k-h_{k+1})}\right)^2.\end{aligned}$$

We will indicate which part of the summation is "non-trivial", i.e. cannot be estimated by using standard tools. For this purpose we will use 3 lemmata which are stated without proof. (Complete proofs can be found in BECK [142].)

Lemma 1.93 *Let $\varphi(n)$ be an arbitrary positive increasing function of n with $\sum_{n\geq 1} 1/\varphi(n) < \infty$. Then for almost all $\underline{\alpha} \in \mathbf{R}^k$, $k \geq 2$,*

$$\sum_{\mathbf{h}\in\mathcal{N}_1} \frac{1}{r(\mathbf{h})\|\mathbf{h}\cdot\underline{\alpha}\|} = \mathcal{O}\left((\log N)^k \varphi(\log\log N)\right) \tag{1.141}$$

holds for all $N \geq 1$, where $\mathcal{N}_1$ denotes the set of $\mathbf{h} = (h_1, \ldots, h_k) \in \mathbf{Z}^k \setminus \{0\}$ which satisfy

$$|h_j| \leq N^2(\log N)^k, \quad 1 \leq j \leq k,$$

and

$$r(\mathbf{h})\|\mathbf{h}\cdot\underline{\alpha}\| \leq (\log N)^{20k}.$$

Lemma 1.94 *For $\mathbf{v} = (v_1, \ldots, v_k), \mathbf{w} = (w_1, \ldots, w_k) \in \mathbf{Z}^k$ let $[\mathbf{v}, \mathbf{w})$ denote the lattice box*

$$[\mathbf{v}, \mathbf{w}) = \{\mathbf{h} \in \mathbf{Z}^k : v_j \leq h_j < w_j, 1 \leq j \leq k\}.$$

Then for almost all $\underline{\alpha} \in \mathbf{R}^k$, $k \geq 2$,

$$\sum_{\mathbf{h}\in[\mathbf{v},\mathbf{w})} \chi_{(0,b(C,\mathbf{h}))}(\{\mathbf{h}\cdot\underline{\alpha}\}) = \sum_{\mathbf{h}\in[\mathbf{v},\mathbf{w})} b(C,\mathbf{h}) + \mathcal{O}\left(C^{3/4+\varepsilon}(\log N)^{k(1/2+\varepsilon)}\right),$$

in which

$$b(C,\mathbf{h}) = \min\left(\frac{1}{2}, \frac{C}{r(\mathbf{h})}\right),$$

holds for all $C = q^4$, $q = 1, 2, \ldots$, and for all lattice boxes $[\mathbf{v}, \mathbf{w})$ with $-\underline{N} \leq \mathbf{v} < \mathbf{w} \leq \underline{N}$, where $\underline{N} = (N, \ldots, N)$ and $\mathbf{v} < \mathbf{w}$ is defined componentwise.

The same assertion holds if we replace the characteristic function $\chi_{(0,b(C,\mathbf{h}))}(\{\mathbf{h}\cdot\underline{\alpha}\})$ by $\chi_{(1-b(C,\mathbf{h}),1)}(\{\mathbf{h}\cdot\underline{\alpha}\})$.

Lemma 1.95 *For almost all $\underline{\alpha} \in \mathbf{R}^k$ the series*

$$\sum_{\mathbf{h}\in\mathbf{Z}^k\setminus\{0\}} \frac{1}{r(\mathbf{h})\|\mathbf{h}\cdot\underline{\alpha}\|\left(\max(1, \log r(\mathbf{h}))\right)^{k+2}} \tag{1.142}$$

converges.

For $\mathbf{h} = (h_1, \ldots, h_k) \in \mathbf{Z}^k$, $h_{k+1} \in \mathbf{Z}$, and $\underline{\alpha} = (\alpha_1, \ldots, \alpha_k) \in \mathbf{R}^k$ let

$$\begin{aligned} A_{\mathbf{h},h_{k+1}}(\underline{\alpha}) &= i^{k-1} \prod_{j=1}^{k} \left(\frac{1 - e(h_j x_j)}{2\pi h_j} \left(\frac{\sin(2\pi \frac{h_j}{N^2})}{2\pi \frac{h_j}{N^2}} \right)^2 \right) \\ &\quad \times \frac{1 - e((\mathbf{h}\cdot\underline{\alpha} - h_{k+1})N)}{2\pi(\mathbf{h}\cdot\underline{\alpha} - h_{k+1})} \left(\frac{\sin(2\pi(\mathbf{h}\cdot\underline{\alpha} - h_{k+1}))}{2\pi(\mathbf{h}\cdot\underline{\alpha} - h_{k+1})} \right)^2 \end{aligned}$$

and

$$S_j = \sum_{(\mathbf{h},h_{k+1})\in H_j} A_{\mathbf{h},h_{k+1}}(\underline{\alpha}) \qquad (1 \leq j \leq 4),$$

where

$$\begin{aligned}
H_1 &= \{(\mathbf{h}, h_{k+1}) \in \mathbf{Z}^{k+1} \;:\; \|\mathbf{h}\|_\infty > N^2(\log N)^k\}, \\
H_2 &= \{(\mathbf{h}, h_{k+1}) \in \mathbf{Z}^{k+1} \;:\; \|\mathbf{h}\|_\infty \le N^2(\log N)^k, |\mathbf{h}\cdot\underline{\alpha} - h_{k+1}| \ge 1/3\}, \\
H_3 &= \{(\mathbf{h}, h_{k+1}) \in \mathbf{Z}^{k+1} \;:\; \|\mathbf{h}\|_\infty \le N^2(\log N)^k, r(\mathbf{h})\|\mathbf{h}\cdot\underline{\alpha}\| \le (\log N)^{20k}, \\
&\qquad h_{k+1} \text{ is the nearest integer to } \mathbf{h}\cdot\underline{\alpha}\}, \\
H_4 &= \{(\mathbf{h}, h_{k+1}) \in \mathbf{Z}^{k+1} \;:\; \|\mathbf{h}\|_\infty \le N^2(\log N)^k, 1/3 > \|\mathbf{h}\cdot\underline{\alpha}\| > (\log N)^{20k}/r(\mathbf{h}), \\
&\qquad h_{k+1} \text{ is the nearest integer to } \mathbf{h}\cdot\underline{\alpha}\}.
\end{aligned}$$

Since $\overline{\Delta}_N(n\underline{\alpha}, \mathbf{x}) \le S_1 + S_2 + S_3 + S_4$ it suffices to estimate S_j, $1 \le j \le 4$. Obviously we have

$$S_1 \ll S_{1.1} + S_{1.2},$$

where

$$S_{1.1} = \sum_{\mathbf{h}\in\mathbf{Z}^k, \|\mathbf{h}\|_\infty > N^2(\log N)^k} \frac{1}{r(\mathbf{h})r(\mathbf{h}/N^2)^2\|\mathbf{h}\cdot\underline{\alpha}\|}$$

and

$$S_{1.2} = \sum_{s=1}^{\infty} \sum_{\mathbf{h}\in\mathbf{Z}^k, \|\mathbf{h}\|_\infty > N^2(\log N)^k} \frac{1}{s^3 r(\mathbf{h})r(\mathbf{h}/N^2)^2},$$

in which s stands for $[|\mathbf{h}\cdot\underline{\alpha} - h_{k+1}| + 1]$. Since

$$\begin{aligned}
S_{1.1} &\ll \sum_{\mathbf{h}\in\mathbf{Z}^k, \|\mathbf{h}\|_\infty > N^2(\log N)^k} \frac{1}{r(\mathbf{h})\|\mathbf{h}\cdot\underline{\alpha}\|(\max(1, \log|h_1\cdots h_k|)^{2k}} \\
&\ll \sum_{\mathbf{h}\in\mathbf{Z}^k\setminus\{0\}} \frac{1}{r(\mathbf{h})\|\mathbf{h}\cdot\underline{\alpha}\|\,(\max(1, \log|h_1\cdots h_k|))^{k+2}}
\end{aligned}$$

Lemma 1.95 implies that $S_{1.1} = \mathcal{O}(1)$ for almost all $\underline{\alpha} \in \mathbf{R}^k$. Furthermore $S_{1.2}$ can easily be estimated by

$$\begin{aligned}
S_{1.2} &\ll \sum_{s=1}^{\infty} \sum_{\mathbf{h}\in\mathbf{Z}^k, \|\mathbf{h}\|_\infty \ge 2} \frac{1}{s^3 r(\mathbf{h})} \prod_{j=1}^{k} (\log|h_j|)^{-2} \\
&\ll \left(\sum_{s=1}^{\infty} \frac{1}{s^3}\right) \cdot \left(\sum_{h=2}^{\infty} \frac{1}{h(\log h)^2}\right)^k = \mathcal{O}(1).
\end{aligned}$$

Thus $S_1 = \mathcal{O}(1)$ for almost all $\underline{\alpha} \in \mathbf{R}^k$.

Next,

$$\begin{aligned}
S_2 &\ll \sum_{\mathbf{h}\in\mathbf{Z}^k, \|\mathbf{h}\|_\infty \le N^2(\log N)^k} \sum_{s=1}^{\infty} \frac{1}{s^3 r(\mathbf{h})} \\
&\ll \left(\sum_{s=1}^{\infty} \frac{1}{s^3}\right) \cdot \prod_{j=1}^{k} \left(\sum_{h_j=1}^{N^3} \frac{1}{h_j}\right) \ll (\log N)^k
\end{aligned}$$

and by Lemma 1.93

$$S_3 \ll (\log N)^k \varphi(\log\log N)$$

for almost all $\underline{\alpha} \in \mathbf{R}^k$.

So it remains to estimate S_4. For every $\underline{\nu} = (\nu_1, \ldots, \nu_k) \in \mathbf{Z}^k$ with $1 \le 2^{\nu_j} \le N^2(\log N)^k$, $1 \le j \le k$, and every $\underline{\varepsilon} = (\varepsilon_1, \ldots, \varepsilon_k) \in \{-1,1\}^k$, let

$$\begin{aligned} T(\underline{\nu}, \underline{\varepsilon}) &= \{\mathbf{h} \in \mathbf{Z}^k : 2^{\nu_j - 1} \le \varepsilon_j h_j \le 2^{\nu_j} \text{ or } h_j = 0 \\ &\qquad \text{if } \nu_j = 0, 1 \le j \le k\}. \end{aligned}$$

Note that $0 \le \nu_j \ll \log N$, $1 \le j \le k$. Furthermore let

$$\begin{aligned} N_1 &= \{\underline{\nu} \in \mathbf{Z}^k : 1 \le 2^{\nu_j} \le N^2(\log N)^k, 1 \le j \le k, \\ &\qquad \text{and } \nu_j \le c^* \log\log N \text{ for some } j\}, \end{aligned}$$

in which $c^* = 20k/\log 2$,

$$\begin{aligned} N_2 &= \{\underline{\nu} \in \mathbf{Z}^k : 1 \le 2^{\nu_j} \le N^2(\log N)^k, 1 \le j \le k, \\ &\qquad \text{and } 2^{\nu_j} \ge N^2/4 \text{ for some } j\}, \end{aligned}$$

and

$$\begin{aligned} S_{4.1} &= \sum_{\underline{\varepsilon} \in \{-1,1\}^k} \left(\sum_{\underline{\nu} \in N_1} \sum_{\mathbf{h} \in T(\underline{\nu}, \underline{\varepsilon}), (\mathbf{h}, h_{k+1}) \in H_4} A_{\mathbf{h}, h_{k+1}}(\underline{\alpha}) \right. \\ &\qquad \left. \sum_{\underline{\nu} \in N_2} \sum_{\mathbf{h} \in T(\underline{\nu}, \underline{\varepsilon}), (\mathbf{h}, h_{k+1}) \in H_4} A_{\mathbf{h}, h_{k+1}}(\underline{\alpha}) \right). \end{aligned}$$

Finally for an integer q with $2^q > (\log N)^{20k} = 2^{c^* \log\log N}$ set

$$T(\underline{\nu}, \underline{\varepsilon}, q) = \{\mathbf{h} \in T(\underline{\nu}, \underline{\varepsilon}) : 2^{q-1} < r(\mathbf{h}) \|\mathbf{h} \cdot \underline{\alpha}\| \le 2^q\}.$$

By Lemma 1.94 it follows that for almost all $\underline{\alpha} \in \mathbf{R}^k$

$$|T(\underline{\nu}, \underline{\varepsilon}, q)| \ll \sum_{\mathbf{h} \in T(\underline{\nu}, \underline{\varepsilon})} \min\left(1, \frac{2^q}{r(\mathbf{h})}\right) + \mathcal{O}\left((2^q)^{\frac{3}{4}+\epsilon} (\log N)^{k(\frac{1}{2}+\epsilon)}\right) \ll 2^q.$$

Furthermore, for every $\mathbf{h} \in T(\underline{\nu}, \underline{\varepsilon})$

$$r(\mathbf{h}) \|\mathbf{h} \cdot \underline{\alpha}\| < (N^2 (\log N)^k)^k,$$

so $2^q \le (N^2(\log N)^k)^k$. Hence

$$\begin{aligned} S_{4.1} &\ll \sum_{\underline{\varepsilon} \in \{-1,1\}^k} \sum_{\underline{\nu} \in N_1 \cup N_2} \sum_{(\log N)^{20k} < 2^q \le (N^2(\log N)^k)^k} \\ &\qquad \sum_{\mathbf{h} \in T(\underline{\nu}, \underline{\varepsilon}, q), (\mathbf{h}, h_{k+1}) \in H_4} \frac{1}{\|\mathbf{h} \cdot \underline{\alpha}\| r(\mathbf{h})} \\ &\ll (\log N)^{k-1} \cdot \log\log N \cdot \sum_{(\log N)^{20k} < 2^q \le (N^2(\log N)^k)^k} \frac{2^q}{2^q} \\ &\ll (\log N)^k \cdot \log\log N. \end{aligned}$$

The real difficulty is to estimate

$$S_{4.2} = \sum_{\underline{\varepsilon} \in \{-1,1\}^k} \sum_{\underline{\nu} \in N_3} \sum_{\mathbf{h} \in T(\underline{\nu},\underline{\varepsilon}), (\mathbf{h}, h_{k+1}) \in H_4} A_{\mathbf{h}, h_{k+1}}(\underline{\alpha}),$$

where

$$N_3 = \{\underline{\nu} \in \mathbf{Z}^k : (\log N)^{20k} < \min_{1 \le j \le k} 2^{\nu_j} \le \max_{1 \le j \le k} 2^{\nu_j} \le N^2/4\}.$$

If we multiply out the product $A_{\mathbf{h},h_{k+1}}(\underline{\alpha})$ we get

$$\begin{aligned} S_{4.2} = \frac{i^{k-1}}{(2\pi)^{k+1}} \Bigg(& \sum_{\mathbf{h} \in N_3} \sum_{h_{k+1} \in N_4(\mathbf{h})} \frac{1}{r(\mathbf{h})(\mathbf{h} \cdot \underline{\alpha} - h_{k+1})} g(\mathbf{h}, h_{k+1}, N, \underline{\alpha}) \\ & + \sum_{\mathbf{s} = (\delta_1, \ldots, \delta_{k+1}) \in \{0,1\}^{k+1} \setminus \{0\}} (-1)^{\delta_1 + \cdots + \delta_{k+1}} \\ & \sum_{\mathbf{h} \in N_3} \sum_{h_{k+1} \in N_4(\mathbf{h})} \frac{e(\mathcal{L}_{\mathbf{s}}(\mathbf{h}, h_{k+1}))}{r(\mathbf{h})(\mathbf{h} \cdot \underline{\alpha} - h_{k+1})} g(\mathbf{h}, h_{k+1}, N, \underline{\alpha}) \Bigg), \end{aligned}$$

in which

$$\begin{aligned} N_4(\mathbf{h}) &= \{h_{k+1} \in \mathbf{Z} : (\log N)^{20k}/r(\mathbf{h}) < |\mathbf{h} \cdot \underline{\alpha} - h_{k+1}| < 1/3\}, \\ g(\mathbf{h}, h_{k+1}, N, \underline{\alpha}) &= \left(\frac{\sin(2\pi(\mathbf{h} \cdot \underline{\alpha} - h_{k+1}))}{2\pi(\mathbf{h} \cdot \underline{\alpha} - h_{k+1})} \right)^2 \cdot \prod_{j=1}^{k} \left(\frac{\sin(2\pi h_j / N^2)}{2\pi h_j / N^2} \right)^2, \quad \text{and} \\ \mathcal{L}_{\mathbf{s}}(\mathbf{h}, h_{k+1}) &= \delta_{k+1} N(\mathbf{h} \cdot \underline{\alpha} - h_{k+1}) + \sum_{j=1}^{k} \delta_j x_j h_j. \end{aligned}$$

Let

$$S_{4.2.1} = \sum_{\mathbf{h} \in N_3} \sum_{h_{k+1} \in N_4(\mathbf{h})} \frac{1}{r(\mathbf{h})(\mathbf{h} \cdot \underline{\alpha} - h_{k+1})} g(\mathbf{h}, h_{k+1}, N, \underline{\alpha}),$$

$\delta = (\log N)^{-2}$, and for every $\underline{\varepsilon} = (\varepsilon_1, \ldots, \varepsilon_{k+1}) \in \{-1,1\}^{k+1}$ and $\mathbf{l} = (l_1, \ldots, l_{k+1}) \in \mathbf{Z}^{k+1}$ with

$$(\log N)^{20k} \le (1+\delta)^{l_j} \le \frac{N^2}{4}, \ 1 \le j \le k, \tag{1.143}$$

and

$$(\log N)^{20k} \le (1+\delta)^{l_{k+1}} \le \left(\frac{N^2}{4} \right)^k. \tag{1.144}$$

Let $U(\mathbf{l}, \underline{\varepsilon})$ be the set of those $\mathbf{h} \in \mathbf{Z}^k$ such that $(1+\delta)^{l_j} \le \varepsilon_j h_j < (1+\delta)^{l_j+1}$, $1 \le j \le k$, and

$$(1+\delta)^{l_{k+1}} \le \varepsilon_{k+1} r(\mathbf{h})(\mathbf{h} \cdot \underline{\alpha} - h_{k+1}) < (1+\delta)^{l_{k+1}+1},$$

in which h_{k+1} is the nearest integer to $\mathbf{h}\cdot\underline{\alpha}$. Obviously, the number of lattice boxes $U(\mathbf{l},\varepsilon)$ is bounded by

$$\ll \left(\frac{\log N}{\delta}\right)^{k+1}. \tag{1.145}$$

Let $\underline{\varepsilon}^+$ and $\underline{\varepsilon}^-$ be two vectors in $\{-1,+1\}^{k+1}$ such that only the $(k+1)$-st coordinates are different. Without loss of generality we may assume that the requirement $\mathbf{h}\in N_3$, $h_{k+1}\in N_4(\mathbf{h})$ can be rewritten as a perfect union of boxes $U(\mathbf{l},\underline{\varepsilon}^+)\cup U(\mathbf{l},\underline{\varepsilon}^-)$. (Otherwise we have to modify the sums $S_1, S_2, S_3, S_{4.1}$ appropriately.) Hence

$$S_{4.2.1} = \sum_{\text{some }(\mathbf{l},\underline{\varepsilon}^\pm)} \left(\sum_{\mathbf{h}\in U(\mathbf{l},\underline{\varepsilon}^+)} \frac{g(\mathbf{h},h_{k+1},N,\underline{\alpha})}{r(\mathbf{h})(\mathbf{h}\cdot\underline{\alpha}-h_{k+1})} + \sum_{\mathbf{h}\in U(\mathbf{l},\underline{\varepsilon}^-)} \frac{g(\mathbf{h},h_{k+1},N,\underline{\alpha})}{r(\mathbf{h})(\mathbf{h}\cdot\underline{\alpha}-h_{k+1})} \right).$$

By Lemma 1.94

$$\begin{aligned} |U(\mathbf{l},\underline{\varepsilon}^\pm)| &= \sum_{\mathbf{h}\in U(\mathbf{l},\underline{\varepsilon}^\pm)} (b(C_1,\mathbf{h})-b(C_2,\mathbf{h})) + \mathcal{O}\left(C_1^{\frac{3}{4}+\varepsilon}(\log N)^{(\frac12+\varepsilon)k}\right) \\ &\ll (C_1-C_2)\delta^k + \mathcal{O}\left(C_1^{\frac34+\varepsilon}(\log N)^{(\frac12+\varepsilon)k}\right), \end{aligned}$$

where $C_1 = (1+\delta)^{l_{k+1}+1}$ and $C_2 = (1+\delta)^{l_{k+1}}$. Hence the leading terms in $S_{4.2.1}$ essentially cancel out:

$$\begin{aligned} &\left| \sum_{\mathbf{h}\in U(\mathbf{l},\underline{\varepsilon}^+)} \frac{g(\mathbf{h},h_{k+1},N,\underline{\alpha})}{r(\mathbf{h})(\mathbf{h}\cdot\alpha-h_{k+1})} + \sum_{\mathbf{h}\in U(\mathbf{l},\underline{\varepsilon}^-)} \frac{g(\mathbf{h},h_{k+1},N,\underline{\alpha})}{r(\mathbf{h})(\mathbf{h}\cdot\underline{\alpha}-h_{k+1})} \right| \\ &\quad \ll \delta\frac{C_1-C_2}{C_1}\delta^k + \mathcal{O}\left(C_1^{-\frac14+\varepsilon}(\log N)^{(\frac12+\varepsilon)k}\right) \\ &\quad \ll \delta^{k+2}, \end{aligned}$$

where we have used that $C_1 > (\log N)^{20k}$. This leads to

$$S_{4.2.1} \ll \left(\frac{\log N}{\delta}\right)^{k+1} \delta^{k+2} = \delta(\log N)^{k+1} = (\log N)^{k-1}.$$

Finally we have to deal with

$$\begin{aligned} S_{4.2.2} &= \sum_{\mathbf{h}\in N_3} \sum_{h_{k+1}\in N_4(\mathbf{h})} \frac{e(\mathcal{L}(\mathbf{h}))}{r(\mathbf{h})(\mathbf{h}\cdot\underline{\alpha}-h_{k+1})} g(\mathbf{h},h_{k+1},N,\underline{\alpha}) \\ &= \sum_{\text{some }(\mathbf{l},\underline{\varepsilon}^\pm)} \left(\sum_{\mathbf{h}\in U(\mathbf{l},\underline{\varepsilon}^+)} \frac{e(\mathcal{L}(\mathbf{h}))}{r(\mathbf{h})(\mathbf{h}\cdot\underline{\alpha}-h_{k+1})} g(\mathbf{h},h_{k+1},N,\underline{\alpha}) \right. \\ &\qquad \left. \sum_{\mathbf{h}\in U(\mathbf{l},\underline{\varepsilon}^-)} \frac{e(\mathcal{L}(\mathbf{h}))}{r(\mathbf{h})(\mathbf{h}\cdot\underline{\alpha}-h_{k+1})} g(\mathbf{h},h_{k+1},N,\underline{\alpha}) \right), \end{aligned}$$

in which $\mathcal{L}(\mathbf{h})$ is any of the $2^{k+1}-1$ linear forms $\mathcal{L}_s(\mathbf{h}, h_{k+1})$, where h_{k+1} is the nearest integer to $\mathbf{h}\cdot\underline{\alpha}$.

We say that $\mathbf{l}=(l_1,\ldots,l_{k+1})\in\mathbf{Z}^{k+1}$ satisfying (1.143) and (1.144) is an $\underline{\varepsilon}$-big vector ($\underline{\varepsilon}\in\{-1,+1\}^k$) if

$$\frac{|U(\mathbf{l},(\underline{\varepsilon},+1))|+|U(\mathbf{l},(\underline{\varepsilon},-1))|}{\log N}\le\Big|\sum_{\mathbf{h}\in U(\mathbf{l},(\underline{\varepsilon},+1))}e(\mathcal{L}(\mathbf{h}))-\sum_{\mathbf{h}\in U(\mathbf{l},(\underline{\varepsilon},-1))}e(\mathcal{L}(\mathbf{h}))\Big|. \tag{1.146}$$

Furthermore for notational convenience we write

$$U^{\pm}(\mathbf{l},\underline{\varepsilon})=U(\mathbf{l},(\underline{\varepsilon},+1))\cup U(\mathbf{l},(\underline{\varepsilon},-1)).$$

Two integral vectors $\mathbf{l}=(l_1,\ldots,l_{k+1})$ and $\mathbf{n}=(n_1,\ldots,n_{k+1})$ satisfying (1.143) and (1.144) will be called $\underline{\varepsilon}$-neighbours if

$$(1+\delta)^{\varepsilon_j(n_j-l_j)} = (\log N)^9,\ 1\le j\le k,\ \text{and} \tag{1.147}$$

$$(1+\delta)^{n_{k+1}-l_{k+1}} = (\log N)^{9(k+1)}. \tag{1.148}$$

The notation $\mathbf{l}\overset{\underline{\varepsilon}}{\to}\mathbf{n}$ means that the ordered pair $\langle\mathbf{l},\mathbf{n}\rangle$ of vectors satisfies (1.147) and (1.148).

Note that, by (1.147) and (1.148), the left sides of

$$\varepsilon_j(n_j-l_j) = \frac{9\log\log N}{\log(1+\delta)},\quad 1\le j\le k,\text{and}$$

$$n_{k+1}-l_{k+1} = (k+1)\cdot\frac{9\log\log N}{\log(1+\delta)}$$

have to be integers, but the right sides are not necessarily integers. However, by slightly modifying the value of $\delta\approx(\log N)^{-2}$, we can easily ensure that $\frac{9\log\log N}{\log(1+\delta)}$ is an integer.

A sequence $N=\langle\mathbf{n}^{(1)},\mathbf{n}^{(2)},\mathbf{n}^{(3)},\ldots\rangle$ of vectors satisfying (1.143) and (1.144) is called a special $\underline{\varepsilon}$-line if $\mathbf{n}^{(1)}\overset{\underline{\varepsilon}}{\to}\mathbf{n}^{(2)}\overset{\underline{\varepsilon}}{\to}\mathbf{n}^{(3)}\overset{\underline{\varepsilon}}{\to}\cdots$, that is, any two consecutive vectors in N are $\underline{\varepsilon}$-neighbours.

The following lemma is the essential step of the proof of Theorem 1.91.

Lemma 1.96 (KEY LEMMA) *For any $\underline{\alpha}\in\mathrm{R}^k$ satisfying Lemma 1.94, every special $\underline{\varepsilon}$-line contains at most one $\underline{\varepsilon}$-big vector.*

Proof. Let $N=\langle\mathbf{n}^{(1)},\mathbf{n}^{(2)},\mathbf{n}^{(3)},\ldots\rangle$ be a special $\underline{\varepsilon}$-line with *two* $\underline{\varepsilon}$-big vectors $\mathbf{n}^{(p)}$ and $\mathbf{n}^{(q)}$, $1\le p<q$. If

$$\|\mathcal{L}(\mathbf{h})\|\le(\log N)^{-2}\ \text{for every}\ \mathbf{h}\in U^{\pm}(\mathbf{n}^{(p)},\underline{\varepsilon}), \tag{1.149}$$

then

$$|1-e(\mathcal{L}(\mathbf{h}))|\ll(\log N)^{-2}\ \text{for every}\ \mathbf{h}\in U^{\pm}(\mathbf{n}^{(p)},\underline{\varepsilon}). \tag{1.150}$$

And so we can repeat the preceding argument (i.e. the cancellation of the main term via Lemma 1.94 which has been used to estimate $S_{4.2.1}$), and by (1.150) we obtain

$$\left| \sum_{\mathbf{h}\in U(\mathbf{n}^{(p)},(\underline{\varepsilon},+1))} e(\mathcal{L}(\mathbf{h})) - \sum_{\mathbf{h}\in U(\mathbf{n}^{(p)},(\underline{\varepsilon},-1))} e(\mathcal{L}(\mathbf{h}))\right|$$

$$\ll \left| |U(\mathbf{n}^{(p)},(\underline{\varepsilon},+1))| - |U(\mathbf{n}^{(p)},(\underline{\varepsilon},-1))| \right| + \frac{|U^{\pm}(\mathbf{n}^{(p)},\underline{\varepsilon})|}{(\log N)^2}$$

$$\ll (\delta + (\log N)^{-2})|U^{\pm}(\mathbf{n}^{(p)},\underline{\varepsilon})| \ll (\log N)^{-2}|U^{\pm}(\mathbf{n}^{(p)},\underline{\varepsilon})|.$$

But this contradicts the assumption that $\mathbf{n}^{(p)}$ is $\underline{\varepsilon}$-big (see (1.146)). The falsity of (1.149) means that there is an $\mathbf{h}^* \in U^{\pm}(\mathbf{n}^{(p)},\underline{\varepsilon})$ such that

$$\|\mathcal{L}(\mathbf{h}^*)\| > (\log N)^{-2}. \tag{1.151}$$

For every $\mathbf{m} \in U^{\pm}(\mathbf{n}^{(q)},\underline{\varepsilon})$ (we recall that $\mathbf{n}^{(q)}$ is *another* $\underline{\varepsilon}$-big vector), consider the arithmetic progression with difference $\mathbf{h}^*$:

$$\mathbf{m} + r\cdot\mathbf{h}^* = (m_1 + r\cdot h_1^*, \ldots, m_{k+1} + r\cdot h_{k+1}^*), \quad r \in \mathbf{Z}. \tag{1.152}$$

We are interested in the number of consecutive members of progression (1.152) which are contained in $U^{\pm}(\mathbf{n}^{(q)},\underline{\varepsilon})$. Since $\mathbf{h}^* \in U^{\pm}(\mathbf{n}^{(p)},\underline{\varepsilon})$ and $\mathbf{m} \in U^{\pm}(\mathbf{n}^{(q)},\underline{\varepsilon})$, we obtain by the definition of $U(\cdot,\underline{\varepsilon})$

$$(1+\delta)^{n_j^{(p)}} \le \varepsilon_j \cdot n_j^* < (1+\delta)^{n_j^{(p)}+1}, \quad 1 \le j \le k, \tag{1.153}$$

$$(1+\delta)^{n_{k+1}^{(p)}} \le r(\mathbf{h}^*)\cdot\|\mathbf{h}^*\cdot\underline{\alpha}\| < (1+\delta)^{n_{k+1}^{(p)}+1}, \tag{1.154}$$

$$(1+\delta)^{n_j^{(q)}} \le \varepsilon_j \cdot m_j < (1+\delta)^{n_j^{(q)}+1}, \; 1 \le j \le k, \tag{1.155}$$

$$(1+\delta)^{n_{k+1}^{(q)}} \le r(\mathbf{m})\cdot\|\mathbf{m}\cdot\underline{\alpha}\| < (1+\delta)^{n_{k+1}^{(q)}+1}. \tag{1.156}$$

We will call $\mathbf{m} \in U^{\pm}(\mathbf{n}^{(q)},\underline{\varepsilon})$ an *inner point* if

$$\begin{aligned}(1+\delta)^{n_j^{(q)}}\left(1+\frac{\delta}{(\log N)^2}\right) &\le \varepsilon_j\cdot m_j \\ &\le \left(1-\frac{\delta}{(\log N)^2}\right)(1+\delta)^{n_j^{(q)}+1}, \; 1\le j\le k,\end{aligned} \tag{1.157}$$

and

$$\begin{aligned}(1+\delta)^{n_{k+1}^{(q)}}\left(1+\frac{\delta}{(\log N)^2}\right) &\le r(\mathbf{m})\cdot\|\mathbf{m}\cdot\underline{\alpha}\| \\ &< \left(1-\frac{\delta}{(\log N)^2}\right)(1+\delta)^{n_{k+1}^{(q)}+1}.\end{aligned} \tag{1.158}$$

The remaining points of $U^{\pm}(\mathbf{n}^{(q)},\underline{\varepsilon})$ are called *border points.*

It follows from (1.147), (1.153) and (1.157) that for every inner point $\mathbf{m} \in U^{\pm}(\mathbf{n}^{(q)}, \underline{\varepsilon})$ and for every $|r| \le (\log N)^4$,

$$\begin{aligned}
(1+\delta)^{n_j^{(q)}} &< \left(1+\frac{\delta}{(\log N)^2}\right)(1+\delta)^{n_j^{(q)}} - (\log N)^4(1+\delta)^{n_j^{(p)}+1} \\
&< \varepsilon_j(m_j + r\cdot h_j^*) \qquad (1.159)\\
&< \left(1-\frac{\delta}{(\log N)^2}\right)(1+\delta)^{n_j^{(q)}+1} + (\log N)^4(1+\delta)^{n_j^{(p)}+1} \\
&< (1+\delta)^{n_j^{(q)}+1}.
\end{aligned}$$

Similarly, from (1.148), (1.154) and (1.158) we obtain that for every inner point $\mathbf{m} \in U^{\pm}(\mathbf{n}^{(q)}, \underline{\varepsilon})$ and for every $|r| \le (\log N)^4$,

$$\begin{aligned}
\|(\mathbf{m}+r\cdot\mathbf{h}^*)\cdot\underline{\alpha}\| &\le \|\mathbf{m}\cdot\underline{\alpha}\| + |r|\cdot\|\mathbf{h}^*\cdot\underline{\alpha}\| \\
&< \frac{(1-\frac{\delta}{(\log N)^2})(1+\delta)^{n_{k+1}^{(q)}+1}}{r(\mathbf{m})} + (\log N)^4\cdot\frac{(1+\delta)^{n_{k+1}^{(p)}+1}}{r(\mathbf{h}^*)} \\
&< \frac{(1+\delta)^{n_{k+1}^{(q)}+1}}{r(\mathbf{m})}. \qquad (1.160)
\end{aligned}$$

The lower bound:

$$\begin{aligned}
\|(\mathbf{m}+r\cdot\mathbf{h}^*)\cdot\underline{\alpha}\| &\ge \|\mathbf{m}\cdot\underline{\alpha}\| - |r|\cdot\|\mathbf{h}^*\cdot\underline{\alpha}\| \\
&> \frac{(1+\frac{\delta}{(\log N)^2})(1+\delta)^{n_{k+1}^{(q)}}}{r(\mathbf{m})} - (\log N)^4\cdot\frac{(1+\delta)^{n_{k+1}^{(q)}+1}}{r(\mathbf{h}^*)} \\
&> \frac{(1+\delta)^{n_{k+1}^{(q)}}}{r(\mathbf{m})}. \qquad (1.161)
\end{aligned}$$

In view of (1.159)-(1.161), for any inner point $\mathbf{m} \in U^{\pm}(\mathbf{n}^{(q)}, \underline{\varepsilon})$, the whole segment $\mathbf{m} + r\mathbf{h}^*$, $r = 0, \pm1, \pm2, \ldots, \pm[(\log N)^4]$ of progression (1.152) is contained in $U^{\pm}(\mathbf{n}^{(q)}, \underline{\varepsilon})$. We can, therefore, decompose $U^{\pm}(\mathbf{n}^{(q)}, \underline{\varepsilon})$ into three parts:

$$U^{\pm}(\mathbf{n}^{(q)}, \underline{\varepsilon}) = AP^+ \cup AP^- \cup BP \qquad (1.162)$$

where $AP^+ \subseteq U(\mathbf{n}^{(q)}, (\underline{\varepsilon}, +1))$ and $AP^- \subseteq U(\mathbf{n}^{(q)}, (\underline{\varepsilon}, -1))$ are two families of disjoint arithmetic progressions $\{\mathbf{m} + r\cdot\mathbf{h}^* \mid 0 \le r \le l-1\}$ with common difference $\mathbf{h}^*$ (see (1.151)) and each length $l \ge (\log N)^4$, and BP is a set of border points of $U^{\pm}(\mathbf{n}^{(q)}, \underline{\varepsilon})$. By using the trivial estimate

$$\left|\sum_{j=0}^{q-1} e(\eta + j\xi)\right| \ll \frac{1}{\|\xi\|},$$

(1.151), and the linearity of $\mathcal{L}$, we obtain

$$\begin{aligned} \left|\sum_{\mathbf{n}\in AP^+} e(\mathcal{L}(\mathbf{h}))\right| &\leq \sum \left|\sum_{r=0}^{l-1} e(\mathcal{L}(\mathbf{m}+r\mathbf{h}^*))\right| \\ &= \sum \left|\sum_{r=0}^{l-1} e(\mathcal{L}(\mathbf{m}) + r\cdot\mathcal{L}(\mathbf{h}^*))\right| \\ &\ll \sum \frac{1}{\|\mathcal{L}(\mathbf{h}^*)\|} < \sum (\log N)^2 \\ &\leq \sum \frac{l}{(\log N)^2} \leq \frac{|U(\mathbf{n}^{(q)},(\underline{\varepsilon},+1))|}{(\log N)^2}, \end{aligned} \tag{1.163}$$

since each length $l \geq (\log N)^4$ and $\|\mathcal{L}(\mathbf{h}^*)\| > (\log N)^{-2}$. Similarly,

$$\left|\sum_{\mathbf{h}\in AP^-} e(\mathcal{L}(\mathbf{h}^*))\right| \ll \frac{|U(\mathbf{n}^{(q)},(\underline{\varepsilon},-1))|}{(\log N)^2}. \tag{1.164}$$

Finally, we estimate the cardinality of the set BP. Obviously $|BP|$ is less or equal the total number of border points of $U^{\pm}(\mathbf{n}^{(q)},\underline{\varepsilon})$. The number of those border points $\mathbf{m} \in U^{\pm}(\mathbf{n}^{(q)},\varepsilon)$ for which for some $j \in \{1,2,\ldots,k\}$, (1.155) holds but (1.157) is violated is cleary $\ll |U^{\pm}(\mathbf{n}^{(q)},\underline{\varepsilon})| \cdot (\log N)^{-2}$. Similarly, by using Lemma 1.94 as we did for estimation of $S_{4.2.1}$, the number of those border points for which (1.156) holds but (1.158) is violated is $\ll |U^{\pm}(\mathbf{n}^{(q)},\underline{\varepsilon})| \cdot (\log N)^{-2}$ for almost every $\underline{\alpha} \in \mathbf{R}^k$. Hence for every $\underline{\alpha} \in \mathbf{R}^k$ satisfying Lemma 1.94, we obtain that the total number of border points of $U^{\pm}(\mathbf{n}^{(q)},\underline{\varepsilon})$ is

$$\ll \frac{|U^{\pm}(\mathbf{n}^{(q)},\underline{\varepsilon})|}{(\log N)^2}.$$

Thus for every $\underline{\alpha} \in \mathbf{R}^k$ satisfying Lemma 1.94, we obtain

$$\left|\sum_{\mathbf{h}\in U(\mathbf{n}^{(q)},(\underline{\varepsilon},+1))} e(\mathcal{L}(\mathbf{h})) - \sum_{\mathbf{h}\in U(\mathbf{n}^{(q)},(\underline{\varepsilon},-1))} e(\mathcal{L}(\mathbf{h}))\right| \ll \frac{|U^{\pm}(\mathbf{n}^{(q)},\underline{\varepsilon})|}{(\log N)^2}$$

which contradicts the assumption that $\mathbf{n}^{(q)}$ is ε-big (see (1.146)). This proves the Key Lemma. □

First we estimate the number of maximal special $\underline{\varepsilon}$-lines $N = \langle \mathbf{n}^{(1)}, \mathbf{n}^{(2)}, \mathbf{n}^{(3)}, \ldots\rangle$ Here $\mathbf{n}^{(1)}$ is the first element of N, that is, if $\mathbf{n}^{(0)} \overset{\varepsilon}{\rightarrow} \mathbf{n}^{(1)}$ holds for some $\mathbf{n}^{(0)}$ then one of the inequalities

$$(1+\delta)^{\varepsilon_j n_j^{(0)}} \geq (\log N)^{20k},\ 1 \leq j \leq k, \text{ and } (1+\delta)^{n_{k+1}^{(0)}} \geq (\log N)^{20k}$$

is violated. Thus by (1.147) and (1.148) one of the inequalities

$$(\log N)^{20k} \leq (1+\delta)^{\varepsilon_j n_j^{(1)}} \leq (\log N)^{20k+9},\ 1 \leq j \leq k, \text{ and} \tag{1.165}$$

$$(\log N)^{20k} \leq (1+\delta)^{n_{k+1}^{(1)}} \leq (\log N)^{20k+9k+9} \tag{1.166}$$

holds. Again by (1.147) and (1.148), for $N = \langle \mathbf{n}^{(1)}, \mathbf{n}^{(2)}, \mathbf{n}^{(3)}, \ldots \rangle$ we have

$$\varepsilon_j(n_j^{(\nu+1)} - n_j^{(\nu)}) = \frac{9 \log\log N}{\log(1+\delta)}, \quad 1 \le j \le k, \text{ and} \tag{1.167}$$

$$n_{k+1}^{(\nu+1)} - n_{k+1}^{(\nu)} = \frac{9(k+1)\log\log N}{\log(1+\delta)}. \tag{1.168}$$

Hence the integral vector

$$\mathbf{w}(N) = \left(n_2^{(\nu)} - n_1^{(\nu)}, n_3^{(\nu)} - n_1^{(\nu)}, \ldots, n_k^{(\nu)} - n_1^{(\nu)}, n_{k+1}^{(\nu)} - (k+1)n_1^{(\nu)}\right) \tag{1.169}$$

is independent of the running index $\nu \ge 1$. Therefore, we can reconstruct the whole line $N = \langle \mathbf{n}^{(1)}, \mathbf{n}^{(2)}, \mathbf{n}^{(3)}, \ldots \rangle$ from the following data: $\underline{\varepsilon}$, $\mathbf{w}(N)$, and any of the $k+1$ coordinates $n_j^{(1)}$. In view of (1.165) and (1.166), for a maximal line N, at least one coordinates $n_j^{(1)}$ of the first element $\mathbf{n}^{(1)}$ of N is restricted to a fixed short interval of const$\cdot \log\log N \cdot \delta^{-1}$ consecutive integers. By (1.143), (1.144), and (1.169) the total number of invariant vectors $\mathbf{w}(N)$ is $\left(\delta^{-1}\log N\right)^k$, thus the total number of maximal special $\underline{\varepsilon}$-lines N is $\ll \log\log N \cdot (\log N)^k \cdot \delta^{-k-1}$. Therefore, by the Key Lemma 1.96, for every $\underline{\alpha} \in \mathbf{R}^k$ satisfying Lemma 1.94, the total number of $\underline{\varepsilon}$-big boxes $U^{\pm}(\mathbf{l}, \underline{\varepsilon})$ is

$$\ll (\log\log N) \cdot (\log N)^k \cdot \delta^{-k-1} \tag{1.170}$$

Now we can estimate $S_{4.2.2} = S_{4.2.2.1} + S_{4.2.2.2}$, in which

$$S_{4.2.2.1} = \sum_{\underline{\varepsilon} \in \{-1,1\}^k} \sum_{\substack{\text{some } (\mathbf{l}, \underline{\varepsilon}\pm) \\ \mathbf{l} \text{ is not } \underline{\varepsilon}-\text{big}}} \sum_{\mathbf{h} \in U^{\pm}(\mathbf{l},\underline{\varepsilon})} \frac{e(\mathcal{L}(\mathbf{h}))}{r(\mathbf{h})(\mathbf{h}\cdot\underline{\alpha} - h_{k+1})} g(\mathbf{h}, h_{k+1}, N, \underline{\alpha}) \tag{1.171}$$

and

$$S_{4.2.2.2} = \sum_{\underline{\varepsilon} \in \{-1,1\}^k} \sum_{\substack{\text{some } (\mathbf{l}, \underline{\varepsilon}\pm) \\ \mathbf{l} \text{ is } \underline{\varepsilon}-\text{big}}} \sum_{\mathbf{h} \in U^{\pm}(\mathbf{l},\underline{\varepsilon})} \frac{e(\mathcal{L}(\mathbf{h})}{r(\mathbf{h})(\mathbf{h}\cdot\underline{\alpha} - h_{k+1})} g(\mathbf{h}, h_{k+1}, N, \underline{\alpha}). \tag{1.172}$$

By definition, for an arbitrary $\mathbf{h} \in U^{\pm}(\mathbf{l}, \underline{\varepsilon})$

$$(1+\delta)^{l_{k+1}} \le r(\mathbf{h})\, \|\mathbf{h}\cdot\underline{\alpha}\| < (1+\delta)^{l_{k+1}+1} \text{ and}$$
$$\|\mathbf{h}\cdot\underline{\alpha}\| = |\mathbf{h}\cdot\underline{\alpha} - h_{k+1}| = (1+\delta)^{l_{k+1}-(l_1+\cdots+l_k)+\mathcal{O}(k)}.$$

Moreover, $\|\mathbf{h}\|_\infty \le N^2/4$ and $|\mathbf{h}\cdot\underline{\alpha} - h_{k+1}| < 1/3$. Hence, by the definition of $S_{4.2.2}$ and (1.171),

$$|S_{4.2.2.1}| \le \sum_{\underline{\varepsilon}} \sum_{\mathbf{l} \text{ is not } \underline{\varepsilon}-\text{big}} (1+\delta)^{-l_{k+1}} |U^{\pm}(\mathbf{l}, \underline{\varepsilon})| \left(\frac{1}{\log N} + \mathcal{O}(\delta)\right)$$

$$\ll \frac{(1+\delta)^{-l_{k+1}}}{\log N} \sum_{\mathbf{l},\underline{\varepsilon}} |U^{\pm}(\mathbf{l}, \underline{\varepsilon})|. \tag{1.173}$$

Again, applying Lemma 1.94 just as we did in estimating $S_{4.2.1}$, for almost all $\underline{\alpha} \in \mathbf{R}^k$, we obtain

$$\begin{aligned} |U^{\pm}(\mathbf{l},\underline{\varepsilon})| &\ll (1+\delta)^{l_{k+1}}\delta^{k+1} + (1+\delta)^{(\frac{3}{4}+\epsilon)l_{k+1}}(\log N)^{(\frac{1}{2}+\epsilon)k} \\ &\ll (1+\delta)^{l_{k+1}}\delta^{k+1}, \end{aligned} \tag{1.174}$$

since by (1.144), $(1+\delta)^{l_{k+1}} \geq (\log N)^{20k}$. In view of (1.145), the number of pairs $\mathbf{l},\underline{\varepsilon}$ is $\ll (\delta^{-1}\log N)^{k+1}$. Thus, by (1.173) and (1.174),

$$\begin{aligned} |S_{4.2.2.1}| &\ll \frac{(1+\delta)^{-l_{k+1}}}{\log N}\left(\frac{\log N}{\delta}\right)^{k+1}(1+\delta)^{l_{k+1}}\delta^{k+1} \\ &= (\log N)^k. \end{aligned}$$

Finally, by (1.170), (1.172), and (1.174), we have

$$\begin{aligned} |S_{4.2.2.2}| &\leq \sum_{\underline{\varepsilon}}\sum_{\mathbf{l}\text{ is }\underline{\varepsilon}-\text{big}} (1+\delta)^{-l_{k+1}}|U^{\pm}(\mathbf{l},\underline{\varepsilon})| \\ &\ll (1+\delta)^{-l_{k+1}}(\log\log N)\cdot(\log N)^k\delta^{-k-1}(1+\delta)^{l_{k+1}}\delta^{k+1} \\ &= (\log N)^k \cdot \log\log N. \end{aligned}$$

Now, if we collect the estimates for $S_{1.1}, S_{1.2}, S_2, S_3, S_{4.1}, S_{4.2.1}, S_{4.2.2.1}$, and $S_{4.2.2.2}$ it follows that

$$\begin{aligned} \overline{\Delta}_N(n\underline{\alpha},\mathbf{x}) &\ll (\log N)^k(\varphi(\log\log N) + \log\log N) \\ &\ll (\log N)^k\varphi(\log\log N) \end{aligned}$$

provided that $\sum_{n=1}^{\infty} 1/\varphi(n) < \infty$. (Note that the hypothesis that $\varphi(n)$ is increasing and $\sum_{n=1}^{\infty} 1/\varphi(n) < \infty$ implies that $\varphi(n) > n$ for sufficiently large n.) Recall that $\overline{\Delta}_N(n\underline{\alpha},\mathbf{x})$ is a sort of "$1/N^2$-neighbourhood average" of $\Delta_N(n\underline{\alpha},\mathbf{x})$. By KHINTCHINE's Theorem 1.72, for almost all $\alpha \in \mathbf{R}$,

$$\|n\alpha\| > \frac{1}{n(\log n)^{1+\epsilon}} > \frac{1}{n^2} \geq \frac{1}{N^2}.$$

Thus $\Delta_N(n\underline{\alpha},\mathbf{x}) = \overline{\Delta}_N(n\underline{\alpha},\mathbf{x}) + \mathcal{O}(1)$, which proves the first part of Theorem 1.91.

The proof of the converse statement (which completes the proof of Theorem 1.91) is much easier. We will not present the details here (compare with BECK [142]). We only want to mention that it essentially relies on the following "standard" lemma.

Lemma 1.97 *Let $\varphi(n)$ be an arbitrary positive increasing function of n with $\sum_{n\geq 1} 1/\varphi(n) = \infty$. Then for almost all $\underline{\alpha} \in \mathbf{R}^k$ there are infinitely many $\mathbf{h} = (h_1,\ldots,h_k) \in \mathbf{Z}^k$, $h_j > 0$, $1 \leq j \leq k$, such that*

$$r(\mathbf{h})(\log r(\mathbf{h}))^k\varphi(\log\log(r(\mathbf{h}^2)))\cdot\|\mathbf{h}\cdot\underline{\alpha}\| < 1. \tag{1.175}$$

1.4.3 Digital Sequences

In the preceding sections we discussed the properties of sequences $(n\alpha)$, where α is an irrational. It is an interesting problem to find integer valued functions $f(n) > 0$ such that $(\alpha f(n))$ is u.d. mod 1 for all irrational numbers α. In this section we will show that there are very prominent examples of this kind which are related to q-ary digital expansions.

Let $q > 1$ be fixed. Then every integer $n \geq 0$ has a unique q-ary representation

$$n = \sum_{k=0}^{\infty} d_k q^k,$$

where $0 \leq d_k = d_k(q;n) \leq q-1$ are the digits.

Definition 1.98 *A function $f : \mathbf{Z}_0^+ \to \mathbf{Z}_0^+$ is called q-additive if*

$$f(n) = \sum_{k=0}^{\infty} f(d_k(q;n)q^k).$$

In addition, if

$$f(n) = \sum_{k=0}^{\infty} f(d_k(q;n))$$

then f is called strongly q-additive.

Note that if f is strongly q-additive then $f(0) = 0$ and $f(1), \ldots, f(q-1)$ determine f. For example, if $L > 0$ is an integer, then the L-th powers of the digits

$$s^{(L)}(q;n) = \sum_{k=0}^{\infty} d_k(q;n)^L.$$

is strongly q-additive.

Theorem 1.99 *Suppose that $f(n)$ is strongly q-additive and that there exists $1 \leq b \leq q-1$ with $f(b) > 0$. Then the sequence $(\alpha f(n))$ is u.d. mod 1 if and only if α is irrational.*

Proof. If α is rational then there exists $h \in \mathbf{Z}^+$ such that $h\alpha \in \mathbf{Z}$. Hence $e(h\alpha f(n)) = 1$ for all n and consequently $(\alpha f(n))$ is not u.d. mod 1 by WEYL's criterion (Theorem 1.19).

Next observe that for every strongly q-additive function $f(n)$ we have

$$\sum_{n<q^k} y^{f(n)} = \left(\sum_{b=0}^{q-1} y^{f(b)} \right)^k.$$

Hence for every positive integer h and real α we obtain

$$\sum_{n<q^k} e(h\alpha f(n)) = \left(\sum_{b=0}^{q-1} e(h\alpha f(b))\right)^k .$$

If α is irrational and if there exist $1 \le b \le q-1$ with $f(b) > 0$ then

$$\left|\sum_{b=0}^{q-1} e(h\alpha f(b))\right| < q,$$

which implies

$$\lim_{k\to\infty} \frac{1}{q^k} \sum_{n<q^k} e(h\alpha f(n)) = 0.$$

Now a non-quantitative application of the following Lemma 1.101 implies

$$\lim_{N\to\infty} \frac{1}{N} \sum_{n<N} e(h\alpha f(n)) = 0.$$

Hence by WEYL's criterion (Theorem 1.19) $(\alpha f(n))$ is u.d. mod 1. □

Next we will discuss the case of a special type of irrationals α in more detail. We consider irrationals α of finite approximation type η, i.e. for every $\varepsilon > 0$ there exists a constant $c(\alpha, \varepsilon) > 0$ such that

$$\|h\alpha\| \ge \frac{c(\alpha,\varepsilon)}{h^{\eta+\varepsilon}}$$

for all positive integers h. (As above $\|x\| = \min(\{x\}, 1 - \{x\})$ denotes the nearest distance to integers.)

Theorem 1.100 *Let α be an irrational of finite approxiation type η and let $f(n)$ be a strongly q-additive function such that there exists $1 \le b \le q-1$ with $f(b) > 0$. Then for every $\varepsilon > 0$ there exists a constant $c(q, \alpha, \varepsilon, f) > 0$ such that*

$$D_N(\alpha f(n)) \le \frac{c(q,\alpha,\varepsilon,f)}{(\log N)^{1/(2\eta)-\varepsilon}}$$

for all $N > 1$. If α is not of approximation type η' for any $\eta' < \eta$ then for every $\varepsilon > 0$ and infinitely many N

$$D_N(\alpha f(n)) \ge \frac{1}{(\log N)^{1/(2\eta)+\varepsilon}}.$$

Remark. The following proof will also show that for every irrational α there exists a constant $c'(q, \alpha, f) > 0$ such that for infinitely many N

$$D_N(\alpha f(n)) > \frac{c'(q,\alpha,f)}{(\log N)^{1/2}}.$$

Note that by the theorem of THUE-SIEGEL-ROTH every irrational real algebraic number α is of approximation type $\eta = 1$. Hence the exponent $1/2$ in this general lower bound cannot be replaced by a larger exponent.

Lemma 1.101 *Let $g : \mathbf{Z}_0^+ \to \mathbf{C}$ be a function such that $g(0) = 1$, $|g(n)| \le 1$, and*

$$g(n) = \prod_{k=0}^{\infty} g(d_k(q;n)q^k) \qquad (n \in \mathbf{Z}^+). \tag{1.176}$$

Suppose that there exists a continuous nondecreasing function $F : [1,\infty) \to (0,\infty)$, which satisfies $F(u) \le u$ and

$$\left| \frac{1}{q^k} \sum_{n=0}^{q^k-1} g(n) \right| \le \frac{1}{F(q^k)} \qquad (k \ge 1).$$

Then

$$\left| \frac{1}{N} \sum_{n=0}^{N-1} g(n) \right| \le \frac{q+1}{F(\sqrt{N})} \qquad (N \ge 1).$$

Proof. Let m be the largest index such that $d_m(q;N) \ne 0$ and set

$$N(j) = \sum_{k=j}^{m} d_k(q;N)q^k.$$

Then

$$\sum_{n=0}^{N-1} g(n) = \sum_{n=0}^{N(m)-1} g(n) + \sum_{j=0}^{m-1} \sum_{n=N(j+1)}^{N(j)-1} g(n),$$

where

$$\sum_{n=0}^{N(m)-1} g(n) = \sum_{l=0}^{d_m-1} \sum_{n=lq^m}^{(l+1)q^m-1} g(n) = \sum_{l=0}^{d_m-1} g(lq^m) \sum_{n=0}^{q^m-1} g(n)$$

and

$$\begin{aligned} \sum_{n=N(j+1)}^{N(j)-1} g(n) &= g(N(j+1)) \sum_{n=0}^{d_j q^j-1} g(n) \\ &= g(N(j+1)) \sum_{l=0}^{d_j-1} g(lq^j) \sum_{n=0}^{q^m-1} g(n). \end{aligned}$$

Hence for every positive integer r we have

$$\begin{aligned} \left| \sum_{n=0}^{N-1} g(n) \right| &\le \sum_{j=0}^{m} \left| \sum_{l=0}^{d_j-1} g(lq^j) \right| \left| \sum_{n=0}^{q^j-1} g(n) \right| \\ &\le \sum_{j=0}^{m} d_j q^j \frac{1}{q^j} \left| \sum_{n=0}^{q^j-1} g(n) \right| \\ &\le \sum_{j=1}^{r} d_j q^j + \sum_{j=r}^{m} d_j q^j \frac{1}{F(q^r)} \le q^r + \frac{N}{F(q^r)}. \end{aligned}$$

Let t be the unique real number such that $(t/q)F(t/q) = N$. Then $t/q \geq N$ because of $F(t/q) \leq t/q$. Choosing r such that $q^{r+1} > t \geq q^r$ we finally obtain

$$\begin{aligned}\left|\sum_{n=0}^{N-1} g(n)\right| &\leq q^r + \frac{N}{F(q^r)} \leq t + \frac{N}{F(t/q)} \\ &= \frac{qN}{F(t/q)} + \frac{N}{F(t/q)} \leq (q+1)\frac{N}{F(\sqrt{N})}.\end{aligned}$$

□

Lemma 1.102 *Let $B = \max_{1 \leq b < q} f(b)$, where f is a function as in the above Theorem. Then*

$$\left|\sum_{j=0}^{q-1} e(\alpha f(j))\right| \leq q - 2\pi\|B\alpha\|^2.$$

Proof. First observe that

$$\begin{aligned}\left|\sum_{j=0}^{q} e(\alpha f(j))\right| &\leq |1 + e(B\alpha)| + q - 2 \\ &= 2|\cos(\pi B\alpha)| + q - 2 = 2\cos(\pi\|B\alpha\|) + q - 2.\end{aligned}$$

Furthermore we have

$$\cos x = 1 - \int_0^x \sin t\, dt \leq 1 - \int_0^x \frac{2}{\pi} t\, dt = 1 - \frac{x^2}{\pi}$$

for $|x| \leq \pi/2$. This proves (1.102). □

Proof of Theorem 1.100. From Lemma 1.102 we directly obtain

$$\left|\sum_{j=0}^{q^k-1} e(h\alpha f(n))\right| = \left|\sum_{j=0}^{q} e(\alpha f(j))\right|^k \leq (q - 4\|hB\alpha\|^2)^k.$$

Hence the assumptions of Lemma 1.101 are satisfied for

$$F(u) = \left(\frac{q}{q - 4\|hB\alpha\|^2}\right)^{\log u/\log q}.$$

By using the inequality $1 - u \leq e^{-u}$ we obtain

$$\begin{aligned}\left|\frac{1}{N}\sum_{n=0}^{N-1} e(h\alpha f(n))\right| &= \frac{q+1}{F(\sqrt{N})} \\ &= (q+1)\left(1 - \frac{4\|hB\alpha\|^2}{q}\right)^{(\log N)/(2\log q)} \\ &\leq (q+1)\exp\left(-\frac{2\|hB\alpha\|^2 \log N}{q\log q}\right).\end{aligned}$$

Now the inequality of ERDŐS-TURÁN-KOKSMA (Theorem 1.21) combined with the assumption $\|hB\alpha\| \ge c_0 h^{-\eta-\varepsilon}$ $(0 < \varepsilon < 1/(4\eta))$ gives

$$D_N(\alpha f(n)) \le 6\left(\frac{1}{H} + \sum_{h=1}^{H} \frac{q+1}{h} \exp\left(-\frac{2c_0^2 h^{-2\eta-2\varepsilon} \log N}{q \log q}\right)\right).$$

Choosing $H = [(\log N)^{1/(2\eta)-\varepsilon}]$ we have for sufficiently large N

$$\begin{aligned}(1+\log H)\exp(-c_1 H^{-2\eta-2\varepsilon}\log N) &\le \frac{1}{2\eta}\log\log N \exp(-c_2(\log N)^{-\varepsilon/\eta+2\varepsilon\eta+2\varepsilon^2})\\ &\le \frac{1}{2\eta}\log\log N \exp(-c_2(\log N)^{-\varepsilon+2\varepsilon+2\varepsilon^2})\\ &\le \frac{c_3}{H}.\end{aligned}$$

Thus

$$D_N(\alpha f(n)) \le \frac{c_4}{H} \le \frac{c(q,\alpha,\varepsilon,f)}{(\log N)^{1/(2\eta)-\varepsilon}}.$$

In order to prove the lower bound set $\kappa = \eta - \varepsilon$ $(\eta > \varepsilon > 0$ arbitrary). Then there are infinitely many positive integers h with $\|h\alpha\| \le ch^{-\kappa} < 1/(2B)$. Since the functions $t \mapsto \cos(2\pi i h t)$ and $t \mapsto \sin(2\pi i h t)$ are of variation $2\pi h$ we obtain from KOKSMA-HLAWKA's inequality (Theorem 1.14)

$$\left|\frac{1}{N}\sum_{n=0}^{N-1} e(h\alpha f(n))\right| \le 4\pi h\, D_N(\alpha f(n)).$$

From $B\|h\alpha\| \le 1/2$ it follows that

$$\begin{aligned}\left|\sum_{j=0}^{q-1} e(h\alpha f(n))\right| &\ge \left|\sum_{j=0}^{q-1} \cos(2\pi h\alpha f(n))\right|\\ &= \left|\sum_{j=0}^{q-1} \cos(2\pi f(n)\|h\alpha\|)\right|\\ &\ge q\cos(2\pi B\|h\alpha\|).\end{aligned}$$

Hence

$$\begin{aligned}\left|\frac{1}{q^k}\sum_{j=0}^{q^k-1} e(h\alpha f(n))\right| &= \left|\frac{1}{q}\sum_{j=0}^{q-1} e(h\alpha f(n))\right|^k\\ &\ge (\cos(2\pi B\|h\alpha\|))^k\\ &\ge \left(1-2\pi^2B^2c^2h^{-2\kappa}\right)^k,\end{aligned}$$

and thus

$$D_{q^k}(\alpha f(n)) \ge \frac{1}{4\pi h}\left(1 - 2\pi^2 B^2 c^2 h^{-2\kappa}\right)^k$$

for infinitely many $h \ge 1$. Choosing $k = [h^{2\kappa}]$ yields

$$D_{q^k}(\alpha f(n)) \ge \frac{1}{4\pi}\left(\frac{\log q}{\log q^k}\right)^{1/(2\kappa)} \exp(-2\pi^2 B^2 c^2)$$

for infinitely many k, which establishes the lower bound of Theorem 1.100.

The general lower bound of the Remark after Theorem 1.100 can be proved along the same lines. By DIRICHLET's approximation theorem (Lemma 1.78) there are infinitely many h with $\|h\alpha\| \le h^{-1}$. Hence we can choose $\kappa = 1$ in the above considerations to obtain the proposed lower bound. □

We now present two variants (generalizations) of Theorem 1.99 which seem to interesting by themselves. The first result is concerned with well distribution; see 1.1.

Theorem 1.103 *Suppose that $f(n)$ is strongly q-additive and that that* $\gcd\{0 < j < q : f(j) > 0\} = 1$. *If a sequence $(x_n)_{n\ge 0}$ is well distributed mod 1 then $(x_{f(n)})$ is well distributed mod 1, too.*

The main ingredience for the proof of Theorem 1.103 is the following lemma due to ODLYZKO and RICHMOND [1388].

Lemma 1.104 *Let $b_0, b_1, \ldots, b_d$ be a finite sequence of non-negative numbers with $b_0 > 0$, $b_d > 0$, and*

$$\gcd\{j : b_j \ne 0\} = 1.$$

Let a_{nk} be defined by

$$\sum_{n\ge 0} a_{nk} x^n = \left(b_0 + b_1 x + \cdots + b_d x^d\right)^k.$$

Then for every $\delta > 0$ there exists a $k_0(\delta)$ such that for every $k \ge k_0(\delta)$

$$a_{nk}^2 \ge a_{n-1,k} a_{n+1,k}\ , \qquad \delta k \le n \le (d-\delta)k. \tag{1.177}$$

Note that the gcd-condition is no real restriction and that (1.177) implies unimodality of the sequence a_{nk}, $k \ge k_0(\delta)$, $\delta k \le n \le (d-\delta)k$, i.e. there exists an n_0 such that a_{nk} is increasing for $n < n_0$ and decreasing for $n > n_0$.

It should be further noticed that this Lemma is strongly related to the central limit theorem for a sum of independent discrete random variables. Set $b = b_0 + b_1 + \cdots + b_d$. Then

$$\frac{a_{nk}}{b^k} = \mathbf{P}[X_1 + X_2 + \cdots + X_k = n],$$

where X_j, $1 \le j \le k$ are independent discrete random variables with

$$\mathbf{P}[X_j = n] = \frac{b_n}{b}.$$

It is well known (see PETROV [1431]) that there is a local limit theorem of the form

$$a_{nk} = \frac{b^k}{\sqrt{2\pi k\sigma^2}}\left(\exp\left(-\frac{(n-k\mu)^2}{2k\sigma^2}\right) + \mathcal{O}(k^{-1/2})\right), \tag{1.178}$$

with

$$\mu = \frac{1}{b}\sum_{j=0}^{d} jb_j \qquad \sigma^2 = \frac{1}{b}\sum_{j=0}^{d}(j-\mu)^2 b_j$$

and exponential tail estimates of the form

$$\sum_{|n-k\mu|\ge x\sqrt{k\sigma^2}} a_{nk} \le e^{-cx^2} q^k$$

for some $c > 0$. Especially, the following properties are satisfied

$$\max_{n\ge 0} a_{nk} = \mathcal{O}\left(\frac{q^k}{\sqrt{k}}\right) \tag{1.179}$$

and for (sufficiently small) $\delta > 0$

$$\sum_{n\le k\delta} a_{nk} + \sum_{n\ge(d-\delta)k} a_{nk} \le q'(\delta)^k, \tag{1.180}$$

where $q'(\delta) < q$.

With help of Lemma 1.104 and using these properties we are able to prove the following lemma.

Lemma 1.105 *Suppose that $f(n)$ and $(x_n)_{n\ge 0}$ satisfy the same assumptions as in Theorem 1.103. Then for every integer $h \ge 1$*

$$\lim_{k\to\infty}\sup_{l\ge 0}\left|\frac{1}{q^k}\sum_{j=0}^{q^k-1} e(hx_{s(j)+l})\right| = 0.$$

Proof. Let

$$\sum_{j=0}^{q^k-1} e(hx_{f(j)+l}) = \sum_{n\ge 0} a_{nk} e(hx_{n+l}).$$

Then

$$a_{nk} = |\{j < q^k \;:\; s(j) = n\}|$$

and

$$\sum_{n\ge 0} a_{nk} x^n = \left(\sum_{j=0}^{q-1} x^{f(j)}\right)^k.$$

By Lemma 1.104, for every $\delta > 0$ there exists a $k_0 = k_0(\delta)$ such that a_{nk}, $\delta k \le n \le (d-\delta)k$, is unimodal ($d = \max\{j < q;\, f(j) > 0\} > 0$). Furthermore (1.179) and (1.180) hold.

Since $(x_n)_{n\ge 0}$ is well distributed it follows from WEYL's criterion (Theorem 1.19) that for every integer $h > 0$ there exists a monotonically decreasing sequence ε_N with $\lim_{N\to\infty} \varepsilon_N = 0$ and

$$\sup_{l\ge 0} \left| \sum_{n=0}^{N-1} e(hx_{n+l}) \right| \le N\varepsilon_N.$$

For any given $\varepsilon > 0$ let $N(\varepsilon)$ be chosen such that $\varepsilon_N \le \varepsilon$ for $N \ge N(\varepsilon)$.

Let $a_{n_0 k} = \max_{n\ge 0} a_{nk}$. Then by use of ABEL' partial summation and (1.179)

$$\begin{aligned}
\left| \sum_{n=n_0}^{(d-\delta)k} a_{nk} e(hx_{n+l}) \right| &\le \sum_{n=n_0}^{(d-\delta)k} a_{nk}((d-\delta)k - n_0 + 1)\varepsilon_{(d-\delta)k-n_0+1} \\
&\quad + \sum_{n=n_0}^{(d-\delta)k-1} (a_{n+1,k} - a_{nk})(n-m+1)\varepsilon_{n-m+1} \\
&\le \sum_{n=n_0}^{(d-\delta)k} a_{nk}((d-\delta)k - n_0 + 1)\varepsilon \\
&\quad + \sum_{n=n_0}^{(d-\delta)k-1} (a_{n+1,k} - a_{nk})(n-m+1)\varepsilon \\
&\quad + \sum_{n=n_0}^{n_0+N(\varepsilon)} (a_{n+1,k} - a_{nk})(n-m+1) \\
&\le \varepsilon \sum_{n=n_0}^{(d-\delta)k} a_{nk} + N(\varepsilon)^2 a_{n_0 k} \\
&\le \varepsilon q^k + \mathcal{O}(N(\varepsilon)^2 q^k/\sqrt{k})
\end{aligned}$$

for sufficiently large k. Similarly we obtain

$$\left| \sum_{n=\delta k}^{n_0-1} a_{nk} e(hx_{n+l}) \right| \le 2\varepsilon q^k.$$

Furthermore, by (1.180)

$$\left| \sum_{n<\delta k} a_{nk} e(hx_{n+l} \sum_{n>(d-\delta)k} a_{nk} e(hx_{n+l} \right| \le q(\delta)^k \le \varepsilon q^k.$$

Hence for every $\varepsilon > 0$ there exists $k(\varepsilon)$ such that for every $k \geq k(\varepsilon)$

$$\left|\sum_{j=0}^{q^k-1} e(hx_{f(j)+l})\right| \leq 6\varepsilon q^k \tag{1.181}$$

uniformly for $l \geq 0$. □

Proof of Theorem 1.103. By Lemma 1.105, for arbitrary $\varepsilon > 0$ there exists $k(\varepsilon)$ such that for every $k \geq k(\varepsilon)$ (1.181) holds for all $l \geq 0$. Suppose that $N \geq q^{k(\varepsilon)}/\varepsilon$ and for every $l \geq 0$ define m_1, m_2 by $(m_1 - 1)q^k \leq l+1 < m_1 q^k$ and by $(m_2-1)q^k \leq l + N < m_2 q^k$. Then

$$\begin{aligned}
\left|\sum_{n=l+1}^{l+N} e(hx_{f(n)})\right| &\leq 2q^k + \sum_{t=m_1}^{m_2-1}\left|\sum_{n=tq^k}^{(t+1)q^k-1} e(hx_{f(n)})\right| \\
&\leq 2q^k + \sum_{t=m_1}^{m_2-1}\left|\sum_{j=0}^{q^k-1} e(hx_{f(t)+f(j)})\right| \\
&\leq 2q^k + (m_2 - m_1)q^k 6\varepsilon \\
&\leq 2N\varepsilon + 6N\varepsilon = 8N\varepsilon.
\end{aligned}$$

This completes the proof of the Theorem 1.103. □

We will say that a function $f(n)$ satisfies a local central limit theorem if there exist sequences μ_N and σ_N with $\lim_{N\to\infty} \sigma_N = \infty$ such that for every interval $[a, b] \subseteq \mathbf{R}$ we have

$$|\{n \leq N \,:\, f(n) = m\}| = \frac{N}{\sqrt{2\pi\sigma_N^2}}\left(\exp\left(\frac{(m-\mu_N)^2}{2\sigma_N^2}\right) + o(1)\right) \tag{1.182}$$

uniformly for all integers $m \in [\mu_N + a\sigma_N, \mu_N + b\sigma_N]$ as $N \to \infty$.

Theorem 1.106 *Suppose that the function $f : \mathbf{N}_0 \to \mathbf{N}_0$ satisfies a local central limit theorem. Then the sequence $(f(n)\alpha)_{n\geq 1}$ is u.d. mod 1 if and only if α is irrational.*

Proof. Set

$$a_{mN} = |\{n \leq N \,:\, f(n) = m\}|.$$

The first step of the proof is to show that

$$\sum_{m\geq 0} |a_{m+1,N} - a_{mN}| = o(N) \tag{1.183}$$

as $N \to \infty$. Let $\varepsilon > 0$ be given and let $T(\varepsilon)$ be defined by

$$\frac{1}{\sqrt{2\pi}}\int_{-T(\varepsilon)}^{T(\varepsilon)} e^{-t^2/2}\, dt = 1 - \varepsilon.$$

Then (1.182) implies

$$\sum_{|m-\mu_N|\le T(\varepsilon)\sigma_N} a_{mN} = N(1-\varepsilon) + o(N).$$

Hence for sufficiently large N we obtain

$$\sum_{|m-\mu_N|\le T(\varepsilon)\sigma_N} a_{mN} < 2\varepsilon N.$$

Furthermore, we have

$$\sum_{|m-\mu_N|\le T(\varepsilon)\sigma_N} |a_{m+1,N} a_{m,N}| \le \frac{2N}{\sqrt{2\pi\sigma_N^2}} + o(N) = o(N).$$

Hence (1.183) follows.

Now suppose that α is irrational and that $h > 0$ is an integer. Then

$$\left|\sum_{n=0}^{N-1} e(h\alpha n)\right| \le \frac{N}{\sin(\pi h\alpha)}.$$

Hence by ABEL summation

$$\begin{aligned}\left|\sum_{n=0}^{N-1} e^{h\alpha f(n)}\right| &= \left|\sum_{m\ge 0} a_{mN} e^{h\alpha m}\right| \\ &\le \sum_{m\ge 0} |a_{m+1,N} - a_{mN}| \frac{N}{\sin(\pi h\alpha)} \\ &= o(N).\end{aligned}$$

By WEYL's criterion (Theorem 1.19) $(f(n)\alpha)$ is u.d. mod 1. □

With help of (1.178) it is an easy exercise to show that strongly q-additive functions satisfy a local central limit theorem (see also SCHMID [1622] for the case $q = 2$). Hence Theorem 1.99 is also a consequence of Theorem 1.106. However, there are other interesting number theoretic functions which satisfy a local central limit theorm.

Corollary 1.107 *Let $\omega(n)$ denote the number of different prime factors of n and $\Omega(n)$ the number of prime factors of n counted with multiplicities. Then the sequences $(\omega(n)\alpha)$ and $(\Omega(n)\alpha)$ are u.d. mod 1 for irrational α.*

This property of $\omega(n)$ and $\Omega(n)$ has already been observed by ERDŐS [561] and DELANGE [449]. We just have to apply a local version of the ERDŐS-KAC theorem (see Elliott [558, 559] or Hwang [857]).

We will finish this section with another concept for producing integer sequences $f(n)$ such that $\alpha f(n)$ is u.d. mod 1 for irrational α, namely sequences with empty spectrum. In order to motivate the subsequent Definition 1.109 we present the following extension of Theorem 1.99.

Theorem 1.108 *Suppose that $f(n)$ is strongly q-additive and that there exists $1 \le b \le q-1$ with $f(b) > 0$. Then the sequence $(\alpha f(n) + \beta n)$ is u.d. mod 1 for every $\alpha \notin \mathbf{Q}$ and for every $\beta \in \mathbf{R}$.*

Proof. Note that $\alpha f(n) + \beta n$ is q-additive. Hence $g(n) = e(h\alpha f(n) + h\beta n)$ satisfies (1.176). Now suppose that $h \in \mathbf{Z} \setminus \{0\}$. From Lemma 1.101 it follows that if

$$\lim_{k\to\infty} \frac{1}{q^k} \sum_{n=0}^{q^k-1} g(n) = 0 \tag{1.184}$$

then

$$\lim_{N\to\infty} \frac{1}{N} \sum_{n=0}^{N-1} g(n) = 0. \tag{1.185}$$

By WEYL's criterion (Theorem 1.19) (1.185) implies that $(\alpha f(n) + \beta n)$ is u.d. mod 1.

Since

$$\sum_{n=0}^{q^k-1} g(n) = \prod_{j=0}^{k-1} \left(\sum_{b=0}^{q-1} e(h\alpha f(b) + h\beta b q^j) \right),$$

it suffices to show that

$$c = \prod_{j=0}^{\infty} \left| \frac{1}{q} \sum_{b=0}^{q-1} e(h\alpha f(b) + h\beta b q^j) \right| = 0.$$

If $c > 0$ then

$$\lim_{j\to\infty} \left| \sum_{b=0}^{q-1} e(h\alpha f(b) + h\beta b q^j) \right| = q.$$

Since $e(h\alpha f(0)) = 1$ it follows that for every $1 \le b \le q-1$

$$\lim_{j\to\infty} h\beta b q^j = -h\alpha f(b) \bmod 1.$$

However, if the sequence xq^j mod 1 converges then its limit is of the form $d/(q-1)$, where d is an integer. But $-h\alpha f(b)$ is irrational for at least one $b > 0$ which leads to a contradiction. Hence $c = 0$. □

Definition 1.109 *Let $(x_n)_{n\ge1}$ be a real sequence. Then the spectrum of (x_n) is the set of those $\beta \in \mathbf{R}$ such that $(x_n + \beta n)$ is u.d. mod 1.*

Theorem 1.108 shows that $(\alpha f(n))$ has an empty spectrum for strongly q-additive f and irrational α. The reason why we are interested in such sequences is that proper subsequences of sequences with empty spectrum are u.d. mod 1, too. For various references on the spectrum of sequences see the Notes of Section 1.6.

Definition 1.110 *A function $p : \mathbf{N}_0 \to \mathbf{C}$ is called almost periodic if for every $\varepsilon > 0$ there exists a trigonometric polyomial*

$$t(x) = \sum_{l=0}^{L} a_l e(\lambda_l x) \qquad (L \ge 0, a_l \in \mathbf{C}, \lambda_l \in \mathbf{R})$$

such that

$$\limsup_{N\to\infty} \frac{1}{N} \sum_{n=0}^{N-1} |p(n) - t(n)| < \varepsilon. \tag{1.186}$$

Furthermore a non-decreasing unbounded sequence (m_n) of positive integers m_n is called almost periodic if the function

$$p(k) = |\{n \ge 1 \,:\, m_n = k\}|$$

is almost periodic. (Sometimes $p(k)$ is called the characteristic function of (m_n).)

Note that by definition for every almost periodic function $p : \mathbf{N}_0 \to \mathbf{C}$ the limit

$$\lim_{N\to\infty} \frac{1}{N} \sum_{n=1}^{N} p(n)$$

exists. (Observe that this limit exists for every trigonometric polynomial.) Furthermore, if (m_n) is an almost periodic sequence then this limit is positive for its characteristic function $p(k)$.

The following two lemmata show that there are non-trivial, non-periodic almost periodic sequences.

Lemma 1.111 *Let $\beta > 0$. Then $m_n = [\beta n]$ is almost periodic.*

Proof. If β is rational then (m_n) is a finite union of arithmetic sequences and thus periodic.

In the case of irrational β we have

$$p(k) = -\left[\left\{-\frac{k}{\alpha}\right\} - \frac{1}{\alpha}\right].$$

Now for an arbitrary $\varepsilon > 0$ let $t(x)$ be a periodic trigonometric polynomial with period 1 such that

$$\sup_{|\{x\}-1/\alpha| \ge \varepsilon/2} |t(x) + [\{x\} - 1/\alpha]| < \frac{\varepsilon}{2}$$

and

$$\sup_{x\in\mathbf{R}} |t(x) + [\{x\} - 1/\alpha]| \le 1.$$

Note that the asymptotic density of those positive integers n with $|\{x\} - 1/\alpha| < \varepsilon/2$ is $\varepsilon/2$. Hence it follows that

$$\limsup_{N\to\infty} \frac{1}{N} \sum_{n=1}^{N} |p(k) - t(\alpha k)| < \varepsilon.$$

Thus $[\beta n]$ is almost periodic. $\square$

Lemma 1.112 *Let E be a subset of the positive integers such that*

$$\sum_{q\in E} \frac{1}{q} < \infty.$$

Let $M(E) = (m_n)_{n\geq 1}$ be the sequence of those positive integers which are not divisible by any $n \in E$. Then $M(E)$ is almost periodic.

Proof. Let p denote the characteristic function of $M(E)$ and for any positive integer A let p_A denote the characteristic function of $M(E\cap[0,A])$. Obviously, p_A is periodic. Furthermore

$$\begin{aligned}
\limsup_{N\to\infty} \frac{1}{N}\sum_{n=1}^{N} |p(n) - p_A(n)| &= \limsup_{N\to\infty} \frac{1}{N} \sum_{n\leq N, n\in M(E\cap[0,A])\setminus M(E)} 1 \\
&\leq \limsup_{N\to\infty} \frac{1}{N} \sum_{n>A, n\in E} \frac{N}{n} \\
&\leq \sum_{n>A,n\in E} \frac{1}{n}.
\end{aligned}$$

By assumption this bound tends to 0 as $N \to \infty$. Hence p is almost periodic. □

Theorem 1.113 *Suppose that $(x_n)_{n\geq 1}$ has empty spectrum. Then $(x_{m_n})_{n\geq 1}$ is u.d. mod 1 for every non-decreasing, unbounded, almost periodic sequence $(m_n)_{n\geq 1}$ of positive integers.*

Proof. Let $(m_n)_{n\geq 1}$ be a non-decreasing, unbounded, almost periodic sequence of non-negative integers and $p : \mathbf{N}_0 \to \mathbf{C}$ its corresponding almost periodic functions defined by $p(k) = |\{n \geq 1 : m_n = k\}|$. Furthermore let $A_N(h)$ denote the WEYL sums

$$A_N(h) = \frac{1}{N}\sum_{n=1}^{N} e(hx_{m_n}) = \frac{m_N}{N}\frac{1}{m_N}\sum_{k=1}^{m_N} p(k)e(hx_k).$$

Since

$$1 = A_N(0) = \frac{m_N}{N}\frac{1}{m_N}\sum_{k=1}^{m_N} p(k),$$

and

$$\frac{1}{m_N}\sum_{k=1}^{m_N} p(k)$$

has a finite positive limit as $N \to \infty$, it follows that

$$\lim_{N\to\infty} \frac{m_N}{N} = B$$

exits. Now let $\varepsilon > 0$ and let

$$t(x) = \sum_{l=0}^{L} a_l e(\lambda_l x)$$

be a trigonometric polynomial which satisfies (1.186). By assumption $(x_k + \beta k)_{k\geq 1}$ is u.d. mod 1 for every $\beta \in \mathbf{R}$. Thus by WEYL's criterion (Theorem 1.19)

$$\lim_{N\to\infty} \frac{1}{m_N} \sum_{k=1}^{m_N} t(k)e(hx_k) = \sum_{l=0}^{L} a_l \lim_{N\to\infty} \frac{1}{m_N} \sum_{k=1}^{m_N} e(\lambda_l k + hx_k) = 0$$

for $h \neq 0$. Therefore, writing

$$A_N(h) = \frac{m_N}{N}\frac{1}{m_N}\sum_{k=1}^{m_N} t(k)e(hx_k) + \frac{m_N}{N}\frac{1}{m_N}\sum_{k=1}^{m_N}(p(k) - t(k))e(hx_k),$$

we immediately obtain

$$\limsup_{N\to\infty} |A_N(h)| \leq B\varepsilon.$$

Since $\varepsilon > 0$ is arbitrary it follows that $\lim_{N\to\infty} A_N(h) = 0$. Thus $(x_{m_n})_{n\geq 1}$ is u.d. mod 1. □

Corollary 1.114 *Suppose that $\alpha \notin \mathbf{Q}$ and $\beta \in \mathbf{R}\setminus\{0\}$. Then the sequence $(\alpha[\beta n]^2)$ is u.d. mod 1.*

Proof. If α is irrational then $(\alpha n^2 + \gamma n)$ is u.d. mod 1 for every $\gamma \in \mathbf{R}$. Hence (αn^2) has empty spectrum. Thus we can apply Theorem 1.113 with $([\beta n])$ as an almost periodic sequence. □

Corollary 1.115 *Let (m_n) denote the sequence of all squarefree positive integers and let $f(n)$ be a (non-trivial) strongly q-multiplicative function. Then $(\alpha f(m_n))_{n\geq 1}$ is u.d. mod 1 if and only if α is irrational.*

Proof. By Theorem 1.108 $f(n)$ has empty spectrum. Furthermore let E denote the set of all integer squares. Then $M(E)$ is the set of all squarefree numbers. □

1.4.4 Normal Numbers

Definition 1.116 *Let $q \geq 2$ be a fixed integer. A real number α with q-adic expansion $\alpha = 0.a_1a_2\ldots$ is said to be* normal to base q *if for all $k \geq 0$ and all blocks $b_1b_2\ldots b_k \in \{0,1,\ldots,q-1\}^k$*

$$\lim_{N\to\infty} \frac{|\{n \leq N : a_na_{n+1}\ldots a_{n+k-1} = b_1b_2\ldots b_k\}|}{N} = q^{-k}.$$

For an introductory survey on normal numbers we refer to KUIPERS and NIEDERREITER [983]. For more recent literature and results see the Notes. It is an easy exercise to prove the following criterion (see [983]).

Theorem 1.117 *α is normal to base $q \geq 2$ if and only if the sequence $x_n = q^n\alpha$ is u.d. mod 1.*

It is also quite easy to see that almost all $\alpha \in [0,1]$ are normal for all bases $q \geq 2$. However, it is a different problem to construct normal numbers explicitly. One method has been observed by DAVENPORT and ERDŐS [430]. They considered numbers of the form

$$\alpha = 0.f(1)f(2)\cdots,$$

where each $f(n)$ is represented in the decimal expansion and the digits of $f(1)$ are succeeded by those of $f(2)$ and so on, and showed that α is normal to base 10 for any polynomial $f(x)$, which values $f(n)$, $n = 1, 2, \ldots$, are positive integers. In the following we will quantify and extend this property (see SCHIFFER [1619] and NAKAI and SHIOKAWA [1254]).

Theorem 1.118 *Let $f(x)$ be a non-constant polynomial with rational coefficients such that $f(x) \geq 1$ for all $x \geq 1$ and let $q \geq 2$ be fixed. Set*

$$a_1a_2\ldots = [f(1)][f(2)]\ldots,$$

where $[f(n)]$ is represented in the q-ary digit expansion, and

$$\alpha = \sum_{k=1}^{\infty} a_k q^{-k} = 0.[f(1)][f(2)]\ldots$$

Then

$$D_N(q^n\alpha) = \mathcal{O}\left(\frac{1}{\log N}\right).$$

Theorem 1.119 *Let $f(x)$ be a linear poylnomial with rational coefficients such that $f(n) \geq 1$ for all $n \geq 1$ and let $q \geq 2$ be fixed. Set*

$$\alpha = 0.[f(1)][f(2)]\ldots,$$

where $[f(n)]$ is represented in the q-ary digit expansion. Then there exists a positive constant c such that

$$D_N(q^n\alpha) \geq \frac{c}{\log N}$$

for infinitely many N.

The crucial step of the proof of Theorem 1.118 is an application of WEYL's inequality

Proposition 1.120 *Suppose that $(a,b) = 1$, $|\alpha - a/b| \leq b^{-2}$, $\phi(x) = \alpha x^k + \alpha_1 x^{k-1} + \ldots + \alpha_{k-1}x + \alpha_k$ and*

$$T(\phi) = \sum_{m=1}^{M} e(\phi(m)).$$

Then

$$T(\phi) \ll M^{1+\varepsilon}(b^{-1} + M^{-1} + bM^{-k})^{1/K},$$

where $K = 2^{k-1}$.

Lemma 1.121 *Suppose that X, Y, α are real numbers with $X \geq 1$, $Y \geq 1$, and $|\alpha - a/b| \leq b^{-2}$ with $(a, b) = 1$. Then*

$$\sum_{m \leq X} \min\left(\frac{XY}{m}, \frac{1}{\|\alpha m\|}\right) \ll XY\left(\frac{1}{b} + \frac{1}{Y} + \frac{b}{XY}\right) \log(2Xb).$$

Proof. We set

$$S = \sum_{m \leq X} \min\left(\frac{XY}{m}, \frac{1}{\|\alpha m\|}\right).$$

Then clearly

$$S \leq \sum_{0 \leq j \leq X/b} \sum_{r=1}^{b} \min\left(\frac{XY}{bj + r}, \frac{1}{\|\alpha(bj + r)\|}\right).$$

For each j let $y_j = [\alpha j b^2]$ and write $\theta = b^2\alpha - ba$. Then

$$\alpha(bj + r) = (y_j + ar)/b + \{\alpha j b^2\} + \theta r b^{-2}.$$

For $j = 0$ and $r \leq b/2$ we have

$$\|\alpha(bj + r)\| \geq \left\|\frac{ar}{b}\right\| - \frac{1}{2b} \geq \frac{1}{2}\left\|\frac{ar}{b}\right\|.$$

Otherwise, for each j there are at most $\mathcal{O}(1)$ values of r for which $\|\alpha(bh + r)\| \geq \frac{1}{2}\|(y_j + ar)/b\|$ fails to hold, and moreover $bj + r \gg b(j + 1)$. Therefore

$$\begin{aligned} S &\ll \sum_{1 \leq r \leq b/2} \frac{1}{\|ar/b\|} + \sum_{0 \leq j \leq X/b} \left(\frac{XY}{b(j+1)} + \sum_{1 \leq r \leq b, \frac{y_j + ar}{b} \notin \mathbf{Z}} \frac{1}{\|(y_j + ar)/b\|}\right) \\ &\ll \frac{XY}{b} \sum_{0 \leq j \leq X} \frac{1}{j+1} + \left(\frac{X}{b} + 1\right) \sum_{0 \leq h \leq b/2} \frac{b}{h} \\ &\ll XY\left(\frac{1}{b} + \frac{1}{Y} + \frac{b}{XY}\right) \log(2Xb), \end{aligned}$$

which proves the lemma. $\square$

Let Δ_j be defined by

$$\begin{aligned} \Delta_1(\phi(x); \beta) &= \phi(x + \beta) - \phi(x) \quad \text{and by} \\ \Delta_{j+1}(\phi(x); \beta_1, \ldots, \beta_{j+1}) &= \Delta_1(\Delta_j(\phi(x); \beta_1, \ldots, \beta_j); \beta_{j+1}) \qquad (j \geq 1), \end{aligned}$$

where $\phi(x)$ denotes a real-valued function.

Lemma 1.122 *Let*

$$T(\phi) = \sum_{m=1}^{M} e(\phi(m)),$$

where ϕ is an arbitrary arithmetical function. Then

$$|T(\phi)|^{2^j} \leq (2M)^{2^j-j-1} \sum_{|h_1|<M} \cdots \sum_{|h_j|<M} |T_j|,$$

with

$$T_j = \sum_{m\in I_j} e\left(\Delta_j(\phi(m); h_1, \ldots, h_j)\right).$$

Additionally, the intervals $I_j = I_j(h_1, \ldots, h_j)$ (possibly empty) satisfy

$$I_j(h_1) \subset [1, M] \quad \text{and} \quad I_j(h_1, \ldots, h_j) \subset I_{j-1}(h_1, \ldots, h_{j-1}).$$

Proof. We proceed by induction on j. For brevity write $\Delta_j(x)$ for $\Delta_j(\phi(x); h_1, \ldots, h_j)$. Obviously

$$\begin{aligned} |T(\phi)|^2 &= \sum_{m=1}^{M} \sum_{h_1=1-m}^{M-m} e(\Delta_1(m)) \\ &= \sum_{h_1=1-M}^{M-1} \sum_{m\in I_1} e(\Delta_1(m)), \end{aligned}$$

where $I_1 = [1, M] \cap [1 - h_1, M - h_1]$.

Now if the conclusion of the lemma is assumed for a particular value of j, then by CAUCHY's inequality,

$$|T(\phi)|^{2^{j+1}} \leq (2M)^{2^{j+1}-2j-2}(2M)^j \sum_{h_1,\ldots,h_j} |T_j|^2.$$

Thus

$$|T_j|^2 = \sum_{|h|<M} \sum_{m\in I_{j+1}} e\left(\Delta_j(m+h) - \Delta_j(m)\right)$$

with $I_{j+1} = I_j \cap \{m : m + h \in I_j\}$. □

Lemma 1.123 *Let $k \geq 2$ be a natural number. Then for every $\varepsilon > 0$ there exists a constant $C(\varepsilon) > 0$ such that for every $h \in \mathbf{N}$*

$$|\{(h_1, \ldots, h_k) \in \mathbf{N}^k \,:\, h_1 \cdots h_k = h\}| \leq C(\varepsilon) h^\varepsilon.$$

Proof. Set $B = 2^{2(k-1)/\varepsilon}$. Then for every $p \geq B$ and $b \geq 1$ we have

$$2^{b+k-1} \leq p^{\varepsilon b}.$$

Furthermore there exists a constant $C > 0$ such that for all subsets J of non-negative integers with $2 \leq \min J \leq \max J \leq B$ and non-negative integers b_j, $j \in J$,

$$\prod_{j\in J} \binom{b_j + k - 1}{k-1} \leq C \prod_{j\in J} j^{\varepsilon b_j}.$$

If h is a natural number and

$$h = p_1^{a_1} \cdots p_r^{a_r}$$

denotes its prime number decomposition ($a_j \ge 1$, $1 \le j \le r$), then

$$|\{(h_1, \ldots, h_k) \in \mathbf{N}^k : h_1 \cdots h_k = h\}| = \prod_{j=1}^{r} \binom{a_j + k - 1}{k - 1}.$$

Since

$$\prod_{p_j < B} \binom{a_j + k - 1}{k - 1} \le C \prod_{p_j < B} p_j^{\varepsilon a_j}$$

and

$$\prod_{p_j \ge B} \binom{a_j + k - 1}{k - 1} \le \prod_{p_j \ge B} 2^{a_j + k - 1} \le \prod_{p_j \ge B} p_j^{\varepsilon a_j},$$

the lemma follows. □

Proof of Proposition 1.120. We will apply Lemma 1.122 with $j = k - 1$. Note that for $\phi(x) = \alpha x^k + \alpha_1 x^{k-1} + \ldots + \alpha_{k-1} x + \alpha_k$ we have

$$\Delta_{k-1}(\phi(x); h_1, \ldots, h_{k-1}) = k!\alpha(x + \frac{1}{2}h_1 + \ldots + \frac{1}{2}h_{k-1}) + (k-1)!\alpha_1.$$

Hence

$$\begin{aligned} |T(\phi)|^K &\le (2M)^{K-k} \\ &\quad \times \underbrace{\sum_{h_1} \cdots \sum_{h_{k-1}}}_{|h_j| \le M} \sum_{m \in I_{k-1}} e\left(h_1, \ldots, h_{k-1} p_{k-1}(m; h_1, \ldots, h_{k-1})\right) \end{aligned}$$

with $p_{k-1}(x; h_1, \ldots, h_{k-1}) = k!\alpha(x + \frac{1}{2}h_1 + \ldots + \frac{1}{2}h_{k-1}) + (k-1)!\alpha_1$. The terms with $h_1, \ldots, h_{k-1} = 0$ contribute $\ll M^{k-1}$. The remaining ones are of the form

$$\sum_{m \in I} e(h\alpha m + \beta),$$

where $|h| \le k!M^{k-1}$ is a non-zero integer, β a real number, and I an interval. Hence, by using

$$\left| \sum_{m \in I} e(h\alpha m + \beta) \right| \ll \frac{1}{\|h\alpha\|}$$

and by applying Lemma 1.123 we obtain

$$\begin{aligned} |T(\phi)|^K &\ll (2M)^{K-k} \left(M^{k-1} + M^{\varepsilon} \sum_{h=1}^{k!M^{k-1}} \min\left(M, \|\alpha h\|^{-1}\right) \right) \\ &\ll M^{K-k+\varepsilon} \left(M^{k-1} + \sum_{h=1}^{k!M^{k-1}} \min\left(M^k h^{-1}, \|\alpha h\|^{-1}\right) \right). \end{aligned}$$

By Lemma 1.121, when $b \le M^k$,this is

$$\ll M^{K+2\varepsilon}\left(b^{-1}+M^{-1}+bM^{-k}\right).$$

The proof is completed by observing that the result is trivial when $b > M^k$. □

For the proof of Theorem 1.118 we will use the following notation. For any function $f : \mathbf{N} \to [1,\infty)$ we consider the q-ary expansion

$$\alpha([f]) = 0.[f(1)][f(2)]\dots[f(n)][f(n+1)]\dots$$

formed by writing the q-ary expansions of the numbers $[f(1)], [f(2)], \dots$ successively at the right side of the radix point. For more convenient notation we separate $[f(n)]$ and $[f(n+1)]$ by commas.

For any positive integers n, l let $T(n)$ denote the sum of the numbers of digits of $[f(1)], \dots, [f(n)]$ and let u_l be the least positive integer x for which the number of digits of $[f(x)]$ is at least l. When f is strictly increasing, $f(u_l - 1) < q^{l-1} \le f(u_l)$ for l sufficiently large.

An example will illustrate these definitions: Let $f(x) = \frac{1}{2}(x^4+6)$ and $q = 10$. Then $\alpha([f]) = 0.3, 11, 43, 131, 315, 651, 1203, \dots$ and $u_1 = 1, u_2 = 2, u_3 = 4, u_4 = 7; T(1) = 1, T(2) = 3, T(3) = 5, T(4) = 8, T(5) = 11, T(6) = 14, T(7) = 18$.

Let

$$\alpha([f]) = \sum_{i=1}^{\infty} a_i q^{-i}$$

be the q-ary representation of $\alpha([f])$ with digits $0 \le a_i \le q-1$. In order to prove Theorem 1.118 it will be necessary to calculate the frequency of occurences of certain blocks of digits as subblocks of the sequence (a_i). Let k be a positive integer, $B = b_1 \cdots b_k$ be a block of digits of length k, and set $\beta(B) = b_1 q^{k-1} + b_2 q^{k-2} + \dots + b_k$. For $1 \le S \le N$ let $\mathcal{N}(B, \alpha, S, N)$ be the number of subblock-occurences of B in $a_S a_{S+1} \cdots a_N$, i.e. the number of i's satisfying $S \le i \le N - k + 1$ and $a_i a_{i+1} \cdots a_{i+k-1} = B$. For $k \le l, u_l \le v < u_{l+1}$ we write $\mathcal{N}(B, v)$ instead of $\mathcal{N}(B, \alpha, T(u_l - 1) + 1, T(v))$. Obviously $\mathcal{N}(B, v)$ is the number of subblock-occurences of B in the sequences of digits formed by $[f(u_l)], [f(u_l + 1)], \dots, [f(v)]$. There are two possibilities for B to occur as subblock of this sequence:

(i) B is a subblock of $[f(u)]$ for a certain u $(u_l \le u \le v)$ and therefore, B does not straddle any comma in $[f(u_l)], \dots, [f(v)]$. Let $\mathcal{N}_1(B, v)$ denote the number of those subblock-occurrences of B;

(ii) B is subblock of $[f(u)], [f(u+1)]$ for a certain u $(u_l \le u \le v)$, straddling the comma between $[f(u)]$ and $[f(u+1)]$. Let $\mathcal{N}_2(B, v)$ denote the number of those subblock-occurrences of B.

Evidently $\mathcal{N}(B, v) = \mathcal{N}_1(B, v) + \mathcal{N}_2(B, v)$ and

$$\mathcal{N}(B, \alpha, S, U) = \mathcal{N}(B, \alpha, S, N) + \mathcal{N}(B, \alpha, N, U) + O(k),$$

when $S \le N \le U$ and B is a block of digits having length k. Let f, α be as in the example given above and let $v = 6, B = 31$. Then $u_3 \le v < u_4, [f(u_3)], \dots, [f(v)] =$

131, 315, 651 and hence $\mathcal{N}_1(B,v)=2, \mathcal{N}_2(B,v)=0, \mathcal{N}(B,v)=2; \mathcal{N}_1(B,\alpha,1,14)=\mathcal{N}_2(B,\alpha,1,14)=2, \mathcal{N}(B,\alpha,1,14)=4$.

Let $d \geq 1; \alpha_0, \alpha_1, \ldots, \alpha_d$ be rational numbers and

$$f(x)=\alpha_d x^d+\ldots+\alpha_1 x+\alpha_0.$$

Without loss of generality we may assume that $f(x)$ is increasing. Then

$$\begin{aligned} & f(x) \gg\ll x^d, f'(x) \gg\ll x^{d-1}; f^{-1}(x) \gg\ll x^{1/d}, (f^{-1})'(x) \gg\ll x^{(1/d)-1} \\ & \text{and} \quad (f^{-1})''(x) \gg\ll x^{(1/d)-2} \quad \text{when} \quad d>1. \end{aligned} \tag{1.187}$$

Thus, for l sufficiently large oviously $u_l < u_{l+1}, f(u_l-1) < q^{l-1} \leq f(u_l)$, the number of digits of $f(x)$ is l if and only if $u_l \leq x < u_{l+1}; T_l \sim l q^{l/d}$.

In order to prove Theorem 1.118 we first show

Lemma 1.124 *Let $u, v, k, l, n \in \mathbf{N}, b \in \mathbf{Z}$ and $0 \leq v$,*

$$q^{l-1} \leq f(u) \leq f(v) < q^l, k \leq n \leq l, 0 \leq b < q^n.$$

Then

$$\underbrace{\sum_{u \leq x < v} \sum_{0 \leq t < q^{n-k}}}_{[f(x)] \equiv b+t(q^n)} 1 \quad = \quad q^{-k}(v-u)+\mathcal{O}\left((v-u)(q^{-n\varepsilon}+q^{n-k-l})\right) \tag{1.188}$$

for an $\varepsilon>0$, where ε and the implicit $\mathcal{O}$-constant do not depend on u, v, k, l, n, b.

Proof. Let $S(u,v)$ denote the expression on the left side of (1.188). When $u_0=[f^{-1}(\frac{1}{2}q^{l-1})]$, obviously $S(u,v)=S(u_0,v)-S(u_0,u)$ and so it suffices to show (1.188) for $u=u_0, q^{l-1} \leq f(v) < q^l$. In this case

$$v-u \quad \sim \quad q^{l/d}. \tag{1.189}$$

The proof of (1.188) splits into two cases.

In the first case we assume that $d=1$ or $n > l(1-1/4d)$. Let

$$U=[(f(u)-b)q^{-n}]+1, V=[(f(v-1)-b)q^{-n}]-1$$

and

$$d(Y)=f^{-1}(Yq^n+b+q^{n-k})-f^{-1}(Yq^n+b) \quad \text{for} \quad U-1 \leq Y \leq V+1.$$

Then

$$\begin{aligned} S(u,v) \quad &= \quad \underbrace{\sum_{Y \in \mathbf{N}} \sum_{u \leq x < v} \sum_{0 \leq t < q^{n-k}}}_{[f(x)-b]=Yq^n+t} 1 \\ &= \quad \sum_{U \leq Y < V} (d(Y)+\mathcal{O}(1))+\mathcal{O}\left(d(U-1)+d(V+1)\right). \end{aligned} \tag{1.190}$$

For Y fixed we have

$$d(Y) = q^{n-k}(f^{-1})'(Yq^n + b) + \frac{1}{2}q^{2n-2k}(f^{-1})''(Yq^n + b + z(Y)),$$

where $z(y) \in [0, q^{n-k}]$. Using (1.187) and $Y \sim q^{l-n}$ we get

$$(f^{-1})''(Yq^n + b + z(Y)) \ll q^{l/d-2l}; d(Y) \ll q^{l/d-l+n-k} \tag{1.191}$$

and $V - U \ll q^{l-n}$. As $d(Y)$ is increasing, we get

$$\begin{aligned}\sum_{U \le Y \le V} d(Y) &= \int_U^V d(y) + \mathcal{O}(d(U)) \\ &= q^{-k}(f^{-1}(Vq^n + b) - f^{-1}(Uq^n + b)) + \mathcal{O}(q^{l/d-l+n-2k}) + \mathcal{O}(d(U)).\end{aligned}$$

As $Uq^n + b = f(u) + \mathcal{O}(q^n)$ we have

$$f^{-1}(Uq^n + b) = u + \mathcal{O}(q^{l/d-l+n})$$

and, similarly

$$f^{-1}(Vq^n + b) = v + \mathcal{O}(q^{l/d-l+n}).$$

With (1.190), (1.191) we obtain

$$S(u, v) = q^{-k}(v - u) + \mathcal{O}(q^{l/d-l+n-k} + q^{l-n}).$$

Now, (1.188) follows from (1.189) and $l - n \le (l/d) - \varepsilon n$ when $d = 1$ or $n > l(1 - 1/(4d))$.

In the second case we assume that $n \le l(1 - 1(4d))$ and $d \ge 2$. We choose $N' \in \mathbf{N}$ in such a way that all coefficients of $N'f$ are integers. Furthermore, let $F = N'f, N = N'q^{n-k}, M = N'q^n$, and $B = N'b$. Thus we get

$$\begin{aligned}S(u, v) &= \underbrace{\sum_{u \le x < v} \sum_{t=0}^{N-1}}_{F(x) \equiv B + t(M)} 1 = \sum_{t=0}^{N-1} \sum_{u \le x < v} \frac{1}{M} \sum_{1 \le h \le M} e\left((F(x) - B - t)\frac{1}{M}\right) \\ &= (v - u)\frac{N}{M} + \frac{R}{M},\end{aligned} \tag{1.192}$$

where

$$R = \sum_{1 \le h \le M} e\left(-B\frac{h}{M}\right) \frac{e(-Nh/M) - 1}{e(-h/M) - 1} \sum_{u \le x < v} e\left(F(x)\frac{h}{M}\right).$$

Using

$$\left|e^{iw} - e^{iz}\right| = 2 \sin\left|\frac{w - z}{2}\right|$$

and writing $E(m,h,t)$ instead of $|\sum_{1\le x\le m} e(F(x+u-1)(h/t)|$ we get

$$\begin{aligned}
R &\ll \sum_{1\le h\le M}\left(\sin\pi\frac{h}{M}\right)^{-1}\left|\sum_{u\le x<v} e\left(F(x)\frac{h}{M}\right)\right| \qquad (1.193)\\
&= \underbrace{\sum_{t|M}\sum_{1\le h<t}}_{(h,t)=1} E(v-u,h,t)\left(\sin\pi\frac{h}{t}\right)^{-1}\\
&= \underbrace{\sum_{t|M}\sum_{1\le h<t}}_{(h,t)=1}\left(\sin\pi\frac{h}{t}\right)^{-1}\left[\frac{v-u}{t}\right]E(t,h,t)\\
&+ \underbrace{\sum_{t|M}\sum_{1\le h<t}}_{(h,t)=1,t<v-u}\left(\sin\pi\frac{h}{t}\right)^{-1}E\left(t\left\{\frac{v-u}{t}\right\},h,t\right)\\
&+ \underbrace{\sum_{t|M}\sum_{1\le h<t}}_{(h,t)=1,t>v-u}\left(\sin\pi\frac{h}{t}\right)^{-1}E(v-u,h,t).
\end{aligned}$$

Let h,t,a,b be integers, $1\le h\le t$, $\gcd(h,t)=1$, $t|M$ and $a/b=\alpha_d N'^h$, $\gcd(a,b)=1$. Then $t(\alpha_d N')^{-1}\le b\le t$ because $\gcd(h,t)=1$.

For $t>v-u$ and l sufficiently large we get

$$\begin{aligned}
(v-u)^{1/8} < t^{1/8} < t(\alpha_d N')^{-1} < b,\\
b\le N'q^n \ll q^{l-l/4d} \ll (v-u)^{d-1/4},
\end{aligned}$$

and therefore $b\le(v-u)^{d-1/8}$. Hence WEYL's inequality (Proposition 1.120) yields

$$\begin{aligned}
E(v-u,h,t) &\ll (v-u)^{1+\varepsilon}((v-u)^{-1}+b^{-1}+b(v-u)^{-d})^{1/K} \qquad (1.194)\\
&\ll (v-u)^{1-\sigma} \quad \text{for} \quad t>v-u \qquad (1.195)
\end{aligned}$$

for an $\sigma=\sigma(d)>0$ and $K=2^{d-1}$. For $t^{1-\sigma}<s<t$ we have

$$s^{1/8}<t<t^{(1-\sigma)(d-1/8)}<s^{d-1/8}$$

(σ can be choosen arbitrarily small) and therefore $E(s,h,t)\ll s^{1-\sigma}$. Thus we have $E(s,h,t)\ll t^{1-\sigma}$ for $s\le t$. Combining this with (1.193) and (1.194) and using

$$\sum_{1\le h<t}\left(\sin\pi\frac{h}{t}\right)^{-1}\ll t\log t,$$

yields

$$\begin{aligned} R &\ll \sum_{t|M}(v-u)t^{1-\sigma}\log t + \sum_{\substack{t|M\\ t<v-u}} tt^{1-\sigma}\log t + \sum_{\substack{t|M\\ t>v-u}} t(v-u)^{1-\sigma}\log t \\ &\ll \log M\left((v-u)M^{1-\sigma} + (v-u)^{1-\sigma}M\log M\right). \end{aligned}$$

As $(v-u)^{-\sigma} \ll q^{-\sigma(l/d)}$ and $n \le l$ we get $R \ll (v-u)q^{n-n\varepsilon}$ for an $\varepsilon > 0$. Now (1.188) follows immediately from (1.192) and the proof of Lemma 1.124 is completed. □

Proof of Theorem 1.118. Let $B = b_1b_2\cdots b_k$ be a block of digits and

$$b - \beta(B) = b_1q^{k-1} + b_2q^{k-2} + \ldots + b_k.$$

When $k < l, u_l \le v < u_{l+1}$, using Lemma 1.124 we obtain

$$\begin{aligned} \mathcal{N}_1(B,v) &= \underbrace{\sum_{0<j<l}\sum_{u_j\le x\le v}\sum_{0\le t\le q^{l-k-j}}}_{[f(x)]\equiv bq^{l-k-j}+t(q^{l-j})} 1 \qquad (1.196)\\ &= q^k(l-k+1)(v-u_l+1) + \sum_{0\le j\le l-k} \mathcal{O}(q^{l/d)-(l-j)\varepsilon} + q^{l/d-k-j}) \\ &= l(v+1-u_l)q^{-k} + \mathcal{O}(q^{l/d)-k\varepsilon}). \end{aligned}$$

Furthermore,

$$\mathcal{N}_2(B,v) \le n_2(B,u_{l+1}-1) \le \sum_{1\le j<k}\sum_{u_l\le x<u_{l+1}} 1 + \mathcal{O}(1),$$

where the last summation is taken only over those x for which $b_1, b_2, \ldots, b_{k-j}$ are the last digits in the q-ary representation of $[f(x)]$ and $b_{k-j+1}, b_{k-j+1}, \ldots, b_k$ are the first digits in the q-ary representation of $[f(x+1)]$. For $b_{k-j+1} \ne 0$ let

$$\begin{aligned} A_j &= [f^{-1}(b_{k-j+1}q^{l-1} + \ldots + b_kq^{l-j} - 1)], \\ B_j &= [f^{-1}(b_{k-j+1}q^{l-1} + \ldots + b_kq^{l-j} + q^{l-j} - 1)]. \end{aligned}$$

Using (1.188) and $B_j - A_j \ll q^{l/d-j}$ we get

$$\begin{aligned} \mathcal{N}_2(B,v) &\ll \sum_{\substack{1\le j<k\\ b_{k-j+1}\ne 0}} \left(\sum_{\substack{A_j\le x\le B_j\\ [f(x)]\equiv b_1q^{k-j-1}+\ldots+b_{k-j}(q^{k-j})}} 1 \right) + \mathcal{O}(1) \qquad (1.197)\\ &= \sum_{\substack{1\le j<k\\ b_{k-j+1}\ne 0}} \left(q^{-(k-j)}(B_j - A_j) + \mathcal{O}\left((B_j-A_j)(q^{-(k-j)\varepsilon} + q^{-l})\right)\right) \\ &\ll q^{l/d-\varepsilon k}. \end{aligned}$$

(1.196) and (1.197) yield

$$\mathcal{N}(B,v) = q^{-k}l(v+1-u_l)+\mathcal{O}(q^{l/d-\varepsilon k}). \tag{1.198}$$

For $N \in \mathbf{N}$ we may choose $n, u \in \mathbf{N}$ in such a way that $k < n/2$, $u_n \le u < u_{n+1}$, $N(u-1) < N \le T(u)$. Then $N = T(u) + \mathcal{O}(n)$, and we obtain

$$\begin{aligned}
\mathcal{N}(B,\alpha,1,N) &= \sum_{1\le l<n} \mathcal{N}(B,u_{l+1}-1)+\mathcal{N}(B,u)+\mathcal{O}(kn)+\mathcal{O}(kT_k) \qquad (1.199)\\
&= \sum_{1\le l<n} (q^{-k}l(u_{l+1}-u_l)+\mathcal{O}(q^{l/d-\varepsilon k}))\\
&\quad +q^{-k}n(u-u_n)+\mathcal{O}(q^{n/d-\varepsilon k}+kn+k^2q^{k/d})\\
&= q^{-k}N+\mathcal{O}(q^{n/d-\varepsilon k}).
\end{aligned}$$

For any interval $I \subset [0,1)$ let

$$\Delta_{N,I} = \frac{1}{N}\sum_{n=0}^{N-1} \chi_I(\{q^n\alpha\}) - \lambda_1(I).$$

Let $k \ge 1$, $c_1, c_2, \ldots, c_k$ be digits, $B = c_1\cdots c_k$, $I = [\gamma, \gamma+q^{-k})$, and

$$\gamma = \beta(B)q^{-k} = \sum_{1\le i\le k} c_i q^{-i}.$$

When $k = \mathcal{O}(\log N)$ we get for sufficiently large N

$$\sum_{n=0}^{N-1}\chi_I(\{q^n\alpha\}) = \sum_{\substack{1\le n\le N\\ c_1\cdots c_k = a_n\cdots a_{n+k-1}}} 1 = \mathcal{N}(B,\alpha,1,N)+\mathcal{O}(k).$$

Since $N \sim nq^{n/d}, n \sim \log N$, (1.199) we obtain

$$|\Delta_{N,[\gamma,\gamma+q^{-k})}| = \left|\frac{1}{N}\mathcal{N}(B,\alpha,1,N) - q^{-k} + \mathcal{O}(k/N)\right| \ll q^{-\varepsilon k}\frac{1}{\log N}. \tag{1.200}$$

Now let $\beta \in [0,1), h \in \mathbf{N}, \beta q^h \in \mathbf{N}, \beta = \sum_{1\le k\le h} b_k q^{-k}$ be the q-ary representation of β and

$$\beta_{k,j} = \sum_{1\le i\le k} b_i q^{-i} + jq^{-j} \quad \text{for} \quad 1 \le k \le h, 0 \le j \le b_k.$$

As $\beta_{1,0} = 0, \beta_{k,b_k} = \beta_{k+1,0}$ for $1 \le k < h$ we get from (1.200) when $h = \mathcal{O}(\log N)$

$$|\Delta_{N,[0,\beta)}| = \sum_{1\le k\le h}\sum_{0\le j\le b_k} |\Delta_{N,[\beta_{k,j},\beta_{k,j+1})}| \ll \sum_{1\le k\le h} q^{-\varepsilon k}\frac{1}{\log N} \ll \frac{1}{\log N}.$$

Finally let γ be any real number in $[0,1)$ and let $N \in \mathbf{N}$. Let $h = [\log\log N]$ and choose α, β in $[0,1)$ with $\alpha \leq \gamma \leq \beta, \beta - \alpha = q^{-h}, \beta q^h \in \mathbf{Z}$. Then

$$\Delta_{N,[0,\gamma)} \leq \frac{1}{N} \sum_{n=0}^{N-1} \chi_{[0,\beta]}(\{q^n \alpha\}) - \alpha = \Delta_{N,[0,\beta)} + q^{-h}.$$

Similarly, we obtain

$$\Delta_N(0,\gamma) \geq \Delta_N(0,\alpha) - q^{-h}.$$

Therefore we have

$$|\Delta_{N,[0,\gamma)}| \leq \max\left\{|\Delta_{N,[0,\alpha)}|, |\Delta_{N,[0,\beta)}\right\} + q^{-h} \ll \frac{1}{\log N} + q^{-h} \ll \frac{1}{\log N}.$$

Thus Theorem 1.118 is proved. □

Lemma 1.125 *Let $\alpha \in [0,1)$ and B_1, B_2 be blocks of digits having same length. If*

$$|\mathcal{N}(B_1,\alpha,1,N) - \mathcal{N}(B_2,\alpha,1,N)| \ > \ K\frac{N}{\log N} \tag{1.201}$$

holds for a constant $K > 0$ and an infinite number of (resp. almost all) $N \in \mathbf{N}$, then

$$D_N(q^n\alpha) > \frac{C}{\log N}$$

for an infinite number of (resp. almost all) $N \in \mathbf{N}$ and a constant $C > 0$. K and C may depend on B_1, B_2, α.

Proof. Let k be the length of B_1 and B_2; $\alpha_i = \beta(B_i q^{-k})$ and $I_i = [\alpha_i, \alpha_i + q^{-k})$ for $i = 1,2$. When $N \geq k$ and N satisfies (1.201) we have

$$\begin{aligned} D_{N-k+1}(q^n\alpha) &\geq \max_{i=1,2}\left|\frac{1}{N}\sum_{j=0}^{N-k} \chi_{I_i}((\{q^j\alpha\}) - q^{-k}\right| \\ &= \max_{i=1,2}\left|\frac{1}{N}\mathcal{N}(B_i,\alpha,1,N) - q^{-k}\right| \\ &> \frac{1}{2N}|\mathcal{N}(B_1,\alpha,1,N) - \mathcal{N}(B_2,\alpha,1,N)| > \frac{C}{\log(N-k+1)} \end{aligned}$$

for a constant $C > 0$. □

Proof of Theorem 1.119. Let C, D be rational numbers, $f(x) = Cx + D, C > 0, C + D > 1$. We choose k and two blocks B_1, B_2 of digits with same length k satisfying the following conditions: $\beta(B_1) = (q[f(1)] + 1)q^h + r$ for some integers h, r with $h \geq 0, 0 \leq r < q^h$ (which means that B has a form like $[f(1)], 1, \ldots$) and $\beta(B_1) = CN$ for an integer N and B_2 is the block consisting of k 0's.

For any block B of k digits and $1 \leq j \leq l-k, q^{j-1} \leq m < q^j$ we define

$$\begin{aligned} U(j,m,B) &= \left[f^{-1}\left(mq^{l-j} + (\beta(B)+1)q^{l-k-j}\right)\right] \\ &\quad -\chi_{\mathbf{N}}\left(f^{-1}\left(mq^{l-j} + (\beta(B)+1)q^{l-k-j}\right)\right) \\ &\quad -\left[f^{-1}\left(mq^{l-j} + \beta(B)q^{l-k-j}\right)\right] \\ &\quad +\chi_{\mathbf{N}}\left(f^{-1}\left(mq^{l-j} + \beta(B)q^{l-k-j}\right)\right), \end{aligned} \tag{1.202}$$

i.e. $U(j,m,B)$ is the number of all x, for which $q^{l-1} \leq [f(x)] < q^l$ and $[f(x)]$ has a q-ary representation of the form $m, B, \ldots$.

Let $v_l = u_{l+1} - 1$ for $l \geq 1$. We deduce

$$\mathcal{N}_1(B, v_l) = \sum_{1 \leq j \leq l-k} \sum_{q^{j-1} \leq m < q^j} U(j,m,B) + \Delta_{0,B,l} + \mathcal{O}(1), \tag{1.203}$$

where

$$\Delta_{0,B,l} = \begin{cases} \frac{1}{C} q^{l-k} & \text{when } \beta(B) \geq q^{k-1} \\ 0 & \text{when } \beta(B) < q^{k-1}. \end{cases}$$

Now let $N \in \mathbf{N}$. There exist $v, n \geq 1$ such that $u_n \leq v < u_{n+1}, N = T(v) + \mathcal{O}(n)$. Let $[f(v)] = q^{n-1}z_1 + q^{n-2}z_2 + \ldots + z_n$ be the q-ary representation of $[f(v)]$ and, for $1 \leq j \leq n-k$:

$$\begin{aligned} A_j &= q^{j-1}z_1 + q^{j-2}z_2 + \ldots + z_j, \\ \gamma_j &= z_{j+1}q^{k-1} + z_{j+2}q^{k-2} + \ldots + z_{j+k}. \end{aligned}$$

Then

$$\mathcal{N}_1(B, v) = \sum_{1 \leq j \leq n-k} \left(\left(\sum_{q^{j-1} \leq m < A_j} U(j,m,B) \right) + \Delta_{j,B} \right) + \Delta_{0,B} + \mathcal{O}(1), \tag{1.204}$$

where $\Delta_{j,B}$ [resp. $\Delta_{0,B}$] denotes the number of all integers x, for which $q^{l-1} \leq [f(x)] \leq v$ and $[f(x)]$ has a q-ary representation $A_j, B, \ldots$ [resp. $B, \ldots$]. Hence

$$\Delta_{j,B} = \begin{cases} U(j, A_j, B) & \text{when } \beta(B) < \gamma_j, \\ v - f^{-1}(A_j q^{n-j} + \beta(B)q^{n-k-j}) + \mathcal{O}(1) & \text{when } \beta(B) = \gamma_j, \\ 0 & \text{when } \beta(B) > \gamma_j \end{cases}$$

and

$$\Delta_{0,B} \geq 0, \quad \Delta_{0,B} = 0 \quad \text{if} \quad \beta(B) < q^{k-1}.$$

Because of $f^{-1}(Y) = (\frac{1}{C})(Y - D)$ we have

$$\Delta_{j,B} < \frac{1}{C} q^{n-k-j} + \mathcal{O}(1) \quad \text{for} \quad 1 \leq j \leq n-k, q^{j-1} \leq m < q^j.$$

$\beta(B_2) < \beta(B_1)$ implies

$$0 \leq \Delta_{j,B_1} \leq \Delta_{j,B_2} + \mathcal{O}(1),$$

and therefore

$$\Delta_{j,B_2} - \Delta_{j,B_1} \leq \frac{1}{C} q^{n-k-j} + \mathcal{O}(1). \tag{1.205}$$

As $\beta(B_1) - \beta(B_2) = CN$, we get

$$U(j,m,B_1) = U(j,m,B_2) \tag{1.206}$$

for $1 \leq j \leq 1-k, q^{j-1} \leq m < q^j$, and (1.203) yields

$$\mathcal{N}_1(B_1,v_1) = \mathcal{N}_1(B_2,v_1) + \frac{1}{C} q^{l-k} + \mathcal{O}(1). \tag{1.207}$$

Obviously $\mathcal{N}_2(B_2,v_1) = 0$. Let $B_1 = b_1 \cdots b_k$ and $i < k$ satisfying $\beta(b_1 \cdots b_k) = [f(1)]$. Let M denote the set of all integers m for which $u_l \leq m \leq v_l = u_{l+1} - 1$ and the q-ary representation of $[f(m)]$ has the form

$$[f(m)] = (b_{i+1} q^{l-1} + \ldots + b_k q^{l-(k-i)}) + \ldots + (b_1 q^{j-1} + \ldots + b_i).$$

For all $x \in \mathbf{N}$

$$[f(1 + (q^i N)x)] = [f(1)] + q^i CNx \equiv [f(1)](q^i),$$

and for sufficiently large l there exists a positive integer t for which

$$[f(1 + tNq^i)] = [f(1)] + CNtq^i \in f(M).$$

Hence M is not empty and contains at least

$$q^i N^{-1} \left(\frac{1}{C} q^{l-(k-i)} + \mathcal{O}(1) \right) + \mathcal{O}(1)$$

elements.

As $\mathcal{N}_2(B_1,v_1) \geq |M| + \mathcal{O}(1)$, we get $\mathcal{N}_2(B_1,v_1) > Kq^l$ for a constant $K > 0$. Hence we obtain from (1.204),(1.205), (1.206), (1.207) and $\Delta_{0,B_1} \geq \Delta_{0,B_2}$

$$\begin{aligned}
\mathcal{N}(B_1,\alpha,1,N) \quad &- \quad \mathcal{N}(B_2,\alpha,1,N) \\
&\geq \sum_{1 \leq l < n} (\mathcal{N}_1(B_1,v_1) - \mathcal{N}_1(B_2,v_1) + \mathcal{N}_2(B_1,v_1)) \\
&\quad + \sum_{1 \leq j \leq n-k} (\Delta_{j,B_1} - \Delta_{j,B_2}) + \mathcal{O}(n) \\
&\geq \sum_{1 \leq l < n} \left(Kq^l + \frac{1}{C} q^{l-k} \right) - \sum_{1 \leq j \leq n-k} \frac{1}{C} q^{n-k-j} + \mathcal{O}(n) > Lq^n
\end{aligned}$$

for a constant $L > 0$ and n sufficiently large.

Thus Theorem 1.119 follows from Lemma 1.125. □

1.4.5 Exponential Sequences

It seems to be a very difficult problem to decide for which real α and $\lambda > 1$ the sequence $(\lambda^n\alpha)$ is u.d. mod 1. Of course, there are special cases, where we know an answer, e.g. if λ is a positive integer and α is a normal number to base λ then $(\lambda^n\alpha)$ is u.d. mod 1. Furthermore there are metric results which are easy to establish: For every $\lambda > 1$ the sequence $(\lambda^n\alpha)$ is u.d. mod 1 for almost all real α and for every $\alpha > 0$ the sequence $(\lambda^n\alpha)$ is u.d. mod 1 for almost all $\lambda > 1$. However, no general method is known to solve this problem for specific α and λ, e.g. it is a famous unsolved problem to decide whether $(3/2)^n$ is u.d. mod 1 or not. In the following we present in detail a construction due to LEVIN (for various references see the Notes).

Definition 1.126 *A real sequence $(x_n)_{n\ge1}$ is called completely uniformly distributed if for every $k \ge 1$ the k-dimensional sequence $(\mathbf{x}_n^{(k)}) = (x_n, x_{n+1}, \ldots, x_{n+k-1})_{n\ge1}$ is u.d. mod 1.*

Let $k(N)$ be a increasing sequence of positive integers with $\lim_{N\to\infty} k(N) = \infty$. Then a real sequence $(x_n)_{n\ge1}$ is called $k(N)$-u.d. mod 1 if

$$\lim_{N\to\infty} D_N(\mathbf{x}_n^{(k(N))}) = 0.$$

Since we always have $D_N(\mathbf{x}_n^{(k')}) \le D_N(\mathbf{x}_n^{(k)})$ for $k' \le k$ it follows that every $k(N)$-u.d. sequence mod 1 is completely u.d. mod 1.

We will concentrate on sequences of the form (λ^n). Theorem 1.188 (with $a_n = n$) says that if $k(N)$ is a increasing sequence of positive integers with $k(N) \le (\log N)^\theta$ for sufficiently large N and some fixed θ satisfying $0 < \theta < 1/2$ then for every $\eta > 0$ and almost all $\lambda > 1$ there exists a constant $C = C(\lambda, \eta) > 0$ such that

$$D_N\left((\lambda^n, \lambda^{n+1}, \ldots, \lambda^{n+k(N)-1})\right) \le C\, N^{-1/2+\eta} \tag{1.208}$$

for all $N \ge 1$. Hence there are $\lambda > 1$ such that (λ^n) is completely u.d. mod 1. The difficult thing is finding a specific one. In what follows we will show how it is possible to construct a real number λ which satisfies (1.208).

Theorem 1.127 *There is an effectively constructable $\lambda > 1$ such that*

$$D_N\left((\lambda^n, \lambda^{n+1}, \ldots, \lambda^{n+k(N)-1})\right) \le C\, N^{-1/2+\eta} \tag{1.209}$$

holds for all increasing sequences $k(N)$ satisfying $k(N) \le (\log N)^\theta$, $0 < \theta < 1/2$, for sufficiently large N.

Let $\rho_1 = 3$, $p_1 = 4$, and for $m \ge 1$

$$t_m = 24m(4(2m+1)^{2m})^{2m+1}, \tag{1.210}$$

$$q_m = \left[\rho_m^{m^{2m}}/t_m + 1\right] t_m(m+1)^4, \tag{1.211}$$

$$p_{m+1} = p_m q_m, \tag{1.212}$$

(1.213)

and

$$\rho_{m+1} = \rho_m + \frac{a_m}{p_m q_m} \tag{1.214}$$

for some non-negative integer $a_m < q_m$ which will be specified in the sequel. Finally we define

$$\lambda = \rho_1 + \sum_{m=1}^{\infty} \frac{a_m}{p_m q_m} = \lim_{m\to\infty} \rho_m. \tag{1.215}$$

For any integral vector $\mathbf{h} = (h_1, \ldots, h_m) \in \mathbf{Z}^m$ consider the polynomial

$$B_m(\mathbf{h}, x) = \sum_{j=1}^{m} h_j x^{j-1}. \tag{1.216}$$

The first property which will be used to select a_m is stated in the following lemma.

Lemma 1.128 *There are at least $\frac{2}{3}q_m$ non-negative integers $a_m < q_m$ such that*

$$\min \left| B_{2m+1}\left(\mathbf{h}, \rho_m + \frac{a_m}{p_m q_m}\right)\right| \ge (p_m t_m)^{-2m}, \tag{1.217}$$

where the minimum is taken over all lattice points $\mathbf{h} \in \mathbf{Z}^{2m+1}$ with

$$0 < \|\mathbf{h}\|_\infty \le (2m+1)^{2m}. \tag{1.218}$$

Proof. The number of sign changes of the polynomials $B_{2m+1}(\mathbf{h}, x)$ and $B'_{2m+1}(\mathbf{h}, x)$ is not greater than $2m$. The amount of numbers x_0 such that the function $B_{2m+1}(\mathbf{h}, \rho_m + \frac{x}{p_m})$ or the function $B'_{2m+1}(\mathbf{h}, \rho_m + \frac{x}{p_m})$ changes the sign is at most

$$4m\left(2(2m+1)^{2m} + 1\right)^{2m+1}$$

at least for one sample of integers $h_1, \ldots h_{2m+1}$ that satisfy (1.218). For every x_0 of this kind there exists a $b(x_0) \in \mathbf{Z}$ with $x_0 \in (\frac{b(x_0)-1}{t_m}, \frac{b(x_0)+1}{t_m})$. Let G_1 be the union of these intervals and $G_2 = [0,1] \backslash G_1$. With (1.210) we have

$$\begin{aligned} \lambda_1(G_1) &\le \frac{2}{t_m} 4m\left(2(2m+1)^{2m} + 1\right)^{2m+1} \le \frac{1}{3} \\ \lambda_1(G_2) &\ge 1 - \lambda_1(G_1) \ge \frac{2}{3}, \end{aligned}$$

(in which λ_1 denotes the LEBESGUE measure). The number of disjoint segments

$$\left[\frac{c}{t_m}, \frac{c+1}{t_m}\right] \subseteq G_2, \qquad c \in \mathbf{Z}, \tag{1.219}$$

is at least $\frac{2}{3}t_m$. Let $h_1, \ldots, h_{2m+1} \in \mathbf{Z}$, satisfying (1.218), and $[\frac{c}{t_m}, \frac{c+1}{t_m}] \subseteq G_2$, $c \in \mathbf{Z}$. Then there exists an $\varepsilon > 0$ such that the functions $B_{2m+1}(\mathbf{h}, \rho_m + \frac{x}{p_m})$ and $B'_{2m+1}(\mathbf{h}, \rho_m + \frac{x}{p_m})$ have the same sign on $(\frac{c}{t_m} - \varepsilon, \frac{c}{t_m} + \varepsilon)$. Therefore $B_{2m+1}(\mathbf{h}, \rho_m +$

$\frac{x}{p_m})$ has the same sign on $(\frac{c}{t_m}-\varepsilon, \frac{c}{t_m}+\varepsilon)$ and is monotone. From (1.216) and (1.218) we get $B_{2m+1} \not\equiv 0$. Therefore $B_{2m+1}(\mathbf{h}, \rho_m + \frac{x}{p_m})$ is monotone and $\neq 0$ on $[\frac{c}{t_m}, \frac{c+1}{t_m}]$. In this way we get for any $h_1, \ldots, h_{2m+1}$, that satisfy (1.218), and for any $c \in \mathbf{Z}$ with $[\frac{c}{t_m}, \frac{c+1}{t_m}] \subseteq G_2$, that $B_{2m+1}(\mathbf{h}, \rho_m + \frac{x}{p_m})$ is monotone and $\not\equiv 0$ on $[\frac{c}{t_m}, \frac{c+1}{t_m}]$. For $c \in \mathbf{Z}$ either $B_{2m+1}(\mathbf{h}, \rho_m + \frac{c}{t_m p_m}) = 0$ or (because of (1.211) - (1.216)) $|B_{2m+1}(\mathbf{h}, \rho_m + \frac{c}{t_m p_m})| \geq (t_m p_m)^{-2m}$ holds. Hence

$$\min\left|B_{2m+1}\left(\mathbf{h}, \rho_m + \frac{x}{p_m}\right)\right| \geq (t_m p_m)^{-2m},$$

with $x \in [\frac{c}{t_m}, \frac{c+1}{t_m}] \subseteq G_2$, $c \in \mathbf{Z}$, and the minimum taken over all $\mathbf{h} = (h_1, \ldots, h_{2m+1})$, that satisfy (1.218). Thus, (1.217) holds for every $a_m \in \mathbf{Z}$, satisfying

$$\frac{a_m}{q_m} \in \left[\frac{c}{t_m}, \frac{c+1}{t_m}\right] \subseteq G_2. \tag{1.220}$$

From (1.211) we get that there are at least $\frac{q_m}{t_m}$ numbers in $[\frac{c}{t_m}, \frac{c+1}{t_m})$ of the form $\frac{a_m}{q_m}$, $a_m \in \mathbf{Z}$, $(m = 1, 2, \ldots)$. Therefore the number of $a_m \in [0, q_m)$ with $\frac{a_m}{q_m} \in G_2$ is at least $\frac{2}{3}q_m$. This completes the proof of Lemma 1.128. □

Corollary 1.129 *For a_m defined as in Lemma 1.128 we get*

$$\min\left|B_{2m}\left(\mathbf{h}, \rho_m + \frac{a_m}{q_m p_m}\right)\right| \geq (p_m t_m)^{-2m}, \tag{1.221}$$

where the minimum is taken over every $\mathbf{h} = (h_1, \ldots, h_{2m}) \in \mathbf{Z}^{2m}$ satisfying $0 < \|\mathbf{h}\|_\infty \leq (2m)^{2m-1}$.

Next set

$$n_1 = 0, \; n_{m+1} = n_m + m^{2m} \; (m \geq 1) \tag{1.222}$$

and for $1 \leq k \leq m$

$$S(c, h_1 \ldots, h_k, h_{m+1}) = \sum_{n=0}^{m^{2m}-1} e\left(\sum_{\nu=1}^{k} h_\nu \left(\rho_m + \frac{c}{p_m q_m}\right)^{n_m+n+\nu-1} + h_{m+1}\frac{n}{m^{2m}}\right),$$

where $h_1, \ldots, h_k, h_{m+1} \in \mathbf{Z}$,

$$D_{m,k}(c) = \sum_{\substack{h_1,\ldots,h_k,h_{m+1}=-m^{m-1} \\ (h_1,\ldots,h_k,h_{m+1})\neq(0,\ldots,0)}}^{m^{m-1}} \frac{|S(c, h_1, \ldots, h_k, h_{m+1})|}{r(h_1, \ldots, h_k, h_{m+1})},$$

and

$$M_{m,k} = \frac{2m}{q_m}\sum_{c=0}^{q_m-1} D_{m,k}(c). \tag{1.223}$$

Now we can specify a_m.

Lemma 1.130 *There are non-negative integers $a_m < q_m$ such that for every integer $1 \le k \le m$*

$$D_{m,k}(a_m) \le M_{m,k}\,. \tag{1.224}$$

Proof. By (1.223) for every k with $1 \le k \le m$ the number of c, $0 \le c < q_m$, with $D_{m,k}(c) > M_{m,k}$ is smaller than $q_m/(2m)$. Hence the number of c, $0 \le c < q_m$, such that there exist k with $D_{m,k}(c) > M_{m,k}$ is smaller than $q_m/2$ and so Lemma 1.130 follows from Lemma 1.128. □

With help of Lemma 1.130 it is possible to construct a_m which satisfy (1.224) inductively and compute λ (defined by (1.215))as precisely as necessary.

Next set for $1 \le k \le m$

$$E(h_1,\ldots,h_k,h_{m+1}) = \frac{1}{q_m}\sum_{c=0}^{q_m-1} |S(c,h_1,\ldots,h_k,h_{m+1})|^2\,. \tag{1.225}$$

Using CAUCHY-SCHWARZ's inequality we obtain

$$\begin{aligned} M_{m,k} &= 2m \sum_{h_1,\ldots,h_k,h_{m+1}=-m^{m-1}}^{m^{m-1}} \frac{1}{r(h_1,\ldots,h_k,h_{m+1})}\frac{1}{q_m}\sum_{c=0}^{q_m-1} |S(c,h_1,\ldots,h_k,h_{m+1})| \\ &\le 2m \sum_{h_1,\ldots,h_k,h_{m+1}=-m^{m-1}}^{m^{m-1}} \frac{E(h_1,\ldots,h_k,h_{m+1})^{1/2}}{r(h_1,\ldots,h_k,h_{m+1})}. \end{aligned}$$

The essential step for the proof of Theorem 1.127 is to estimate $E(h_1,\ldots,h_k,h_{m+1})$.

Proposition 1.131 *For $m \ge m_0, 1 \le k \le m$ and $0 < \max\{|h_1|,\ldots,|h_k|,|h_{m+1}|\} \le m^{m-1}$ we have*

$$E(h_1,\ldots,h_k,h_{m+1}) < 16m^{2m}.$$

The proof of Proposition 1.131 requires rather technical estimates. We postpone its proof to the end of this section and first show how Theorem 1.127 can be deduced from it.

Proof of Theorem 1.127. From Lemma 1.130 and Proposition 1.131 we obtain

$$D_{m,k}(a_m) \le M_{m,k} \le 8m^{m+1}(3+2\log(m^{m-1}))^{k+1}$$

or

$$D_{m,k}(a_m) = \mathcal{O}(m^{m+1}(\log m^m)^{k+1})\,,$$

where the $\mathcal{O}$-constant does not depend on k. For fixed N let l_0 be defined by $n_{l_0} \le N < n_{l_0+1}$ and set $R = N - n_{l_0}$. For an interval $I \subseteq [0,1)^k$ let $A(I,N_1,N_2,\lambda^n)$ denote the number of integers n satisfying $N_1 < n \le N_2$ and

$$(\{\lambda^{n+\nu-1}\})_{1\le\nu\le k} \in I.$$

Then the discrepancy can be estimated by

$$D_N\left((\lambda^n,\lambda^{n+1},\ldots,\lambda^{n+k-1})\right) \le \frac{1}{N}\sum_{m=1}^{l_0-1}\sup_I |A(I,n_m,n_{m+1},\lambda^n)-m^{2m}\lambda_k(I)| + \frac{1}{N}\sup_I\left|A(I,n_{l_0},N\lambda^n)-R\lambda_k(I)\right|, \qquad (1.226)$$

(where λ_k denotes the k-dimensional LEBESGUE measure. By ERDŐS-TURÁN-KOKSMA's inequality (Theorem 1.21) we have for $k\le m$

$$|A(I,n_m,n_m+R,\lambda^n)-Rm^{2m}m^{-2m}\lambda_k(I)| \le \left(\frac{3}{2}\right)^k \times$$

$$\times\left(2m^{m+1}+\sum_{0<\max\limits_{1\le\nu\le k}(|h_\nu|,|h_{m+1}|)\le m^{m-1}} L(h_1,\ldots,h_k,h_{m+1})\right),$$

uniformly for all $R\le m^{2m}$, where

$$L(h_1,\ldots,h_k,h_{m+1}) = \frac{\left|\sum\limits_{n=0}^{m^{2m}-1} e\left(\sum\limits_{\nu=1}^{k} h_\nu\lambda^{n_m+n+\nu-1}+h_{m+1}\frac{n}{m^{2m}}\right)\right|}{r(h_1)\cdots r(h_k)r(h_{m+1})}.$$

Note that we have used the $(k+1)$-dimensional sequence

$$(\lambda^{n_m+n},\lambda^{n_m+n+1},\ldots,\lambda^{n_m+n+k},nm^{-2m}),\quad 0\le n<m^{2m},$$

instead of the original k-dimensional sequence $(\lambda^{n_m+n},\lambda^{n_m+n+1},\ldots,\lambda^{n_m+n+k})$, $0\le n<m^{2m}$. Hence the number $A(I,n_m,n_m+R,\lambda^n)$, $R\le m^{2m}$, is exactly the number of elements of $(\lambda^{n_m+n},\lambda^{n_m+n+1},\ldots,\lambda^{n_m+n+k},nm^{-2m})\in I\times[0,Rm^{-2m})$, $0\le n<m^{2m}$.

For $\max_{1\le\nu\le s}|h_\nu|\le m^{m-1}$, $s\le m$ let us estimate the difference

$$U_m=\sum_{n=0}^{m^{2m}-1} e\left((\sum_{\nu=1}^{s} h_\nu\lambda^{n_m+n+\nu-1}+h_{m+1}\frac{n}{m^{2m}})\right)-S(a_m,h_1,\ldots,h_s,h_{m+1}). \qquad (1.227)$$

With (1.223) and the inequality $|e(\theta_1)-e(\theta_2)|\le 2\pi|\theta_1-\theta_2|$ we have

$$|U_m|\le 2\pi\sum_{n=0}^{m^{2m}-1}|\sum_{\nu=1}^{s} h_\nu(\lambda^{n_m+n+\nu-1}-\rho_{m+1}^{n_m+n+\nu-1})|.$$

From (1.222), (1.230), (1.231), (1.232), and from $s\le m$ we get

$$|U_m| \ll m^m\frac{m^{2m}}{p_{m+1}}\sum_{n=0}^{m^{2m}-1}\rho_{m+1}^{n_m+n+m-1} \ll$$

$$\ll \frac{m^{3m}\rho_{m+1}^{n_{m+1}+m}}{((m+1)!)^4\rho_{m+1}^{n_{m+1}}(\rho_{m+1}-1)} \ll 1$$

for $m \to \infty$. Hence, for $k \le m$

$$\left|A(I, n_m, n_m + R, \lambda^n) - R\lambda_k(I)\right| \le C(3/2)^k m^{m+k+2}(\log m)^{k+1}. \tag{1.228}$$

Combining (1.228) with the trivial estimate for $k > m$ and with (1.226) we conclude

$$\begin{aligned} D_N((\lambda^n, \dots, \lambda^{n+k-1})) &\le \frac{1}{N}\left(2k^{2k} + C(3/2)^k \sum_{m=1}^{l_0} m^{m+k+2}(\log m)^{k+1}\right) \\ &\le \frac{2k^{2k}}{N} + C_1 3^k \frac{(\log N)^{k+3}}{\sqrt{N}}, \end{aligned}$$

where C_1 does not depend on N and k. Hence we immediately obtain for $k = k(N) \le (\log N)^\theta$ $(0 < \theta < 1)$ and every $\eta > 0$

$$D_N\left((\lambda^n, \dots, \lambda^{n+k(N)-1})\right) = \mathcal{O}(N^{-1/2+\eta})$$

as proposed. □

For the proof of Proposition 1.131 we need a series of lemmata.

Lemma 1.132 *With the notations defined above we have:*

- *For every $m \ge 1$ we have*

$$m^{2m} \le n_{m+1} \le 2m^{2m}. \tag{1.229}$$

- *For every m, i, z, $1 \le i \le m$, $0 \le z \le 1$, there exist $\theta(m,i,z) \in [0,1]$ with*

$$3 \le \rho_m + \frac{z}{p_m} = \rho_i + \frac{2\theta(m,i,z)}{p_i} < 4. \tag{1.230}$$

- *For every $m \ge 1$ we have*

$$p_m \gg\ll (m!)^4 \rho_m^{n_m}. \tag{1.231}$$

- *For every $m \ge 1$ we have*

$$\left(\frac{\lambda}{\rho_{m+1}}\right)^{n_{m+1}+m} = 1 + \mathcal{O}\left(\frac{m^{2m}}{p_{m+1}}\right). \tag{1.232}$$

- *For every $m \ge 2$ we have*

$$\left(p_i t_i\left(\rho_{i+1} + \frac{2}{p_{i+1}}\right)\right)^{2i} m^{3m+2} = o(p_{i+1}), \tag{1.233}$$

where $i = [m/2]$.

- *Set*

$$A_m = \rho_m^{m^{m-1}-2^m}. \tag{1.234}$$

Then for every $m \geq 2$ we have

$$(p_i t_i)^{2i} = o(A_m), \tag{1.235}$$

where $i = [m/2]$.

- *There exists a constant c_1 such that for every $m \geq c_1$*

$$24m^{2m} p_m A_m \rho_m^{-n_m+1} \ll \rho_m^{m^{m-1}} \tag{1.236}$$

and

$$\left(\rho_m + \frac{1}{p_m}\right)^{n_{m+1}+m-1} m^m (n_{m+1} + m) \leq \frac{p_{m+1}}{4}. \tag{1.237}$$

Proof. (1.229) and (1.230) follow immediately from definition (1.222) resp. (1.214). By (1.211) and (1.222)

$$\begin{aligned} \frac{p_m}{(m!)^4}\rho_m^{-n_m} &= 4\rho_m^{-n_m} \prod_{k=1}^{m-1} \left[\rho_k^{k^{2k}} t_k^{-1} + 1\right] t_k = \\ &= 4\prod_{k=1}^{m-1} \left(\frac{\rho_m}{\rho_k}\right)^{-k^{2k}} \prod_{k=1}^{m-1} \left(\frac{\theta_k t_k}{\rho_k^{k^{2k}}} + 1\right) \end{aligned} \tag{1.238}$$

with $0 \leq \theta_k \leq 1$. Furthermore, by (1.210) $t_k = \mathcal{O}(3^{k^3})$ as $k \to \infty$. Hence

$$\prod_{k=1}^{\infty} \left(\frac{\theta_k t_k}{\rho_k^{k^{2k}}} + 1\right) \tag{1.239}$$

is convergent and

$$1 \leq \frac{\rho_m}{\rho_k} = 1 + \frac{2\theta(m,k,0)}{p_k \rho_k} < 1 + \frac{1}{p_k}.$$

From $p_k > (k!)^4$ it follows that

$$\frac{k^{2k}}{p_k} = \mathcal{O}\left(\frac{1}{k^{2k}}\right)$$

and that

$$\left(1 + \frac{1}{p_k}\right)^{k^{2k}} = 1 + \mathcal{O}\left(\frac{k^{2k}}{p_k}\right).$$

Thus

$$\prod_{k=1}^{\infty} \left(1 + \frac{1}{p_k}\right)^{k^{2k}}$$

is convergent. So we get

$$1 \geq \prod_{k=1}^{m-1} \left(\frac{\rho_m}{\rho_k}\right)^{-k^{2k}} \geq \prod_{k=1}^{\infty} \left(1+\frac{1}{p_k}\right)^{-k^{2k}}, \tag{1.240}$$

and a combination of (1.238), (1.239), and (1.240) proves (1.231).

By(1.215) and (1.230)

$$\lambda = \rho_{m+1} + \frac{2\theta(m+1)}{p_{m+1}}$$

with $0 \leq \theta(m+1) \leq 1$. Furthermore, by (1.229) and (1.231) $n_{m+1} + m = o(p_{m+1})$ which gives

$$\begin{aligned}(\frac{\lambda}{\rho_{m+1}})^{n_{m+1}+m} - 1 &= (1+\frac{2\theta(m+1)}{p_{m+1}\rho_{m+1}})^{n_{m+1}+m} - 1 \\ &\ll \frac{n_{m+1}+m}{p_{m+1}} \ll \frac{m^{2m}}{p_{m+1}}\end{aligned}$$

and proves (1.232).

Again, by (1.229) and (1.231) we get

$$\log_{\rho_i}(p_{i+1}) \gg n_{i+1} \gg i^{2i}.$$

Combining this with (1.230) and the estimate $t_k = \mathcal{O}\left(3^{k^3}\right)$ yields

$$\begin{aligned}\log_{\rho_i}\left(p_i t_i(\rho_{i+1} + \frac{2}{p_{i+1}})\right)^{2i} m^{3m+2}) &\ll 2i(\log_{\rho_i} p_i + \log_{\rho_i} t_i + 1) \\ &+ (3(2i+1)+2)\log_{\rho_i}(2i+1) \\ &\ll i^4 + in_i \\ &\ll i^{2i-1} = o(i^{2i}).\end{aligned} \tag{1.241}$$

Taking (1.231) into account this proves (1.233).

From (1.230) and (1.241) it follows that

$$\log_{\rho_i}(p_i t_i) \ll i^{2i-1},$$

whereas (1.235) gives

$$\log_{\rho_i} A_m \gg (2i)^{2i-1} - 2^{2i+1} \gg 2^i i^{2i-1}.$$

Thus (1.234) follows.

From (1.230) , (1.231) and (1.234) we get

$$\begin{aligned}24m^{2m} p_m A_m \rho_m^{-n_m+1} &\ll m^{2m}(m!)^4 \rho_m^{n_m} \rho_m^{m^{m-1}-2^m} \rho_m^{-n_m+1} \\ &\ll m^{2m}(m!)^4 3^{-2^m} \rho_m^{m^{m-1}} = o(\rho_m^{m^{m-1}})\end{aligned} \tag{1.242}$$

Finally, from $n_{m+1} + m - 1 = \mathcal{O}(p_m)$ it follows that

$$\left(1 + \frac{1}{p_m \rho_m}\right)^{n_{m+1}+m-1} = \mathcal{O}(1).$$

By using this and $n_{m+1} + m \ll m^{2m}$ we obtain

$$\left(\rho_m + \frac{1}{p_m}\right)^{n_{m+1}+m-1} m^m (n_{m+1} + m) \ll \rho_m^{n_{m+1}} 4^m m^{3m} = o(\rho_{m+1}^{n_{m+1}} ((m+1)!)^4).$$

which leads to (1.237) via (1.231). □

Similarly to $B_m(\mathbf{h}, x)$ we define the polynomial $C_m(\mathbf{h}, x)$, $\mathbf{h} = (h_1, \ldots, h_m) \in \mathbf{Z}^k$, by

$$C_m(\mathbf{h}, x) = \sum_{\nu=1}^{m} h_\nu (\nu - 1) x^{\nu-1} \quad (m = 1, 2, \ldots) \tag{1.243}$$

Lemma 1.133 *Let $0 \le u \le 1$, $\mathbf{h} = (h_1, \ldots, h_m) \in \mathbf{Z}^m$, $0 < \|\mathbf{h}\| \le m^{m-1}$. There exists a $C_2 > 1$ such that for every $m > C_2$,*

$$|B_m(\mathbf{h}, \rho_m)| > A_m^{-1}, \tag{1.244}$$

$$\begin{aligned} B_m\left(\mathbf{h}, \rho_m + \frac{u}{q_m p_m}\right) &= B_m(\mathbf{h}, \rho_m)\left(1 + \frac{\theta_1}{10 m^{2m}}\right), \\ C_m\left(\mathbf{h}, \rho_m + \frac{u}{q_m p_m}\right) &= C_m(\mathbf{h}, \rho_m) + \frac{\theta_2}{10} B_m(\mathbf{h}, \rho_m), \end{aligned} \tag{1.245}$$

with $|\theta_1| \le 1$ and $|\theta_2| \le 1$.

Proof. By combining (1.217) and (1.221) we obtain

$$|B_m(\rho_{i+1})| = |B_m(\rho_i + \frac{a_i}{q_i p_i})| \ge (p_i t_i)^{-2i}, \tag{1.246}$$

where $i = [\frac{m}{2}]$. From the mean-value-theorem and from (1.230) it follows that

$$B_m(\mathbf{h}, \rho_m) - B_m(\mathbf{h}, \rho_{i+1}) = \frac{2\theta(m, i+1, 0)}{p_{i+1}} B_m'(\mathbf{h}, \rho_{i+1} + \frac{2\theta_3}{p_{i+1}}) \tag{1.247}$$

with $0 \le \theta_3 \le 1$. By using (1.233) this leads to

$$\left| \frac{B_m'(\mathbf{h}, \rho_{i+1} + \frac{2\theta_3}{p_{i+1}})}{B_m(\mathbf{h}, \rho_{i+1})} \right| \le \left(p_i t_i (\rho_{i+1} + \frac{2}{p_{i+1}})\right)^{2i} m^{m+1} = o(p_{i+1}).$$

Hence

$$B_m(\mathbf{h}, \rho_m) = B_m(\mathbf{h}, \rho_{m+1})(1 + \mathcal{O}(1))$$

as $m \to \infty$. According to (1.235) we have proved (1.244). (1.245) can be proved the same way. □

Lemma 1.134 *Let the function $\varphi(u)$ be differentiable on $[0,N]$ such that $\frac{1}{2} \geq |\varphi'(u)| \geq L > 0$, $u \in [0,N]$, for some constant $L > 0$ and that $\varphi'(u)$ has n intervals of monotonicity on $[0,N]$. Then*

$$\left|\sum_{k=0}^{N-1} e(\varphi(k))\right| \leq \frac{4n}{L}.$$

Proof. According to $\varphi(k+1) - \varphi(k) = \varphi'(\xi_k) \neq 0$, where $\xi_k \in (k, k+1)$, $k = 0, 1, \ldots, N-1$ we get

$$\frac{1}{2} \geq |\varphi(k+1) - \varphi(k)| \geq L > 0. \tag{1.248}$$

Therefore

$$2|\varphi(k+1) - \varphi(k)| \leq |\sin(\pi(\varphi(k+1) - \varphi(k)))| \leq \pi|\varphi(k+1) - \varphi(k)|. \tag{1.249}$$

Setting

$$\begin{aligned} a_k &= (1 - e(\varphi(k+1) - \varphi(k)))^{-1} \\ b_k &= e(\varphi(k)) - e(\varphi(k+1)) \end{aligned} \tag{1.250}$$

we obtain by ABEL's summation

$$\begin{aligned} \sum_{k=0}^{N-1} e(\varphi(k)) &= \sum_{k=0}^{N-1} a_k b_k = \\ &= a_{N-1} \sum_{j=0}^{N-1} b_j - \sum_{k=0}^{N-2} (a_{k+1} - a_k) \sum_{j=0}^{k} b_j. \end{aligned} \tag{1.251}$$

Because of (1.250) we get

$$\left|\sum_{j=0}^{k} b_j\right| \leq 2 \qquad (0 \leq k \leq N-1). \tag{1.252}$$

The differences $a_{k+1} - a_k$, $k = 0, 1, \ldots, N-2$, can be estimated with help of (1.248) and (1.249):

$$\begin{aligned} |a_{k+1} - a_k| &\leq \frac{|\sin\pi(\varphi(k+2) - 2\varphi(k+1) + \varphi(k))|}{2|\sin\pi(\varphi(k+2) - \varphi(k+1))| \cdot |\sin\pi(\varphi(k+1) - \varphi(k))|} \\ &\leq \frac{\pi}{8} \frac{|\varphi(k+2) - 2\varphi(k+1) + \varphi(k)|}{|\varphi(k+2) - \varphi(k+1)| \cdot |\varphi(k+1) - \varphi(k))|} = \\ &= \frac{\pi}{8} \left| \frac{1}{\varphi'(\xi_{k+1})} - \frac{1}{\varphi'(\xi_k)} \right|. \end{aligned} \tag{1.253}$$

Furthermore,

$$|a_{N-1}| \leq \frac{1}{2|\sin(\pi(\varphi(N) - \varphi(N-1)))|} \leq \frac{1}{4|\varphi(N) - \varphi(N-1)|} \leq \frac{1}{4L} \tag{1.254}$$

By applying (1.252)–(1.254) to (1.251) we obtain

$$\left|\sum_{k=0}^{N-1} e(\varphi(k))\right| \le \frac{1}{2L} + \frac{\pi}{4}\sum_{k=0}^{N-2}\left|\frac{1}{\varphi'(\xi_{k+1})} - \frac{1}{\varphi'(\xi_k)}\right|. \tag{1.255}$$

If ξ_k, $k = i, \dots, i+j$, $j \ge 1$, are contained in an interval of monotonicity of $\varphi'(n)$, then

$$\sum_{k=i}^{i+j-1}\left|\frac{1}{\varphi'(\xi_{k+1})} - \frac{1}{\varphi'(\xi_k)}\right| \le \frac{2}{L}. \tag{1.256}$$

Obviously (1.256) holds for $j = 1$, too, if ξ_i, ξ_{i+1} are in different intervals of monotonicity of $\varphi'(u)$. Therefore

$$\left|\sum_{k=0}^{N-1} e\left(\varphi(k)\right)\right| \le \frac{1}{2L} + \frac{\pi n}{2L} + \frac{\pi(n-1)}{2L} \le \frac{4n}{L}$$

which proves the lemma. □

The following notation will be used in the sequel:

$$f(m,u,x) = \sum_{\nu=1}^{m} h_\nu \left(\rho_m + \frac{u}{q_m p_m}\right)^{n_m+x+\nu-1},$$

$$T_m(x,y) = \frac{1}{q_m}\sum_{c=0}^{q_m-1} e\left(\left(f(m,c,x) - f(m,c,y)\right)\right), \tag{1.257}$$

and

$$x_m(z) = -n_m - \frac{C_m(\mathbf{h},z)}{B_m(\mathbf{h},z)} \qquad (\text{if } B_m(\mathbf{h},z) \ne 0). \tag{1.258}$$

Furthermore, let Δ_m be the interval $[x_m(\rho_m) - 2, x_m(\rho_m) + 2]$.

Lemma 1.135 *Let $m > C_2$, $0 \le u \le q_m$, $0 < \|\mathbf{h}\|_\infty \le m^{m-1}$. $x, y \in \mathbf{Z}$, $x \in \Delta_m$, and $m^{2m} > x > y \ge 0$. Then*

$$\left|\frac{\partial f}{\partial u}(m,u,x) - \frac{\partial f}{\partial u}(m,u,y)\right| \ge \frac{1}{3}\left(p_m q_m A_m\right)^{-1} \rho_m^{n_m+x-1}.$$

Proof. From (1.257), (1.216) and (1.243) we obtain

$$\begin{aligned} p_m q_m \frac{\partial f}{\partial u}(m,u,x) &= \left(\rho_m + \frac{u}{q_m p_m}\right)^{n_m+x-1} \\ &\quad \times \left((n_m+x)B_m\left(\mathbf{h}, \rho_m + \frac{u}{q_m p_m}\right) + C_m\left(\mathbf{h}, \rho_m + \frac{u}{q_m p_m}\right)\right) \\ &= \left(\rho_m + \frac{u}{q_m p_m}\right)^{n_m+x-1} B_m(\mathbf{h}, \rho_m)\left(x - x_m(\rho_m) + \frac{\theta(x)}{2}\right) \end{aligned}$$

with $|\theta| \le 1$. Next Lemma 1.133, (1.258), and (1.229) imply

$$q_m p_m \left(\frac{\partial f}{\partial u}(m,u,x) - \frac{\partial f}{\partial u}(m,u,y)\right) = (\rho_m + \frac{u}{q_m p_m})^{n_m+x-1} B_m(\mathbf{h},\rho_m)$$
$$\times\left(\left(1-\left(\rho_m+\frac{u}{p_m q_m}\right)^{y-x}\right)\left(x-x_m(\rho_m)+\frac{\theta(x)}{2}\right)\right.$$
$$\left.+\left(x-y+\frac{\theta(x)-\theta(y)}{2}\right)\left(\rho_m+\frac{u}{q_m p_m}\right)^{y-x}\right). \tag{1.259}$$

Furthermore, according to the assumption we have

$$\left|x-x_m(\rho_m)+\frac{\theta(x)}{2}\right| \ge \frac{3}{2} \tag{1.260}$$

Clearly

$$\left(\rho_m+\frac{u}{q_m p_m}\right)^{y-x} \le 3^{y-x} \le \frac{1}{3} \tag{1.261}$$

and

$$(x-y+1)3^{y-x} \le \frac{2}{3}, \tag{1.262}$$

in which $x, y \in \mathbf{Z}$, $x > y$. Applying (1.260) - (1.262) to (1.259), we obtain

$$q_m p_m \left|\frac{\partial f}{\partial u}(m,u,x) - \frac{\partial f}{\partial u}(m,u,y)\right| \ge \frac{1}{3}|B_m(\mathbf{h},\rho_m)|\rho_m^{n_m+x-1}.$$

Combining this with (1.244) proves the lemma. □

Lemma 1.136 *Let* $m > \max(C_1, C_2)$, $0 < \|\mathbf{h}\|_\infty \le m^{m-1}$, $x, y \in \mathbf{Z}$, $m^{2m} > x > y \ge 0$, *and* $x \notin \Delta_m$. *Then*

$$|T_m(x,y)| \le \rho_m^{m^{m-1}-x}.$$

Proof. Lemma 1.135 applied to $\varphi(u) = f(m,u,x) - f(m,u,y)$ (with $u \in [0, q_m]$) yields

$$|\varphi'(u)| \ge \frac{1}{3 q_m p_m A_m}\rho_m^{n_m+x-1}.$$

Conversely, from (1.257), (1.211), (1.222), and (1.237) (with $u \in [0, q_m]$) it follows that

$$|\varphi'(u)| \le \frac{2}{p_{m+1}}(n_m+x+m-1)m\cdot m^{m-1}\left(\rho_m+\frac{u}{q_m p_m}\right)^{n_m+x+m-2} \le \frac{1}{2}$$

Since

$$\varphi''(u) = \left(\rho_m+\frac{u}{q_m p_m}\right)^{n_m-2} P(u),$$

with $P(u)$ being a polynomial of degree $< m^{2m} + m - 1 < 2m^{2m} - 1$, it follows that $\varphi'(u)$ has at most $2m^{2m}$ intervals of monotonicity on $[0, q_m]$. Applying Lemma 1.134 to (1.257), we get

$$|T_m(x,y)| \;\le\; 24m^{2m} p_m A_m \rho_m^{-n_m - x + 1}$$

which proves the lemma via (1.236). □

Proof of Proposition 1.131. By (1.223) and (1.257)

$$S(c, h_1, \ldots, h_k, h_{m+1}) = \sum_{x=0}^{m^{2m}-1} e\left(f(m,c,x) + h_{m+1} x m^{-2m}\right)$$

which implies

$$E(h_1, \ldots, h_k, h_{m+1}) = \sum_{x,y=0}^{m^{2m}-1} T_m(x,y) e\left(h_{m+1}(x-y)m^{-2m}\right). \tag{1.263}$$

If $h_1, = \ldots = h_k = 0$, then $h_{m+1} \not\equiv 0 \; (m^{2m})$. Consequently (1.223) and (1.225) yield

$$E(0, \ldots, 0, h_{m+1}) = \sum_{x,y=0}^{m^{2m}-1} e\left(h_{m+1} \frac{x-y}{m^{2m}} \right) = 0. \tag{1.264}$$

Now suppose that $0 < \max_{1 \le \nu \le k} |h_\nu| \le m^{m-1}$. If we use (1.263) and Lemma 1.136 (where we set $h_{k+1} = \ldots = h_m = 0$ for $k < m$) and the trivial inequality $|T_m(x,y)| \le 1$, then we get

$$\begin{aligned} E(h_1, \ldots, h_k, h_{m+1}) \;&\le\; \sum_{m^{2m} > x, y \ge 0} |T_m(x,y)| \\ &\le\; \sum_{m^{2m} > x = y > 0} 1 + 2 \sum_{m^{m-1} \ge x > y \ge 0} 1 + \\ &\quad +2 \sum_{m^{2m} > x > y \ge 0, x \in \Delta_m} 1 + 2 \sum_{m^{2m} > x > y \ge 0, x \notin \Delta_m, x > m^{m-1}} \rho_m^{m^{m-1} - x} \end{aligned}$$

Substituting $z = x - m^{m-1}$ and considering (1.258) and (1.230) we finally obtain

$$\begin{aligned} E(h_1, \ldots, h_k, h_{m+1}) \;&<\; 15m^{2m} + 2 \sum_{y=0}^{m^{2m}-1} \sum_{z=1}^{\infty} \rho_m^{-z} = \\ &=\; 15m^{2m} + 2m^{2m}\left(\frac{1}{\rho_m - 1}\right) \le 16m^{2m} \end{aligned} \tag{1.265}$$

which completes the proof of Proposition 1.131. □

Notes

There is a vast literature on the $(n\alpha)$-sequence. Miklavc [1195] gave an elementary proof for the u.d. of this sequence. Dupain [504] has precisely estimated the involved constant in the discrepancy bound when α is the golden ratio. Dupain and Sós [508] showed that this constant is minimal for $\alpha = \sqrt{2}$. The problem of estimating the discrepancy of the $(n\alpha)$-sequence was first tackled by Behnke [155, 156], Ostrowski [1398], Hardy and Littlewood [735] and Hecke [759]. More recently, it was taken up by Niederreiter [1278], Lesca [1060] and Ramshaw [1500]. The characterization of all α such that $ND_N(n\alpha) = \mathcal{O}(\log N)$ is already contained in the work of Behnke. Schoißengeier [1655] established an explicit formula for $D_N(n\alpha)$ which yields fast computations of the discrepancy for special $\alpha's$ (e.g. $\alpha = \pi$); see also [1656]. A very special result and applications to the phyllotaxis of plants are due to van Ravenstein [1521]. Schoißengeier [1658] computed the $\mathcal{O}$-constant in the discrepancy bound for a wide class of numbers α generalizing the earlier results of Dupain and Sós. Bounds for sums of Bernoulli polynomials $\sum_{n\leq N} B_1(n\alpha)$ were given by Schoißengeier [1657], where estimates for Dedekind sums are used; see also [1613]. For more informations on Dedekind sums in this context we refer to Dieter [462]. In Schoißengeier [1659] one-sided discrepancy estimates are proved. Baxa and Schoißengeier [113] also gave an explicit formula for $\limsup_{N\to\infty} ND_N^*(n\alpha)/\log N$ for all α for which the discrepancy is $\mathcal{O}(\log N/N)$. Baxa [112] studied the image of the map $\alpha \to v(\alpha) = \limsup_{N\to\infty} ND_N^*(n\alpha)/\log N$ showing that $v(\mathbf{R}\backslash\mathbf{Q}) = [v(\sqrt{2}), \infty]$. For further contributions on the $(n\alpha)$-sequence see Brown and Shiue [303]. Oskolkov [1396] investigated the convergence of the arithmetic mean of $f(n\alpha)$ to the integral for unbounded functions. A lower bound for such arithmetic means in the case of piecewise Lipschitz functions is due to Perelli and Zannier [1424].

The dispersion of the sequence $(n\alpha)$ and the corresponding dispersion spectrum was studied by Drobot [493], Kopetzky and Schnitzer [952], Jager and De Jonge [869], Tripathi [1876] and Ji and Lu [874]. Schoißengeier [1660] proved explicit formulas for the dispersion of $(n\alpha)$-sequences. Lambert [1010] constructed a low-dispersion sequence in the unit square. Mitchell [1197] determined the dispersion of Hammersley and Halton sequences; see also Peart [1422] and Section 3.1. For further results on the dispersion of special sequences (e.g. of good lattice point sequences) see Larcher [1019, 1022]. Related investigations to the one-dimensional dispersion are due to Ramshaw [1499].

Steinhaus conjectured that the set of differences $s_{j+1} - s_j$, $j = 1, \ldots, N-1$, where s_j denote the first N elements of the $(n\alpha)$-sequence in increasing order, consists of at most 3 elements. For proofs of this assertion see Sós [1724], Swierczkowski [1766] and Slater [1695]. Chung and Graham [366] established a generalization to m sequences $(n_i\alpha + \beta_i)$, $n_i \leq N_i$, $1 \leq i \leq m$. For further extensions of the Steinhaus problem see Mayer [1159] and Lohöfer, Mayer [1093], Geelen and Simpson [649] and Fried and Sós [625]. Cardin [325] investigated the structure of subsequences of the $(n\alpha)$-sequence. For a gap problem for the $(n^2\alpha)$-sequence we refer to Drobot [494]. An optimal bound for the L^2-discrepancy of the two-dimensional point set $(n/N, n\alpha)$, $1 \leq n \leq N$ was shown by Sós and Zaremba [1727].

Kesten [920] proved for the "local" discrepancy function $g(N, I) = A(N, I, n\alpha) - N\lambda(I)$ the following theorem: The sequence $g(N, I)$, $N = 1, 2, \ldots$ is bounded if and only if $\lambda(I) = h\alpha \bmod 1$, $h \in \mathbf{Z}$.

An extension and a new proof of Kesten's theorem is due to Shapiro [1668] and Stewart [1742]; see also Veech [1907]. For various relations to topological dynamics we refer to Furstenberg, Keynes and Shapiro [636]. Dupain [501, 502, 503] and Dupain and Sós [507] constructed intervals I such that the sequence $g(N, I)$ is bounded from above but not from below, or unbounded in both directions (e.g. for the golden ratio and $\lambda(I) \neq h\alpha \bmod 1$). In [507] necessary and sufficient conditions for the one-sided boundedness are given in the case that α has bounded partial quotients. A strong irregularity result for the $(n\alpha)$-sequence is due to Sós [1726]. Motivated by Kesten's theorem, a subset B of a compact metric space is called a bounded remainder set for a sequence (x_n) if there exists $a \in [0, 1)$ such that $A(N, B, x_n) - Na$ is bounded. If B is an interval Kesten's result gives a characterization of B to be a bounded remainder set for the $(n\alpha)$-sequence. Generalizations were given by Petersen [1428] and Oren [1395]. Bounded remainder sets for the $(n\alpha)$-sequence were also studied by Rauzy [1519]. Liardet [1085] proved an analogon of Kesten's theorem in the k-dimensional case, and he shows that there exist no (non-trivial) intervals which are bounded remainder sets for polynomial sequences (of degree > 1) with irrational leading coefficient. Schmidt [1631] has shown that in arbitrary dimension and for arbitrary sequences in the k-dimensional hypercube there are only countably many possible values for the Lebesgue measure of I (interval) provided that the corresponding discrepancy function $g(N, I)$ is bounded. In the case of the $(n\alpha)$-sequence Tijdeman and Voorhoeve [1859] established an algorithm for computing this countable set. Hellekalek [763, 764] determined explicitly the bounded remainder sets for generalized Halton sequences (for definitions see Section 3.1). For further results on bounded remainder sets we refer to Hellekalek and Larcher [771]. Hellekalek [767, 766, 768, 769, 770] considered the boundedness of Weyl sums over irrational rotations.

For a geometric interpretation of u.d. we refer to Deshouillers [456] and Dekking and Mendès France [447]. For a survey on u.d. see Jacobs [867]. Boyd and Steele [289] study the length of longest increasing subsequences of the $(n\alpha)$-sequence. In a series of papers Berend [172, 173, 175, 176, 177] investigated the denseness of sequences $(a_n x)$, where (a_n) is an additive semigroup generated by a strictly increasing sequence of integers; i.e. an IP-set. In [177] for various classes of sequences of the type $(a_n x)$ characterizations are shown such that these sequences are dense or u.d. for all irrational x. This problem is solved for instance for linear recurring sequences and divisibility sequences. Berend and Peres [182] studied asymptotically dense dilations of sets on the circle. Bel'nov [163] gave a sufficient condition for the existence of a non-zero limiting point mod 1 of such sequences, where (a_n) is an arbitrary unbounded sequence. Forrest [615] has shown that the Weyl sums of $(\alpha n^2 + xn)$ are dense in the complex plane for certain transcendental numbers α if x is chosen from a dense G_δ subset in [0,1]; the proof is based on ideas from ergodic theory and it is also shown that the associated cocycle is topologically transitive. For more on ergodicity and cocycles over irrational rotations we refer to Lemanczyk and Mauduit [1059].

The multi-dimensional version of the $(n\alpha)$-sequence is called Kronecker sequence. Larcher [1025] proved that the essentially best possible lower bound for the isotropic discrepancy of the Kronecker sequence is of the order $N^{-1/k}$; see also [1023, 1028] and [1029] for the cubic discrepancy. Larcher and Niederreiter [1037] investigated nonarchimedian Kronecker-type sequences; these investigations were motivated by the study of net sequences and applications in Quasi-Monte Carlo methods. Larcher [1034] established a metric discrepancy estimate for such sequences. Moshchevitin [1207] showed that there exist positive (absolute) constants C, δ such that for infinitely many N the discrepancy of the Kronecker sequence is less than $CN^{-\delta}$ provided that the coefficients α_i have "no good" common approximation. Miles and Thomas [1196] show that the Kronecker sequence (with rationally independent algebraic coefficients) is u.d. "polynomially fast" in every closed subset with positive measure whose boundary is Lipschitz. Ostrowski [1399, 1400, 1401] continued his classical work and established also results on Kronecker sequences. Wang and Yu [1943] established metric results concerning the Kronecker sequence.

A famous conjecture of Littlewood says that for an arbitrary given $c > 0$ and real numbers $\alpha_1, \ldots, \alpha_k$, $k \geq 2$ the diophantine inequality

$$|h| \, \|h\alpha_1\| \cdots \|h\alpha_k\| \geq c$$

has no integral solution $h \neq 0$; see for instance A. Baker [62] and Schmidt [1637]. Skubenko [1694] studied exceptional sets related to this conjecture. A. Baker [62] stated the conjecture that, for any $\varepsilon > 0$, the inequality $q^{1+\varepsilon}\|q\theta\| \ldots \|q\theta^n\| < 1$ has only finitely many integer solutions q for almost all real θ. This conjecture has been proved for $n = 2$ by Sprindzhuk [1730]. Yu [1979, 1978] sharpened Sprindzhuk's result and extended it to complex θ in the cases $n = 2, n = 3$. Furthermore for $n = 2$ the Hausdorff dimension of the exceptional set was considered.

A classical theorem of Khintchine says that the Lebesgue measure $\lambda_1(X)$ of the set $X = \{x \in [0,1] : \|q_n x\| < q_n f(q_n)$ for infinitely many $q_n\}$ is 1 provided that $\sum q_n f(q_n) = \infty$, where q_n are pairwise distinct positive integers and $q_n^2 f(q_n)$ is decreasing. Duffin and Schaeffer [495] conjectured that $\lambda_1(X) = 1$ provided that $\sum \varphi(q_n) f(q_n) = \infty$, where φ denotes Euler's φ-function. Quite recent contributions to this conjecture are due to Harman [744] and Nakada and Wagner [1251]; the last authors considered a certain complex version of the conjecture. Pollington and Vaughan [1449, 1448] established a proof for a k-dimensional version of this conjecture and gave further contributions to the case $k = 1$. Strauch [1748] introduced "uniformly quick" sequences and proved some sufficient conditions for the Duffin-Schaeffer conjecture. Strauch [1749] investigated the uniform quickness of the $(n\alpha)$-sequence. This work is continued in [1750, 1751] where also distribution properties of the Farey sequence are considered.

Reversat [1529, 1530, 1531] investigates so-called eutaxic sequences. (u_n) in [0,1) is said to be eutaxic if for all decreasing real numbers $\varepsilon_n > 0$ with divergent $\sum \varepsilon_n$ the inequality $\|x - u_n\| < \varepsilon_n$ has infinitely many solutions for almost all x. This concept was introduced in the thesis of Lesca and later studied by de Mathan [1130, 1131]. So for instance, the $(n\alpha)$-sequence is eutaxic if and only if $\limsup(n\|n\alpha\|)^{-1}$ is finite. Reversat also considers eutaxic sequences in higher dimensions. A related concept is the notion of "uniform quickness" introduced by Strauch.

Sequences $([\theta n + \gamma])$ $\theta \geq 1$ are called Beatty sequences; for a survey on the investigation of that sequences we refer to Porta and Stolarsky [1451]. An application of Vaaler's method [1887] to Beatty sequences is due to Abercrombie [1]. Fraenkel and Holzman [618] investigated gap problems and the intersection of two Beatty sequences; for further results on such sequences we refer to Bowman [284] and Komatsu [944, 945, 946]. Higher dimensional generalizations were studied by Niederreiter [1275, 1277]. Various contributions on Beatty sequences and related topics are due to Fraenkel [617] and Eggleton, Fraenkel and Simpson [513]. Carlson [326] proved that sequences $([n\beta]\alpha)$ are u.d. mod 1 if $1, \beta, \alpha\beta$ are linearly independent over the rationals. Extensions of this result to polynomial sequences are due to Haland [714, 715] and Haland and Knuth [716]. A distribution property for linear combinations of the integer parts of polynomials was discussed by Mendès France [1168]. Sequences of the type $a_n := (-1)^{[\alpha n]}$ were considered by Borwein and Gawronski [268] and Bundschuh [312]; they showed, for instance, that $\sum a_n/n$ is convergent for special quadratic irrationals α. Fraenkel [616] investigated the distribution of $([n\alpha])$ and complementary sets. Related investigations are due to Brown [302], Ito and Yasutomi [866] and Crisp, Moran, Pollington and Shiue [422]. Pollington [1446] proved metric results involving lacunary sequences of integers and Beatty sequences. Fiorito [598] studied "periodically monotone" sequences.

A classical theorem of Heilbronn [760] says that for any $\varepsilon > 0$, the diophantine inequality $\|n^2\alpha\| < n^{-\frac{1}{2}+\varepsilon}$ has infinitely many integer solutions, where α is an arbitrary real number. Recently, Zaharescu [1982] could establish a remarkable improvement by obtaining Heilbronn's theorem with exponent $-\frac{2}{3} + \varepsilon$. For previous results on that problem we also refer to Huxley [856], Friedlander and Iwaniec [626] and Li [1081]. For results on the fractional parts of $(n^2\alpha)$ we refer to Heath-Brown [757]. Excellent surveys on diophantine inequalities of the Heilbronn type and extensions are due to Baker [87] and Schmidt [1636]. In a series of papers Baker [75, 77, 80, 85, 84, 81, 82, 90] proved upper bounds for $\min_{1\leq n\leq N} \|f(n)\|$, where $f(n)$ denote certain polynomials; see also Baker [73], Harman [748] and Baker and Brüdern [91]. Related results are due to Baker, Brüdern and Harman [92] who considered fractional parts of αn^k for square-free n. Baker and Harman [93, 83] proved diophantine inequalities for quadratic forms, Balog and Perelli [102] inequalities for the approximation by square free numbers. Balog and Friedlander [101] considered simultaneous diophantine approximation using prime numbers. Results on small solutions of quadratic forms are also due to Schinzel, Schlickewei and Schmidt [1620]. An upper bound for $\|\alpha n^2 + \beta n\|$ was shown by Schmidt [1635], upper bounds for exponential sums over polynomials were proved by Schmidt [1638, 1639]. Berend [174] investigated diophantine inequalities for smooth functions and Chong and Liu [352] considered sums of polynomials. In a series of papers Cook [377, 378] proved various diophantine inequalities involving polynomials in one and more variables. For instance, in [376] quadratic forms are considered and it is proved that for any $\varepsilon > 0, N > 1$ and $n \geq n_0(\varepsilon)$ there exist integers $x_1, \ldots, x_n$ satisfying $\max_{1\leq j\leq n} |x_j| \leq N$ and

$$\|Q(x_1, \ldots, x_n)\| < C(n, \varepsilon)N^{-2+\varepsilon};$$

see also [1620], Schaeffer [1616] and Baker and Schaeffer [97].

Pétermann [1426] studied the oscillation of the so-called Golomb sequence. Niederreiter [1279] investigated the distribution of the Farey points; see also Meijer and Niederreiter [1162], Codeca [372] and Codeca and Perelli [373]. Schoißengeier [1650] and Niederreiter and Schoißengeier [1269] considered special u.d. sequences using almost periodic functions (in the sense of Bohr). Boshernitzan [273] investigated the u.d. of special sequences $f(n)$ related to Hardy fields.

In a series of papers Faure [580, 581, 582, 583, 584, 585, 586, 587, 590] investigated generalized Van der Corput sequences; see also Béjian [157], and Béjian and Faure [158, 159]. In [584] he proved that a generalized Van der Corput sequence $\gamma_q(n)$ in $[0, 1]$ (in base q) has a bounded remainder function $A(J, N, x_n) - N\lambda_1(J)$ with respect to a given interval $J = [0, \alpha)$ if and only if α is q-adic, i.e. $\alpha = k/q^m$. This extends results of Schmidt [1629] and Hellekalek [763]. For a further generalization see Hellekalek [764]. In the important contribution [582] Faure considered a class of sequences of the net type $[0,1]^k$ and proved a discrepancy bound $N D_N \leq c_k(\log N)^k + O_k((\log N)^{k-1})$, where c_k tends to 0 super-exponentially (as $k \to \infty$). In [590] it is proved that for every base r there exists a permutation of the digits such that the corresponding Van der Corput sequence satisfies $\limsup_{N\to\infty} ND_N^* / \log N \leq 1/\log 2$. Later sequences of the net type were systematically investigated by Niederreiter, see Section 3.1. Chaix and Faure [333, 334, 335, 336, 594, 595] systematically investigated the discrepancy and the diophany of generalized Van der Corput sequences (with respect to given bases and permutations of the digits). The constants $\limsup_{N\to\infty} ND_N / \log N$ can be determined explicitly for sequences of this type. This extends results of Atanassov [54], Proinov and Atanassov [1482], Proinov and Grozdanov [1483, 1484] and Xiao [1973]; see also a series of papers by Proinov [1460, 1459, 1462, 1463, 1465, 1479, 1472, 1470], Thomas [1806] and Xiao [1973]. These authors mainly consider the Halton-sequence and the symmetrization of the Van der Corput-Halton sequence or the $(n\alpha)$-sequence and its symmetrization. Sharp bounds for the L^2-discrepancy and for the diophany were proved; see also Grozdanov [706]. Larcher [1020] estimated the isotropic discrepancy of Hammersley and Halton sequences. More detais on such kinds of sequences see Section 3.1. For related low discrepancy sequences produced by iteration of functions we refer to Borel [264], Lapeyre and Pages [1011] and Lapeyre, Pages and Sab [1012]; see also Pages [1403]. Lambert [1007] showed how low discrepancy sequences can arise from ergodic transformations. Borel [256, 258, 262, 265] studied in detail "self-similar" sequences, extending the classical Van der Corput sequence; he described these sequences using piecewise monotonic transformations and obtained discrepancy bounds. Rauzy [1515] constructed a low discrepancy sequence which is completely u.d. (by a modification of the Van der Corput sequence). White [1954, 1955] estimated the L^2-discrepancy of special sequences such as the Hammersley sequence.

There is a vast literature on normal numbers. Ito and Shiokawa [865] extend Champernowne's construction [337] to β-normality with respect to the β-expansion $\sum_{n=0}^{\infty} w_n(x)\beta^{-n-1}$, $w_n(x) = [T_\beta^n x]$. Here $T_\beta = \beta x - [\beta x]$ ($\beta > 1$, $0 < x < 1$) is the so called β-shift, see Parry [1405] and Renyi [1528]. For recent contributions to the β-expansion we refer to Frougny and Solomyak [628], Schweiger [1664], Solomyak [1717], Flatto [605], Loraud [1095] and Keane, Smorodinsky and Solomyak [916]. Shiokawa and Uchiyama [1674] analyzed carefully the distribution of the digits of dyadic Champernowne numbers. A slightly different approach is due to Grabner [676]. Ito [864] considered ergodic properties of some maps which are useful for estimating the discrepancy of $(n\alpha)$, see also Sós [1725]. Kamae [889, 890] studied normal numbers from an ergodic point of view, especially properties of subsequences. Pearce and Keane [1421] present a new (probabilistic) proof of a classical theorem of Schmidt saying that for positive integers r, s with $r^m \neq s^n$ (for all integers m, n) there exist r-normal numbers that are not s-normal. For another proof of this theorem and various further results on normal numbers we refer to Nagasaka [1232, 1233, 1234] and Nagasaka and Batut [1238]. Brown and Moran [298] could prove for two commuting ergodic $n \times n$-matrices S, T that either $S^q = T^p$ for natural numbers p and q or there are uncountably many points in $\mathbf{R}^n$ which are normal in base S but not normal in base T (and conversely); for the basic facts concerning normality with respect to matrices see Schmidt [1623]. Kulikova [995] and Postnikova [1456, 1457, 1455] provide constructions of normal numbers, see also Korobov [956], Kano [896], Zhu [1993] and Martinelli [1127]. Galambos [645, 646] considered u.d. mod 1 in connection with Cantor representation of numbers. Further investigations related to Cantor representations are due to Borel [263], Lacroix [998] and Salát [1592]. Grekos and Volkmann [697] proved relations between asymptotic densities of subsets of a set A of positive integers and the lengths of the gaps of A. For further properties of normal numbers we refer to Rauzy [1507, 1508], Broglio and Liardet [296] and Mauduit [1147, 1150, 1151]. For metric results concerning normal numbers see Slivka and Severo [1696], Ki and Linton [923]. For normal numbers and and entropy we refer to Host [849]. Constructions of normal numbers using transducer automata are due to Dumont and Thomas [499]; see also Blanchard, Dumont and Thomas [227], Dumont and Thomas [498] and Blanchard [226]. They obtain a new approach (using substitutions) to the results of Schiffer [1619] and Schoißengeier [1651], Nakai and Shiokawa [1252, 1253, 1254] and Shiokawa [1672]. Discrepancy results related to normal numbers are due to Moran and Pearce [1200]. For further constructions of normal numbers, results on distribution measures of sequences $(g^n x)$ and related results we refer to Volkmann [1914, 1915, 1916], Szüsz and Volkmann [1918, 1770, 1771], Dumont and Thomas [497], Bertrand-Mathis [196], Bertrand-Mathis and Volkmann [209] and Grabner

[676]. In a recent contribution Nakai and Shiokawa [1255] proved that the number

$$0.[f(2)][f(3)][f(5)][f(7)][f(11)][f(13)]\dots$$

(in base q) is normal, where $[m]$ denotes the q-adic digital representations of m and f is nonconstant polynomial taking positive integral values at all positive integers.

In a series of papers Mauduit [1145, 1146, 1148, 1149] discussed automatic aspects of the theory of normal sets (in the sense of Mèndes France [1164]). Here a set M of real numbers is called normal, if there exists a sequence (u_n) of real numbers such that (xu_n) is u.d. mod 1 if and only if $x \in M$. Coquet [394, 395, 397] investigated properties of normal sets in connection with completely u.d. sequences; see also [396] and Zame [1983]. For further contributions on normal sets and self-similar measures see Borel [255, 254, 260, 257, 259]. Moran and Pollington [1201] considered normality of real numbers with respect to distinct non-integer bases; see also Brown [297] and Brown, Moran and Pollington [300]. Wagner [1931] proved a general result on normal numbers from which a construction of numbers x follows such that, for any non-constant polynomial $q \in \mathbf{Z}[x]$, the number $q(x)$ is normal to base 3 but not normal to base 15. For related results we refer to Kano [895] and Kano and Shiokawa [897]. Brown, Moran and Pearce [299] established (by combining Fourier analysis, Riess products and transcendence methods) a generalization of a result of Schmidt: If $s > 1$ is a positive integer, then every real number can be expressed as the sum of four numbers none of which is normal to base s but all of which are normal for all bases r multiplicatively independent of s, so that the set of all real numbers which are normal to every base r multiplicatively independent of s but not (simply) normal to base s has the cardinality of the continuum. Stoneham [1744, 1745] studied an extension of normal numbers, the so-called (j,ε)-normality. Briefly, a rational fraction is (j,ε)-normal when represented to base b if each block of j digits that actually occurs in the b-expansion of the fraction occurs to "almost" the expected normal frequency; see also Stoneham [1746, 1747]. Wagon [1932] presented a brief report on computational results on the digits of π. "Weighted normal numbers" were studied by Choe [351].

Adler, Keane and Smorodinski [3] established a construction of normal numbers with respect to Gauss measure. Levin [1077] constructed Markov-normal sequences of discrepancy $\mathcal{O}(N^{-1/2}(\log N)^2)$; for further constructions of normal numbers and related aspects see Levin [1072, 1075, 1076].

Berend and Boshernitzan [178] established a further improvement of the result of Mahler [1119] on digital properties of the $(n\alpha)$-sequence; see also Alon and Peres [33] and Volkmann and Szüsz [1917]. Further contributions in this direction are due to Berend and Boshernitzan [179]. Blecksmith, Filaseta and Nicol [229] proved that the number of digital blocks in the b-ary expansion of a^n ($a, b \geq 2$ integers) tends to infinity (as $n \to \infty$) provided that $\log a/\log b$ is irrational; here a digital block is a successive sequence of equal digits of maximal length. For a somewhat related result see Mercer [1187].

Johnson and Newman [876] investigated when 0 belongs to the convex hull of a^m, b^m, c^m for some positive integer m and three unimodular complex numbers a, b, c. Shiue [1677] fills a gap in [876]. Myerson [1225] studied how small the sum of N-th roots of unity can be in absolute value. Pinch [1438] considered a special sequence in the unit square. A conjecture of Zaremba [1987] says that for every integer $m \geq 2$ there exists a reduced fraction $0 < a/m < 1$ such that all the partial quotients in the continued fraction expansion of a/m are ≤ 5. Niederreiter [1311] proved this conjecture for the special case where m is a power of 2, 3 or 5. Weaker bounds (dependent on m) are due to Zaremba [1987] and Cusick [426]; see also Sander [1598], who proved bounds for the number of integers $m \leq x$ such that there exists a positive integer $a < m$, coprime to m, satisfying that the partial quotients of a/m are bounded by some constant. Recently, Larcher [1013] obtained various metric diophantine approximation results related to optimal coefficients and good lattice points (see Section 3.1); in this paper also aspects of Zaremba's conjecture are discussed. For a general metric inequality and applications see also Larcher [1026].

Del Junco and Steele [879] establish some results for the number of elements of a largest increasing subsequence (upto a given index) of a u.d. sequence. Garifullina [648] considered u.d. matrix exponential functions. Haight [713] studied multiples of increasing sequences (b_n) with irrational quotients b_m/b_n. Erdős and Hall considered the distribution of $(\log d)$ mod 1, where d runs through the divisors of an integer n. For various extensions see Hall [720, 721, 722, 723, 724, 725], Dupain, Hall and Tennenbaum [505], Tennenbaum [1797], and Hall and Tennenbaum [726]. Erdős and Sheng [567] investigated the distribution of rational numbers in short intervals; see also Isbell and Schanuel

[860]. Another class of sequences of rational numbers was studied by Porubsky, Salát and Strauch [1453]. Myerson [1226] studied the u.d. of a special sequence connected with Dedekind sums. Vardi [1901] established a connection of special Kloosterman sums and Dedekind sums with applications to u.d. For a related result see also Zheng [1992].

Uniform distribution properties of the zeros of the Riemann zeta function were investigated by Hlawka [802], Griso [698], Schoißengeier [1652, 1648, 1649] and Fujii [630, 631, 632]. Kaczorowski [880, 881, 882, 883] investigated u.d. of the zeros of the Riemann zeta function with respect to a related matrix summation method and obtained a discrepancy bound. Discrepancy estimates for the value distribution of the Riemann zeta function are due to Matsumoto [1138, 1139, 1140] and Harman and Matsumoto [752]. Reich [1523, 1524] proved the convergence of a discrete mean value to a continuous mean value for Euler products using u.d. sequences. A special class of sequences in the Gaussian number field was studied by Hlawka [830].

In a series of papers Rauzy [1502, 1504, 1505] investigated the u.d. mod 1 of sequences $(f(n))$, where f is an entire function with real Taylor coefficients and satisfying suitable growth conditions. Improvements and extensions are due to Baker [86, 89, 88]. Bobrovskij [244] considered special sequences $(f(n))$ satisfying certain conditions on the first and second derivatives of f. For distribution properties of analytically described sequences we also refer to Cater, Crittenden and van den Eynden [328]. Rhin [1535, 1537, 1538] investigated the u.d. of $(f(p))$, where f is an entire function and p runs through the primes. For further results on the u.d. of sequences $(f(n))$ and $(f(p))$, where f satisfies suitable analytic conditions we refer to Toffin [1863, 1862, 1864]. Borel [246, 249] investigated distribution properties of sequences related to Beurling prime numbers; for more on prime number sequences see the Notes of Section 1.2. Schoißengeier [1654] proved sharp bounds for the discrepancy of (αn^σ), $(\alpha > 0, 0 < \sigma < 1)$, in particular $\mathcal{O}(N^{-1/2})$ for $\sigma = 1/2$, where the $\mathcal{O}$-constant is precisely determined when the square of α is irrational. The remaining case is much more complicated and was recently settled by Baxa and Schoißengeier [114]. Beck [141] established a central limit theorem for the sequence $n\sqrt{2}$. For distribution properties of oscillating sequences we refer to Berend and Kolesnik [181]. Goto and Kano [670], Fiorito, Musmeci and Strano [599] considered special uniform distribution problems and Bukovska and Salát [308] established topological properties for sequences $(n^k x)$ and applications to trigonometric series.

Katai [908, 909, 910] established results on the distribution mod 1 of additive functions, see also Mauclaire [1141], Bedin and Deutsch [153], Ruzsa [1582], Tolenov and Fainlaib [1865], Manstavicius [1120], and for multiplicative and additive functions Kubilius [966]. Summation of multiplicative functions on semigroups is studied by Manstavicius and Skarabutenas [1122]. Various results on distribution properties of arithmetic functions can be found in Alladi [14]; for special arithmetic functions see also Katai [911], Sarkőzy [1601] and Shirokov [1675]. Indlekofer, Jarai and Katai [859] investigated related dynamical properties of iterated function systems. Nagasaka, Shiue and Yu [1241] studied regularity properties of multiplicative functions. Recent contributions on the distribution of complex valued multiplicative functions are due to Laurincikas [1046]. For a result on divisors and Brownian motion we refer to Manstavicius [1121].

Kawai investigated digital sequences with respect to the Ostrowski expansion generalizing a result of Coquet [400]; see also Kopecek, Larcher, Tichy and Turnwald [950] for discrepancy bounds. In a series of papers Coquet [384, 387, 393, 390, 401, 402, 404, 406] and Coquet, Rhin and Toffin [413] proved the u.d. (or in some cases even the well distribution) of sequences of the types $P(n, s_q(n))$, $(xs_q(n) + ys(n))$, or of certain "digital" subsequences of polynomials, where $s(n)$ is the sum of digits function (or a related digital function) with respect to a given base sequence, P is a polynomial (in two variables) with at least one irrational coefficient ($\neq P(0,0)$), $s_q(n)$ is the usual q-ary sum of digits function, and x, y are real numbers not both rational; see also [403] and Liardet [1086]. Later these results were complemented by discrepancy bounds; see Tichy and Turnwald [1846, 1848], Larcher [1024, 1032], Larcher and Tichy [1042, 1043]. Grabner and Tichy [684] considered digital expansions with respect to (not necessarily finite) linear recurrences and obtained similar bounds for the discrepancy and also for the discrepancy with respect to well-distribution. For further results on digital sequences we refer to Bassily [110], Grabner [673, 677], Grabner and Tichy [682], Larcher [1027], Olivier [1393, 1394] and Radoux [1496]. Horbowicz [846] constructs a new class of sequences having the "substitution property" with respect to well distribution; this extends an earlier result of Coquet [393]. Mauduit and Rivat [1152] investigated the distribution of q-multiplicative functions in the sequence $([n^c]), c > 1$. For ergodic aspects of digital expansions see Grabner, Liardet and

Tichy [680], Mauduit and Mosse [1153] and Vershik and Sidorov [1909]. Digital problems related to diophantine equations and linear recurrences were studied by Pethö and Tichy [1430]. Barat [104] studied quasi-periodicity and automaticity of some quite general digital sequences. Allouche and Salon [29] considered polynomial subsequences of automatic sequences. Mendès France [1165, 1166] investigated sequences with empty spectrum; q-multiplicative functions and sequences with empty spectrum were studied by Coquet [379] and Coquet and Mendès France [411]; see also [402, 405]. For more on spectral properties of sequences we refer to the Notes of Section 1.6. Lesigne, Mauduit and Mosse [1061] studied ergodic properties of q-multiplicative sequences. Barat and Grabner [105] investigated the existence of an asymptotic distribution of q-additive sequences. Bertrand-Mathis [204, 205, 206] considered digital sequences related to Parry's β-expansion of real numbers. For automatic aspects concerning Parry's β-expansion we mention Berend and Frougny [180]. For results on digital expansions and Hausdorff dimension see to Bisbas [223], Jäger [871] and Pollicott and Simon [1441]. A special metric result concerning q-adic expansions is due to Erdős, M. Joo and I. Joo [564].

In a series of papers Levin [1067, 1069, 1068, 1070, 1071, 1074] investigated sequences of the type (ax^n). He mainly constructed such sequences, which are u.d. or even completely u.d. including proofs of discrepancy bounds. For instance, in [1074] he constructed a sequence (x^n) such that its discrepancy is $\mathcal{O}((\log N)^4 N^{-1/2})$ which is almost as good as the corresponding metric results; see Section 1.6. For related results we also refer to Levin and Shparlinskij [1078], Shparlinskij [1679] and Muhutdinov [1215]. A special distribution property of exponentially growing sequences was shown by Strzelecki [1758]. Bass [108] gave a construction of completely u.d. sequences which are used in the theory of pseudorandom functions. Strauch [1753] improved an integral inequality of Koksma involving sequences of the type $(a_n x)$, where (a_n) is an integer sequence.

Of special interest are sequences of the type (θ^n) for specific values of θ. For instance it is well known that such a sequence cannot be dense modulo 1 if θ is a Pisot number. We recall that a real algebraic integer θ is called a Pisot number if its absolute value is larger than 1 and all its conjugates are of absolute value less than 1; θ is called a Salem number if some of the conjugates may lie on the unit circle. For recent contributions on Pisot numbers and on Salem numbers (and generalizations) see Bertin [200, 201], Bertin and Boyd [197], Bertin and Pathiaux-Delefosse [203], Boyd [286, 291, 292] and Boyd and Parry [288]. For investigations of Pisot E-sequences ($e_n = N(e_{n-1}^2/e_{n-2})$) and related sequences of positive integers we refer to Boyd [290, 285, 287], Cantor [323] and Bertin [199, 198]. A quite recent monograph on Pisot and Salem numbers is due to Bertin, Decomps-Guilloux, Grandet-Hugot, Pathiaux-Delefosse and Schreiber [202]. Kőrnyei [955] considered linear recurring sequences with characteristic roots not contained in the interior of the unit disc and investigated the problem when the distance to the nearest integer tends to 0. Borel [261] studied certain properties of polynomials and explained the links with u.d. Kahane [885] reminds on the work of Pisot. Mendès France [1171] investigated relations between statistical independence of sequences and Pisot numbers. Mendès France [1170] and Coquet [383, 381] gave a characterization of Pisot numbers in terms of the u.d. of certain sequences, and in [1180] the first author established diophantine inequalities relating polynomials and exponentially growing sequences. Coquet and Mendès France [411] and Mendès France [1182] established diophantine inequalities related to certain statements on Pisot numbers. Mendès France [1168] investigated several questions concerning the set of all x such that $(x\theta^n)$ is u.d. mod 1. Drobot and McDonald considered polynomials p with integer coefficients which are bounded in absolute value by some constant M; it is shown that the set of values $(p(t))$ for Pisot numbers t with $1 < t < N+1$ is discrete.

It is a famous open problem to decide whether the sequence $\left(\left(\frac{3}{2}\right)^n\right)$ is u.d. mod 1. One open question concerning the $g(k)$ in Waring's problem can be reformulated in the following way: Can $\|\left(\frac{3}{2}\right)^k\| > \left(\frac{3}{4}\right)^k$ be false for $k \geq 5$? Beukers [216, 217] proved the lower bound $\|\left(\frac{3}{2}\right)^k\| > 2^{-\frac{9}{10}k}$ for $k > 5000$. For classical work on this problem we refer to Mahler [1118], Tijdeman [1852] and Baker and Coates [63]; for improvements see Bennett [165]. In a series of papers Choquet [356, 354, 353, 357, 358, 355, 359, 360] established further results on the distribution properties of sequences $\left(\left(\frac{3}{2}\right)^n\right)$ and $(k\theta^n)$, where $k > 0, \theta > 1$ are given constants. Mignotte [1192] proved that $\|\theta^n\| \leq \frac{1}{(\theta+2)^2}$ implies that θ is an integer. Pollington [1447] gave an effective procedure for the construction of numbers k with $\|k\left(\frac{3}{2}\right)^n\| > 0.108$ for all n; see also [1445]. It is also shown that the set of all real numbers k for which $(k\theta^n)$ is not dense mod 1 has Hausdorff dimension at least $\frac{1}{2}$. Flatto, Lagarias and Pollington [606] proved the remarkable inequality $\limsup_{n\to\infty}\{x(p/q)^n\} - \liminf_{n\to\infty}\{x(p/q)^n\} \geq 1/p$, where $p > q \geq 2$ are relatively prime integers and x is any positive real number. This implies that no interval of length $< 1/p$ can contain all but finitely many of the limit points mod 1 of the sequence $(x(p/q)^n)$. Related results are due to Bennett [166], who established effective lower bounds for the fractional parts of powers of a dense set of rationals.

1.5 Distribution Functions

1.5.1 Basic Concepts and Maldistribution

In this section we want to discuss some recent results on asymptotic distribution functions.

Definition 1.137 *The sequence* $(\mathbf{x}_n)_{n\geq 1}$ *in* $\mathbf{R}^k$ *is said to have the asymptotic distribution function mod 1 (abbreviated a.d.f. mod 1)* $g(\mathbf{x})$ *if*

$$\lim_{N\to\infty} \frac{A([0,\mathbf{x}),N,\mathbf{x}_n)}{N} = g(\mathbf{x}),$$

for every $\mathbf{x} \in [0,1)^k$.

The most important case is $k = 1$. Evidently, the function g on $[0,1]$ is non-decreasing with $g(0) = 0, g(1) = 1$. An arbitrary sequence (x_n) need not have an a.d.f. mod 1. But in any case the limits

$$\liminf_{N\to\infty} \frac{A([0,x),N,x_n)}{N} = \varphi(x),$$

$$\limsup_{N\to\infty} \frac{A([0,x),N,x_n)}{N} = \Phi(x),$$

for $0 \leq x \leq 1$ can be considered. The functions φ and Φ are non-decreasing with $\varphi(0) = \Phi(0) = 0, \varphi(1) = \Phi(1) = 1$ and $0 \leq \varphi(x) \leq \Phi(x) \leq 1$ for $0 \leq x \leq 1$. The function φ is called lower distribution function mod 1 of (x_n) and Φ upper distribution function mod 1. If $\varphi \equiv \Phi$, then the sequence has the a.d.f. mod 1 φ. If $\varphi(x) = \Phi(x) = x$ for $0 \leq x \leq 1$, then the sequence (x_n) is uniformly distributed mod 1.

Definition 1.138 *Let* $(\mathbf{x}_n)_{n\geq 1}$ *be a sequence in* $\mathbf{R}^k$ *and let* $N_1 < N_2 < \ldots$ *be an increasing sequence of positive integers such that*

$$\lim_{i\to\infty} \frac{A([0,\mathbf{x}),N_i,\mathbf{x}_n)}{N_i} = f(\mathbf{x})$$

for $\mathbf{x} \in [0,1]^k$. *Then* $f(\mathbf{x})$ *is called a distribution function mod 1 of* $(\mathbf{x}_n)_{n\geq 1}$. *If*

$$\lim_{i\to\infty} \frac{A([0,\mathbf{x}),N_i,\mathbf{x}_n)}{N_i} = f(\mathbf{x})$$

holds with $f(\mathbf{x}) = x^{(1)} \cdots x^{(k)}$ $(\mathbf{x} = (x^{(1)}, \ldots, x^{(k)}))$ *then the sequence* $(\mathbf{x}_n)_{n\geq 1}$ *is called almost uniformly distributed mod 1.*

Remark. It follows immediately from HELLY's selection principle that an arbitrary sequence (x_n) of real numbers has at least one distribution function mod 1., see KUIPERS and NIEDERREITER [983, p. 54] for the one-dimensional case and LOÈVE [1092].

Theorem 1.139 *(WEYL's Criterion) A sequence* $(\mathbf{x}_n)$ *has the continuous a.d.f. mod 1* $g(\mathbf{x})$ *if and only if for every real-valued continuous function* f *on* $[0,1]^k$,

$$\lim_{N\to\infty} \sum_{n=1}^{N} f(\{\mathbf{x}_n\}) = \int_{[0,1]^k} f(\mathbf{x}) dg(\mathbf{x}).$$

An equivalent condition is

$$\lim_{N\to\infty} \sum_{n=1}^{N} e^{2\pi i (\mathbf{h}, \mathbf{x}_n)} = \int_{[0,1]^k} e^{2\pi i (\mathbf{h}, \mathbf{x})} dg(\mathbf{x})$$

for all integer vectors $\mathbf{h} \neq \mathbf{0}$.

The one-dimensional case is proved in [983]. The proof of the general case runs along the same lines. Using a k-dimensional version of the HELLY-BRAY-lemma (see LOÈVE [1092]). The existence of a continuous a.d.f. mod 1 is answered by the famous WIENER-SCHOENBERG Theorem. In the following we sketch a k-dimensional version.

Theorem 1.140 *The sequence* $(\mathbf{x}_n)$ *has a continuous a.d.f. mod 1 if and only if for every* $\mathbf{h} \in \mathbf{Z}^k \setminus \{\mathbf{0}\}$ *the limit*

$$\omega_{\mathbf{h}} = \lim_{N\to\infty} \frac{1}{N} \sum_{n=1}^{N} e(\mathbf{h} \cdot \mathbf{x}_n)$$

exists, and in addition

$$\lim_{H_1,\ldots,H_k\to\infty} \frac{1}{H_1 \ldots H_k} \sum_{h_1=1}^{H_1} \cdots \sum_{h_k=1}^{H_k} |\omega_{\mathbf{h}}|^2 = 0.$$

Sketch Proof. The existence of the limits $\omega_{\mathbf{h}}$ is clearly necessary. By the WEYL criterion (Theorem 1.139) we have

$$\omega_{\mathbf{h}} = \int_{[0,1]^k} e(\mathbf{h} \cdot \mathbf{x}) \, dg(\mathbf{x}),$$

thus

$$\lim_{H_1,\ldots,H_k\to\infty} \frac{1}{H_1 \ldots H_k} \sum_{h_1=1}^{H_1} \cdots \sum_{h_k=1}^{H_k} |\omega_{\mathbf{h}}|^2 =$$

$$= \lim_{H_1,\ldots,H_k\to\infty} \frac{1}{H_1\ldots H_k} \sum_{h_1=1}^{H_1} \cdots \sum_{h_k=1}^{H_k} \int_{[0,1]^k} \int_{[0,1]^k} e(\mathbf{h}\cdot(\mathbf{x}-\mathbf{y}))\,dg(\mathbf{x})\,dg(\mathbf{y}) =$$

$$= \iint_{\{(\mathbf{x},\mathbf{y})\in[0,1]^{2k}\,|\,\mathbf{x}-\mathbf{y}\in\mathbf{Z}^k\}} dg(\mathbf{x})dg(\mathbf{y}).$$

The last integral is zero if and only if g is continuous. In particular, if $(\mathbf{x}_n)$ has a continuous a.d.f. mod 1, then the relation

$$\lim_{H_1,\ldots,H_k\to\infty} \frac{1}{H_1\ldots H_k} \sum_{h_1=1}^{H_1} \cdots \sum_{h_k=1}^{H_k} |w_{\mathbf{h}}|^2 = 0$$

follows. The converse direction follows as in the one-dimensional case from the RIESZ representation theorem, see [983, 1575]. □

Several properties of the set $G((\mathbf{x}_n))$ of distribution functions of a given sequence $(\mathbf{x}_n)$ were investigated in a series of fundamental papers [1891, 1892, 1893, 1894, 1895, 1896, 1897, 1898] by VAN DER CORPUT. More recently, MYERSON [1229] and STRAUCH [1756] obtained further results in that direction, especially on non-continuous distribution functions. These investigations were completed by GRABNER, TICHY and STRAUCH [681]. Let us mention three types of sequences $(\mathbf{x}_n)$, which were considered in [681] and [1229]. The first class is given by

$$G((\mathbf{x}_n)) = \{\chi_{\underline{\alpha}}(\mathbf{x}) : \underline{\alpha} \in [0,1]^k\}, \tag{1.266}$$

where $\chi_\alpha(\mathbf{x}) = \chi_{[\underline{\alpha},\mathbf{1}]}(\mathbf{x})$; $\mathbf{1} = (1,\ldots,1)$. The second class is defined by

$$\begin{aligned} G((\mathbf{x}_n)) = \{t\chi_{\underline{\alpha}}(\mathbf{x}) + (1-t)\chi_{\underline{\beta}}(\mathbf{x}) : t \in [0,1], \\ \underline{\alpha},\underline{\beta} \in [0,1]^k, \alpha_i \neq \beta_i \Longrightarrow \alpha_i = 1, \beta_i = 0, 1 \le i \le k\}. \end{aligned} \tag{1.267}$$

The third class was considered by MYERSON [1229]. It can be characterized by

$$G((\mathbf{x}_n)) \supset \{\chi_{\underline{\alpha}}(\mathbf{x}) : \mathbf{x} \in [0,1]^k\}. \tag{1.268}$$

MYERSON calls a sequence (x_n) of real numbers *maldistributed* if for any interval $J \subset [0,1)$

$$\limsup_{N\to\infty} \frac{A(J,N,x_n)}{N} = 1,$$

which is equivalent to (1.268).

Proposition 1.141 *Let $(M_i)_{i\ge1}$ be a sequence of positive numbers satisfying*

$$\lim_{L\to\infty} \sum_{i=1}^{L-1} \frac{M_i}{M_L} = 0.$$

For a given sequence $(\mathbf{y}_m)_{m\geq 1}$ *in* $[0,1)^k$, *we construct the sequence* $(\mathbf{x}_n)_{n\geq 1}$ *in* $[0,1)^k$ *by* $\mathbf{x}_n = \mathbf{y}_m$ *for*

$$\sum_{i=1}^{m-1} M_i \leq n < \sum_{i=1}^{m} M_i.$$

Finally, let $H_{(\mathbf{y}_m)} \subseteq [0,1]^k \times [0,1]^k$ *denote the set of all limit points of the sequence* $((\mathbf{y}_{m-1}, \mathbf{y}_m))_{m\geq 2}$. *Then*

$$G((\mathbf{x}_n)) = \{t\chi_{\underline{\alpha}}(\mathbf{x}) + (1-t)\chi_{\underline{\beta}}(\mathbf{x}) \mid t \in [0,1], (\underline{\alpha}, \underline{\beta}) \in H_\sigma\}.$$

Proof. For

$$N = \sum_{l=1}^{L-1} M_l + \theta_L M_L, \qquad 0 \leq \theta_L < 1,$$

we have

$$A([\mathbf{0},\mathbf{x}), N, \mathbf{x}_n) = o(M_{L-1}) + \begin{cases} M_{L-1} & \text{if } \mathbf{y}_{L-1} \in [\mathbf{0},\mathbf{x}) \text{ and } \mathbf{y}_L \notin [\mathbf{0},\mathbf{x}), \\ \theta_L M_L & \text{if } \mathbf{y}_{L-1} \notin [\mathbf{0},\mathbf{x}) \text{ and } \mathbf{y}_L \in [\mathbf{0},\mathbf{x}), \\ M_{L-1} + \theta_L M_L & \text{if } \mathbf{y}_{L-1}, \mathbf{y}_L \in [\mathbf{0},\mathbf{x}). \end{cases}$$

Assuming that $A([\mathbf{0},\mathbf{x})N_i, \mathbf{x}_n)/N_i \to g(\mathbf{x})$ for selected sequences of indices

$$N_i = \sum_{l=1}^{L_i-1} M_l + \theta_L M_L,$$

$L_i = L(N_i)$ we can choose a subsequence $(N_{i_j})_{j\geq 1}$ such that

$$\lim_{j\to\infty} (\mathbf{y}_{L_{i_j}-1}, \mathbf{y}_{L_{i_j}}) = (\underline{\alpha}, \underline{\beta})$$

and

$$\lim_{j\to\infty} \frac{M_{L_{i_j}-1}}{M_{L_{i_j}-1} + \theta_{L_{i_j}} M_{L_{i_j}}} = t$$

for some $\underline{\alpha}, \underline{\beta} \in H_{(\mathbf{y}_m)}$ and $t \in [0,1]$. Thus

$$g(\mathbf{x}) = t\chi_{\underline{\alpha}}(\mathbf{x}) + (1-t)\chi_{\underline{\beta}}(\mathbf{x}).$$

Conversely, we can easily construct an index sequence

$$N_i = \sum_{l=1}^{L_i-1} M_l + \theta_{L_i} M_{L_i}$$

satisfying

$$\lim_{i\to\infty} \frac{M_{L_i-1}}{M_{L_i-1} + \theta_{L_i} M_{L_i}} = t \in [0,1],$$

provided that $(\mathbf{y}_{L_i-1}, \mathbf{y}_{L_i}) \to (\underline{\alpha}, \underline{\beta})$. Thus the proof is complete. □

Corollary 1.142 *For a given set $H \subseteq [0,1]^k$ we suppose that there exists a sequence $(\mathbf{y}_n)_{n\geq 1}$ in $[0,1]^k$ such that*
(i) H coincides with the set of limit points of $(\mathbf{y}_n)_{n\geq 1}$,
(ii) $\lim_{n\to\infty}(\mathbf{y}_n - \mathbf{y}_{n-1}) = \mathbf{0}$.

Then there exists a sequence $(\mathbf{x}_n)_{n\geq 1}$ in $[0,1)^k$ such that

$$G((\mathbf{x}_n)) = \{\chi_{\underline{\alpha}}(\mathbf{x}) : \underline{\alpha} \in H\}.$$

Further properties on the set $G((\mathbf{x}_n))$ can be found in GRABNER, TICHY and STRAUCH [681]. In the following subsection we discuss recent results on distribution functions and linear sequences.

1.5.2 Distribution Functions and Recurring Sequences

First we discuss distribution properties of the quotients of successive terms of second order linear recurring sequences; in prinicipal we follow here KISS and TICHY [934]. Let us note some facts concerning the zero terms of a second order linear recurring sequence $(u_n)_{n\geq 0}$, $u_n = Au_{n-1}+Bu_{n-2}$, with negative discriminant $D = A^2+4B < 0$. Obviously, u_n is given by

$$u_n = aw_1{}^n + bw_2{}^n, \tag{1.269}$$

where w_1, w_2, and a, b are complex conjugate numbers. We have

$$w_1 = re^{i\pi\vartheta}, \quad w_2 = re^{-i\pi\vartheta} \tag{1.270}$$

and

$$a = r_1 e^{i\pi\omega}, \quad b = r_1 e^{-i\pi\omega}, \tag{1.271}$$

with

$$0 < \vartheta = \frac{1}{\pi}\arctan\frac{\sqrt{-D}}{A} < 1,$$

$$\omega = \frac{1}{\pi}\arctan\frac{Au_0 - 2u_1}{u_0\sqrt{-D}},$$

and r, r_1 are positive numbers. We say (u_n) is a non-degenerate sequence if w_1/w_2 is not a root of unity (i.e. if ϑ is an irrational number).

If $u_n = u_m = 0$ for some $n < m$, then by (1.269) we obtain

$$\omega + n\pi\vartheta = \frac{2k_1+1}{2}$$

and

$$\omega + m\pi\vartheta = \frac{2k_2+1}{2}$$

with some integers k_1 and k_2: Since these equations do not hold simultaneously for an irrational number ϑ, a non-degenerate sequence (u_n) contains at most one zero term. Thus there exists an integer $c = c(u_n)$ depending on the sequence (u_n) such that $u_n \neq 0$ for $n \geq c$. The elements with indices $n < c$ are not relevant for the asymptotic distribution behaviour of $(\{u_{n+1}/u_n\})$; hence without loss of generality we assume in the following that $u_n \neq 0$ for $n > 0$.

Theorem 1.143 *Let $(u_n)_{n\geq 0}$ be a linear recurring sequence defined by $u_n = Au_{n-1} + Bu_{n-2}$ with non-zero real coefficients A, B, real initial values u_0, u_1 and negative discriminant $D = A^2 + 4B$. If the number*

$$\vartheta = \frac{1}{\pi}\arctan\frac{\sqrt{-D}}{A}$$

is irrational, then the asymptotic distribution function modulo 1 of the sequence $(u_{n+1}/u_n)_{n=1}^{\infty}$ is given by $F(x) = F_1(x - \{A/2\}) - F_1(-\{A/2\})$ with

$$F_1(x) = x + \frac{1}{\pi}\arctan\frac{\sin 2\pi x}{\exp(\pi\sqrt{-D}) - \cos 2\pi x}. \tag{1.272}$$

Furthermore the following estimate holds (for sufficiently large N):

$$\sup_{0\leq x\leq 1}\left|\frac{1}{N}\sum_{n=1}^{N}\chi_{[0,x)}(\{u_{n+1}/u_n\}) - F(x)\right| \leq 2\sqrt{2}(-D)^{1/4}\cdot\sqrt{D_N} + 6D_N, \tag{1.273}$$

where $D_N = D_N(\vartheta n)$ denotes the discrepancy of the sequence $(\vartheta n)_{n\geq 1}$.

Remark. An easy calculation shows that the derivative of $F(x)$ is given by ($E = \exp(\pi\sqrt{-D})$):

$$F'(x) = 1 + 2\frac{E\cos 2\pi(x - \{A/2\}) - 1}{E^2 - 2E\cos 2\pi(x - \{A/2\}) + 1} \tag{1.274}$$

Thus the graph of $F(x)$ is steepest at $x = \{A/2\}$ corresponding to $\{u_{n+1}/u_n\}$ favouring values near this point.

Since the asymptotic distribution function $F(x)$ in Theorem 1.143 is strictly increasing and different from x, we obtain

Corollary 1.144 *Let $(u_n)_{n\geq 0}$ be a second order linear recurring sequence as considered in Theorem 1.143. If ϑ is an irrational number, then the sequence of fractional parts $(\{u_{n+1}/u_n\})_{n\geq 1}$ is everywhere dense in $[0,1]$ but not uniformly distributed.*

Using Corollary 1.64 another consequence follows immediately.

Corollary 1.145 *Let $G = (u_n)_{n=0}^{\infty}$ be a second order linear recurring sequence as considered in Theorem 1.143. If ϑ is an irrational number such that the partial quotients of its continued fraction expansion are bounded by K, then*

$$\sup_{0\leq x\leq 1}\left|\frac{1}{N}\sum_{n=1}^{N}\chi_{[0,x)}(\{u_{n+1}/u_n\}) - F(x)\right| \leq C\sqrt{\frac{\log N}{N}}$$

with a constant C only depending on K and the discriminant D of G.

Let us consider sequences G defined by rational constants A, B, u_0 and u_1. Using the above notations, PETHŐ [1429] has shown that in this case ϑ is a rational number (or w_1/w_2 is a root of unity) if and only if $A^2 = -kB$, where $k = 1, 2, 3$ or 4. Therefore by Theorem 1.143 we derive:

Corollary 1.146 *Let $(u_n)_{n=0}^{\infty}$ be a second order linear recurring sequence defined by non-zero rational constants A, B, u_0 and u_1 with $D = A^2 + 4B < 0$. If $A^2 \neq -kB$ for $k = 1, 2, 3$ and 4, then the asymptotic distribution function modulo 1 of the sequence $(u_{n+1}/u_n)_{n=1}^{\infty}$ is given by 1.272 and fulfills the estimate 1.273.*

Proof of Theorem 1.143 Let (u_n) be a second order linear recurring sequence satisfying the conditions of the Theorem 1.143. As we have seen above, we may suppose that $u_n \neq 0$ for $n > 0$. Because of (1.269), (1.270) and (1.271) we obtain for all $n \geq 1$:

$$\begin{aligned}
\frac{u_{n+1}}{u_n} &= \frac{r_1 r^{n+1}\exp(i\pi(\omega+(n+1)\vartheta)) + r_1 r^{n+1}\exp(-i\pi(\omega+(n+1)\vartheta))}{r_1 r^n \exp(i\pi(\omega+(n\vartheta)) + r_1 r^n \exp(-i\pi(\omega+n\vartheta))} \\
&= r\cdot\frac{\cos\pi(\omega+(n+1)\vartheta)}{\cos\pi(\omega+n\vartheta)} = r(\cos\pi\vartheta - \sin\pi\vartheta\tan\pi(n\vartheta+\omega)) \\
&= c + d\cdot\tan(\pi(n\vartheta+\omega)),
\end{aligned}$$

where $c = A/2 \neq 0$ and $d = -\frac{1}{2}\sqrt{-D} < 0$ are real numbers not depending on n. Now we have for an arbitrary ε with $0 < \varepsilon < \frac{1}{2}$:

$$\Phi(y) = \begin{cases} \tan\pi y & \text{if } \{y\} \leq \frac{1}{2} - \varepsilon \text{ or } \{y\} \geq \frac{1}{2} + \varepsilon \\ -c/d & \text{if } \frac{1}{2} - \varepsilon < \{y\} < \frac{1}{2} + \varepsilon. \end{cases} \tag{1.275}$$

Furthermore we put

$$\begin{aligned}
\psi_1(x,y) &= \chi_{[0,x)}(\{c + d\tan(\pi(y+\omega))\}), \\
\psi_2(x,y) &= \chi_{[0,x)}(\{c + d\Phi(\pi(y+\omega))\}).
\end{aligned} \tag{1.276}$$

Then we get

$$\frac{1}{N}\sum_{n=1}^{N}(\psi_1(x,\vartheta n) - \psi_2(x,\vartheta n)) = 2\delta_1\varepsilon + 2\delta_2 D_N, \tag{1.277}$$

where $D_N = D_N(\vartheta n)$ is the discrepancy of the sequence $\{\vartheta n\}$ and $-1 \leq \delta_1 \leq 0$, $|\delta_2| \leq 1$. From (1.275) and (1.276) we also obtain

$$\int_0^1 (\psi_1(x,y) - \psi_2(x,y))dy = 2\delta_3\varepsilon \tag{1.278}$$

with $0 \leq \delta_3 \leq 1$. Hence we get

$$\begin{aligned}
\frac{1}{N}\sum_{n=1}^{N}\psi_1(x,n\vartheta) - \int_0^1 \psi_1(x,y)dy &= \frac{1}{N}\sum_{n=1}^{N}\psi_2(x,n\vartheta) - \int_0^1 \psi_2(x,y)dy \\
&+ 2\delta_4\varepsilon + 2\delta_2 D_N,
\end{aligned} \tag{1.279}$$

where $|\delta_4| \leq 1$. The function $y \mapsto \psi_2(x,y)$ (for fixed x) is of bounded variation V on $[0,1]$ satisfying the estimate

$$V \leq 4\left(\frac{|d|}{2\varepsilon} + 1\right). \tag{1.280}$$

This estimate follows from the fact that ψ_2 is constant (0 or 1) with the exception of a finite number of discontinuity points, and the number of discontinuity points is bounded by

$$4\left(1+|d|\tan\left(\frac{\pi}{2}-\varepsilon\pi\right)\right)\leq 4\left(1+\frac{|d|}{\sin\pi\varepsilon}\right)\leq 4\left(1+\frac{|d|}{2\varepsilon}\right).$$

An application of KOKSMA-HLAWKA's inequality (Theorem 1.14) yields

$$\left|\frac{1}{N}\sum_{n=1}^{N}\psi_2(x,n\vartheta)-\int_0^1\psi_2(x,y)dy\right|\leq V\cdot D_N\leq 4D_N\left(\frac{|d|}{2\varepsilon}+1\right). \tag{1.281}$$

Combining (1.279) and (1.281) we derive for any ε $(0<\varepsilon<\frac{1}{2})$

$$\left|\frac{1}{N}\sum_{n=1}^{N}\psi_1(x,n\vartheta)-\int_0^1\psi_1(x,y)dy\right|\leq D_N\left(\frac{2|d|}{\varepsilon}+6\right)+2\varepsilon. \tag{1.282}$$

Now we set $\varepsilon=(|d|D_N)^{1/2}$; since $\{n\vartheta\}$ is uniformly distributed, D_N tends to 0 and so we can assume N so large that $\varepsilon<\frac{1}{2}$. Hence we have because of $|d|=\sqrt{-D/2}$

$$\left|\frac{1}{N}\sum_{n=1}^{N}\psi_1(x,n\vartheta)-\int_0^1\psi_1(x,y)dy\right|\leq\frac{4(-D)^{1/4}}{\sqrt{2}}D_N^{1/2}+6D_N. \tag{1.283}$$

As D_N tends to 0 (for $N\to\infty$) the distribution function of $\{u_{n+1}/u_n\}$ is given by

$$F(x)=\int_0^1\psi_1(x,y)dy=\int_{-1/2}^{1/2}\chi_{[0,x)}(\{c+d\tan(\pi y)\})dy. \tag{1.284}$$

In the following we shall compute the integral (1.284) in the case $c=0$. By the substitution $u=d\tan(\pi y)$ we get

$$F_1(x)=\frac{|d|}{\pi}\int_{-\infty}^{\infty}\frac{\chi_{[0,x)}(u)}{d^2+u^2}du. \tag{1.285}$$

We use the FOURIER series expansion of the characteristic function

$$\chi_{[0,x)}(u)=x+\frac{1}{\pi}\sum_{m=1}^{\infty}\frac{\sin 2\pi mx}{m}\cos 2\pi mu+\frac{1}{\pi}\sum_{m=1}^{\infty}\frac{1-\cos 2\pi mx}{m}\sin 2\pi mu$$

and swap summation and integration applying LEBESGUE's theorem on dominated convergence. By the integral forumulae

$$\begin{aligned}\int_{-\infty}^{\infty}\frac{\cos 2\pi mu}{d^2+u^2}du &= \frac{\pi}{|d|}\cdot e^{-2\pi m|d|},\\ \int_{-\infty}^{\infty}\frac{\sin 2\pi mu}{d^2+u^2}du &= 0\end{aligned}$$

we obtain

$$F_1(x) = x + \frac{1}{\pi} \sum_{m=1}^{\infty} \frac{\sin 2\pi m x}{m} e^{2\pi m d} = x + \frac{1}{\pi} \cdot \mathrm{Im} \sum_{m=1}^{\infty} \frac{w^m}{m}$$

with $w = \exp(2\pi(d + ix))$. Since $d < 0$ we have $|w| < 1$ and $Re(1 - w) > 0$, so

$$\sum_{m=1}^{\infty} \frac{w^m}{m} = -\log(1 - w).$$

From $\Re(1 - w) = 1 + \exp(2\pi d) \cdot \cos(2\pi x)$ and $\Im(1 - w) = \exp(2\pi d) \cdot \sin(2\pi x)$ it immediately follows that

$$F_1(x) = x + \frac{1}{\pi} \arctan \frac{e^{2\pi d} \sin(2\pi x)}{1 - e^{2\pi d} \cos(2\pi x)}.$$

Since $F(x) = F_1(x - c) - F_1(-c)$, the proof is complete. □

We now turn to the more general case of linear recurring sequences of higher order. Let $(u_n)_{n \geq 0}$ be a k-th order linear recurring sequence with real coefficients c_i ($c_0 \neq 0$) defined by

$$u_{n+k} = c_{k-1} u_{n+k-1} + \cdots + c_0 u_n\,, \quad n \geq 0 \tag{1.286}$$

and real initial values $u_0, \ldots, u_{k-1}$ (not all equal to 0). Furthermore assume that $u_n \neq 0$ for sufficiently large n and that none of the quotients w_i/w_j of the distinct roots of the characteristic polynomial

$$z^k - c_{k-1} z^{k-1} - \cdots - c_0 = \prod_{\ell=1}^{s} (z - w_\ell)^{\sigma_\ell} \tag{1.287}$$

is a root of unity. It is well known that there is a unique representation of u_n in the form

$$u_n = \sum_{\ell=1}^{s} P_\ell(n) w_\ell^n, \tag{1.288}$$

where $P_\ell(n)$ is a polynomial of degree $\leq \sigma_\ell - 1$. Without proof we state the following generalization of Theorem 1.143; for details we refer to GOLDSTERN, TICHY and TURNWALD [665].

Theorem 1.147 *Assume that (u_n) satisfies a linear recurrence of type (1.286), but no linear recurrence of order $< k$. Without loss of generality we assume $|w_1| = \ldots = |w_q| = r$, $\sigma_1 = \cdots = \sigma_q = \sigma$ and $|w_j| < r$ or $|w_j| = r$ and $\sigma_j < \sigma$ for $j > q$. Let a_ℓ denote the leading coefficient of $P_\ell(n)$ in (1.288) and define*

$$u'_n = \sum_{\ell=1}^{q} a_\ell w_\ell^n.$$

Furthermore we suppose that $\bar{w}_1 = w_{p+1}, \ldots, \bar{w}_p = w_{2p}$ *and that* w_ℓ *is real for* $2p < \ell \le q$. *Assume that* $1, \vartheta_1, \ldots, \vartheta_p$ *are linearly independent over the rationals, where* ϑ_ℓ *is defined by* $w_\ell = r\exp(2\pi i \vartheta_\ell)$, $0 \le \vartheta_\ell < 1$. *If* $u_n, v_n \ne 0$ *for sufficiently large* n, *then* $x_n = \frac{u_{n+1}}{u_n}$ *and* $x'_n = \frac{u'_{n+1}}{u'_n}$ *have the same asymptotic distribution function* $F(t)$, *which is given by*

$$F(t) =$$

$$\frac{1}{2}\int_{[0,1]^p} \chi_{[0,t)}\left(\left\{r\frac{\sum_{j=1}^p 2|a_j|\cos(2\pi x_j) + b + c}{\sum_{j=1}^p 2|a_j|\cos(2\pi(x_j - \vartheta_j)) - b + c}\right\}\right) dx_1 \cdots dx_p +$$

$$\frac{1}{2}\int_{[0,1]^p} \chi_{[0,t)}\left(\left\{r\frac{\sum_{j=1}^p 2|a_j|\cos(2\pi(x_j + \vartheta_j)) - b + c}{\sum_{j=1}^p 2|a_j|\cos(2\pi x_j) + b + c}\right\}\right) dx_1 \cdots dx_p, \quad (1.289)$$

where b, c *are defined via*

$$\frac{u'_n}{r^n} = 2|a_j|\cos(2\pi(n\vartheta_j + \alpha_j)) + b(-1)^n + c,$$

in which α_j *is given by* $a_j = |a_j|e(\alpha_j)$.

For a proof and estimates for the error see GOLDSTERN, TICHY and TURNWALD [665].

In the sequel we want to consider more general linear recurring sequences (v_n) of the form

$$v_{n+k} = c_{k-1}(n)v_{n+k-1} + \cdots + c_0(n)v_n \qquad (n \ge 0), \qquad (1.290)$$

where the coefficients are functions of n obtained by slight perturbations of constants:

$$c_j(n) = c_j + \mathcal{O}\left(n^{-\lambda}\right) \qquad (1.291)$$

with $\lambda > 1$. As above we assume that $v_n \ne 0$ for sufficiently large n. In [1843] TICHY proved that for second order linear recurring sequences the distribution function $F(t)$ is stable under such perturbations, i.e. in the case of recurrences of type (1.290) (with $k = 2$) the distribution function is the same as in the case of constant coefficients c_j. The general case is tackled in GRABNER, TICHY and WINKLER [686]. The main tool for the proof is a recent approach of KOOMAN [948] which was originally used for generalizing classical results of PERRON and POINCARÉ; see also TIJDEMAN and KOOMAN [949]. In the following we give a sketch for the proof of this general stability result concerning k-th order linear recurring sequences.

Theorem 1.148 *Let*

$$v_{n+k} = c_{k-1}(n)v_{n+k-1} + \cdots + c_0(n)v_n \qquad (n \ge 0)$$

be a linear recurring sequence as above and $c_j(n) = c_j + \mathcal{O}(n^{-\lambda})$ *for* $j = 0, \ldots, k-1$. *Without loss of generality we assume* $|w_1| = \ldots = |w_q| = r$, $\sigma_1 = \cdots = \sigma_q = \sigma$ *and* $|w_j| < r$ *or* $|w_j| = r$ *and* $\sigma_j < \sigma$ *for* $j > q$. *Furthermore we suppose that* $\bar{w}_1 =$

$w_{p+1}, \dots, \bar{w}_p = w_{2p}$ and that w_ℓ is real for $2p < \ell \le q$. Assume that $1, \vartheta_1, \dots, \vartheta_p$ are linearly independent over the rationals, where ϑ_ℓ is defined by $w_\ell = re(\vartheta_\ell)$, $0 \le \vartheta_\ell < 1$.

If $\lambda > \sigma$, $v_n \ne 0$ for sufficiently large n, and if $v_n = \Omega(r^n \sigma^{n-1})$ then $\frac{v_{n+1}}{v_n}$ has the asymptotic distribution function $F(t)$ stated in Theorem 1.147.

Without loss of generality we assume that $v_n \ne 0$ for $n \ge 0$ as well as $u_n \ne 0$ for $n \ge 0$, where u_n is the corresponding recurring sequence with constant coefficients and suitable initial values (precisely defined in the proof). As mentioned above the main tool is the following proposition due to KOOMAN [948].

Proposition 1.149 *Suppose that (v_n) is a linear recurring sequence as above, where the coefficients $c_j(n)$ of the recurrence (1.290) satisfy (1.291) with $c_0 \ne 0$ and $\lambda > \sigma$. Let α be a characteristic root of (1.287) with multiplicity λ. Then (1.290) has σ linearly independent solutions $v_n^{(1)}, \dots, v_n^{(\sigma)}$ such that*

$$\lim_{n\to\infty} \frac{v_n^{(i)}}{\alpha^n n^{i-1}} = 1$$

for $i = 1, \dots, \sigma$.

The basic idea of the proof of Theorem 1.148 is to show that the sequence (v_{n+1}/v_n) does not differ too much from the sequence $(\frac{u'_{n+1}}{u'_n})$ except for some (few) points.

Lemma 1.150 *Let v_n be a solution of (1.290) with $v_n = \Omega(r^n n^{\sigma-1})$. Then there exists a solution u_n of (1.286) such that $v_n - u_n = \mathcal{O}(r^n n^{2\sigma-\lambda-1})$, where $\lambda > \sigma$ is defined in Theorem 1.148.*

Proof. Notice first that Proposition 1.149 yields the existence of a set of linearly independent solutions of (1.290) which are asymptotically equivalent to the solutions $w_j^n n^t$, $t = 0, \dots, \sigma_j - 1$ of (1.286). Therefore it suffices to prove an estimate for the difference of two solutions.

Let now $d_n = v_n - u_n$. It is an easy observation that d_n satisfies a recurrence

$$d_n = c_{k-1} d_{n-1} + \dots + c_0 d_{n-k} + f(n), \tag{1.292}$$

where $f(n) = (c_{k-1}(n) - c_{k-1}) v_{n-1} + \dots + (c_0(n) - c_0) v_{n-k} = \mathcal{O}(r^n n^{\sigma-\lambda-1})$. We have to prove the existence of a solution $d_n = \mathcal{O}(r^n n^{2\sigma-\lambda-1})$. Observe now that one partial solution of (1.292) is given by a suitable linear combination of terms

$$g_{j,t}(n) = \sum_{\ell=0}^{n} (n-\ell)^t w_j^{n-\ell} f(\ell), \tag{1.293}$$

which can again be written as a linear combination of terms

$$w_j^n n^{t-\kappa} \sum_{\ell=0}^{n} \ell^\kappa w_j^{-\ell} f(\ell).$$

By our assumptions on λ we obtain that this last sum converges for $n \to \infty$, $|w_j| = r$ and $\kappa < \lambda - \sigma$ yielding a term $C_{j,t} w_j^n n^t + \mathcal{O}(r^n n^{t+\sigma-\lambda})$. For $|w_j| = r$ and values of κ, for which the infinite sum does not converge, it yields a $\mathcal{O}(r^n n^{t+\sigma-\lambda})$ contribution. For all the other roots $|w_j| < r$ the sum is a $\mathcal{O}(r^n n^{\sigma-\lambda})$. Thus we have found a partial solution of (1.292) whose main term is a linear combination of $w_j^n n^{\sigma-1}$ and whose other terms are $O(r^n n^{2\sigma-\lambda-1})$. Observing that the main terms are solutions of the homogeneous equation yields the proof of the lemma. □

Remark The proof of Lemma 1.150 shows that more asymptotic information for v_n can be obtained, if λ is large. It is also clear how the result has to be changed, if $v_n = \mathcal{O}(r^n n^\tau)$ with $\tau < \sigma - 1$ (this is the case of a non-generic solution of the recurrence).

We will also use the following two observations whose proofs can be found in GOLDSTERN, TICHY and TURNWALD [665]

Lemma 1.151 *Let v_n be a solution of (1.290) with $v_n = \Omega(r^n n^{\sigma-1})$ and u_n a solution of (1.286) whose existence is stated in Lemma 1.150. Then the sequence $U(n) = u_n r^{-n} n^{-\sigma+1}$ has an a.d.f. mod 1 $G(t)$ which satisfies a* LIPSCHITZ *condition of the form $|G(t_2) - G(t_1)| \le L|t_2 - t_1|^{1/2}$.*

Lemma 1.152 *Under the same assumptions as in Theorem 1.148. Then the function $F(t)$ (given in (1.289)) satisfies a* LIPSCHITZ *condition of the form $|F(t_2) - F(t_1)| \le K|t_2 - t_1|^{1/5}$.*

Lemma 1.153 *The sequences (v_{n+1}/v_n) and (u_{n+1}/u_n) have the same a.d.f., where u_n is the sequence, whose existence was stated in Lemma 1.150*

Proof. Let d_n be the sequence used in the proof of Lemma 1.150. Setting $\alpha = 2\sigma - \lambda - 1$ we have

$$\left|\frac{v_{n+1}}{v_n} - \frac{u_{n+1}}{u_n}\right| = \left|\frac{v_{n+1} d_n - v_n d_{n+1}}{v_n u_n}\right| =$$

$$r\left|\frac{\left(U(n+1)\frac{(n+1)^{\sigma-1}}{n^{\alpha}} + D(n+1)\right) D(n) - \left(U(n)n^{\lambda-\sigma} + D(n)\right) D(n+1)\left(\frac{n+1}{n}\right)^{\alpha}}{\left(U(n)n^{\lambda-\sigma} + D(n)\right) U(n)}\right|$$

$$\le \frac{K}{|U(n)n^{\lambda-\sigma} + D(n)| \cdot |U(n)|}, \tag{1.294}$$

where $U(n)$ and $D(n)$ are bounded functions given by $d_n = D(n) r^n n^{2\sigma-\lambda-1}$ and $u_n = U(n) r^n n^{\sigma-1}$. We can write

$$\delta_n = \left|\frac{v_{n+1}}{v_n} - \frac{u_{n+1}}{u_n}\right| = n^{\sigma-\lambda} \frac{K(n)}{|(U(n) + D(n)n^{\sigma-\lambda})U(n)|},$$

in which $|U(n)|$ and $|K(n)|$ are bounded by K. This shows that v_{n+1}/v_n and u_{n+1}/u_n have the same distribution function, if $U(n)$ takes "small" values only on a "thin" subset of the integers.

It remains to specify what "small" and "thin" should mean. For this purpose we analyze the quantity

$$\begin{aligned}\Phi(N,\delta) &= |\{1 \le n \le N \mid \delta_n \ge \delta\}| \\ &= \left|\left\{1 \le n \le N \;\middle|\; |U(n)| \cdot |U(n) + D(n)n^{\sigma-\lambda}| \le \frac{K(n)n^{\sigma-\lambda}}{\delta}\right\}\right|.\end{aligned}$$

Since $|U(n)/2| > |D(n)|n^{\sigma-\lambda}$ and $U(n)^2/2 > |K(n)|n^{\sigma-\lambda}/\delta$ imply

$$|U(n)| \cdot |U(n) + D(n)n^{\sigma-\lambda}| > \left|\frac{U(n)}{2}\right| \cdot |U(n)| > \frac{K(n)n^{\sigma-\lambda}}{\delta},$$

we obtain for $0 < \delta < \min(\delta, 1/(2K))$ and for any $0 < \gamma < 1$

$$\begin{aligned}&\Phi(N,\delta) \\ &\le \left|\left\{1 \le n \le N : \left|\frac{U(n)}{2}\right| \le |D(n)|n^{\sigma-\lambda} \text{ or } \left|\frac{U(n)}{2}\right| \cdot |U(n)| \le \frac{|K(n)|n^{\sigma-\lambda}}{\delta}\right\}\right| \\ &\le \left|\left\{1 \le n \le N : |U(n)| \le \sqrt{\frac{2Kn^{\sigma-\lambda}}{\delta}}\right\}\right| \\ &\le N^\gamma + \left|\left\{1 \le n \le N : |U(n)| \le \sqrt{\frac{2K}{\delta N^{\gamma(\lambda-\sigma)}}}\right\}\right|. \qquad (1.295)\end{aligned}$$

By Lemma 1.151

$$\left|\left\{1 \le n \le N : |U(n)| \le \sqrt{\frac{2K}{\delta N^{\gamma(\lambda-\sigma)}}}\right\}\right| \le N\left(2D_N'' + L\left(\frac{2C}{\delta N^{\gamma(\lambda-\sigma)}}\right)^{\frac{1}{4}}\right) \qquad (1.296)$$

for some constant $L > 0$, where D_N'' denotes the discrepancy of the sequence $u_n n^{-\sigma+1} r^{-n}$ with respect to its distribution function $G(t)$. By Lemma 1.151 this discrepancy tends to 0.

We again use information on the distribution of (u_{n+1}/u_n) to derive

$$\left|A\left(N, \left(\frac{v_{n+1}}{v_n}\right), t\right) - A\left(N, \left(\frac{u_{n+1}}{u_n}\right), t\right)\right| \le \Phi(N,\delta) + 2ND_N' + 2NK\delta^{\frac{1}{5}}, \quad (1.297)$$

where D_N' denotes the discrepancy of (u_{n+1}/u_n) with respect to its distribution function $G(t)$ and the last term originates from the LIPSCHITZ condition $|F(t_1) - F(t_2)| \le K|t_1 - t_2|^{\frac{1}{5}}$ (cf. Lemma 1.152). Combining (1.295), (1.296) and (1.297) yields

$$\begin{aligned}&\left|A\left(N, \left(\frac{v_{n+1}}{v_n}\right), t\right) - A\left(N, \left(\frac{u_{n+1}}{u_n}\right), t\right)\right| \\ &\qquad \le N^\gamma + 2ND_N'' + NL\left(\frac{2C}{\delta N^{\gamma(\lambda-\sigma)}}\right)^{\frac{1}{4}} + 2ND_N' + 2KN\delta^{\frac{1}{5}}.\end{aligned}$$

Choosing $\delta = N^{-\frac{5(\lambda-\sigma)}{9+\lambda-\sigma}}$ and $\gamma = \dfrac{9}{9+\lambda-\sigma}$ yields

$$\left| A\left(N, \left(\frac{v_{n+1}}{v_n}\right), t\right) - A\left(N, \left(\frac{u_{n+1}}{u_n}\right), t\right) \right| \\ \leq (1 + L(2C)^{\frac{1}{4}} + 2K)N^{\gamma} + 2N\left(D_N'' + D_N'\right).$$

Since the right hand side is $o(N)$ the desired result is proved. □

1.5.3 Subdivisions

Let $(R_n)_{n\geq 1}$ and $(x_n)_{n\geq 1}$ be increasing and unbounded sequences of real numbers with $R_1 = 0$. Let (i_n), $n = 1, 2, \dots$ be the sequence of positive integers defined by

$$R_{i_n - 1} \leq x_n < R_{i_n}.$$

Define a sequence $(\Delta_R x_n)_{n\geq 1}$ by

$$\Delta_R x_n = \frac{x_n - R_{i_n-1}}{R_{i_n} - R_{i_n-1}}.$$

The sequence (x_n) is said to be uniformly distributed modulo (R_n) if the sequence $\Delta_R x_n$ is uniformly distributed mod 1, i.e.

$$\lim_{N\to\infty} \frac{A([0,x), N, \Delta_R x_n)}{N} = x$$

for $0 \leq x \leq 1$. In the case $R_n = n$ we get the usual uniform distribution. The uniform distribution modulo a subdivision (R_n) of the interval $[0, \infty)$ was introduced by LE VEQUE [1065]. Later on the sequence $(x_n) = (n\theta)$, θ a positive irrational was investigated modulo slowly increasing sequences (R_n), c.f. DAVENPORT and LEVEQUE [433] and DAVENPORT and ERDŐS [431]. Another type of results was obtained by KISS [930] and by KISS and TICHY [936]. A function $F : [0,1] \to [0.1]$ is called a distribution function (d.f.) of x_n mod (R_n) if $(0 \leq x \leq 1)$

$$\lim_{N_i\to\infty} \frac{A(x, N_i, \Delta_R x_n)}{N_i} = F(x),$$

where N_i runs through an infinite subsequence $N_1 < N_2 < N_3 < \dots$ of the positive integers. KISS [930] proved that the sequence $(x_n) = (n\theta)$, θ a positive irrational, always has a distribution function modulo an increasing linear recurring sequence (R_n). In KISS and TICHY [936] this result was generalized to sequences (x_n) of polynomial growth modulo subdivisions (R_n) of exponential growth.

As for the usual subdivision $R_n = n$, one can define lower and upper asymptotic distribution functions modulo arbitrary subdivisions:

$$\varphi(x) = \liminf_{N\to\infty} \frac{A([0,x), N, \Delta_R x_n)}{N},$$

$$\Phi(x) = \limsup_{N\to\infty} \frac{A([0,x), N, \Delta_R x_n)}{N}.$$

If $\varphi(x) = \Phi(x)$ for $0 \le x \le 1$ the sequence (x_n) has a uniquely determined d.f. F and (x_n) is called asymptotically distributed mod (R_n); in the case $F(x) = x$ (x_n) is called uniformly distributed mod (R_n). If the distribution function is not uniquely determined one says as usual that the sequence (x_n) is almost asymptotically distributed modulo (R_n).

In the following we compute lower and upper asymptoic d.f. in the case of the sequence $(x_n) = (n\theta)$, where θ is an arbitrary positive real number, modulo exponentially growing sequences (R_n).

Theorem 1.154 *Let θ be an arbitrary positive number and let (R_n) be a subdivision satisfying*

$$\lim_{n\to\infty} \frac{R_{n+1}}{R_n} = q > 1.$$

Then lower and upper asymptotic distribution functions of $(n\theta)$ modulo (R_n) are given by

$$\varphi(x) = x \qquad \text{and} \qquad \Phi(x) = \frac{qx}{(q-1)x+1}.$$

Proof. Since

$$A([0,x), N, \Delta_R x_n) = \sum_{m=0}^{M-1} \sum_{R_m \le \theta k < R_{m+1}} \chi_{[0,x)}(\Delta_R \theta k) + \sum_{R_M \le \theta k < x_N} \chi_{[0,x)}(\Delta_R \theta k),$$

where M is a positive integer satisfying $R_M \le \theta N < R_{M+1}$. Setting $\Delta R_n = R_{n+1} - R_n$ we obtain

$$\begin{aligned}
\sum_{0\le m<M} \sum_{R_m \le \theta k < R_{m+1}} \chi_{[0,x)}(\Delta_R \theta k) &= \sum_{0\le m<M} \sum_{R_m \le \theta k < R_m + x\Delta R_m} 1 \\
&= \sum_{0\le m<M} \left(\frac{x}{\theta}\Delta R_m + \mathcal{O}(1)\right) \\
&= \frac{x}{\theta} R_M + \mathcal{O}(M).
\end{aligned}$$

Furthermore we have

$$\begin{aligned}
\sum_{R_M \le \theta k < \theta N} \chi_{[0,x)}(\Delta_R \theta k) &= \min\left(N - \frac{R_M}{\theta}, x\frac{\Delta R_M}{\theta}\right) + \mathcal{O}(1) \\
&= \frac{\Delta R_M}{\theta} \min(\Delta_R \theta N, x) + \mathcal{O}(1).
\end{aligned}$$

Thus

$$\frac{A([0,x), N, \Delta_R x_n)}{N} = \min\left(\frac{xR_{M+1}}{\theta N}, \frac{\theta N - (1-x)R_M}{\theta N}\right) + \mathcal{O}(\frac{M}{N}).$$

Since the mapping $N \mapsto (\theta N - (1-x)R_M)/(\theta N)$ is increasing and $N \mapsto xR_{M+1}/(\theta N)$ is decreasing for $R_M/\theta \le N < R_{M+1}/\theta$ it follows that the minimal value is obtained for $N = -[-R_M/\theta]$ or for $N = [R_{M+1}/\theta]$, i.e.

$$\lim_{M\to\infty} \min_{R_M/\theta \le N < R_{M+1}/\theta} \frac{A([0,x), N, \Delta_R x_n)}{N} = x.$$

Similarly the maximal value is attained if N is approximately $(R_M + x\Delta R_M)/\theta$ providing that

$$\lim_{M\to\infty} \max_{R_M/\theta \le N < R_{M+1}/\theta} \frac{A([0,x), N, \Delta_R x_n)}{N} = \frac{qx}{(q-1)x+1}.$$

This proves the theorem. □

Notes

Hlawka [784] systematically investigated asymptotic distribution of sequences with respect to general distribution measures on compact spaces. Later an ergodic approach is due to Denker, Grillenberger, and Sigmund [450]. Recently, Winkler [1961] obtained a characterization of the structure of weak limit points of sequences of measures on locally compact spaces. Strauch [1755, 1754, 1757] studied various problems concerning the existence of distribution functions; especially related moment problems are considered. In [1750, 1751] Strauch considered some applications of Franel's integral. Molnar gave a characterization of all functions f such that for every pair F, G of distribution functions there exists a sequence (x_n) of real numbers mod 1 such that (x_n) has distribution function F and the image sequence $f(x_n)$ has distribution function G.

Goto and Kano [666, 669] established a Wiener-Schoenberg theorem for weighted means and well distribution. Kano [898] gave a survey on u.d. and distribution functions; see also Goto and Kano [671]. Brown and Duncan [301] showed certain convergence theorems for distribution functions; for some special results on almost u.d. we mention Chen [340]. Ruzsa [1579] gave a survey on u.d. and almost u.d. of sequences $(a_n x)$. Nakajima and Ohkubo [1256] obtained an extension of LeVeque's inequality to general distribution functions.

Strauch [1752] proved that a sequence (x_n) in the unit interval is u.d. if and only if $(|x_m - x_n|)$ $(1 \le m, n \le N)$ is asymptotically distributed to the function $2x - x^2$.

Kiss and Molnar considered the distribution mod 1 of (xG_n), where x is a real number and G_n a linear recurring sequence with rational coefficients. For diophantine approximation using linear recurring sequences we refer to Kiss [932], Kiss and Sinka [933], and Grabner, Kiss and Tichy [679]. Kiss and Tichy [934] computed the distribution function of the quotients G_{n+1}/G_n modulo 1 for second order linear recurring sequences G_n. These investigations were extended to more general recurrences by Goldstern, Tichy, and Turnwald [665]. Later Grabner, Tichy, and Winkler [686] could also identify the distribution function for a wide class of linear recurrences with non-constant coefficients extending Tichy [1843]. For further results on the asymptotic distribution of recurring sequences we refer to Tichy [1830, 1828, 1825, 1842].

Van de Lune [1112] showed that $(\log n/g_n)$ is not u.d. mod (an) for any a, where g_n denotes the largest prime dividing n.

Kiss and Tichy [936] investigated u.d. with respect to subdivisions generated by linear recurring sequences, and more generally with respect to sequences of exponential growth.

Sun [1764] has shown that every u.d. sequence mod 1 can be decomposed into finitely many sequences with given asymptotic densities and asymptotic distribution functions. Pastéka [1407] studied transformations which preserve distribution functions. Schatte [1615] investigated transformations of distribution functions on the unit interval.

Proinov [1476, 1477, 1478, 1480] proved extensions of the Erdős-Turán inequality (in one dimension) to general distribution functions obtaining improved versions of Berry-Esseen type inequalities.

For similar results and applications see Fainleib [577] and Elliott [557]; see also the Notes of Section 1.2, especially Niederreiter and Philipp [1352]. Niederreiter [1354] studied the distribution of values of Kloosterman sums; for further information in this direction see the fundamental monograph by Katz [912].

1.6 Metric Theory

1.6.1 ERDŐS-GÁL-KOKSMA's Method

In this section we want to develop some main tools of the metric theory of uniform distribution. We start with a description of the classical approach of ERDŐS-GÁL-KOKSMA. First of all we apply a well-known method from the theory of orthogonal series to establish a general metric upper bound for the discrepancy and related functions.

Theorem 1.155 *Let $F(M,N,x)$ be non-negative functions in $L^p[0,1]$ $(p \geq 1)$ for $M,N = 0,1,\ldots$ such that $F(M,0,x) = 0$ and*

$$F(M,N,x) \leq F(M,N',x) + F(M+N',N-N',x) \tag{1.298}$$

for all $M = 0,1,\ldots$ and all $0 \leq N' \leq N$. Suppose further

$$\int_0^1 F(M,N,x)^p dx = \mathcal{O}(\Psi(N))$$

uniformly in $M = 0,1,\ldots$, where $\Psi(N)/N$ is a non-decreasing function. Then for arbitrary $\varepsilon > 0$

$$F(0,N,x)^p = \mathcal{O}\left(\Psi(N)(\log N)^{p+1+\varepsilon}\right)$$

for almost all $x \in [0,1]$.

The proof of this theorem is based on the following two lemmata.

Lemma 1.156 *Let $a_n(x) \geq 0$, $n = 1,2,\ldots$ be integrable functions on $[0,1]$ such that*

$$\sum_{n=1}^{\infty} \int_0^1 a_n(x)dx < \infty.$$

Then $a_n(x) = o(1)$ almost everywhere.

Proof. By a standard theorem from the theory of infinite series there exists an increasing sequence of positive numbers $\alpha(n) \to \infty$, as $n \to \infty$ such that

$$\sum_{n=1}^{\infty} \int_0^1 \alpha(n)a_n(x)dx < \infty.$$

Let E be the set of all points in $[0,1]$ with $\limsup_{n\to\infty} \alpha(n)a_n(x) \geq 1$. Furthermore let E_n be the sets of points x such that $2\alpha(n)a_n(x) > 1$. Let ε be an arbitrary positive number. Then there exists an $n_0 = n_0(\varepsilon)$ such that

$$\sum_{n=n_0}^{\infty} \int_0^1 \alpha(n)a_n(x)dx < \frac{\varepsilon}{2}.$$

From the definition of the sets E_n we obtain

$$\mu(E) \leq \sum_{n=n_0}^{\infty} \mu(E_n) \leq 2\sum_{n=n_0}^{\infty} \int_{E_n} \alpha(n)a_n(x)dx \leq 2\sum_{n=n_0}^{\infty} \int_0^1 \alpha(n)a_n(x)dx.$$

This yields $\mu(E) = 0$. Thus we have $\limsup_{n\to\infty} \alpha(n)\cdot a_n(x) < 1$ for almost all $x \in [0,1]$. Since $\alpha(n) \to \infty$, the proof of the lemma is complete. □

Lemma 1.157 *Let $p > 1$ and let $t(n,k) \geq 1$ for $n = 1,2\ldots,$ $k = 1,2,\ldots,n$ and define*

$$K(N) = \begin{cases} 1 & \text{for } p = 1, \\ \max\limits_{2^\nu \leq N} \left(1 + \sum\limits_{k=1}^{\nu} t(\nu,k)^{-\frac{p}{p-1}}\right)^{p-1} & \text{for } p > 1. \end{cases}$$

Let $F(M,N,x)$ be non-negative functions in $L^p(0,1)$ $(p \geq 1)$ as in Theorem 1.155 and set

$$H(n,p,x) = \left(F(0,2^n,x)^p + \sum_{k=1}^{n} t(n,k)^p \sum_{j_k=0}^{2^{n-k}-1} F(2^n + j_k 2^k, 2^{k-1}, x)^p\right)^{1/p}.$$

Then we have

$$F(0,N,x) \leq K(N)^{1/p} H(n(N),p,x)$$

for all $x \in [0,1]$, where $n = n(N)$ is defined by $2^{n(N)} \leq N < 2^{n(N)+1}$ (for $N \geq 2$).

Proof. Let $N \geq 2$ and set

$$N = 2^n + \varepsilon_{n-1}2^{n-1} + \ldots + \varepsilon_1 \cdot 2 + \varepsilon_0$$

in binary expansion, where $n = n(N) \geq 1$ and $\varepsilon_i = 0,1$. Then we have

$$\begin{aligned} F(0,N,x) &\leq F(0,2^n,x) + F(2^n, \varepsilon_{n-1}2^{n-1} + \ldots + \varepsilon_0, x) \\ &\leq F(0,2^n,x) + F(2^n, \varepsilon_{n-1}2^{n-1}, x) \\ &\quad + F(2^n + \varepsilon_{n-1}2^{n-1}, \varepsilon_{n-2}2^{n-2} + \ldots + \varepsilon_0, x) \\ &\leq F(0,2^n,x) + F(2^n, \varepsilon_{n-1}2^{n-1}, x) \\ &\quad + F(2^n + \varepsilon_{n-1}2^{n-1}, \varepsilon_{n-2}2^{n-2}, x) \end{aligned}$$

$$\begin{aligned}
&+F(2^n+\varepsilon_{n-1}2^{n-1}+\varepsilon_{n-2}2^{n-2},\varepsilon_{n-3}2^{n-3}+\ldots+\varepsilon_0,x)\\
\leq\ & F(0,2^n,x)+F(2^n,\varepsilon_{n-1}2^{n-1},x)\\
&+\sum_{k=1}^{n-1}F(2^n+\varepsilon_{n-1}2^{n-1}+\ldots+\varepsilon_k2^k,\varepsilon_{k-1}2^{k-1},x).
\end{aligned}$$

Set $m_n=0$ and $m_k=\varepsilon_k+\varepsilon_{k+1}2+\ldots+\varepsilon_{n-1}2^{n-k-1}$ for $k=1,2,\ldots n-1$. From the above inequality we derive

$$F(0,N,x)\leq F(0,2^n,x)+\sum_{k=1}^{n}F(2^n+m_k2^k,\varepsilon_{k-1}2^{k-1},x).$$

Assuming $F\geq 0$ and $F(M,0,x)=0$, we have

$$F(2^n+m_k2^k,\varepsilon_{k-1}2^{k-1},x)\leq F(2^n+m_k2^k,2^{k-1},x).$$

Thus

$$F(0,N,x)\leq F(0,2^n,x)+\sum_{k=1}^{n}F(2^n+m_k2^k,2^{k-1},x)$$

for $N\geq 2$.

Next we consider the special case $p=1$, i.e. $K(N)=1$. Since $t(n,k)\geq 1$, we obtain

$$F(0,N,x)\leq F(0,2^n,x)+\sum_{k=1}^{n}t(n,k)F(2^n+m_k2^k,2^{k-1},x).$$

Hence

$$\begin{aligned}
F(0,N,x)\ &\leq\ F(0,2^n,x)+\sum_{k=1}^{n}t(n,k)\sum_{m_k=0}^{2^{n-k}-1}F(2^n+m_k2^k,2^{k-1},x)\\
&=\ H(n,1,x)=K(N)^{1/1}H(n(N),1,x),
\end{aligned}$$

which yields the lemma for $p=1$. In the case $p>1$ we apply Hölder's inequality and obtain from above

$$\begin{aligned}
F(0,N,x)\ &\leq\ F(0,2^n,x)+\sum_{k=1}^{n}t(n,k)^{-1}(t(n,k)F(2^n+m_k2^k,2^{k-1},x))\\
&\leq\ K(N)^{1/p}\cdot(F(0,2^n,x)^p+\sum_{k=1}^{n}t(n,k)^pF(2^n+m_k2^k,2^{k-1},x)^p)^{1/p}\\
&\leq\ K(N)^{1/p}H(n,p,x).
\end{aligned}$$

Thus the proof of the lemma is complete. □

Proof of Theorem 1.155. In order to prove the theorem we set $t(n,k)=1$, $\varphi(N)=8\Psi(N)\log^{2+\varepsilon}N$ and $a_n(x)=H(n,p,x)^p\varphi(2^n)^{-1}$. Obviously, we have

$K(N) = \mathcal{O}(\log^{p-1} N)$. Since $\Psi(N)/N$ is non-decreasing, we obtain $\Psi(2^k) \le 2^{k-n}\Psi(2^n)$ for $1 \le k \le n$. Hence we can estimate

$$\sum_{n=1}^{\infty} \int_0^1 a_n(x)dx \ll \sum_{n=1}^{\infty} \left(\Psi(2^n)\varphi(2^n)^{-1} + \sum_{k=1}^{n} \sum_{m_k=0}^{2^{n-k}-1} \Psi(2^k)\varphi(2^n)^{-1} \right)$$
$$\le \sum_{n=1}^{\infty} \frac{1}{n^{1+\varepsilon}} < \infty.$$

Lemma 1.156 yields

$$a_n(x) = \frac{H(n,p,x)^p}{\varphi(2^n)} = o(1)$$

for almost all $x \in [0,1]$. Setting $n = n(N)$ with $2^n \le N < 2^{n+1}$ and applying Lemma 1.157 completes the proof of the theorem. □

Remark. Of course, the above results remain true for general probability spaces (instead of the unit interval).

Using the above general theorem a metric discrepancy bound can be proved for sequences $(f_n(x))$, since the discrepancy $F(M,N,x) = ND_N(f_{n+M}(x))$ of shifted sequences satisfies the relation (1.298).

Theorem 1.158 *Let $(f_n(x))_{n\ge1}$ be a sequence of continuously differentiable functions, defined for $x \in [a,b]$, such that for $n_1 \ne n_2$*

$$f'_{n_1}(x) - f'_{n_2}(x)$$

is either a non-decreasing or a non-increasing function of x on $[a,b]$ and that

$$\min_{a\le x\le b} |f'_{n_1}(x) - f'_{n_2}(x)| \ge \delta,$$

where δ denotes a positive number which does not depend on n_1, n_2 or x. Then for almost all x the discrepancy $D_N(f_n(x))$ of the sequence $(f_n(x))_{n\ge1}$ satisfies the inequality

$$D_N(x) = \mathcal{O}(N^{-\frac{1}{2}}(\log N)^{\frac{5}{2}+\varepsilon}) \qquad (\varepsilon > 0).$$

Proof. We use the ERDŐS-TURÁN-KOKSMA inequality (see Theorem 1.21) for $k = 1$ and set $M = [\sqrt{n}]$. Thus we obtain

$$N^2 D_N^2(x) \ll \left(N + 2\sqrt{N} \sum_{h=1}^{[\sqrt{N}]} \frac{1}{h} \left| \sum_{n=1}^{N} e(hf_n(x)) \right| \right)$$
$$+ \sum_{h=1}^{[\sqrt{N}]} \sum_{k=1}^{[\sqrt{N}]} \frac{1}{hk} \left| \sum_{n=1}^{N} e(hf_n(x)) \right| \cdot \left| \sum_{n=1}^{N} e(kf_n(x)) \right| .$$

Integrating this inequality yields

$$\int_a^b N^2 D_N^2(x)\,dx \ll \left(N(b-a) + 2\sqrt{N} \sum_{h=1}^{[\sqrt{N}]} \frac{1}{h} \int_a^b \left| \sum_{n=1}^{N} e(hf_n(x)) \right| dx \right.$$
$$\left. + \sum_{h=1}^{[\sqrt{N}]} \sum_{k=1}^{[\sqrt{N}]} \frac{1}{hk} \int_a^b \left| \sum_{n=1}^{N} e(hf_n(x)) \right| \left| \sum_{n=1}^{N} e(kf_n(x)) \right| dx. \right)$$

Applying the CAUCHY-SCHWARZ inequality we derive

$$\int_a^b N^2 D_N^2(x)\,dx \ll \left(N(b-a) + 2\sqrt{N} \sum_{h=1}^{[\sqrt{N}]} \frac{1}{h} \left(\int_a^b 1^2 dx\, I_h(N) \right)^{\frac{1}{2}} \right.$$
$$\left. + \sum_{h=1}^{[\sqrt{N}]} \sum_{k=1}^{[\sqrt{N}]} \frac{1}{hk} \left(I_h(N) \cdot I_k(N) \right)^{\frac{1}{2}}, \right)$$

where

$$I_h(N) = \int_a^b \left| \sum_{n=1}^{N} e(hf_n(x)) \right|^2 dx.$$

In order to estimate $I_h(N)$ we proceed as in KUIPERS and NIEDERREITER [983, Theorem 4.3, p.34]. Since

$$\begin{aligned} I_h(N) &= \sum_{n_1,n_2=1}^{N} \int_a^b e(h(f_{n_1}(x) - f_{n_2}(x)))\,dx \\ &\le \sum_{n_1,n_2=1}^{N} \left| \int_a^b e(h(f_{n_1}(x) - f_{n_2}(x)))\,dx \right| \\ &= (b-a)N + 2 \sum_{1\le n_1<n_2\le N} \left| \int_a^b e(h(f_{n_1}(x) - f_{n_2}(x)))\,dx \right|, \end{aligned}$$

and (with some $x_0', x_0'' \in [a,b]$)

$$\left| \int_a^b e(h(f_{n_1}(x) - f_{n_2}(x)))\,dx \right| = \left| \frac{1}{2\pi ih} \int_a^b \frac{de(h(f_{n_1}(x) - f_{n_2}(x)))}{f_{n_1}'(x) - f_{n_2}'(x)} \right|$$
$$\le \left| \frac{1}{2\pi ih} \left(\frac{1}{f_{n_1}'(a) - f_{n_2}'(a)} \int_a^{x_0'} d\cos(2\pi h(f_{n_1}(x) - f_{n_2}(x))) \right. \right.$$
$$\left. \left. + \frac{1}{f_{n_1}'(b) - f_{n_2}'(b)} \int_{x_0'}^b d\cos(2\pi h(f_{n_1}(x) - f_{n_2}(x))) \right) \right|$$
$$+ \left| \frac{1}{2\pi ih} \left(\frac{1}{f_{n_1}'(a) - f_{n_2}'(a)} \int_a^{x_0''} d\sin(2\pi h(f_{n_1}(x) - f_{n_2}(x))) \right. \right.$$

$$+\frac{1}{f'_{n_1}(b)-f'_{n_2}(b)}\int_{x''_0}^{b} d\cos(2\pi h(f_{n_1}(x)-f_{n_2}(x)))\Bigg|$$
$$\leq \frac{1}{h\pi}\left(\frac{2}{|f'_{n_1}(a)-f'_{n_2}(a)|}+\frac{2}{|f'_{n_1}(b)-f'_{n_2}(b)|}\right)$$

we immediately obtain

$$I_h(N)\leq (b-a)N+\frac{4}{\pi h}\sum_{n_2=2}^{N}\sum_{n_1=1}^{n_2-1}\left(\frac{1}{|f'_{n_1}(a)-f'_{n_2}(a)|}+\frac{1}{|f'_{n_1}(b)-f'_{n_2}(b)|}\right).$$

For fixed $2\leq n_2\leq N$, we can order the numbers $f'_1(a), f'_2(a), \ldots, f'_{n_2}(a)$ according to their magnitude. In the new ordering, the difference of any two consecutive numbers will be $\geq \delta$. Hence

$$\sum_{n_1=1}^{n_2-1}\frac{1}{|f'_{n_1}(a)-f'_{n_2}(a)|}\ll \sum_{n=1}^{N}\frac{1}{n\delta}\ll \frac{1}{\delta}\log N.$$

Therefore we finally obtain

$$I_h(N)\ll N+\frac{N}{h}\log N.$$

Inserting this in the above inequality and applying Theorem 1.155 (due to GÁL and KOKSMA) yields the desired result. □

The above discrepancy bound can be applied to various classes of special sequences, e.g. to sequences of the form $(a_n x)$, where a_n is an increasing sequence of positive integers, or (x^n) for $x>1$. In the case of lacunary sequences the exponent of the logarithmic term can be improved by a slightly different method of ERDŐS and KOKSMA.

Theorem 1.159 *Let $[a,b]$ be a given real interval and $\delta>0$. Let $f_n(x)$ denote a sequence of real functions defined on $[a,b]$ such that*

$$f'_{n+1}(x)\geq (1+\delta)f'_n(x)>0,\quad f''_{n+1}(x)\geq (1+\delta)f''_n(x)\geq 0$$

for all $x\in[a,b]$ and $n\geq 1$. Let $\varphi(n)$ denote an arbitrary increasing function of n such that $\varphi(n)\to\infty$ as $n\to\infty$.

Then for almost all $x\in[a,b]$ the discrepancy of the sequence $(f_n(x))$ satisfies the estimate

$$D_N(f_n(x)) = O(N^{\frac{1}{2}}(\log N)^{\frac{3}{2}}(\log\log N)^{\frac{1}{2}}\ \varphi(N)).$$

Remark. We omit a proof of this theorem here and refer to ERDŐS and KOKSMA [565]. The main idea of the proof is to introduce L^{2p}-norms instead of L^2-norms as above. Then the $2p-th$ powers are computed explicitly by combinatorial arguments. Again the ERDŐS-TURÁN-KOKSMA (Theorem 1.21) inequality is applied and the exponential sums are estimated via the second mean value theorem.

Obviously, this Theorem 1.159 can be applied to sequences of the type $(a_n x)$, where a_n is lacunary, i.e. $a_{n+1} \geq (1+\delta)a_n$ for some $\delta > 0$. Furthermore, this estimate can also be applied to (x^n) for $x > 1$. Various generalizations of these sequences are discussed in the Notes.

In the following we present a different approach due to BAKER [79], which yields an improvement of the exponent $\frac{5}{2}$ in the metric discrepancy bound of $(a_n x)$, where a_n is a strictly increasing sequence of positive integers (not necessarily lacunary). The method depends heavily on the following deep inequality of CARLSON-HUNT from FOURIER analysis; see for instance MOZZOCHI [1213].

Lemma 1.160 *Let f be in $L^2[0,1]$ (extended with period one) and let $s_r(f,x)$ denote the r-th partial sum of the* FOURIER *series of f. Then*

$$\left\| \max_{r\geq 1} |s_r(f,x)| \right\|_2 \leq C_2 \|f\|_2,$$

where $C_2 > 0$ is a constant and $\|\cdot\|_2$ denotes the L^2-norm on $[0,1]$.

Corollary 1.161 *Set*

$$S_h(M,N,x) = \sum_{j=M+1}^{M+N} e(ha_j x).$$

Then

$$\left\| \max_{1\leq r\leq N} |S_h(0,r,x)| \right\|_2 \leq C_2 N^{1/2}. \tag{1.299}$$

Theorem 1.162 *Let (a_n) be a strictly increasing sequence of positive integers. Then for almost all x and arbitrary $\varepsilon > 0$*

$$D_N(a_n x) = \mathcal{O}(N^{-\frac{1}{2}}(\log N)^{\frac{3}{2}+\varepsilon})$$

Proof. We set

$$\psi(N) = N^{1/2}(\log N)^{3/2+\varepsilon} \quad (N \geq 1)$$

and note that

$$\psi(4^{k+1}) < 3\psi(4^k)$$

for $k \geq k_0$. In the following we use the notations of the above lemmata and set $D(N,x) = D_N(a_n x)$. Now let

$$E = \{x \in [0,1] : ND(N,x) > \psi(N) \text{ for infinitely many } N\}.$$

Then we have

$$E \subset \bigcap_{r=1}^{\infty} \bigcup_{k=r}^{\infty} B_k,$$

where

$$B_k = \left\{ x \in [0,1] : \max_{1\leq N\leq 4^k} ND(N,x) > \frac{1}{3}\psi(4^k) \right\}.$$

This follows, since $x \in E$ implies that there are arbitrarily large integers k such that, for some integer N in $[4^{k-1}, 4^k)$,

$$ND(N,x) > \psi(N) \geq \psi(4^{k-1}) > \frac{1}{3}\psi(4^k).$$

Now by the inequality of ERDŐS-TURÁN-KOKSMA (Theorem 1.21) we obtain

$$ND(N,x) \leq 6\left(N^{1/2} + \sum_{h=1}^{[N^{1/2}]} h^{-1}|S_h(0,N,x)|\right).$$

Thus we have

$$\max_{1\leq N\leq 4^k} ND(N,x) \leq 6\left(2^k + \sum_{h=1}^{2^k} h^{-1} \max_{1\leq r\leq 4^k} |S_h(0,r,x)|\right).$$

From MINKOWSKI's inequality and Corollary 1.161 we derive

$$\begin{aligned}\left\|\max_{1\leq N\leq 4^k} ND(N,x)\right\|_2 &\leq 6.2^k + 6\sum_{h=1}^{2^k} h^{-1}\left\|\max_{1\leq r\leq 4^k} |S_h(0,r,x)|\right\|_2 \\ &\leq 6.2^k + 6C_2\sum_{h=1}^{2^k} h^{-1}2^k \leq C_3 k 2^k,\end{aligned}$$

where $C_3 > 0$ is again a constant. Hence the LEBESGUE measure $\lambda_1(B_k)$ of B_k satisfies

$$\lambda_1(B_k)\left(\frac{1}{3}\psi(4^k)\right)^2 \leq C_3^2 k^2 4^k.$$

Thus

$$\sum_{k=1}^{\infty} \lambda_1(B_k) \leq 9C_3^2 \sum_{k=1}^{\infty} k^{-1-2\varepsilon} < \infty.$$

The theorem now follows immediatly by the BOREL-CANTELLI lemma. □

1.6.2 Lacunary Sequences

In the following we want to establish a law of iterated logarithm for the discrepancy function $D_N(a_n x)$ provided that (a_n) is a lacunary sequence, i.e. $a_{n+1} \geq q\, a_n$ with $q = 1+\delta > 1$. For exponential sums such laws of iterated logarithm were established by GÁL and GÁL [642]. In the following we present a result of PHILIPP [1434] obtaining a law of iterated logarithm for the discrepancy function.

Theorem 1.163 *Let (a_n) be a lacunary sequence of positive integers satisfying $a_{n+1} \geq qa_n$ with $q > 1$. Then for almost all x*

$$\frac{1}{4} \leq \limsup_{N\to\infty} \frac{ND_N(a_n x)}{\sqrt{N \log\log N}} \leq C,$$

where

$$C \leq 166 + 664(q^{1/2} - 1)^{-1}.$$

As an immediate consequence of this theorem and KOKSMA-HLAWKA's inequality (Theorem 1.14) we obtain

Corollary 1.164 *Let $f(x)$ be a function of bounded variation on $[0,1]$ with*

$$f(x+1) = f(x) \quad \text{and} \quad \int_0^1 f(x)\,dx = 0. \tag{1.300}$$

Then for any lacunary sequence (a_n) of integers satisfying $a_{n+1} \geq qa_n$ the relation

$$\limsup_{N\to\infty} \frac{\left|\sum\limits_{n\leq N} f(a_n x)\right|}{\sqrt{N \log\log N}} \ll 1$$

holds for almost all x, where the constant implied by $\ll$ depends on f and q.

For Lipschitz functions satisfying (1.300) the corollary is due to TAKAHASHI [1772]. In fact a stronger result than the above corollary can be proved:

Corollary 1.165 *Let $\mathcal{F}$ be the class of functions of bounded variation $V(F)$ on $[0,1]$ not exceeding V and satisfying (1.300). Then for any lacunary sequence (a_n) of integers the relation*

$$\frac{1}{8}V \leq \limsup_{N\to\infty} \frac{\sup\limits_{f\in\mathcal{F}} \left|\sum\limits_{n\leq N} f(a_n x)\right|}{\sqrt{N \log\log N}} \leq CV$$

holds for almost all x, where a bound for C is given in Theorem 1.163.

In the following we present the proof of the Theorem 1.163 which is based on a combination of the ERDŐS-GÁL-KOKSMA method and an approach by TAKAHASHI [1772, 1778].

Proposition 1.166 *Let $M \geq 0, N \geq 1$ be integers and let $R \geq 2$. Suppose that $f(x)$ satisfies (1.300) and $\|f\|_2 \gg N^{-1/4}$ and has variation $V(f) \leq 2$. Then*

$$\lambda_1 \left\{ x : \left| \sum_{k=M+1}^{M+N} f(a_k x) \right| \geq (1+2\eta_0) C_1 R \|f\|_2^{1/4} (N \log\log N)^{1/2} \right\}$$
$$\ll \exp(-(1+\eta_0)\|f\|_2^{-1/2}(R-2)\log\log N) + R^{-2} N^{-3/4},$$

where

$$C_1 = \frac{1}{2} + 2(q^{1/2}-1)^{-1}, \tag{1.301}$$

$\eta_0 > 0$ *is a (universal) constant, and the constant implied by* $\ll$ *depends on* q.

We expand $f(x)$ in a FOURIER series with partial sums

$$f_n(x) = \sum_{|j|<n} c_j e(jx).$$

Obviously, $c_0 = 0$ and

$$c_j = \int_0^1 f(x)e(-jx)\,dx = \frac{-1}{2\pi j}\int_0^1 f(x)\,de(-jx) = \frac{1}{2\pi j}\int_0^1 e(-jx)\,df(x)$$

provides

$$|c_j| \le \frac{V}{2\pi j} \le \frac{1}{\pi j} \quad (1 \le |j| \le n). \tag{1.302}$$

We set

$$\varphi_n(x) = f(x) - f_n(x),$$

$$\psi(x,m,n) = \sum_{m\le|j|<n} c_j e(jx),$$

and

$$\Phi_N(x,m,n) = \sum_{k\le N} \psi(a_k x, m, n).$$

For proving Proposition 1.166 we will use the following lemmata.

Lemma 1.167 *Let* $m \le n$ *and* N *be positive integers. Then*

$$\|\Phi_N(\cdot,m,n)\|_2^2 \ll \frac{N}{m}\log(n/m).$$

Proof. Let $\delta_{x,y}$ denote the KRONECKER symbol; i.e. $\delta_{x,x} = 1$ and $\delta_{x,y} = 0$ for $x \ne y$. By the definition of Φ_N we have

$$\begin{aligned}
\|\Phi_N(\cdot,m,n)\|_2^2 &= \int_0^1 \left(\sum_{k\le N}\sum_{m\le|j|<n} c_j e(ja_k x)\right)^2 dx \\
&= \sum_{k,l\le N}\sum_{m\le|i|,|j|<n} c_i\bar{c}_j \delta_{ia_k,ja_l} \\
&\ll \sum_{k,l\le N}\sum_{m\le|i|,|j|<n} (|c_i|^2+|c_j|^2)\delta_{ia_k,ja_l} \\
&\ll \sum_{k\le N}\sum_{m\le|i|<n} |c_i|^2 \sum_{l\le N, m\le|j|<n} \delta_{ia_k,ja_l}.
\end{aligned} \tag{1.303}$$

But for fixed k and i with $m \le |i| < n$ the number of solutions of the equation

$$ia_k = ja_l, \quad l \le N, m \le |j| < n \tag{1.304}$$

does not exceed a constant times $\log(n/m)$. For showing this, we may assume w.l.o.g. that $i > 0$ and $j > 0$. Let l_1 be the smallest index l so that a_l is a solution of (1.304) and let l_2 be the largest index with that property. Then

$$j_1 a_{l_1} = j_2 a_{l_2}, \quad m \le j_2 \le j_1 < n$$

and

$$q^{l_2 - l_1} \le a_{l_2}/a_{l_1} = j_1/j_2 \le n/m.$$

Hence

$$l_2 - l_1 \ll \log(n/m).$$

Consequently, the inner sum in (1.303) is bounded by that quantity and thus by (1.302)

$$\|\Phi_N\|_2^2 \ll N \log(n/m) \sum_{|i| \ge m} |c_i|^2 \ll \frac{N}{m} \log(n/m). \qquad \square$$

Lemma 1.168 *For any positive integer T we have*

$$\int_0^1 \Big(\sum_{k \le N} \varphi_T(a_k x) \Big)^2 dx \ll NT^{-1}.$$

Proof. Let h_0 be the smallest integer with $T \le 2^{h_0}$. We apply Lemma 1.167 with $m = T, n = 2^{h_0}$ and $m = 2^h, n = 2^{h+1}$, $h \ge h_0$, and obtain

$$\begin{aligned} \Big\| \sum_{k \le N} \varphi_T(a_k x) \Big\|_2 &\le \Big\| \sum_{k \le N} \psi(a_k x, T, 2^{h_0}) \Big\|_2 + \sum_{h \ge h_0} \Big\| \sum_{k \le N} \psi(a_k x, 2^h, 2^{h+1}) \Big\|_2 \\ &\ll N^{1/2} \left(T^{-1/2} + \sum_{h \ge h_0} 2^{-\frac{1}{2}h} \right) \ll N^{1/2} T^{-1/2}. \qquad \square \end{aligned}$$

In the following let H be any integer with

$$q^H > 3H^6. \tag{1.305}$$

We put

$$g(x) = \sum_{1 \le |j| \le H^6} c_j e(jx) \quad \text{and} \quad U_m(x) = \sum_{k=Hm+1}^{H(m+1)} g(a_k x) \tag{1.306}$$

and observe by (1.302)

$$\sum_{|j| \ge n} |c_j|^2 \le n^{-1}.$$

Lemma 1.169 *Let $\kappa > 0$ be with $\kappa H^{3/2} < 1$. Then there exists $\eta_0 > 0$ such that for any positive integer P*

$$\int_0^1 \exp\left(\kappa \sum_{m=0}^{P-1} U_{2m}(x)\right) dx \le \exp\left((1+\eta_0)C_1\kappa^2\|f\|_2 HP\right)$$

and

$$\int_0^1 \exp\left(\kappa \sum_{m=1}^{P} U_{2m-1}(x)\right) dx \le \exp\left((1+\eta_0)C_1\kappa^2\|f\|_2 HP\right),$$

where C_1 is defined in the Proposition 1.166.

Proof. Since $\|g\|_\infty = \mathcal{O}(\log H)$ and $\kappa H^{3/2} < 1$ we have $\kappa U_{2m}(x) \le z_0$ for a constant z_0. Hence, for $|z| \le z_0$ we can apply the inequality $e^z \le 1 + z + \frac{1}{2}(1+\eta_0)z^2$ and obtain

$$\int_0^1 \exp\left(\sum_{m=0}^{P-1} U_{2m}(x)\right) dx \le \int_0^1 \prod_{m=0}^{P-1}\left(1 + \kappa U_{2m}(x) + \frac{1}{2}\kappa^2(1+\eta_0)U_{2m}^2(x)\right) dx. \tag{1.307}$$

Since $U_{2m}(x)$ is real we have $|U_{2m}^2(x)| = U_{2m}^2(x)$. Thus

$$\begin{aligned} U_{2m}^2(x) &= \sum_{l,j=H(2m)+1}^{H(2m+1)} \sum_{1\le|r|,|s|\le H^6} c_r\bar{c}_s e((a_jr - a_ls)x) \\ &= W_{2m}(x) + H\sum_{|k|\le H^6}|c_k|^2 \\ &\quad +2\sum_{l=H(2m)+1}^{H(2m+1)-1}\ \sum_{j=l+1}^{H(2m+1)}\ \sum_{\substack{1\le|r|,|s|\le H^6\\ |a_jr-a_ls|<a_{H(2m)}}} c_r\bar{c}_s e((a_ls - a_jr)x), \end{aligned} \tag{1.308}$$

where $W_{2m}(x)$ is the sum of trigonometric functions whose frequencies are between a_{H2m} and $2H^6 a_{H(2m+1)} < q^H a_{H(2m+1)} \le a_{H2(m+1)}$. Let $V_{2m}(x)$ denote the last term of the right hand side of (1.308). Observe that for any choice of j, l with $j > l > H(2m)$ then for any $r \ne 0$ there are at most 2 possible $s \ne 0$ which satisfy $|a_jr - a_ls| < a_{H(2m)}$. Especially those s satisfy $|s| \ge a_j/a_l$. Hence

$$\begin{aligned} |V_{2m}(x)| &\le 2\sum_{l=H(2m)+1}^{H(2m+1)-1}\ \sum_{j=l+1}^{H(2m+1)}\ \sum_{\substack{1\le|r|,|s|\le H^6\\ |a_ls-a_jr|<a_{H(2m)}}} |c_rc_s| \le \\ &\le 4\sum_{l=H(2m)}^{H(2m+1)}\sum_{j>l}\left(\sum_{|r|\ge1}|c_r|^2\right)^{1/2}\left(\sum_{|s|\ge a_j/a_l}|c_s|^2\right)^{1/2} \\ &\le 4\|f\|_2 \sum_{l=H(2m)}^{H(2m+1)}\sum_{j>l}\left(\frac{a_l}{a_j}\right)^{1/2} \end{aligned}$$

$$\le 4\|f\|_2 \sum_{l=H2m}^{H(2m+1)} \sum_{j>l} q^{(j-l)/2}$$
$$\le 4\|f\|_2 \frac{H}{q^{1/2}-1}.$$

From this we get $U_{2m}^2(x) \le 2C_1\|f\|_2 H + W_{2m}(x)$, and thus we obtain the following upper bound for the integral (1.307):

$$\int_0^1 \prod_{m=0}^{P-1} \left(1 + \kappa^2(1+\eta_0)HC_1\|f\|_2 + \kappa U_{2m}(t) + \frac{\kappa^2}{2}(1+\eta_0)W_{2m}(t)\right) dt.$$

Furthermore, since the frequencies of $U_{2m}(x)$ and $W_{2m}(x)$ are disjoint for different m this upper bound reduces to

$$\prod_{m=0}^{P-1} (1 + \kappa^2(1+\eta_0)HC_1\|f\|_2) \le \exp\left(\kappa^2(1+\eta_0)HC_1\|f\|_2 P\right)$$

and proves the first inequality. The second one can be proved along the same lines. □

Proof of Proposition 1.166. It is sufficient to prove the proposition for $M = 0$ only, since the shifted sequence $(a_{M+k})_{k\ge1}$ is lacunary with the same factor q.

Let $N \ge N_0$ be given. Put $H = [N^{1/6}]$. If N_0 is sufficiently large then (1.305) is satisfied. With this choice of H we define the functions $g(x)$ and $U_m(x)$ by (1.306). For any $Q > 0$ we have

$$\lambda_1\left(\left\{x : \sum_{k\le N} f(a_k x) \ge (1+2\eta_0)Q\right\}\right) \le \lambda_1(B_1) + \lambda_1(B_2), \tag{1.309}$$

where

$$B_1 = \left\{x : \sum_{k\le N} g(a_k x) \ge (1+\eta_0)Q\right\}, \quad B_2 = \left\{x : \sum_{k\le N} \varphi_T(a_k x) \ge \eta_0 Q\right\}$$

with

$$T = [N^{1/6}]^6. \tag{1.310}$$

We put $Q = C_1\|f\|_2^{1/4} R(N\log\log N)^{1/2}$ and apply the inequality

$$\lambda_1(\{x : h(x) \ge (1+\eta_0)Q\}) \le e^{-(1+\eta_0)Q\kappa} \cdot \int_0^1 e^{\kappa h(x)} dx \tag{1.311}$$

with

$$h(x) = \sum_{k\le N} g(a_k x) \quad \text{and} \quad \kappa = (\|f\|_2^{-3/2} N^{-1} \log\log N)^{1/2}.$$

To estimate the integral in (1.311) we define the integer P by

$$2HP \leq N < 2H(P+1). \tag{1.312}$$

Then from above and the hypothesis on $\|f\|_2$ we obtain

$$\kappa \left| \sum_{k\leq N} g(a_k x) - \sum_{m=0}^{2P-1} U_m(x) \right| \leq \kappa \sum_{k=2HP+1}^{N} |g(a_k x)| = O(\kappa H \log H) = o(1).$$

Hence CAUCHY-SCHWARZ's inequality, Lemma 1.169 and (1.312)

$$\int_0^1 e^{\kappa h(x)}\,dx \ll \int_0^1 \exp\left\{\kappa \sum_{m=0}^{2P-1} U_m(x)\right\} dx$$

$$\leq \left(\int_0^1 \exp\left\{2\kappa \sum_{m=0}^{P-1} U_{2m}(x)\right\} dx\right)^{1/2} \left(\int_0^1 \exp\left\{2\kappa \sum_{m=1}^{P} U_{2m-1}(x)\right\} dx\right)^{1/2}$$

$$\ll e^{2(1+\eta_0)C_1\kappa^2\|f\|_2 N}.$$

Thus (1.311) yields

$$\lambda_1(B_1) \ll \exp(-(1+\eta_0)C_1(R-2)\|f\|_2^{-1/2} \log\log N). \tag{1.313}$$

Furthermore Lemma 1.168 provides

$$\lambda_1(B_2) \ll \frac{1}{\eta_0^2 Q^2} \int_0^1 \left(\sum_{k\leq N} \varphi_T(a_k x)\right)^2 dx \ll Q^{-2} \ll R^{-2}N^{-3/4}. \tag{1.314}$$

Hence (1.309), (1.313) and (1.314) give

$$\lambda_1 \left\{ x : \sum_{k=1}^{N} f(a_k x) \geq (1+2\eta_0)C_1 R \|f\|_2^{1/4} (N \log\log N)^{1/2} \right\}$$

$$\ll \exp(-(1+\eta_0)\|f\|_2^{-1/2}(R-2)\log\log N) + R^{-2}N^{-3/4}.$$

The same estimate holds for $-f$. Thus the proof of Proposition 1.166 is finished. □

Let $\delta_1, \delta_2 > 0$ and let $N \geq N_0(\delta_1, \delta_2)$ be given. Put

$$H = [\log N / \log 4] + 1. \tag{1.315}$$

Any $0 \leq \alpha \leq 1$ can be written in dyadic expansion:

$$\alpha = \sum_{j=1}^{\infty} 2^{-j}\varepsilon_j, \quad \varepsilon_j \in \{0,1\}.$$

Obviously, we have

$$\sum_{j=1}^{H} 2^{-j}\varepsilon_j \le \alpha \le \sum_{j=1}^{H} 2^{-j}\varepsilon_j + 2^{-H}.$$

We define for $1 \le h \le H$ the intervals

$$I(h) = \left[\sum_{j=1}^{h} 2^{-j}\varepsilon_j, \sum_{j=1}^{h+1} 2^{-j}\varepsilon_j\right) \quad \text{and} \quad J(H) = \left[\sum_{j=1}^{H} 2^{-j}\varepsilon_j, \sum_{j=1}^{H} 2^{-j}\varepsilon_j + 2^{-H}\right),$$

and set

$$\begin{aligned} \varrho_h(x) &= \varrho_h(x,\alpha) = \tilde{\chi}_{I(h)} \quad (1 \le h < H), \\ \sigma_H(x) &= \sigma_H(x,\alpha) = \tilde{\chi}_{J(H)}\,. \end{aligned}$$

Here $\tilde{\chi}_I$ denotes the characteristic function of an interval I extended with period 1. Then

$$\sum_{h=1}^{H-1} \varrho_h(x) \le \tilde{\chi}_{[0,\alpha]}(x) \le \sum_{h=1}^{H-1} \varrho_h(x) + \sigma_H(x). \tag{1.316}$$

For fixed h there are only 2^h different functions ϱ_h and there are only 2^H different functions σ_H as α varies between 0 and 1. We denote these functions by $\varrho_h^{(j)} (1 \le j \le 2^h)$ and $\sigma_H^{(j)} (1 \le j \le 2^H)$. Moreover, these functions have the same structure. Thus we define

$$\varphi_h^{(j)} = \begin{cases} \varrho_h^{(j)}, & 1 \le j \le 2^h \quad \text{if} \quad 1 \le h < H, \\ \sigma_H^{(j)}, & 1 \le j \le 2^H \quad \text{if} \quad h = H. \end{cases} \tag{1.317}$$

For integers $1 \le j \le 2^h, 1 \le h \le H, N \ge 1, M \ge 0$ we set

$$F(M,N,j,h,x) = \left| \sum_{k=M+1}^{M+N} \left(\varphi_h^{(j)}(a_k x) - \int_0^1 \varphi_h^{(j)}(x)\,dx \right) \right|. \tag{1.318}$$

Lemma 1.170 *Define n by $2^n \le N < 2^{n+1}$. Then there are integers m_l with $0 \le m_l < 2^{n-l} (1 \le l \le n)$ such that*

$$F(0,N,j,h,x) \le F(0,2^n,j,h,x) + \sum_{\frac{1}{3}n \le l \le n} F(2^n + m_l 2^l, 2^{l-1}, j, h, x) + N^{1/3}.$$

Proof. For notational convenience we set $F(M,N) = F(M,N,j,h,x)$. We have

$$F(0,N) \le F(0,2^n) + F(2^n, N-2^n).$$

Next we expand N as a dyadic integer obtaining

$$N = 2^n + e_{n-1}2^{n-1} + e_{n-2}2^{n-2} + \ldots + e_0,$$

with digits $e_i \in \{0,1\}$. Since

$$F(M,N) \le F(M,N') + F(M+N', N-N') \text{ for } N' < N,$$

we get

$$\begin{aligned} F(2^n, N-2^n) &= F(2^n, e_{n-1}2^{n-1} + \ldots + e_0) \\ &\le F(2^n, e_{n-1}2^{n-1} + \ldots + e_1 \cdot 2) \\ &\quad + F(2^n + e_{n-1}2^{n-1} + \ldots + e_1 \cdot 2, e_0) \\ &\le F(2^n, e_{n-1}2^{n-1}) \\ &\quad + \sum_{l=1}^{n-1} F(2^n + e_{n-1}2^{n-1} + \ldots + e_l 2^l, e_{l-1}2^{l-1}). \end{aligned}$$

Next we set $m_n = 0$, $m_l = e_{n-1}2^{n-1-l} + \ldots + e_l$ for $l \ge 1$. Then $0 \le m_l \le 2^{n-l} - 1$. Furthermore

$$\begin{aligned} F(2^n + e_{n-1}2^{n-1} + \ldots + \ldots + e_l 2^l, e_{l-1}2^{l-1}) &= F(2^n + m_l 2^l, e_{l-1}2^{l-1}) \\ &\le F(2^n + m_l 2^l, 2^{l-1}). \end{aligned}$$

This yields the result. □

Next we set

$$\phi(N) = 2(1 + 2\eta_0)C_1(N \log\log N)^{1/2} \tag{1.319}$$

where C_1 is given in (1.301) and define the sets

$$G(n,j,h) = \left\{x : \; F(0, 2^n, j, h) \ge 2^{-h/8}\phi(2^n)\right\},$$

$$H(n,j,h,l,m) = \left\{x : F(2^n + m2^l, 2^{l-1}, j, h) \ge 2^{-h/8}2^{(l-n-3)/6}\phi(2^n)\right\},$$

$$G_n = \bigcup_{h \le H} \bigcup_{j \le 2^h} G(n,j,h),$$

and

$$H_n = \bigcup_{h \le H} \bigcup_{j \le 2^h} \bigcup_{\frac{1}{3}n \le l \le n} \bigcup_{m \le 2^{n-l}} H(n,j,h,l,m).$$

Lemma 1.171 *There is an $n_0 = n_0(\delta_1, \delta_2)$ such that*

$$\lambda_1\left(\bigcup_{n \ge n_0} (G_n \cup H_n)\right) < \delta_1.$$

Proof. We observe that by (1.315)

$$N^{-1/2} \ll 2^{-h-1} \le \left\| \varphi_h^{(j)} - \int_0^1 \varphi_h^{(j)} \right\|_2^2 \le 2^{-h} \tag{1.320}$$

for $1 \le j \le 2^h, 1 \le h \le H$. Now Proposition 1.166 with $M = 0, N = 2^n$ and $R = 3$ gives

$$\lambda_1(G(n,j,h)) \ll \exp(-(1+\delta_2)2^{h/4}\log n) + N^{-3/4},$$

and so

$$\lambda_1(G_n) \ll n^{-1-\frac{1}{2}\delta_2} \tag{1.321}$$

if N is sufficiently large. A similar application of Proposition 1.166 with $M = 2^n + m2^l, N = 2^{l-1}$ and $R = 2^{(n-l)/3}$ yields

$$\lambda_1(H(n,j,h,l,m)) \ll \exp(-(1+\delta_2)2^{h/4}R\log n) + 2^{2(l-n)/3}N^{-3/4}.$$

Hence

$$\lambda_1(H_n) \ll n^{-1-\frac{1}{2}\delta_2}, \tag{1.322}$$

and the proof of the lemma is complete. □

Proof of the upper bound of Theorem 1.163. By using the above notation we obtain for all $\beta \in [0,1]$

$$\begin{aligned}
\left|\sum_{k\le N} \tilde{\chi}_{[0,\beta]}(a_k x) - N\beta\right| &\le \sum_{h=1}^{H}\left|\sum_{k\le N}\varphi_h^{(j)}(a_k x) - N\int_0^1 \varphi_h^{(j)}(x)\,dx\right| + 2^{-H}N \\
&\le \sum_{h\le H}\left(F(0,2^n,j,h,x) + \sum_{\frac{1}{3}n\le l\le n} F(2^n + m_l 2^{l-1}, j, h, x)\right) + 2N^{1/2} \\
&\le \phi(2^n)\sum_{h\le H} 2^{-h/8}\left(1 + \sum_{\frac{1}{3}n\le l\le n} 2^{(l-n-3)/6}\right) + 2N^{1/2} \\
&\le \phi(N)(2^{1/8}-1)^{-1}(1 + 2^{-1/2}(1-2^{-1/6})^{-1}) + 2N^{1/2} \\
&\le (1+4\delta_2)(83 + 332(q^{1/2}-1)^{-1})(N\log\log N)^{1/2}
\end{aligned}$$

for all x except a set of measure less than δ_1. Hence for those x

$$\begin{aligned}
&\left|\sum_{k\le N} \tilde{\chi}_{[\beta_1,\beta_2]}(a_k x) - N(\beta_1 - \beta_2)\right| \\
&\qquad \le (1+4\delta_2)(166 + 664(q^{1/2}-1)^{-1})(N\log\log N)^{1/2}.
\end{aligned}$$

Since $\delta_1 > 0$ and $\delta_2 > 0$ can be chosen arbitrarily small the upper bound in the theorem follows immediately. □

The proof of the lower bound of Theorem 1.163 relies on KOKSMA-HLAWKA's inequality (Theorem 1.14), which implies

$$\left|\sum_{k\le N} e(a_k x)\right| \le 4ND_N(a_n x), \tag{1.323}$$

and on the following proposition.

Proposition 1.172 *For every $\varepsilon > 0$ and $\eta > 0$ and for every integer $N > 0$ there exists a finite sequence of integers $N < N_1 < N_2 < \cdots < N_k$ such that*

$$\max_{1\le\nu\le k} \frac{\left|\sum_{k\le N_\nu} e(a_k x)\right|}{\sqrt{N\log\log N}} \ge 1-\varepsilon$$

for every $x \in [0,1]$ except possibly a set of measure $\le \eta$.

Obviously, Proposition 1.172 and (1.323) provide the lower bound of Theorem 1.163.

However, the proof of Proposition 1.172 requires some preliminaries. First we have to estimate the number of solutions of the equation

$$L(\mathbf{x},\mathbf{y}) = (x_1+x_2+\cdots+x_p)-(y_1+y_2+\cdots+y_p) = 0,$$

in which $x_1, x_2, \ldots, x_p$ and $y_1, y_2, \ldots, y_p$ are restricted to values of the set $\{a_1, a_2, \ldots, a_N\}$.

Proposition 1.173 *Let $1 \le p \le N$ and $L(\mathbf{x},\mathbf{y}) = (x_1+x_2+\ldots+x_p) - (y_1+y_2+\ldots+y_p)$, where $x_1,\ldots,x_p$ and $y_1,\ldots,y_p$ are restricted to the values $a_1 = 1, a_2, \ldots, a_N$. Then there exists a $c = c(q) > 0$ independent of p,s and the sequence $a_1 = 1, a_2, \ldots, a_N$ such that*

$$p!^2\binom{N}{p} \le \sum_{L(\mathbf{x},\mathbf{y})=0} 1 \le p!^2\binom{N}{p} + (cp)^{3p}N^{p-1}, \tag{1.324}$$

and for every real s

$$0 \le \sum_{s-\frac12\le L(\mathbf{x},\mathbf{y})\le s+\frac12,\ L(\mathbf{x},\mathbf{y})\ne 0} 1 \le (cp)^{3p}N^{p-1}. \tag{1.325}$$

The proof of Proposition 1.173 requires several steps. First we prove the following estimate for the number of a_ν's lying in a given interval:

Lemma 1.174 *If $0 < \alpha < \beta$ then*

$$\sum_{\alpha\le a_\nu\le\beta} 1 \le \frac{\log((\beta/\alpha)q)}{\log q}, \tag{1.326}$$

and if α is arbitrary real then

$$\sum_{\alpha\le a_\nu\le\alpha+1} 1 \le \frac{\log(2q)}{\log q}. \tag{1.327}$$

Proof. In order to prove (1.326) let ν_0 be defined by the inequality $a_{\nu_0} < \alpha \le a_{\nu_0+1}$ ($a_0 = 0$) and $i \ge 0$ be defined by the inequality $a_{\nu_0+i} \le \beta < a_{\nu_0+i+1}$. If $i = 0$ then (1.326) is true. If $i \ge 1$ then we have

$$\beta \ge a_{\nu_0+i} \ge q^{i-1}a_{\nu_0+1} \ge q^{i-1}\alpha.$$

Hence $\beta q/\alpha \geq q^i$ and (1.326) follows immediately. Now we prove (1.327): Since $a_1 = 1$ (1.327) is trivial for negative values of α. If $0 \leq \alpha \leq 1$ we have

$$\sum_{\alpha \leq a_\nu \leq \alpha+1} 1 \leq \sum_{1 \leq a_\nu \leq 2} 1,$$

whence we obtain (1.327) by using (1.326) with $\alpha = 1, \beta = 2$. If $\alpha > 1$ then (1.327) is an immediate consequence of (1.326). □

Next we want to prove that the number of $a_k \neq a_l$ pairs satisfying the inequality $s - \frac{1}{2} \leq a_k - a_l \leq s + \frac{1}{2}$ is uniformly bounded for every real s. More precisely we prove the following:

Lemma 1.175 *Let s be an arbitrary real number and let $A(l, N, s)$ denote the number of those $a_k \neq a_l$ pairs which satisfy the inequality*

$$s - \frac{1}{2} \leq a_k - a_l \leq s + \frac{1}{2}.$$

Then there exists a positive constant $c=c(q)$ independent of s and the choice of the sequence $a_1 = 1, a_2, \ldots, a_N$ such that

$$0 \leq A(l, N, s) \leq c. \tag{1.328}$$

Remark. In fact we shall prove that (1.328) holds for any $c = c(q)$ satisfying

$$c \geq 2\frac{\log(2q)}{\log q} \cdot \frac{\log(2q^2/(q-1))}{\log q}, \tag{1.329}$$

whence the independence is obvious.

Proof of Lemma 1.175. First of all $A(l, N, s) = A(l, N, -s)$, hence we may assume that $s \geq 0$. If $0 \leq s \leq \frac{1}{2}$ then $\frac{1}{2}A(l, N, s)$ is not more than the number of a_k, a_l pairs satisfying $0 < a_k - a_l \leq l$. Since $k \geq l + 1$ we get

$$1 \geq a_k - a_l \geq a_k(1 - q^{-1}),$$

and on the other hand $a_k \geq 1$. Therefore the possible a_k's satisfy the inequality $1 \leq a_k \leq q/(q-1)$ and their number can be estimated by (1.326):

$$\sum_{a_k} 1 \leq \frac{\log(q^2/(q-1))}{\log q}.$$

For a fixed value of a_k the number of possible a_l's can be estimated by (1.327). Namely, we have $a_k - 1 \leq a_l \leq a_k$, and so by (1.327)

$$\sum_{a_l} 1 \leq \frac{\log(2q)}{\log q}.$$

Consequently, we have for $0 \leq s \leq \frac{1}{2}$

$$A(l, N, s) \leq 2\frac{\log(2q)}{\log q} \cdot \frac{\log(q^2/(q-1))}{\log q}.$$

Now let $s \geq \frac{1}{2}$. We have $a_k > a_l$ and so

$$s + \frac{1}{2} \geq a_k - a_l \geq a_k(1 - q^{-1}).$$

On the other hand $a_k \geq \max(1, s - \frac{1}{2})$. Hence if $s \geq \frac{3}{2}$ we have the inequality

$$s - \frac{1}{2} \leq a_k \leq \frac{2(s - \frac{1}{2})}{(1 - q^{-1})},$$

and if $\frac{1}{2} \leq s \leq \frac{3}{2}$ then

$$1 \leq a_k \leq \frac{2}{(1 - q^{-1})}.$$

In the first case we may use (1.326) with $\alpha = s - \frac{1}{2}, \beta = 2(s - \frac{1}{2})q/(q-1)$ and in the second case with $\alpha = 1, \beta = 2q/(q-1)$. Consequently in both cases

$$\sum_{a_k} 1 \leq \frac{\log(2q^2/(q-1))}{\log q}.$$

For a fixed value of a_k the number of possible a_l's can be estimated by (1.327):

$$\sum_{a_l} 1 \leq \frac{\log(2q)}{\log q}.$$

Therefore we have for $s \geq \frac{1}{2}$

$$A(l, N, s) \leq \frac{\log(2q)}{\log q} \cdot \frac{\log 2q^2/(q-1)}{\log q}.$$

This completes the proof of (1.328) and (1.329). □

Now we consider the inequality

$$s - \frac{1}{2} \leq L(\mathbf{x}, \mathbf{y}) \leq s + \frac{1}{2}.$$

We want to prove that the number of distinct (x_p, y_p) pairs which occur among the solutions is at most $\mathcal{O}(N)$ uniformly in s. (Here the restriction $L(\mathbf{x}, \mathbf{y}) \neq 0$ is omitted. Hence choosing $s = 0$ we obtain a similar result for the solution of $L(\mathbf{x}, \mathbf{y}) = 0$.)

Lemma 1.176 *Let $p \geq 1$ and s arbitrary real. Let $B_p(s)$ denote the number of distinct $x_p = a_k, y_p = a_l$ pairs which occur among the solutions of*

$$s - \frac{1}{2} \leq L(\mathbf{x}, \mathbf{y}) \leq s + \frac{1}{2} \tag{1.330}$$

where the x's and y's take the values $a_1 = 1, a_2, \ldots, a_N$ and are restricted to the conditions $x_1 \leq x_2 \leq \ldots \leq x_p$ and $y_1 \leq y_2 \leq \ldots \leq y_p$. Then we have

$$B_p(s) \leq 8pN \frac{\log(1+q)}{\log q}. \tag{1.331}$$

Proof. Since $B_p(s) = B_p(-s)$ we may assume that $s \geq 0$. We must distinguish between two types of solutions; 1) those for which $x_p \leq 2s+1$ and 2) those for which $x_p > 2s+1$. Let us consider solutions of the first kind; let $x_p = a_k$ be a possibility. Then using (1.330) we get $(s - \frac{1}{2}) \leq pa_k$, and so

$$\frac{s-\frac{1}{2}}{p} \leq a_k \leq 2s+1.$$

From this we conclude for $s \geq \frac{3}{2}$

$$\frac{s-\frac{1}{2}}{p} \leq a_k \leq 4(s - \frac{1}{2}),$$

and for $0 \leq s \leq \frac{3}{2}$ we obtain $1 \leq a_k \leq 4$. Hence we may use (1.326) in both cases and get in either case

$$\sum_{x_p \leq 2s+1} 1 \leq \frac{\log(4pq)}{\log q} < 4p \frac{\log(1+q)}{\log q}.$$

For a fixed value of $x_p = a_k$ there are at most N choices for y_p (namely $a_1, a_2, \ldots, a_N$) and so

$$\sum_{(x_p, y_p),\ x_p \leq 2s+1} 1 \leq 4pN \frac{\log(1+q)}{\log q}.$$

Next we consider the solutions of the second kind. Here we first estimate the number of possible y_p's. Let $y_p = a_l$ be a possibility. Then using $x_p > 2s+1$ we obtain from (1.330)

$$x_p \leq py_p + s + \frac{1}{2} < py_p + \frac{x_p}{2},$$

that is to say $x_p/2p < y_p$. On the other hand we obtain again from (1.330)

$$y_p \leq px_p + \frac{1}{2} - s \leq px_p + \frac{1}{2} < 2px_p.$$

Hence $x_p/2p \leq y_p = a_l \leq 2px_p$. Consequently (1.326) can be used, and it follows that if $x_p > 2s+1$ then

$$\sum_{y_p} 1 \leq \frac{\log 4p^2 q}{\log q} < 4p \frac{\log(1+q)}{\log q}.$$

There are at most N possibilities for $x_p > 2s+1$, hence

$$\sum_{(x_p, y_p),\ x_p > 2s+1} 1 \leq 4pN \frac{\log(1+q)}{\log q}.$$

This establishes the inequality (1.331). □

Now we are able to complete the proof of Proposition 1.173.

Proof of Proposition 1.173. Let p; $1 \leq p \leq N$ be a positive integer and let $D(p, N)$ denote the number of solutions of the diophantine equation $L(\mathbf{x}, \mathbf{y}) = 0$ and $D(p, N, s)$

($s \in \mathbf{R}$) the number of those solutions of the inequalitiy $s - \frac{1}{2} \leq L(\mathbf{x}, \mathbf{y}) \leq s + \frac{1}{2}$ for which $L(\mathbf{x}, \mathbf{y}) \neq 0$ with the restrictions that in both cases $x_1 \leq x_2 \leq \ldots \leq x_p$ and $y_1 \leq y_2 \leq \ldots \leq y_p$ and that $x_1, \ldots, x_p$ and $y_1, \ldots, y_p$ can take the values $a_1 = 1, a_2, \ldots, a_N$. We will show that there exists a positive constant $c = c(q)$ independent of p,s and the choice of the sequence $a_1 = 1, a_2, \ldots, a_N$ such that

$$\binom{N}{p} \leq D(p, N) \leq \binom{N}{p} + (cp)^p N^{p-1} \tag{1.332}$$

and

$$0 \leq D(p, N, s) \leq (cp)^p N^{p-1} \tag{1.333}$$

for every $1 \leq p \leq N$ and real s. Obviously, Proposition 1.173 follows from (1.332) and (1.333).

We will prove (1.332) and (1.333) by induction on p. First let $p = 1$. In this case $D(1, N) = N$ and so (1.332) is obviously true. For $p = 1$ (1.333) had been established in Lemma 1.175, inequality (1.328). Hence $c(q) > 0$ must satisfy (1.329).

Let us assume now that (1.333) is true for $1, 2, \ldots, (p-1)$ and let us prove (1.332) for $p \geq 2$. Let ν ($2 \leq \nu \leq p$) be fixed and let $D^{(\nu)}$ denote the number of those solutions of $L(\mathbf{x}, \mathbf{y}) = 0$ which satisfy $x_1 \leq x_2 \leq \ldots \leq x_p$ and $y_1 \leq y_2 \leq \ldots \leq y_p$ and also the additional condition

$$x_\nu \neq y_\nu \quad \text{and} \quad x_{\nu+1} = y_{\nu+1}, x_{\nu+2} = y_{\nu+2}, \ldots, x_p = y_p. \tag{1.334}$$

Using Lemma 1.176 we can estimate the number of possible x_ν, y_ν pairs. For, $L(\mathbf{x}, \mathbf{y}) = 0$ and (1.334) imply

$$-\frac{1}{2} < L(\mathbf{x}, \mathbf{y}) = (x_1 + \ldots + x_\nu) - (y_1 + \ldots + y_\nu) < \frac{1}{2},$$

thus (1.330) is satisfied for every solution and so by (1.331)

$$\sum_{(x_\nu, y_\nu)} 1 \leq 8\nu N \frac{\log(1+q)}{\log q}. \tag{1.335}$$

Now we fix one possible x_ν, y_ν pair; $x_\nu = a_k$ and $y_\nu = a_l$, say. Let $D^{(\nu)}(a_k, a_l)$ denote the number of those solutions of $L(\mathbf{x}, \mathbf{y}) = 0$ and (1.334) for which $x_\nu = a_k$ and $y_\nu = a_l$. Obviously

$$D^{(\nu)} = \sum_{x_\nu = a_k,\ y_\nu = a_l} D^{(\nu)}(a_k, a_l). \tag{1.336}$$

In order to estimate $D^{(\nu)}(a_k, a_l)$ we consider the system of equations

1. $x_1 + x_2 + \ldots + x_{\nu-1} = y_1 + y_2 + \ldots + y_{\nu-1} + (a_l - a_k)$
2. $x_1 \leq x_2 \leq \ldots \leq x_{\nu-1}$; $y_1 \leq y_2 \leq \ldots \leq y_{\nu-1}$
3. $x_{\nu+i} = y_{\nu+i}$ ($i = 1, 2, \ldots, p - \nu$).

It is obvious that $D^{(\nu)}(a_k, a_l)$ is majorized by the number of solutions of this system. However $a_l \neq a_k$, and so the number of solutions of 1. subject to the condition 2. is at most $D(\nu-1, N, a_l - a_k)$. The number of solutions of 3. is exactly $\binom{N}{p-\nu}$. Hence using our assumption (1.333) we get

$$D^{(\nu)}(a_k, a_l) \leq D(\nu-1, N, a_l - a_k)\binom{N}{p-\nu} \leq c^{\nu-1}\nu^{\nu-1}N^{p-2}.$$

Now we use (1.335) and (1.336) and obtain the estimate

$$D^{(\nu)} \leq c^{\nu-1}\nu^{\nu-1}N^{p-2} \sum_{x_\nu = a_k,\ y_\nu = a_l} 1 \leq 8\frac{\log(1+q)}{\log q}c^{\nu-1}\nu^{\nu}N^{p-1}.$$

Consequently $D^{(\nu)} \leq \frac{1}{4}(c\nu)^{\nu}N^{p-1}$, provided $c = c(q)$ satisfies

$$c \geq 32\frac{\log(1+q)}{\log q}. \tag{1.337}$$

Finally we sum up with respect to $\nu = 2, 3, \ldots, p$:

$$D' = \sum_{\nu=2}^{p} D^{(\nu)} \leq \frac{1}{4}c^p N^{p-1}\sum_{\nu=2}^{p}\nu^{\nu} < \frac{1}{2}(cp)^p N^{p-1}.$$

Here D' denotes those solutions of $L(\mathbf{x}, \mathbf{y}) = 0$ for which at least one $x_\nu \neq y_\nu$. The number of remaining solutions is between $\binom{N}{p}$ and $\binom{N}{p} + N^{p-1}p^p$, hence in fact

$$\binom{N}{p} \leq D(p, N) \leq \binom{N}{p} + (cp)^p N^{p-1}.$$

We assume again that (1.333) is true for $1, 2, \ldots, (p-1)$ and we prove (1.333) for $p \geq 2$. We are interested in those solutions of

$$s - \frac{1}{2} \leq L(\mathbf{x}, \mathbf{y}) = (x_1 + x_2 + \ldots + x_p) - (y_1 + y_2 + \ldots + y_p) \leq s + \frac{1}{2}$$

for which $L(\mathbf{x}, \mathbf{y}) \neq 0$ and $x_1 \leq x_2 \leq \ldots \leq x_p$ and $y_1 \leq y_2 \leq \ldots \leq y_p$. We must distinguish between two types of solutions: 1. those for which $x_1 = y_1$ and 2. those for which $x_1 \neq y_1$.

If $x_1 = y_1$ is fixed we have

$$s - \frac{1}{2} \leq (x_2 + \ldots + x_p) - (y_2 + \ldots + y_p) \leq s + \frac{1}{2}$$

and $x_2 + \ldots + x_p \neq y_2 + \ldots + y_p$. Hence using our assumption, for fixed $x_1 = y_1$ the number of solutions is at most $D(p-1, N, s) \leq c^{p-1}p^p N^{p-2}$. There are N possibilities for $x_1 = y_1$, hence the number of solutions of the first type is

$$D_1 \leq c^{p-1}p^p N^{p-1}.$$

Now we consider the solutions of the second type: If $x_{\nu+1}, \dots, x_p$ and $y_{\nu+1}, \dots, y_p$ are fixed then the number of possible x_ν, y_ν pairs can be estimated by Lemma 1.176, inequality (1.331). Hence using (1.337) the number of choices for $x_2, x_3, \dots, x_p$ and $y_2, y_3, \dots, y_p$ can be estimated by

$$\sum_{x_1,\dots,x_p,\ y_2,\dots,y_p} 1 \le \left(8\frac{\log(1+q)}{\log q}\right)^{p-1} p^{p-1}N^{p-1} \le \frac{1}{4}c^{p-1}p^pN^{p-1}.$$

For fixed $x_2, \dots, x_p$ and $y_2, \dots, x_p$ the number of possible $x_1 \ne y_1$ pairs can be estimated by Lemma 1.175, inequality (1.328). Hence the number of solutions of the second type is at most

$$D_2 \le \frac{1}{4}c^p p^p N^{p-1}.$$

Finally $D(p, N, s) = D_1 + D_2 \le (cp)^p N^{p-1}$. This completes the proof of Proposition 1.173. □

The next step is to find proper estimates for the function

$$\phi(t) = \left(\{x | \alpha \le x \le \beta;\ F(N; x) \ge \sqrt{tN \log\log N}\}\right), \tag{1.338}$$

where

$$F(N; x) = \left|\sum_{n=1}^{N} e(a_n x)\right|$$

and $0 < a_1 < \dots < a_N$ is a fixed lacunary sequence satisfying $a_{\nu+1}/a_\nu \ge q > 1$, $1 \le \nu < N$. (Notice that the condition $a_1 = 1$ has been dropped.) For this purpose we give an asymptotic expression for the value of the integral

$$I = \int_\alpha^\beta |F(N; x)|^{2p} dx, \tag{1.339}$$

when $p = \mathcal{O}(\log\log N)$, $\beta - \alpha \ge 1/(a_1\sqrt{N})$ and $N \to \infty$.

Lemma 1.177 *Let α, β be real and such that $\beta - \alpha \ge 1/(a_1\sqrt{N})$. Furthermore let p be a positive integer satisfying $1 \le p \le 3 \log\log N$. Then*

$$|I - (\beta - \alpha)p! N^p| \le (\beta - \alpha)N^{p-\frac{1}{2}} \tag{1.340}$$

for every $N \ge N_0(q)$, where $N_0(q)$ is independent of α, β, p and the sequence $a_1, a_2, \dots, a_N$.

Proof. We have from (1.339)

$$I = \sum_{1 \le k_\nu, l_\nu \le N} \int_\alpha^\beta e\left(\sum_{\nu=1}^{p}(a_{k_\nu} - a_{l_\nu})x\right) dx.$$

Hence introducing the notations of the previous section we have

$$I = (\beta-\alpha)\sum_{L(\mathbf{x},\mathbf{y})=0} 1 + \sum_{\substack{-(a_1/2)\le L(\mathbf{x},\mathbf{y})\le a_1/2\\ L(\mathbf{x},\mathbf{y})\ne 0}} \int_\alpha^\beta e\left(L(\mathbf{x},\mathbf{y})\xi\right)\,d\xi$$

$$+\sum_{s\ne 0}{}' \sum_{\substack{a_1(s-\frac12)\le L(\mathbf{x},\mathbf{y})\le a_1(s+\frac12)\\ L(\mathbf{x},\mathbf{y})\ne 0}} \int_\alpha^\beta e\left(L(\mathbf{x},\mathbf{y})\xi\right)\,d\xi.$$

The dash in $\sum'$ indicates that there are at most N^{2p} distinct values for $s=\pm1,\pm2,\dots$ for which the contribution is not zero.

In the first sum we estimate the integral by

$$\left|\int_\alpha^\beta e\left(L(\mathbf{x},\mathbf{y})\xi\right)\,d\xi\right| \le \beta-\alpha$$

and in the second sum we use the inequality

$$\left|\int_\alpha^\beta e\left(L(\mathbf{x},\mathbf{y})\xi\right)d\xi\right| \le \frac{1}{\pi|L(\mathbf{x},\mathbf{y})|}.$$

Hence applying the inequalities (1.324) and (1.325) of Lemma 1.173 we obtain

$$\begin{aligned}\left|I-(\beta-\alpha)p!^2\binom{N}{p}\right| &\le (\beta-\alpha)(cp)^{3p}N^{p-1}+(\beta-\alpha)\sum_{-(a_1/2)\le L(\mathbf{x},\mathbf{y})\le a_1/2} 1\\ &\quad+\sum_{s\ne0}{}'\sum_{a_1(s-\frac12)\le L(\mathbf{x},\mathbf{y})\le a_1(s+\frac12)} L(\mathbf{x},\mathbf{y})^{-1}\\ &\le 2(\beta-\alpha)(cp)3pN^{p-1}+\frac{(cp)^{3p}N^{p-\frac12}}{a_1\sqrt{N}}\sum_{s\ne0}{}'\frac{1}{|s-\frac12|}.\end{aligned}$$

Since there are at most N^{2p} distinct choices for $s=\pm1,\pm2,\dots$ in $\sum'$ we get from above

$$\left|I-(\beta-\alpha)p!^2\binom{N}{p}\right| \le 6(\beta-\alpha)(cp)^{3p+1}N^{p-\frac12}\log N,$$

provided that $\beta-\alpha\ge 1/a_1\sqrt{N}$. If p satisfies the inequality $1\le p\le 3\log\log N$ then the right hand side is less then $\frac12(\beta-\alpha)N^{p-\frac12}$ for $N\ge N_0(c)=N_0(q)$, and $p!^2\binom{N}{p}$ can be replaced by $p!N^p$ which yields an error less than $\frac12 N^{p-\frac12}$. Hence (1.340) follows. □

The following upper bounds for the measure $\phi(t)$ $(0\le t\le N)$ (defined in (1.338)) is a little bit sharper than that of Proposition 1.166.

Lemma 1.178 *We have*

$$\phi(t)\le\begin{cases}(\beta-\alpha)\frac{18\log\log N}{(\log N)^t} & \text{for } 0\le t\le 3, \text{ and}\\ (\beta-\alpha)\frac{6\log\log N}{t^{2\log\log N}} & \text{for } 3\le t\le N\end{cases}\tag{1.341}$$

provided that $\beta - \alpha \geq 1/(a_1\sqrt{N})$ *and* $N \geq N_0(q)$ *where* $N_0(q)$ *is independent of* α, β *and the sequence* $a_1, a_2, \ldots, a_N$.

Proof. Obviously we have

$$\phi(t) \leq \int\limits_{F(N;x)^2 \geq t \log\log N} \left(\frac{F(N;x)}{\sqrt{tN \log\log N}}\right)^{2p} dx \leq \frac{I}{(tN \log\log N)^p}$$

for any $t > 0$, $p = 1, 2, \ldots$. Hence using (1.340) and replacing $p!$ by $p(p/e)^p$ it follows that

$$\phi(t) \leq 2(\beta - \alpha)p\left(\frac{p}{et \log\log N}\right)^p,$$

provided that $N \geq N_0(q)$, $p_0 \leq p \leq 3 \log\log N$ and $\beta - \alpha \geq 1/(a_1\sqrt{N})$.

If $0 < t \leq 3$ we choose $p = [t \log\log N]$ and obtain

$$\phi(t) \leq 6(\beta - \alpha)(\log\log N)e^{-[t \log\log N]} < \frac{18(\beta - \alpha)\log\log N}{(\log N)^t}.$$

If $t \geq 3$ we choose $p = [e \log\log N]$ and get

$$\phi(t) < 6(\beta - \alpha)(\log\log N)t^{-[e \log\log N]} < \frac{6(\beta - \alpha)\log\log N}{t^{2 \log\log N}}.$$

This proves Lemma 1.178. □

Using (1.340) and (1.341) we can even find a lower bound for $\phi(t)$; $0 < t < 1$.

Lemma 1.179 *Let* ε, $0 < \varepsilon < 1$, *be arbitrary. Then*

$$\phi(1 - \varepsilon) > \frac{\beta - \alpha}{(\log N)^{1-4\varepsilon^2}}. \tag{1.342}$$

for any α, β *satisfying* $\beta - \alpha \geq 1/(a_1\sqrt{N})$ *and every* $N \geq N_0(q, \varepsilon)$. *This bound* $N_0(q, \varepsilon)$ *is independent of* α *and* β.

Proof. For the sake of simplicity let

$$R(x) = F(N;x)^2/N \log\log N.$$

Furthermore let us introduce the following subsets of the interval $\alpha \leq x \leq \beta$:

$$\begin{aligned}
E &= \{x \in [a,b] \,|\, 1 - \varepsilon \leq R(x) \leq 1\} \\
E_1 &= \{x \in [a,b] \,|\, 0 < R(x) < 1 - \varepsilon\} \\
E_2 &= \{x \in [a,b] \,|\, 1 < R(x) \leq 3\} \\
E_3 &= \{x \in [a,b] \,|\, 3 < R(x) \leq N\}.
\end{aligned}$$

According to Lemma 1.177

$$\int_\alpha^\beta R(x)^p dx > (\beta - \alpha)\left(\frac{p}{e \log\log N}\right)^p$$

for $1 \le p \le 3 \log\log N, \beta - \alpha \ge 1/(a_1\sqrt{N})$ and $N \ge N_0(q)$. Hence

$$\begin{aligned} \phi(1-\varepsilon) &\ge \lambda_1(E) \ge \int_E R(x)^p dx \ge \\ &\ge (\beta-\alpha)\left(\frac{p}{e\log\log N}\right)^p - \left(\int_{E_1} + \int_{E_2} + \int_{E_3}\right) R(x)^p dx, \end{aligned}$$

or equivalently

$$\frac{\phi(1-\varepsilon)}{\beta-\alpha} \ge \left(\frac{p}{e\log\log N}\right)^p - (I_1+I_2+I_3), \tag{1.343}$$

where

$$I_i = \frac{1}{\beta-\alpha}\int_{E_i} R(x)^p dx \qquad (i=1,2,3).$$

We chose $p = [(1-(\varepsilon/2))\log\log N]$ and estimate I_1, I_2 and I_3 from above. First of all, by using Lemma 1.178 we obtain

$$\begin{aligned} I_1 &= -\int_0^{1-\varepsilon} t^p d\varphi(t) \le \frac{2p}{\beta-\alpha}\int_0^{1-\varepsilon} t^{p-1}\phi(t)dt \\ &\le 2p\int_0^{1-\varepsilon} t^{p-1}\frac{18\log\log N}{(\log N)^t}dt \\ &= 36\, p(\log\log N)^{1-p}\int_0^{(1-\varepsilon)\log\log N} u^{p-1}e^{-u}du. \end{aligned}$$

Since $u^{p-1}e^{-u}$ has its maximum at $u = p-1$ and $(1-\varepsilon)\log\log N \le p-1$, we get for $N \ge N_0(q,\varepsilon)$

$$\begin{aligned} I_1 &< 36(\log\log N)^2(1-\varepsilon)^p e^{-(1-\varepsilon)\log\log N} \\ &< 72(\log\log N)^2(1-\varepsilon)^{(1-(\varepsilon/2))\log\log N}(\log N)^{-(1-\varepsilon)}. \end{aligned}$$

Finally for $N \ge N_0(q,\varepsilon)$

$$I_1 \le \frac{72(\log\log N)^2}{(\log N)^\theta}, \tag{1.344}$$

where

$$\theta = 1-\varepsilon-\left(1-\frac{\varepsilon}{2}\right)\log(1-\varepsilon).$$

Next we estimate I_2 by using the same procedure:

$$I_2 < 36(\log\log N)^{1-p}p\int_{\log\log N}^{3\log\log N} u^{p-1}e^{-u}du.$$

Since $p-1 < u = \log\log N$ we obtain

$$I_2 < \frac{72(\log\log N)^2}{\log N} \tag{1.345}$$

for all $N \geq N_0(q,\varepsilon)$.

In order to estimate I_3 we proceed in a simliar way, but we must apply the second, weaker estimate of Lemma 1.178:

$$\begin{aligned} I_3 &\leq 2p \int_3^N t^{p-1} \frac{6\log\log N}{t^{2\log\log N}} dt \leq 12(\log\log N) t^{p-2\log\log N}|_{t=N}^3 \\ &< 12(\log\log N) e^{-\log\log N} = \frac{12\log\log N}{\log N}. \end{aligned} \tag{1.346}$$

Now we combine the inequalities (1.343), (1.344), (1.345) and (1.346).
Since $p = [(1-(\varepsilon/2))\log\log N]$, we have for $N \geq N_0(q,\varepsilon)$

$$\begin{aligned} \left(\frac{p}{e\log\log N}\right)^p &\geq \left(1 - \frac{2}{\log\log N}\right)^p \left(\frac{1-(\varepsilon/2)}{e}\right)^p \\ &\geq \left(1 - \frac{2}{\log\log N}\right)^{\log\log N} \left(\frac{1-(\varepsilon/2)}{e}\right)^{(1-(\varepsilon/2))\log\log N}. \end{aligned}$$

Hence

$$\left(\frac{p}{e\log\log N}\right)^p > \frac{1}{9(\log N)^{\vartheta}},$$

where

$$\vartheta = \left(1 - \frac{\varepsilon}{2}\right) - \left(1 - \frac{\varepsilon}{2}\right)\log\left(1 - \frac{\varepsilon}{2}\right).$$

An easy computation shows that $\vartheta < 1 - (\varepsilon^2/4)$ for $0 < \varepsilon < 1$, hence we have from (1.345) and (1.346)

$$\frac{1}{2}\left(\frac{p}{e\log\log N}\right)^p > I_2 + I_3$$

for $N \geq N_0(q,\varepsilon)$. Consequently we obtain from (1.343) and (1.344)

$$\frac{\phi(1-\varepsilon)}{\beta - \alpha} \geq \frac{1}{18(\log N)^{\vartheta}} - \frac{72(\log\log N)^2}{(\log N)^{\theta}}.$$

In order to establish the statement of the lemma it is sufficient to show that $\vartheta < \theta$, i.e.

$$\frac{\varepsilon}{2} < \left(1 - \frac{\varepsilon}{2}\right)\log\frac{1-(\varepsilon/2)}{1-\varepsilon},$$

where $0 < \varepsilon < 1$. This last inequality clearly holds, as can be seen by expanding each side of the inequality into a power series. This completes the proof of Lemma 1.179. □

Proof of Proposition 1.172. We will use the notations

$$\psi(N) = \sqrt{N \log\log N}$$

and for $M \geq 0, N \geq 1$

$$F(M,N;x) = \left| \sum_{n=M+1}^{M+N} e(a_n x) \right|.$$

Furthermore we will make use of the observation that the intersection $(I_1 \cup I_2 \cup \cdots \cup I_m) \cap (J_1 \cup J_2 \cup \cdots \cup J_n)$ consists of at most $m+n$ intervals on the real line if $I_1, I_2, \ldots, I_m$ and $J_1, J_2, \ldots, J_n$ are arbitrary intervals on the real line.

Now let $\varepsilon > 0$ be given and let $a \geq a_0(\varepsilon), u \geq 1$ be arbitrary integers the exact value of which will be determined at the end of the sequel. At present the condition $a \geq a_0(\varepsilon)$ assures only that Lemma 1.179 can be applied to any of the sums

$$F_k(x) = F(a^u + a^{u+1} + \ldots + a^{u+k-1}, a^{u+k}, x) \tag{1.347}$$

with $k = 1, 2, 3, \ldots$.

For the sake of shortness we will also use the abbreviation $\psi_k = \psi(a^{u+k})$ and let $I = [0,1]$. We define the set

$$E_1 = \left\{ x \in I \,|\, F_1(x) \geq \left(1 - \frac{\varepsilon}{2}\right) \psi_1 \right\},$$

and inductively

$$E_k = \left\{ x \in I \setminus (E_1 \cup E_2 \cup \cdots \cup E_{k-1}) \,|\, F_k(x) \geq \left(1 - \frac{\varepsilon}{2}\right) \psi_k \right\} \tag{1.348}$$

for $k = 1, 2, 3 \ldots$ (E_0 denotes the empty set). Our aim is to obtain an upper estimate for the measure

$$\lambda_1(I \setminus (E_1 \cup E_2 \cup \cdots \cup E_k)).$$

For this purpose we estimate $\lambda_1(E_k)$ from below.

Set $m_k = a_{a^u + a^{u+1} + \ldots + a^{u+k}}$. Then $F_k(x)^2$ is a trigonometric polynomial of degree $2m_k$, and the set

$$\left\{ x \in I \,|\, F_k(x) < \left(1 - \frac{\varepsilon}{2}\right) \psi_k \right\}$$

consists of at most $4m_k$ intervals. In particular $I - E_1$ consists of ϱ_1 intervals where $\varrho_1 < 4m_1$. In general the set $I \setminus (E_1 \cup E_2 \cup \cdots \cup E_k)$ consists of ϱ_k intervals where

$$\varrho_k < 4(m_1 + m_2 + \ldots + m_k) < 4km_k.$$

For, according to the definition of $E_1, E_2, \ldots, E_k$ in (1.348) we have

$$\begin{aligned} &I \setminus (E_1 \cup E_2 \cup \cdots \cup E_k) \\ &\quad = (I \setminus (E_1 \cup E_2 \cup \cdots \cup E_{k-1})) \cap \left\{ x | x \in I;\ F_k(x) < \left(1 - \frac{\varepsilon}{2}\right) \psi_k \right\}. \end{aligned}$$

Hence $\varrho_k < \varrho_{k-1} + 4m_k$, which proves the above estimate.

Now let e_{k+1} $(k \geq 1)$ be the union of those intervals of

$$I \setminus (E_1 \cup E_2 \cup \cdots \cup E_k),$$

the length of which is less than $\delta_k = m_k^{-1} a^{-(u+k+1)/2}$. Then we have

$$\lambda_1(e_{k+1}) \leq \varrho_k \delta_k < 4k m_k \delta_k = 4k a^{-(u+k+1)/2}. \tag{1.349}$$

The set $I \setminus (E_1 \cup E_2 \cup \cdots \cup E_k) - e_{k+1}$ consists of intervals the lengths of which is at least

$$\delta_k > \left(n(1 + a^u + a^{u+1} + \ldots + a^{u+k})\sqrt{a^{u+k+1}}\right)^{-1}.$$

Hence the condition $\beta - \alpha \geq 1/(a_1\sqrt{N})$ of Lemma 1.179 is satisfied for every interval of the set $I \setminus (E_1 \cup E_2 \cup \cdots \cup E_k) - e_{k-1}$. Since $a \geq a_0(\varepsilon)$ we may use Lemma 1.179 in order to estimate $\lambda_1(E_{k+1})$:

$$\begin{aligned} \lambda_1(E_{k+1}) &\geq \frac{\lambda_1[I \setminus (E_1 \cup E_2 \cup \cdots \cup E_k) - e_{k-1}]}{(u+k+1)\log a} \\ &\geq \frac{\lambda_1[I \setminus (E_1 \cup E_2 \cup \cdots \cup E_k)]}{(u+k+1)\log a} - \lambda_1(e_{k+1}). \end{aligned}$$

According to Lemma 1.179 we also have

$$\lambda_1(E_1) \geq \frac{1}{(u+1)\log a} = \frac{\lambda_1(I)}{(u+1)\log a}.$$

From these last inequalities we obtain by induction on k;

$$\lambda_1[I \setminus (E_1 \cup E_2 \cup \cdots \cup E_k)] \leq \prod_{\nu=1}^{k} \left(1 - \frac{1}{(u+\nu)\log a}\right) + \sum_{\nu=2}^{k} \lambda_1(e_\nu).$$

By (1.349),

$$\sum_{\nu=2}^{k} \lambda_1(e_\nu) \leq 4a^{-u/2} \sum_{k=1}^{\infty} k a^{-(k+1)/2} = \mathcal{O}(a^{-u/2}).$$

Finally it follows that

$$\lambda_1\left(I \setminus (E_1 \cup E_2 \cup \cdots \cup E_k)\right) \leq \prod_{\nu=1}^{k} \left(1 - \frac{1}{(u+\nu)\log a}\right) + \mathcal{O}(a^{-u/2}).$$

Let us introduce the notation

$$N_k = a^u + a^{u+1} + \ldots + a^{u+k} \ (k \geq 0)$$

and let us define the sets E_k', $k \geq 1$, by

$$E_k' = \{x \in I \mid F(0, N_{k-1}; x) \geq \sqrt{2}\phi(N_{k-1})\}. \tag{1.350}$$

The measure $\lambda_1(E'_k)$ can be estimated by Lemma 1.178,

$$\lambda_1(E'_k) < \frac{18 \log\log N_{k-1}}{(\log N_{k-1})^2} < \frac{1}{2(u+k-1)^{3/2}}$$

provided that $a \geq a_0$. Hence

$$\sum_{\nu=1}^{k} \lambda_1(E'_\nu) < \frac{1}{2}\left(\frac{1}{u^{3/2}} + \frac{1}{(u+1)^{3/2}} + \ldots\right) < \frac{1}{\sqrt{u-1}}.$$

Using our previous estimates we see that

$$\lambda_1[I \setminus (E_1 \cup E_2 \cup \cdots \cup E_k)] + \lambda_1(E'_1 \cup E'_2 \cup \ldots \cup E'_k) \tag{1.351}$$
$$< \prod_{\nu=1}^{k} \left(1 - \frac{1}{(u+\nu)\log a}\right) + \frac{1}{\sqrt{u-1}} + \mathcal{O}(a^{-u/2}).$$

With the help of this inequality the proof of Proposition 1.172 can be finished.

According to the definition of $F(M, N; x) \geq 0$ we have

$$\frac{F(0, N_\nu; x)}{\psi(N_\nu)} \geq \frac{F_\nu(x)}{\psi_\nu} \cdot \frac{\psi_\nu}{\psi(N_\nu)} - \frac{F(0, N_{\nu-1}; x)}{\sqrt{2}\psi(N_{\nu-1})} \cdot \frac{\sqrt{2}\psi(N_{\nu-1})}{\psi(N_\nu)}.$$

An elementary computation shows that

$$\frac{\psi_\nu}{\psi(N_\nu)} \geq \frac{\log u}{(1+(2/a))\log(u+1)} > \frac{1}{(1+(2/a))(1+(1/u))}$$

for any $\nu = 1, 2, 3, \ldots$; $a \geq 2$ and $u \geq 3$. Similarly one obtains $\psi(N_{\nu-1})/\psi(N_\nu) < \sqrt{2/(a-1)}$ for any $\nu = 1, 2, 3, \ldots$; $a \geq 2$ and $u \geq 0$. Hence we have

$$\frac{F(0, N_\nu; x)}{\psi(N_\nu)} \geq \frac{F_\nu(x)}{\psi_\nu} \cdot \frac{1}{(1+(2/a))(1+(1/u))} - \frac{F(0, N_{\nu-1}; x)}{\sqrt{2}\psi(N_{\nu-1})} \cdot \sqrt{\frac{4}{(a-1)}}$$

for any $\nu = 1, 2, 3, \ldots$; $a \geq 2$ and $u \geq 3$.

If we restrict ourselves to those x which belong to the set

$$E = (E_1 \cup E_2 \cup \cdots \cup E_k) \cap (I \setminus (E'_1 \cup E'_2 \cup \ldots \cup E'_k))$$

then by (1.350) $F(0, N_{\nu-1}; x) < \sqrt{2}\psi(N_{\nu-1})$ for every $\nu = 1, 2, \ldots, k$ and by (1.348) $F_\nu(x)/\psi_\nu \geq 1 - \varepsilon/2$ for a suitable $\nu = \nu(x) \leq k$. Hence on the set E we have

$$\max_{1 \leq \nu \leq k} \frac{F(0, N_\nu; x)}{\psi(N_\nu)} \geq \left(1 - \frac{\varepsilon}{2}\right) \frac{1}{(1+(2/a))(1+(1/u))} - \sqrt{\frac{4}{a-1}}.$$

Moreover according to (1.351) we have

$$\lambda_1(E) \geq 1 - \prod_{\nu=1}^{k} \left(1 - \frac{1}{(u+\nu)\log a}\right) - \frac{1}{\sqrt{u-1}} + \mathcal{O}(a^{-u/2}). \tag{1.352}$$

Now it is clear how to complete the proof of Proposition 1.172. Given $\varepsilon > 0, \eta > 0$ and N, first we choose $a = a(\varepsilon)$ such that $a \geq a_0(\varepsilon)$, $a > N$,

$$\sqrt{\frac{4}{a-1}} \leq \frac{\varepsilon}{2}, \quad \text{and} \quad \frac{1}{1+(2/a)} \geq 1 - \frac{\varepsilon}{4}.$$

Then $N_\nu > N$ ($\nu \geq 1$) is satisfied. Next we choose $u = u(q,\varepsilon,\eta) \geq 3$ such that $(1+(1/u))^{-1} \geq 1 - \frac{\varepsilon}{4}$, and the sum of the last two terms in (1.352) is numerically less than $\eta/2$.

Finally we choose $k = k(q,\varepsilon,\eta)$ such that

$$\prod_{\nu=1}^{k} \left(1 - \frac{1}{(u+\nu)\log a}\right) \leq \frac{\eta}{2}.$$

Hence the statement of Proposition 1.172 holds for the set E and $\lambda_1(E) \geq 1 - \eta$. □

In his Nijmrode lecture in 1962 ERDŐS [562] conjectured for arbitrary increasing sequences (a_k) of positive integers that

$$D_N(a_n x) = \mathcal{O}\left(N^{-\frac{1}{2}} (\log\log N)^\gamma\right) \tag{1.353}$$

holds for almost all x with some constant $\gamma \geq \frac{1}{2}$. This, of course, would also improve BAKER's theorem. Conjecture (1.353) was reformulated in a personal communication by BAKER provided that a growth condition

$$a_{k+1}/a_k \geq 1 + ck^{-\alpha} \tag{1.354}$$

for some $c > 0$ and $\alpha > 0$ is satisfied. If $0 < \alpha < \frac{1}{2}$ TAKAHASHI [1778] and BERKES [186] could prove a law of iterated logarithm again. However, for $\alpha \geq \frac{1}{2}$ the conjecture has been disproved by BERKES and PHILIPP [187] who could show the following result.

Theorem 1.180 *Let f be a non-decreasing function with*

$$\sup_{k\geq 1} \frac{f(k^2)}{f(k)} < \infty \quad \textit{and} \quad \sum_{k=1}^{\infty} \frac{1}{k(f(k))^2} = \infty. \tag{1.355}$$

Then there exists an increasing sequence (a_k) of integers such that

$$\limsup_{N\to\infty} \frac{\left|\sum_{k\leq N} \cos(2\pi a_k x)\right|}{\sqrt{N} f(N)} = \infty$$

for almost all x.

Remark. The sequence (a_k) can be constructed in a way that

$$\frac{a_{k+1}}{a_k} \geq 1 + \frac{1}{k^{\frac{1}{2}} (\log k)^\beta}$$

for some $\beta > 0$ ($k \geq k_0$). From KOKSMA-HLAWKA's inequality (Theorem 1.14) and Theorem 1.180 with $f(k) = \sqrt{\log k}$, we immediately derive

$$\limsup_{N\to\infty} \frac{N\,D_N(x)}{\sqrt{N \log N}} = \infty$$

for almost all x, which disproves the conjecture of ERDŐS.In the following we present a sketch of the proof of the above theorem. From (1.355) we immediately obtain by iteration

$$\sup_{k\geq 1} \frac{f(k^p)}{f(k)} < \infty$$

for any $p > 0$. Hence there is a constant $c_1 > 0$ with $f(k^3) \leq c_1 f(k)$ (for $k \geq 1$). This implies

$$f(k) \ll (\log k)^C \quad \text{with} \quad C = \frac{\log c_1}{\log 3}.$$

We set $m_k = [k \cdot f(k)^2]$ and obtain

$$\sum \frac{1}{m_k} = \infty\,, \qquad m_k \ll k(\log k)^{2C}.$$

Let $b_k = 2^{k^2-k}$ and $I_k = \{2b_k, 4b_k, 6b_k, \ldots, 2m_k b_k\}$. Clearly, the sets I_k are disjoint for $k \geq k_0$. The (monotone) sequence of integers a_k is now defined.

$$\{a_l\} = \bigcup_{k\geq 1} I_k.$$

In the following we consider the unit interval $[0,1]$ as a probability space (with LEBESGUE measure). Furthermore let $\mathcal{F}_k$ denote the σ-field generated by the dyadic intervals

$$U_\nu = \left[\nu 2^{-k^2}, (\nu+1)2^{-k^2}\right), \quad (0 \leq \nu < 2^{k^2}).$$

Then we introduce the following random variables:

$$\begin{aligned} \xi_j(x) &= 2^{k^2} \int_{U_\nu} \cos(2\pi j t)\, dt \qquad (x \in U_\nu), \\ X_k(x) &= \sum_{j\in I_k} \cos(2\pi j x), \\ Z_k(x) &= \sum_{j\in I_k} \xi_j(x). \end{aligned}$$

Obviously, ξ_j is $\mathcal{F}_k$-measurable if $j \in I_k$. Moreover, we have for $j \in I_k$ and $0 \leq x < 1$

$$|\xi_j(x) - \cos 2\pi j x| \leq 2\pi j 2^{-k^2} \ll k^2 2^{-k},$$

and

$$\dot{X}_k(x) - Z_k(x) \ll m_k k^2 2^{-k} \ll k^4 2^{-k}.$$

By standard tools from probability theory (characteristic functions) it can be proved that $Z_1, Z_2, \ldots$ are independent (details see in BERKES and PHILIPP [187]). The next lemma is crucial.

Lemma 1.181 *For almost all $x \in [0,1)$ we have*

$$\limsup_{k\to\infty} \frac{X_k}{m_k} \geq \frac{2}{\pi}.$$

Proof. We need to show that for almost all x infinitely many of the events $\{X_k \geq 2m_k/\pi\}$ occur. By the formula

$$X_k = \frac{\sin(2\pi(2m_k+1)b_k x)}{\sin 2\pi b_k x} - \frac{1}{2}$$

we have

$$\lambda_1\left(\left\{x:\ 0 \leq x < 1, X_k \geq (2m_k+1)/\pi - \frac{1}{2}\right\}\right)$$
$$= \lambda_1\left(\left\{x:\ 0 \leq x < 1: \frac{\sin((2m_k+1)\pi x)}{2\sin tx} \geq \frac{2m_k+1}{\pi}\right\}\right) \geq \frac{1}{4m_k+2}.$$

This yields

$$\sum_k \lambda_1\left(\left\{x: 0 \leq x < 1, Z_k \geq \frac{2m_k^{-1}}{\pi}\right\}\right) = \infty,$$

and the result follows from the independence of the Z_k and the BOREL-CANTELLI lemma. □

Now the theorem follows easily. Setting

$$S_N(x) = \sum_{j\leq N} \cos 2\pi a_j x \qquad \text{and} \qquad M_k = \sum_{i\leq k} m_i$$

yields

$$S_{M_k}(x) = \sum_{j\leq k} X_j(x).$$

Thus by Lemma 1.181 we derive

$$\limsup_{k\to\infty} \frac{|S_{M_k}(x)|}{m_k} \geq \frac{1}{\pi}$$

for almost all x. Applying the estimates

$$M_k \leq k^2 f(k)^2 \leq k^3$$

we obtain $M_k^{1/2} f(M_k) \leq k f(k) f(k^3) \leq c_1 M_k$. Thus

$$\limsup_{k\to\infty} \frac{|S_{M_k}(x)|}{M_k^{1/2} f(M_k)} \geq \frac{1}{\pi c_1}$$

for almost all x. Now one just has to take a function $g(k)$ with $g(k) \to \infty$ (as $k \to \infty$) such that conditions (1.355) remain valid for $f(k)g(k)$ replacing $f(k)$. This completes the proof of the theorem. □

We conclude this section by stating (without proof) a result of PHILIPP [1436] concerning a law of iterated logarithm for sequences $(a_n x)$, where the sequence (a_n) is given by

$$\{a_n\}_{n=1}^{\infty} = \{q_1^{\alpha_1} \dots q_\tau^{\alpha_\tau} : \alpha_i \in \mathbf{Z}, \alpha_i \geq 0\} \tag{1.356}$$

in increasing order, where $\{q_1 \dots, q_\tau\}$ is a finite set of coprime integers. (a_n) is not lacunary, it only satisfies a growth condition

$$\frac{a_{n+1}}{a_n} \geq 1 + n^{-\alpha_n},$$

where α_n is in the average $1 - \frac{1}{\tau} > \frac{1}{2}$ (for $\tau > 2$). So a law of the iterated logarithm need not be expected. However, a combination of strong approximation theorems for independent random variables, martingale inequalities and a theorem of TIJDEMAN [1854] on S-unit equations yield the following theorem.

Theorem 1.182 *There exists a constant $C > 0$ depending on the total number of primes occuring in the prime factorization of $q_1, \dots, q_\tau$ such that for almost all x*

$$\frac{1}{4} \leq \limsup_{N \to \infty} \frac{N\, D_N(a_n x)}{\sqrt{N \log \log N}} \leq C,$$

where a_n is given in (1.356).

1.6.3 Completely Uniformly Distributed Sequences

Let us recall here the basic definitions of completely uniformly distributed (for short cpl.u.d.) sequences. A sequence (x_n) of numbers in $[0, 1)$ is cpl.u.d. if for all $k \geq 1$ the sequence of points $x_n^{(k)} = (x_n, x_{n+1}, \dots, x_{n+k-1})_{n \geq 1}$ is u.d. mod 1 in $\mathbf{R}^k$. FRANKLIN [620] proved that the sequence (x^n) is cpl.u.d. mod 1 for almost all $x > 1$; for sequences in higher dimensions see also FRANKLIN [621]. Of course, the uniform distribution of this sequence (including discrepancy bounds of the type $\mathcal{O}(N^{-1/2}(\log N)^{3/2+\varepsilon})$ follow immediately by the ERDŐS-GÁL-KOKSMA method. KNUTH [940, section 3.5] introduces several distribution concepts for sequences which define a sequence (x_n) in $[0, 1)$ to be "random". His definition $R4$ requires that for every sequence b_n of distinct positive integers obtained by an effective algorithm, the sequence $(\{x_{b_n}\})$ is cpl.u.d. He raised the problem whether (x^n) satisfies this definition $R4$ of randomness for almost all $x > 1$. Since there are only countably many effective algorithms for specifying integral sequences (b_n), the problem of KNUTH is settled in the affirmative by the following result of NIEDERREITER and TICHY [1364]. For various notions of randomness we refer to Section 3.4.

Theorem 1.183 *If a_n is any sequence of distinct positive integers, then the sequence (x^{a_n}) is cpl.u.d. mod.1 for almost all $x > 1$.*

The following three lemmata are the key ingredients of the proof:

Lemma 1.184 *(DAVENPORT, ERDŐS and LEVEQUE [432]). Let (u_n) be a sequence of LEBESGUE-measurable functions on the interval $[a,b]$ and put*

$$I(N) = \int_a^b \left| \frac{1}{N} \sum_{n=1}^{N} e(u_n(x)) \right|^2 dx.$$

If $\sum_{N=1}^{\infty} I(N)$ converges, then

$$\lim_{N\to\infty} \frac{1}{N} \sum_{n=1}^{N} e(u_n(x)) = 0 \textit{ for almost all } x \in [a,b].$$

(Compare with Lemma 1.156)

Lemma 1.185 *(VAN DER CORPUT). Let the function u on $[a,b]$ have a continuous and monotone first derivative that satisfies $|u'(x)| \geq L > 0$ for all $x \in [a,b]$. Then*

$$\left| \int_a^b e(u(x))dx \right| \leq \frac{1}{L}$$

(Compare with KUIPERS and NIEDERREITER [983, p. 15] or with the proof of Theorem 1.158.)

Lemma 1.186 *If P is a monic polynomial of degree $d \geq 1$, then for the supremum norm $\|P\|_\infty$ on $[a,b]$ we have*

$$\|P\|_\infty \geq 2^{1-2d}(b-a)^d.$$

The lower bound of Lemma 1.186 is a standard result of approximation theory; cf. NATANSON [1264, p.39]. It is essentially best possible. In NIEDERREITER and TICHY [1365] the above theorem was extended to sequences (a_n) of positve numbers satisfying $\inf_{m\neq n} |a_m - a_n| > 0$. The main idea of the proof of this more general result is to replace the lower bound of Lemma 1.186 by the following one. Instead of polynomials MÜNTZ polynomials

$$P(x) = x^{\lambda_s} + \sum_{j=1}^{s-1} c_j x^{\lambda_j}$$

with $0 \leq \lambda_1 < \lambda_2 < \ldots < \lambda_s, s \geq 2$ occur.

Lemma 1.187 *For any monic MÜNTZ polynomial P and any interval $[a,b]$ with $0 < a < b$ we have*

$$\|P\|_\infty \geq c_s a^{\lambda_s} \prod_{j=1}^{s-1} \frac{\lambda_s - \lambda_j}{\lambda_j - \lambda_1 + (1/\tau)},$$

where $\tau = \frac{1}{2}\log\frac{b}{a}$ and $c_s > 0$ depends only on s.

The proof of this lemma essentially depends on a result by KOREVAAR and LUXEMBURG [1113] on MÜNTZ-SZASZ approximation and entire functions and on a PALEY-WIENER argument. For an introduction to MÜNTZ-SZASZ approximation we refer to RUDIN [1575].

The methods we have discussed until now only yield quantitative results. The following theorem is due to TICHY [1832] and gives a generalization as well as a discrepancy bound. Let $k = k(N)$ be a sequence of positive integers and denote by $D_N(k(N), x_n)$ the discrepancy of the sequence $(x_n, x_{n+1}, \ldots, x_{n+k(N)-1})$ in $\mathbf{R}^{k(N)}$ modulo 1.

Theorem 1.188 *Let $k = k(N)$ be a sequence of positive integers satisfying $k(N) \leq (\log N)^\theta$ with $0 < \theta < \frac{1}{2}$, let $(a_n)_{n=1}^\infty$ be a sequence of positive numbers such that $|a_m - a_n| \geq \delta > 0$ (for $m \neq n$) and let $\eta > 0$ be arbitrary. Then for almost all $x < 1$ and all N the bound*

$$D_N(k(N), x^{a_n}) \ll N^{-\frac{1}{2}+\eta}$$

holds, where the implied constant may depend on x, θ, δ, and η.

Remark. Our proof gives further

$$D_N(k, x^{a_n}) \ll N^{-\frac{1}{2}}(\log N)^{k+3/2+\eta} ,$$

if $k = k(N)$ is a constant sequence.

Lemma 1.189 *Let $F(x) = c_1 e^{\gamma_1 x} + \ldots + c_s e^{\gamma_s x}$ be defined for $0 < \alpha \leq x \leq \beta$ and let $s \geq 2$, where c_j, γ_j are real numbers with $|c_1| \geq 1$, $\sum_{j=2}^s |c_j| \geq H$ $(H \geq 2)$ and $\gamma_1 \geq \gamma_2 \geq \ldots \geq \gamma_s \geq 0, \gamma_1 - \gamma_2 \geq \delta$ with $0 < \delta < 1$. Then for any function $\psi(N)$ with $\psi(N) \to \infty$ (as $N \to \infty$) and all sufficiently large $N \geq N_0(\alpha, \beta, \delta, \psi)$ we have*

$$\left|\frac{d^M F}{dx^M}(x)\right| = |F^{(M)}(x)| \geq \frac{1}{2} e^{\delta\alpha} H^{(\log \delta)\gamma_1 \log \psi(N)}$$

for all $x \in [\alpha, \beta]$, where $M = [\gamma_1 \log H \log \psi(N)]$. ($N_0$ does not depend on H; for $s = 1$ the bound is trivially satisfied with $M = [\log \psi(N)]$).

Proof. Since $\dfrac{\gamma_j}{\gamma_1} \leq 1 - \dfrac{\delta}{\gamma_1}$, $1 \leq j \leq s$, we have

$$\left|F^{(M)}(x)\right| \leq \gamma_1^M e^{\gamma_1 x} \left|1 - H\left(1 - \frac{\delta}{\gamma_1}\right)^M\right|$$

for all $x \in [\alpha, \beta]$. From

$$\left(1 - \frac{\delta}{\gamma_1}\right)^{\gamma_1} \leq e^{-\delta} < 1$$

we obtain

$$\left|F^{(M)}(x)\right| \geq \frac{e^{\delta x}}{2} H^{(\log \delta)\gamma_1 \log \psi(N)}$$

for all $x \in [\alpha, \beta]$ and $N \geq N_0(\alpha, \beta, \delta, \psi)$. □

Lemma 1.190 *Let F be a function on $[\alpha, \beta]$ with continuous derivatives $F^{(k)}(x)$, $1 \leq k \leq M$, up to the M-th order. If for all $x \in [\alpha, \beta]$*

$$|F^{(M)}(x)| \geq A > 0,$$

then for every $\varepsilon > 0$ there exist at most $M(M+1)/2+1$ intervals $I_j = I_j(\varepsilon)$, $1 \leq j \leq M(M+1)/2+1$ with

$$\lambda_1 \left(\bigcup_j I_j \right) \geq \beta - \alpha - 2M(M+1)\varepsilon,$$

such that all derivatives $F^{(M-k)}$, $0 \leq k \leq M$ are strictly monotone on each interval I_j. Furthermore for all $x \in \bigcup_j I_j$ the estimate

$$|F^{(M-k)}|(x) \geq A\varepsilon^k$$

holds.

Proof. ROLLE's theorem combined with the assumptions implies that $F^{(M-k)}$, $0 \leq k \leq M$, has at most k zeros in the interval $[\alpha, \beta]$. We write $p = p(k) \leq k$ for the number of zeros $a_{k,h}$, $h = 1, \ldots, p$ and assume that they are arranged in increasing order. Put $J = [a_{1,1} - \varepsilon, a_{1,1} + \varepsilon]$ for $p(1) = 1$. The complement of J consists of at most $\frac{1 \cdot 2}{2} + 1 = 2$ intervals I_1, I_2 and for each $x \in I_1 \cup I_2$ we have from the mean-value theorem

$$|F^{(M-1)}(x)| = |F^{(M)}(\xi)|\,|x - a_{1,1}| \geq A \cdot \varepsilon.$$

Now we use induction on k. Let us suppose that $q = q(k) \leq k(k+1)/2$ intervals $J_1, \ldots, J_{q(k)}$ with

$$\lambda_1 \left(\bigcup_{i=1}^{q} J_i \right) \leq 2k(k+1)\varepsilon \qquad \text{and} \qquad |F^{(M-k=}(x)| \geq A \cdot \varepsilon^k$$

exist for all x in the complement of $\bigcup_{i=1}^q J_i$.

We consider the intervals $J'_h = [a_{k+1,h} - \varepsilon, a_{k+1,h} + \varepsilon]$, for $1 \leq h \leq p(k+1) \leq k+1$. From the intervals J_i and J'_h we construct a family of $q(k+1) \leq (k+1)(k+2)/2$ intervals $J''_1, \ldots, J''_{q(k+1)}$, by putting all overlapping intervals together, and by enlargening all the intervals containing a zero of $F^{(M-k)}$ by the length ε in both sides.

Now we have

$$\begin{aligned} \lambda_1 \left(\bigcup_{i=1}^{q(k+1)} J''_i \right) &\leq \lambda_1 \left(\bigcup_{i=1}^{q(k)} J_i \right) + \lambda_1 \left(\bigcup_{h=1}^{p(k+1)} J'_h \right) + 2k\varepsilon \\ &\leq 2k(k+1)\varepsilon + 2(k+1)\varepsilon + 2k\varepsilon \leq 2(k+1)(k+2)\varepsilon. \end{aligned}$$

The complement of $\bigcup_{i=1}^{q(k+1)} J''_i$ consists of at most $(k+1)(k+2)/2+1$ intervals. Let $I = [c, d]$ be such an interval. For two consecutive zeros of $F^{(M-k)}$, $a_{k,h} < c$ and $a_{k,h+1} > d$ the following two cases are possible:

(i) there is no zero of $F^{(M-k-1)}$ in $(a_{k,h}, a_{k,h+1})$;

(ii) there is exactly one zero a of $F^{(M-k-1)}$ in $(a_{k,h}, a_{k,h+1})$.

We assume now that $F^{(M-k-1)}$ is strictly increasing in the interval $[a_{k,h}, a_{k,h+1}]$.

If in case (i) $F^{(M-k-1)}$ is positive in $[a_{k,h}, a_{k,h+1}]$, then the mean-value theorem yields for all $x \in (c, d)$ and an appropriate ξ with $c - \varepsilon < \xi < x$:

$$F^{(M-k-1)}(x) = F^{(M-k-1)}(c-\varepsilon) + F^{(M-k)}(\xi)(x - c + \varepsilon) \geq A\varepsilon^{k+1},$$

while $c - \varepsilon$ lies in the interior of $\bigcup_{i=1}^{q(k)} J_i$. For a negative $F^{(M-k-1)}$ we may use the analogous argument.

Assuming $F^{(M-k-1)}$ to be increasing in the case (ii), there are just two possible subcases:

(ii.1) $a_{k,h} < c < d < a < a_{k,h+1}$ and $F^{(M-k-1)}(x) < 0$ for all $x \in (a_{k,h}, a)$.

(ii.2) $a_{k,h} < a < c < d < a_{k,h+1}$ and $F^{(M-k-1)}(x) > 0$ for all $x \in (a, a_{k,h+1})$.

In the case (ii.1) we obtain from the mean-value theorem for all $x \in (c, d)$ and an appropriate ξ with $x < \xi < d + \varepsilon$:

$$F^{(M-k-1)}(x) = F^{(M-k-1)}(d+\varepsilon) + F^{(M-k)}(\xi)(x - d - \varepsilon) \leq -A\varepsilon^{k+1}.$$

Similarly we obtain in the case (ii.2) for all $x \in (c, d)$ and an appropriate ξ with $c - \varepsilon < \xi < x$:

$$F^{(M-k-1)}(x) = F^{(M-k-1)}(c-\varepsilon) + F^{(M-k)}(\xi)(x - c + \varepsilon) \geq A\varepsilon^{k+1}.$$

Replacing F by $(-F)$, we obtain the same estimates for decreasing $F^{(M-k-1)}$ on the interval $[a_{k,j}, a_{k,j+1}]$. In this way we get for all x contained in the complement of $\bigcup_{i=1}^{q(k+1)} J_i''$ the estimate

$$|F^{(M-k-1)}(x)| \geq A\varepsilon^{k+1}.$$

Thus the proof of the lemma is complete. □

In the following we take a sufficiently large N, a sequence $k = k(N) \leq (\log N)^\theta$ (for $N \geq N^*$) and a fixed lattice point $\mathbf{h} = (h_0, \dots, h_{k-1})$ with integral coordinates satisfying $0 < \|\mathbf{h}\| \leq H = [\sqrt{N}]$. Then we estimate for almost all x in a fixed interval $[\alpha, \beta]$ $(0 < \alpha < \beta)$ the WEYL sum

$$\frac{1}{N}\sum_{\nu=1}^{N} e(u_\nu(\mathbf{h}, x)), \tag{1.357}$$

where $u_\nu(\mathbf{h}, x) = u_\nu(x) = \sum_{j=0}^{k-1} h_j e^{x a_{\nu+j}}$. We set $\Omega = \{j \in \mathbf{Z} : 0 \leq j \leq k-1, h_j \neq 0\}$ and $\rho = |\Omega|$; furthermore let $\tilde{b_1}(\nu) > \tilde{b_2}(\nu) > \dots > \tilde{b_\rho}(\nu)$ be the $a_{\nu+j}$, $j \in \Omega$, in decreasing order. Now we have

$$u_\nu(x) = \sum_{j=1}^{\rho} \tilde{g_j}(\nu) e^{x\tilde{b_j}(\nu)},$$

where $\tilde{g}_j(\nu)$, $1 \le j \le \rho$ is a permutation of h_j, $j \in \Omega$. Furthermore

$$\sum_{j=1}^{\rho} |\tilde{g}_j(\nu)| = \sum_{j\in\Omega} |h_j| =: H \le \sqrt{N}\log N.$$

In order to estimate the above exponential sum (1.357) we provide an upper bound for

$$I(N) = \int_{\alpha}^{\beta} \left| \frac{1}{N}\sum_{\nu=1}^{N} e(u_\nu(x)) \right|^{2w} dx, \tag{1.358}$$

where $w = w(N) = [k(N)\cdot \log\log N]$. Clearly

$$\left| \sum_{\nu=1}^{N} e(u_\nu(x)) \right|^{2w} = \sum_{\mathbf{n}} e(f_{\mathbf{n}}(x)),$$

where $\mathbf{n} = (n_1, \ldots, n_{2w})$, $1 \le n_l \le N$ and

$$f_{\mathbf{n}}(x) = u_{n_1}(x) - u_{n_2}(x) + \ldots + u_{n_{2w-1}}(x) - u_{n_{2w}}(x).$$

Lemma 1.191 *Let $\{\bar{b}_i(\mathbf{n}), 1 \le i \le \bar{r}\}$ denote an enumeration of $\tilde{b}_j(n_l)$ ($j = 1,\ldots,\rho; l = 1,\ldots,2w$) in decreasing order (with $\bar{r} \le 2\rho w$) and let X_N be the set of all $\mathbf{n} = (n_1,\ldots,n_{2w})$ satisfying that each $\bar{b}_i(\mathbf{n})$, which occurs just once among all $\tilde{b}_j(n_l)$, is smaller than $\sqrt{N}$. Then*

$$|X_N| \le 2\left(\frac{8w}{\delta}\right)^{2w} N^w.$$

Proof. Obviously

$$|X_N| \le (2w)!\, |\{\mathbf{n} \in X_N : n_1 \le n_2 \le \ldots \le n_{2w}\}|.$$

Furthermore, each $\mathbf{n}$ induces a partition on $\{1,\ldots,2w\}$ with connected blocks; we write $\lambda(\mathbf{n})$ for the number of partition blocks containing one element. For each such block $\{l\}$ we have: $n_l \le \sqrt{N}$ or $l > 1$ and $n_{l-1} < n_l \le n_{l-1} + k - 1$ (with $k = k(N)$). Let $\mu(\mathbf{n})$ ($\le \lambda(\mathbf{n})$) be the number of the blocks $\{l\}$ defined above, such that $n_l \le \sqrt{N}$. Then we have

$$|X_N| \le (2w)! \sum_{\lambda=0}^{2w} \sum_{\mu=0}^{\lambda} \sum \binom{\lambda}{\mu} N^{w-(\lambda/2)} \left(\frac{\sqrt{N}}{\delta}\right)^{\lambda}, \tag{1.359}$$

and the most interior sum is taken over all $\mathbf{n}$ with $\lambda(\mathbf{n}) = \lambda$ and $\mu(\mathbf{n}) = \mu$. (Note that the number of blocks containing more than one element is $\le \frac{1}{2}(2w - \lambda)$ and $k(N) \le (\log N)^\theta \le \sqrt{N}$.) Finally, we can estimate the number $\mathbf{n}$ with $\lambda(\mathbf{n}) = \lambda$ and $\mu(\mathbf{n}) = \mu$ by the number of all partitions with contiguous blocks and therefore by

2^{2w}. This combinatorial consideration can be refined, see GOLDSTERN [661]. In that way we obtain

$$|X_N| \leq \left(\frac{2}{\delta}\right)^{2w} (2w)! N^w \cdot 2^{2w+1} \leq 2 \cdot \left(\frac{8w}{\delta}\right)^{2w} N^w. \tag{1.360}$$

□

Proof of Theorem 1.188. Lemma 1.191 implies

$$I(N) \leq \frac{1}{N^{2w}} \sum_{\mathbf{n} \notin X_N} \left| \int_\alpha^\beta e(f_{\mathbf{n}}(x))dx \right| + \mathcal{O}((8w/\delta)^{2w} N^{-w}). \tag{1.361}$$

For $L = 0, 1, 2$ and $\mathbf{n} \notin X_N$ we set

$$f_{\mathbf{n}}^{(L)} = \frac{d^L}{dx} f_{\mathbf{n}}(x) = \sum_{i=1}^{r} b_i^L(\mathbf{n}) g_i(\mathbf{n}) e^{x b_i(\mathbf{n})}, \tag{1.362}$$

where $b_i(\mathbf{n})$ $(i = 1, \dots, r; r \leq \bar{r} \leq 2\rho w \leq 2k(N)w(N))$ is a strictly decreasing ordered subset of $\tilde{b}_i(n_l)$ and $g_i(\mathbf{n})$ are linear combinations of $\tilde{g}_i(n_l)$. Because of $\mathbf{n} \notin X_N$ we have $b(\mathbf{n}) := b_1(\mathbf{n}) \geq \sqrt{N}$ and $|g_1(\mathbf{n})| \geq 1$; Furthermore

$$\sum_{i=1}^{r} |g_i(\mathbf{n})| \leq 2wH.$$

Now we set, for $j \geq 2$, $\sigma(j) = \frac{1}{4} - \frac{1}{\log(j+2)}$ and distinguish the following three cases:

(i) r=1;

(ii) $r \geq 2$ and $b(\mathbf{n}) - b_2(\mathbf{n}) > N^{\sigma(2)}$;

(iii) $r \geq 2$ and there exists a maximal s, $2 \leq s \leq r$ with

$$b(\mathbf{n}) - b_s(\mathbf{n}) \leq N^{\sigma(s)}.$$

In case (ii) we get for all sufficiently large $N \geq N_0$:

$$\begin{aligned} e^{-xb(\mathbf{n})} \cdot \left| \frac{f_{\mathbf{n}}^{(L)}}{b^L(\mathbf{n})} \right| &= |g_1(\mathbf{n})| + \sum_{i=2}^{r} \frac{b_i^L(\mathbf{n})}{b^L(\mathbf{n})} g_i(\mathbf{n}) e^{x(b_i(\mathbf{n}) - b(\mathbf{n}))} \\ &\geq |g_1(\mathbf{n})| - \sum_{i=1}^{r} |g_i(\mathbf{n})| e^{-xN^{\sigma(2)}} \\ &\geq 1 - 2wHe^{-\alpha N^{\sigma(2)}} \geq \frac{1}{2}. \end{aligned} \tag{1.363}$$

In case (i) the estimate (1.363) trivially holds. Because of $b(\mathbf{n}) \geq \sqrt{N}$ we conclude from (1.362) and (1.363) (with $L = 1$) for $f'_{\mathbf{n}} = f_{\mathbf{n}}^{(1)}$ and all sufficiently large $N \geq N_1$:

$$|f_{\mathbf{n}}^{(1)}(x)| \geq \frac{1}{2} b(\mathbf{n}) e^{xb(\mathbf{n})} \geq e^{\alpha\sqrt{N}}. \tag{1.364}$$

From (1.362) and (1.363) (with $L = 2$) we get an analogous estimate for $f''_{\mathbf{n}} = f^{(2)}_{\mathbf{n}}$. Note that $f_{\mathbf{n}}$ is monotone on $[\alpha,\beta]$. Therefore, Lemma 1.185 yields in the cases (i) and (ii)

$$\left| \int_\alpha^\beta e(f_{\mathbf{n}}(x))\,dx \right| \le e^{-\alpha\sqrt{N}} \quad \text{for} \quad N \ge N_1. \tag{1.365}$$

In case (iii) we write

$$\sum_{i=1}^{s} \frac{b_i^L(\mathbf{n})}{b^L(\mathbf{n})} g_i(\mathbf{n}) e^{x(b_i(\mathbf{n})-b(\mathbf{n}))} + \sum_{i=s+1}^{r} \frac{b_i^L(\mathbf{n})}{b^L(\mathbf{n})} g_i(\mathbf{n}) e^{x(b_i(\mathbf{n})-b(\mathbf{n}))} =: U_{L,s} + V_{L,s}. \tag{1.366}$$

If $s \le r-1$, then $b(\mathbf{n}) - b_{s+1}(\mathbf{n}) > N^{\sigma(s+1)}$ (because of the definition of s). Thus we have

$$V_{L,s}(x) \le e^{-xN^{\sigma(s+1)}} \sum_{i=s+1}^{r} |g_i(\mathbf{n})| \le 2wHe^{-\alpha N^{\sigma(s+1)}}. \tag{1.367}$$

Because of $V_{L,s}(x) = 0$ for $s = r$, (1.367) is valid in this case as well. Now we set $d = b(\mathbf{n}) - b_s(\mathbf{n})$ and (for $N \ge N_2$)

$$\begin{aligned} U_{L,s}(x) &= \frac{b_s^L(\mathbf{n})e^{-dx}}{b^L(\mathbf{n})} \sum_{i=1}^{s} \frac{b_i^L(\mathbf{n})}{b_s^L(\mathbf{n})} g_i(\mathbf{n}) e^{x(b_i(\mathbf{n})-b_s(\mathbf{n}))} \\ &= \frac{b_s^L(\mathbf{n})e^{-dx}}{b^L(\mathbf{n})} F_{\mathbf{n}}(x) \ge \frac{e^{-dx}}{4} F_{\mathbf{n}}(x). \end{aligned} \tag{1.368}$$

Here we have

$$F_{\mathbf{n}}(x) = F_{L,\mathbf{n}}(x) = \sum_{i=1}^{s} c_i e^{x\gamma_i} \quad \text{with} \quad c_i = \frac{b_i^L(\mathbf{n})}{b_s^L(\mathbf{n})} g_i(\mathbf{n});$$

$\gamma_i = b_i(\mathbf{n}) - b_s(\mathbf{n})$ is a function as in Lemma 1.189.

Now all necessary conditions of Lemma 1.189 are fulfilled uniformly in $\mathbf{n}$ and L. Hence we obtain with $M = [\gamma_i \log(2Hw) \log\log N]$ that

$$|F_{L,\mathbf{n}}^{(M)}| \ge \frac{e^{\delta\alpha}}{2} (2wH)^{(\log\delta)(\log\log N)\gamma_1} \tag{1.369}$$

for all sufficiently large $N \ge N_3$. It follows from Lemma 1.190 and (1.369) that $[\alpha,\beta]$ can be divided into a union of at most $M(M+1)/2+1$ intervals I_j such that

$$|F_{L,\mathbf{n}}| \ge \frac{e^{\delta\alpha}}{2} (2wH)^{(\log\delta)(\log\log N)\gamma_1} \varepsilon^M \tag{1.370}$$

is valid on each such interval I_j and for any $\varepsilon > 0$. Now we set $\varepsilon = 1/N^{2w}$ and use $2wH \le N$, $\gamma_1 = d \le N^{\sigma(s)}$ (for sufficiently large N). Then for all $x \in W = \bigcup_j I_j$ and all sufficiently large $N \geq N_4$ we have

$$\begin{aligned}
&|U_{L,s}(x) - V_{L,s}(x)| \\
&\geq |e^{-\beta\delta}\frac{e^{\delta\alpha}}{8}N^{(\log\delta)(\log\log N)\gamma_1 - (\log N)^4 N^{\sigma(s)}}| - |Ne^{-\alpha N^{\sigma(s+1)}}| \qquad (1.371)\\
&\geq e^{-N^{3/8}},
\end{aligned}$$

because $s \leq 2wk \leq 2(\log N)^{2\theta}\log\log N$. In this way we obtain from (1.361) and (1.362) and using $b(\mathbf{n}) \geq \sqrt{N}$ that for all $x \in W$, $L = 1, 2$ and all sufficiently large $N \geq N_5$

$$|f_{\mathbf{n}}^{(L)}(x)| \geq \sqrt{N}e^{\alpha\sqrt{N}}e^{-N^{3/8}} \geq e^{(\alpha/2)\sqrt{N}}. \qquad (1.372)$$

Therefore $f_{\mathbf{n}}$ is strictly monotone on each of the intervals I_j, and we get from Lemma 1.185

$$\left|\int_{I_j} e(f_{\mathbf{n}}(x))dx\right| \leq e^{-(\alpha/2)\sqrt{N}}. \qquad (1.373)$$

Moreover, for all sufficiently large $N \geq N_5$ and all $\mathbf{n} \notin X_N$ we obtain

$$\left|\int_W e(f_{\mathbf{n}}(x))dx\right| \leq \left(\frac{M(M+1)}{2} + 1\right)e^{-(\alpha/2)\sqrt{N}} \leq e^{-(\alpha/4)\sqrt{N}}.$$

Hence

$$\begin{aligned}
\left|\int_\alpha^\beta e(f_{\mathbf{n}}(x))\,dx\right| &\leq \left|\int_{[a,b]\setminus W} e(f_{\mathbf{n}}(x))\,dx\right| + \left|\int_W e(f_{\mathbf{n}}(x))\,dx\right| \\
&\leq \frac{(\log N)^2}{N^{2w}} + e^{-(\alpha/4)\sqrt{N}} \leq \frac{1}{N^w} \qquad (1.374)
\end{aligned}$$

for all sufficiently large $N \geq N_6$ and all $\mathbf{n} \notin X_N$. Combining this with (1.361) yields

$$I(N) = \mathcal{O}\left(\left(\frac{8w}{\delta}\right)^{2w} N^{-w}\right). \qquad (1.375)$$

Now fix some $\eta > 0$ and consider the set

$$\mathcal{M}(N; \mathbf{h}) = \left\{x \in [\alpha, \beta] : |\sum_{\nu=1}^N e(u_\nu(x))| \geq \frac{8w}{\delta}N^{1/2}N^\eta\right\}. \qquad (1.376)$$

By (1.375) the Lebesgue measure of this set satisfies

$$\lambda_1(\mathcal{M}(N, \mathbf{h})) \leq N^{-\eta w}.$$

Since $w(N) = [k(N)\log\log N]$ we obtain

$$\lambda_1(\mathcal{M}(N)) \leq (2\sqrt{N} + 1)^{k(N)}N^{-\eta w} \leq 1/N^2 \qquad (1.377)$$

for all sufficiently large $N \geq N_7$, where

$$\mathcal{M}(N) = \bigcup_{\|\mathbf{h}\| \leq \sqrt{N}} \mathcal{M}(N, \mathbf{h}).$$

Thus the series $\sum_N \lambda_1(\mathcal{M}(N))$ converges and it follows from the BOREL-CANTELLI lemma, that for almost all $x \in [\alpha, \beta]$ there is an integer $\tilde{N}(x) > 0$, such that for all $N > \tilde{N}(x)$ and all lattice points $\mathbf{h}$ with $0 \leq \|\mathbf{h}\| \leq \sqrt{N}$ we have

$$\left| \sum_{\nu=1}^{N} e(u_\nu(\mathbf{h}, x)) \right| = \mathcal{O}(N^{1/2+\eta}). \tag{1.378}$$

Now the ERDŐS-TURÁN-KOKSMA inequality (Theorem 1.21) with $H = [\sqrt{N}]$ may be used, and we obtain the desired result after replacing x by $\log x$. □

1.6.4 PHILIPP's Theorem

In this final section we will state a general theorem of the iterated logarithm due to PHILIPP [1432], which is uniform in a special parameter and applies even for weakly dependent random variables; for a contribution concerning a functional law for weakly dependent random variables see PHILIPP [1435]. We will also give a first application. Anotherone is provided in Section 2.3. For a detailed discussion, a proof of the following general theorem, and various applications we refer to [1432].

Theorem 1.192 *Let $\mathcal{A} = \{X_n(\gamma) \mid n \geq 1, \gamma \in \Gamma\}$ be a family of random variables uniformly bounded by 1. Let*

$$S_N^2(\gamma) = S_N^2(X_n(\gamma)) = \mathbf{E}\left(\sum_{n=1}^{N} X_n(\gamma) - \mathbf{E}X_n(\gamma) \right)^2 \tag{1.379}$$

be the variance of the partial sum, and assume that

$$\lim_{N\to\infty} \frac{1}{N} S_N^2(\gamma) = \sigma^2(\gamma) < 1$$

uniformly for $\gamma \in \Gamma$.

Suppose that for each non-negative integer θ there is a subclass $\Gamma_\theta \subseteq \Gamma$ which defines a subfamily $\mathcal{A}_\theta \subseteq \mathcal{A}$ whose elements we denote by $\{X_n(\theta) \mid n \geq 1\}$. Suppose further that for each θ the familiy $\mathcal{A}_\theta$ satisfies the following conditions

1. *$1 \leq |\mathcal{A}_\theta| \leq e^{c_1\theta}$ for some constant $c_1 > 0$.*
2. *Each process $\{X_n(\theta) \mid n \geq 1\} \in \mathcal{A}_\theta$ satisfies the mixing condition*

$$|\mathbf{P}(A \cap B) - \mathbf{P}(A)\mathbf{P}(B)| \leq \mathbf{P}(A)\mathbf{P}(B)\psi(M) \tag{1.380}$$

with $\psi(M) = e^{-\lambda M}$ for $M > 1$ ($\lambda > 0$ a fixed constant), for any event $A \in \mathcal{M}_{0,L}$ and any event $B \in \mathcal{M}_{L+M,\infty}$, where $\mathcal{M}_{S,T}$ denotes the σ-field generated by the variables X_n with $S \leq n \leq T$.

3. *For each random variable* $X_n(\gamma)$ *there are two sequences of random variables* $\overline{X}_n(0), \overline{X}_n(1), \ldots,$ *and* $X_n(0), X_n(1), \ldots,$ *such that for any* $s \geq 0$

$$\overline{Y}_n(s) = \sum_{\theta=0}^{s} \overline{X}_n(\theta) \leq X_n(\gamma) \leq \sum_{\theta=0}^{s} X_n(\theta) = Y_n(s)$$

and

$$0 \leq \mathbf{E}(Y_n(s) - \overline{Y}_n(s)) \leq e^{-c_2 s}$$

with c_2 *a positive constant. For each* θ *both* $\{\overline{X}_n(\theta) \,|\, n \geq 1\}$ *and* $\{X_n(\theta) \,|\, n \geq 1\}$ *are members of* $\mathcal{A}_\theta$.

4. *Using notation (1.379)*

$$S_N^2(\theta) = \mathcal{O}(\theta^{-6} N)$$

uniformly in $\theta < \log N$.

5. *Each process* $\{\overline{Y}_n(s) \,|\, n \geq 1\}$, *and* $\{Y_n(s) \,|\, n \geq 1\}$ *satisfies the mixing condition (1.380) with* $\psi(M)$ *defined in 2.*

6. $S_N^2(\overline{Y}_n(s)) = N\sigma^2(\overline{Y}_n(s)) + \mathcal{O}(1)$ *and* $S_N^2(Y_n(s)) = N\sigma^2(Y_n(s)) + \mathcal{O}(1)$ *uniformly in* s *and* N.

Then, for any $\varepsilon > 0$, *there almost always is a* $N_0(\varepsilon)$ *such that*

$$\frac{\sum_{n=1}^{N}(X_n(\gamma) - \mathbf{E}X_n(\gamma))}{\sqrt{2N \log\log N}} < \sigma(\gamma) + \varepsilon, \quad \textit{for all } \gamma \in \Gamma,\ N > N_0. \tag{1.381}$$

Furthermore

$$\mathbf{P}\left[\limsup_{N\to\infty} \frac{\sum_{n=1}^{N}(X_n(\gamma) - \mathbf{E}X_n(\gamma))}{\sqrt{2N \log\log N}} \geq \sigma(\gamma)\right] = 1, \quad \textit{for all } \gamma \in \Gamma. \tag{1.382}$$

As a corollary we get a law of the iterated logarithm for the disrepancy in the modulo 1 case.

Theorem 1.193 *Let* $k \geq 1$. *Then*

$$\limsup_{N\to\infty} \frac{\sqrt{2N} D_N(\mathbf{x}_n)}{\sqrt{\log\log N}} = 1$$

for almost all sequences $(x_n)_{n\geq 1}$, $x_n \in \mathbf{R}^k$.

Notes

By the individual ergodic theorem one immediately gets for a fixed irrational α and any fixed Lebesgue measureable set $I \subset [0,1)$ of measure $\lambda_1(I)$ that for almost all x

$$\lim_{N\to\infty} \frac{A(I, N, \{x + n\alpha\})}{N} = \lambda_1(I).$$

In a series of papers Bourgain [276, 278, 279, 277, 281, 280, 283, 282] developed a powerful method to obtain similar results for subsequences, eg.

$$\lim_{N\to\infty} \frac{A(I, N, \{x+n^2\alpha\})}{N} = \lambda_1(I),$$

$$\lim_{N\to\infty} \frac{A(I, N, \{x+p_n\alpha\})}{N} = \lambda_1(I)$$

for almost all x (p_n denoting the n-th prime number). Bourgain's method is a combination of various deep tools from number theory, Fourier analysis and ergodic theory. In a recent article by Rosenblatt and Wierdl [1563] a very good survey on Bourgain's approach and related results is given, also more general subsequences (a_n) (instead of the squares or primes) are considered. For a further detailed discussion of these results see Thouvenot [1807].

By the individual ergodic theorem one immediately gets for any Lebesgue integrable function

$$\lim_{N\to\infty} \frac{1}{N} \sum_{n=1}^{N} f(\{x+n\alpha\}) = \int_0^1 f(x)dx$$

for almost all x. It is not known whether this is true for subsequences like n^2 or p_n, but it holds for every $f \in L^p, p > 1$. This concepts can be generalized to arbitrary probability spaces (with measure μ) and ergodic transformations T (instead of irrational rotations on the torus). Subsequences (a_n) of the positive integers with the property

$$\lim_{N\to\infty} \frac{1}{N} \sum_{n=1}^{N} f(T^{a_n}(x)) = \int f d\mu$$

for μ-almost all x (and an arbitrary integrable function f) are called "universal good". Several classes of such sequences are discussed in Bellow and Losert [161], see also [162]. For special results concerning the metric theory of continued fractions we refer to Nair [1248]. Further results on subsequence ergodic theorems see Jones, Olsen and Wierdl [877]. Negative results are of great interest in this context, too. Bellow [160] has shown that for every irrational x there exists a characteristic function f for which

$$\frac{1}{N} \sum_{n=1}^{N} f(\{x+2^n\alpha\})$$

diverges for almost every x. Thus the sequence (2^n) is called "universal bad". Bourgain [277] has developed the so-called entropy method which can be applied for proving almost everywhere divergence results. One main point is to construct non-lacunary universal bad sequences. Further results on universal bad sequences can be found in Rosenblatt [1562].

Nair [1247] investigated polynomials in primes from an ergodic point of view. Nakada [1250] described connections between diophantine approximations and ergodicity of geodesic flows. For further results on ergodic averages over subsequences we refer to Furstenberg [635].

Let (X, B, μ) be a probability space and $T : X \to X$ a measure preserving mapping; then (X, B, μ, T) is called a dynamical system; see Walters [1935]. Bergelson [184] considered weakly mixing dynamical systems $(X, \mathcal{B}, \mu, T)$ and distinct polynomials $p_1, \dots, p_k$. Then for essentially bounded functions $f_1, \dots, f_k$ the averages

$$\frac{1}{N-M} \sum_{n=M}^{N-1} T^{p_1(n)} f_1 T^{p_2(n)} f_2 \cdots T^{p_k(n)} f_k$$

converge to the product of the integrals $\int f_i \, d\mu$. A recent contribution on the convergence of such averages generated by three ergodic transformations is due to Zhang [1990].

A subset $P \subseteq \mathbb{N}$ is called a Poincaré set if there exists a dynamical system, a set A of positive measure and a number $n \in P$ such that $\mu(T^{-n}A \cap A) > 0$. A subset $H \subseteq N$ is called an intersective set if for every $S \subset \mathbb{N}$ of positive density $H \cap (S-S) \neq \emptyset$, where $S - S = \{a - b : a, b \in S, a \neq b\}$.

Bertrand-Mathis [207] proved that Poincaré sets and intersective sets are equal; see also [208]. Ruzsa [1577, 1578] has shown that in the definition of intersective sets one can take subsets S of strictly positive upper density or of strictly positive Banach density; see also Ruzsa [1580]. The result of Bertrand-Mathis has independently been proved by Liardet, Méla and Katznelson. Ruzsa [1578] conjectured that Poincaré sets (or intersective sets) are equivalent to Van der Corput sets. Van der Corput sets are strongly related with Van der Corput's difference theorem: A subset $H \subseteq \mathbf{N}$ is a Van der Corput set if and only if a sequence $(u_n)_{n\geq 0}$ of real numbers is uniformly distributed provided that the difference sequences $(u_{n+h} - u_n)_{n\geq 0}$ are uniformly distributed for all $h \in H$. (Note: By the difference theorem the set of natural numbers is a Van der Corput set.) Equivalent formulations can be given in terms of continuity of measures: H is a Van der Corput set if and only if all positive measures μ on $[0, 2\pi]$ with Fourier coefficients $\int_0^{2\pi} e^{-ihx}\, d\mu = 0$ (for every $h \in H$) are continuous at 0. Van der Corput sets were extensively studied by Kamae and Mendès France [892] and Mendès France [1174, 1173]; for Van der Corput sets in groups and summation methods we refer to Peres [1425]. For further results and an excellent survey on problems of this kind see Furstenberg [634]. Up to now the conjecture of Rusza seems to be open. More results on difference sequences can be found in Ruzsa [1576] and Stewart and Tijdeman [1741]. Further contributions to Van der Corput's difference theorem and generalization are due to Taschner [1784, 1783, 1785, 1791, 1793]. In Bertrand-Mathis [207] various results in the spirit of Kamae and Mendès France [892] are obtained and they are applied to the uniform distribution of special sequences of the type $(x\theta^n)_{n\geq 0}$. Boshernitzan [270] investigated homogeneously distributed sequences and Poincaré sequences of subexponential growth.

Ergodic aspects of normal numbers and generalizations can be found in Kamae [889, 890, 891], in Kamae and Weiss [893] and in Sigmund [1690]. For more results on ergodic theory and uniform distribution we refer to Coffey [374], Couot [416] and to Helmberg [774]. For ergodic sequences of measures in connection with additive number theory see Niederreiter [1294]. Ferenczi, Kwiatkowski and Mauduit [596] considered ergodic problems in connection with arithmetic functions.

Halasz [717] proved that the remainder in the individual ergodic theorem cannot be improved for any non-atomic system. Several papers of Hellekalek [765], Hellekalek and Larcher [772, 771] and Larcher [1033] are devoted to ergodic problems of cylinder flows and the connection with Weyl sums and skew products over irrational rotations. For related investigations see Rauzy [1510, 1512, 1511], Merill [1189], Oren [1395], Bagett [61], Choe [349], and Liardet [1082, 1084]. Rauzy [1514, 1516] and Keane and Rauzy [915] studied ergodic properties of interval exchange transformations (generalizing the irrational rotations). Liardet [1087] gave an extensive discussion of metric properties of subsequences, especially on combinations of statistical independence and u.d. subsequences, see also the Notes of Section 2.4. Recently, Dumont, Kamae and Takahashi [496] gave a detailed study of minimal cocycles and substitutions. Liardet and Volny [1089] investigated the speed of convergence in the ergodic theorem and related questions, mainly for irrational rotations on the torus. As an important survey on ergodic problems of this type we refer to Gabriel, Lemanczyk and Liardet [641]. Furthermore Liardet and Volny [1088] constructed smooth cocycles over irrational rotations.

For applications of ergodic methods to combinatorial number theory we refer to Furstenberg's monograph [634]. Recently, Boshernitzan [272] gave an elementary proof of the following theorem by Furstenberg [633]: *Let S be a non-lacunary multiplicative semigroup of positive integers and α be an irrational; then $S\alpha$ is dense mod 1.* Related results are due to Berend [170, 171, 172, 173, 174, 175, 176]; for simultaneous diophantine approximation and so-called IP-sets see Furstenberg and Weiss [637]. Boshernitzan [273] investigated the denseness (and u.d. mod 1) of sequences $f(n)$ under certain regularity conditions on the function f.

Maxones, Richards, Rindler and Schoißengeier [1155] precisely determined the asymptotics (as $N \to \infty$) for the integral (extended over all sequences in $[0, 1]^\infty$) of the discrepancy. Improvements on the error term are due to Schoißengeier [1653]. Niederreiter [1281] established formulae for the (infinite) product measure of the set of sequences ω satisfying $D_N(\omega) \leq a$.

In a series of papers the spectral measure of special sequences was investigated, see for instance Coquet and Mendès France [411] and Coquet, Kamae and Mendès France [408]. A complex sequence (u_n) belongs to the space S of Wiener if

$$\gamma(n) = \lim_{N\to\infty} \frac{1}{N} \sum_{0\leq n<N} u_{n+k}\bar{u}_k$$

exits for every $n \geq 0$. According to the Bochner-Herglotz representation theorem there exists a bounded positive measure Λ such that (for $(u_n) \in S$)

$$\gamma(n) = \int_{\mathbf{R}/\mathbf{Z}} e(nx)\, d\Lambda(x).$$

Λ is called the spectral measure of the sequence (u_n), see Wiener [1958, 1959], Mahler [1117] and Kahane [884]. We mention here also that the Wiener space is strongly connected with so-called "pseudo-random" sequences, c.f. Bass, [106, 107, 108, 109], Bertrandias [210] and Coquet [380, 385, 382, 392, 391]. An interesting concept related to pseudorandomness is the statistical independence of sequences, see Mendès France [1171], Coquet [396, 389], Coquet,Rhin and Toffin [413, 398, 414] and Coquet and Liardet [409, 410]; see also Grabner and Tichy [685]. Bésineau [212, 213, 214, 215] studied statistical independence and almost periodicity of sequences. Almost periodicity and spectral properties were studied by Daboussi and Mendès France [428] and Mendès France [1167].

In Coquet, Kamae and Mendès France [408] q-multiplicative sequences are investigated in detail from a spectral theoretic point of view. An explicit formula is established and several general properties of spectral measures are shown. It is proved that in case of a q-multiplicative sequence the spectral measure Λ is either purely singular or purely atomic. Furthermore a special sequence of Kakutani [886], [887] is shown to be p^2-multiplicative ($p \geq 3$ a prime number) and to be pseudo-random with a singular spectral measure. Recurrence formulas in connection with correlation products and spectral measures can be found in Queffelec [1493] and [1494]. For further results on spectral properties of sequences see Dekking and Mendès France [428], Coquet [388, 386, 399], Coquet, Rhin and Toffin [412], Prunner [1489], Mendès France [1165], and Choe [350]. In Mendès France [1172] statistical independence and spectral properties are investigated.

Niederreiter [1294] considered ergodic sequences of measures and relations to additive number theory; see also Niederreiter [1284]. Erdös and Taylor [568] have proved that there exist increasing sequences $S = (n_i)$ of positive integers such that the set $E(S)$ of points x for which (xn_i) is not uniformly distributed mod 1 is not countable; furthermore the sequence can be chosen such that $1 \leq n_{i+1} - n_i \leq 100$. Baker [67] has shown a refinement of this theorem; see also de Mathan [1132, 1133, 1134], where it is shown that in the theorem of Erdős and Taylor "not u.d." can be replaced by "not dense", and that the Hausdorff dimension of the corresponding set of x is 1. Related results and new proofs are due to Fan [579]. Pollington [1443, 1444] has shown that the Hausdorff dimension of the set of all x such that sequences of the types (x^n), $x > 1$, or $(a_n x)$, (a_n) a lacunary sequence of positive integers, are not dense mod 1 is one. On the other hand Ajtai, Havas and Komlós [9] proved that for any given sequence $\varepsilon_n > 0$ tending to 0 there is a sequence of positive integers (a_n) with $a_{n+1}/a_n > 1 + \varepsilon_n$ such that $(a_n x)$ is u.d. mod 1 for all irrationals x. Salát [1593, 1595] studied the Hausdorff dimension of sets related to sequences of the type $(a_n x)$ (for increasing integers a_n) and related to subsequences of a given sequence; for similar metric results see Salát [1591] and Bukovska and Salát [308].

In a series of papers Baker considers the distribution mod 1 of sequences $(\lambda_n(x))$ where x is from a locally compact abelian group G and $\lambda_1, \lambda_2, \ldots$ is a sequence in the dual group $\hat{G}$. Metric results analogous to the Erdős-Gál-Koksma theorems are proved and the exceptional sets are studied, c.f. [64, 65, 66, 68]. In Baker [71] sequences $(g_i(x))$ of differentiable functions on the interval $[a, b]$ are considered such that $g_m'(x) - g_n'(x)$ is monotone and uniformly bounded away from 0 for $m \neq n$; see also Rousseau and Egele [1572]. For an arbitrary function in $L^2[0, 1]$ with period 1 it is shown that

$$\lim_{n \to \infty} \sum_{k=1}^{n} f(g_k(x) - y) = \int_0^1 f(t)dt$$

for almost all (x, y) in $[a, b] \times [0, 1]$. In Baker [72] a similar result is proved provided that the spacing condition is weakened to $g_m'(x) - g_n'(x) \geq C$ whenever $m \geq n + cn^a$ with $0 \leq a < 1$ and $C, c > 0$. A metric bound for the discrepancy is proved and the Hausdorff dimension of the exceptional set is estimated. For further results in that direction we mention [76, 78] and Yavid [1976]. A metric discrepancy bound for sequences with repetitions is due to Topuzoglu [1871].

Several results on Hausdorff dimension and density mod 1 can be found in Pollington [1442] and Boshernitzan [271]; see also Borel [251, 252] and Thomas [1805]. Turán [1878] established metric results on partitions and normal numbers. For results concerning the Hausdorff dimension and simultaneous diophantine approximation we refer to Dickinson [460, 461] and Rynne [1585].

As a general reference on ergodic theory and metric number theory we mention the monograph by Schweiger [1663].

Khintchine [921] conjectured that

$$\lim_{N\to\infty}\frac{A(E,N,\{n\alpha\})}{N}=\lambda_1(E)$$

for any measurable set $E\subseteq[0,1)$ of Lebesgue measure $\lambda_1(E)$. This conjecture was disproved by Marstrand [1125]. Baker [69] gave a positive result if E is a suitable set. For further results in this direction we refer to Nair [1245]. It follows already from Weyl [1953] that $(\lambda_n\cdot x)$ is uniformly distributed mod 1 for almost all $x\in\mathbf{R}$ if $\lambda_1\leq\lambda_2\leq\ldots$ is an increasing sequence of integers provided that at most $\mathcal{O}(n(\log n^{-1-\varepsilon}))$ consecutive terms coincide. Baker [70] proved that apart from ε in the exponent this result is best possible. Harman [743] showed several metric results on the distribution of sequences (xa_n+b_n), where $(a_n),(b_n)$ are two sequences of real numbers, a_n satisfying suitable growth conditions; see also [742, 741, 747]. In [747] it is proved that if (a_n) is an increasing sequence of reals satisfying certain (fairly mild) conditions, then the inequality $||xa_n||<(n\log\log 9n)^{-1}$ has infinitely many solutions for almost all real x.

For results on approximating Lebesgue integrals by Riemann sums we refer to Revesz and Ruzsa [1532]. Furthermore we mention here Baker [74], where special Riemann sums are used for the approximation of Lebesgue integrals; see also Nair [1249]. Schneider [1645] obtained results on the capacity of sets related to u.d. sequences.

Bernik [188] established a metric bound for linearly independent polynomials; see also [189, 190], Bernik and Kovalevskaya [191] and Kovalevskaya [964]. Aumayr [57] proved metric results on the existence of uniformly distributed sequences; this is the continuation of the work of Taschner [1792].

An excellent presentation of the (qualitative) metric theory of uniform distribution is due to Rauzy [1509]; another survey on the metric theory of u.d. was given by Rhin [1534]. The metric theory of Erdős-Gál-Koksma was systematically developed in a series of papers [563, 565, 566, 642, 643, 644]. For various other metric results concerning sequences (xa^n) we refer to Moskvin [1209, 1210, 1211, 1212]. A special example is due to Kano [899].

Laws of iterated logarithms for the discrepancy of lacunary sequences are due to Philipp [1433, 1434] and Stackelberg [1734]. In our presentation we followed Philipp [1434] who used results of Takahashi [1772, 1778] on lacunary Fourier series in his original proof. However there is a more probabilistic approach given by Philipp and Stout [1437]. For further contributions of Takahashi on lacunary Fourier series we refer to [1773, 1774, 1776, 1775, 1777]. A survey on laws of iterated logarithms in number theory is given in the monograph by Philipp [1432]; see also Philipp [1435]. A law of iterated logarithm for Sobol's range discrepancy is established by Grabner [678].

LeVeque extended metric discrepancy estimates to trigonometric and complex valued sequences, see [1063, 1064]. Tichy [1811] proved that for almost all quaternions z of norm >1 the sequence of powers (z^n) is uniformly distributed in $\mathbf{R}^4$. Metric discrepancy bounds for quaternions and more general hypercomplex systems are due to Nowak [1377] and Nowak and Tichy [1382]. Further results on the metric theory of uniform distribution in hypercomplex systems can be found in Tichy [1813] and Tichy and Winkler [1850]; for related results see also Nowak [1375, 1376, 1379] and Nair [1244, 1246]. In a series of papers Losert, Nowak, and Tichy developed the Erdős-Gál- Koksma technique for power sequences of a given matrix, see [1381, 1380, 1105, 1383]. A typical result of that papers is the following:

Theorem 1.194 *Let $(p(n))_{n\geq1}$ be a strictly increasing sequence of positive integers. Then for almost all (in the sense of the s^2-dimensional Lebesgue measure) real $s\times s$-matrices A with at least one eigenvalue of modulus larger than 1 there exists a constant $C(A)$ such that the discrepancy $D(N,A)$ of the sequence $(A^{p(n)})_{n=1}^N$ (considered as a sequence in $\mathbf{R}^{s^2}$) can be estimated by*

$$D(N,A)\leq C(A)N^{-1/2}(\log N)^{s^2+1}.$$

Lacroix and Thomas [999] established a metric theorem on completely u.d. sequences related to general digital expansions. A metric discrepancy bound for "homothetic" sequences is due to Borel [253].

Completely uniformly distributed sequences of matrices were studied by Drmota,Tichy and Winkler [491]. This is a combination of the approach used for proving metric theorems on uniform distribution of power sequences of matrices and the method developed for solving Knuth's problem which was extensively discussed in this section. For different approaches solving Knuth's problem and quantitative refinements see Niederreiter and Tichy [1364, 1365], Tichy [1832] and Goldstern [661]. Petersen [1427] investigated sequences (a_n) with the strong Weyl property, i.e. for every rearrangement $\pi(n)$ of the positive integers the sequence $(xa_{\pi(n)})$ is uniformly distributed mod 1 for almost every $x \in \mathbf{R}$ (with respect to Lebesgue measure). A continuation of this investigations is due to Winkler [1964].

Metric properties of subsequences of uniformly distributed sequences and of multi-sequences are due to Tichy [1812] and Losert and Tichy [1108], see also [1102], [1816]. Further results in the metric theory of uniform distribution can be found in Loynes [1109] and in Niederreiter [1280, 1289]. In [1289] Koksma's metric theorem is extended to u.d. with respect to matrix summation methods (see Section 2.2).

Chapter 2

General Concepts of Uniform Distribution

2.1 Geometric Concepts of Discrepancy

2.1.1 Discrepancy Systems

The notion of uniform distribution can be easily generalized to compact spaces X (See Section 2.4).

Definition 2.1 *Let X be a compact (*HAUSDORFF*) space and μ a positive regular normalized* BOREL *measure on X. A sequence $(x_n)_{n\geq 1}$, $x_n \in X$, is called* uniformly distributed with respect to μ *(μ-u.d.) if*

$$\lim_{N\to\infty} \frac{1}{N} \sum_{n=1}^{N} f(x_n) = \int_X f\, d\mu \tag{2.1}$$

holds for all continuous functions $f : X \to \mathbf{R}$.

But this concept cannot be used to generalize the notion of discrepancy. First we need a property where we can replace the continuous functions f in (2.1) by characteristic functions χ_M of special subsets $M \subseteq X$. One possibility is using μ-continuity sets.

Definition 2.2 *A* BOREL *set $M \subseteq X$ is called* μ-continuity set *if $\mu(\partial M) = 0$, where ∂M denotes the boundary of M.*

Theorem 2.3 KUIPERS *and* NIEDERREITER [983, p. 178] *A sequence $(x_n)_{n\geq 1}$, $x_n \in X$, is μ-u.d. if and only if*

$$\lim_{N\to\infty} \frac{1}{N} \sum_{n=1}^{N} \chi_M(x_n) = \mu(M) \tag{2.2}$$

holds for all μ-continuity sets $M \subseteq X$.

This leads to the definition

Definition 2.4 *A system $\mathcal{D}$ of μ-continuity sets of X is called* discrepancy system *if*

$$\lim_{N\to\infty} \sup_{M\in\mathcal{D}} \left| \frac{1}{N} \sum_{n=1}^{N} \chi_M(x_n) - \mu(M) \right| = 0 \tag{2.3}$$

holds if and only if $(x_n)_{n\geq 1}$ is μ-u.d. For a discrepancy system $\mathcal{D}$ the supremum

$$D_N^{(\mathcal{D})}(x_n) = \sup_{M\in\mathcal{D}} \left| \frac{1}{N} \sum_{n=1}^{N} \chi_M(x_n) - \mu(M) \right| \tag{2.4}$$

is called discrepancy *of $(x_n)_{n\geq 1}$ with respect to $\mathcal{D}$.*

There are many cases where we have already shown that a system $\mathcal{D}$ of subsets $M \subseteq X$ is a discrepancy system. E.g. in Section 1.1.2. we have shown that the system of cubes with sides parallel to the axes and the system of all convex subsets contained in the torus are discrepancy systems. Therefore each system of convex subsets containing all squares with sides parallel to the axes are discrepancy systems. It is clear that we should suppose that every continuous function $f : X \to \mathbf{R}$ can be uniformly approximated by a finite linear combination of characteristic funcions of sets contained in $\mathcal{D}$. This can be checked by the following condition.

Proposition 2.5 *Suppose that $\mathcal{D}$ is a system of closed μ-continuity sets $M \subseteq X$ with $X \in \mathcal{D}$ that satisfies the following separation condition:*

> *For any two different points $x, y \in X$ there exist $M_0, M_1 \in \mathcal{D}$ with $x \in M_0^\circ \subseteq M_0 \subset M_1^\circ$ and $y \notin M_1$ and an* URYSOHN *function $u : X \to [0,1]$ with $u|_{M_0} = 0$, $u|_{X\setminus M_1^\circ} = 1$, and $M_\lambda = \{z \in X : u(z) \leq \lambda\} \in \mathcal{D}$ for all $\lambda \in [0,1]$.*

Then every continuous function $f : X \to \mathbf{R}$ can be approximated uniformly by a finite linear combination of characteristic funcions χ_M of sets $M \in \mathcal{D}$.

Proof. By STONE-WEIERSTRASS' approximation theorem finite linear combinations of the point-separating URYSOHN functions $u : X \to [0,1]$ and χ_X can approximate any continuous funcion $f : X \to \mathbf{R}$ uniformly. Furthermore any URYSOHN function $u : X \to [0,1]$ with $M_\lambda \in \mathcal{D}$ satisfies

$$\sup_{x\in X} \left| u(x) - \frac{1}{n} \sum_{k=1}^{n} (1 - \chi_{M_{k/n}}(x)) \right| \leq \frac{1}{n}.$$

This proves Proposition 2.5. □

It is quite practical to restrict the discussion to compact (HAUSDORFF) spaces satisfying the second axiom of countability. This additional assumption leads to the following simplifications. First we need not assume regularity for μ. Every positive normalized BOREL measure is automatically regular, see HALMOS [727]. Furthermore by URYSOHN's metrization theorem X is metrizable. The next proposition gives a sufficient condition for a system $\mathcal{D}$ to be a discrepancy system.

Proposition 2.6 *Let (X,d) be a compact metric space and μ a positive normalized BOREL measure on X, let $\mathcal{D}$ be a system of measurable μ-continuity sets of X, and suppose that every continuous function $f: X \to \mathbf{R}$ can be approximated uniformly by a finite linear combination of characteristic funcions χ_M of sets $M \in \mathcal{D}$. Suppose that $\varphi(t)$, $t>0$, is a positive real monotone function with $\lim_{t\to 0+}\varphi(t)=0$ and assume that for every $\varepsilon>0$ and for every set $M\in\mathcal{D}$ there exist an open set $V\subseteq X$ and a compact set $K\subseteq X$ with $K\subseteq M\subseteq V$, $\mu(V\setminus K)<\varphi(\varepsilon)$,*

$$\overline{d}(K, X\setminus M)\geq\varepsilon \quad or\ K=\emptyset \tag{2.5}$$

and

$$\overline{d}(M, X\setminus V)\geq\varepsilon \quad or\ V=X, \tag{2.6}$$

where $\overline{d}(A,B)=\inf\{d(a,b): a\in A, b\in B\}$. Then $\mathcal{D}$ is a discrepancy system.

Proof. Since every continuous function $f: X\to\mathbf{R}$ can be approximated uniformly by a finite linear combination of characteristic funcions χ_M of sets $M\in\mathcal{D}$ it immediately follows from $\lim_{N\to\infty} D_N^{(\mathcal{D})}(x_n)=0$ that $(x_n)_{n\geq 1}$ is μ-u.d.

Now suppose that $(x_n)_{n\geq1}$ is μ-u.d. Then we have by Theorem 2.3 that (2.2) holds for any μ-continuity set $M\subseteq X$, especially for all sets $M\in\mathcal{D}$. Therefore we just have to prove that the convergence in (2.2) is uniform for all $M\in\mathcal{D}$. For this purpose we consider for any $\varepsilon>0$ a finite system of closed balls $A_i=\overline{B}(x_i,r_i)=\{x\in X: d(x,x_i)\leq r_i\}$ $(i\in I)$ covering X such that $r_i<\frac{\varepsilon}{2}$ and $\mu(\partial A_i)=0$. This is always possible since X is compact and there are at most countably many radii $r>0$ such that $\mu(\partial(\overline{B}(x_i,r)))>0$. Next set

$$\mathcal{B}=\left\{\bigcup_{i\in J}A_i : J\subseteq I\right\}.$$

Since we always have $\partial(C\cup D)\subseteq\partial C\cup\partial D$ it follows that $\mathcal{B}$ consists of finitely many μ-continuity sets. Hence there exists some positive integer $N_0(\varepsilon)$ such that

$$\left|\frac{1}{N}\sum_{n=1}^{N}\chi_B(x_n)-\mu(B)\right|<\varepsilon$$

holds for all $B\in\mathcal{B}$ and all $N\geq N_0(\varepsilon)$. By assumptions for any set $M\in\mathcal{D}$ there exists a compact set $K\subseteq M$ satisfying (2.5) and an open set $V\supseteq M$ satisfying (2.6) such that $\mu(V\setminus K)<\varphi(\varepsilon)$. Obviously

$$B_1=\bigcup_{A_i\subseteq M}A_i \quad\text{and}\quad B_2=\bigcup_{A_i\subseteq V}A_i$$

satisfy $K \subseteq B_1 \subseteq M \subseteq B_2 \subseteq V$. Consequently

$$\begin{aligned}\frac{1}{N}\sum_{n=1}^{N}\chi_{B_1}(x_n) - \mu(B_1) + \mu(B_1) - \mu(M) \\ \le \frac{1}{N}\sum_{n=1}^{N}\chi_M(x_n) - \mu(M) \\ \le \frac{1}{N}\sum_{n=1}^{N}\chi_{B_2}(x_n) - \mu(B_2) + \mu(B_2) - \mu(B_1)\end{aligned}$$

implies

$$\left|\frac{1}{N}\sum_{n=1}^{N}\chi_M(x_n) - \mu(M)\right| \le 2\varepsilon + \mu(V \setminus K) \le 2\varepsilon + \varphi(\varepsilon)$$

for $N \ge N_0(\varepsilon)$ uniformly for all $M \in \mathcal{D}$. □

Using these two propositions it is easy to verify directly that the above mentioned systems (i.e. systems of convex subsets containing all cubes with sides parallel to the axes) and similar systems are discrepancy systems. It also helps us to identify the system of all (closed) balls as a discrepancy system in a very general setting.

Proposition 2.7 *Let (X, d) be a compact metric space and μ a positive normalized* BOREL *measure on X. Furthermore, suppose that there exists a real positive monotone function $\varphi(t)$, $t > 0$, with $\lim_{t\to 0+}\varphi(t) = 0$ and*

$$\mu(\overline{B}(x, r+\varepsilon)) - \mu(\overline{B}(x,r)) < \varphi(\varepsilon)$$

uniformly for $x \in X$ and $r \ge 0$. Then the system $\mathcal{D} = \{\overline{B}(x,r) : x \in X, r \ge 0\}$ of all (closed) balls is a discrepancy system.

Proof. From

$$\left|d(x,z) - \frac{1}{n}\sum_{k\ge 1}(1 - \chi_{\overline{B}(x,k/n)}(z))\right| \le \frac{1}{n}$$

it follows that for every fixed $x \in X$ the (continuous) function $f_x(z) = d(x,z)$ can be uniformly approximated by a linear combination of characteristic functions of (closed) balls. Consequently, by the STONE-WEIERSTRASS theorem every continuous function $f : X \to \mathbf{R}$ can be approximated uniformly by a finite linear combination of characteristic functions χ_M of sets $M \in \mathcal{D}$.

Next, let $M = \overline{B}(x,r)$ be a (closed) ball. Suppose that $\varepsilon > 0$ is given and set $V = B(x, r+\varepsilon)$ and $K = \overline{B}(x, r-\varepsilon)$ (i.e. $K = \emptyset$ if $r < \varepsilon$). Then V is open, K is compact, $K \subseteq M \subseteq V$, $\mu(V \setminus K) < \varphi(2\varepsilon)$ and the relations (2.5) and (2.6) are satisfied.

Thus, we can apply Proposition 2.6 and conclude that $\mathcal{D}$ is indeed a discrepancy system. □

Before we start with the euclidean and the spherical case (Sections 2.1.2 and 2.1.3) we will discuss the possibility to generalize the notion of L^p-discrepancy to metric spaces. (Compare with Section 1.1.2) In this context we will use the system of all (closed) balls as a (discrepancy) system $\mathcal{D}$. We want to present a general variant of the results given in Section 1.1.2.

Let (X, d) denote a compact metric space and let λ and μ denote two probability (BOREL) measures on X such that μ satisfies the following additional (smoothness) conditions:

$$|\mu(\overline{B}(x,r_1)) - \mu(\overline{B}(x,r_2))| \leq L_1|r_2 - r_1|^\beta, \tag{2.7}$$
$$|\mu(\overline{B}(x_2,r)) - \mu(\overline{B}(x_1,r))| \leq L_2 d(x_1,x_2)^\beta, \tag{2.8}$$
$$\mu(\overline{B}(x,r)) \geq L_0 r^k, \tag{2.9}$$

where L_1, L_0, β, s are positive (absolute) constants and $L_2 \geq 0$. In the case of an $k-$dimensional closed, compact, connected manifold X with surface measure μ the LIPTSCHITZ conditions (2.7) and (2.8) are satisfied with $\beta = 1$; d denotes the geodesic metric. In the special case of the sphere (or more general in the case of homogeneous manifolds) (2.8) holds with $L_2 = 0$.

Theorem 2.8 *Let (X, d) be a compact metric space and let λ, μ be (BOREL) probability measures on X such that μ satiesfies the additional properties (2.7), (2.8), (2.9). For $y \in X$ and $r \geq 0$ let $D(y,r) = \lambda(\overline{B}(y,r)) - \mu(\overline{B}(y,r))$ and let φ be a nondecreasing function on $[0,1]$. Then*

$$\int_X \int_0^\theta \varphi(|D(y,r)|)\, dr\, d\mu(y) \geq c\|D\|_\infty^{\frac{k+1}{\beta}} \varphi\left(\frac{1}{6}\|D\|_\infty\right),$$

where c is a positive constant only depending on the space X and on L_0, L_1, L_2, β and k; θ denotes the diameter of X.

The most important special case is the case of L^p-discrepancy of sequences on closed, connected, compact metric manifolds X. Let (x_n), $n = 1, 2, \ldots, N$ be a sequence on X. Furthermore, let μ be the surface measure on X and

$$\mu(E) = \frac{1}{N}\sum_{n=1}^{N} \chi_E(x_n). \tag{2.10}$$

Then for $p \geq 1$ the L^p-discrepancy $D_N^{(p)}(x_n)$ of the point sequence (x_n) is defined by

$$D_N^{(p)}(x_n) = \left(\int_X \int_0^\theta |D(y,r)|^p dr\, d\mu(y)\right)^{1/p}. \tag{2.11}$$

We will compare it with the discrepancy

$$D_N(x_n) = \sup_{y \in X, r \geq 0} |D(y,r)|. \tag{2.12}$$

Inserting $\varphi(t) = t^p, \beta = 1$ and $k = \dim X$ in Theorem 2.8 yields the bound

$$\delta D_N(x_n)^{\frac{k+p+1}{p}} \leq D_N^{(p)}(x_n) \leq \theta D_N(x_n) \tag{2.13}$$

(with a positive constant $\delta > 0$).

Hence, if $\mathcal{D}$, the system of all balls, is a discrepancy system, we can describe all μ-u.d. sequences by considering any L^p-discrepancy.

Proof of Theorem 2.8. We use the notation $D(x,r) = G(x,r) - F(x,r)$, in which $G(x,r) = \lambda(B(x,r))$ and $F(x,r) = \mu(B(x,r))$. Then there exists a pair $(x_0, r_1) \in (X, \mathbf{R}_+)$ such that $|D(x_0, r_1)| > \|D\|_\infty - \varepsilon$ (for arbitrary $\varepsilon > 0$). We set $\Theta(x_0) := \sup\{d(x_0, y) : y \in X\}$. In the following we will show for arbitrary a with $0 < a < \frac{1}{2}\Theta(x_0)$:

$$m(a) := \sup_{r \in [a, \Theta(x_0) - a]} |D(x_0, r)| \geq \frac{1}{3}\|D\|_\infty - \frac{2}{3}L_1 a^\beta. \tag{2.14}$$

Setting $A(x_0) = \{y : \Theta(x_0) - a \leq d(x_0, y) \leq \Theta(x_0)\}$ and $r' = \min(r, \Theta(x_0) - a)$ we obtain

$$|D(x_0, r) - D(x_0, r')| \leq \max(\lambda(A(x_0)), \mu(A(x_0))). \tag{2.15}$$

By the LIPTSCHITZ condition (2.7)) we have

$$\mu(A(x_0)) \leq L_1 a^\beta$$

and

$$\mu(A(x_0)) = 1 - \mu(\overline{B}(x_0, \Theta(x_0) - a)) - D(x_0, \Theta(x_0) - a) \leq L_1 a^\beta + m(a).$$

Thus

$$|D(x_0, r)| \leq |D(x_0, r')| + L_1 a^\beta + m(a). \tag{2.16}$$

Using the notations $\overline{B}(x_0) = \overline{B}(x_0, a)$ and $r'' = \max(r', a)$ we have

$$|D(x_0, r') - D(x_0, r'')| \leq \max(\lambda(\overline{B}(x_0)), \mu(\overline{B}(x_0))). \tag{2.17}$$

By the LIPTSCHITZ condition (2.7) we get as above $\mu(\overline{B}(x_0)) \leq L_1 a^\beta$ and

$$\mu(\overline{B}(x_0)) \leq \mu(\overline{B}(x_0)) + |D(x_0, a)| \leq L_1 a^\beta + m(a).$$

Inserting these estimates in (2.17) and combining with (2.16) yields

$$|D(x_0, r)| \leq 2L_1 a^\beta + 3m(a) \tag{2.18}$$

for arbitrary $r \geq 0$. Thus the assertion (2.14) is proved. We set

$$a = \min\left\{\left(\frac{1}{2}\Theta(x_0)\right)^{1/\beta}, 6^{-1/\beta}\left(\frac{2}{3}L_1 + \frac{L_1 + L_2}{2^\beta}\right)^{-1/\beta}, \frac{1}{2}\Theta(x_0)\|D\|_\infty^{1/\beta}\right\}$$

and choose an $r_0 \in [a, \Theta(x_0) - a]$ such that

$$|D(x_0, r_0)| \geq \frac{1}{3}\|D\|_\infty - \frac{2}{3}L_1 a^\beta - \varepsilon \tag{2.19}$$

(for arbitrary $\varepsilon > 0$).

We distinguish the cases $D(x_0, r_0) \geq 0$ and $D(x_0, r_0) < 0$ and set $b = \frac{1}{4}a$. Then we have for every $y \in \overline{B}(x_0, b)$ and every r with $r_0 + 2b \leq r \leq r_0 + a$:

$$\overline{B}(y, r) \supseteq \overline{B}(x_0, r_0). \tag{2.20}$$

In the case $D(x_0, r_0) \geq 0$ we obtain for $y \in \overline{B}(x_0, b)$ and $r \in [r_0 + 2b, r_0 + a]$:

$$\begin{aligned} G(y,r) - F(y,r) &\geq G(x_0, r_0) - F(y, r_0) - L_1(r - r_0)^\beta \\ &\geq G(x_0, r_0) - F(x_0, r_0) - L_1(r - r_0)^\beta - L_2 b^\beta \\ &\geq G(x_0, r_0) - F(x_0, r_0) - (L_1 + L_2) b^\beta \end{aligned}$$

(because of the conditions (2.7) and (2.8)). Hence we derive by (2.19)

$$|G(y,r) - F(y,r)| \geq \frac{1}{3}\|D\|_\infty - \frac{2}{3}L_1 a^\beta - \frac{L_1 + L_2}{2^\beta}a^\beta - \varepsilon \geq \frac{1}{6}\|D\|_\infty. \tag{2.21}$$

Thus we have

$$\begin{aligned} \int_{y\in X}\int_{r\geq 0} \varphi(|D(y,r)|)d\mu(y)dr &\geq \int_{y\in\overline{B}(x_0,b) r_0+2b} \varphi\left(\frac{1}{6}\|D\|_\infty\right) d\mu(y)dr \\ &\geq L_0 b^{k+1} \varphi\left(\frac{1}{6}\|D\|_\infty\right), \end{aligned} \tag{2.22}$$

where we have used (2.9) in the last step. Since $\inf\{\Theta(x_0) : x_0 \in X\} > 0$ we obtain

$$\int_{y\in X}\int_{r\geq 0} \varphi(|D(y,r)|)d\mu(y)dr \geq c\|D\|_\infty^{\frac{k+1}{\beta}} \varphi\left(\frac{1}{6}\|D\|_\infty\right) \tag{2.23}$$

with a positive constant c (only depending on the space X and the constants L_0, L_1, L_2, β).

Next we consider the case $D(x_0, r_0) < 0$. Here we obtain for $y \in \overline{B}(x_0, b)$ and $r \in [r_0 - 2b, r_0 - b]$:

$$\begin{aligned} F(y,r) - G(y,r) &\geq F(y, r_0) - G(x_0, r_0) - L_1(r_0 - r)^\beta \\ &\geq F(x_0, r_0) - G(x_0, r_0) - L_1(r_0 - r)^\beta - L_2 b^\beta \\ &\geq F(x_0, r_0) - G(x_0, r_0) - (L_1 + L_2) b^\beta \end{aligned}$$

(because of the conditions (2.7) and (2.8)). Hence we derive as in (2.21)

$$|G(y,r) - F(y,r)| \geq \frac{1}{6}\|D\|_\infty,$$

and, from this inequality, (2.23) follows as above also in the case $D(x_0, r_0) < 0$. Thus Theorem 2.8 is proved. □

We will close this section with a non-trivial example of a fractal compact space. We want to describe the most important ideas for the classical SIERPINSKI gasket, a well-known planar fractal set introduced by SIERPINSKI [1689]. Let A_0 be a closed equilateral triangle of unit sides e_1, e_2, e_3 with vertices $P_1\left(\frac{1}{2}, \frac{\sqrt{3}}{2}\right), P_2(0,0), P_3(1,0)$. Let A_1 be the set obtained by deleting the open equilateral triangle whose vertices are the midpoints of the edges of A_0. Thus A_1 consists of 3 equilateral triangles with side $\frac{1}{2}$. Repeating this procedure we obtain successively $A_2, A_3, \ldots A_n$ consists of 3^n equilateral triangles of side 2^{-n}, which are called *elementary triangles of level n*. Furthermore we denote the set of all vertices of A_n by V_n and the boundary of A_n by E_n. Thus $F_n = (V_n, E_n)$ is defining a finite graph. The set $G = \bigcap_{n=0}^{\infty} A_n$ is called the (bounded) SIERPINSKI gasket.

Remark. Any point $p \in G$ can be represented by the triple (k_1, k_2, k_3) with $k_1 + k_2 + k_3 = 2$, where

$$k_i = k_i(p) = \sum_{l=1}^{\infty} \frac{\varepsilon_l^{(i)}}{2^l}$$

and $\varepsilon_l^{(1)} + \varepsilon_l^{(2)} + \varepsilon_l^{(3)} = 2$ for all $l \geq 1$. Note that $(1 - k_i)\frac{\sqrt{3}}{2}$ is just the distance of p to the side e_i.

By standard techniques as described in FALCONER [578] it is easy to see that the SIERPINSKI gasket has HAUSDORFF dimension $\alpha = \frac{\log 3}{\log 2}$ and finite positive HAUSDORFF measure. Let μ denote the (normalized) HAUSDORFF measure of dimension α on G.

In the following we give a short introduction into the main ideas of uniformly distributed sequences on the gasket. This theory was systematically developed in GRABNER and TICHY [675]. Many of the concepts can be generalized to more general fractals.

G is a compact space the topology of which is induced by the following metric d. Any two points a and b in G are contained in elementary triangles of level k, $\Delta_k(a), \Delta_k(b)$, respectively. Let a_k, b_k be the lower vertices of $\Delta_k(a), \Delta_k(b)$ respectively, and observe that a_k and b_k are vertices of the finite graph F_k. We set

$$d(a,b) = \lim_{k \to \infty} 2^{-k} d_k(a_k, b_k),$$

where d_k is the minimal length of a chain connecting a_k and b_k. Obviously $d(a,b)$ is the geodesic distance of a and b, i.e. the length of the shortest continuous curve in G connecting a and b. Without proof we state the following explicit formula for this distance, for details see GRABNER and TICHY [675].

Proposition 2.9 *Let a and b be two points in G given by their digital representation $a = (\varepsilon_l^{(i)}), b = (\delta_l^{(i)}), i = 1,2,3, l = 1,2,\ldots$ Let L be the first index such that the triples $\varepsilon_L^{(i)})$ and $\delta_L^{(i)} = 0$ and $\delta_L^{(j)} = 0$. Then the distance of a and b is given by*

$$\sum_{l=L}^{\infty} 2^{-l}\left(\varepsilon_l^{(j)} + \delta_l^{(i)} - 1\right).$$

(The formula does not depend on different representations of the same points.)

Remark. Let $p \in G$ be a point different from P_i, $i = 1, 2, 3$ and let $\varepsilon > 0$ be sifficiently small. Then the ε-ball $B(p, \varepsilon) = \{x \in G | d(x, p) < \varepsilon\}$ consists of two congruent equilateral triangles (intersected with G) with one common vertex. Obviously, $B(P_i, \varepsilon)$ consists of one triangle. Thus the metric d induces the topology of the gasket.

Let us consider the discrepancy system $\mathcal{B}$ of all balls in G, and let $D_N^{(\mathcal{B})}(x_n)$ denote the discrepancy with respect to $\mathcal{B}$. Furthermore we can introduce a L^p-discrepancy $D_N^{(\mathcal{B}),p}(x_n)$ by

$$D_N^{(\mathcal{B},p)}(x_n) = \left(\int_G \int_0^1 \left| \sum_{n=1}^N \chi_{B(x,r)} - \mu(B(x,r)) \right|^p dr \, d\mu(x) \right)^{1/p}.$$

Then Theorem 2.8 gives

$$D_N^{(\mathcal{B},p)}(x_n) \gg \left(D_N^{(\mathcal{B})}(x_n) \right)^{(\alpha+p+1)/p}.$$

Furthermore VAN DER CORPUT type sequences (x_n) can be considered and the upper bound

$$D_N^{(\mathcal{B})}(x_n) = \mathcal{O}\left(\frac{1}{N^{\frac{\alpha-1}{\alpha}}} \right)$$

is shown for these sequences; see GRABNER and TICHY [675]. In this paper also a probabilistic construction of a finite sequence $(x_n)_{n=1}^N$ of N points is established such that

$$D_N^{(\mathcal{B})}(x_n) \ll N^{\frac{1}{2\alpha}-1} (\log N)^{1/2}.$$

Finally we want to mention here that also continuous uniform distribution of curves (see Section 2.3) can be extended to the gasket. For a metric result in this direction we again refer to GRABNER and TICHY [675].

2.1.2 Euclidean Discrepancy Systems

The most important discrepancy systems in the k-dimensional euclidean space $\mathbf{R}^k$ and in the k-dimensional torus $\mathbf{R}^k/\mathbf{Z}^k$ are similar or homothetic convex sets.

In the following we will concentrate on two kinds of problems. First, let $X \subseteq \mathbf{R}^k$ be a compact subset with $\mu_k(X) = 1$ and non-empty interior and let μ be the restriction of the k-dimensional LEBESGUE measure λ_k on X. Hence X is a compact metric space and μ a normalized BOREL measure on X. Fix any compact body $A \subseteq \mathbf{R}^k$. Then by Propositions 2.5 and 2.6 it follows that

$$\mathcal{D}_s(A, X) = \{(\rho(\xi A) + \mathbf{y}) \cap X : 0 < \rho \leq 1, \xi \in SO(k), \mathbf{y} \in \mathbf{R}^k\}, \tag{2.24}$$

where $SO(k)$ denotes the set of all rotations in $\mathbf{R}^k$, and

$$\mathcal{D}_h(A, X) = \{(\rho A + \mathbf{y}) \cap X : 0 < \rho \leq 1, \mathbf{y} \in \mathbf{R}^k\} \tag{2.25}$$

are discrepancy systems for sequences in X. In these cases (almost) optimal upper and lower bounds for the corresponding discrepancies are known. In the following,

$r(A)$ will denote the radius of the largest ball $B \subseteq A$, $\mathrm{diam}(A)$ the diameter of A, and $\sigma(\partial A)$ the surface measure of the boundary ∂A.

Theorem 2.10 *Let $k \geq 2$ and $X, A \subseteq \mathbf{R}^k$ be defined as above. Then we have for any finite sequence $\mathbf{x}_1, \ldots, \mathbf{x}_N \in X$ $(N \geq r(A)^{-k})$*

$$D_N^{(\mathcal{D}_s(A,X))}(\mathbf{x}_n) \gg \frac{(\sigma(\partial A))^{\frac{1}{2}}}{(\mathrm{diam}(A) + (\mathrm{diam}(X))^k} N^{-\frac{1}{2}-\frac{1}{2k}}, \tag{2.26}$$

where the constant implied by $\gg$ only depends on the dimension k. On the other hand, if X is convex, then for any $N \geq 1$ there exists a finite sequence $\mathbf{x}_1, \ldots, \mathbf{x}_N \in X$ such that

$$D_N^{(\mathcal{D}_s(A,X))}(\mathbf{x}_n) \ll N^{-\frac{1}{2}-\frac{1}{2k}} (\log N)^{\frac{1}{2}}, \tag{2.27}$$

where the constants implied by $\ll$ depends on the dimension k, on $\mathrm{diam}(X)$, *on* $\mathrm{diam}(A)$, *and on $r(A)$.*

It is obvious that such a general theorem cannot hold for the system of homothetic sets $\mathcal{D}_h(A, X)$ without any further assumptions. If A is a ball then $\mathcal{D}_h(A, X) = \mathcal{D}_s(A, X)$ and we get exactly the same estimates. On the other hand, if A is a cube then we have completely different results. (Compare with Theorems 1.40 and 3.13). Hence it might be conjectured that smoothness conditions on the boundary of A are essential for the behaviour of $D_N^{(\mathcal{D}_h(A,X))}$. The next theorems give a partial answer of this question.

Theorem 2.11 *Let $k \geq 2$ and $A \subseteq \mathbf{R}^k$ be a convex body such that the boundary ∂A can be mapped (uniquely) onto the unit sphere (using parallel pairs of support hyperplanes) by a bijective $6k$-times continuously differentiable function and assume that the product $L = r_1 \cdots r_k$ of all radii of the main curvature is uniformly bounded below by a positive constant, i.e.* $\inf L > 0$. *Then we have for any finite sequence $\mathbf{x}_1, \ldots, \mathbf{x}_N \in X$*

$$D_N^{(\mathcal{D}_h(A,X))}(\mathbf{x}_n) \gg N^{-\frac{1}{2}-\frac{1}{2k}}. \tag{2.28}$$

On the other hand, if X is convex there exists a finite sequence $\mathbf{x}_1, \ldots, \mathbf{x}_N \in X$ for any $N > 1$ such that

$$D_N^{(\mathcal{D}_h(A,X))}(\mathbf{x}_n) \ll N^{-\frac{1}{2}-\frac{1}{2k}} (\log N)^{\frac{1}{2}}. \tag{2.29}$$

The constants implied by $\gg$ and $\ll$ depend on the dimension k, on X, and on A.

Theorem 2.12 *Let $k \geq 2$ and $A \subseteq \mathbf{R}^k$ a convex simple polytope. Then we have for any finite sequence $\mathbf{x}_1, \ldots, \mathbf{x}_N \in X$*

$$D_N^{(\mathcal{D}_h(A,X))}(\mathbf{x}_n) \gg N^{-1} (\log N)^{\frac{1}{2}(k-1)}. \tag{2.30}$$

On the other hand, if X is convex there exists a finite sequence $\mathbf{x}_1, \ldots, \mathbf{x}_N \in X$ for any $N > 1$ such that

$$D_N^{(\mathcal{D}_h(A,X))}(\mathbf{x}_n) \ll N^{-1} (\log N)^{\max\left(\frac{3}{2}k+\varepsilon, 2k-1\right)}. \tag{2.31}$$

The constants implied by $\gg$ and $\ll$ depend on the dimension k, on X, and on A.

A convex polytope $P \subseteq \mathbf{R}^k$ is called simple if every vertex has exactly k adjacent vertices.

It should be mentioned that these three theorems remain true in the modulo 1 case. Let us consider a convex body $A \subseteq \mathbf{R}^k$ with $\operatorname{diam}(A) < 1$. Obviously we can interprete the reduction A mod 1 as a proper "convex" subset of $\mathbf{R}^k/\mathbf{Z}^k$.

Theorem 2.13 *Let $A \subseteq \mathbf{R}^k$ a convex body with* $\operatorname{diam}(A) < 1$ *and set*

$$\mathcal{D}_s(A, \mathbf{R}^k/\mathbf{Z}^k) = \{(\mu(\xi A) \bmod 1) + \mathbf{y} : 0 < \mu \le 1, \xi \in SO(k), \mathbf{y} \in \mathbf{R}^k/\mathbf{Z}^k\},$$

and

$$\mathcal{D}_h(A, \mathbf{R}^k/\mathbf{Z}^k) = \{(\mu A \bmod 1) + \mathbf{y}) \cap X : 0 < \mu \le 1, \mathbf{y} \in \mathbf{R}^k/\mathbf{Z}^k\}.$$

Then $\mathcal{D}_s(A, \mathbf{R}^k/\mathbf{Z}^k)$ and $\mathcal{D}_h(A, \mathbf{R}^k/\mathbf{Z}^k)$ are discrepancy systems on $\mathbf{R}^k/\mathbf{Z}^k$ and the corresponding discrepancies related to finite sequences $\mathbf{x}_1, \dots, \mathbf{x}_N \in \mathbf{R}^k/\mathbf{Z}^k$ satisfy exactly the same estimates as stated in Theorems 2.10, 2.11, and 2.12 for the euclidean case.

It seems to be a very difficult problem to decide which properties of the compact body A determine the irregularity behaviour of the discrepancy with respect to $\mathcal{D}_h(A, X)$. A partial answer for the case $k = 2$ is given in BECK and CHEN [143, p. 173 ff].

First we will prove lower and upper bound of Theorem 2.10. The methods to be used are essentially due to BECK. The proofs of Theorem 2.11 and 2.12 use related but different methods. BECK's FOURIER transform approach is always applied. However, some additional considerations have to be used. For example, the proof of Theorem 2.12 is essentially a combination of ROTH's and BECK's method.

For the proof of the lower bounds (2.26) in Theorem 2.10 we will make use of an estimate by BECK [143, p. 170].

Lemma 2.14 *Let $k \ge 2$, $A \subseteq \mathbf{R}^k$ a convex body with $r(A) \ge 1$. Set $Q_m = [-100 \cdot 2^{-m+1}, 100 \cdot 2^{-m+1}] \setminus [-100 \cdot 2^{-m}, 100 \cdot 2^{-m}]$, where m is an arbitrary positive integer. Then*

$$\min_{\mathbf{t} \in Q_m} \int_{SO(k)} \int_0^1 |\hat{\chi}_{\mu(\xi A)}(\mathbf{t})|^2 \, d\mu \, d\xi \gg \min\{\lambda_l(A)^2, m^2 \sigma(\partial A)\}, \tag{2.32}$$

where $d\xi$ denotes the Haar measure on $SO(k)$, σ the surface measure and the constant implied by $\gg$ only depends on the dimension k.

Proof of (2.26). Let $X' = N^{\frac{1}{k}} \cdot X$, $\mathbf{y}_n = N^{\frac{1}{k}} \cdot \mathbf{x}_n$, and $A' = N^{\frac{1}{k}} \cdot A$. We will show that there exist $\mathbf{x} \in \mathbf{R}^k$, $\xi \in SO(k)$, $\rho \in (0, 1]$ such that

$$\begin{aligned} &\left| \sum_{n=1}^{N} \chi_{\rho(\xi A') + \mathbf{x}}(\mathbf{y}_n) - \lambda_k((\rho(\xi A') + \mathbf{x}) \cap X') \right| \\ &\qquad \gg (\operatorname{diam}(A) + \operatorname{diam}(X))^k (\sigma(\partial A'))^{\frac{1}{2}} \qquad (2.33) \\ &\qquad = (\operatorname{diam}(A) + \operatorname{diam}(X))^k (\sigma(\partial A))^{\frac{1}{2}} N^{\frac{1}{2} - \frac{1}{2k}} \end{aligned}$$

for $N \geq r(A)^{-k}$. Obviously, (2.33) is equivalent to (2.26). For this purpose let $F_{\rho,\xi} = \chi_{-\rho(\xi A')} * d\nu'$, where $\nu' = \sum_{n=1}^{N} \delta_{\mathbf{y}_n} - \mu'$ and μ' is the restriction of λ_k to X'. More explicitly,

$$F_{\rho,\xi}(\mathbf{x}) = \sum_{n=1}^{N} \chi_{\rho(\xi A')-\mathbf{x}}(\mathbf{x}_n) - \lambda_k\left((\rho(\xi A') - \mathbf{x}) \cap X'\right). \tag{2.34}$$

By PARSEVAL's identity and Lemma 2.14 we get for any positive integer m and $N \geq r(A)^{-k}$ (which is equivalent to $r(A') \geq 1$)

$$\begin{aligned}
\int_{SO(k)} \int_0^1 \int_{\mathbf{R}^k} |F_{\rho,\xi}(\mathbf{x})|^2 \, d\mathbf{x} \, d\rho \, d\xi &= \int_{\mathbf{R}^k} \left(\int_{SO(k)} \int_0^1 |\hat{\chi}_{\rho(\xi A')}(\mathbf{t})|^2 \, d\rho \, d\xi \right) \cdot |\hat{\nu}'(\mathbf{t})|^2 \, d\mathbf{t} \\
&\geq \int_{Q_m} \left(\int_{SO(k)} \int_0^1 |\hat{\chi}_{\rho(\xi A')}(\mathbf{t})|^2 \, d\rho \, d\xi \right) \cdot |\hat{\nu}'(\mathbf{t})|^2 \, d\mathbf{t} \\
&\gg \min\{\lambda_k(A')^2, m^2 \sigma(\partial A')\} \cdot \int_{Q_m} |\hat{\nu}'(\mathbf{t})|^2 \, d\mathbf{t}.
\end{aligned}$$

From $r(A') \geq 1$ we get $\lambda_k(A') \gg \sigma(\partial A')$ and hence $\lambda_k(A')^2 \gg \sigma(\partial A')(\log N)^2$. Therefore, since $F_{\rho,\xi}(\mathbf{x})$ is supported in a set of volume $\ll N(\mathrm{diam}(X) + \mathrm{diam}(A))^k$ it suffices to show that there is a positive integer $m \ll \log N$ such that

$$\int_{Q_m} |\hat{\nu}'(\mathbf{t})|^2 \, d\mathbf{t} \gg N. \tag{2.35}$$

First we will show that

$$\int_{[-100,100]^k} |\hat{\nu}'(\mathbf{t})|^2 \, d\mathbf{t} \gg N. \tag{2.36}$$

Let

$$f(\mathbf{x}) = \prod_{j=1}^{k} \left(\frac{2\sin(50x_j)}{\sqrt{2\pi} x_j} \right)^2$$

($\mathbf{x} = (x_1, \ldots, x_k)$). Since

$$\hat{f}(\mathbf{t}) = \prod_{j=1}^{k} (100 - |t_j|)^+,$$

where $\mathbf{t} = (t_1, \ldots, t_k)$ and $y^+ = \max\{y, 0\}$, we get by PARSEVAL's identity

$$\int_{\mathbf{R}^k} |(f * d\nu')(\mathbf{x})|^2 \, d\mathbf{x} = \int_{\mathbf{R}^k} |\hat{f}(\mathbf{t})|^2 |\hat{\nu}'(\mathbf{t})|^2 \, d\mathbf{t} \ll \int_{[-100,100]^k} |\hat{\nu}'(\mathbf{t})|^2 \, d\mathbf{t}.$$

From
$$f(\mathbf{x}-\mathbf{x}_n) - \int_{\mathbf{R}^k} f(\mathbf{x}-\mathbf{y})\,d\mu'(\mathbf{y}) \gg 1$$
for $\mathbf{x}-\mathbf{x}_n \in \left[-\frac{1}{50}, \frac{1}{50}\right]^k$ it follows
$$\begin{aligned} |(f * d\nu')(\mathbf{x})| &= \left|\sum_{n=1}^{N} f(\mathbf{x}-\mathbf{x}_n) - \int_{\mathbf{R}^k} f(\mathbf{x}-\mathbf{y})\,d\mu'(\mathbf{y})\right| \\ &\gg \sum_{n=1}^{N} \chi_{\left[-\frac{1}{50},\frac{1}{50}\right]^k+\mathbf{x}}(\mathbf{x}_n), \end{aligned}$$
and consequently
$$\begin{aligned} \int_{\mathbf{R}^k} |(f * d\nu')(\mathbf{x})|^2\,d\mathbf{x} &\gg \int_{\mathbf{R}^k} \left(\sum_{n=1}^{N} \chi_{\left[-\frac{1}{50},\frac{1}{50}\right]^k+\mathbf{x}}(\mathbf{x}_n)\right)^2 d\mathbf{x} \\ &\gg \int_{\mathbf{R}^k} \left(\sum_{n=1}^{N} \chi_{\left[-\frac{1}{50},\frac{1}{50}\right]^k+\mathbf{x}}(\mathbf{x}_n)\right) d\mathbf{x} \\ &\gg N, \end{aligned}$$
which proves (2.36).

Since $|\hat{\nu}'(\mathbf{t})| \ll N$ there exists a constant $c > 0$ such that
$$\int_{[-cN^{-1/k},cN^{-1/k}]^k} |\hat{\nu}'(\mathbf{t})|\,d\mathbf{t} < \frac{1}{2} \int_{[-100,100]^k} |\hat{\nu}'(\mathbf{t})|\,d\mathbf{t}.$$
Thus there exists some positive integer m satisfying (2.35).□

Remark. The proof for the corresponding inequality in the modulo 1 case runs along the same lines. Let $m = N^{\frac{1}{k}}$, identify $\mathbf{R}^k/\mathbf{Z}^k$ with the cube $[0,m)^k$, tranform the finite sequence $\mathbf{x}_1,\ldots,\mathbf{x}_N \in \mathbf{R}^k/\mathbf{Z}^k \cong [0,1)^k$ to the corresponding sequence $\mathbf{y}_n = m\mathbf{x}_n \in [0,m)^k$, and set $A' = mA$. Next, let M be an integral multiple of m to be specified in the sequel. Set $N' = 2^{M/m}N$ and extend the finite sequence $\mathbf{y}_1,\ldots,\mathbf{y}_N$ periodically with period m to $\mathbf{z}_1,\ldots,\mathbf{z}_{N'} \in [-M,M]^k$. Set $F_{\rho,\xi} = \chi_{-\rho(\xi A')} * d\nu'$, where $\nu' = \sum_{n=1}^{N'} \delta_{\mathbf{z}_n} - \mu'$ and μ' is the restriction of λ_k to $[-M,M]^k$. Hence we get by the same calculations as above ($N \geq r(A)^{-k}$)
$$\begin{aligned} \int_{\mathbf{R}^k}\int_{SO(k)}\int_0^1 |F_{\rho,\xi}(\mathbf{x})|^2, d\rho\,d\xi\,d\mathbf{x} &= \int_{Q_1}\int_{SO(k)}\int_0^1 |F_{\rho,\xi}(\mathbf{x})|^2\,d\rho\,d\xi\,d\mathbf{x} \\ &\gg \min\left\{\sigma(\partial A'), \frac{\lambda_k(A')^2}{(\log M)^2}\right\} M^k, \end{aligned}$$

in which $Q_1 = [-M - \mathrm{diam}(A), M + \mathrm{diam}(A)]^k$. Hence either

$$\int_{Q_2}\int_{SO(k)}\int_0^1 |F_{\rho,\xi}(\mathbf{x})|^2, d\rho\, d\xi\, d\mathbf{x} \gg \min\left\{\sigma(\partial A'), \frac{\lambda_k(A')^2}{(\log M)^2}\right\} M^k, \tag{2.37}$$

where $Q_2 = [-M + \mathrm{diam}(A), M - \mathrm{diam}(A)]^k$, or

$$\int_{Q_1\setminus Q_1}\int_{SO(k)}\int_0^1 |F_{\rho,\xi}(\mathbf{x})|^2, d\rho\, d\xi\, d\mathbf{x} \gg \min\left\{\sigma(\partial A'), \frac{\lambda_k(A')^2}{(\log M)^2}\right\} M^k. \tag{2.38}$$

Now let M be an integral multiple of m satisfying

$$\frac{C}{2}(\mathrm{diam}(A'))^{2k+2}M \geq 2C(\mathrm{diam}(A'))^{2k+2},$$

where $C > 0$ will be specified in the sequel.

Suppose that (2.38) holds. Then there exist $\mathbf{x} \in \mathbf{R}^k$, $\rho \in (0,1]$, $\xi \in SO(k)$ such that

$$|F_{\rho,\xi}(\mathbf{x})|^2 \gg \frac{M}{\mathrm{diam}(A')}\min\left\{\sigma(\partial A'), \frac{\lambda_k(A')^2}{(\log M)^2}\right\}.$$

Since $r(A') \geq 1$, we have

$$\lambda_k(A') \gg \sigma(\partial A') \gg \mathrm{diam}(A') \gg M^{\frac{1}{2k+2}},$$

and hence

$$|F_{\rho,\xi}(\mathbf{x})|^2 \gg \frac{M}{\mathrm{diam}(A')} \gg \mathrm{diam}(A')(\lambda_k(A'))^2 \geq (\lambda_k(A'))^2.$$

Choosing C sufficiently large, we obtain $|F_{\rho,\xi}(\mathbf{x})| > 2\lambda_k(A')$. Thus

$$\sum_{n=1}^{N'} \chi_{\rho(\xi A')-\mathbf{x}}(\mathbf{z_n}) \geq 2\lambda_k(A').$$

Keeping in mind that $\rho(\xi A') - \mathbf{x}$ corresponds to a set $S_1 = (\frac{1}{m}((\rho(\xi A') - \mathbf{x}) \cap [-M, M]^k) \bmod 1$ which is contained in $S_2 = (\frac{1}{m}(\rho(\xi A') - \mathbf{x}) \bmod 1 \in \mathcal{D}_s(A, \mathbf{R}^k/\mathbf{Z}^k)$, we immediately get

$$\begin{aligned}\sum_{n=1}^{N} \chi_{S_2}(\mathbf{x}_n) - N\lambda_k(S_2) &\geq \sum_{n=1}^{N} \chi_{S_1}(\mathbf{x}_n) - N\lambda_k(S_2)\\ &> \lambda_k(A')\\ &\gg (\sigma(\partial A'))^{\frac{1}{2}}.\end{aligned}$$

Finally suppose that (2.37) holds. It follows that there exist $\mathbf{x} \in \mathbf{R}^k$, $\rho \in (0,1]$, $\xi \in SO(k)$ such that

$$|F_{\rho,\xi}(\mathbf{x})|^2 \gg \min\left\{\sigma(\partial A'), \frac{\lambda_k(A')^2}{(\log M)^2}\right\} \gg \sigma(\partial A')$$

and $\rho(\xi A') - \mathbf{x} \subseteq [-M, M]^k$. Hence there is a direct relation to $\mathbf{R}^k/\mathbf{Z}^k$ and the proof of the lower bound in the modulo 1 case is complete. □

In order to prove the upper bound (2.27) we will make use of a probabilistic estimate.

Lemma 2.15 *Let $X_1, \ldots, X_m$ be independent random variables with $|X_i| \leq 1$ for $i = 1, \ldots, m$. Then*

$$\mathbf{P}\left[\left|\sum_{i=1}^{m}(X_i - \mathbf{E}X_i)\right| \geq \gamma\right] \leq 2e^{-\gamma^2/(4m)}. \tag{2.39}$$

As ususal, $\mathbf{E}X$ denotes the expexted value of the random variable X, and $\mathbf{P}$ the probability.

Proof. Set $Y = \sum_{i=1}^m (X_i - \mathbf{E}X_i)$. Then we have

$$\mathbf{P}[Y \geq \gamma] = \mathbf{P}[e^{tY} \geq e^{t\gamma}] \leq e^{-t\gamma}\mathbf{E}e^{tY} = \leq e^{-t\gamma}\prod_{i=1}^{m}\mathbf{E}e^{t(X_i - \mathbf{E}X_i)},$$

where the real parameter t will be fixed later. Since the function

$$\varphi(t) = \log\left(Ee^{tX_i}\right)$$

satisfies $\varphi(0) = 0$, $\varphi'(0) = \mathbf{E}X_i$, and

$$\begin{aligned} \varphi''(t) &= \frac{\mathbf{E}[X_i^2 e^{tX_i}]\mathbf{E}e^{tX_i} - \left(\mathbf{E}[X_i e^{tX_i}]\right)^2}{(\mathbf{E}e^{tX_i})^2} \\ &\leq \frac{\mathbf{E}[X_i^2 e^{tX_i}]}{\mathbf{E}e^{tX_i}} \leq 1, \end{aligned}$$

we get by TAYLOR's theorem

$$\varphi(t) \leq t \cdot \mathbf{E}X_i + \frac{t^2}{2}.$$

Hence we obtain for $t = \frac{\gamma}{m}$

$$\mathbf{P}[Y \geq \gamma] \leq \left(-t\gamma + m\frac{t^2}{2}\right) = \exp\left(-\frac{\gamma^2}{2m}\right).$$

Repeating the same calculations for $\mathbf{P}[Y \leq -\gamma]$, we obtain the same upper bounds. Hence we can estimate $\mathbf{P}[|Y| \geq \gamma]$ by $\mathbf{P}[Y \geq \gamma] + \mathbf{P}[Y \leq -\gamma]$. This completes the proof of (2.39). □

Proof of (2.27). Let $N \ge 1$ be given and set $\varepsilon = N^{-\frac{k+1}{2k}}$. Consider a minimal set $Z = \{\mathbf{z}_1, \ldots, \mathbf{z}_L\}$ such that the balls $B(\mathbf{z}_i, \varepsilon) = \{\mathbf{x} \in \mathbf{R}^k : |\mathbf{x} - \mathbf{z}| < \varepsilon\}$ cover $X + A$. Clearly $L \ll (\mathrm{diam}(X) + \mathrm{diam}(A))^k \varepsilon^{-k} \ll \varepsilon^{-k}$. Furthermore let R be a (minimial) system of rotations $\xi \in SO(k)$ such that for any $\xi' \in SO(k)$ there exist a $\xi \in SO(k)$ satisfying $\max_{\mathbf{x}\in A} |\xi\mathbf{x} - \xi'\mathbf{x}| \le \varepsilon$. Since we can assume w.l.o.g. that $(0, \ldots, 0) \in A$, we have $|R| \ll (\mathrm{diam}(A))^{k-1}\varepsilon^{1-k} \ll \varepsilon^{1-k}$. Now let $\mathcal{A}_N$ be the system of sets $(\rho(\xi A) + \mathbf{z}) \cap X$, where $\mathbf{z} \in Z$, $\xi \in SO(k)$, and $\rho = j\frac{2\varepsilon}{r(A)}$ for $j = 0, 1, \ldots, \frac{r(A)}{2\varepsilon} + 1$.

This system of sets $\mathcal{A}_N$ has cardinality $|\mathcal{A}_N| \ll \varepsilon^{-2k}$ and satisfies the condition that for any set $S \in \mathcal{D}_s(A, X)$ there exist $S_1, S_2 \in \mathcal{A}_N$ such that $S_1 \subseteq S \subseteq S_2$ and

$$\lambda_k(S_2 \setminus S_1) \ll \frac{(\mathrm{diam}(A))^k}{r(A)} \varepsilon \ll \varepsilon.$$

Hence it suffices to show that there exists a finite sequence $\mathbf{x}_1, \ldots, \mathbf{x}_N \in X$ such that

$$\max_{S\in\mathcal{A}_N} \left| \frac{1}{N} \sum_{n=1}^N \chi_S(\mathbf{x}_n) - \lambda_k(S) \right| \ll N^{-\frac{1}{2}-\frac{1}{2k}}.$$

Now consider the system of cubes $\mathcal{Q}$ of sidelengths $N^{-\frac{1}{k}}$ with vertices contained in the lattice $L = N^{-\frac{1}{k}} \cdot \mathbf{Z}^k \subseteq \mathbf{R}^k$. Let $Q_1, \ldots, Q_M$ be those cubes of $\mathcal{Q}$ which are contained in X. Obviously $N - M \ll \sigma(\partial X) N^{1-\frac{1}{k}}$. Furthermore, let $Q_{M+1}, \ldots, Q_N$ be an arbitrary partition of $X \setminus (Q_1 \cup \cdots \cup Q_M)$ such that $\lambda_k(Q_{M+1}) = \cdots = \lambda_k(Q_N) = N^{-\frac{1}{k}}$.

We will use a probabilistic estimate. For this purpose let Z_n, $1 \le n \le N$, be independent random variables such that Z_n is uniformly distributed on Q_n. For any $S \in \mathcal{A}_N$ let $X_n(S) = \chi_S(Z_n)$. Hence $\mathbf{E}X_n(S) = N\lambda_k(S \cap Q_n)$ and we have $X_n(S) \equiv \mathbf{E}X_n(S)$ if $Q_n \subseteq S$ or if $Q_n \subseteq X \setminus S$. Thus if we are interested in the sum

$$\sum_{n=1}^N X_n(S) - \mathbf{E}X_n(S) = \sum_{n=1}^N \chi_S(X_n) - N\lambda_k(S)$$

we just have to take into account those n satisfying $\emptyset \ne S \cap Q_n \ne Q_n$ into account. Since this number is $\ll \sigma(\partial A) N^{1-\frac{1}{k}} + (N - M) \ll N^{1-\frac{1}{k}}$, we immediately derive Lemma 2.15

$$\mathbf{P}\left[\left| \sum_{n=1}^N X_n(S) - \mathbf{E}X_n(S) \right| \ge \gamma \right] \le 2\exp\left(-c_k \gamma^2 N^{\frac{1}{k}-1}\right).$$

Now, set $\gamma = cN^{\frac{1}{2}-\frac{1}{k}}(\log N)^{\frac{1}{2}}$ and choose c large enough such that

$$2\exp\left(-c_k\gamma^2 N^{\frac{1}{k}-1}\right) \le \frac{1}{2|\mathcal{A}_N|}.$$

Hence

$$\mathbf{P}\left[\left| \frac{1}{N} \sum_{n=1}^N \chi_S(X_n) - \lambda_k(S) \right| \ge cN^{\frac{1}{2}-\frac{1}{k}}(\log N)^{\frac{1}{2}} \text{for some } S \in \mathcal{A}_N \right] \le \frac{1}{2}$$

implies that there is a finite sequence $\mathbf{x}_1, \ldots, \mathbf{x}_N$ such that

$$D_N^{(\mathcal{D}_*(A,X))}(\mathbf{x}_n) \le C N^{\frac{1}{2}-\frac{1}{k}} (\log N)^{\frac{1}{2}},$$

where C only depends on the dimension k, on $\mathrm{diam}(X)$, on $\mathrm{diam}(A)$, and on $r(A)$. □

Remark. Apparently the upper bound in the modulo 1 case follows from (2.27). Let $X = [0,1]^k$ and $\mathbf{x}_1, \ldots, \mathbf{x}_N \in [0,1)^k$.

$$D_N^{(\mathcal{D}_*(A,\mathbf{R}^k/\mathbf{Z}^k))}(\mathbf{x}_n) \le 2^k D_N^{(\mathcal{D}_*(A,X))}(\mathbf{x}_n).$$

Now we have finished the proof of Theorem 2.10 and its analogue in the modulo 1 case. Since the upper bound (2.29) of Theorem 2.11 follows from the upper bound (2.27) of Theorem 2.10 the next step is to prove the lower bound (2.28) in Theorem 2.11 (and its modulo 1 analogue).

For this purpose we will use an asymptotic fomula by HLAWKA [780, 781].

Lemma 2.16 *Let $k \ge 2$ and $\mathbf{0} \in A \subseteq \mathbf{R}^k/\mathbf{Z}^k$ a convex body satisfying the the same assumptions as in Theorem 1. Let $H(\mathbf{u})$ denote the support function*

$$H(\mathbf{u}) = \sup_{\mathbf{x}\in A} \mathbf{u}\cdot\mathbf{x}$$

and $L(\mathbf{u}) = L = r_1 \cdots r_k$ the product of the radii of the main curvature corresponding to the the point $\mathbf{u}$ on the unit sphere $|\mathbf{u}| = 1$. Then we have

$$\begin{aligned}\int_B e^{i\mathbf{t}\mathbf{x}}\, d\mathbf{x} &= \frac{(2\pi)^{\frac{k-1}{2}}}{|\mathbf{t}|^{\frac{k+1}{2}}}\left(\frac{\sqrt{L(\frac{\mathbf{t}}{|\mathbf{t}|})}}{H(\frac{\mathbf{t}}{|\mathbf{t}|})}\exp\left(i|\mathbf{t}|H(\frac{\mathbf{t}}{|\mathbf{t}|}) - i\frac{(k+1)\pi}{4}\right)\right. \\ &\left. + \frac{\sqrt{L(-\frac{\mathbf{t}}{|\mathbf{t}|})}}{H(-\frac{\mathbf{t}}{|\mathbf{t}|})}\exp\left(-i|\mathbf{t}|H(-\frac{\mathbf{t}}{|\mathbf{t}|}) + i\frac{(k+1)\pi}{4}\right) + \mathcal{O}\left(|\mathbf{t}|^{-\frac{1}{2}}\right)\right)\end{aligned} \tag{2.40}$$

uniformly for $\mathbf{t} \in \mathbf{R}^k$, $|\mathbf{t}| \to \infty$.

Now set

$$\varphi_q(\mathbf{t}) = \frac{1}{q}\int_q^{2q} |\hat{\chi}_{rA}(\mathbf{t})|^2\, dr, \tag{2.41}$$

where $\hat{\chi}_{rA}$ is the FOURIER transform of the characteristic function of rA. Lemma 2.16 enables us to prove

Lemma 2.17 *Let $0 < p < q$. Then we have*

$$\frac{\varphi_q(\mathbf{t})}{\varphi_p(\mathbf{t})} \gg \left(\frac{q}{p}\right)^{k-1} \tag{2.42}$$

uniformly for $\mathbf{t} \in \mathbf{R}^k$.

Proof. It is sufficient to show that there is some constant $C > 0$ such that

$$\frac{1}{y}\int_y^{2y} |\hat{\chi}_{rA}(\mathbf{t})|^2\,dt \gg\ll \begin{cases} y^{k-1}|\mathbf{t}|^{-k-1} & \text{for } y|\mathbf{t}| \geq C \\ y^{2k} & \text{for } y|\mathbf{t}| < C. \end{cases} \tag{2.43}$$

Set

$$f(\mathbf{t}) = \frac{1}{(2\pi)^{\frac{k}{2}}}\int_A e^{-i\mathbf{tx}}. \tag{2.44}$$

Then we have

$$\begin{aligned} \hat{\chi}_{rA}(\mathbf{t}) &= r^k f(r\mathbf{t}) \\ &= \frac{1}{(2\pi)^{\frac{1}{2}}}\frac{r^{\frac{k-1}{2}}}{|\mathbf{t}|^{\frac{k+1}{2}}}\left(\frac{\sqrt{L(\frac{\mathbf{t}}{|\mathbf{t}|})}}{H(\frac{\mathbf{t}}{|\mathbf{t}|})}\exp\left(-ir|\mathbf{t}|H(\frac{\mathbf{t}}{|\mathbf{t}|}) + i\frac{(k+1)\pi}{4}\right)\right. \\ &\qquad \left. + \frac{\sqrt{L(-\frac{\mathbf{t}}{|\mathbf{t}|})}}{H(-\frac{\mathbf{t}}{|\mathbf{t}|})}\exp\left(ir|\mathbf{t}|H(-\frac{\mathbf{t}}{|\mathbf{t}|}) - i\frac{(k+1)\pi}{4}\right) + \mathcal{O}\left((r|\mathbf{t}|)^{-\frac{1}{2}}\right)\right) \\ &= \frac{1}{(2\pi)^{\frac{1}{2}}}\frac{r^{\frac{k-1}{2}}}{|\mathbf{t}|^{\frac{k+1}{2}}}\left(g(r,\mathbf{t}) + \mathcal{O}\left((r|\mathbf{t}|)^{-\frac{1}{2}}\right)\right). \end{aligned} \tag{2.45}$$

Let $C_1 > 0$ be sufficiently small such that

$$f(r\mathbf{t}) \gg\ll \lambda_k(A) \gg\ll 1$$

for $r|\mathbf{t}| \leq C_1/2$. Hence

$$\frac{1}{y}\int_y^{2y} |\hat{\chi}_{rA}(\mathbf{t})|^2\,dt \gg\ll y^{2k}$$

for $y|\mathbf{t}| \leq C_1$. Furthermore it is an easy exercise to obtain

$$\frac{1}{y}\int_y^{2y} |g(r,\mathbf{t})|^2\,dr \gg\ll 1$$

uniformly for $\mathbf{t} \in \mathbf{R}^k$ since $L(\mathbf{u})$ and $H(\mathbf{u})$ are uniformly bounded by $0 < L_1 \leq L(\mathbf{u}) \leq L_2$ and by $0 < H_1 \leq H(\mathbf{u}) \leq H_2$, where L_1, L_2, H_1, H_2 depend on A. Thus

$$\frac{1}{y}\int_y^{2y} \left|g(r,\mathbf{t}) + \mathcal{O}\left((r|\mathbf{t}|)^{-\frac{1}{2}}\right)\right|^2 dr \gg\ll 1$$

uniformly for $y|\mathbf{t}| \geq C_2$, where $C_2 > 0$ is sufficiently large. Hence

$$\frac{1}{y}\int_y^{2y} |\hat{\chi}_{rA}(\mathbf{t})|^2\,dt \gg\ll \frac{r^{k-1}}{|\mathbf{t}|^{k+1}}$$

for $y|\mathbf{t}| \geq C_2$. Finally, by continuity we have

$$\min_{C_1 \leq y|\mathbf{t}| \leq C_2} \frac{1}{y}\int_y^{2y} |f(r\mathbf{t})|^2\,dr = \min_{C_1 \leq V \leq C_2,\,|\mathbf{u}|=1} \frac{1}{V}\int_V^{2V} |f(v\mathbf{u})|^2\,dv > 0$$

and so we get (2.43) for $C = C_2$. □

Next we use the notation $\nu = \frac{1}{N}\sum_{n=1}^{N} \delta_{\mathbf{x}_n} - \lambda_k$ and introduce the functions

$$\begin{aligned} F_\rho(\mathbf{x}) &= (\chi_{-\rho A} * (d\nu))(\mathbf{x}) \qquad (2.46) \\ &= \frac{1}{N}\sum_{n=1}^{N} \chi_{\rho(A)-\mathbf{x}}(\mathbf{x}_n) - \lambda_k\left((\rho A - \mathbf{x}) \cap X\right) \end{aligned}$$

and

$$\begin{aligned} \Phi(q) &= \frac{1}{q}\int_q^{2q}\int_{\mathbf{R}^k} |F_\rho(\mathbf{x})|^2\, d\mathbf{x}\, d\rho \\ &= \frac{1}{q}\int_q^{2q}\int_{\mathbf{R}^k} |\hat{F}_\rho(\mathbf{t})|^2\, d\mathbf{t}\, d\rho \qquad (2.47) \\ &= \int_{\mathbf{R}^k} \varphi_q(\mathbf{t})|\hat{\nu}(\mathbf{t})|^2\, d\mathbf{t}. \end{aligned}$$

It is easy to obtain a trivial bound for $\Phi(cN^{-1/k})$.

Lemma 2.18 *Let $X \subseteq \mathbf{R}^k$ a compact set with non-empty interior and* LEBESGUE *measure $\lambda_k(X) = 1$. Then there exists some constant $c > 0$ such that*

$$\Phi\left(cN^{-\frac{1}{k}}\right) \gg N^{-2} \qquad (N \geq 1). \qquad (2.48)$$

Proof. Assume that $q \leq \rho \leq 2q$, where $q = \frac{1}{2}(2\lambda_k(A)N)^{-1/k}$. Then

$$\frac{1}{2^{k+1}N} \leq \lambda_k(\rho A) \leq \frac{1}{2N}$$

implies

$$\left|\sum_{n=1}^{N} \chi_{\rho A-\mathbf{x}}(\mathbf{x}_n) - N\lambda_k\left((\rho A - \mathbf{x}) \cap X\right)\right| \geq \frac{1}{2}\sum_{n=1}^{N} \chi_{\rho A-\mathbf{x}}(\mathbf{x}_n),$$

and

$$\begin{aligned} &\int_{\mathbf{R}^k} \left|\frac{1}{N}\sum_{n=1}^{N} \chi_{\rho A-\mathbf{x}}(\mathbf{x}_n) - \lambda_k\left((\rho A - \mathbf{x}) \cap X\right)\right|^2 d\mathbf{x} \\ &\geq \frac{1}{4N^2}\int_{\mathbf{R}^k} \left(\sum_{n=1}^{N} \chi_{\rho A-\mathbf{x}}(\mathbf{x}_n)\right)^2 d\mathbf{x} \\ &\geq \frac{1}{4N^2}\int_{\mathbf{R}^k} \left(\sum_{n=1}^{N} \chi_{\rho A-\mathbf{x}}(\mathbf{x}_n)\right) d\mathbf{x} \\ &= \frac{1}{4N} N\lambda_k(A) \\ &\geq \frac{1}{2^{k+3}N^2}, \end{aligned}$$

which proves $\Phi(q) \gg N^{-2}$. □

Proof of Theorem 2.11. Now we can apply Lemma 2.17 with $p = cN^{-1/k}$ and $q = \frac{1}{2}$ to obtain

$$\begin{aligned}\Phi\left(\frac{1}{2}\right) &= \int_{\mathbf{R}^k} \varphi_q(\mathbf{t})|\hat{\nu}(\mathbf{t})|^2\,d\mathbf{t} \\ &\gg N^{1-\frac{1}{k}}\Phi(p) \qquad (2.49)\\ &\gg N^{-1-\frac{1}{k}}.\end{aligned}$$

Hence there exists some ρ with $\frac{1}{2} \le \rho \le 1$ and some $\mathbf{x} \in \mathbf{R}^k$ such that

$$|F_\rho(\mathbf{x})| \ge N^{-\frac{1}{2}-\frac{1}{2k}}$$

which proves the lower bound of Theorem 2.11 in the euclidean case.

The lower bound for the modulo 1 case follows by identifying $\mathbf{R}^k/\mathbf{Z}^k$ with $[0,1)^k$, extending periodically to $[-M,M]^k$, where M is chosen sufficiently large, and using averaging arguments similar to the the proof of Theorem 2.10. Finally, as mentioned above, the upper bounds are immediate consequences of Theorem 2.10. □

We now turn to the proof of the lower bound of Theorem 2.12. (The proof of the upper bound is due to KÁROLYI [906] and will not be presented here.) As mentioned above the proof uses a combination of ROTH's and BECK's method. In order to be more precise we have to introduce some notations. Let $A = P$ be a simple convex polytope, which means that every vertex $\mathbf{x}_0$ has exactly k adjacent ones. Furthermore, locally around a vertex $\mathbf{x}_0$ of P, P can be represented as the intersection $\bigcap_i H_i$ of k closed halfspaces H_i. Let $\{\mathbf{x}_1,\ldots,\mathbf{x}_k\} = \Gamma(\mathbf{x}_0)$ be the vertices adjacent to $\mathbf{x}_0$ and L_i, $1 \le i \le k$, be hyperplanes

$$L_i = \left\{\mathbf{x}_0 + \sum_{j\ne i}\lambda_j(\mathbf{x}_j - \mathbf{x}_0)\ :\ \lambda_j \in \mathbf{R}\right\},$$

which support the halfspaces H_i, $1 \le i \le k$. Then the set

$$\tilde{\mathcal{P}} = \left\{\tilde{P}(\rho_1,\ldots,\rho_k) = P \cap \bigcap_{i=1}^{k}(H_i + \rho_i(\mathbf{x}_i - \mathbf{x}_0))\ :\ 0 \le \rho_i \le \varepsilon,\ 1 \le i \le k\right\} \qquad (2.50)$$

is such a system of polytopes of the same "shape" as P if $\varepsilon > 0$ is chosen sufficiently small.

The following proposition is essentially due to KÁROLYI [906] and is in fact a generalization of ROTH's Theorem 1.40.

Proposition 2.19 *Suppose that $P \subseteq \mathbf{R}^k$ is a simple convex polytop. Then there exists $\varepsilon > 0$ such that for every finite set of points $\mathbf{x}_1,\ldots,\mathbf{x}_N \in X$*

$$\int_{[0,\varepsilon)^k}\int_{\mathbf{R}^k}\left|\frac{1}{N}\sum_{n=1}^{N}\chi_{\tilde{P}(\rho_1,\ldots,\rho_k)+\mathbf{y}}(\mathbf{x}_n) - \lambda_k((\tilde{P}(\rho_1,\ldots,\rho_k)+\mathbf{y})\cap X)\right|^2 d\mathbf{y}\,d\rho_1\cdots d\rho_k$$
$$\gg N^{-2}(\log N)^{k-1}.$$

Proof. Let $\mathbf{y} \in \mathbf{R}^k$ and set

$$D(\underline{\rho}, \mathbf{y}) = \frac{1}{N} \sum_{n=1}^{N} \chi_{\tilde{P}(\rho_1,\ldots,\rho_k)+\mathbf{y}}(\mathbf{x}_n) - \lambda_k((\tilde{P}(\rho_1,\ldots,\rho_k) + \mathbf{y}) \cap X),$$

where $\underline{\rho} = (\rho_1, \ldots, \rho_k) \in [0, \varepsilon]^k$. The idea of the proof is to find a function $F(\underline{\rho}, \mathbf{y})$ such that

$$\int_{\mathbf{R}^k} \int_{[0,\varepsilon]^k} F(\underline{\rho}, \mathbf{y}) D(\underline{\rho}, \mathbf{y}) \, d\underline{\rho} \, d\mathbf{y} \gg N^{-1} (\log N)^{k-1} \tag{2.51}$$

and

$$\int_{\mathbf{R}^k} \int_{[0,\varepsilon]^k} F(\underline{\rho}, \mathbf{y})^2 \, d\underline{\rho} \, d\mathbf{y} \ll (\log N)^{k-1}. \tag{2.52}$$

Obviously, (2.51) and (2.52) (combined with CAUCHY-SCHWARZ's inequality) imply Proposition 2.19.

As in the proof of Theorem 1.58 we use RADEMACHER functions

$$R_{\mathbf{r}}(\mathbf{x}) = \prod_{j=1}^{k} R_{r_j}(x_j);$$

$\mathbf{r} = (r_1, \ldots, r_k)$ is a vector of positive integers. If $(\tilde{P}(\rho_1, \ldots, \rho_k) + \mathbf{y}) \cap X \neq \emptyset$ we set

$$F(\underline{\rho}, \mathbf{y}) = \sum_{\|\mathbf{r}\|_1 = n} f_{\mathbf{r}}(\underline{\rho}),$$

where n is defined by $2N \le 2^n < 4N$ and on every $\mathbf{r}$-box $I = I_1 \times \cdots \times I_k$ (i.e. I_j are of the form $I_j = [m_j 2^{-r_j}, (m_j+1)2^{-r_j})$, where $0 \le m_j < 2^{r_j}$, see Section 1.3) we have

$$f_{\mathbf{r}}(\underline{\rho}) = \pm R_{\mathbf{r}}\left(\frac{\rho_1}{\varepsilon}, \ldots, \frac{\rho_k}{\varepsilon}\right),$$

where the sign is chosen such that (for every $\mathbf{r}$-box I)

$$\int_I f_{\mathbf{r}}(\underline{\rho}) D(\underline{\rho}, \mathbf{y}) \, d\underline{\rho} \ge 0. \tag{2.53}$$

Whereas, if $(\tilde{P}(\rho_1, \ldots, \rho_k) + \mathbf{y}) \cap X = \emptyset$ then we set $F(\underline{\rho}, \mathbf{y}) = 0$. It is well known (and easy to check, see BECK and CHEN [143]) that the system of functions $f_{\mathbf{r}}$, $\|\mathbf{r}\|_1 = n$, are orthogonal (for every fixed n). Thus (2.52) is automatically satisfied.

Now consider an $\mathbf{r}$-box $I = I_1 \times \cdots I_k$, $I_j = [m_j 2^{-r_j}, (m_j+1)2^{-r_j})$, such that

$$\tilde{I} = \left\{ \mathbf{x}_0 + \sum_{j=1}^{k} \rho_j : \frac{\rho_j}{\varepsilon} \in I \right\}$$

does not contain a point $\mathbf{x}_1, \ldots, \mathbf{x}_N$. By the definition of the RADEMACHER functions we directly get

$$\int_I R_{\mathbf{r}}(\underline{\rho}) D(\varepsilon\underline{\rho}, \mathbf{y})\, d\underline{\rho}$$

$$= \int_{I'} \sum_{i_1=0}^{1} \cdots \sum_{i_k=0}^{1} D(\varepsilon(\rho_1 + i_1 2^{-r_1-1}), \ldots, \varepsilon(\rho_1 + i_1 2^{-r_1-1}), \mathbf{y})\, d\underline{\rho},$$

where $I' = I'_1 \times \cdots \times I'_k$ and $I'_j = [m_j 2^{-r_j}, (m_j + 1/2) 2^{-r_j})$. For $\underline{\rho} \in I'$ set

$$\tilde{B}(\underline{\rho}) = \left\{ \mathbf{x}_0 + \sum_{j=1}^{k} (\rho_j + \kappa_j) : 0 \le \kappa_j < 2^{-r_j-1},\ 1 \le j \le k \right\}$$

and

$$\tilde{D}(\underline{\rho}, \mathbf{y}) = \frac{1}{N} \sum_{n=1}^{N} \chi_{\tilde{B}(\underline{\rho})+\mathbf{y}} - \lambda_k((\tilde{B}(\underline{\rho}) + \mathbf{y}) \cap X).$$

Then

$$\sum_{i_1=0}^{1} \cdots \sum_{i_k=0}^{1} D(\varepsilon(\rho_1 + i_1 2^{-r_1-1}), \ldots, \varepsilon(\rho_1 + i_1 2^{-r_1-1})) = \tilde{D}(\underline{\rho}).$$

Furthermore, since $\tilde{B}(\underline{\rho}) \subseteq \tilde{I}$ for $\underline{\rho} \in I'$ we directly obtain

$$\int_I R_{\mathbf{r}}(\underline{\rho}) D(\varepsilon\underline{\rho}, \mathbf{y})\, d\underline{\rho} = (-1)^k \int_{I'} (-\lambda_k((\tilde{B}(\underline{\rho}) + \mathbf{y}) \cap X)). \tag{2.54}$$

Note that for every $\mathbf{r}$ with $\|\mathbf{r}\|_1 = n$ there are exactly 2^n $\mathbf{r}$-boxes and that $N \le \frac{1}{2} 2^n$. Hence, (2.53) and (2.54) imply

$$\int_{\mathbf{R}^k} \int_I f_{\mathbf{r}}(\underline{\rho}) D(\varepsilon\underline{\rho}, \mathbf{y})\, d\underline{\rho}\, d\mathbf{y} \ge c' 2^{-2\|\mathbf{r}\|_1 - 2k},$$

and consequently (2.51). □

Proposition 2.19 and the following observation immediately provide the lower bound of Theorem 2.12.

Proposition 2.20 *Let $P \subseteq \mathbf{R}^k$ be a simple convex polytop. Then there exists $\varepsilon > 0$ such that for every finite set of points $\mathbf{x}_1, \ldots, \mathbf{x}_N \in X$*

$$\int_{[0,\varepsilon)^k} \int_{\mathbf{R}^k} \left| \sum_{n=1}^{N} \chi_{\tilde{P}(\rho_1,\ldots,\rho_k)+\mathbf{y}}(\mathbf{x}_n) - N\lambda_k((\tilde{P}(\rho_1, \ldots, \rho_k) + \mathbf{y}) \cap X) \right|^2 d\mathbf{y}\, d\rho_1 \cdots d\rho_k$$

$$\ll \int_0^1 \int_{\mathbf{R}^k} \left| \sum_{n=1}^{N} \chi_{\rho P+\mathbf{y}}(\mathbf{x}_n) - N\lambda_k((\rho P + \mathbf{y}) \cap X) \right|^2 d\mathbf{y}\, d\rho.$$

The proof of Proposition 2.20 is quite similar to that of Proposition 1.47. Essentially it relies of the following asymptotic relation.

Lemma 2.21 *Let $k \geq 1$ and $P \subseteq \mathbf{R}^k$ be a k-dimensional convex simple polytope with vertex set $V(P)$. For $\mathbf{x} \in V(P)$ let $\Gamma(\mathbf{x})$ denote the set set of those vertices in $V(P)$ which are adjacent to $\mathbf{x}$. Then*

$$|\hat{\chi}_P(\mathbf{t})|^2 \ll \sum_{\mathbf{x}\in V(P)} \prod_{\mathbf{y}\in\Gamma(\mathbf{x})} \frac{1}{(1+|\mathbf{t}\cdot(\mathbf{x}-\mathbf{y})|)^2} \tag{2.55}$$

and

$$\int_0^1 |\hat{\chi}_{\rho P}(\mathbf{t})|^2\, d\rho \gg \sum_{\mathbf{x}\in V(P)} \prod_{\mathbf{y}\in\Gamma(\mathbf{x})} \frac{1}{(1+|\mathbf{t}\cdot(\mathbf{x}-\mathbf{y})|)^2} \tag{2.56}$$

uniformly for all $\mathbf{t} \in \mathbf{R}^k$, where the constants implied by $\ll$ and $\gg$ only depend on P.

Proof. First we show that the FOURIER transform of the characteristic function χ_P is given by

$$\hat{\chi}_P(\mathbf{t}) = (2\pi)^{-\frac{k}{2}} \sum_{\mathbf{x}\in V(P)} \frac{|\det(\mathbf{x}-\mathbf{y})_{\mathbf{y}\in\Gamma(\mathbf{x})}|}{\prod_{\mathbf{y}\in\Gamma(\mathbf{x})}(-i\mathbf{t}\cdot(\mathbf{x}-\mathbf{y}))} e^{-i\mathbf{t}\cdot\mathbf{x}}. \tag{2.57}$$

It is an easy exercise to verify (2.57) for k-dimensional simplices. Next, observe that $\hat{\chi}_P(\mathbf{t})$ depends only on the vertices of P and that it has the form

$$\hat{\chi}_P(\mathbf{t}) = \sum_{\mathbf{x}\in V(P)} c_{\mathbf{x}}(\mathbf{t}) e^{-i\mathbf{t}\mathbf{x}}.$$

Now choose any vertex $\mathbf{x}_0 \in V(P)$ and let $P = P' \cup S$, where S is the simplex obtained by taking the convex hull of $\mathbf{x}_0$ and all vertices $\Gamma(\mathbf{x}_0)$ adjacent to $\mathbf{x}_0$ and $P' = \overline{P \setminus S}$. It is clear that

$$\hat{\chi}_P(\mathbf{t}) = \hat{\chi}_{P'}(\mathbf{t}) + \hat{\chi}_S(\mathbf{t})$$

and that $\hat{\chi}_{P'}(\mathbf{t})$ has no influence on the coefficient $c_{\mathbf{x}_0}(\mathbf{t})$. Hence $c_{\mathbf{x}_0}(\mathbf{t})$ is given by

$$c_{\mathbf{x}_0}(\mathbf{t}) = (2\pi)^{-\frac{k}{2}} \frac{|\det(\mathbf{x}_0-\mathbf{y})_{\mathbf{y}\in\Gamma(\mathbf{x}_0)}|}{\prod_{\mathbf{y}\in\Gamma(\mathbf{x}_0)}(-i\mathbf{t}\cdot(\mathbf{x}_0-\mathbf{y}))},$$

and (2.57) follows.

Next, we will prove the upper bound (2.55). As usual, a facet F of P is the intersection of P with a supporting hyperplane, e.g. the 0-dimensional facets are the vertices $V(P)$ and the 1-dimensional facets are exactly the edges joining adjacent vertices.

(2.55) has two trivial cases. If $|\mathbf{t}| < C$ for some constant $C > 0$ then (2.55) is satisfied since

$$|\hat{\chi}_P(\mathbf{t})| \leq (2\pi)^{-\frac{k}{2}} \lambda_k(P).$$

The second one occurs if $|\mathbf{t}\cdot(\mathbf{x}-\mathbf{y})| \ge \varepsilon$ for all adjacent vertices $\mathbf{x},\mathbf{y} \in V(P)$ and some fixed $\varepsilon > 0$. Here we have

$$\begin{aligned}|\hat{\chi}_P(\mathbf{t})| &\ll \sum_{\mathbf{x}\in V(P)} \prod_{\mathbf{y}\in\Gamma(\mathbf{x})} \frac{1}{|\mathbf{t}\cdot(\mathbf{x}-\mathbf{y})|} \\ &\le \left(\frac{2}{\varepsilon}\right)^k \sum_{\mathbf{x}\in V(P)} \prod_{\mathbf{y}\in\Gamma(\mathbf{x})} \frac{1}{1+|\mathbf{t}\cdot(\mathbf{x}-\mathbf{y})|},\end{aligned}$$

and consequently (2.55).

In what follows we will need a (fixed) sequence $\varepsilon = \varepsilon_1 < \varepsilon_2 < \cdots < \varepsilon_k$ such that ε_k and the fractions $\varepsilon_l/\varepsilon_{l+1}$, $1 \le l < k$, are sufficiently small. We assume now that $|\mathbf{t}| \ge C$ for some sufficiently large constant C and that there exist two adjacent vertices $\mathbf{x},\mathbf{y} \in V(P)$ with $|\mathbf{t}\cdot(\mathbf{x}-\mathbf{y})| < \varepsilon$. We partition $V(P)$ into vertex sets $V(P) = \bigcup_{j\in J} V_j$ such that the following conditions hold. For any $j \in J$ there exists a facet F_j of P of dimension l_j such that $F_j \cap P = V_j$ and there exists $\mathbf{x}_j \in V_j$ such that $|\mathbf{t}\cdot(\mathbf{x}_j-\mathbf{y})| < \varepsilon_{l_j}$ for all $\mathbf{y} \in \Gamma(\mathbf{x}_j)\cap V_j$ and $|\mathbf{t}\cdot(\mathbf{x}_j-\mathbf{y})| \ge \varepsilon_{l_j+1}$ for all $\mathbf{y} \in \Gamma(\mathbf{x}_j)\setminus V_j$. Note that $|\Gamma(\mathbf{x}_j)\cap V_j| = l_j$ and that there exists at least one set V_j with $|V_j| \ge 2$. Furthermore $|\mathbf{t}| \ge C$ ensures that $l_j < k$.

Let $\mathbf{y} \in \Gamma(\mathbf{x}_j)\setminus V_j$. Then for any vertex $\mathbf{x}' \in V_j$ there exists $\mathbf{y}' \in \Gamma(\mathbf{x}')$ corresponding to $\mathbf{y}$. This means that $\{\mathbf{x}-\mathbf{x}'\} \cup \{\mathbf{y}-\mathbf{x}_j\}$ generate the same subspace as $\{\mathbf{x}-\mathbf{x}'\} \cup \{\mathbf{y}'-\mathbf{x}'\}$. For technical reasons we replace any $\mathbf{y}' \in \Gamma(\mathbf{x}')$ by $\mathbf{x}' + \rho(\mathbf{y}'-\mathbf{x}')$, in which $\rho > 0$, such that $(\mathbf{x}'+\rho(\mathbf{y}'-\mathbf{x}')) - \mathbf{y}$ is parallel to $\{\mathbf{x}-\mathbf{x}'\}$. For brevity we will denote $\mathbf{x}'+\rho(\mathbf{y}'-\mathbf{x}')$ again by $\mathbf{y}'$. Hence there exists $\mu_{\mathbf{y}}$ such that $\mathbf{x}'-\mathbf{y}' = \mathbf{x}_j - \mathbf{y} + \mu_{\mathbf{y}}(\mathbf{x}'-\mathbf{x}_j)$ for all $\mathbf{y}'$ corresponding to $\mathbf{y}$. Note that the coefficients $c_{\mathbf{x}'}(\mathbf{t})$ do not change during this procedure.

Our aim is to show that

$$\left|\sum_{\mathbf{x}\in V_j} c_{\mathbf{x}}(\mathbf{t}) e^{-i\mathbf{t}\cdot\mathbf{x}}\right| \ll \prod_{\mathbf{y}\in\Gamma(\mathbf{x}_j)\setminus V_j} \frac{1}{|\mathbf{t}\cdot(\mathbf{x}_j-\mathbf{y})|}. \tag{2.58}$$

It is clear that (2.58) implies (2.55) since $|\mathbf{t}\cdot(\mathbf{x}'-\mathbf{x}'')| = \mathcal{O}(\varepsilon_{l_j})$ for any two adjacent vertices $\mathbf{x}',\mathbf{x}'' \in V_j$ and

$$\prod_{\mathbf{y}'\in\Gamma(\mathbf{x}')\setminus V_j} (-i\mathbf{t}\cdot(\mathbf{x}'-\mathbf{y}')) = \prod_{\mathbf{y}\in\Gamma(\mathbf{x}_j)\setminus V_j} (-i\mathbf{t}\cdot(\mathbf{x}_j-\mathbf{y}))\left(1+\mathcal{O}\left(\frac{\varepsilon_{l_j}}{\varepsilon_{l_j+1}}\right)\right)$$

for any $\mathbf{x}' \in V_j$.

Next, consider a triangulation $\mathcal{T}$ of F_j such that for any simplex $S \in \mathcal{T}$ the vertices $V(S)$ of S are contained in V_j. For $\mathbf{x} \in V(S)$ let $\Gamma_S(\mathbf{x}) = (V(S)\setminus\{\mathbf{x}\}) \cup (\Gamma(\mathbf{x})\setminus V_j)$ and set

$$c_S(\mathbf{t}) = (2\pi)^{-\frac{k}{2}} \sum_{\mathbf{x}\in V(S)} \frac{|\det(\mathbf{x}-\mathbf{y})_{\mathbf{y}\in\Gamma_S(\mathbf{x})}|}{\prod_{\mathbf{y}\in\Gamma_S(\mathbf{x})}(-i\mathbf{t}\cdot(\mathbf{x}-\mathbf{y}))} e^{-i\mathbf{t}\cdot\mathbf{x}}.$$

Then by considering the convex hull of $S \cup \bigcup_{\mathbf{x}\in V(S)} (\Gamma(\mathbf{x}) \setminus V_j)$, and by using (2.57) it follows that

$$\sum_{S\in\mathcal{T}} c_S(\mathbf{t}) = \sum_{\mathbf{x}\in V_j} c_{\mathbf{x}}(\mathbf{t})e^{-i\mathbf{t}\cdot\mathbf{x}}.$$

Therefore it remains to show that

$$|c_S(\mathbf{t})| \ll \prod_{\mathbf{y}\in\Gamma(\mathbf{x}_0)\setminus V_j} \frac{1}{|\mathbf{t}\cdot(\mathbf{x}_0-\mathbf{y})|}, \tag{2.59}$$

where $\mathbf{x}_0$ is any fixed vertex of S.

Let $S \in \mathcal{T}$ and denote by $\mathbf{x}_0, \mathbf{x}_1, \dots, \mathbf{x}_{l_j}$ the vertices of S. Furthermore set

$$F(u) = \frac{e^{-iu}}{\prod_{\mathbf{y}\in\Gamma(\mathbf{x}_0)\setminus V_j}(-i\mathbf{t}\cdot(\mathbf{x}_0-\mathbf{y}) - \mu_{\mathbf{y}}(iu - i\mathbf{t}\cdot\mathbf{x}_0))},$$

$$f_1(u_0,u_1) = \frac{F(u_1)-F(u_0)}{u_1-u_0},$$

and inductively ($l \geq 1$)

$$f_{l+1}(u_0,u_1,\dots,u_{l+1}) = \frac{f_l(u_0,u_1\dots,u_{l-1},u_{l+1}) - f_l(u_0,u_1,\dots,u_{l-1},u_l)}{u_{l+1}-u_l}.$$

Then

$$f_l(u_0,\dots,u_l) = \sum_{n=0}^{l} \frac{F(u_n)}{\prod_{m\neq n}(u_n-u_m)},$$

and consequently

$$c_S(\mathbf{t}) = (2\pi)^{-\frac{k}{2}}(-i)^{-l_j} f_{l_j}(\mathbf{t}\cdot\mathbf{x}_0,\dots,\mathbf{t}\cdot\mathbf{x}_{l_j}).$$

On the other hand it follows by induction and by TAYLOR's theorem that

$$f_l(u_0,\dots,u_l) = \frac{1}{l!}\left(F_1^{(l)}(v_1) + iF_2^{(l)}(v_2)\right),$$

where $F_1(u) = \Re(F(u))$, $F_2(u) = \Im(F(u))$, and $v_1, v_2 \in [\min u_n, \max u_n]$. Since $\mathbf{t}\cdot(\mathbf{x}_n-\mathbf{x}_m) = \mathcal{O}(\varepsilon_{l_j})$ and $|\mathbf{t}\cdot(\mathbf{x}_n-\mathbf{y})| > \varepsilon_{l_j+1} - \mathcal{O}(\varepsilon_{l_j})$ for $\mathbf{y}\in\Gamma(\mathbf{x}_n)\setminus V_j$ we obtain

$$c_S(\mathbf{t}) = (2\pi)^{-\frac{k}{2}}\lambda_k(\tilde{S})(-i)^{l_j}\left(F_1^{(l_j)}(\mathbf{t}\cdot\mathbf{x}_0 + \mathcal{O}(\varepsilon_{l_j})) + F_1^{(l_j)}(\mathbf{t}\cdot\mathbf{x}_0 + \mathcal{O}(\varepsilon_{l_j}))\right),$$

where $\tilde{S}$ denotes the convex hull of $S \cup (\Gamma(\mathbf{x}_0)\setminus V_j)$. Since

$$\begin{aligned}&(-i)^{l_j}F^{(l_j)}(u) \\ &= e^{-iu}\sum_{n=0}^{l_j}\frac{l_j!}{(l_j-n)!}\sum_{\sum m_{\mathbf{y}}=n}\prod_{\mathbf{y}\in\Gamma(\mathbf{x}_0)\setminus V_j}\frac{\mu_{\mathbf{y}}^{m_{\mathbf{y}}}}{(-i\mathbf{t}\cdot(\mathbf{x}_0-\mathbf{y}) - \mu_{\mathbf{y}}(iu - i\mathbf{t}\cdot\mathbf{x}_0))^{m_{\mathbf{y}}+1}}\end{aligned} \tag{2.60}$$

and $|\mathbf{t}\cdot(\mathbf{x}_0-\mathbf{y})| \gg \varepsilon_{l_j+1}$, $\mathbf{y}\in\Gamma(\mathbf{x}_0)\setminus V_j$, we finally get

$$c_S(\mathbf{t}) = (2\pi)^{-\frac{k}{2}}\lambda_k(\tilde{S})e^{-i\mathbf{t}\cdot\mathbf{x}_0}$$
$$\times\sum_{n=0}^{l_j}\frac{l_j!}{(l_j-n)!}\sum_{\sum m_{\mathbf{y}}=n}\prod_{\mathbf{y}\in\Gamma(\mathbf{x}_0)\setminus V_j}\frac{\mu_{\mathbf{y}}^{m_{\mathbf{y}}}}{(-i\mathbf{t}\cdot(\mathbf{x}_0-\mathbf{y}))^{m_{\mathbf{y}}+1}}\left(1+\mathcal{O}\left(\frac{\varepsilon_{l_j}}{\varepsilon_{l_j+1}}\right)\right),$$

which proves (2.59) and consequently (2.55).

The proof of the lower bound (2.56) uses a similar procedure but is more refined. First, the case $|\mathbf{t}|<C$ is trivial since

$$\min_{|\mathbf{t}|<C}\int_0^1 |\hat{\chi}_{\rho P}(\mathbf{t})|^2\,d\rho > 0.$$

Note that $\hat{\chi}_{\rho P}(\mathbf{t})$ has the representation

$$\hat{\chi}_{\rho P}(\mathbf{t}) = \rho^k\hat{\chi}_P(\rho\mathbf{t}) = \sum_{\mathbf{x}\in V(P)} c_{\mathbf{x}}(\mathbf{t})e^{-i\rho\mathbf{t}\cdot\mathbf{x}}.$$

In order to give a flavour of the method let us consider the easiest case, in which $|\mathbf{t}\cdot(\mathbf{x}-\mathbf{y})|>\varepsilon_1$ for all adjacent vertices $\mathbf{x},\mathbf{y}\in V(P)$. Set $v=|V(P)|$. By elementary considerations it follows that there exists $\delta\in[\frac{1}{2v},\frac{1}{v}]$ such that

$$\left|e^{-i\delta\mathbf{t}\cdot\mathbf{x}}-e^{-i\delta\mathbf{t}\cdot\mathbf{y}}\right| \ge \eta = \min\left(\frac{\delta\varepsilon_1}{2},\frac{1}{2v^2}\right)$$

for any pair of adjacent vertices $\mathbf{x},\mathbf{y}\in V(P)$. Then the matrix

$$M(\tau) = \left(e^{-i(\delta m+\tau)\mathbf{t}\cdot\mathbf{x}}\right)_{0\le m<v,\mathbf{x}\in V(P)}$$

is regular and its determinant can be estimated by

$$|\det M(\tau)| = \prod_{(\mathbf{x},\mathbf{y})\in E(P)}\left|e^{-i\delta\mathbf{t}\cdot\mathbf{x}}-e^{-i\delta\mathbf{t}\cdot\mathbf{y}}\right| \ge \eta^{\frac{1}{2}v(v-1)},$$

where $E(P)$ denotes the set of edges of P. Next, consider the vectors $\mathbf{c} = (c_{\mathbf{x}})_{\mathbf{x}\in V(P)}(\mathbf{t})$ and $\mathbf{v}(\tau) = \left(\hat{\chi}_{(\delta m+\tau)P}(\mathbf{t})\right)_{0\le m<v}$. From $\mathbf{c}=M(\tau)^{-1}\mathbf{v}(\tau)$ we get

$$\begin{aligned}\|\mathbf{v}(\tau)\|_\infty &= \max_{0\le m<v}|\hat{\chi}_{(\delta m+\tau)P}(\mathbf{t})|\\ &\ge \frac{1}{\|M(\tau)^{-1}\|}\|\mathbf{c}\|_\infty\\ &\ge \frac{\eta^{\frac{1}{2}v(v-1)}}{v!}\|\mathbf{c}\|_\infty,\end{aligned}$$

and finally

$$\int_0^1|\hat{\chi}_{\rho P}(\mathbf{t})|^2\,d\rho \gg \int_0^\delta \|\mathbf{v}(\tau)\|_\infty^2\,d\tau \gg \|\mathbf{c}\|_\infty^2,$$

which proves (2.56).

If there exist adjacent vertices $\mathbf{x}, \mathbf{y} \in V(P)$ with $|\mathbf{t} \cdot (\mathbf{x} - \mathbf{y})| \le \varepsilon_1$ then we use the same partition $V(P) = \bigcup_{j\in J} V_j$ as in the proof for the upper bound (2.55). Furthermore, for any $j \in J$ we get by exactly the same methods as above

$$\sum_{\mathbf{x}\in V_j} c_{\mathbf{x}}(\mathbf{t}) e^{-i\rho\mathbf{t}\cdot\mathbf{x}} = (2\pi)^{-\frac{k}{2}} \lambda_k(\tilde{V}_j) e^{-i\rho\mathbf{t}\cdot\mathbf{x}_j} \times \tag{2.61}$$

$$\sum_{n=0}^{l_j} \rho^{l_j-n} \frac{l_j!}{(l_j-n)!} \sum_{\sum m_{\mathbf{y}}=n} \prod_{\mathbf{y}\in\Gamma(\mathbf{x}_0)\setminus V_j} \frac{\mu_{\mathbf{y}}^{m_{\mathbf{y}}}}{(-i\mathbf{t}\cdot(\mathbf{x}_j - \mathbf{y}))^{m_{\mathbf{y}}+1}} \left(1 + \mathcal{O}\left(\frac{\varepsilon_{l_j}}{\varepsilon_{l_j+1}}\right)\right),$$

where $\mathbf{x}_j \in V_j$. Now, let $v' = \sum_j (l_j + 1)$ and choose $\delta \in [\frac{1}{2v'}, \frac{1}{v'}]$ such that

$$\left|e^{-i\delta\mathbf{t}\cdot\mathbf{x}_j} - e^{-i\delta\mathbf{t}\cdot\mathbf{x}_{j'}}\right| \ge \eta' = \min\left(\frac{\delta\varepsilon_1}{2}, \frac{1}{2v'^2}\right)$$

for $j \ne j'$. Again the determinant of the matrix

$$M(\tau) = \left(e^{-i(\delta m+\tau)\mathbf{t}\cdot\mathbf{x}_j} (\delta m + \tau)^{l_j - n}\right)_{0\le m<v', (j\in J, 0\le n\le l_j)}$$

can be bounded below by

$$\begin{aligned} |\det M(\tau)| &= \delta^{\sum_{j\in J} l_j} \left(\prod_{j\in J}\prod_{n=0}^{l_j} \frac{1}{n!}\right) \prod_{j\ne j'\in J} \left|e^{-i\delta\mathbf{t}\cdot\mathbf{x}_j} - e^{-i\delta\mathbf{t}\cdot\mathbf{x}_{j'}}\right|^{(l_j+1)(l_{j'}+1)} \\ &\ge \delta^{\sum_{j\in J} l_j} \eta'^{\sum_{j\ne j'\in J}(l_j+1)(l_{j'}+1)} \prod_{j\in J}\prod_{n=0}^{l_j} \frac{1}{n!}. \end{aligned}$$

Finally consider the vectors $\mathbf{c} = (c_{jn})_{(j\in J, 0\le n\le l_j)} = M(\tau)^{-1}\mathbf{v}(\tau)$ and $\mathbf{v}(\tau) = \left(\hat{\chi}_{(\delta m+\tau)P}(\mathbf{t})\right)_{0\le m<v}$. By (2.61) c_{jn} is given by

$$c_{jn} = \frac{l_j!}{(l_j-n)!} \sum_{\sum m_{\mathbf{y}}=n} \prod_{\mathbf{y}\in\Gamma(\mathbf{x}_0)\setminus V_j} \frac{\mu_{\mathbf{y}}^{m_{\mathbf{y}}}}{(-i\mathbf{t}\cdot(\mathbf{x}_j - \mathbf{y}))^{m_{\mathbf{y}}+1}} \left(1 + \mathcal{O}\left(\frac{\varepsilon_{l_j}}{\varepsilon_{l_j+1}}\right)\right).$$

As above they satisfy

$$\|\mathbf{v}(\tau)\|_\infty \gg \|\mathbf{c}\|_\infty \ge \max_{j\in J} |c_{j0}|.$$

Consequently, we obtain (2.56) in any case. □

Proof of Propositon 2.20. Set

$$F_\rho(\mathbf{y}) = \sum_{n=1}^{N} \chi_{\rho P+\mathbf{y}}(\mathbf{x}_n) - N\lambda_k((\rho P + \mathbf{y}) \cap X).$$

and

$$F_{\rho_1,\ldots,\rho_k}(\mathbf{y}) = \sum_{n=1}^{N} \chi_{P(\rho_1,\ldots,\rho_k)+\mathbf{y}}(\mathbf{x}_n) - N\lambda_k((\rho P + \mathbf{y}) \cap X).$$

Then $\hat{F}_\rho(\mathbf{t}) = \hat{\chi}_{-\rho P} \cdot \widehat{d\nu_N}$ and $\hat{F}_{\rho_1,\ldots,\rho_k}(\mathbf{t}) = \hat{\chi}_{-P(\rho_1,\ldots,\rho_k)} \cdot \widehat{d\nu_N}$, where ν_N denotes the signed measure

$$\nu_N(S) = \sum_{n=1}^{N} \chi_S(\mathbf{x}_n) - N\lambda_k(S \cap X).$$

Furthermore, observe that the estimates (2.55) and (2.56) proposed in Lemma 2.21 can be made uniform for a system of parallel polytopes $\mathcal{P}$ (see (2.50) if $\varepsilon > 0$ is sufficiently small. Hence, we obtain

$$\begin{aligned} \int_0^1 \int_{\mathbf{R}^k} |F_\rho(\mathbf{y})|^2 \, d\mathbf{y}\, d\rho &= \int_{\mathbf{R}^k} \int_0^1 |\hat{\chi}_{-\rho P}(\mathbf{t})|^2 \, d\rho \, |\widehat{d\nu_N}|^2 \, d\mathbf{t} \\ &\gg \int_{\mathbf{R}^k} \int_{[0,1)^k} |\hat{\chi}_{-P(\rho_1,\ldots,\rho_k)}(\mathbf{t})|^2 \, d\rho_1 \cdots d\rho_k \, |\widehat{d\nu_N}|^2 \, d\mathbf{t} \\ &= \int_0^1 \int_{\mathbf{R}^k} |F_\rho(\mathbf{y})|^2 \, d\mathbf{y}\, d\rho_1 \cdots d\rho_k. \end{aligned}$$

This completes the proof of Proposition 2.20. □

2.1.3 Spherical Discrepancy Systems

Let $\mathbf{S}^k$ denote the k-dimensional unit sphere $\mathbf{S}^k = \{\mathbf{x} \in \mathbf{R}^{k+1} : |\mathbf{x}| = 1\}$, d the geodesic distance, and σ_k resp. σ_k^* the surface resp. the normalized surface measure on $\mathbf{S}^k$. In the following we will discuss two different discrepancy systems, first the system of spherical caps

$$\mathcal{D}(C, \mathbf{S}^k) = \{C(\mathbf{x}, r) : \mathbf{x} \in \mathbf{S}^k,\ -1 \le r \le 1\}, \tag{2.62}$$

where $C(\mathbf{x}, r) = \{\mathbf{z} \in \mathbf{S}^k : \mathbf{x} \cdot \mathbf{z} \ge r\}$, and then the system of spherical slices

$$\mathcal{D}(S, \mathbf{S}^k) = \{S(\mathbf{x}, \mathbf{y}) : \mathbf{x}, \mathbf{y} \in \mathbf{S}^k\}, \tag{2.63}$$

where $S(\mathbf{x}, \mathbf{y}) = \{\mathbf{z} \in \mathbf{S}^k : \mathbf{x} \cdot \mathbf{z} \ge 0,\ \mathbf{y} \cdot \mathbf{z} \ge 0\}$. Such a slice is the intersection of two half-spheres. By Propositions 2.5 and 2.6 it is clear that $\mathcal{D}(C, \mathbf{S}^k)$ is a discrepancy system. In the case of slices it is not so easy to prove that $\mathcal{D}(S, \mathbf{S}^k)$ is a discrepancy system. We will give a proof of this assertion at the end of this section.

For both systems we get estimates of the same kind.

Theorem 2.22 *Let $k \ge 2$ and $\mathcal{D}$ the system of all sherical caps $\mathcal{D}(C, \mathbf{S}^k)$ or the system of all sperical slices $\mathcal{D}(S, \mathbf{S}^k)$. If $\mathbf{x}_1, \ldots, \mathbf{x}_N \in \mathbf{S}^k$ is any finite sequence on $\mathbf{S}^k$ then we have*

$$D_N^{(\mathcal{D})}(\mathbf{x}_n) \gg N^{-\frac{1}{2}-\frac{1}{2k}}. \tag{2.64}$$

On the other hand, for every $N > 1$ there exists a finite sequence $\mathbf{x}_1, \ldots, \mathbf{x}_N \in \mathbf{S}^k$ such that

$$D_N^{(\mathcal{D})}(\mathbf{x}_n) \ll N^{-\frac{1}{2}-\frac{1}{2k}}(\log N)^{\frac{1}{2}}. \tag{2.65}$$

The constants implied by $\gg$ and $\ll$ only depend on the dimension k.

First, let us discuss the case of caps.

Proof of Theorem 2.22 for caps. Let $B(r) = \{\mathbf{x} \in \mathbf{R}^{k+1} : |\mathbf{x}| \le r\}$ denote the closed ball with radius r and set $F_r = \chi_{B(r)} * d\nu : \mathbf{R}^{k+1} \to \mathbf{R}$, where

$$\nu = \frac{1}{N}\sum_{n=1}^{N} \delta_{\mathbf{x}_n} - \mu,$$

$\mathbf{x}_1, \ldots, \mathbf{x}_N \in \mathbf{S}^k \subseteq \mathbf{R}^{k+1}$, and μ is defined by $\mu(A) = \sigma_k^*(A \cap \mathbf{S}^k)$. By PARSEVAL's identity we get

$$\begin{aligned}\Phi(q) &= \frac{1}{q}\int_q^{2q}\int_{\mathbf{R}^{k+1}} |F_r(\mathbf{x})|^2\, d\mathbf{x}\, dr \\ &= \int_{\mathbf{R}^{k+1}} \varphi_q(\mathbf{t})|\hat{\nu}(\mathbf{t})|^2\, d\mathbf{t}.\end{aligned}$$

Recall that Lemma 2.17 gives

$$\frac{\varphi_q(\mathbf{t})}{\varphi_p(\mathbf{t})} \gg \left(\frac{q}{p}\right)^k \tag{2.66}$$

uniformly for $\mathbf{t} \in \mathbf{R}^{k+1}$ and $0 < p < q$. (It should be noted that the FOURIER transform $\hat{\chi}_{B(r)}(\mathbf{t})$ is (despite of a constant factor) a BESSEL function. For BESSEL functions asymptotic expansions of the kind (2.40) are classical. So we can prove (2.66) without using Lemma 2.16.

Let p be defined in a way such that $\frac{1}{4N} \le \max_{\mathbf{x}\in\mathbf{R}^{k+1}} \sigma_k^*(B(p) + \mathbf{x}) \le \frac{1}{2N}$. This may be done by setting $p = c_k N^{-\frac{1}{k}}$ and choosing c_k appropriately. Hence we get

$$\left|\frac{1}{N}\sum_{n=1}^{N} \chi_{B(r)+\mathbf{x}}(\mathbf{x}_n) - \sigma_k^*((B(r)+\mathbf{x}) \cap \mathbf{S}^k)\right| \ge \sigma_k^*((B(r)+\mathbf{x}) \cap \mathbf{S}^k) \gg N^{-1}$$

uniformly for all r and $\mathbf{x}$ satisfying $\frac{1}{2}p \le r \le p$ and $1 - \frac{p}{2}|\mathbf{x}| \le 1$. This gives the trivial lower bound

$$\Phi(p) \gg N^{-2-\frac{1}{k}}.$$

Now (2.66) implies

$$\Phi(1) \gg N\,\Phi(p) \gg N^{-1-\frac{1}{k}},$$

which completes the proof of the lower bound (2.64).

In order to prove the upper bound (2.65) we can use a similar probabilistic approach to the proof of the upper bound (2.27) in Theorem 2.10. Let $N \ge 1$ be given and set $\varepsilon = N^{-\frac{k+1}{2k}}$. Then we have to construct a finite system $\mathcal{A}_N \subseteq \mathcal{D}(C, \mathbf{S}^k)$ such that for any $C \in \mathcal{D}(C, \mathbf{S}^k)$ there exist $C_1, C_2 \in \mathcal{A}_N$ with $C_1 \subseteq C \subseteq C_2$ and

$$\sigma_k^*(C_2 \setminus C_1) \ll \varepsilon.$$

Clearly, such a system exists with cardinality $|\mathcal{A}_N| \ll \varepsilon^{-k-1}$. Next we have to find a partition $Q_1, \ldots, Q_N$ of $\mathbf{S}^k$ such that $\sigma_k^*(Q_n) = \frac{1}{N}$ $(1 \leq n \leq N)$ and $\max_{1 \leq n \leq N} \operatorname{diam}(Q_n) \ll N^{-\frac{1}{k}}$. (It is an easy exercise to verify that such a system exists.) Now the existence of a finite sequence $\mathbf{x}_1, \ldots, \mathbf{x}_N \in \mathbf{S}^k$ satisfying (2.65) follows by exactly the same methods as in the proof of (2.27). □

In order to prove the lower bound for slices, we have to introduce a new concept, namely FOURIER series on the homogeneous space $\mathbf{S}^k$ (cf. MÜLLER [1218] and VILENKIN [1910]).

We identify $\mathbf{S}^k$ with the left coset space $SO(k+1)/SO(k)$, $SO(k)$ being the subgroup of $SO(k+1)$ which leaves $(1, 0, \ldots, 0)$ fixed.

The adjoint map of the projection $\flat$

$$C(SO(k+1)) \mapsto C(\mathbf{S}^k), \quad f \mapsto f^\flat : \quad f^\flat(\dot{\xi}) = \int_{SO(k)} f(\xi\eta)\, d\eta$$

($d\eta$ being the Haar measure of $SO(k)$ and $\dot{\xi}$ the image of ξ under the canonical projection $\phi : SO(k+1) \to \mathbf{S}^k$) is the map $\sharp$

$$\mathcal{M}(\mathbf{S}^k) \mapsto \mathcal{M}(SO(k+1)), \quad \nu \to \nu^\sharp$$

of the space of RADON measures on $\mathbf{S}^k$, $\mathcal{M}(\mathbf{S}^k)$ onto the subspace of measures in $\mathcal{M}(SO(k+1))$ which are invariant under the right action of $SO(k)$.

We set

$$\nu = \frac{1}{N} \sum_{n=1}^{N} \delta_{\mathbf{x}_n} - \sigma_k^*,$$

(δ_x being the Dirac measure at x) and define for a BOREL subset A of $\mathbf{S}^k$ the mean square A-discrepancy by averaging over all sets $\xi(A)$, where $\xi(A)$ is the image of A under the action of $\xi \in SO(k+1)$:

$$\Delta_N(A, 2)^2 = \int_{SO(k+1)} (\nu(\xi(A))^2\, d\xi = \int_{SO(k+1)} \left(\int_{\mathbf{S}^k} \chi_A(\xi^{-1}(y))\, d\nu(y) \right)^2 d\xi.$$

It is an easy consequence of the regularity of ν that $\nu(A) = \nu^\sharp(\phi^{-1}(A))$ and we have

$$\int_{\mathbf{S}^k} \chi_A(\xi^{-1}(y))\, d\nu(y) = \int_{SO(k+1)} \chi_{\phi^{-1}(A)}(\xi^{-1}\eta)\, d\nu^\sharp(\eta) = \nu^\sharp * \tilde{\chi}_{\phi^{-1}(A)}(\xi),$$

where $\tilde{\chi}_{\phi^{-1}(A)}(\xi) = \chi_{\phi^{-1}(A)}(\xi^{-1})$. As $\tilde{\chi}_{\phi^{-1}(A)}$ is bounded $\nu^\sharp * \tilde{\chi}_{\phi^{-1}(A)}$ is in $L^2(SO(k+1))$ and we obtain for the L^2-norm $\|\cdot\|$:

$$\Delta_N(A,2) = \|\nu^\sharp * \tilde{\chi}_{\phi^{-1}(A)}\|.$$

Since $\hat{\nu}^\sharp$ is the FOURIER-STIELTJES transform of a measure which is invariant under the right action of $SO(k)$ it has nonvanishing elements only in the column containing the trivial representation of $SO(k)$. $\chi_{\phi^{-1}(A)}$ is a function which is invariant under the right action of $SO(k)$, hence $\tilde{\chi}_{\phi^{-1}(A)}$ is invariant under the left action of $SO(k)$. Furthermore, all elements of the row containing the trivial representation of $SO(k)$ vanish in $\hat{\tilde{\chi}}_{\phi^{-1}(A)}(l)$, which is the FOURIER transform of $\tilde{\chi}_{\phi^{-1}(A)}$ at l. This follows from that fact that for $f \in L^1(SO(k+1))$ we have $\hat{\tilde{f}}(l) = \overline{\hat{f}(l)}^t$ and the observation that $\hat{\tilde{\chi}}_{\phi^{-1}(A)}$ is a representation of class 1, i.e. a representation of $SO(k+1)$ containing the trivial representation of $SO(k)$. Therefore the HILBERT-SCHMIDT norm $\|\cdot\|_2$ of the product of these two transforms is equal to the product of the norms:

$$\|(\nu^\sharp * \tilde{\chi}_{\phi^{-1}(A)})^\wedge(l)\|_2 = \|\hat{\nu}^\sharp(l)\hat{\tilde{\chi}}_{\phi^{-1}(A)}(l)\|_2 = \|\hat{\nu}^\sharp(l)\|_2\|\hat{\chi}_{\phi^{-1}(A)}(l)\|_2.$$

Since $\nu^\sharp * \tilde{\chi}_{\phi^{-1}(A)}$ is an absolutely continuous measure with bounded density we may identify it with a function in $L^2(SO(k+1))$ and it follows from PARSEVAL's formula that the function $\nu^\sharp * \tilde{\chi}_{\phi^{-1}(A)}$ with FOURIER expansion

$$(\nu^\sharp * \tilde{\chi}_{\phi^{-1}(A)})(\xi) = \sum_l d_l \operatorname{tr}((\nu^\sharp * \tilde{\chi}_{\phi^{-1}(A)})^\wedge(l) U_l^*(\xi))$$

has L^2-norm

$$\|\nu^\sharp * \tilde{\chi}_{\phi^{-1}(A)}\| = \left(\sum_l d_l \|\hat{\nu}^\sharp(l)\|_2^2 \|\hat{\tilde{\chi}}_{\phi^{-1}(A)}(l)\|_2^2\right)^{1/2},$$

where $U_l(\xi)$ is the unitary matrix representing the action of $\xi \in SO(k+1)$ on the d_l-dimensional space of harmonic polynomials in $n+1$ variables and degree l. Since $\nu^\sharp$ is a right $SO(k)$-invariant measure the sum may be restricted to representations of class 1, which may be labeled by $l = 0, 1, 2, \ldots$.

For A we consider slices $S_\varphi = S((0,\ldots,0,1,0),(0,\ldots,\cos\varphi,\sin\varphi))$ and spherical caps $C_r = C((1,0,\ldots,0),r)$ centered at the unit coset of $\mathbf{S}^k$ and define

$$D_N(C,2) = \left(\int_{-1}^{1} \Delta_N(\tau_{C_r},2)^2\,dr\right)^{1/2}, \qquad D_N(S,2) = \left(\int_0^\pi \Delta_N(S_\varphi,2)^2\,d\varphi\right)^{1/2}.$$

First, we observe

Lemma 2.23 *Let $k \geq 2$. Then we have for any finite sequence $\mathbf{x}_1,\ldots,\mathbf{x}_N \in \mathbf{S}^k$*

$$D_N(C,2) \gg N^{-\frac{1}{2}-\frac{1}{2k}}. \tag{2.67}$$

Proof. Obviously, $\Phi(1)$ can be transformed to an integral of the form

$$\Phi(1) = \int_{-1}^{1} \int_{SO(k+1)} \left(\int_{\mathbf{S}^k} \chi_{C_r}(\xi^{-1}(y))\, d\nu(y) \right)^2 d\xi\, f_k(r)\, dr,$$

where $f_k(r) \ll 1$ is a bounded weight function (compare with BECK and CHEN [143, p. 226 f]. Hence we get

$$D_N(C,2)^2 \gg \Phi(1) \gg N^{-1-\frac{1}{k}},$$

which proves (2.23). □

In view of this lemma it suffices to show $D_N(S,2) \gg D_N(C,2)$.

Proof of (2.64) for slices. It follows from the considerations above that we obtain for $\nu = \frac{1}{N}\sum_{n=1}^{N} \delta_{\mathbf{x}_n} - \sigma_k^*$

$$\begin{aligned} D_N(S,2) &= \left(\sum_l d_l \|\hat{\nu}^\sharp(l)\|_2^2 \int_0^{\pi} \|\hat{\tilde{\chi}}_{\phi^{-1}(S_\varphi)}(l)\|_2^2\, d\varphi \right)^{1/2}, \\ D_N(C,2) &= \left(\sum_l d_l \|\hat{\nu}^\sharp(l)\|_2^2 \int_{-1}^{1} \|\hat{\tilde{\chi}}_{\phi^{-1}(C_r)}(l)\|_2^2\, d\varphi \right)^{1/2}. \end{aligned}$$

$D_N(S,2) \gg D_N(C,2)$ will follow if we show that

$$\int_0^{\pi} \|\hat{\chi}_{\phi^{-1}(S_\varphi)}(l)\|_2^2\, d\varphi \geq \lambda_k \int_{-1}^{1} \|\hat{\chi}_{\phi^{-1}(C_r)}(l)\|_2^2\, dr, \quad \lambda_k > 0.$$

It will be more convenient to compute the $L^2(\mathbf{S}^k, \sigma_k)$-coefficients $c_{l,J}(\chi_{S_\varphi})$ (resp. $c_{l,J}(\chi_{C_r})$) of χ_{S_φ} (resp. χ_{C_r}) with respect to a basis of spherical harmonics. Especially, we have

$$d_l \sigma_k(S_k) \|\hat{f}(l)\|_2^2 = \sum_J |c_{l,J}(f^\flat)|^2$$

($(d\sigma_k^*)^\sharp$ is just the normalized HAAR measure $d\eta$ on $SO(k+1)$).

We introduce polar coordinates on $\mathbf{S}^k$

$$\begin{aligned} x_k &= \sin\theta_1 \ldots \sin\theta_{k-1} \sin\varphi \\ x_{k-1} &= \sin\theta_1 \ldots \sin\theta_{k-1} \cos\varphi \\ &\vdots \\ x_1 &= \sin\theta_1 \cos\theta_2 \\ x_0 &= \cos\theta_1, \end{aligned}$$

and set $t_i = \cos\theta_i$, $i = 1, \ldots, k-1$, $t_i \in [-1,1]$, $\varphi \in [0, 2\pi)$, so that the measure σ_k on $\mathbf{S}^k$ is given by

$$\prod_{i=1}^{k-1} (1 - t_i^2)^{\frac{k-i-1}{2}}\, dt_i\, d\varphi.$$

An orthonormal basis for the space of spherical harmonics on $\mathbf{S}^k$ is given by the functions

$$\Xi_{l,J} = \frac{1}{\sqrt{2\pi}} A_{l,j_1}(k+1,t_1)A_{j_1,j_2}(k,t_2)\cdots A_{j_{k-2},j_{k-1}}(3,t_{k-1})e^{ij_k\varphi},$$

where $l \geq 0$, $J = (j_0, j_1, \ldots, j_k)$ with $l = j_0 \geq j_1 \geq j_2 \geq \cdots \geq j_{k-1} \geq 0$, $j_k = \pm j_{k-1}$, and the functions

$$A_{i,j}(m,t) = a_{i,j,m}(1-t^2)^{\frac{j}{2}} C_{i-j}^{\frac{m-2}{2}+j}(t)$$

are associate LEGENDRE functions of degree i, order j and dimension m, normalized with respect to the weight $(1-t^2)^{(m-3)/2}$ with C_l^p denoting the GEGENBAUER polynomial

$$C_l^p(t) = \frac{(-1)^l\Gamma(p+\frac{1}{2})\Gamma(l+2p)(1-t^2)^{\frac{1}{2}-p}}{2^l\Gamma(l+p+\frac{1}{2})\Gamma(2p)l!}\frac{d^l}{dt^l}(1-t^2)^{l+p-\frac{1}{2}}.$$

The zonal harmonic polynomial of degree l then corresponds to $J_0 = (l, 0, 0, \ldots, 0)$.

First we will prove that the coefficients $c_{l,J}(\chi_{C_r})$, $l > 0$ satisfy

$$\int_{-1}^{1} \sum_J |c_{l,J}(\chi_{C_r})|^2 \, dr \leq l^{-2}.$$

Since χ_{C_r} is a zonal spherical function all coefficients of χ_{C_r} vanish except $c_{l,J_0}(\chi_{C_r})$, corresponding to the normalized zonal harmonic polynomial of degree l

$$\Xi_{l,J_0}(t) = \sqrt{\frac{2^{k-2}l!(l+\frac{k-1}{2})\Gamma^2(\frac{k-1}{2})}{\sigma_{k-1}(S_{k-1})\pi\Gamma(l+k-1)}}\, C_l^{\frac{k-1}{2}}(t).$$

Since

$$\begin{aligned}
&\int_r^1 C_l^{\frac{k-1}{2}}(t)(1-t^2)^{\frac{k-2}{2}}\, dt = \\
= \quad & \left(\frac{-1}{2}\right)^l \frac{\Gamma(\frac{k}{2})\Gamma(l+k-1)}{\Gamma(l+\frac{k}{2})\Gamma(k-1)l!} \int_r^1 \left(\frac{d}{dt}\right)^l (1-t^2)^{l+(k-2)/2}\, dt \\
= \quad & -\frac{k-1}{l(l+k-1)}(1-r^2)^{k/2} C_{l-1}^{\frac{k+1}{2}}(r),
\end{aligned}$$

we get using

$$\int_{-1}^{1} [C_l^p(t)]^2 (1-t^2)^{p-\frac{1}{2}}\, dt = \frac{\pi\Gamma(2p+l)}{2^{2p-1}l!(l+p)\Gamma^2(p)}$$

(cf. VILENKIN [1910, p. 462]) the estimate

$$\int_{-1}^{1} |c_{l,J_0}(\chi_{C_r})|^2\, dr \quad = \quad K(l,k)\int_{-1}^{1} \left(C_{l-1}^{\frac{k+1}{2}}(r)\right)^2 (1-r^2)^k\, dr$$

$$
\begin{aligned}
&\le K(l,k)\int_{-1}^{1}\left(C_{l-1}^{\frac{k+1}{2}}(r)\right)^2(1-r^2)^{k/2}\,dr \\
&= K(l,k)\frac{\pi\Gamma(k+l)}{2^k(l-1)!(l-1+\frac{k+1}{2})\Gamma^2(\frac{k+1}{2})} \\
&= \frac{1}{l(l+k-1)}\le l^{-2},
\end{aligned}
$$

where

$$
K(l,k)=\frac{2^{k-2}l!(l+\frac{k-1}{2})\Gamma^2(\frac{k-1}{2})}{\pi\Gamma(l+k-1)}\frac{(k-1)^2}{l^2(l+k-1)^2}.
$$

In order to complete the proof we will finally show

$$
\int_0^\pi \sum_J |c_{l,J}(\chi s_\varphi)|^2\,d\varphi \ge \begin{cases} c_k l^{-2} & \text{for } l>0 \\ c_k & \text{for } l=0 \end{cases} \tag{2.68}
$$

for some $\lambda_k > 0$.

The associated LEGENDRE functions satisfy the differential equation

$$
\left[\frac{d}{dt}(1-t^2)^{\frac{m-1}{2}}\frac{d}{dt}+(1-t^2)^{\frac{m-3}{2}}i(i+m-2)\right.
$$
$$
\left.-(1-t^2)^{\frac{m-5}{2}}j(j+m-3)\right]A_{i,j}(m,t)=0.
$$

Multiplying this equation with $A_{i,j'}(m,t)$ and subtracting the same equation with j and j' interchanged yields after integration by parts over the interval $[-1,1]$ for fixed i,m the orthonormality of the functions $b_{i,j,m}A_{i,j}(m,t)$, $0<j\le i$ with respect to the weight $(1-t^2)^{\frac{m-5}{2}}$ and constants $b_{i,j,m}$. Since $A_{i,j}(m,t)$ is normalized with respect to the weight $(1-t^2)^{\frac{m-3}{2}}$ it follows that $b_{i,j,m}\le 1$. Since $A_{i,0}(3,t)\notin \mathrm{L}^2\left([-1,1],(1-t^2)^{-1}\,dt\right)$ we have to single out the case $j=0$.

For fixed $i>0$ we have

$$
\sum_{j=0}^{i}\left|\int_{-1}^{1}A_{i,j}(m,t)(1-t^2)^{\frac{m-3}{2}}\,dt\right|^2\ge
$$
$$
\ge\sum_{j=1}^{i}\left|\int_{-1}^{1}b_{i,j,m}A_{i,j}(m,t)(1-t^2)(1-t^2)^{\frac{m-5}{2}}\,dt\right|^2.
$$

The right hand side of this inequality is just the square of the norm of the orthogonal projection of the function $1-t^2$ on the subspace of $\mathrm{L}^2\left([-1,1],(1-t^2)^{\frac{m-5}{2}}dt\right)$ spanned by the orthonormal basis

$$
\{b_{i,j,m}A_{i,j}(m,t):j=0,1,\dots,i\}.
$$

It follows from the definition of the associated LEGENDRE functions $A_{i,j}$ and the above orthogonality, that for i odd, the $(i+1)/2$ functions $A_{i,1}(m,t), A_{i,3}(m,t), \dots, A_{i,i}(m,t)$ span the space of functions of the form $(1-t^2)^{1/2}P(t)$, with $P(t)$ an even polynomial of degree $\le i-1$, which clearly contains the function $(1-t^2)^{1/2}$. We hence can estimate the norm of the orthogonal projection of $1-t^2$ on this space by the norm of the orthogonal projection on the onedimensional subspace spanned by the function $(1-t^2)^{1/2}$ in $\mathrm{L}^2\left([-1,1], (1-t^2)^{\frac{m-5}{2}}\,dt\right)$:

$$\sum_{j=1}^{i}\left|\int_{-1}^{1} b_{i,j,m}A_{i,j}(m,t)(1-t^2)(1-t^2)^{\frac{m-5}{2}}\,dt\right|^2 \ge \frac{\left|\int_{-1}^{1}(1-t^2)^{\frac{m-2}{2}}\,dt\right|^2}{\int_{-1}^{1}(1-t^2)^{\frac{m-3}{2}}\,dt}.$$

For $i>0$, i even, the functions $A_{i,2}(m,t), A_{i,4}(m,t), \dots, A_{i,i}(m,t)$ constitute a set of $i/2$ linearly independent even polynomials of degree i which factor through $1-t^2$. Hence these functions span the space of all even polynomials of degree $\le i$ with zeros in ± 1 which contains the polynomial $1-t^2$. It follows by the CAUCHY-SCHWARZ inequality that for $i>0$ even

$$\sum_{j=2}^{i}\left|\int_{-1}^{1} b_{i,j,m}A_{i,j}(m,t)(1-t^2)(1-t^2)^{\frac{m-5}{2}}\,dt\right|^2 = \|(1-t^2)\|_2^2 =$$

$$= \int_{-1}^{1}(1-t^2)^{\frac{m-1}{2}}\,dt \ge \frac{\left|\int_{-1}^{1}(1-t^2)^{\frac{m-2}{2}}\,dt\right|^2}{\int_{-1}^{1}(1-t^2)^{\frac{m-3}{2}}\,dt}.$$

For $i=0$ we have

$$\left|\int_{-1}^{1} A_{0,0}(m,t)(1-t^2)^{\frac{m-3}{2}}\,dt\right|^2 = \int_{-1}^{1}(1-t^2)^{\frac{m-3}{2}}\,dt.$$

Combining the above inequalities we obtain for $i \ge 0$, $m \ge 3$

$$\sum_{j=0}^{i}\left|\int_{-1}^{1} A_{i,j}(m,t)(1-t^2)^{\frac{m-3}{2}}\,dt\right|^2 \ge \alpha(m) > 0.$$

Hence everything follows applying

$$l^{-2} \le \frac{1}{2\pi}\int_0^{\pi}\left|\int_0^{\varphi} e^{ij_k\alpha}\,d\alpha\right|^2 d\varphi = \begin{cases} j_k^{-2} & |j_k|>0 \\ 2\pi^2/3 & j_k=0 \end{cases}.$$

We repeat the above estimate for $m=3,4,\dots,k+1$ for the FOURIER coefficient $c_{l,J}(\chi_{S_\varphi}) = \int_{S_\varphi}\Xi_{l,J}(x)\,dx$ of χ_{S_φ} which is

$$\frac{1}{\sqrt{2\pi}}\int_{-1}^{1}\cdots\int_{-1}^{1}\int_0^{\varphi} A_{j_0,j_1}(k+1,t_1)\cdots A_{j_{k-2},j_{k-1}}(3,t_{k-1})e^{ij_k\alpha}$$
$$(1-t_1^2)^{\frac{k-2}{2}}\cdots(1-t_{k-2}^2)^{\frac{1}{2}}\,d\alpha\,dt_{k-1}\cdots dt_1$$

and obtain

$$\begin{aligned}\int_0^\pi \sum_J |c_{l,J}(\chi_{S_\varphi})|^2 \, d\varphi = \\ = \sum_{j_1=0}^{l} \left(\left| \int_{-1}^{1} A_{j_0,j_1}(k+1,t_1)(1-t_1^2)^{\frac{k-2}{2}} \, dt_1 \right|^2 \cdot \right. \\ \sum_{j_2=0}^{j_1} \left(\left| \int_{-1}^{1} A_{j_1,j_2}(k,t_2)(1-t_2^2)^{\frac{k-3}{2}} \, dt_2 \right|^2 \cdot \right. \\ \vdots \qquad \vdots \\ \sum_{j_{k-1}=0}^{j_{k-2}} \left(\left| \int_{-1}^{1} A_{j_{k-2},j_{k-1}}(3,t_{k-1}) \, dt_{k-1} \right|^2 \cdot \right. \\ \left. \left. \left. \sum_{j_k=\pm j_{k-1}} \frac{1}{2\pi} \int_0^\pi \left| \int_0^\varphi e^{ij_k\alpha} \, d\alpha \right|^2 d\varphi \right) \right) \cdots \right) \\ \geq l^{-2} \prod_{m=3}^{k+1} \alpha(m).\end{aligned}$$

□

Proof of (2.65) for slices. Again we will use a probabilistic approach similar to the proof of (2.27). But the system $\mathcal{A}_N$ cannot consist of slices.

Let $N \geq 1$ be given and set $\varepsilon = N^{-\frac{k+1}{2}k}$. Let $Z = \{\mathbf{z}_1, \ldots, \mathbf{z}_L\} \subseteq \mathbf{S}^k$ be a minimal set such that the balls $B(\mathbf{z}_l, \varepsilon) \subseteq \mathbf{R}^{k+1}$ $(1 \leq l \leq L)$ cover $\mathbf{S}^k$. Clearly $L \ll \varepsilon^{-k}$. For any pair of points $\mathbf{z}', \mathbf{z}'' \in Z$ consider the generalized slices

$$\begin{aligned} S_\varepsilon(\mathbf{z}',\mathbf{z}'') &= \{\mathbf{x} \in \mathbf{S}^k : \mathbf{z}' \cdot \mathbf{x} \geq \varepsilon, \mathbf{z}'' \cdot \mathbf{x} \geq \varepsilon\} \\ S_{-\varepsilon}(\mathbf{z}',\mathbf{z}'') &= \{\mathbf{x} \in \mathbf{S}^k : \mathbf{z}' \cdot \mathbf{x} \geq -\varepsilon, \mathbf{z}'' \cdot \mathbf{x} \geq -\varepsilon\} \end{aligned}$$

and set

$$\mathcal{A}_N = \{S_{\pm\varepsilon}(\mathbf{z}',\mathbf{z}'') : \mathbf{z}', \mathbf{z}'' \in Z\}.$$

Let $S(\mathbf{x},\mathbf{y})$ be an arbitrary slice, and let $\mathbf{z}', \mathbf{z}'' \in Z$ with $|\mathbf{x} - \mathbf{z}'| \leq \varepsilon$ and $|\mathbf{y} - \mathbf{z}''| \leq \varepsilon$. Since

$$\mathbf{z}' \cdot \mathbf{z} - \varepsilon \leq \mathbf{x} \cdot \mathbf{z} \leq \mathbf{z}' \cdot \mathbf{z} + \varepsilon$$

it follows that

$$S_\varepsilon(\mathbf{z}',\mathbf{z}'') \subseteq S(\mathbf{x},\mathbf{y}) \subseteq S_{-\varepsilon}(\mathbf{z}',\mathbf{z}'').$$

Furthermore

$$\sigma_k^*(S_{-\varepsilon}(\mathbf{z}',\mathbf{z}'') \setminus S_\varepsilon(\mathbf{z}',\mathbf{z}'')) \ll \varepsilon.$$

Therefore, if we can prove that there exists a finite sequence $\mathbf{x}_1, \ldots, \mathbf{x}_N \in \mathbf{S}^k$ satisfying

$$\max_{S' \in \mathcal{A}_N} \left| \frac{1}{N} \sum_{n=1}^{N} \chi_{S'}(\mathbf{x}_n) - \sigma_k^*(S') \right| \ll N^{-\frac{1}{2}-\frac{1}{2k}} (\log N)^{\frac{1}{2}}, \tag{2.69}$$

it follows that

$$D_N^{(\mathcal{D}(S,\mathbf{S}^k))}(\mathbf{x}_n) \ll N^{-\frac{1}{2}-\frac{1}{2k}}(\log N)^{\frac{1}{2}}.$$

In order to prove (2.65) we just have to use a partition $Q_1, \ldots, Q_N$ of $\mathbf{S}^k$ such that $\sigma_k^*(Q_n) = \frac{1}{N}$ $(1 \le n \le N)$ and $\max_{1\le n\le N} \operatorname{diam}(Q_n) \ll N^{-\frac{1}{k}}$. Then we have to take independent uniformly distributed random variables Z_n on Q_n. Now (2.65) follows along the same lines as in the proof of (2.27). □

Finally we will show that the system of spherical slices $\mathcal{D}(S, \mathbf{S}^k)$ is indeed a discrepancy system. By Proposition 2.6 it only follows that any u.d. sequence $(x_n)_{n\ge1}$ on $\mathbf{S}^k$ satisfies

$$\lim_{N\to\infty} D_N^{\mathcal{D}(Sl,\mathbf{S}^k)}(x_n) = 0. \tag{2.70}$$

In order to prove the converse statement it suffices to show that (2.70) implies

$$\lim_{N\to\infty} D_N^{\mathcal{D}(C,\mathbf{S}^k)}(x_n) = 0 \tag{2.71}$$

since $\mathcal{D}(C, \mathbf{S}^k)$ is a discrepancy system.

From (2.70) and

$$D_N(S,2) \gg D_N(C,2)$$

(cf. the proof of the lower bound for slices) it follows that

$$\lim_{N\to\infty} D_N(C,2) = 0. \tag{2.72}$$

Now assume that for some subsequence (N_i)

$$\lim_{i\to\infty} \frac{1}{N_i}\sum_{n=1}^{N_i} \chi_{C(\mathbf{y}_i,r_i)}(\mathbf{x}_n) - \sigma_k^*(C(\mathbf{y}_i,r_i)) = 2\alpha > 0.$$

By compactness of $\mathbf{S}^k \times [-1,1]$ we may assume $\mathbf{y}_i \to \mathbf{y}_0$ and $r_i \to r_0$. For sufficiently small ε we have $C(\mathbf{y}, r_0 - 2\varepsilon) \supset C(\mathbf{y}_i, r_i)$ and

$$\sigma_n^*(C(x, r_0 - 3\varepsilon) \setminus C(x_i, r_i)) < \alpha$$

for $i \ge i_0$ and $d(x, x_0) < \varepsilon$. It follows that

$$\frac{1}{N_i}\sum_{n=1}^{N_1} \chi_{C(x,r_0-s)}(\mathbf{x}_n) - \sigma_k^*(C(\mathbf{y}, r_0 - s)) > \alpha$$

for $s \in [2\varepsilon, 3\varepsilon]$. Therefore,

$$D_N(C,2) \ge \alpha\sqrt{\varepsilon\sigma_k^*(\{x : d(x,x_0) < \varepsilon\})}$$

contradicts (2.72). For $\alpha < 0$ the proof is similar. Thus the system of slices indeed is a discrepancy system.

2.1.4 ALEXANDER's Method

The previous section is devoted to spherical discrepancy systems. The most prominent system in this context is that of spherical caps $\mathcal{D}(C, \mathbf{S}^k)$. In the proof for the lower bound $N^{-\frac{1}{2}-\frac{1}{2k}}$ we have used the fact that $\mathcal{D}(C, \mathbf{S}^k)$ can be obtained by intersecting balls $B(\mathbf{x}, r) \in \mathbf{R}^{k+1}$ with $\mathbf{S}^k$. On the other hand we get the same system $\mathcal{D}(C, \mathbf{S}^k)$ by intersecting half spaces H with $\mathbf{S}^k$. ALEXANDER's method [12, 13] to be described below works exactly with discrepancy systems of this kind.

Theorem 2.24 *Suppose that M is a convex k-dimensional hypersurface embedded in $\mathbf{R}^{k+1}$. Let $X \subseteq M$ be a measurable subset with unit surface measure and $\mathbf{x}_1, \ldots, \mathbf{x}_N \in X$ a finite sequence in X. Then we have*

$$D_N^{\mathcal{D}(H,X)}(\mathbf{x}_n) \gg (\operatorname{diam}(X))^{-\frac{1}{2}} N^{-\frac{1}{2}-\frac{1}{2k}}, \tag{2.73}$$

where $\mathcal{D}(H, X)$ is the system of all halfspaces $H \subseteq \mathbf{R}^{k+1}$ intersected with X and the constant implied by $\gg$ only depends on the dimension k.

On the other hand, if X is a convex set on M, then there exists a finte sequence $\mathbf{x}_1, \ldots, \mathbf{x}_N \in X$ for any fixed $N > 1$ such that

$$D_N^{\mathcal{D}(H,X)}(\mathbf{x}_n) \ll N^{-\frac{1}{2}-\frac{1}{2k}} (\log N)^{\frac{1}{2}}. \tag{2.74}$$

It should be noted that $\mathcal{D}(H, X)$ is indeed a discrepancy system. You only have to apply Propositions 2.5 and 2.6.

Theorem 2.24 covers not only the case of sperical caps, where it gives the same lower bound as by BECK's method, but also ROTH's disc segment problem, where X is a dics of unit area in the plane ($M = \mathbf{R}^2 \subseteq \mathbf{R}^3$). BECK [120] has given a lower bound of the kind $N^{-\frac{3}{4}}(\log N)^{-\frac{7}{2}}$. R. ALEXANDER's method sharpens this bound to $N^{-\frac{3}{4}}$.

It should be also mentioned that ALEXANDER's theorem [13] is a little bit more general. It just requires that $\mathbf{x}_1, \ldots, \mathbf{x}_N \in M$. But in the present context we are only interested in the case $\mathbf{x}_1, \ldots, \mathbf{x}_N \in X$.

In the sequel we will use the following notations. Let $\nu = \nu^+ - \nu^-$ be a signed BOREL measure on X and $I(\nu)$ the functional

$$I(\nu) = \int_X \int_X |\mathbf{x} - \mathbf{y}| \, d\nu(\mathbf{x}) \, d\nu(\mathbf{y}). \tag{2.75}$$

Of course, we are mainly interested in the signed measure

$$\nu = \nu^+ - \nu^- = \frac{1}{N} \sum_{n=1}^{N} \delta_{\mathbf{x}_n} - \sigma_X, \tag{2.76}$$

where σ_X denotes the restriction of the surface measure of M to X.

The relation between $I(\nu)$ and the discrepancy is stated in

Proposition 2.25 *Let $\nu = \nu^+ - \nu^-$ be a bounded signed measure on X (i.e. $|\mu| = \nu^+ + \nu^-$ is bounded) satisfying $\nu(X) = 0$. Then we have*

$$0 \le -I(\nu) \ll (\mathrm{diam}(X)) \left(\sup_{H \subseteq \mathbf{R}^{k+1}} |\nu(H \cap X)| \right), \tag{2.77}$$

where $H \subseteq \mathbf{R}^{k+1}$ denotes a halfspace and the constant implied by $\ll$ only depends on the dimension k.

Proof. First, let us introduce a BOREL measure on the the hyperplanes $h \subseteq \mathbf{R}^{k+1}$. (It will be clear from the description below which systems of hyperplanes are BOREL sets.) For $\mathbf{u} \in \mathbf{R}^{k+1}$ with $|\mathbf{u}| = 1$ let $l(\mathbf{u}) = \{\rho\mathbf{u} : \rho \in \mathbf{R}\}$ be the line containing $\mathbf{u}$ and let $\lambda_{1,\mathbf{u}}$ be the LEBESGUE measure on $l(\mathbf{u})$. Furthermore, let σ_k denote the surface measure on $\mathbf{S}_k$.

Now, let $T \in \mathcal{H}$ be a ("measurable") system of hyperplanes $h \in \mathbf{R}^{k+1}$ and $T(\mathbf{u}) = \{h \in T : h \perp \mathbf{u}\}$. Then the measure $\mu'(T)$ of T is defined by

$$\mu'(T) = \int_{\mathbf{S}^k} \lambda_{1,\mathbf{u}}(T(\mathbf{u}) \cap l(\mathbf{u}))\, d\sigma_k(\mathbf{u}).$$

This measure has a remarkable relation to the euclidean distance. Set

$$\begin{aligned} d(\mathbf{x},\mathbf{y}) &= \mu'\left(\{h : h \cap \overline{\mathbf{x},\mathbf{y}} \ne \emptyset\}\right) \\ &= \int_{\mathbf{S}^k} |(\mathbf{x}-\mathbf{y}) \cdot \mathbf{u}|\, d\sigma_k(\mathbf{u}), \end{aligned}$$

where $\overline{\mathbf{x},\mathbf{y}}$ denotes the open segment connection but not containing $\mathbf{x},\mathbf{y}$. Then we have $d(\rho\mathbf{x},\mathbf{0}) = |\rho| d(\mathbf{x},\mathbf{0})$ ($\rho \in \mathbf{R}$) and $d(\xi\mathbf{x},\mathbf{0}) = d(\mathbf{x},\mathbf{0})$ ($\xi \in SO(k+1)$). Hence $d(\mathbf{x},\mathbf{y}) = C|\mathbf{x}-\mathbf{y}|$ for some constant $C > 0$. Thus, for $\mu = \frac{2}{C}\mu'$ we have

$$|\mathbf{x}-\mathbf{y}| = \frac{1}{2}\mu\left(\{h : h \cap \overline{\mathbf{x},\mathbf{y}} \ne \emptyset\}\right).$$

Set $A(h^+) = \nu(h^+)$ and $B(h^-) = \nu(h^-)$, where h^+ and h^- represent the open halfspaces $\subseteq \mathbf{R}^{k+1}$ determined by h. Then we obtain

$$\begin{aligned} I(\nu) &= \frac{1}{2}\int_X \int_X \int_{\mathcal{H}} \chi(h \cap \overline{\mathbf{x},\mathbf{y}})\, d\nu(\mathbf{x})\, d\nu(\mathbf{y})\, d\mu(h) \\ &= \frac{1}{2}\int_{\mathcal{H}} \int_X \int_X \chi(h \cap \overline{\mathbf{x},\mathbf{y}})\, d\mu(h)\, d\nu(\mathbf{x})\, d\nu(\mathbf{y}) \\ &= \frac{1}{2}\int_{\mathcal{H}} \int A(h) \times B(h) \cup B(h) \times A(h) d(\nu \times \nu)\, d\mu(h) \\ &= \int_{\mathcal{H}} A(h)B(h)\, d\mu(h), \end{aligned}$$

where $\chi(h \cap \overline{\mathbf{x},\mathbf{y}}) = 1$ if $h \cap \overline{\mathbf{x},\mathbf{y}} \ne \emptyset$ and $\chi(h \cap \overline{\mathbf{x},\mathbf{y}}) = 0$ otherwise. Since $A(h) + B(h) + \nu(h) = 0$ and $\nu(h) = 0$ μ-a.e., we have $A(h) = -B(h)$ μ-a.e. and hence

$$I(\nu) = -\int_{\mathcal{H}} A(h)^2\, d\mu(h) \le 0.$$

Furthermore, since

$$\int_{\mathcal{H}} \chi(X \cap h)\, d\mu(h) \ll \operatorname{diam}(X),$$

where $\chi(X \cap h) = 1$ if $X \cap h \neq \emptyset$ and $\chi(X \cap h) = 0$ otherwise, the assertion (2.77) follows immediately. □

Now Theorem 2.24 follows from

Proposition 2.26 *Let ν be given by (2.76). Then*

$$-I(\nu) \gg N^{-1-\frac{1}{k}}, \tag{2.78}$$

where the constant implied by $\gg$ only depends on the dimension k.

For the proof of Proposition 2.26 we need a series of Lemmata.

Lemma 2.27 *Given $n+2$ distinct points $r_0, r_1, \ldots, r_{n+1} \in \mathbf{R}$ there is a non-zero signed measure ϕ with $\phi(\mathbf{R}) = 0$ that is supported by these points and has its first n moments vanishing. Furthermore*

$$I^{2k}(\phi) = \int_{\mathbf{R}} \int_{\mathbf{R}} |x-y|^{2k}\, d\phi(x)\, d\phi(y) = 0$$

for $1 \leq k \leq n$.

Proof. W.l.o.g. we can assume that $r_0 = 0$ and $\phi(r_0) = 1$. This leads to the system of $n+1$ linear equations

$$\sum_{i=1}^{n+1} r_i^m \eta_i = -1 \qquad (0 \leq m \leq n),$$

where $\eta_i = \phi(r_i)$. Since the associated VANDERMONDE matrix is non-singular there is a unique solution.

Furthermore, by the binomial theorem and FUBINI's theorem

$$I^{2k}(\phi) = \sum_{j=0}^{2k} \binom{2k}{j} (-1)^j \int_{\mathbf{R}} x^{2k-j}\, d\phi(x) \int_{\mathbf{R}} y^j\, d\phi(y) = 0$$

for $1 \leq k \leq n$. □

Lemma 2.28 *Let ϕ be a signed discrete measure with finite support and ψ a signed bounded measure with compact support and $\psi(\mathbf{R}^k) = 0$. Then*

$$-I(\phi * \psi) \leq -|\phi|^2 I(\psi).$$

Proof. Set

$$J(\nu_1,\nu_2) = \int_{\mathbf{R}^k} |x-y|\, d\nu_1(x)\, d\nu_2(y). \tag{2.79}$$

Suppose that ν_1, ν_2 are signed measures with $\nu_1(\mathbf{R}^k) = \nu_2(\mathbf{R}^k) = 0$. Then by Proposition 2.25 the quadratic form

$$Q(\alpha,\beta) = -I(\alpha\nu_1+\beta\nu_2) = -I(\nu_1)\alpha^2 - 2J(\nu_1,\nu_2)\alpha\beta - I(\nu_2)\beta^2$$

is positive semidefinite. Hence the discriminant is non negative, resp.

$$J(\nu_1,\nu_2)^2 \le I(\nu_1)I(\nu_2). \tag{2.80}$$

Furthermore, by (2.80) we get a triangle inequality

$$(-I(\nu_1,\nu_2))^{\frac{1}{2}} \le (-I(\nu_1))^{\frac{1}{2}} + (-I(\nu_2))^{\frac{1}{2}}. \tag{2.81}$$

Now suppose that ϕ is concentrated on the points r_i. Then

$$\phi * \psi = \sum \phi(r_i)\psi_{-r_i},$$

where ψ_x denotes the measure defined by $\psi_x(M) = \psi(M+x)$. Since $I(\psi_x) = I(\psi)$, we get by (2.81)

$$\begin{aligned} -I(\phi*\psi) &= -I\left(\sum \phi(r_i)\psi_{-r_i}\right) \\ &\le \left(\sum (-I(\phi(r_i)\psi_{-r_i}))^{\frac{1}{2}}\right)^2 \\ &= \left(\sum |\psi(r_i)|(-I(\psi))^{\frac{1}{2}}\right)^2 \\ &= -|\phi|^2 I(\psi). \end{aligned}$$

□

Lemma 2.29 *Let ϕ be a signed measure with finite support contained on the x_{k+2}-axis of $\mathbf{R}^{k+2}$ such that $\phi(\mathbf{R}^{k+2}) = 0$. For any $a \in \mathbf{R}^{k+2}$ orthogonal to the x_{k+2}-axis (i.e. a can be considered as an element of $\mathbf{R}^{k+1} \subseteq \mathbf{R}^{k+2}$) let ϕ_a be the shifted measure defined by $\phi_a(M) = \phi(M-a)$ Then the functional $J(\phi,\phi_a)$ (see (2.80)) depends on $y = |a|$. Furthermore $-J(\phi,\phi_a)$ is a strictly decreasing positive function of $y = |a|$.*

Proof. Let ϕ be concentrated on the points r_i and set $\phi(r_i) = \eta_i$. Since

$$J(\phi,\phi_a) = \sum_{i,j} \left(|r_i - r_j|^2 + |a|^2\right)^{\frac{1}{2}} \eta_i\eta_j,$$

the functional $J(\phi,\phi_a)$ only depends on $y = |a|$. Now define $c > 0$ by

$$c^{-1} = \int_0^\infty \left(1 - e^{-s^2}\right) s^{-2}\, ds.$$

Hence

$$z = c\int_0^\infty \left(1 - e^{-z^2s^2}\right) s^{-2}\, ds$$

and $\sum \eta_i = 0$ imply

$$\begin{aligned} J(\phi,\phi_a) &= \sum_{i,j} c\int_0^\infty \left(1 - e^{-(|r_i-r_j|+y^2)^2s^2}\right) s^{-2}\, ds \\ &= -c\int_0^\infty e^{-(ys)^2}\left(\sum_{i,j} e^{-|r_i-r_j|^2s^2}\eta_i\eta_j\right) s^{-2}\, ds. \end{aligned}$$

Since

$$\begin{aligned} \sum_{i,j} e^{-|r_i-r_j|^2s^2}\eta_i\eta_j &= \sum_{i,j}\eta_i\eta_j\frac{1}{2\pi}\int_{-\infty}^{\infty} e^{-t^2/2}e^{i\sqrt{2}(r_i-r_j)st}\, dt \\ &= \frac{1}{2\pi}\int_{-\infty}^{\infty} e^{-t^2/2}\left|\sum_i e^{i\sqrt{2}r_ist}\right| dt > 0, \end{aligned}$$

we immediately get $-J(\phi,\phi_a) > 0$. Furthermore, since $e^{-(ys)^2}$ is a strictly decreasing function of y, the functional $-J(\phi,\phi_a)$ is stricly decreasing, too. □

Lemma 2.30 *Let ϕ be a signed measure with finite support contained in $\left[-\frac{h}{2},\frac{h}{2}\right]$ ($0 < h \le \frac{1}{2}$) which is identified by the corresponding interval of the x_{k+2}-axis of $\mathbf{R}^{k+2}$ such that $\phi(\mathbf{R}^{k+2}) = 0$, $|\phi| = 1$, and assume that the first n moments vanish. Then*

$$-J(\phi,\phi_a) < h^{2n}|a|^{-2n-1} \qquad \textit{for } |a| \ge 2.$$

Proof. Let $c_l = \binom{1/2}{l}$ denote the coefficients of the binomial series

$$(1+x^2)^{\frac{1}{2}} = \sum_{l=0}^{\infty} c_l x^{2l}.$$

Obviously $c_0 = 1$ and $|c_l| < 1$ for $l \ge 1$. Now suppose that ϕ is supported on r_i and set $\phi(r_i) = \eta_i$. Then we get

$$\begin{aligned} -J(\phi,\phi_a) &= -\sum_{i,j}\left(|r_i-r_j|^2 + |a|^2\right)^{\frac{1}{2}}\eta_i\eta_j \\ &= -|a|\sum_{i,j}\left(1 + \sum_{l=0}^{\infty} c_l|r_i-r_j|^{2l}|a|^{-2l}\right)\eta_i\eta_j \\ &< |a|\sum_{l>n} h^{2l}|a|^{-2l} \\ &= |a|\frac{(h^2|a|^{-2})^{n+1}}{1-h^2|a|^{-2}} \\ &< h^{2n}|a|^{-2n-1}, \end{aligned}$$

since $h^2|a|^{-2} \le \frac{1}{10}$, $1 < (1-h^2|a|^{-2})^{-1} \le \frac{16}{15}$, and $h^2 \le \frac{1}{4}$. □

Lemma 2.31 *Let ψ_1, ψ_2 be bounded signed measures on $\mathbf{R}^{k+1}$ and ϕ a signed measure on $\mathbf{R}$ with finite support. Then*

$$J(\psi_1 \times \phi, \psi_2 \times \phi) = \int_{\mathbf{R}^{k+1}} \int_{\mathbf{R}^{k+1}} J(\phi_p, \phi_q)\, d\psi_1(p)\, d\psi_2(q).$$

Proof. Suppose that ϕ is concentrated on the points r_i and set $\phi(r_i) = \eta_i$. Then we get by FUBINI's theorem and by the relation $J(\phi_p, \phi_q) = J(\phi, \phi_{p-q})$

$$\begin{aligned} J(\psi_1 \times \phi, \psi_2 \times \phi) &= \int_{\mathbf{R}^{k+1}} \int_{\mathbf{R}^{k+1}} \int_{\mathbf{R}} \int_{\mathbf{R}} \left(|r-s|^2 + |p-q|^2\right)^{\frac{1}{2}} d\phi(r)\, d\phi(s)\, d\psi(p)\, d\psi(q) \\ &= \int_{\mathbf{R}^{k+1}} \int_{\mathbf{R}^{k+1}} \sum_{i,j} \left(|r_i - r_j|^2 + |p-q|^2\right)^{\frac{1}{2}} \eta_i \eta_j\, d\psi(p)\, d\psi(q) \\ &= \int_{\mathbf{R}^{k+1}} \int_{\mathbf{R}^{k+1}} J(\phi_p, \phi_q)\, d\psi(p)\, d\psi(q). \end{aligned}$$

□

Proof of Proposition 2.26 Let $K > 0$ be a parameter to be specified in the sequel. Instead of X and the finite sequence $\mathbf{x}_1, \ldots, \mathbf{x}_N \in X$ we will now consider the similar set $Y = K^{\frac{1}{k}} X$ and the sequence $\mathbf{y}_n = K^{\frac{1}{k}} \mathbf{x}_n$. Notice that $\sigma(Y) = K$. Furthermore, if we define $\overline{\nu}$ by

$$\overline{\nu} = \overline{\nu}^+ - \overline{\nu}^- = \frac{K}{N} \sum_{n=1}^{N} \delta_{\mathbf{y}_n} - \sigma_Y,$$

where σ_Y denotes the surface measure restricted on Y, then we get

$$I(\overline{\nu}) = K^{2+\frac{1}{k}} I(\nu).$$

In the following we will estimate $-I(\overline{\nu})$. Let ϕ be a signed measure with finite support contained in $\left[-\frac{1}{4}, \frac{1}{4}\right]$ (which will be identified with the corresponding interval on the x_{k+2}-axis of $\mathbf{R}^{k+2}$) such that $\phi(\mathbf{R})0$, $|\phi| = 1$, and assume that the first n moments vanish. By Lemma 2.27 such a measure always exists. Hence by Lemma 2.28

$$\begin{aligned} -I(\overline{\nu}) &\geq -I(\overline{\nu} * \phi) \\ &= -I(\overline{\nu} \times \phi) \\ &= -I(\overline{\nu}^+ \times \phi) + 2J(\overline{\nu}^+ \times \phi, \overline{\nu}^- \times \phi) - I(\overline{\nu}^- \times \phi). \end{aligned}$$

Since $(\overline{\nu}^+ \times \phi)(\mathbf{R}^{k+2}) = 0$ we have $-I(\overline{\nu}^+ \times \phi) \geq 0$.

Next consider $-I(\overline{\nu}^- \times \phi)$. Since $\overline{\nu}^+(\mathbf{y}_n) = \overline{\nu}(\mathbf{y}_n) > 0$ we get by Lemmata 2.31 and 2.29

$$-I(\overline{\nu}^- \times \phi) = -\sum_{n,m=1}^{N} J(\phi_{\mathbf{y}_n}, \phi_{\mathbf{y}_m}) \overline{\nu}(\mathbf{y}_n) \overline{\nu}(\mathbf{y}_m)$$

$$\begin{aligned}
&= -I(\phi)\sum_{n=1}^{N}\overline{\nu}(\mathbf{y}_n) - 2\sum_{1\le n<m\le N} J(\phi,\phi_{\mathbf{y}_n-\mathbf{y}_m})\overline{\nu}(\mathbf{y}_n)\overline{\nu}(\mathbf{y}_m)\\
&\ge -I(\phi)\sum_{n=1}^{N}\overline{\nu}(\mathbf{y}_n)\\
&= c_1\frac{K^2}{N}. \qquad (2.82)
\end{aligned}$$

Finally we have to discuss $J(\overline{\nu}^+\times\phi,\overline{\nu}^-\times\phi)$. Applying Lemma 2.31 gives

$$\begin{aligned}
J(\overline{\nu}^+\times\phi,\overline{\nu}^-\times\phi) &= \int_{\mathbf{R}^{k+1}}\int_{\mathbf{R}^{k+1}} J(\phi,\phi_{p-q})\,d\overline{\nu}^+(p)\,d\overline{\nu}^-(q)\\
&= \sum_{n=1}^{N}\overline{\nu}^+(\mathbf{y}_n)\int_{\mathbf{R}^{k+1}} J(\phi,\phi_{p-q})\,d\overline{\nu}^-(q).
\end{aligned}$$

By Lemma 2.29

$$-\int_{|\mathbf{y}_n-q|\le 2} J(\phi,\phi_{p-q})\,d\overline{\nu}^-(q) \le -I(\phi)\int_{|\mathbf{y}_n-q|\le 2} d\overline{\nu}^-(q) \le -I(\phi)2^k\sigma(\mathbf{S}^k).$$

Furthermore,

$$\begin{aligned}
-\int_{|\mathbf{y}_n-q|>2} J(\phi,\phi_{p-q})\,d\overline{\nu}^-(q) &= -\sum_{l=1}^{\infty}\int_{2^l<|\mathbf{y}_n-q|\le 2^{l+1}} J(\phi,\phi_{p-q})\,d\overline{\nu}^-(q)\\
&\quad -\sum_{l=1}^{\infty} J_{2^l}\int_{2^l<|\mathbf{y}_n-q|\le 2^{l+1}} d\overline{\nu}^-(q)\\
&\le \sigma(\mathbf{S}^k)\sum_{l=1}^{\infty} J_{2^l}2^{k(l+1)},
\end{aligned}$$

where J_{2^l} denotes the value of $J(\phi,\phi_{\mathbf{y}_n-q})$ with $|\mathbf{y}_n-q| = 2^l$. Since ϕ is supported on $\left[-\frac{1}{2},\frac{1}{2}\right]$ and the first n moments vanish, Lemma 2.30 gives

$$-J_{2^l} \le 2^{-2(k+1)}2^{-l(k+1)}2^{-l} < 2^{-(l+1)(k+1)}.$$

Hence

$$-\sum_{l=1}^{\infty} J_{2^l}2^{k(l+1)} < \sum_{l=1}^{\infty} 2^{-(l+1)} = \frac{1}{2}$$

and consequently

$$\begin{aligned}
2\,J(\overline{\nu}^+\times\phi,\overline{\nu}^-\times\phi) &> -2\sigma(\mathbf{S}^k)\left(-I(\phi)2^k+\frac{1}{2}\right)\sum_{n=1}^{N}\overline{\nu}^+(\mathbf{y}_n)\\
&= -c_2K. \qquad (2.83)
\end{aligned}$$

Combining (2.82), (2.83) and choosing $K = 2\frac{c_2}{c_1}N$ immediately gives

$$-I(\overline{\nu}) > c_1\frac{K^2}{N} - c_2K = 2c_1c_2^2N.$$

Since $I(\nu) = K^{-2-\frac{1}{k}}I(\overline{\nu})$ we have finally proved

$$-I(\nu) \gg N^{-1-\frac{1}{k}},$$

where the constant implied by $\gg$ only depends on the dimension k. □

Notes

Schmidt [1625, 1626, 1627, 1628] was the first who considered irregularities of more general notions of discrepancies, e.g. he considered rectangles in arbitrary position, balls and sherical caps. By using his integral equation method he showed that a lower bound for the corresponding discrepancies is given by $N^{-1/2-1/(2k)-\varepsilon}$ (with an arbitrary $\varepsilon > 0$). This means that even if the usual discrepancy $D_N(\mathbf{x}_n)$ with respect to rectangles with sides parallel to the axes is rather small, e.g. $D_N(\mathbf{x}_n) = \mathcal{O}((\log N)^{k-1}/N)$, then there exists a rotated rectangle with much larger discrepancy.

Zaremba [1986] and Schmidt [1632] provided lower bounds for the isotropic discrepancy, e.g. Schmidt proved that $J_N(\mathbf{x}_n) \ge c_k N^{-2/(k+1)}$. Although Schmidt's proof is rather short and simple his bound is optimal (despite of logarithmic factor). Stute [1759] showed that in the case $k = 3$ almost all sequences satisfy $J_N(\mathbf{x}_n) = \mathcal{O}(N^{-1/2}(\log N)^{3/2})$ and for $k \ge 4$ almost all sequences satisfy $J_N(\mathbf{x}_n) = \mathcal{O}(N^{-2/(k+1)}(\log N)^{2/(k+1)})$. Finally Beck [134] settled the two dimensional case $k = 2$. There exists a sequence with $J_N(\mathbf{x}_n) = \mathcal{O}(N^{-2/3}(\log N)^4)$. A discrepancy inequaltiy between the isotropic and the discrepancy with respect to balls is due to Smyth [1699]. Steele [1736] estimated the length of the shortest path through a point sequence in terms of the discrepancy.

In 1983 Beck [120] introduced a new concept in the theory of irregularities of distribution, a Fourier transform approach. He solved a problem of Roth concerning the irregularities of point distributions on a disc with respect to segments, i.e. intersections of halfplanes with the disc. With help of this Fourier transform approach Beck reproved and widely generalized various theorems in the theory of irregularities of distribution, e.g. Roth's result Theorem 1.40 (see Theorem 1.41, [143]) and theorems that Schmidt [1625, 1626, 1627, 1628] was able to prove with his integral equation method (see [122, 128] and [143]). He also obtained upper bounds in order to show that his lower bounds are (almost) best possible [124, 143]. The most general theorem is the following one [128, 143].

Theorem 2.32 *Let $A \subseteq \mathbf{R}^k$ $(k \ge 2)$ a convex body such that there exists a ball with radius 1 which is contained in A and set $\mathcal{D}_s(A, \mathbf{R}^k) = \{\lambda(\xi A) + \mathbf{y} \,|\, 0 \le \lambda \le 1, \xi \in SO(k), \mathbf{y} \in \mathbf{R}^k\}$. Then for every sequence $(\mathbf{x}_n)_{n\ge1}$, $\mathbf{x}_n \in \mathbf{R}^k$, we have*

$$D^{\mathcal{D}_s(A,\mathbf{R}^k)}(\mathbf{x}_n) = \sup_{A' \in \mathcal{D}_s(A,\mathbf{R}^k)} \left| \sum_{n\ge1} \chi_{A'}(\mathbf{x}_n) - \lambda_k(A') \right| \ge c_k(\sigma(\partial A))^{1/2}.$$

Conversely there exists a sequence $(\mathbf{x}_n)_{n\ge1}$ with

$$D^{\mathcal{D}_s(A,\mathbf{R}^k)}(\mathbf{x}_n) \le c_k'(\sigma(\partial A))^{1/2}(\log \sigma(\partial A))^{1/2}.$$

($c_k > 0$ and $c_k > 0$ are constants just depending on the dimension k.)

Obviously, Theorem 2.10 and the first part of Theorem 2.13 are related to Theorem 2.32. In fact the proof of Theorem 2.10 essentially relies on the methods presented in Beck and Chen [143] and the first part of Theorem 2.13 is a corollary of Theorem 2.32. All these lower bounds are optimal despite of a logarithmic factor $(\log)^{1/2}$. It is impossible to sharpen the lower bounds by using quadratic means since the there are sequences with L^2-discrepancy of the same order of magnitude as the lower bound (see Beck and Chen [143]). Similar upper and lower bounds hold for the corresponding L^p-discrepancies for $p > 0$ resp. for $p \geq 2i$. The upper bounds are obtained by using probabilistic methods; see Beck and Chen [145, 147] and Chen [347]. The situation may change drastically if the L^1-discrepancy is considered. In [149, 148] Beck and Chen provided two examples in which the L^1-discrepancy is of order $(\log N)^2/N$ whereas the lower bound $N^{-3/4}$ holds for the L^2-discrepancy.

With help of his Fourier transform method Beck could also solve a problem of Erdős [129] (a related problem is discussed in [133]) and a lattice point problem of Moser [131, 132]. Beck and Chen [145] showed that there is no Roth phenomenon for the discrepancy with respect to balls. In [136] Beck provided lower bounds for the one-sided discrepancy with respect to balls.

Beck [130] also introduced the discrepancy with respect to homothetic sets $\mathcal{D}_h(A, \mathbf{R}^k) = \{\lambda A + \mathbf{y} \mid 0 \leq \lambda \leq 1, \mathbf{y} \in \mathbf{R}^k\}$ in which $A \subseteq \mathbf{R}^k$ is a convex body. He observed (in the case $k = 2$) that general lower bounds for the discrepancy with respect to $\mathcal{D}_h(A, \mathbf{R}^2)$ essentially depend on a special notion of approximability of A by polygons. Especially if the boundary of A is twice continously cifferentiable then the lower bound is (almost) the same as that for the discrepancy with respect to $\mathcal{D}_s(A, \mathbf{R}^2)$. On the other hand, if A is a convex polygone then we obtain exactly the same lower bound as the classical one by Roth [1567], e.g. in the modulo 1 case $(\log N)^{1/2}/N$ is lower bound for the corresponding L^2-discrepancy. Beck and Chen [144] showed that this L^2-bound is optimal for convex polygons. Theorems 2.11 and 2.12 (see Drmota [485, 476]) show that there are similar phenomena in arbitrary dimension k. In the case of polygons (polytopes) Beck and Chen [146] and Karolyi [906] obtained lower bounds for the discrepancy with respect to polygons (polytopes) where each sides is parallel to a member of a fixed finite system of hyperplanes; see also Karolyi [907]. Baire category results in the theory of irregularities of distributions (for discrepancies with respect to convex sets) are due to Beck [138].

The first (non-trivial) lower bound ($N^{-1/2-1/(2k)-\epsilon}$) for the spherical cap discrepancy on $\mathbf{S}_k$ was obtained by Schmidt [1626] with help of his integral equation method. Beck [125] sharpened this bound to $N^{-1/2-1/(2k)}$ which is again (almost) optimal (see Theorem 2.22). Alexander [12, 13] (Theorem 2.24) gave a third proof of this lower bound in a much more general setting. His method especially works in the case of the "separation discrepancy", i.e. the discrepancy with respect to half spaces; see also Rogers [1560] The second part of Theorem 2.22 is due to Blümlinger [236] who generalized Beck's Fourier transform technique to the homogenous space $\mathbf{S}^k$. Similarly Drmota [483] adapted Beck's technique to the hyperbolic plane. It should be further mentioned that those sequences with (almost) optimal cap discrepancy are not effectively coumputable. The existence proof is mainly probabilistic. A group theoretic construction of sequences on the two dimensional sphere is due to Lubotzky, Phillips, and Sarnak [1110, 1111]; see the Notes of Section 3.2.

2.2 Summation Methods

2.2.1 Weighted Means

Usually, a sequence $(\mathbf{x}_n)_{n\geq 1}$, $\mathbf{x}_n \in \mathbf{R}^k$, is said to be u.d. if

$$\lim_{N\to\infty} \frac{1}{N} \sum_{n=1}^{N} f(\mathbf{x}_n) = \int_{[0,1)^k} f\, d\lambda_k$$

holds for a special system of functions, for instance for all continuous functions. But the left hand side is exactly the first CESARO limit of the sequence $f(\mathbf{x}_n)$. The CESARO mean is a special case of a matix limitation method.

Definition 2.33 *Let $A = (a_{Nn})_{N,n\geq 1}$ be an infinite matrix with real elements. Then a real sequence $(x_n)_{n\geq 1}$ is said to be A-limitable to x if*

$$\lim_{N\to\infty} \sum_{n=1}^{\infty} a_{Nn} x_n = x.$$

In this case we write

$$A\text{–}\lim x_n = x.$$

It is an important problem to decide whether a limitation method is regular. For matrix limitation methods this problem is solved by TOEPLITZ's theorem (see [1989]).

Theorem 2.34 *A matrix limitation method A is regular, i.e. $\lim x_n = x$ implies $A\text{–}\lim x_n = x$, if and only if the following three conditions hold:*

$$\text{(i)} \quad \lim_{N\to\infty} \textstyle\sum_{n=1}^{\infty} a_{Nn} = 1, \tag{2.84}$$

$$\text{(ii)} \quad \sup_{N\geq 1} \sum_{n=1}^{\infty} |a_{Nn}| < \infty, \tag{2.85}$$

$$\text{(iii)} \quad \lim_{n\to\infty} a_{Nn} = 0 \qquad \text{for all } N \geq 1. \tag{2.86}$$

In the following sections we will mainly concentrate on weighted means which are special cases of matrix limitation methods.

Definition 2.35 *Let $P = (p_n)_{n\geq 1}$ be a sequence of non-negative numbers and set*

$$P_N = \sum_{n=1}^{N} p_n.$$

Then the matrix limitation method $A = (a_{Nn})_{N,n\geq 1}$ defined by

$$a_{Nn} = \begin{cases} \frac{p_n}{P_N} & \text{for } n \leq N \\ 0 & \text{for } n > N \end{cases}$$

is called weighted mean M_P, i.e. a real sequence $(x_n)_{n\geq 1}$ is M_P-limitable to x if

$$\lim_{N\to\infty} \frac{1}{P_N} \sum_{n=1}^{N} p_n x_n = x.$$

By TOEPLITZ's theorem a weighted mean M_P is regular if and only if

$$\lim_{N\to\infty} P_N = \infty. \tag{2.87}$$

In the sequel we will always assume that (2.87) is satisfied.

As indicated above we can generalize the notion of uniform distribution by using a matrix limitation method.

Definition 2.36 *Let $A = (a_{Nn})_{N,n\geq 1}$ be a matrix limitation method with $a_{Nn} \geq 0$. Then a sequence $\mathbf{x}_n \in \mathbf{R}^k$ is said to be uniformly distributed modulo 1 with respect to A (for short A-u.d. mod 1) if*

$$A\text{–}\lim \chi_I(\{\mathbf{x}_n\}) = \lambda_k(I)$$

holds for every interval $I \subseteq \mathbf{R}^k/\mathbf{Z}^k$. Furthermore the A-discrepancy $A\text{–}D_N(\mathbf{x}_n)$ is defined by

$$A\text{–}D_N(\mathbf{x}_n) = \sup_{I\subseteq \mathbf{R}^k/\mathbf{Z}^k} \left| \sum_{n=1}^{\infty} a_{Nn}\chi_I(\{\mathbf{x}_n\}) - \lambda_k(I) \right|.$$

Similarly the star discrepancy $A\text{–}D_N^*(\mathbf{x}_n)$ is defined by taking the supremum over all rectangles of the form $I = [0, \mathbf{x})$, where $\mathbf{x} \in [0,1]^k$.

Remark. It should be noted that we can also consider A-u.d. sequences in a compact metric space X with respect to a positive normalized BOREL measure μ on X. We will call a sequence $(x_n)_{n\geq 1}$, $x_n \in X$ (A,μ)-u.d. if

$$A\text{–}\lim f(x_n) = \int_X f\, d\mu$$

holds for all continuous functions $f : X \to \mathbf{R}$.

As usual, the discrepancy characterizes A-u.d. sequences mod 1, i.e. a sequence $(\mathbf{x}_n)_{n\geq 1}$ is A-u.d. mod 1 if and only if

$$\lim_{N\to\infty} A\text{–}D_N(\mathbf{x}_n) = 0.$$

Furthermore we can use continuous functions or RIEMANN integrable functions instead of characteristic functions in the definition of A-u.d. sequences mod 1, especially we get

Theorem 2.37 (WEYL's criterion) *A sequence of points $(\mathbf{x}_n)_{n\geq 1}$ in the k-dimensional space $\mathbf{R}^k$ is A-u.d. mod 1 if and only if*

$$A\text{–}\lim e(\mathbf{h}\cdot \mathbf{x}_n) = 0$$

holds for all non-zero interal lattice points $\mathbf{h} \in \mathbf{Z}^k \setminus \{\mathbf{0}\}$.

Moreover there are analoga to the KOKSMA-HLAWKA and the ERDŐS-TURAN-KOKSMA inequality. The proof runs along the same lines as the usual ones.

Theorem 2.38 *Let $A = (a_{Nn})_{N,n\geq 1}$ be a matrix limitation method with $a_{Nn} \geq 0$ and let f be a function of bounded variation on $[0,1]^k$ in the sense of* HARDY *and* KRAUSE. *Let $(\mathbf{x}_n^{(i_1,\ldots,i_l)})_{n=1}^N = (\mathbf{x}_n^{(F)})_{n=1}^N$ denote the projection of the sequence $(\mathbf{x}_n)_{n=1}^N$, $\mathbf{x}_n \in [0,1]^k$, on the $(k-l)$-dimensional face F of $[0,1]^k$ defined by $x_{i_1} = \cdots = x_{i_l} = 1$. Then we have*

$$\left| \sum_{n=1}^{\infty} a_{Nn} f(\mathbf{x}_n) - \int_{[0,1]^k} f(\mathbf{x})\, d\mathbf{x} \right| \leq \sum_{l=0}^{k-1} \sum_{F_l} A\text{–}D_N^*(\mathbf{x}_n^{(F_l)}) V^{(k-l)}(f^{(F_l)}), \qquad (2.88)$$

where the second sum is extended over all $(k-l)$-dimensional faces F_l of the form $x_{i_1} = \cdots = x_{i_l} = 1$. The discrepancy $A\text{–}D_N^*(\mathbf{x}_n^{(F_l)})$ is computed in the face of $[0,1]^k$ in which $(\mathbf{x}_n^{(F_l)})_{n=1}^N$ is contained.

Theorem 2.39 *Let $A = (a_{Nn})_{N,n\geq 1}$ be a matrix limitation method with $a_{Nn} \geq 0$ and let $\mathbf{x}_1, \ldots, \mathbf{x}_N$ be points in the k-dimensional space $\mathbf{R}^k$ and H an arbitrary positive integer. Then*

$$A\text{–}D_N(\mathbf{x}_n) \leq \left(\frac{3}{2}\right)^k \left(\frac{2}{H+1} + \sum_{0<\|\mathbf{h}\|_\infty \leq H} \frac{1}{r(\mathbf{h})} \left|\sum_{n=1}^{\infty} a_{Nn} e(\mathbf{h}\cdot\mathbf{x}_n)\right|\right), \tag{2.89}$$

where $r(\mathbf{h}) = \prod_{i=1}^k \max\{1, |h_i|\}$ for $\mathbf{h} = (h_1, \ldots, h_k) \in \mathbf{Z}^k$.

As a non-trivial example we will consider the sequence

$$x_n = \alpha n + \beta \log n \tag{2.90}$$

for real α, β with $\beta \neq 0$. Such sequences need not be u.d. in the usual sence, e.g. the sequence $x_n = \log n$ is not uniformly distributed (mod 1). However, TSUJI [1877] proved that $(\log n)$ is M_P-u.d. for the logarithmic weighted mean $p_n = \frac{1}{n}$. We will say that it is logarithmically uniformly distributed (mod 1). HLAWKA [812] extended this result to the case $x_n = \alpha n + \beta \log n$, where $\beta \neq 0$. In contrast to the usual discrepancy D_N (with optimal order of magnitude $(\log N)/N$) the logarithmic discrepancy $D_N^{\log}(x_n) = M_P\text{–}D_N(x_n)$ is surely bounded below by $1/(2\log N)$ This follows from the observation that

$$\sum_{n=1}^N \frac{1}{n}\chi_{[0,y)}(x_n) - y\sum_{n=1}^N \frac{1}{n}$$

has a jump of at least 1 at $y = x_1$, unless $x_1 = 0$ in which case we have

$$\lim_{y\to 0}\left(\sum_{n=1}^N \frac{1}{n}\chi_{[0,y)}(x_n) - y\sum_{n=1}^N \frac{1}{n}\right) = \sum_{x_n=0} \frac{1}{n} \geq 1.$$

It is remarkable that this (trivial) bound is best possible. TICHY and TURNWALD [1847] showed that

$$D_N^{\log}(x_n) < C(\alpha,\beta)\frac{1}{\log N}$$

for the sequence (2.90) whenever α is not a LIOUVILLE number and if $\beta > 0$; the constant $C(\alpha, \beta)$ depends only on α adn β. A few years later BAKER and HARMAN [94]) proved a stronger theorem.

Theorem 2.40 *Let α, β be real numbers with $\beta \neq 0$. Then the logarithmic discrepancy of the sequence (2.90) is bounded by*

$$D_N^{\log}(x_n) < C(\beta)\frac{1}{\log N},$$

where the constant $C(\beta)$ depends only on β.

Theorem 2.40 is a special case ($c = K = |\beta|, \delta = H = 1$) of the following more general one.

A real valued function f defined on $[1, \infty)$ is said to be of class H if there are real numbers $1 = x_0 < x_1 < \ldots < x_H$ such that f is monotone in each of the intervals $[x_{j-1}, x_j]$, $1 \leq j \leq H$, and $[x_H, \infty)$.

Theorem 2.41 *Let f be a real valued twice differentiable function on $[1, \infty)$. Suppose that there are positive constants c, K, δ and H with the following properties:*

(i) $x(f'(x) - \lambda)$ is of class H for every real λ;

(ii) f' is bounded on bounded intervals;

(iii) For $x \geq 1$

$$cx^{-2} \leq f''(x) \leq Kx^{-1-\delta} \quad or \quad cx^{-2} \leq -f''(x) \leq Kx^{-1-\delta}.$$

Then the logarithmic discrepancy of the sequence $x_n = f(n)$ satisfies

$$D_N^{\log}(x_n) < C(c, K, \delta, H)\frac{1}{\log N}.$$

We remark that $f(x) = \alpha x + \beta \log x$ is extremal in the sense that $f''(x) = -\beta x^{-2}$. (Of course, $x(f'(x) - \lambda)$ is linear.)

The weighted ERDŐS-TURÁN-KOKSMA inequality (see Theorem 2.39) seems not to be sufficient to prove Theorem 2.40. Therefore we have to provide an improved estimate (see BAKER and HARMAN [94]).

Lemma 2.42 *Let $0 < \sigma \leq 1$. Then the logarithmic discrepancy of a real sequence (x_n) satisfies*

$$\log N \cdot D_N^{\log}(x_n) < C_1(\sigma) + C_2(\sigma) \sum_{1 \leq h \leq N^\sigma} \frac{1}{h} \max_{A \geq h^{1/\sigma}} \left| \sum_{n=A}^{N} \frac{1}{n} e(hx_n) \right|.$$

In order to apply Lemma 2.42 we must estimate exponential sums.

Lemma 2.43 *Suppose that f satisfies the hypotheses of Theorem 2.41. Let h and ν be integers with $h \geq 1$ and $B > A \geq 1$ real numbers. Then*

$$\left| \int_A^B \frac{e(hf(x) - \nu x)}{x} dx \right| < C(c, H)h^{-\frac{1}{2}}.$$

Proof. Since

$$\int_A^B x^{-1}e(-hf(x)+\nu x)dx = \overline{\int_A^B x^{-1}e(hf(x)-\nu x)dx},$$

we can replace f by $-f$ (if necessary) and may assume henceforth that $f''(x) \geq cx^{-2}$.

Let $g(x) = f(x) - \nu x/h$. Then $[A,B]$ may be partitioned into at most $H+1$ intervals in each of which $xg'(x)$ is monotone. It therefore suffices to show that

$$\left|\int_A^B x^{-1}e(hg(x))dx\right| < C(c)h^{-\frac{1}{2}} \tag{2.91}$$

under the hypotheses that

$$g''(x) \geq cx^{-2}, \tag{2.92}$$

$$xg'(x) \text{ is monotone in } [A,B]. \tag{2.93}$$

Since g' is strictly increasing it has at most one zero in $[1,\infty)$.
Suppose first that g' has a zero in $[A,B]$, at a, say. Let

$$a' = \max\left(A, \frac{a}{2}\right), \quad a'' = \min(2a, B),$$

$$I_1 = [A, a'), \quad I_2 = [a', a''], \quad I_3 = (a'', B].$$

Suppose that I_1 is non-empty. Then $(a' = \frac{1}{2}a)$ and

$$\int_{I_1} \frac{e(hg(x))\,dx}{x} = \int_A^{a'} \frac{d(e(hg(x)))}{2\pi i h g'(x)x} = \left[\frac{e(hg(x))}{2\pi i h g'(x)x}\right]_A^{a'} - \int_A^{a'} \frac{e(hg(x))}{2\pi i h} d\left(\frac{1}{xg'(x)}\right)$$

with

$$g'(x) = -\int_x^a g''(t)dt \leq -\int_x^{2x} \frac{c}{t^2}dt = -\frac{c}{2x}. \quad (x \in I_1)$$

Thus $xg'(x) \leq -\frac{1}{2}c$. From (2.93), we have

$$\left|\int_A^{a'} d\left(\frac{1}{xg'(x)}\right)\right| = \int_A^{a'} \left|d\left(\frac{1}{xg'(x)}\right)\right| < \frac{2}{c}.$$

In combination with the above integration by parts, this yields

$$\left|\int_A^{a'} \frac{e(hg(x))}{x}dx\right| < \frac{3}{\pi ch}. \tag{2.94}$$

Similarly we get

$$\left|\int_{a''}^B \frac{e(hg(x))}{x}dx\right| < \frac{3}{2\pi ch}. \tag{2.95}$$

For y in I_2, we define

$$J(y) = \int_{a'}^{y} e(hg(x))dx.$$

Then we have $g''(x) \geq c/y^2 > 0$ for $x \in [a', y]$. First, assume that $a' + \delta \leq a \leq y - \delta$, where

$$\delta = \frac{2y}{\sqrt{hc}},$$

and set

$$I_{21} = [a', a-\delta), \quad I_{22} = [a-\delta, a+\delta], \quad I_{23} = (a+\delta, y].$$

In I_{23}

$$g'(x) = \int_a^x g''(t)\,dt \geq \frac{c}{y^2}(x-a) \geq \frac{c\delta}{y^2}.$$

Hence, by using the method of the proof of Theorem 2.79

$$\left| \int_{I_{23}} e(hg(x))\,dx \right| \leq \frac{4y^2}{hc\delta}.$$

Similarly we can treat I_{21}. Hence, by using the trivial bound 2δ for I_{22} we obtain

$$|J(y)| \leq \frac{8y^2}{hc\delta} + 2\delta = \frac{8y}{\sqrt{hc}}.$$

If $a < a' + \delta$ or $a\ y - \delta$ we get the same upper bound. Consequently, the modulus of the integral

$$\int_{I_2} \frac{e(hg(x))}{x}\,dx = \left[\frac{J(y)}{y}\right]_{a'}^{a''} + \int_{a'}^{a''} \frac{J(y)}{y^2}\,dy$$

is bounded above by

$$\frac{16}{\sqrt{hc}} + \frac{8}{\sqrt{hc}} \int_{\frac{1}{2}a}^{2a} \frac{1}{y}dy < \frac{30}{\sqrt{hc}}. \tag{2.96}$$

The desired inequality (2.91) follows combining (2.94)-(2.96).

Now suppose that g' does not vanish in $[A, B]$. There are two possibilities: either $g'(x)$ is positive on $[A, B]$ or $g'(x)$ is negative on $[A, B]$. In the first case we obtain

$$\left| \int_A^{\min(2A,B)} \frac{e(hg(x))}{x}\,dx \right| \leq \frac{30}{\sqrt{hc}}$$

by the method used for the proof of (2.96), while for $2A < x \leq B$ we have

$$g'(x) = g'(A) + \int_A^x g''(t)dt > g'(A) + \int_{\frac{1}{2}x}^{x} \frac{c}{t^2}dt = g'(A) + \frac{c}{x} \geq \frac{c}{x}.$$

Then we can proceed as we did to obtain (2.94).

Finally, in the second case (if $g'(x)$ is negative on $[A, B]$) we apply an analogue of (2.96) for $[\max(A, \frac{B}{2}, B]$ and and analogue of (2.94) to $[A, \max(A, \frac{B}{2})]$.

In all cases we obtain (2.91) and the proof of Lemma 2.43 is complete. □

The following lemma can also be found in TITCHMARSH [1861, p. 76] and will be used to complete the proof of Theorem 2.41.

Lemma 2.44 *Let $f, g : [a,b] \to \mathbf{R}$ be two twice continuously differentiable real functions such that $f'(x)$ and $|g'(x)|$ are decreasing and $g(x) > 0$. Then*

$$\begin{aligned}\sum_{a<n\le b} g(n)e(f(n)) &= \sum_{\alpha-\eta<\nu<\beta+\nu} \int_a^b g(x)e(f(x)-\nu x)\,dx \qquad (2.97)\\ &\quad +\mathcal{O}\big(g(a)\log(\beta-\alpha+2)\big)+\mathcal{O}\big(|g'(a)|\big),\end{aligned}$$

where $\alpha = f'(b)$, $\beta = f'(a)$, and ν is any positive constant less than 1.

Proof. First observe that without loss of generality we may assume that $\eta - 1 < \alpha \le \eta$. For, if k is the integer such that $\eta-1<\alpha-k\le\eta$ and if we set $\overline{f}(x) = f(x)-kx$ then (2.97) is equivalent to

$$\begin{aligned}\sum_{a<n\le b} g(n)e(\overline{f}(n)) &= \sum_{\alpha'-\eta<\nu-k<\beta'+\nu} \int_a^b g(x)e(\overline{f}(x)-(\nu-k)x)\,dx\\ &\quad +\mathcal{O}(g(a)\log(\beta'-\alpha'+2))+\mathcal{O}(|g'(a)|),\end{aligned}$$

in which $\alpha' = \alpha - k$ and $\beta' = \beta - k$.

By EULER's formula

$$\begin{aligned}\sum_{a<n\le b} g(n)e(f(n)) &= \int_a^b g(x)e(f(x))\,dx\\ &\quad + \int_a^b \left(x-[x]-\frac12\right)(g'(x)+2\pi i g(x)f'(x))\,e(f(x))\,dx + \mathcal{O}(1).\end{aligned}$$

Hence, by using the FOURIER series

$$x-[x]-\frac12 = -\frac1\pi\sum_{\nu=1}^\infty \frac{\sin(2\nu\pi x)}{\nu}$$

we obtain

$$\begin{aligned}&\int_a^b \left(x-[x]-\frac12\right)(g'(x)+2\pi i g(x)f'(x))\,e(f(x))\,dx\\ &= -2\pi\sum_{\nu=1}^\infty \int_a^b \frac{\sin(2\nu\pi x)}{\nu}(g'(x)+2\pi i g(x)f'(x))\,e(f(x))\,dx\\ &= \sum_{\nu=1}^\infty \frac1\nu \int_a^b (e(-\nu x)-e(\nu x)\,(g'(x)+2\pi i g(x)f'(x))\,e(f(x))\,dx\\ &= \sum_{\nu=1}^\infty \frac1\nu\frac{1}{2\pi i}\int_a^b \frac{g'(x)}{f'(x)-\nu}\,d(e(f(x)-\nu x))\\ &\quad - \sum_{\nu=1}^\infty \frac1\nu\frac{1}{2\pi i}\int_a^b \frac{g'(x)}{f'(x)+\nu}\,d(e(f(x)+\nu x))\end{aligned}$$

$$+\sum_{\nu=1}^{\infty}\frac{1}{\nu}\frac{1}{2\pi i}\int_a^b g(x)\frac{f'(x)}{f'(x)-\nu}\,d(e(f(x)-\nu x))$$
$$-\sum_{\nu=1}^{\infty}\frac{1}{\nu}\frac{1}{2\pi i}\int_a^b g(x)\frac{f'(x)}{f'(x)+\nu}\,d(e(f(x)+\nu x)).$$

Since $f'(x)$ is decreasing it follows that $g(x)f'(x)/(f'(x)+\nu)$ is decreasing, too; and the latter is bounded by $g(a)\beta/(\beta+\nu)$. Thus, by integration by parts

$$\begin{aligned}\left|\int_a^b g(x)\frac{f'(x)}{f'(x)+\nu}\,d(e(f(x)+\nu x))\right| &\leq \frac{f'(b)}{f'(b)+\nu}+\frac{f'(a)}{f'(a)+\nu}e(f(a)+\nu a)\\ &+\int_a^b\left|d\left(g(x)\frac{f'(x)}{f'(x)+\nu}\right)\right|\,dx \qquad (2.98)\\ &\leq 4g(a)\frac{\beta}{\beta+\nu}.\end{aligned}$$

Consequently,

$$\begin{aligned}\sum_{\nu=1}^{\infty}\frac{1}{\nu}\frac{1}{2\pi i}\int_a^b g(x)\frac{f'(x)}{f'(x)+\nu}\,d(e(f(x)+\nu x)) &= \mathcal{O}\left(\sum_{\nu=1}^{\infty}\frac{\beta}{\nu(\beta+\nu)}\right)\\ &= \mathcal{O}(\log(\beta+2))+\mathcal{O}(1).\end{aligned}$$

Similarly, we obtain

$$\begin{aligned}\sum_{\nu\geq\beta+\nu}\frac{1}{\nu}\frac{1}{2\pi i}\int_a^b g(x)\frac{f'(x)}{f'(x)-\nu}\,d(e(f(x)-\nu x)) &= \mathcal{O}\left(\sum_{\nu\geq\beta+nu}\frac{\beta}{\nu(\beta-\nu)}\right)\\ &= \mathcal{O}(\log(\beta+2))+\mathcal{O}(1).\end{aligned}$$

Next we have to estimate

$$\sum_{1\leq\nu<\beta+\nu}\frac{1}{\nu}\int_a^b g(x)f'(x)e(f(x)-\nu x)\,dx.$$

Since

$$\begin{aligned}&\int_a^b g(x)f'(x)e(f(x)-\nu x)\,dx\\ &=\int_a^b g(x)(f'(x)-\nu x)e(f(x)-\nu x)\,dx+\nu\int_a^b g(x)e(g(x)-\nu x)\,dx\\ &=\frac{1}{2\pi i}(g(b)e(f(b)-\nu b)-g(a)e(f(a)-\nu a))-\frac{1}{2\pi i}\int_a^b g'(x)e(f(x)-\nu x)\,dx\\ &\quad+\nu\int_a^b g(x)e(f(x)-\nu x)\,dx,\end{aligned}$$

we get

$$\int_a^b g(x)f'(x)e(f(x)-\nu x)\,dx = \mathcal{O}\left(g(a)+g(a)\frac{\nu}{\beta-\nu}\right),$$

where we have used the inequality

$$\left|\int_a^b g(x)e(f(x)-\nu x)\,dx\right| \le \frac{4g(a)}{\beta-\nu},$$

which can be derived in the same way as (2.98). This gives

$$\sum_{1\le\nu<\beta+\nu}\frac{1}{\nu}\int_a^b g(x)f'(x)e(f(x)-\nu x)\,dx = \mathcal{O}(\log(\beta+2))+\mathcal{O}(1).$$

Finally, the integrals

$$\int_a^b \frac{g'(x)}{f'(x)\pm\nu}\,d(e(f(x)-\nu x))$$

can be estimated in a similar way. □

Lemma 2.45 *Suppose that f satisfies the hypotheses of Theorem 2.41. Then, for any natural numbers h, A, B we have*

$$\left|\sum_{n=A}^{B}\frac{e(hf(n))}{n}\right| < C\cdot(h^{\frac{1}{2}}A^{-\delta}+h^{-\frac{1}{2}})$$

with a constant $C=C(c,K,\delta,H)$ depending on c, K, and H.

Proof. We apply Lemma 2.44 with $g(x)=x^{-1}$ and $hf(x)$ in place of $f(x)$. This gives

$$\begin{aligned}\sum_{n=A}^{B}\frac{e(hf(n))}{n} &= \sum_{hf'(B)-\frac{1}{2}<\nu<hf'(A)+\frac{1}{2}}\int_A^B\frac{1}{x}e(hf(x)-\nu x)dx\\ &\quad+\mathcal{O}(A^{-1}\log(h|f'(A)-f'(B)|+2)),\end{aligned} \tag{2.99}$$

where the implied constant is absolute. (The upper and lower bounds in the summation over ν are to be reversed if $f''>0$.)

Next observe that

$$|f'(A)-f'(B)| = \int_A^B |f''(x)|dx \le \frac{KA^{-\delta}}{\delta}\le\frac{K}{\delta}.$$

Thus, the $\mathcal{O}$-term in (2.99) is bounded by $C'A^{-1}\log h$, where $C'=C'(K,\delta)$ depends on K and δ.

By Lemma 2.43 each term of the ν-sum of the right hand side of (2.99) is bounded by $Ch^{-\frac{1}{2}}$. Since the sum consists of at most $C''(hA^{-\delta}+1)$ terms, where $C'' = C''(K,\delta)$ depends on K and δ, we immediately derive Lemma 2.45. □

Proof of Theorem 2.41. By Lemma 2.42 and Lemma 2.45 we have

$$\log N \cdot D_N^{\log}(x_n) < C_1(\delta) + C_2(\delta) \sum_{1\le h\le N^\delta} \frac{1}{h}.2C_9 h^{-\frac{1}{2}} = C_4(c,K,\delta,H),$$

which proves Theorem 2.41. □

2.2.2 Well Distribution

In Section 1.1 we already introduced the notion of well distributed sequences mod 1. This notion is motivated by the observation that every shifted sequence $(\mathbf{x}_{n+\nu})_{n\ge1}$, $\nu \ge 0$, is u.d. mod 1 if $(\mathbf{x}_n)_{n\ge1}$ is a u.d. sequence mod 1. However, the convergence of

$$\lim_{N\to\infty} \frac{1}{N} \sum_{n=1}^{N} \chi_I(\{\mathbf{x}_{n+\nu}\}) = \lambda_k(I)$$

need not be uniform for $\nu \ge 0$. (For convenience we reall the second part of Definition 1.1 of Section 1.1.)

Definition 2.46 *A sequence $(\mathbf{x}_n)_{n\ge1}$, $\mathbf{x}_n \in \mathbf{R}^k$ is said to be well distributed modulo 1 (for short w.d. mod 1) if*

$$\lim_{N\to\infty} \sup_{\nu\ge0} \left| \frac{1}{N} \sum_{n=1}^{N} \chi_I(\{\mathbf{x}_{n+\nu}\}) - \lambda_k(I) \right| = 0$$

holds for all intervals $I \subseteq \mathbf{R}^k/\mathbf{Z}^k$. The uniform discrepancy $\tilde{D}_N(\mathbf{x}_n)$ is given by

$$\tilde{D}_N(\mathbf{x}_n) = \sup_{I\subseteq\mathbf{R}^k/\mathbf{Z}^k} \sup_{\nu\ge0} \left| \frac{1}{N} \sum_{n=1}^{N} \chi_I(\{\mathbf{x}_{n+\nu}\}) - \lambda_k(I) \right|.$$

Remark. Clearly we can consider w.d. sequences in compact metric spaces, too. It should be further noted that every w.d. sequence is trivially u.d. but the contrary is not true. For example, it can be shown (in a very general setting) that almost all sequences are u.d. (see Section 2.4) but almost no sequences are w.d. (see KUIPERS and NIEDERREITER [983]).

The situation changes if we want to define w.d. sequences with respect to a limitation method A. Here it need not be true that all shifted sequence $(\mathbf{x}_{n+\nu})_{n\ge1}$ are A-u.d. if $(\mathbf{x}_n)_{n\ge1}$ is A-u.d., even if A is a weighted mean M_P. We need additional regularity conditions. It is an easy exercise to prove

Lemma 2.47 *Let M_P, $P=(p_n)_{n\ge1}$ be a weighted mean such that p_n is monotone, $P_N\to\infty$, and $\lim_{n\to\infty} p_n/P_n = 0$. Then every shifted sequence $(\mathbf{x}_{n+\nu})_{n\ge1}$ is M_P-u.d. mod 1 if $(\mathbf{x}_n)_{n\ge1}$ is M_P-u.d.*

This gives rise to the following definition

Definition 2.48 *Let M_P, $P = (p_n)_{n\geq 1}$, be a weighted mean. A sequence $(\mathbf{x}_n)_{n\geq 1}$, $\mathbf{x}_n \in \mathbf{R}^k$ is said to be M_P-w.d. mod 1 of type (i) if*

$$\lim_{N\to\infty} \sup_{\nu\geq 0} \left| \frac{1}{P_N} \sum_{n=1}^{N} p_n \chi_I(\{\mathbf{x}_{n+\nu}\}) - \lambda_k(I) \right| = 0$$

holds for all intervals $I \subseteq \mathbf{R}^k/\mathbf{Z}^k$.

But as the following theorem shows, this definition does not really depend on the weighted mean.

Theorem 2.49 *Let M_P, $P = (p_n)_{n\geq 1}$, be a weighted mean such that p_n is monotone, $P_N \to \infty$, and $\lim_{n\to\infty} p_n/P_n = 0$. Then a sequence $(\mathbf{x}_n)_{n\geq 1}$, $\mathbf{x}_n \in \mathbf{R}^k$ is M_P-w.d. mod 1 of type (i) if and only if it is w.d. mod 1.*

Proof. According to Theorem 3 by LORENTZ [1096], for a regular matrix method $A = (a_{Nn})$ the relation

$$\sup_{\nu\geq 0} \left| \sum_{n=1}^{\infty} a_{Nn}(\chi_{[0,x)}(\{x_{n+\nu}\}) - x) \right| \to 0$$

(as $N \to \infty$) is equivalent to the definition of well-distribution provided that

$$\sum_{n=1}^{\infty} |a_{Nn} - a_{N,n+1}| \to 0 \tag{2.100}$$

(as $N \to \infty$). The last condition (2.100) is called LORENTZ condition, which is satisfied in our case. □

Furthermore, if we define w.d. sequences with respect to a weighted mean in the following way then we have a similar result.

Definition 2.50 *Let M_P, $P = (p_n)_{n\geq 1}$, be a weighted mean. A sequence $(\mathbf{x}_n)_{n\geq 1}$, $\mathbf{x}_n \in \mathbf{R}^k$ is said to be M_P-w.d. mod 1 of type (ii) if*

$$\lim_{N\to\infty} \sup_{\nu\geq 0} \left| \frac{1}{P_{N+\nu} - P_\nu} \sum_{n=\nu+1}^{\nu+N} p_n \chi_I(\{\mathbf{x}_n\}) - \lambda_k(I) \right| = 0 \tag{2.101}$$

holds for all intervals $I \subseteq \mathbf{R}^k/\mathbf{Z}^k$.

Theorem 2.51 *Let M_P, $P = (p_n)_{n\geq 1}$, be a weighted mean such that p_{n+1}/p_n is non-decreasing, $P_N \to \infty$, and $\lim_{n\to\infty} p_n/P_n = 0$. Then a sequence $(\mathbf{x}_n)_{n\geq 1}$, $\mathbf{x}_n \in \mathbf{R}^k$ is M_P-w.d. mod 1 of type (ii) if and only if it is w.d. mod 1.*

Proof. Let the sequence $\mathbf{x}_n$ be well-distributed of type (i) and set

$$A_n(\mathbf{x}) = \chi_{[0,\mathbf{x})}(\{\mathbf{x}_n\}) - x^{(1)} \dots x^{(k)}$$

for $\mathbf{x} = (x^{(1)}, \dots, x^{(k)})$ and $[0, \mathbf{x}) = [0, x^{(1)}) \times \cdots \times [0, x^{(k)})$. Then given $\varepsilon > 0$ there exists an N_1 such that

$$\left| \sum_{n=\nu+1}^{\nu+N} A_n(\mathbf{x}) \right| \le \varepsilon N \tag{2.102}$$

for $N \ge N_1$ and all $\nu, \mathbf{x}$. By partial summation

$$\sum_{n=\nu+1}^{\nu+N} p_n A_n(\mathbf{x}) = p_{\nu+N} \sum_{n=\nu+1}^{\nu+N} A_n(\mathbf{x}) + \sum_{n=\nu+2}^{\nu+N} (p_{n-1} - p_n) \sum_{m=\nu+1}^{n-1} A_m(\mathbf{x}).$$

Hence, for $N \ge N_1$

$$\begin{aligned} \left| \sum_{n=\nu+1}^{\nu+N} p_n A_n(\mathbf{x}) \right| &\le \varepsilon p_{\nu+N} N \sum_{n=\nu+2}^{\nu+N_1} (p_{n-1} - p_n)(n - \nu - 1) \\ &\quad + \varepsilon \sum_{n=\nu+2}^{\nu+N} (p_{n-1} - p_n)(n - \nu - 1) \\ &\le \sum_{n=\nu+1}^{\nu+N_1} p_n + \varepsilon \sum_{n=\nu+1}^{\nu+N} p_n. \end{aligned}$$

Furthermore,

$$\frac{p_n}{p_m} = \frac{p_n}{p_{n-1}} \cdot \frac{p_{n-1}}{p_{n-2}} \cdots \frac{p_{m+1}}{p_m} \le \frac{p_{n+\nu}}{p_{m+\nu}}$$

for $\nu \ge 0$, $n \ge m$ implies

$$\sum_{m=1}^{N_1} p_{\nu+m} \sum_{n=1}^{N} p_n \le \sum_{m=1}^{N_1} p_m \sum_{n=1}^{N} p_{\nu+n}.$$

Thus

$$\sum_{n=\nu+1}^{\nu+N_1} p_n \Big/ \sum_{n=\nu+1}^{\nu+N} p_n \le \sum_{n=1}^{N_1} p_n \Big/ \sum_{n=1}^{N} p_n \to 0,$$

as $N \to \infty$ uniformly in ν. Hence, it follows that the sequence is w.d. of type (ii).

Conversely, suppose that $(\mathbf{x}_n)$ is w.d. of type (ii). Then we introduce

$$q_1(\nu) = 1/p_{\nu+1} \quad \text{and} \quad q_n(\nu) = 1/p_{\nu+n} - 1/p_{\nu+n-1}$$

for $n > 1$. We have $q_n(\nu) \ge 0$ and

$$p_{\nu+n} \sum_{m=1}^{n} q_m(\nu) = 1. \tag{2.103}$$

From (2.101) it follows that for given $\varepsilon > 0$ there exists an N_1 such that

$$\left| \sum_{n=\nu+1}^{\nu+N} p_n A_n(\mathbf{x}) \right| \leq \varepsilon \sum_{n=\nu+1}^{\nu+N} p_n \tag{2.104}$$

for $N \geq N_1$ and all ν. This implies

$$\sum_{m=1}^{N-N_1} q_m(\nu) \left| \sum_{n=m}^{N} p_{\nu+n} A_{\nu+n} \right| \leq \varepsilon \sum_{m=1}^{N-N_1} q_m(\nu) \sum_{n=m}^{N} p_{\nu+n}.$$

Furthermore we have

$$\begin{aligned} \left| \sum_{m=N-N_1+1}^{N} q_m(\nu) \sum_{n=m}^{N} p_{\nu+n} A_{\nu+n} \right| &\leq \sum_{m=N-N_1+1}^{N} q_m(\nu) \sum_{n=m}^{N} p_{\nu+n} \\ &= \sum_{n=N-N_1+1}^{N} p_{\nu+n} \sum_{m=N-N_1+1}^{n} q_m(\nu) \leq N_1, \end{aligned}$$

where we applied (2.103). Thus, we obtain

$$\begin{aligned} \left| \sum_{n=1}^{N} A_{\nu+n} \right| &= \left| \sum_{m=1}^{N} q_m(\nu) \sum_{n=m}^{N} p_{\nu+m} A_{\nu+n} \right| \\ &\leq N_1 + \varepsilon \sum_{m=1}^{N} q_m(\nu) \sum_{n=m}^{N} p_{\nu+n} \\ &= N_1 + \varepsilon N. \end{aligned}$$

Hence the sequence is also w.d. of type (i). □

However, there exists a third natural definition for w.d. sequences with respect to weighted means, which is due to SCHATTE.

Definition 2.52 *Let M_P, $P = (p_n)_{n\geq 1}$, be a weighted mean and let $L(\nu, N)$ be defined by*

$$\sum_{n=\nu+1}^{L(\nu,N)} p_n \leq P_N < \sum_{n=\nu+1}^{L(\nu,N)+1} p_n.$$

A sequence $(\mathbf{x}_n)_{n\geq 1}$, $\mathbf{x}_n \in \mathbf{R}^k$ is said to be M_P-w.d. mod 1 of type (iii) if

$$\lim_{N\to\infty} \sup_{\nu\geq 0} \left| \frac{1}{P_N} \sum_{n=\nu+1}^{L(\nu,N)} p_n \chi_I(\{\mathbf{x}_n\}) - \lambda_k(I) \right| = 0$$

holds for all intervals $I \subseteq \mathbf{R}^k/\mathbf{Z}^k$. Furthermore the uniform discrepancy M_P–$\tilde{D}_N(\mathbf{x}_n)$ is given by

$$M_P\text{–}\tilde{D}_N(\mathbf{x}_n) = \sup_{I\subseteq \mathbf{R}^k/\mathbf{Z}^k} \sup_{\nu\geq 0} \left| \frac{1}{P_N} \sum_{n=\nu+1}^{L(\nu,N)} p_n \chi_I(\{\mathbf{x}_n\}) - \lambda_k(I) \right|.$$

As usual, a sequence is M_P-w.d. mod 1 of type (iii) if and only if $\lim_{N\to\infty} M_P$–$\tilde{D}_N(\mathbf{x}_n) = 0$. Furthermore we can check M_P-well distribution by a WEYL criterion.

Theorem 2.53 *Let M_P, $P = (p_n)_{n\geq 1}$, be a weighted mean. Then a sequence $(\mathbf{x}_n)_{n\geq 1}$, $\mathbf{x}_n \in \mathbf{R}^k$ is M_P-w.d. mod 1 of type (iii) if and only if*

$$\lim_{N\to\infty} \sup_{\nu\geq 0} \left| \frac{1}{P_N} \sum_{n=\nu+1}^{L(\nu,N)} p_n e(\mathbf{h}\cdot\mathbf{x}_n) \right| = 0$$

holds for all non-zero interal lattice points $\mathbf{h} \in \mathbf{Z}^k \setminus \{0\}$.

It is not obvious that there always exists a M_P-w.d. sequence of type (iii). The following theorems give a first insight and explicit examples. Since the third notion of w.d. is the most interesting one we will omit "type (iii)" in the following.

Theorem 2.54 *If there exists a M_P-w.d. sequence mod 1 then the sequence $P = (p_n)_{n\geq 1}$ is bounded.*

If the sequence $P = (p_n)_{n\geq 1}$ is convergent, then there exists a M_P-w.d. sequence mod 1.

Proof. First we suppose that there is an infinite sequence of weights (p_{n_i}) with $\lim_{i\to\infty} p_{n_i} = \infty$. Then $L(n_i - 1, N) = n_i - 1$ for all sufficiently large i, which proves the first assertion.

Now assume $\lim_{n\to\infty} p_n = c > 0$. Since

$$\lim_{N\to\infty} \frac{P_N}{cN} = 1,$$

we obtain

$$\left| \sum_{n=\nu+1}^{L(\nu,N)} p_n - cN \right| \leq \left| \sum_{n=\nu+1}^{L(\nu,N)} p_n - P_N \right| + |P_N - cN| \leq \max_{n\geq 1} p_n + o(N).$$

Thus, in this case a sequence is (P, μ) - w.d. if and only if it is well-distributed in the usual sense, and the existence of w.d. sequences is guaranteed.

In the case $\lim_{n\to\infty} p_n = 0$ define m_l by $m_1 = 1$ and by

$$\sum_{n=1}^{n_l - 1} p_n \leq l < \sum_{n=1}^{m_l} p_n \quad \text{for} \quad l > 1.$$

Let $(\mathbf{y}_n)$ denote a w.d. sequence in the usual sense. Then we define a sequence $(\mathbf{x}_n)$ by $\mathbf{x}_n = \mathbf{y}_l$ for $m_l \leq n < m_{l+1}$. Furthermore, set $h(n) = \max\{m_l : m_l \leq n\}$ and

$l(n) = \max\{l : m_l \leq n\}$. Hence, for every $\varepsilon > 0$ we obtain

$$\begin{aligned}
&\left| \sum_{n=\nu+1}^{L(\nu,N)} p_n e(\mathbf{h}\cdot\mathbf{x}_n) - \sum_{j=l(\nu+1)}^{l(\nu+1)+[P_N]-1} e(\mathbf{h}\cdot\mathbf{y}_j) \right| \\
&\qquad \leq \left| \sum_{n=\nu+1}^{L(\nu,N)} p_n e(\mathbf{h}\cdot\mathbf{x}_n) - \sum_{n=h(\nu+1)}^{m_{l(\nu+1)+[P_N]}-1} e(\mathbf{h}\cdot\mathbf{x}_n) \right| \\
&\qquad + \left| \sum_{n=h(\nu+1)}^{m_{l(\nu+1)+[P_N]}-1} e(\mathbf{h}\cdot\mathbf{x}_n) - \sum_{j=l(\nu+1)}^{l(\nu+1)+[P_N]-1} e(\mathbf{h}\cdot\mathbf{y}_j) \right| \\
&\qquad \leq 2 + \varepsilon P_N + c(\varepsilon)
\end{aligned}$$

for some constant $c(\varepsilon) > 0$ only depending on ε. Thus, $(\mathbf{x}_n)$ is M_P-w.d. □

We now want to point out how to prove that the sequence $([p_n]\alpha)$ is M_P-w.d. for irrational α (under suitable conditions on the sequence p_n).

The following proposition is a direct generalization of a result of OHKUBO [1391] (see [489]). The proof can easily be given by using EULER's summation formula, ABEL summation, and some elementary asymptotic arguments.

Proposition 2.55 *Let $g(t)$ be twice continuously differentiable and $p(t)$ a continuously differentiable, positive, and non-increasing function satisfying the following conditions ($1 \leq t < \infty$):*

$$\begin{aligned}
&g(t) \to \infty \qquad \text{as } t \to \infty, \\
&g'(t) \to \text{const.} < 1 \qquad \text{monotonically as } t \to \infty, \\
&g'(t)/p(t) \qquad \text{is monotone for } t \geq 1.
\end{aligned}$$

Then we have for integers $h \neq 0$, $\nu \geq 0$ and irrational α

$$\begin{aligned}
&\left| \sum_{n=\nu+1}^{L(\nu,N)} p(n) e(h[g(n)]\alpha) \right| \\
&\qquad \leq \frac{1}{\|h\alpha\|} \left(3 \frac{p(L(\nu,N))}{g'(L(\nu,N))} + \int_{\nu+1}^{L(\nu,N)} p(t) g'(t)\, dt + \mathcal{O}(1) \right),
\end{aligned}$$

where the $\mathcal{O}$-constants are independent of h, ν, and α. ($\|x\| = \min(\{x\}, 1 - \{x\})$).

Theorem 2.56 *If in addition to the assumptions of Proposition 2.55*

$$\lim_{t\to\infty} g'(t) = 0 \quad \text{and} \quad \frac{p(t)}{g'(t)} = \mathcal{O}(1) \quad \text{as } t \to \infty$$

then the sequence $([g(n)]\alpha)_{n\geq 1}$ is M_P-w.d. mod 1. Furthermore, if the irrational number α is of approximation type $\eta > 1$ then for arbitrary $\varepsilon > 0$

$$M_P\text{–}\tilde{D}_N([g(n)]\alpha) \ll G(N)^{\varepsilon - \frac{1}{\eta}},$$

and if $\eta = 1$ *then we have*

$$M_P\text{–}\tilde{D}_N([g(n)]\alpha) \ll \frac{(\log G(N))^2}{G(N)},$$

where

$$G(N) = P_N \left(\int_1^N p(t)g'(t)\, dt \right)^{-1}.$$

Remark. Note that all assumptions are satisfied in the important case

$$\begin{aligned} g'(t) &= p(t) \qquad \text{monotonically as } t \to \infty \text{ and} \\ g(t) &\to \infty \qquad \text{as } t \to \infty. \end{aligned}$$

Proof. For a fixed integer $N \geq 1$ we define a continuously differentiable function $L(\tau)$ by

$$\int_\tau^{L(\tau)} p(t)\, dt = P_N.$$

It is clear that $L(\tau)$ satisfies $L'(\tau) = p(\tau)/p(L(\tau))$. Hence, a simple computation of derivatives shows that the mapping

$$\tau \mapsto \int_\tau^{L(\tau)} p(t)g'(t)\, dt$$

is non-increasing. Thus, we obtain by Proposition 2.55

$$\sum_{n=\nu+1}^{L(\nu,N)} p(n)e(h[g(n)]\alpha) = \mathcal{O}\left(1 + \int_1^N p(t)g'(t)\, dt\right) = o(P_N)$$

as $N \to \infty$ uniformly in $\nu \geq 0$, since $g'(t) \to 0$. Therefore, by Theorem 2.53 the sequence $([g(n)]\alpha)_{n\geq 1}$ is M_P-w.d. mod 1.

By using the same arguments as above, Theorem 2.39, and the asymptotic relations

$$\sum_{h=1}^{m} \frac{1}{h\|h\alpha\|} = \begin{cases} \mathcal{O}(m^{\eta-1+\varepsilon}) & \text{for } \eta > 1, \\ \mathcal{O}(\log^2 m) & \text{for } \eta = 1 \end{cases},$$

the estimates for the M_P-discrepancy of $([g(n)]\alpha)$ follow immediately. □

It has already been mentioned that almost no sequences are w.d. mod 1 in the usual sense. It is interesting to observe that the situation changes for general weighted means if p_n tends to zero too fast.

Theorem 2.57 *Let* M_P, $P = (p_n)_{n\geq 1}$, *be a weighted mean such that*

$$\limsup_{n\to\infty} p_n \log n = \infty.$$

Then almost no sequences $(\mathbf{x}_n)_{n\geq 1}$, $\mathbf{x}_n \in [0,1]^k$ are M_P-w.d. mod 1.

On the other hand, if

$$\limsup_{n\to\infty} p_n \log n < \infty \qquad \text{and} \qquad \lim_{N\to\infty} \frac{P_N}{\log N} = \infty \tag{2.105}$$

or if p_n is non-increasing and for some $\varepsilon > 0$

$$\lim_{n\to\infty} p_n \sqrt{n} \log n (\log\log n)^{\frac{3}{2}+\varepsilon} = 0, \tag{2.106}$$

then almost all sequences $(\mathbf{x}_n)_{n\geq 1}$, $\mathbf{x}_n \in [0,1]^k$ are M_P-w.d. mod 1.

In addition, in the case (2.106) there are two absolute constants $C_1, C_2 > 0$ such that almost all sequences $(\mathbf{x}_n)_{n\geq 1}$ satisfy

$$\frac{C_1}{P_N} \leq M_P\text{–}\tilde{D}_N(\mathbf{x}_n) \leq \frac{C_2}{P_N}.$$

Remark. This theorem shows that in the cases $P = (n^\alpha)$, $-1 \leq \alpha < 0$, and $P = ((\log(n+1))^\beta)$, $\beta < -1$, almost all sequences $(\mathbf{x}_n)_{n\geq 1}$ are M_P-w.d. mod 1 whereas in the case $P = ((\log(n+1))^\beta)$, $-1 \leq \beta < 0$ almost no sequences $(\mathbf{x}_n)_{n\geq 1}$ are M_P-w.d. mod 1.

Proof. First let us indicate the proof of the "almost no" statement. We define a sequence $\nu_i(N)$, $i \geq 0$, of integers by $\nu_0(N) = 0$ and by $\nu_{i+1}(N) = L(\nu_i(N), N)$. Furthermore, fix an interval $I \subseteq \mathbf{R}^k/\mathbf{Z}^k$ with $0 < \alpha = \lambda_k(I) < 1$ and set

$$W_N = \bigcap_{i=0}^{\infty} \bigcup_{n'=\nu_i(N)+1}^{\nu_{i+1}(N)} \{(\mathbf{x}_n)_{n\geq 1} : \mathbf{x}_{n'} \in I\}.$$

By definition, for every M_P-w.d. sequence $(\mathbf{x}_n)_{n\geq 1}$ there exists N such that $(\mathbf{x}_n)_{n\geq 1} \in W_N$. Therefore, it is sufficient to prove that $\lambda_\infty(W_N) = 0$ for all $N \geq 1$, where λ_∞ denotes the infinite product measure generated by the LEBESGUE measure.

From $\lim_{n\to\infty} p_n \log n$ it follows that there exists a constant $C > 0$ such that $\log \nu_i(N) \leq C \log i$. Now choose i_0 in a way that

$$p_n \log n \geq P_N C \log \frac{1}{1-\alpha}$$

for $n \geq \nu_{i_0}(N)$. Hence,

$$P_N \geq \sum_{n=\nu_{i-1}(N)-1}^{\nu_{i+1}(N)} p_n \geq P_N C \log \frac{1}{1-\alpha} (\nu_i(N) - \nu_{i-1}(N)) \frac{1}{\log \nu_i(N)}$$

implies

$$\nu_i(N) - \nu_{i-1}(N) \leq \frac{\log i}{\log \frac{1}{1-\alpha}}$$

for $i \ge i_0$. Consequently

$$(1-\alpha)^{\nu_i(N)-\nu_{i-1}(N)} \ge \frac{1}{i}$$

for $i \ge i_0$. Thus

$$\sum_{i\ge 1}(1-\alpha)^{\nu_i(N)-\nu_{i-1}(N)} = 0$$

implies

$$\lambda_\infty(W_N) = \prod_{i\ge 1}\left(1-(1-\alpha)^{\nu_i(N)-\nu_{i-1}(N)}\right) = 0.$$

This proves the "almost no" part of Theorem 2.57.

Now we turn to the "almost all" part of Theorem 2.57. For a fixed function $f_{\mathbf{h}}(\mathbf{x}) = e(\mathbf{h}\cdot\mathbf{x})$, $(\mathbf{h} \in \mathbf{Z}^k \setminus \{\mathbf{0}\})$ let $B_{N,\nu}(\eta)$, $\eta > 0$ denote the set

$$B_{N,\nu}(\eta) = \left\{(\mathbf{x}_n) \in X^\infty : \left|\frac{1}{P_N}\cdot\sum_{n=\nu+1}^{L(\nu,N)} p_n f_{\mathbf{h}}(\mathbf{x}_n)\right| \le \eta\right\},$$

where $X = [0,1]^k$. Then by Lemma 2.15 we have for the complement $B_{N,\nu}(\eta)^c$

$$\lambda_\infty(B_{N,\nu}(\eta)^c) \le 2\exp\left(-\eta^2 \Big/ \left(\frac{2}{P_N^2}\sum_{n=\nu+1}^{L(\nu,N)} p_n^2\right)\right). \tag{2.107}$$

Set $M(\eta) = \bigcup_{N'=1}^{\infty}\bigcap_{N=N'}^{\infty}\bigcap_{\nu=0}^{\infty} B_{N,\nu}(\eta)$ and note that (2.105) implies

$$\sum_{N=N_0(\delta)}\sum_{\nu=0}^{\infty}\exp\left(-\delta\Big/\left(\frac{1}{P_N^2}\sum_{n=\nu+1}^{L(\nu,N)} p_n^2\right)\right) < \infty \qquad (\delta > 0).$$

Hence, a simple application of the BOREL-CANTELLI-lemma yields

$$\lambda_\infty(M(\eta)) \ge 1-\varepsilon$$

for arbitrary $\varepsilon > 0$. Thus $\lambda_\infty(M(\eta)) = 1$ and $\lambda_\infty\left(\bigcap_{r=1}^{\infty} M\left(\frac{1}{r}\right)\right) = 1$. Taking the intersection over all functions $f_{\mathbf{h}}$, $\mathbf{h} \in \mathbf{Z}^k \setminus \{\mathbf{0}\}$ proves the first part of the "almost all" assertion.

Now we assume (2.106). Since the sum $\sum_{n=1}^{N} f(\mathbf{x}_n)$ may be interpreted as a sum of independent random variables we have

$$\sum_{n=1}^{N} f_h(\mathbf{x}_n) = \mathcal{O}(\sqrt{N\log\log N}) \qquad (N \ge N_0) \tag{2.108}$$

for almost all sequences $(\mathbf{x}_n)$. As above, by partial summation we obtain

$$\sum_{n=1}^{N} p_n f(\mathbf{x}_n) = \mathcal{O}\left(1 + p_N \sqrt{N} (\log\log N)^{1/2} + \sum_{n=N_0}^{N-1} (p_n - p_{n+1}) \sqrt{n} (\log\log n)^{1/2}\right)$$

$$= \mathcal{O}\left(1 + \sum_{n=N_0}^{N} p_n \sqrt{\frac{\log\log n}{n}}\right) = \mathcal{O}(1).$$

Hence, by (2.106)

$$\sum_{n=\nu+1}^{L(\nu,N)} p_n f(\mathbf{x}_n) = \mathcal{O}(1)$$

for almost all sequences $(\mathbf{x}_n)$, where the $\mathcal{O}$-constant is uniformly in $\nu = 0, 1, 2, \ldots$. Thus the proof of the "almost all" assertion is complete.

The proof of the upper discrepancy bound runs along the same lines. Instead of $f_{\mathbf{h}}(\mathbf{x}) = e(\mathbf{h} \cdot \mathbf{x})$ we have to use $f(\mathbf{x}) = \chi_{[\mathbf{a},\mathbf{b})}(\mathbf{x}) - \lambda_k([\mathbf{a},\mathbf{b}))$ and the uniform law of the iterated logarithm (Theorem 1.193). The lower discrepancy bound is trivial. □

2.2.3 Abel's Summation Method

A very important limitation method that is not a matrix method is the following one related to power series.

Definition 2.58 *A real sequence $(x_n)_{n\ge 0}$ is said to be* Abel *limitable to x if*

$$\lim_{r\to 1-} (1-r) \sum_{n=0}^{\infty} x_n r^n = x.$$

By Abel's theorem on power series, the above limitation method is regular. Furthermore, bounded sequences are Abel limitable if and only if they are Cesaro limitable. Therefore, sequences $(\mathbf{x}_n)_{n\ge 1}$, $\mathbf{x}_n \in \mathbf{R}^k$ are u.d. mod 1 if and only if

$$\lim_{r\to 1-} (1-r) \sum_{n=0}^{\infty} \chi_I(\{\mathbf{x}_n\}) r^n = \lambda_k(I)$$

holds for all intervals $I \subseteq \mathbf{R}^k/\mathbf{Z}^k$. Therefore it is natural to introduce the notion of Abel discrepancy of Hlawka [793] and Niederreiter [1285].

Definition 2.59 *Let $(\mathbf{x}_n)_{n\ge 0}$ be a sequence in the k-dimensional space $\mathbf{R}^k$. Then, for $0 < r < 1$, the* Abel *discrepancy $D_r(\mathbf{x}_n)$ is defined by*

$$D_r(\mathbf{x}_n) = \sup_{I\subseteq \mathbf{R}^k/\mathbf{Z}^k} \left| (1-r) \sum_{n=0}^{\infty} \chi_I(\{\mathbf{x}_n\}) r^n - \lambda_k(I) \right|,$$

and for a positive integer N, the tructated Abel *discrepancy is given by*

$$D_{r,N}(\mathbf{x}_n) = \sup_{I\subseteq \mathbf{R}^k/\mathbf{Z}^k} \left| \frac{(1-r)}{1-r^N} \sum_{n=0}^{N-1} \chi_I(\{\mathbf{x}_n\}) r^n - \lambda_k(I) \right|.$$

Remark. Note that we now consider sequences $(\mathbf{x}_n)_{n\geq 0}$ instead of $(\mathbf{x}_n)_{n\geq 1}$. This is more natural in connection with power series. We also make the convention that $D_N(\mathbf{x}_n)$ should denote $D_N(\mathbf{x}_0,\ldots,\mathbf{x}_{N-1})$ in this section.

The connection between D_r and $D_{r,N}$ is the following one. Note that in our definition the supremum ranges over all intervals. (In NIEDERREITER [1285] only intervals of the type $[0,x)$ are allowed.)

Lemma 2.60 *For any sequence $(\mathbf{x}_n)_{n\geq 1}$, $\mathbf{x}_n \in \mathbf{R}^k$, we have*

$$|D_r(\mathbf{x}_n) - D_{r,N}(\mathbf{x}_n)| \leq r^N.$$

Hence, $\lim_{N\to\infty} D_{r,N}(\mathbf{x}_n) = D_r(\mathbf{x}_n)$ *for all* $0 < r < 1$.

The proof follows from simple estimates (see NIEDERREITER [1285]).

As already mentioned, $\lim_{N\to\infty} D_N(\mathbf{x}_n) = 0$ if and only if $\lim_{r\to 1-} D_r(\mathbf{x}_n) = 0$. But it is not trivial to derive bounds between these two kinds of discrepancies.

Theorem 2.61 *Let $(\mathbf{x}_n)_{n\geq 0}$, $\mathbf{x}_n \in \mathbf{R}^k$, be an arbitrary sequence. Then we have*

$$D_r(\mathbf{x}_n) \leq 4 \sup_{N\geq (1-r)^{-1/2}} D_N(\mathbf{x}_n) \tag{2.109}$$

and

$$D_N(\mathbf{x}_n) \leq c(-\log(D_r(\mathbf{x}_n)))^{-1}, \tag{2.110}$$

where

$$r = N^{-\frac{1}{N}},$$

and $c > 0$ is an absolute real constant.

Proof of (2.109). Let $f(\mathbf{x}) = \chi_{[\mathbf{0},\mathbf{y})}(\mathbf{x})$ and set

$$\delta(r,f,\mathbf{x}_n) = \left|(1-r)\sum_{n=0}^{\infty} f(\mathbf{x}_n)r^n - \int_{I^k} f(\mathbf{x})d\mathbf{x}\right|;$$

$$\Delta_N(f) = \frac{1}{N+1}\sum_{n=0}^{N} f(\mathbf{x}_n) - \int_{I^k} f(\mathbf{x})d\mathbf{x}.$$

A simple application of the KOKSMA-HLAWKA inequality (Theorem 1.14) shows for an arbitrary positive integers M

$$\delta(r,f,\mathbf{x}_n)(1-r)^{-1} \leq 2M + V(f)(M + (1-r)^{-1})D_M^*,$$

where $V(f) = 1$ is the total variation of f and

$$D_M^* = \sup_{n\geq M} D_N^*(\mathbf{x}_n).$$

Setting $M = [(1-r)^{-1/2}]$ immediately yields the first estimate (2.109). □

The proof of the second inequality (2.110) is much more involved. It makes use of the following approximation lemma (which can be found in GANELIUS [647]).

Lemma 2.62 *Let $h : [0,1] \to \mathbf{R}$ be a function of bounded variation. Then there exist constants c_1, c_2 only depending on $h(x)$ such that for every positive sufficiently large integer d there are polynomials $p_1(x)$, $p_2(x)$, $q(x)$ of degree d satisfying*

$$p_1(x) = \sum_{m=0}^{d} b_m^{(1)} x^m \le h(x) \le p_2(x) = \sum_{m=0}^{d} b_m^{(2)} x^m \quad \text{and} \quad q(x) = \frac{p_2(x) - p_1(x)}{x(1-x)},$$

where

$$\sum_{n=m}^{g} \left(|b_m^{(1)}| + |b_m^{(2)}| \right) \le e^{c_1 d}$$

and

$$\int_0^1 q(x)\, dx \le \frac{c_2}{d}.$$

The following lemma will be essential in the proof of (2.110).

Lemma 2.63 *Suppose that a_n is a real bounded sequence, say $|a_n| \le 1$, and that there exists a decreasing function $\varphi(x) > 0$, $x > 0$, with $\lim_{x\to 0+} \varphi(x) = 0$ such that*

$$\left| (1-r) \sum_{n=0}^{\infty} a_n r^n \right| \le \varphi(1-r)$$

for $0 < r < 1$. Then for every $N \ge d \ge 1$

$$\left| \frac{1}{N} \sum_{n=0}^{N-1} a_n \right| \le c_1' e^{c_2' d} \varphi\left(c_3' \frac{d}{N} \right) + c_4' \frac{e^{c_2' d}}{N} + \frac{c_5'}{d}$$

with absolute constants $c_1', c_2', c_3', c_4', c_5' > 0$.

Proof. Set $h(x) = \chi_{[1/2,1]}$ and $N(x) = [\log 2 / \log(1/x)]$. Then

$$(1-r) \sum_{n\ge 0} a_n h(r^n) \sim \frac{\log 2}{N(r)} \sum_{n=0}^{N(r)} a_n.$$

Since $h(x)$ is of bounded variation we can apply Lemma 2.62 to estimate

$$\begin{aligned}
(1-r)\sum_{n\ge 0} a_n h(r^n) &= (1-r)\sum_{n\ge 0} a_n p_1(r^n) + (1-r)\sum_{n\ge 0} a_n (h(r^n) - p_1(r^n)) \\
&= \sum_{m=1}^{d} b_m^{(1)} (1-r) \sum_{n\ge 0} a_n r^{mn} + (1-r) \sum_{n\ge 0} a_n (h(r^n) - p_1(r^n)) \\
&\le \varphi(1-r^d) \sum_{m=1}^{d} |b_m^{(1)}| + \sum_{n\ge 0} (1-r) r^n q_1(r^n) \\
&\le e^{c_1 d} \varphi(1-r^d) + \frac{c_2}{d} + c_3 (1-r) e^{c_1 d},
\end{aligned}$$

where we have used that

$$q_1(x) = (1-x)q(x) = \sum_{m=1}^{d} (b_m^{(2)} - b_m^{(1)})x^{m-1} = \sum_{m=0}^{d-1} b_m^{(3)} x^m$$

satisfies

$$\begin{aligned}
\sum_{n\geq 0}(1-r)r^n q_1(r^n) &= \sum_{m=0}^{d-1} b_m^{(3)} \frac{1-r}{1-r^{m+1}} \\
&= \sum_{m=0}^{d-1} \frac{b_m^{(3)}}{m+1} + \mathcal{O}\left((1-r)\sum_{m=0}^{d-1} |b_m^{(3)}|\right) \\
&= \int_0^1 q_1(x)\,dx + \mathcal{O}\left((1-r)e^{c_1 d}\right) \\
&\leq \frac{c_2}{d} + c_(1-r)e^{c_1 d}.
\end{aligned}$$

In the same manner we obain a lower bound. Since

$$r = 2^{-1/(N(r)+\mathcal{O}(1))} = 1 - \frac{\log 2}{N(r)} + \mathcal{O}\left(\frac{1}{N(r)^2}\right)$$

the result follows immediately. □

Proof of (2.110). Note that $D_r^*(\mathbf{x}_n) \geq 1 - r$ (see Theorem 2.67). Hence we immediately obtain

$$D_N(\mathbf{x}_n) \leq c_1' e^{c_2' d} D_r^*(\mathbf{x}_n) + c_4' \frac{e^{c_2' d}}{N} + \frac{c_5'}{d} \leq C_1 e^{c_2' d} D_r^*(\mathbf{x}_n) + \frac{C_2}{d},$$

where $r = 2^{-d/(N+1)}$.

Then we set $d = \left[\frac{1}{2c_2'} \log \frac{1}{D_r^*}\right] - 1$

and obtain

$$D_N(\mathbf{x}_N) \ll \left(\log \frac{1}{D_r^*(\mathbf{x}_n)}\right)^{-1},$$

which proves (2.110). □

Furthermore there are analoga to the KOKSMA-HLAWKA (Theorem 1.14) and the ERDŐS-TURÁN-KOKSMA (Theorem 1.21) inequalities.

Theorem 2.64 *Let $(\mathbf{x}_n)_{n\geq 1}$ be a sequence in the k-dimensional space $\mathbf{R}^k$ and let f be a function that is of bounded variation $V(f)$ on $[0,1]^k$ in the sense of* HARDY *and* KRAUSE. *Then, for any $0 < r < 1$ we have*

$$\left|(1-r)\sum_{n=0}^{\infty} f(\{\mathbf{x}_n\})r^n - \int_{[0,1]^k} f(\mathbf{x})\,d\mathbf{x}\right| \leq V(f) D_r(\mathbf{x}_n).$$

Theorem 2.65 *Let $(\mathbf{x}_n)_{n\geq 1}$ be a sequence in the k-dimensional space $\mathbf{R}^k$ and H an arbitrary positive integer. Then for any $0 < r < 1$*

$$D_r(\mathbf{x}_n) \leq \left(\frac{3}{2}\right)^k \left(\frac{2}{H+1} + (1-r) \sum_{0<\|\mathbf{h}\|_\infty \leq H} \frac{1}{r(\mathbf{h})} \left| \sum_{n=1}^{\infty} e(\mathbf{h}\cdot\mathbf{x}_n) r^n \right| \right), \quad (2.111)$$

where $r(\mathbf{h}) = \prod_{i=1}^{k} \max\{1, |h_i|\}$ for $\mathbf{h} = (h_1, \ldots, h_k) \in \mathbf{Z}^k$.

Next we want to discuss special sequences of the form $(n\alpha)_{n\geq 1}$, where α is irrational.

Theorem 2.66 *Let α be of finite approximation type $\eta \geq 1$. Then the* Abel *discrepancy satisfies*

$$D_r(n\alpha) = \mathcal{O}\left((1-r)^{\frac{1}{\eta}-\varepsilon}\right)$$

and

$$D_r(n\alpha) = \Omega\left((1-r)^{\frac{1}{\eta}+\varepsilon}\right)$$

for any $\varepsilon > 0$.

Sketch Proof. The upper bound follows as in the classical case from the Erdős-Turán-Koksma inequality Theorem 2.65 (compare with Kuipers and Niederreiter [983]).

The lower bound is trivial for $\eta = 1$, we assume $\eta > 1$ for the rest of the proof. For given $\varepsilon > 0$, choose a real number δ with

$$0 < \delta < \min\left(\eta - 1, \frac{\varepsilon\eta^2}{\varepsilon\eta + 1}\right).$$

Then determine $\gamma > 0$ from the equation

$$\frac{1+\gamma}{\eta - \delta} = \frac{1}{\eta} + \varepsilon.$$

There exist infinitely many positive integers q and corresponding integers p such that

$$\left|\alpha - \frac{p}{q}\right| < q^{-1-\eta+(\delta/2)};$$

see Kuipers and Niederreiter [983, chapter 2].

Choose one such q, and set $N = [q^{\eta-\delta}]$ and $r = 1 - N^{-1/(1+\gamma)}$. By writing

$$\alpha = \frac{p}{q} + \theta q^{-1-\eta+(\delta/2)} \quad \text{with} \quad |\theta| < 1,$$

we obtain for $1 \leq n \leq N$,

$$n\alpha = \frac{np}{q} + \theta_n \quad \text{with} \quad |\theta_n| < q^{-1-(\delta/2)}.$$

Thus none of the numbers $0, \{\alpha\}, \{2\alpha\}, \ldots, \{N\alpha\}$ lies in the interval

$$J = [q^{-1-(\delta/2)}, q^{-1} - q^{-1-(\delta/2)}).$$

Therefore,

$$D_r^*(n\alpha) \geq \lambda_1(J) - (1-r)\sum_{n=0}^{\infty} \chi_J(\{n\alpha\})r^n \geq \lambda_1(J) - (1-r)\sum_{n=N+1}^{\infty} r^n = \lambda_1(J) - r^{N+1}.$$

For sufficiently large q, we have $\lambda_1(J) \geq 1/2q$. Moreover,

$$q^{\eta-\delta} \leq 2N = 2(1-r)^{-1-\gamma},$$

and so

$$q \leq 2^{1/(\eta-\delta)}(1-r)^{-(1+\gamma)/(\eta-\delta)} < 2(1-r)^{-(1/\eta)-\varepsilon}.$$

Thus

$$D_r^*(n\alpha) > \frac{1}{4}(1-r)^{(1/\eta)+\varepsilon} - r^{(1-r)^{-1-\gamma}}$$

for an infinite sequence of values of r tending to 1.
From this the lower bound of the theorem follows easily. □

It is worth mentioning that in the case $k = 1$ there exists a similar optimal lower bound for the ABEL discrepancy $D_r(x_n)$ as it is $\frac{1}{N}$ for the usual discrepancy $D_N(x_n)$.

Theorem 2.67 *Any real sequence* $(x_n)_{n\geq 1}$ *satisfies*

$$D_r(x_n) \geq 1 - r$$

for all $0 < r < 1$.

On the other hand for every $0 < r < 1$ *there exists a sequence* $(x_n)_{n\geq 1}$ *depending on* r *such that*

$$D_r(x_n) = 1 - r.$$

Sketch Proof. The lower bound follows immediately by considering small intervals containing the first point of the sequence. The second assertion follows from the example

$$x_n = \frac{2 - r^n - r^{n+1}}{2}.$$

□

Finally it should be mentioned that it is possible to introduce a weighted ABEL method by

$$\lim_{s\to 0+} s\sum_{n=1}^{\infty} p_n x_n e^{-sP_n} = x.$$

Again we have a similar property to the above one.

Theorem 2.68 *Let M_P, $P = (p_n)_{n\geq 1}$, be a weighted mean such that $P_N \to \infty$ and $\lim_{n\to\infty} p_n/P_n = 0$. Then the weighted* ABEL *method is regular and is equivalent to M_P for bounded sequences.*

Proof. First we show

$$\lim_{s\to 0+} s \sum_{n=1}^{\infty} p_n e^{-sP_n} = 1. \tag{2.112}$$

Evidently (2.112) implies that the weighted ABEL method is regular.

Since $\lim_{n\to\infty} P_n/P_{n-1} = 1$ we have $P_n \leq P_{n-1}(1+\varepsilon)$ for $n \geq N(\varepsilon)$. Thus

$$\int_{P_{n-1}}^{P_n} e^{-(1+\varepsilon)st}\,dt \leq p_n e^{-(1+\varepsilon)sP_{n-1}} \leq p_n e^{-sP_n} \leq \int_{P_{n-1}}^{P_n} e^{-st}\,dt$$

yields

$$\frac{1}{1+\varepsilon} + sK(\varepsilon) \leq s \sum_{n=1}^{\infty} p_n e^{-sP_n} \leq 1$$

for some constant $K(\varepsilon)$. Hence (2.112) follows.

Next suppose that a bounded sequence $(x_n)_{n\geq 1}$ satisfies M_P–$\lim x_n = x$. Since x_n is bounded it is no loss of generality to assume that $x_n > 0$. Set $y_n = \sum_{k=1}^{n} p_k x_k$. Then $y_n \sim xP_n$. As above it follows that

$$\lim_{s\to 0+} s^2 \sum_{n=1}^{\infty} p_n P_n e^{-sP_n} = 1,$$

which implies

$$\lim_{s\to 0+} s^2 \sum_{n=1}^{\infty} p_n y_n e^{-sP_n} = x. \tag{2.113}$$

Now use

$$\sum_{k=1}^{\infty} p_k y_k e^{-sP_k} = \sum_{n=1}^{\infty} p_n x_n \sum_{k=n}^{\infty} p_k e^{-sP_k},$$

and (for $n \geq N(\varepsilon)$)

$$\begin{aligned}
\frac{e^{-s(1+\varepsilon)P_{n-1}}}{s(1+\varepsilon)} &= \int_{P_{n-1}}^{\infty} e^{-s(1+\varepsilon)t}\,dt \leq \sum_{k=n}^{\infty} p_k e^{-sP_k} \\
&\leq \int_{P_n}^{\infty} e^{-st}\,dt + p_n e^{-sP_n} \\
&= \frac{e^{-sP_n}}{s} + p_n e^{-sP_n}
\end{aligned}$$

to obtain

$$sK(\varepsilon) + \frac{s}{1+\varepsilon}\sum_{n=1}^{\infty} p_n x_n e^{-s(1+\varepsilon)P_n} \le s^2 \sum_{k=1}^{\infty} p_k y_k e^{-sP_k}$$

$$\le s \sum_{n=1}^{\infty} p_n x_n e^{-sP_n} + s^2 \sum_{n=1}^{\infty} p_n^2 x_n e^{-sP_n}.$$

Since $p_n = o(P_n)$ we have

$$\lim_{s\to 0+} s^2 \sum_{n=1}^{\infty} p_n^2 x_n e^{-sP_n} = 0.$$

Thus (x_n) is limitable with respect to the weighted ABEL method with limit x.

Finally, suppose that $x_n > 0$ and

$$\lim_{s\to 0+} s \sum_{n=1}^{\infty} p_n x_n e^{-sP_n} = x. \tag{2.114}$$

By a variation of KARAMATA's method we show that (2.114) implies M_P–$\lim x_n = x$. Obviously, it follows from (2.114) that for $f(x) = x^k$, $k \ge 0$,

$$\lim_{s\to 0+} s \sum_{n=1}^{\infty} p_n x_n e^{-sP_n} f(e^{-sP_n}) = x \int_0^1 f(t)\, dt. \tag{2.115}$$

Thus (2.115) holds for all RIEMANN integrable functions $f(x)$. Especially if we set $f(x) = 0$ for $0 \le x < e^{-1}$ and $f(x) = 1/x$ for $e^{-1} \le x \le 1$ and $s = 1/P_N$ we obtain

$$\lim_{N\to\infty} \frac{1}{P_N} \sum_{P_n \le P_N} p_n x_n = x \int_{e^{-1}}^1 \frac{dx}{x} = x.$$

This completes the proof of Theorem 2.68. □

Notes

Tichy [1809] considered double sequences with respect to special summation methods were considered. Special triangular arrays were studied by Nowak [1373, 1374]. For further results on double sequences and c.u.d. functions we refer to Taschner [1786]. In Tichy [1810, 1814] the discrepancy of chains is estimated, answering a problem of Hlawka [801]; see also Hlawka [809]. General lower bounds for the discrepancy with respect to weighted means and metric results can be found in Tichy [1819, 1818, 1817]. Improved bounds are due to Niederreiter and Tichy [1363]. Among other results a necessary condition of Topuzoglu [1868] (see also Sinnadurai [1691]) for the u.d. of sequences $(a_n x)$ is generalized to weighted means. Niederreiter [1304] investigated the distribution mod 1 of monotone sequences with respect to general weights. Among other results he gave a necessary and sufficient condition for the existence of u.d. sequences with respect to weighted means. For further results on u.d. with respect to weighted means see Tichy [1823, 1827, 1821]; see also Ohkubo [1390, 1391] and for some general results Nakajima and Ohkubo [1256]. Dowidar [472, 473] showed the u.d. of $(n^2\alpha)$

with respect to special weighted means; in [471] the same author considered u.d. of integer sequences with respect to matrix summation methods.

Losert [1104] gave a necessary and sufficient condition on a weighted mean P such that almost all sequences are P-u.d. Note that the classical Hill-condition (see Kuipers and Niederreiter [983]) is only sufficient but in general (for nonmonotonic weights) not necessary.

Saffari and Vaughan [1587, 1588, 1589] proved limit theorems for sequences of rational numbers with respect to special weighted means and studied applications to Dirichlet's divisor problem. Vaaler [1884, 1885] proved Tauberian theorems with applications to u.d. of sequences mod 1 with respect to weighted means.

Results on weighted exponential sums related to Koksma's inequality are due to Horbowicz and Niederreiter [847]. Hlawka [812] investigated special sequences of the type $(\alpha n + \beta \log n)$ with respect to the harmonic mean and relations to the convergence of power series at the boundary of the convergence circle. Tichy [1820] presented quantitative formulations of these results. He obtained some upper estimates of the weighted discrepancies of the sequences contained in some wide class, but unfortunately, the upper estimate of the logarithmic discrepancy of the sequence $(\alpha n + \beta \log n)$, namely

$$D_N^{\log}(\alpha n + \beta \log n) \leq \frac{c(\beta)}{\log N}$$

does not follow from the general upper bounds. In Tichy and Turnwald [1847] these investigations were continued and a final solution is due to Baker and Harman [94]. A generalization to sequences and weights satisfying some analytic properties is due to Ohkubo [1392]. Well distribution with respect to weighted means was studied by Drmota and Tichy [489] and Schatte [1609, 1611, 1607]. For further remarks on well distributed sequences we refer to Sobol [1703]. A weighted diophany was studied by Lev [1062].

Kemperman [918] studied the distribution of slowly changing sequences, i.e. $x_{n+1} - x_n$ tending to 0, with respect to general summation methods.

Benford's law gives rise to an important field of applications of u.d. sequences with respect to the limitation method H_∞. Benford's law is the phenomenon that the occurence of the first digit of the mantissa in a random sequence is higher than that of the remaining ones, see Benford [164], Diaconis [458]. For various contributions to Benford's law (mainly concerned with special sequences) we refer to Boyle [293], Deakin [443], Flehinger [607], Giuliano Antonioni [44], Goto [667], Jager and Liardet [870], Hill [777, 778], Kanemitsu, Nagasaka, Rauzy and Shiue [894], Konheim [947], Nagasaka, Kanemitsu and Shiue [1239], Shiue and Nagasaka [1617], Newcomb [1268], Pavlov [1420], Pinkham [1439], Raimi [1498], Filipponi and Menicocci [597], Schatte [1602, 1604, 1605, 1603, 1606, 1610, 1608, 1612, 1613, 1614, 1616], Sentance [1665], Tichy [1834], Washington [1946], and Webb [1949]. In several of these papers discrepancy estimates with respect to the limitation method H_∞ are proved; see also [1827, 1825]. Further results on digital representations of numbers are discussed in Tichy [1836]. Kunoff [996] established Benford's law for the sequence $n!$; Hlawka [812] provided a discrepancy bound. Fuchs and Letta [629] investigated Benford's law in connection with lower and upper densities of integer sequences. For related sequences we refer to Fiorito, Musmeci and Strano [599], Sury [1765] and Too [1867]. Jech [872] studied the logarithmic distribution of the leading digits and connections to finitely additive measures.

Akita, Iseka, and Kano [11] studied the convergence of series of the type $\sum n^{-\alpha} e(\theta n^\beta)$ in connection with u.d.. The systematic study of uniform distribution with respect to Abel's summation method was initiated by Hlawka [793] and continued by Niederreiter [1285]. In Burg, Drmota and Tichy [316] similar results for the Le Roi summation method are developed, see also Drmota [484].

In Hlawka [798, 799, 797] a concept of discrepancy with respect to polynomials instead of characteristic functions of intervals was introduced, Schmidt [1640] proved the optimality of the discrepancy bounds given there. Tichy [1822] extended Hlawka's bounds relating the polynomial discrepancy and the usual discrepancy to weighted means. Klinger and Tichy [938] gave generalizations to the spherical cap discrepancy.

2.3 Continuous Uniform Distribution

2.3.1 Basic Results

Let $\mathbf{x} : [0,\infty) \to \mathbf{R}^k$ be a continuous function, which may be interpreted as a motion $\mathbf{x}(t)$ in time $t \ge 0$. As for sequences we can reduce $\mathbf{x}(t)$ modulo 1, i.e. $\{\mathbf{x}(t)\}$, which is nothing else than a continuous motion on $\mathbf{R}^k/\mathbf{Z}^k$. Instead of counting the number of points in a (k-dimensional) interval I, we will measure the time of stay of $\mathbf{x}(t)$ in I. This leads to the definition

Definition 2.69 *A continuous function* $\mathbf{x} : [0,\infty) \to \mathbf{R}^k$ *is said to be* continuously uniformly distributed modulo 1 *(for short c.u.d. mod 1) if for every interval* $I \subseteq \mathbf{R}^k/\mathbf{Z}^k$ *we have*

$$\lim_{T\to\infty} \frac{1}{T} \int_0^T \chi_I(\mathbf{x}(t))\,dt = \lambda_k(I). \tag{2.116}$$

It is clear that any shifted function $\mathbf{x}_\tau(t) = \mathbf{x}(t+\tau)$ ($\tau \ge 0$) of a c.u.d. function is again c.u.d. But the convergence in (2.116) need not be uniform. Therefore we can introduce a stronger concept.

Definition 2.70 *A continuous function* $\mathbf{x} : [0,\infty) \to \mathbf{R}^k$ *is said to be* continuously well distributed modulo 1 *(for short c.w.d. mod 1) if for every interval* $I \subseteq \mathbf{R}^k/\mathbf{Z}^k$ *the convergence in*

$$\lim_{T\to\infty} \frac{1}{T} \int_0^T \chi_I(\mathbf{x}(t+\tau))\,dt = \lambda_k(I) \tag{2.117}$$

is uniform for $\tau \ge 0$.

As in the case for sequences there are several criteria to check whether a function $\mathbf{x}(t)$ is c.u.d. (resp. c.w.d.) mod 1 or not.

Theorem 2.71 (Criterion A) *A continuous function* $\mathbf{x} : [0,\infty) \to \mathbf{R}^k$ *is c.u.d. mod 1 if and only if*

$$\lim_{T\to\infty} \frac{1}{T} \int_0^T f(\{\mathbf{x}(t)\})\,dt = \int_{[0,1]^k} f\,d\lambda_k \tag{2.118}$$

holds for all RIEMANN *integrable functions* $f : [0,1]^k \to \mathbf{R}$ *and it is c.w.d. mod 1 if and only if*

$$\lim_{T\to\infty} \frac{1}{T} \int_0^T f(\{\mathbf{x}(t+\tau)\})\,dt = \int_{[0,1]^k} f\,d\lambda_k \tag{2.119}$$

holds for all RIEMANN *integrable functions* $f : [0,1]^k \to \mathbf{R}$ *uniformly for* $\tau \ge 0$.

Theorem 2.72 (Criterion B) *A continuous function* $\mathbf{x} : [0,\infty) \to \mathbf{R}^k$ *is c.u.d. (resp. c.w.d.) mod 1 if and only if (2.118) (resp. (2.119)) holds for all continuous functions* $f : [0,1]^k \to \mathbf{R}$ *(uniformly for* $\tau \ge 0$*).*

Theorem 2.73 (Criterion C, WEYL's criterion) *A continuous function* $\mathbf{x} : [0,\infty) \to \mathbf{R}^k$ *is c.u.d. mod 1 if and only if*

$$\lim_{T\to\infty} \frac{1}{T} \int_0^T e(\mathbf{h}\cdot\mathbf{x}(t))\,dt = 0 \tag{2.120}$$

holds for all non-zero integral lattice points $\mathbf{h} \in \mathbf{Z}^k \setminus \{(0,\ldots,0)\}$ *and it is c.w.d. mod 1 if and only if*

$$\lim_{T\to\infty} \frac{1}{T} \int_0^T e(\mathbf{h}\cdot\mathbf{x}(t+\tau))\,dt = 0 \tag{2.121}$$

holds for all non-zero integral lattice points $\mathbf{h} \in \mathbf{Z}^k \setminus \{(0,\ldots,0)\}$ *uniformly for* $\tau \geq 0$.

It should be noted that WEYL has already discussed a special type of c.u.d. funtions mod 1 in his groundbreaking paper [1953] in connection with a problem in statistical mechanics. He used linear functions $\mathbf{x}(t) = \underline{\alpha}\,t$, where $\underline{\alpha} = (\alpha_1,\ldots,\alpha_k) \in \mathbf{R}^k$. By WEYL's criterion it turns out that $\mathbf{x}(t) = \underline{\alpha}\,t$ is c.u.d. mod 1 if and only if $\alpha_1,\ldots,\alpha_k$ are linearly independent over $\mathbf{Q}$. (This is of course a weaker condition than that for the sequence $\mathbf{x}_n = \underline{\alpha}\,n$, where the linear independence of $1,\alpha_1,\ldots,\alpha_k$ is required.) Moreover it is easily seen that $\{\underline{\alpha}\,t\}$ is dense in $\mathbf{R}^k/\mathbf{Z}^k$ if and only if $\mathbf{x}(t) = \underline{\alpha}\,t$ is c.u.d. mod 1.

WEYL's criterion enables us to reduce the higher dimensional case to the case dimensional one. It immediately follows that $\mathbf{x} : [0,\infty) \to \mathbf{R}^k$ is c.u.d. (resp. c.w.d.) mod 1 if and only if the one dimensional functions $\mathbf{h}\cdot\mathbf{x}(t)$ are c.u.d. (resp. c.w.d.) mod 1 for all non-zero lattice points $\mathbf{h} \in \mathbf{Z}^k \setminus \{(0,\ldots,0)\}$. This justifies that most of the following general theorems only concern the one dimensional case.

Furthermore we can introduce a notion of discrepancy.

Definition 2.74 *Let* $\mathbf{x} : [0,T] \to \mathbf{R}^k$ *be a continuous function. Then the number*

$$D_T(\mathbf{x}(t)) = \sup_{I\subseteq \mathbf{R}^k/\mathbf{Z}^k} \left| \frac{1}{T}\int_0^T \chi_I(\mathbf{x}(t))\,dt - \lambda_k(I)\right| \tag{2.122}$$

is called discrepancy of $\mathbf{x}(t)$ *and the number*

$$D_T^*(\mathbf{x}(t)) = \sup_{\mathbf{y}\in[0,1]^k} \left| \frac{1}{T}\int_0^T \chi_{[0,\mathbf{y})}(\mathbf{x}(t)) - \lambda_k([0,\mathbf{y}))\right| \tag{2.123}$$

is called star discrepancy of $\mathbf{x}(t)$. *For a continuous function* $\mathbf{x} : [0,\infty) \to \mathbf{R}^k$, $D_T(\mathbf{x}(t))$ *(resp.* $D_T^*(\mathbf{x}(t))$*) should denote the (star) discrepancy of the restriction of* $\mathbf{x}$ *to* $[0,T]$ *and is called (star) discrepancy, too.*

Similarly to the discrete case, the notion of uniform distribution can be characterized with the help of discrepancy.

Theorem 2.75 *A continuous function* $\mathbf{x} : [0,\infty) \to \mathbf{R}^k$ *is c.u.d. mod 1 if and only if*

$$\lim_{T\to\infty} D_T(\mathbf{x}(t)) = 0. \tag{2.124}$$

Since $D_T^*(\mathbf{x}(t)) \le D_T(\mathbf{x}(t)) \le 4^k D_T^*(\mathbf{x}(t))$ the previous theorem holds for the star discrepancy, too.

Moreover, we can generalize KOKSMA-HLAWKA's (Theorem 1.14) inequality and ERDŐS-TURÁN-KOKSMA's (Theorem 1.21) inequality to the continuous case.

Theorem 2.76 *Let f be of bounded variation on $[0,1]^k$ in the sense of* HARDY *and* KRAUSE. *Let $\mathbf{x}^{(i_1,\ldots,i_l)} = \mathbf{x}^{(F)}$ denote the projection of the (measurable) function $\mathbf{x} : [0,T] \to [0,1]^k$, on the $(k-l)$-dimensional face F of $[0,1]^k$ given by $x_{i_1} = \cdots = x_{i_l} = 1$. Then we have*

$$\left| \frac{1}{T} \int_0^T f(\mathbf{x}(t))\, dt - \int_{[0,1]^k} f(\mathbf{y})\, d\mathbf{y} \right| \le \sum_{l=0}^{k-1} \sum_{F_l} D_T^*(\mathbf{x}^{(F_l)}(t)) V^{(k-l)}(f^{(F_l)}), \qquad (2.125)$$

where the second sum is extended over all $(k-l)$-dimensional faces F_l of the form $x_{i_1} = \cdots = x_{i_l} = 1$. The discrepancy $D_N^*(\mathbf{x}^{(F_l)}(t))$ is computed in the face of $[0,1]^k$ in which $\mathbf{x}^{(F_l)}(t)$ is contained.

Theorem 2.77 *Let $\mathbf{x} : [0,T] \to \mathbf{R}^k$ be a continuous funtion and H an arbitrary positive integer. Then*

$$D_T(\mathbf{x}(t)) \le \left(\frac{3}{2}\right)^k \left(\frac{2}{H+1} + \sum_{0<\|\mathbf{h}\|_\infty \le H} \frac{1}{r(\mathbf{h})} \left| \frac{1}{T} \int_0^T e(\mathbf{h}\cdot\mathbf{x}(t))\, dt \right| \right), \qquad (2.126)$$

where $r(\mathbf{h}) = \prod_{i=1}^{k} \max\{1, |h_i|\}$ for $\mathbf{h} = (h_1, \ldots, h_k) \in \mathbf{Z}^k$.

First we will list some elementary properties and difference theorems.

Theorem 2.78 *Let $\mathbf{x} : [0,\infty) \to \mathbf{R}$ be convex and strictly increasing for $t > t_0$. Then $\mathbf{x}(t)$ is c.u.d. mod 1. Especially, we have*

$$D_T(\mathbf{x}(t)) = \mathcal{O}\left(\frac{1}{T}\right).$$

Proof. W.l.o.g. we may assume that $\mathbf{x}(t)$ is convex and strictly increasing for $t > 0$ and that $\mathbf{x}(0) = 0$. Let $\mathbf{y}(s)$ denote the inverse function of $\mathbf{x}(t)$ which is obviously concave. Hence for every T, for which $\mathbf{x}(T)$ is a positive integer, and for every $x \in [0,1]$ we have

$$\begin{aligned} \int_0^T \chi_{[0,x)}(\mathbf{x}(t)) - xT &= \sum_{n=0}^{\mathbf{x}(T)-1} (\mathbf{y}(n+x) - \mathbf{y}(n+1)) - xT \\ &= \sum_{n=0}^{\mathbf{x}(T)-1} (\mathbf{y}(n+x) - x\mathbf{y}(n) - (1-x)\mathbf{y}(n+1)) \\ &= x(\mathbf{y}(x) - \mathbf{y}(0)) + (1-x)(\mathbf{y}(T-1+x) - \mathbf{y}(T)) \end{aligned}$$

$$+ \sum_{n=1}^{\mathbf{x}(T)-1} ((1-x)\mathbf{y}(n-1+x) + x\mathbf{y}(n+x) - \mathbf{y}(n))$$
$$\leq \mathbf{y}(1).$$

Thus $T\,D_T(\mathbf{x}(t)) = \mathcal{O}(1)$. □

Theorem 2.79 *Suppose that* $\mathbf{x} : [0,\infty) \to \mathbf{R}$ *is twice continuously differentiable such that* $\operatorname{sgn}(\mathbf{x}''(t))$ *is constant for* $t > t_0$. *If* $|\mathbf{x}'(t)| \geq C$ $(t > t_0)$ *for some positive constant then* $\mathbf{x}(t)$ *is c.w.d. mod 1. If* $\lim_{t\to\infty} |t\mathbf{x}'(t)| = \infty$ *then* $\mathbf{x}(t)$ *is c.u.d. mod 1.*

Proof. Let $\tau \geq 0$ and $T \geq t_0$. Then we have

$$\left| \int_{t_0}^{T} e(h\mathbf{x}(t+\tau))\,dt \right|$$
$$= \left| \int_{t_0}^{T} e^{2\pi i h\mathbf{x}(t+\tau)} \frac{2\pi i h\mathbf{x}'(t+\tau)}{2\pi i h\mathbf{x}'(t+\tau)}\,dt \right|$$
$$\leq \left| e^{2\pi i h\mathbf{x}(t+\tau)} \frac{1}{2\pi i h\mathbf{x}'(t+\tau)} \right|_{t=t_0}^{T} + \frac{1}{2\pi|h|} \int_{t_0}^{T} \left| \frac{\mathbf{x}''(t+\tau)}{\mathbf{x}'(t+\tau)^2} \right| dt$$
$$\leq \frac{1}{2\pi|h|} \left(\left| \frac{1}{\mathbf{x}'(t_0+\tau)} \right| + \left| \frac{1}{\mathbf{x}'(T+\tau)} \right| + \left| \int_{t_0}^{T} \frac{d}{dt} \left(\frac{1}{\mathbf{x}'(t+\tau)} \right) dt \right| \right)$$
$$\leq \frac{1}{\pi|h|} \left(\left| \frac{1}{\mathbf{x}'(t_0+\tau)} \right| + \left| \frac{1}{\mathbf{x}'(T+\tau)} \right| \right).$$

Hence, we get in the first case

$$\left| \frac{1}{T} \int_0^T e(h\mathbf{x}(t+\tau))\,dt \right| \leq \frac{1}{T} \left(t_0 + \frac{2}{\pi C|h|} \right),$$

which implies that $\mathbf{x}(t)$ is c.w.d. mod 1. In the second case,

$$\left| \frac{1}{T} \int_0^T e(h\mathbf{x}(t))\,dt \right| \leq \frac{1}{T} \left(t_0 + \frac{1}{\pi|h\mathbf{x}'(t_0)|} + \frac{1}{\pi|h\mathbf{x}'(T)|} \right)$$

ensures that $\mathbf{x}(t)$ is c.u.d. mod 1.□

The next theorem is an analogon to the difference theorem of VAN DER CORPUT.

Theorem 2.80 *Let* $\mathbf{x} : [0,\infty) \to \mathbf{R}$ *be given and suppose that for all positive integers* R *the function* $\mathbf{x}(t+R) - \mathbf{x}(t)$ *is c.u.d. (resp. c.w.d.) mod 1. Then* $\mathbf{x}(t)$ *is c.u.d. (resp. c.w.d.) mod 1.*

This theorem is an immediate consequence of an continuous analogon to the inequality of VAN DER CORPUT which can be proved as in the discrete case (see KUIPERS and NIEDERREITER [983] or HLAWKA [815]).

Proposition 2.81 *Let $f : [0,T] \to \mathbf{R}$ be a measurable function and suppose that $R \le T$. Then*

$$\left|\frac{1}{T}\int_0^T e(f(t))\,dt\right|$$
$$\le \left(\frac{1}{R}+\frac{1}{T}-\frac{1}{RT}\right)\left(1+2\sum_{r=1}^{R}\left(1-\frac{r}{R}\right)\left|\frac{1}{T}\int_0^{T-r} e(f(t+r)-f(t))\,dt\right|\right).$$

The next theorem is interesting since it is only true for c.w.d. functions mod 1.

Theorem 2.82 *Suppose that $\mathbf{x} : [0,\infty) \to \mathbf{R}$ and $\mathbf{y} : [0,\infty) \to \mathbf{R}$ are continuously differentiable such that*

$$\lim_{t\to\infty}(\mathbf{x}'(t)-\mathbf{y}'(t)) = 0.$$

If $\mathbf{x}(t)$ is c.w.d. mod 1 then $\mathbf{y}(t)$ is c.w.d. mod 1, too.

Proof. Set $f(t) = \mathbf{x}'(t) - \mathbf{y}'(t)$ and suppose that $|f(t)| \le M$ for $t \in [a,b]$. Then

$$\mathbf{y}(s) = -\int_a^s f(t)\,dt + \mathbf{x}(s) + \mathbf{y}(a) - \mathbf{x}(a)$$

implies

$$\left|\int_a^b e(h\mathbf{y}(t))\,dt\right| - \left|\int_a^b e(h\mathbf{x}(t))\,dt\right|$$
$$\le \left|\int_a^b e(h\mathbf{y}(t))\,dt - e(h(\mathbf{y}(a)-\mathbf{x}(a)))\int_a^b e(h\mathbf{x}(t))\,dt\right|$$
$$= \left|\int_a^b \left(e\left(-h\int_a^s f(t)\,dt\right)-1\right) e(h(\mathbf{y}(a)-\mathbf{x}(a)))e(h\mathbf{x}(s))\,ds\right|$$
$$\le \int_a^b \left|2\pi h\int_a^s f(t)\,dt\right|\,ds$$
$$\le \pi(b-a)^2 M|h|.$$

Now, suppose that $\mathbf{x}(t)$ is c.w.d. mod 1, i.e. for every integer $h \ne 0$ and every $\varepsilon > 0$ there exists $T_0 \ge 0$ such that

$$\left|\frac{1}{T}\int_0^T e(h\mathbf{x}(t+\tau))\,dt\right| \le \varepsilon$$

for $T \ge T_0/\varepsilon$ and $\tau \ge 0$. Furthermore, suppose that $|\mathbf{x}'(t)-\mathbf{y}'(t)| \le \varepsilon/T_0$ for $t \ge \tau_0$. Hence we get for $T \ge T_0/\varepsilon$ and $\tau \ge \tau_0$

$$\left|\frac{1}{T}\int_0^T e(h\mathbf{y}(t+\tau))\,dt\right| \le \left|\frac{1}{T}\sum_{q=1}^{[T/T_0]}\int_{(q-1)T_0}^{qT_0} e(h\mathbf{y}(t+\tau))\,dt\right| + \frac{T_0}{T}$$

$$\leq \frac{1}{T}\left[\frac{T}{T_0}\right]\left(\varepsilon T_0 + \pi T_0^2 \frac{\varepsilon}{T_0}|h|\right) + \frac{T_0}{T}$$
$$\leq (2+\pi|h|)\varepsilon + \frac{T_0}{T}.$$

Thus we have for $\tau < \tau_0$

$$\left|\frac{1}{T}\int_0^T e(h\mathbf{y}(t+\tau))\,dt\right| = \left|\frac{1}{T}\left(\int_\tau^{\tau_0} + \int_{\tau_0}^{T+\tau_0} - \int_{T+\tau}^{T+\tau_0}\right) e(h\mathbf{y}(t)\,dt\right|$$
$$\leq \frac{2\tau_0}{T} + (2+\pi|h|)\varepsilon + \frac{T_0}{T},$$

which implies that $\mathbf{y}(t)$ is c.w.d. mod 1, too.□

The above theorems provide a lot of examples of c.u.d. and c.w.d. functions mod 1. For example, $\mathbf{x}(t) = p(t)$, where $p(t)$ is a non-constant polynomial and $\mathbf{x}(t) = t^\alpha$ ($\alpha > 1$) are c.w.d. mod 1 and therefore c.u.d. mod 1. It is interesting to see that $\mathbf{x}(t) = t^\alpha$ ($0 < \alpha < 1$) is c.u.d. mod 1 (Theorem 2.79) but not c.w.d. mod 1 (Theorem 2.82). Note that all these theorems need assumptions on the sign of the second derivative of $\mathbf{x}(t)$. The following theorem on entire functions of very small growth does not require an assumption of this kind. For similar results concerning sequences we refer to the Notes of Section 1.2.

Theorem 2.83 *Let $f(z)$ be a (non-constant) entire function satisfying*

$$\limsup_{r\to\infty} \frac{\log\log M(r)}{\log\log r} < \frac{3}{2} \tag{2.127}$$

(where $M(r) = \max_{|z|\leq r}|f(z)|$) such that all TAYLOR *coefficients of $f(z)$ are real. Then $\mathbf{x}(t) = f(t)$ ($t \geq 0$) is c.u.d. mod 1.*

By using Theorem 2.83 and WEYL's criterion (Theorem 2.73) we obtain as a corollary

Theorem 2.84 *Let $f(z)$ be an entire function satisfying (2.127) such that either the quotient $\Re(a_n)/\Im(a_n)$ is irrational for some* TAYLOR *coefficient a_n with $n \geq 1$ or there are two* TAYLOR *coefficients a_n, a_m ($1 \leq m < n$) with $\Re(a_n)\Im(a_m) \neq \Re(a_m)\Im(a_n)$. Then $\mathbf{x}(t) = (\Re(f(t)), \Im(f(t)))$ ($t \geq 0$) is c.u.d. mod 1.*

One basic tool for the proof is the following estimate.

Lemma 2.85 *Let $p(t) = at^N + a_1t^{N-1} + \ldots + a_{N-1}t + a_N$ be a polynomial of degree N with real coefficients. Then for all $A < B$*

$$\left|\int_A^B e(p(t))\,dt\right| < \frac{28}{|a|^{1/N}}. \tag{2.128}$$

Proof. Using the substitution $u = t(N|a|)^{1/N}$ for $N \geq 2$ we have to prove

$$\left| \int_\alpha^\beta e(q(u))\,du \right| \leq 28 \tag{2.129}$$

for any α, β and a polynomial $q(u) = \frac{1}{N}u^N + b_1u^{N-1} + \cdots + b_N$. Applying Theorem 3.4.1 of BOAS [243] we have $|q'(u)| \leq K$ for $u \in U$, where U is the union of at most $N-1$ intervals with LEBESGUE measure $\lambda_1(U) \leq 12K^{\frac{1}{N-1}}$. Therefore there are at most N subintervals contained in $[\alpha, \beta]$, where $|q'(u)| > K$. Moreover, since $q''(u)$ has at most $N-2$ zeroes there are at most $2N-2$ subintervals $I_j \subseteq [\alpha, \beta]$ such that in each of them $q''(u)$ has constant sign and $|q'(u)| > K$. As in the proof of Theorem 2.79 we obtain for such an interval

$$\left| \int_{I_j} e(q(u))\,du \right| \leq \frac{4}{\pi K} < \frac{2}{K}.$$

Combining this with the trivial bound

$$\left| \int_U e(q(u))\,du \right| \leq 12K^{\frac{1}{N-1}},$$

and choosing $K = N$ yields (2.129). Thus the proof is finished since the case $N = 1$ is trivial. □

Let

$$f(z) = \sum_{n=0}^{\infty} a_n z^n$$

be the TAYLOR expansion of $f(z)$. Since non-constant polynomials are always c.u.d. mod 1 (see Theorem 2.78 or Lemma 2.85) it suffices to discuss the case, where infinitely many TAYLOR coefficients a_n are non-zero.

Basically we will use the fact that (2.127) implies that there is some constant $c > 3$ such that

$$\lim_{n\to\infty} \frac{\log(1/|a_n|)}{n^c} = \infty. \tag{2.130}$$

By (2.127) there exist $\varepsilon > 0$ and $r_0 > 0$ such that

$$\log M(r) < (\log r)^{\frac{3}{2}-\varepsilon} \qquad \text{for } r \geq r_0.$$

Furthermore, by CAUCHY's inequality

$$|a_n| \leq \frac{M(r)}{r^n}.$$

Choosing $r \geq r_0$ (for sufficiently large n) such that $(\log r)^{\frac{1}{2}-\varepsilon} = 2n$ we get

$$\log \frac{1}{|a_n|} \geq n \log r - (\log r)^{\frac{3}{2}-\varepsilon} \gg n^{3+4\varepsilon},$$

which proves (2.130) for some c satisfying $3 < c < 3 + 4\varepsilon$.

Next we will consider the NEWTON polygon of the points $(n, \log \frac{1}{|a_n|})$. Let $n_0 = \min\{n : a_n \neq 0\}$. If n_r is defined set $l_r = \log(1/|a_{n_r}|)$ and

$$m_r = \sup\{m : \log(1/|a_n|) \geq l_r + m(n - n_r) \text{ for all } n \geq n_r\}$$

and define

$$n_{r+1} = \max\{n > n_r : \log(1/|a_n|) = l_r + m_r(n - n_r)\}.$$

Furthermore we set

$$p_r = \frac{n_r l_r}{n_r^2 - 1}, \quad P_r = e^{p_r}, \quad Q_r = e^{m_r} 4^{-n_r - 1}.$$

Lemma 2.86 *The quantities defined above satisfy the following properties:*

$$P_{r+1} < Q_r \quad (r \geq r_0), \tag{2.131}$$

$$\lim_{r\to\infty} 2^{n_r+2} e^{-m_r} |a_{n_r}| P_r^{n_r+1} = 0, \tag{2.132}$$

and

$$\lim_{r\to\infty} \frac{1}{P_r} \left(\frac{2}{|a_{n_r}|}\right)^{\frac{1}{n_r}} = 0. \tag{2.133}$$

Proof. In order to prove (2.131) we will show

$$\lim_{r\to\infty} \frac{m_r - p_{r+1}}{n_{r+1}^{c-2}} = \infty. \tag{2.134}$$

Obviously (2.134) implies (2.131).

Since $\lim_{r\to\infty} l_r/(n_r - 2) = \infty$, there exists some $r = r_1$ such that

$$\frac{l_r}{n_r - 2} \leq \frac{l_{r+1}}{n_{r+1} - 2}.$$

Since $l_{r+1} = l_r + m_r(n_{r+1} - n_r)$ we get

$$m_r - \frac{l_{r+1}}{n_{r+1} - 2} = \frac{n_r - 2}{n_{r+1} - n_r}\left(\frac{l_{r+1}}{n_{r+1} - 2} - \frac{l_r}{n_r - 2}\right) \geq 0,$$

and hence

$$\begin{aligned}
l_{r+2} &= l_{r+1} + m_{r+1}(n_{r+2} - n_{r+1}) \\
&\geq l_{r+1} + m_r(n_{r+2} - n_{r+1}) \\
&\geq l_{r+1} + \frac{l_{r+1}}{n_{r+1} - 2}(n_{r+2} - n_{r+1}) \\
&= \frac{n_{r+2} - 2}{n_{r+1} - 2} l_{r+1},
\end{aligned}$$

which gives by induction

$$\frac{l_r}{n_r-2} \le \frac{l_{r+1}}{n_{r+1}-2} \le m_r \qquad \text{for } r \ge r_1.$$

Consequently we get

$$\frac{m_r - p_{r+1}}{n_{r+1}^{c-2}} \ge \frac{l_{r+1}}{n_{r+1}^{c-2}} \left(\frac{1}{n_{r+1}-2} - \frac{1}{n_{r+1}-1} \right) \gg \frac{l_{r+1}}{n_{r+1}^{c}} \to \infty,$$

which proves (2.134).

Similarly we obtain

$$\begin{aligned} \frac{m_r + l_r - (n_r+1)p_r}{n_r^{c-2}} &= \frac{1}{n_r^{c-2}} \left(m_r - \frac{l_r}{n_r-1} \right) \\ &\ge \frac{l_r}{n_r^{c-2}} \left(\frac{1}{n_r-2} - \frac{1}{n_r-1} \right) \\ &\gg \frac{l_r}{n_r^{c}} \to \infty \end{aligned}$$

which gives (2.132).

Finally

$$\frac{p_r - l_r/n_r}{n_r^{c-3}} = \frac{l_r}{n_r^{c-3}} \left(\frac{n_r}{n_r^2-1} - \frac{1}{n_r} \right) \gg \frac{l_r}{n_r^{c}} \to \infty,$$

proving (2.133).□

Proof of Theorem 2.83. Let h be a positive integer and set

$$S(T) = \int_0^T e(hf(t))\,dt.$$

Furthermore, for any $r \ge 0$ we use the abbreviations

$$g_r(t) = \sum_{j=0}^{n_r} \frac{(t-T)^j}{j!} f^{(j)}(T),$$

$\alpha_r = f^{(n_r)}(T)/n_r!$, and

$$S'(T) = \int_T^{T+P_r} e(hg_r(t))\,dt.$$

Since $\log(1/|a_n|) \ge l_r + m_r(n - n_r)$ for all $n \ge n_r$, we obtain

$$|a_n| \le |a_{n_r}| e^{m_r(n_r - n)},$$

and therefore

$$\left| \frac{f^{(n_r+1)}(t)}{(n_r+1)!} \right| \le \left| \sum_{n=n_r+1}^{\infty} \binom{n}{n_r+1} a_n t^{n-(n_r+1)} \right| \le \frac{|a_{n_r}|}{e^{m_r}} \left(\frac{1}{1 - Q_r e^{-m_r}} \right)^{n_r+2}$$

for $0 \le t \le Q_r$. Hence, by TAYLOR's formula and $Q_r \le \frac{1}{2}e^{m_r}$ we derive

$$|f(t) - g_r(t)| \le 2^{n_r+2}|a_{n_r}|e^{-m_r}P_r^{n_r+1} \tag{2.135}$$

for $T \le t \le T + P_r$ and $T \le Q_r - P_r$. Furthermore

$$|\alpha_r - a_{n_r}| = \frac{f^{(n_r)}(T) - f^{(n_r)}(0)}{n_r!} \le (n_r + 1)2^{n_r+1}|a_{n_r}|Q_r e^{-m_r} \le \frac{|a_{n_r}|}{2}.$$

Therefore we get by Lemma 2.85, Lemma 2.86, and (2.135)

$$\begin{aligned} |S(T+P_r) - S(T)| &\le |S(T+P_r) - S(T) - S'(T)| + |S'(T)| \\ &\le \left(2^{n_r+2}e^{-m_r}|a_{n_r}|P_r^{n_r+1} + \frac{26}{P_r}\left(\frac{2}{|a_{n_r}|}\right)^{\frac{1}{n_r}}\right)P_r \\ &\le \varepsilon P_r \end{aligned} \tag{2.136}$$

for $T \le Q_r - P_r$ and $r \ge r_0(\varepsilon)$, where $\varepsilon > 0$ is an arbitrary positive given number.

The final step of the proof is to show that (2.136) implies

$$|S(T)| \le \varepsilon T + C \tag{2.137}$$

for $T \ge 0$. We will proceed by induction. Suppose that (2.137) holds for $0 \le T \le Q_{r-1}$. (Choosing $C = C(\varepsilon) = Q_{r_0(\varepsilon)}$ it follows that (2.137) is satisfied for $r = r_0(\varepsilon)$.) If $0 \le T \le Q_r$ then $T = \left[\frac{T}{P_r}\right]P_r + R$ with $0 \le R < P_r < Q_{r-1}$, and consequently (2.136) combined with (2.137) for $0 \le T \le Q_{r-1}$ yields

$$|S(T)| \le |S(T) - S(R)| + |S(R)| < \varepsilon \left[\frac{T}{P_r}\right] P_r + \varepsilon R + C = \varepsilon T + C.$$

Since $\lim_{r\to\infty} Q_r = \infty$ we get for every $\varepsilon > 0$

$$\limsup_{T\to\infty} \left|\frac{S(T)}{T}\right| \le \varepsilon,$$

hence by Theorem 2.73, $\mathbf{x}(t) = f(t)$ is c.u.d. mod 1. □

We will conclude this section by introducing a general notion of continuous uniform distribution on compact spaces.

Definition 2.87 *Let (X, d) be a arcwisely connected compact metric space and μ a positive normalized* BOREL *measure on X. A continuous funtion $\mathbf{x} : [0, \infty) \to X$ is said to be* continuously uniformly distributed with respect to *μ (for short μ-c.u.d.) if for all continuous functions $f : X \to \mathbf{R}$*

$$\lim_{T\to\infty} \frac{1}{T} \int_0^T f(\mathbf{x}(t))\, dt = \int_X f\, d\mu. \tag{2.138}$$

It is said to be continuously well distributed with respect to μ *(for short μ-c.w.d.) if for all continuous functions $f : X \to \mathbf{R}$*

$$\lim_{T\to\infty} \frac{1}{T}\int_0^T f(\mathbf{x}(t+\tau))\,dt = \int_X f\,d\mu \tag{2.139}$$

holds uniformly for $\tau \geq 0$.

Theorem 2.88 *Let (X,d) be a compact metric arcwisely connected space and μ a positive normalized* BOREL *measure on X. Then there exists a μ-c.w.d. function $\mathbf{x} : [0,\infty) \to X$. Obviously, this function is μ-c.u.d., too.*

Proof. By BAAYEN and HEDRLÍN [58] there exists a μ-w.d. sequence $(x_n)_{n\geq 1}$ on X. This means that for every continuous function $f : X \to \mathbf{R}$

$$\lim_{N\to\infty} \frac{1}{N}\sum_{n=\nu+1}^{N+\nu} f(x_n) = \int_X f\,d\mu$$

holds uniformly for $\nu \geq 0$. Now set

$$\mathbf{x}(t) = \begin{cases} x_1 & \text{for} \quad 0 \leq t \leq \frac{3}{2}, \\ x_n & \text{for} \quad n \leq t \leq n+1-2^{-n} \quad (n \geq 2), \\ \mathbf{y}_n(t) & \text{for} \quad n - 2^{-n+1} \leq t \leq n \quad (n \geq 2), \end{cases}$$

where $\mathbf{y}_n : [n-2^{-n+1}, n] \to X$ are continuous functions with $\mathbf{y}_n(n-2^{-n+1}) = x_{n-1}$ and $\mathbf{y}_n(n) = x_n$. Thus, $\mathbf{x}(t)$ is continuous and satisfies

$$\left| \frac{1}{T}\int_\tau^{T+\tau} f(\mathbf{x}(t))\,dt - \frac{1}{[T]}\sum_{n=[\tau]+1}^{[T]+[\tau]} f(x_n) \right| \leq \frac{5}{T}\max_{x\in X}|f(x)|.$$

Hence $\mathbf{x}(t)$ is μ-c.w.d. and μ-c.u.d. □

In the same way as for sequences we can introduce the *discrepancy* of $\mathbf{x}(t)$ with respect to a (discrepancy) system $\mathcal{D}$ by

$$D_T^{(\mathcal{D})}(\mathbf{x}(t)) = \sup_{M\in\mathcal{D}} \left| \frac{1}{T}\int_0^T \chi_M(\mathbf{x}(t)) - \mu(M) \right|. \tag{2.140}$$

We will call such a system $\mathcal{D}$ *discrepancy system* if $\mathbf{x}(t)$ is μ-c.u.d. if and only if

$$\lim_{T\to\infty} D_T^{(\mathcal{D})}(\mathbf{x}(t)) = 0.$$

As in the case for sequences, Proposition 2.6 gives a sufficient condition for $\mathcal{D}$ to be a discrepancy system.

2.3.2 Discrepancy Bounds

We will now consider the question how fast the discrepancy $D_T^{(\mathcal{D})}(\mathbf{x}(t))$ can converge to 0. It turns out that the most interesting parameter in this context is the arclength $s(T)$ of the curve $\mathbf{x}(t)$ $(0 \le t \le T)$. We will first discuss the modulo 1 case and then some other geometric discrepancy systems.

The only case which is completely understood is the one dimensional case mod 1 (see TASCHNER [1786]).

Theorem 2.89 *For any non-increasing function $\psi : [0,\infty) \to [0,\infty)$ there exists $T_0 \ge 0$ and a c.u.d. function $\mathbf{x} : [0,\infty) \to \mathbf{R}$ mod 1 such that*

$$D_T(\mathbf{x}(t)) \le \psi(T)$$

for $T \ge T_0$. This means that the discrepancy can tend to 0 arbitrarily fast.

Proof. Suppose that $\psi(t) \le 1$ for $t \ge N_0 \ge 1$. Set $a_N = N$ for $0 \le N \le N_0$ and $a_{N+1} = a_N + ([N\psi(N+1)])^{-1}$ for $N \ge N_0$. Furthermore set $T_n = N + (n - a_N)/(a_{N+1} - a_N)$ for $a_N < n \le a_{N+1}$. Now the function

$$\mathbf{x}(t) = n + \frac{t - T_n}{T_{n+1} - T_n} \qquad \text{for } T_n \le t < T_{n+1}$$

satisfies $D_{T_n}(\mathbf{x}(t)) = 0$ and

$$D_T(\mathbf{x}(t)) \le \frac{1}{T}(T_{n+1} - T_n) \le \frac{1}{N} N\psi(N+1) < \psi(T)$$

for $N_0 \le N \le T_n < T < T_{n+1} \le N + 1$. □

Note that the function $\mathbf{x}(t)$ contructed in the preceding proof is monotone and therefore of bounded variation. In the following we will always use the assumption that the *arclength*

$$s(T) = \int_0^T d(|\mathbf{x}(t)|) = \sup_{0=a_0<a_1<\cdots<a_n=T} \sum_{i=1}^{N} |\mathbf{x}(a_i) - \mathbf{x}(a_{i-1})| \tag{2.141}$$

is finite for all $T > 0$. The reason why the discrepancy can tend to 0 arbitrarily fast in the one dimensional case is the fact that $D_T(\mathbf{x}(t))$ can be 0 for a discrete set of $T > 0$. In general we can only prove

Theorem 2.90 *Suppose that (X, d) is a compact metric arcwisely connected space, μ a positive normalized BOREL measure on X, and let $\mathcal{D}$ be a system of measurable subsets of X. If there exists a continuous function $\mathbf{x} : [0,\infty) \to X$ with $\lim_{T\to\infty} D_T^{(\mathcal{D})}(\mathbf{x}(t)) = 0$, then for every $\varepsilon > 0$ there exists a continuous function $\mathbf{x}_\varepsilon : [0,\infty) \to X$ satisfying*

$$\limsup_{T\to\infty} T\, D_T^{(\mathcal{D})}(\mathbf{x}_\varepsilon(t)) \le \varepsilon. \tag{2.142}$$

Proof. Let $\varepsilon > 0$ be fixed and set $T_0 = 0$. If T_n is defined then choose T_{n+1} such that

$$D_{T_{n+1}-T_n}(\mathbf{x}(t+T_n)) \le 2^{-n-2}.$$

We now set

$$\mathbf{x}_\varepsilon(t) = \mathbf{x}(T_n + (2t/\varepsilon - n)(T_{n+1} - T_n))$$

for $n\varepsilon/2 \le t < (n+1)\varepsilon/2$, $n \ge 0$. Thus

$$D_{\varepsilon/2}(\mathbf{x}_\varepsilon(t + n\varepsilon/2)) = D_{T_{n+1}-T_n}(\mathbf{x}(t+T_n)) \le 2^{-n-2}$$

implies

$$\begin{aligned} D_T(\mathbf{x}_\varepsilon(t)) &\le \frac{1}{T}\left(\sum_{n=0}^{[2T/\varepsilon]-1} \frac{\varepsilon}{2} D_{\varepsilon/2}(\mathbf{x}_\varepsilon(t+n\varepsilon/2)) + \frac{\varepsilon}{2}\right) \\ &\le \frac{1}{T}\left(\frac{\varepsilon}{2} + \frac{\varepsilon}{2}\right) = \frac{\varepsilon}{T}. \end{aligned}$$

This proves the theorem. □

It seems that (2.142) is best possible in the following sense

Conjecture. *Let $k \ge 2$ and $\mathbf{x} : [0,\infty) \to \mathbf{R}^k$ such that the arclength $s(T)$ is finite for any $T \ge 0$. Is it true that*

$$\limsup_{T\to\infty} T\, D_T(\mathbf{x}(t)) > 0\ ? \tag{2.143}$$

It is clear that it would be sufficient to prove this conjecture for $k = 2$ and that a verification would settle the problem how fast the discrepancy can tend to 0 in terms of T in the (classical) modulo 1 case.

As already mentioned the arclength $s(T)$ is also a very interesting parameter in this context. Again, the case $k = 1$ is completely solved.

Theorem 2.91 *Let $\varphi : [0,\infty) \to [0,\infty)$ be a non-increasing function satisfying*

$$\int_0^\infty \varphi(t)\, dt < \infty \tag{2.144}$$

and let $\mathbf{x} : [0,\infty) \to \mathbf{R}$ be a continuous function such that the arclength $s(T)$ is finite for all $T > 0$ and assume that $\lim_{T\to\infty} s(T) = \infty$. Then, for every $T' > 0$ there exists $T > T'$ such that

$$D_T(\mathbf{x}(t)) > \varphi(s(T)). \tag{2.145}$$

Conversely, if

$$\int_0^\infty \varphi(t)\, dt = \infty \tag{2.146}$$

then there exists a continuous function $\mathbf{x} : [0, \infty) \to \mathbf{R}$ *(such that the arclength* $s(T)$ *is finite for all* $T > 0$ *and* $\lim_{T\to\infty} s(T) = \infty$*) and* $T_0 > 0$ *satisfying*

$$D_T(\mathbf{x}(t)) \leq \varphi(s(T)). \tag{2.147}$$

for $T \geq T_0$.

Proof. Since $s(T)$ is continuous and $s(T) \to \infty$ we can consider a monotone unbounded sequence $(T_n)_{n\geq 1}$ satisfying $s(T_n) = \frac{n}{2}$. First we show that

$$\sum_{n=1}^{\infty} \frac{T_{n+1} - T_n}{T_{n+1}} = \infty. \tag{2.148}$$

If $(T_{n+1} - T_n)/T_{n+1} > \frac{1}{2}$ for infinitely many n then (2.148) is obviously satisfied. Conversely, if $(T_{n+1} - T_n)/T_{n+1} \leq \frac{1}{2}$ for almost all n then we can use the inequality

$$\log T_{n+1} - \log T_n = -\log\left(1 - \frac{T_{n+1} - T_n}{T_{n+1}}\right) \leq 2\frac{T_{n+1} - T_n}{T_{n+1}}$$

which is valid for $(T_{n+1} - T_n)/T_{n+1} \leq \frac{1}{2}$, and (2.148) follows again. Furthermore, (2.144) implies that

$$\sum_{n=1}^{\infty} \varphi\left(\frac{n}{2}\right) < \infty.$$

Hence there are infinitely many n such that

$$\frac{T_{n+1} - T_n}{T_{n+1}} > 8\varphi\left(\frac{n}{2}\right).$$

If $D_{T_n}(\mathbf{x}(t)) > \varphi(s(T_n)) = \varphi\left(\frac{n}{2}\right)$ for such an n then there is nothing to show. In the other case we have

$$\begin{aligned} D_{T_{n+1}}(\mathbf{x}(t)) &\geq \frac{(T_{n+1} - T_n)D_{T_{n+1}-T_n}(\mathbf{x}(t + T_n)) - T_n D_{T_n}(\mathbf{x}(t))}{T_{n+1}} \\ &> 8\varphi\left(\frac{n}{2}\right)\frac{1}{4} - \varphi\left(\frac{n}{2}\right) \\ &= \varphi\left(\frac{n}{2}\right) \geq \varphi(s(T_{n+1})), \end{aligned}$$

since $s(T_{n+1}) - s(T_n) = \frac{1}{2}$ implies $D_{T_{n+1}-T_n}(\mathbf{x}(t + T_n)) \geq \frac{1}{4}$. This proves (2.145).

Now suppose that (2.146) holds and set $T_1 = 0$ and

$$T_{n+1} = \exp\left(\frac{1}{2}\int_0^{n+1} \varphi(t)\,dt\right)$$

for $n > 1$. Then the continuous function

$$\mathbf{x}(t) = n - 1 + \frac{t - T_n}{T_{n+1} - T_n} \qquad \text{for } T_n \leq t < T_{n+1}$$

satisfies $s(T_n) = n - 1$, $D_{T_n}(\mathbf{x}(t)) = 0$, and

$$\begin{aligned} D_T(\mathbf{x}(t)) &\le \frac{T_{n+1} - T_n}{T_n} \\ &= \exp\left(\frac{1}{2}\int_n^{n+1} \varphi(t)\,dt\right) - 1 \\ &\le \exp\left(\frac{\varphi(n)}{2}\right) - 1 \\ &\le \varphi(n) = \varphi(s(T_{n+1})) \le \varphi(s(T)) \end{aligned}$$

for $T_n \le T \le T_{n+1}$, where $T_n > T'$ is chosen such that $\exp(\varphi(s(T'))/2) - 1 \le \varphi(s(T'))$.
□

Remark. It should be noted that the lower bound (2.145) can be generalized to any "reasonable" discrepancy system on a compact space (see DRMOTA [482]). Especially all of the following examples will satisfy (2.145). But it seems that (2.145) is only optimal in the one dimensional case. In fact, for dimensions $k \ge 3$ we will prove even better bounds.

Now we will change our point of view. In the following we will not discuss a fixed function $\mathbf{x}(t)$ and the behaviour of $D_T(\mathbf{x}(t))$ for $T \to \infty$, but we will fix some positive number s and will be looking for those functions $x(t)$ $(0 \le t \le 1)$ with $s(T) = s$ and smallest possible discrepancy.

Since linear transformations do not affect the discrepancy, from now on we will always consider continuous functions $\mathbf{x} : [0,1] \to X$, and we will denote the arclength by s and the discrepancy by $D(\mathbf{x}(t))$.

The first theorem is not only remarkable due to its simple proof but it is (almost) best possible in the modulo 1 case.

Theorem 2.92 *Let (X, d) be a compact metric arcwisely connected space and μ a positive normalized* BOREL *measure on X satisfying*

$$\mu(B(x,r)) \le \alpha r^k \tag{2.149}$$

for all open balls $B(x,r)$ with center $x \in X$ and radius $r > 0$, where $\alpha > 0$ and $k > 1$ are real numbers. Suppose that $\mathcal{D}$ is a system of subsets of X such that for every open ball $B(x,r)$ there exists a set $M \in \mathcal{D}$ with $B(x,r) \subseteq M \subseteq B(x,\beta r)$, where $\beta \ge 1$ is a real constant. Let $\mathbf{x} : [0,1] \to X$ be a continuous function with finite arclength

$$s = \sup_{0=a_0<a_1<\cdots<a_N=T} \sum_{i=1}^{N} d(\mathbf{x}(a_i), \mathbf{x}(a_{i-1})) \ge 1.$$

Then there exists a set $M \in \mathcal{D}$ such that

$$\left|\int_0^1 \chi_M(\mathbf{x}(t))\,dt - \mu(M)\right| \ge c\, s^{-1-\frac{1}{k-1}}, \tag{2.150}$$

where the constant $c > 0$ is given by

$$c = \frac{1}{6}\left(\beta^k \max\{1, 3\alpha\}\right)^{-\frac{1}{k-1}}.$$

Proof. Obviously, for every $r > 0$ there exists a subinterval $[a, b] \subseteq [0, 1]$ satisfying $\mathbf{x}([a, b]) \subseteq B(\mathbf{x}(a), r)$ and

$$b - a \geq \left(\left[\frac{s}{r}\right] + 1\right)^{-1}.$$

We only have to subdivide the arc $\mathbf{x}([0, 1])$ into $\left[\frac{s}{r}\right] + 1$ subarcs of length $< r$. By assumption there is a set $M \in \mathcal{D}$ such that $B(x, r) \subseteq M \subseteq B(x, \beta r)$. Hence we get

$$\begin{aligned}
\left|\int_0^1 \chi_M(\mathbf{x}(t))\,dt - \mu(M)\right| &\geq \int_0^1 \chi_M(\mathbf{x}(t))\,dt - \mu(M) \\
&\geq \int_0^1 \chi_{B(\mathbf{x}(a), r)}(\mathbf{x}(t))\,dt - \mu(B(\mathbf{x}(a), \beta r)) \\
&\geq \left(\left[\frac{s}{r}\right] + 1\right)^{-1} - \alpha\beta^k r^k.
\end{aligned}$$

Choosing $r = (s\beta^k \max\{1, 3\alpha\})^{-1/(k-1)}$ we immediately obtain (2.150). □

It is obvious that Theorem 2.92 can be applied for all discrepancy systems already discussed with only two exceptions: the case of spherical slices and the halfspaces used in ALEXANDER's method. However, in these and several other cases there are better lower bounds known than provided by Theorem 2.92. Nevertheless in the "classical" modulo 1 case it is (almost) optimal.

Theorem 2.93 *Let $k \geq 2$ and $s \geq 2$. Then there exists a continuous function $\mathbf{x} : [0, 1] \to \mathbf{R}^k$ such that*

$$D(\mathbf{x}(t)) \ll s^{-1-\frac{1}{k-1}}(\log s)^{k-1},$$

where the constant implied by $\ll$ only depends on the dimension k.

The proof of this theorem is a direct combination of a general principle (Proposition 2.97) and the fact that for every $N \geq 1$ there exists a sequence $\mathbf{x}_1, \ldots, \mathbf{x}_N \in [0, 1]^k$ with $D_N(\mathbf{x}_n) \ll N^{-1}(\log N)^{k-1}$.

In the two and three dimensional case we can give more precise examples.

Theorem 2.94 *Let p, q be two coprime integers and set $\mathbf{x}(t) = (p, q)t$ $(0 \leq t \leq 1)$. Then we have*

$$D(\mathbf{x}(t)) \ll \frac{1}{|pq|}.$$

Especially, if $q = p + 1$, we get

$$D(\mathbf{x}(t)) \ll s^{-2}.$$

Proof. We will use Theorem 2.77. For this purpose we have to calculate ($\mathbf{h} = (h_1, h_2)$)

$$\int_0^1 e(\mathbf{h}\cdot\mathbf{x}(t))\,dt = \begin{cases} 1 & \text{if } h_1p + h_2q = 0, \\ 0 & \text{if } h_1p + h_2q \neq 0. \end{cases}$$

Since all solutions of $h_1p + h_2q = 0$ are given by $h_1 = mq$, $h_2 = -mp$ ($m \in \mathbf{Z}$) we have

$$\begin{aligned} D(\mathbf{x}(t)) &\ll \sum_{\mathbf{h}\neq 0} \frac{1}{r(\mathbf{h})}\left|\int_0^1 e(\mathbf{h}\cdot\mathbf{x}(t))\,dt\right| \\ &= \frac{2}{|pq|}\sum_{m=1}^{\infty}\frac{1}{m^2} \ll \frac{1}{|pq|}, \end{aligned}$$

which proves Theorem 2.94. □

Theorem 2.95 *Let F_n be the* FIBONACCI *numbers defined by $F_0 = 0$, $F_1 = 1$, and by $F_{n+1} = F_n + F_{n-1}$ for $n \geq 1$ and set $\mathbf{x}(t) = (F_n, F_{n+1}, F_{n+1}+1)t$ $(0 \leq t \leq 1)$, where $n \equiv 1 \bmod 6$. Then*

$$D(\mathbf{x}(t)) \ll s^{-\frac{3}{4}}.$$

Sketch Proof. Again by Theorem 2.73 we can estimate the discrepancy $D(\mathbf{x}(t))$ by

$$D(\mathbf{x}(t)) \ll \sum_{\mathbf{a}\cdot\mathbf{h}=0,\mathbf{h}\neq 0} \frac{1}{r(\mathbf{h})},$$

where $\mathbf{a} = (F_n, F_{n+1}, F_{n+1}+1)$ and $\mathbf{h} = (h_1, h_2, h_3) \in \mathbf{Z}^3$. Using the fact that the partial quotients of $F_n/(F_{n+1}+1)$ are bounded by 5 (F_n and F_{n+1} are coprime for $n \equiv 1 \bmod 6$, see DRMOTA [486]) and applying approximation properties of principal convergents yields

$$\sum_{\mathbf{a}\cdot\mathbf{h}=0,\mathbf{h}\neq 0} \frac{1}{r(\mathbf{h})} \ll F_n^{-\frac{3}{2}} \ll s^{-\frac{3}{4}}.$$

□

As in the case of sequences the situation drastically changes if rotation is allowed. The next theorem gives a list of "large" discrepancy bounds in s which are optimal despite of a logarithmic factor. Our notation is similar to that used for sequences. If $\mathcal{D}$ is a system of (measurable) subsets of X then

$$D^{(\mathcal{D})}(x(t)) = \sup_{D\in\mathcal{D}}\left|\int_0^1 \chi_D(x(t)) - \mu(D)\right|$$

denotes the discrepancy of $x(t)$, $0 \leq t \leq 1$, with respect to $\mathcal{D}$ and the probability measure μ on X.

Theorem 2.96 *Assume that X and $\mathcal{D}$ denote one of the examples in the following list.*

- $X \subseteq \mathbf{R}^k$ $(k \geq 2)$ *is a convex body with* $\lambda_k(X) = 1$ *and* $\mathcal{D} = \mathcal{D}_s(A, X)$, *where* $A \subseteq \mathbf{R}^k$ *is an arbitrary compact body.*
- $X = \mathbf{R}^k/\mathbf{Z}^k$ $(k \geq 2)$ *and* $\mathcal{D} = \mathcal{D}_s(A, \mathbf{R}^k/\mathbf{Z}^k)$, *where* $A \subseteq \mathbf{R}^k$ *is an arbitrary compact body with* $\mathrm{diam}(A) < 1$.
- $X \subseteq \mathbf{R}^k$ $(k \geq 2)$ *is a convex body with* $\lambda_k(X) = 1$ *and* $\mathcal{D} = \mathcal{D}_h(A, X)$, *where* $A \subseteq \mathbf{R}^k$ *is a compact body satisfying the same differentiability conditions for the boundary* ∂A *as in Theorem 2.11.*
- $X = \mathbf{R}^k/\mathbf{Z}^k$ $(k \geq 2)$ *and* $\mathcal{D} = \mathcal{D}_h(A, \mathbf{R}^k/\mathbf{Z}^k)$, *where* $A \subseteq \mathbf{R}^k$ *is a compact body satisfying the same differentiability conditions for the boundary* ∂A *as in Theorem 2.11.*
- $X = \mathbf{S}^k$ $(k \geq 2)$ *and* $\mathcal{D} = \mathcal{D}(C, \mathbf{S}^k)$ *is the system of all spherical caps.*
- $X = \mathbf{S}^k$ $(k \geq 2)$ *and* $\mathcal{D} = \mathcal{D}(S, \mathbf{S}^k)$ *is the system of all spherical slices.*

Then, for any continuous function $\mathbf{x} : [0,1] \to X$ *with finite arclength* $s \geq 2$

$$D^{(\mathcal{D})}(\mathbf{x}(t)) \gg s^{-\frac{1}{2}-\frac{1}{k-1}}. \tag{2.151}$$

On the other hand, for any given number $s \geq 2$ *there exists a continuous function* $\mathbf{x} : [0,1] \to X$ *such that*

$$D^{(\mathcal{D})}(\mathbf{x}(t)) \ll s^{-\frac{1}{2}-\frac{1}{k-1}}(\log s)^{\frac{1}{2}}. \tag{2.152}$$

Proof of (2.151). The proof of (2.151) in the case $\mathcal{D} = \mathcal{D}_s(A, X)$, where $X \subseteq \mathbf{R}^k$ is a convex body withe $\lambda_k(X) = 1$, runs along the same line as the proof of Theorem 2.10. We set $X' = T^{1/k}X$, $A' = T^{1/k}A$, and $\mathbf{y}(t) = T^{1/k}\mathbf{x}(t/T)$, $0 \leq t \leq T$, where $T = c\, s^{k/(k-1)}$ for some constant $c > 0$ which will be chosen in the sequel. Note that $\mathbf{y}(t)$, $0 \leq t \leq T$, has arc length $s' = T^{1/k}s$. We will show that there exists $\mathbf{x} \in \mathbf{R}^k$, $\xi \in SO(k)$, and $\rho \in (0,1]$ such that

$$\left| \int_0^T \chi_{\rho(\xi A')+\mathbf{x}}(\mathbf{y}(t))\, dt - \lambda_k((\rho(\xi A') + \mathbf{x}) \cap X') \right| \gg (\sigma(\partial A'))^{\frac{1}{2}} \tag{2.153}$$

for sufficiently large T. Since

$$(\sigma(\partial A'))^{\frac{1}{2}} = (\sigma(\partial A))^{\frac{1}{2}} T^{\frac{1}{2}-\frac{1}{2k}} = (\sigma(\partial A))^{\frac{1}{2}} T s^{-\frac{1}{2}-\frac{1}{k-1}} c^{-\frac{1}{2}-\frac{1}{k-1}}$$

(2.153) is equivalent to (2.151).

Let ν' be defined by

$$\nu'(M) = \int_0^T \chi_M(\mathbf{y}(t))\, dt - \lambda_k(M \cap X').$$

The main goal is to show that

$$\int_{\mathbf{R}^k} |(f * \nu')(\mathbf{x})|^2\, d\mathbf{x} \gg T, \tag{2.154}$$

where

$$f(\mathbf{x}) = \prod_{j=1}^{k} \left(\frac{2\sin(50x_j)}{\sqrt{2\pi}\, x_j} \right)^2 \qquad (\mathbf{x} = (x_1, \ldots, x_k)).$$

By using precisely the same arguments as in the proof of Theorem 2.10, (2.154) implies (2.153).

In order to prove (2.154) we will show that there exists a constant $c > 0$ ($T = c\, s^{k/(k-1)}$) such that

$$\int_{\mathbf{R}^k} |(f * \nu')(\mathbf{x})| \, d\mathbf{x} \gg \int_{\mathbf{R}^k} v(\mathbf{x})^2 \, d\mathbf{x} \tag{2.155}$$

and

$$\int_{\mathbf{R}^k} v(\mathbf{x})^2 \, d\mathbf{x} \gg T, \tag{2.156}$$

where $v(\mathbf{x})$ stands for

$$v(\mathbf{x}) = \int_0^T \chi_{Q+\mathbf{x}}(y(t)) \, dt \,,$$

and $Q = [-\frac{1}{100}, \frac{1}{100}]^k$.

Let t_i ($i = 0, \ldots, [100\, s']$) be defined by

$$\int_0^{t_i} |d\mathbf{y}(t)| = \frac{i}{100} \,.$$

Therefore $s, t \in [t_i, t_{i+1}]$ satisfy $|y(t) - y(s)| \le \frac{1}{100}$. Since

$$v(\mathbf{x})^2 = \int_0^T \int_0^T \chi_{Q+\mathbf{x}}(\mathbf{y}(t)) \, \chi_{Q+\mathbf{x}}(\mathbf{y}(s)) \, ds \, dt \,,$$

we obtain by applying CAUCHY-SCHWARZ's inequality

$$\begin{aligned}
\int_{\mathbf{R}^k} v(\mathbf{x})^2 \, dx &= \int_0^T \int_0^T \int_{\mathbf{R}^k} \chi_{Q+\mathbf{x}}(y(t)) \, \chi_{Q+\mathbf{x}}(y(s)) \, dx \, ds \, dt \\
&\ge \underbrace{\int_0^T \int_0^T}_{|y(t)-y(s)| \le \frac{1}{100}} 50^{-k} \, ds \, dt \\
&\ge 50^{-k} \sum_{i=1}^{[100\, s']} (t_i - t_{i-1})^2 + 50^{-k} (T - t_{[100\, s']})^2 \\
&\ge 50^{-k} \frac{T^2}{100\, s' + 1} \gg \frac{T^2}{T^{1/k}\, s} = c^{\frac{k-1}{k}} T.
\end{aligned}$$

This proves (2.156).

Now set

$$I = \int_{\mathbf{R}^k} f(\mathbf{x})\, d\mathbf{x} = 100^k .$$

Then we obtain

$$\underline{f} := \inf_{\mathbf{x}\in Q} g(\mathbf{x}) = \left(\frac{2}{\pi}\right)^k \left(100 \sin\left(\frac{1}{2}\right)\right)^{2k} \geq 10\, I\, .$$

Furthermore $c_1 = I/(10\, I - 1)$ satisfies $\frac{1}{10} < c_1 < \frac{1}{9}$. Let R_1, R_2 denote the sets $R_1 = \{\mathbf{x} \in \mathbf{R}^k : v(\mathbf{x}) \geq c_1\}$ and $R_2 = \mathbf{R}^k \setminus R_1$, then we have for $\mathbf{x} \in R_1$

$$\begin{aligned} (f * d\nu')(\mathbf{x}) &= \int_0^T f(\mathbf{x} - \mathbf{y}(t))\, dt - \int_{X'} f(\mathbf{x} - \mathbf{y})\, d\mathbf{y} \\ &\geq \int_0^T \underline{f} \chi_{Q+\mathbf{x}}(\mathbf{y}(t))\, dt - I = \underline{f}\, v(\mathbf{x}) - I \\ &\geq 10\, I\, v(\mathbf{x}) - I \geq v(\mathbf{x}) + (10\, I - 1)c_1 - I \geq v(\mathbf{x})\, . \end{aligned}$$

Since $\int_{R_2} v(\mathbf{x})^2\, d\mathbf{x} \ll T$ we can choose c in a way that

$$\int_{R_2} v(\mathbf{x})^2\, d\mathbf{x} \leq \frac{1}{2} \int_{\mathbf{R}^k} v(\mathbf{x})^2\, d\mathbf{x}\, .$$

Thus

$$\int_{\mathbf{R}^k} (f * d\nu')(\mathbf{x})^2\, d\mathbf{x} \geq \int_{R_1} (f * d\nu')(\mathbf{x})^2\, d\mathbf{x} \geq \int_{R_1} v(\mathbf{x})^2\, d\mathbf{x} \gg \int_{\mathbf{R}^k} v(\mathbf{x})^2\, d\mathbf{x}\, ,$$

if c is chosen in this way. Thus we have proved (2.155). This completes the proof of (2.151) in the case $\mathcal{D} = \mathcal{D}_s(A, X)$, where X is a convex body.

If $\mathcal{D} = \mathcal{D}_s(A, \mathbf{R}^k/\mathbf{Z}^k)$ then the proof of (2.151) is quite similar to the above one. One only has to follow the modifications which are indicated in the Remark following the proof of Theorem 2.10.

Next consider the case $\mathcal{D} = \mathcal{D}_h(A, X)$, where the boundary of A satisfies proper differentiability conditions. Here the proof is similar to that of the second part of Theorem 2.11. We set

$$\nu(M) = \int_0^1 \chi_M(\mathbf{x}(t))\, dt - \lambda_k(M \cap X),$$

$F_\rho = \chi_{-\rho A} * (d\nu)$, and

$$\Phi(q) = \frac{1}{q} \int_q^{2q} \int_{\mathbf{R}^k} |F_\rho(\mathbf{x})|^2\, d\mathbf{x}\, d\rho = \int_{\mathbf{R}^k} \varphi_q(\mathbf{t}) |\hat{\nu}(\mathbf{t})|^2\, d\mathbf{t}.$$

Instead of Lemma 2.18 we have to show

$$\Phi\left(cs^{-\frac{1}{k-1}}\right) \gg s^{-\frac{2k}{k-1}} \tag{2.157}$$

for some constant $c > 0$. Obviously Lemma 2.17 and (2.157) imply

$$\Phi(1) \gg \left(s^{\frac{1}{k-1}}\right)^{k-1} \Phi\left(cs^{-\frac{1}{k-1}}\right) \gg s^{-1-\frac{2}{k-1}},$$

which proves (2.151).

For the proof of (2.157) let us consider a k-dimensional cube U contained in X of side length $d > 0$ and subdivide U into N^k subcubes U_m, $1 \le m \le N^k$ of side length d/N. (N will be fixed in a moment.) Furthermore subdivide the interval $[0,1]$ into $[(N+1)s/d]$ intervals $[t_j, t_{j+1}]$ ($j = 0, \ldots, [(N+1)s/d] - 1$), $0 = t_0 < t_1 < \cdots < t_{[s(N+1)/d]} = 1$, such that the arclength $\int_{t_j}^{t_{j+1}} d|\mathbf{x}(t)| < d/N$ ($j = 0, \ldots, [(N+1)s/d] - 1$). Trivially, the number of cubes U_m, $1 \le m \le N^k$ with $\mathbf{x}([t_j, t_{j+1}]) \cap Q_m \ne \emptyset$ is less or equal 2^k. Therefore there are at most $2^k[(N+1)s/d]$ cubes U_m such that $x([0,1]) \cap U_m \ne \emptyset$. Now choose the minimal N satisfying $2^k(N+1)s/d \le N^k/2$. Hence there are at least $N^k/2$ cubes U_m such that $x([t_j, t_{j+1}]) \cap U_m) = \emptyset$. Now consider subcubes C_m of U_m with the same center and side length $d/(2N)$. If $\rho < d(4Nk^{1/2})^{-1}$ and $\mathbf{x} \in C_m$ we have

$$F_\rho(\mathbf{x}) = -(2\rho)^k,$$

and therefore, if $q < d(8Nk^{1/2})^{-1}$, we obtain

$$\Phi(q) \gg q^{2k}.$$

Since $N \gg\ll s^{1/(k-1)}$, this proves (2.157).

The proof of (2.151) in the case $\mathcal{D} = \mathcal{D}_h(A, \mathbf{R}^k/\mathbf{Z}^k)$, where the boundary of A satisfies proper differentiability conditions, runs along similar lines. (Compare with the Remark following the proof of Theorem 2.10.)

Finally, if $X = \mathbf{S}^k$ ($k \ge 2$) and $\mathcal{D} = \mathcal{D}(C, \mathbf{S}^k)$ is the system of all spherical caps then we can follow the ideas used in the proof of Theorem 2.22.

Let $B(\rho) = \{\mathbf{x} \in \mathbf{R}^{k+1} : |\mathbf{x}| \le \rho\}$ denote the closed ball with radius ρ and set $F_\rho = \chi_{B(\rho)} * d\nu : \mathbf{R}^{k+1} \to \mathbf{R}$, where

$$\nu(M) = \int_0^1 \chi_M(\mathbf{x}(t))\, dt - \sigma_k^*(M \cap \mathbf{S}^k).$$

By PLANCHEREL's identity we get

$$\begin{aligned} \Phi(q) &= \frac{1}{q}\int_q^{2q} \int_{\mathbf{R}^{k+1}} |F_r(\mathbf{x})|^2\, d\mathbf{x}\, dr \\ &= \int_{\mathbf{R}^{k+1}} \varphi_q(\mathbf{t}) |\hat{\nu}(\mathbf{t})|^2\, d\mathbf{t}. \end{aligned}$$

Thus (2.151) follows from Lemma 2.17, which says that

$$\frac{\varphi_q(\mathbf{t})}{\varphi_p(\mathbf{t})} \gg \left(\frac{q}{p}\right)^k$$

uniformly for $\mathbf{t} \in \mathbf{R}^{k+1}$ and $0 < p < q$; here we also have to use a bound of the form

$$\Phi(c\, s^{-1/(k-1)}) \gg s^{-\frac{2k+1}{k-1}} \tag{2.158}$$

for some constant $c > 0$. Namely, if we set $q = c\, s^{-1/(k-1)}$ we obtain

$$\Phi(1) \gg s^{\frac{k}{k-1}} \Phi(q) \gg s^{-1-\frac{2}{k-1}}.$$

Therefore, it remains to prove (2.158).

We represent $\mathbf{S}^k$ by polar coordiantes. Let $Y : (0, \infty) \times [0, \pi]^{k-1} \times [0, 2\pi) \to \mathbf{R}^{k+1}$ be the mapping

$$\begin{aligned}
x_1 &= r \sin\varphi_1 \sin\varphi_2 \cdots \sin\varphi_{k-2} \sin\varphi_{k-1} \sin\varphi_k, \\
x_2 &= r \sin\varphi_1 \sin\varphi_2 \cdots \sin\varphi_{k-2} \sin\varphi_{k-1} \cos\varphi_k, \\
x_3 &= r \sin\varphi_1 \sin\varphi_2 \cdots \sin\varphi_{k-2} \cos\varphi_{k-1}, \\
&\vdots \\
x_k &= r \sin\varphi_1 \cos\varphi - 2, \\
x_{k+1} &= r \cos\varphi_1.
\end{aligned}$$

Then $\mathbf{S}^k = Y(\{1\} \times \times [0, \pi]^{k-1} \times [0, 2\pi))$. Now consider the images $R_{(m_i)} = Y(\{1\} \times B_{(m_i)})$ of the $4N^k$ cubes

$$B_{(m_i)} = \prod_{j=1}^{k-1} \left[\frac{\pi}{4}\left(1 + 2\frac{m_j}{N}\right), \frac{\pi}{4}\left(1 + 2\frac{m_j + 1}{N}\right)\right) \times \left[\pi\frac{m_k}{2N}, \pi\frac{m_k + 1}{2N}\right),$$

where $0 \le m_1, \ldots, m_{k-1} < N$ and $0 \le m_k < 4N$.

As above it is possible to choose $N \gg\ll s^{1/(k-1)}$ such that for at least $2N^k$ sets $R_{(m_i)}$ we have $\mathbf{x}([0,1]) \cap R_{(m_i)} = \emptyset$. For these $R_{(m_i)}$ consider the images $\overline{R}_{(m_i)} = Y(C_{(m_i)})$ of the rectangles

$$\begin{aligned}
C_{(m_i)} &= \left[1, 1 + \frac{1}{8N}\right] \times \prod_{j=1}^{k-1} \left[\frac{\pi}{4}\left(1 + 2\frac{m_j + \frac{1}{4}}{N}\right), \frac{\pi}{4}\left(1 + 2\frac{m_j + \frac{3}{4}}{N}\right)\right) \\
&\quad \times \left[\frac{\pi(m_k + 1/4)}{2N}, \frac{\pi(m_k + 3/4)}{2N}\right).
\end{aligned}$$

If $\rho \in \left[\frac{\pi}{16N}, \frac{\pi}{8N}\right]$ and $\mathbf{x} \in C_{(m_i)}$ (for such $R_{(m_i)}$) we have

$$|F_\rho(\mathbf{x})| \gg N^{-k}.$$

This implies (2.158). (Note that we are working in $\mathbf{R}^{k+1}$.)

In the case of spherical slices (2.158) follows from the corresponding estimate for the L^2-discrepancy with respect to spherical caps and an comparision of FOURIER coefficients as in the proof of Theorem 2.22. □

For the proof of (2.152) we just have to apply the following principle and the corresponding results for sequences.

Proposition 2.97 *Let X be either a convex body $X \subseteq \mathbf{R}^k$ with $\lambda_k(X) = 1$ or $X = \mathbf{R}^k/\mathbf{Z}^k$ or $X = \mathbf{S}^k$. If $X \subseteq \mathbf{R}^k$ or $X = \mathbf{R}^k/\mathbf{Z}^k$ then let μ denote λ_k and if $X = \mathbf{S}^k$ then let μ denote σ_k^*. For a system $\mathcal{D}$ of measurable subsets of X we set*

$$\Omega^{\mathcal{D}}(N) = \inf_{\mathbf{x}_1,\ldots,\mathbf{x}_N \in X} \sup_{D\in\mathcal{D}} \left| \frac{1}{N}\sum_{n=1}^{N} \chi_D(\mathbf{x}_n) - \mu(D) \right|$$

and

$$\Delta^{\mathcal{D}}(s) = \inf_{\mathbf{x}\in C_s} \sup_{D\in\mathcal{D}} \left| \int_0^1 \chi_D(\mathbf{x}(t))\,dt - \mu(D) \right|,$$

where C_s denotes the set of all continuous functions $\mathbf{x} : [0,1] \to X$ with arclength s. Then there exists a constant $c > 0$ such that

$$\Delta^{\mathcal{D}}(s) \le \Omega^{\mathcal{D}}(N) \tag{2.159}$$

for $s = cN^{1-1/k}$.

Proof. The first step of the proof is to show that if $\mathbf{x}_1,\ldots,\mathbf{x}_N$ are N points in X then there is a permutation π of $\{1,\ldots,N\}$ such that

$$\sum_{i=1}^{N-1} |\mathbf{x}_{\pi(i+1)} - \mathbf{x}_{\pi(i)}| \ll N^{1-1/k}. \tag{2.160}$$

First assume that $X \subseteq \mathbf{R}^k$. W.l.o.g. we may assume that X is contained in a ball B centered at $\mathbf{0}$ and $\mathrm{diam}(B) \ll \mathrm{diam}(X)$. Let X_1 be the intersection of X with the hyperplane H_1 given by $x_1 = 0$. Set $M = [N^{1/k}]$ and let $Z \subseteq X_1$ be the set of all points $\mathbf{z} \in X_1$ such that $M\mathbf{z}$ is an integer point. Trivially $|Z| \ll M^{k-1} \ll N^{1-1/k}$. For every $i = 1,\ldots,N$ let $\mathbf{z}(\mathbf{x}_i) \in Z$ be chosen such that $|\mathbf{x}_i - \mathbf{z}(\mathbf{x}_i)|$ is minimal. Let $\mathbf{z}_1,\ldots,\mathbf{z}_{|Z|}$ be a numeration of Z, I_j $(1 \le j \le |Z|)$ the set of indices $i \in \{1,\ldots,N\}$ with $\mathbf{z}(x_i) = \mathbf{z}_j$, and $k_j = |I_j|$. Now define a permutation π of $\{1,\ldots,N\}$ by

$$\left\{ \pi(i) : \sum_{l<j} k_l < i \le \sum_{l\le j} k_l \right\}, = I_j$$

and that the first coordinates of $\mathbf{x}_{\pi(i)}$, $\sum_{l<j} k_l \le i \le \sum_{l\le j} k_l$ are ordered. Thus

$$\begin{aligned}
\sum_{i=1}^{N-1} |\mathbf{x}_{\pi(i+1)} - \mathbf{x}_{\pi(i)}| &= \sum_{j=1}^{|Z|} \sum_{\sum_{l<j} k_l \le i \le \sum_{l\le j} k_l} |\mathbf{x}_{\pi(i+1)} - \mathbf{x}_{\pi(i)}| \\
&\le |Z|\,\mathrm{diam}\,(X) + \sum_{j=1}^{|Z|} \left(\mathrm{diam}\,(X) + k_j \frac{\sqrt{k-1}}{M} \right) \\
&\ll N^{1-1/k}.
\end{aligned}$$

Obviously, if $X = \mathbf{R}^k/\mathbf{Z}^k$ or if $X = \mathbf{S}^k$ we can use exactly the same considerations. In any case we obtain (2.160).

Next it is an easy exercise to show that for every $\varepsilon > 0$ there exists a continuous function $\mathbf{x} : [0,1] \to X$ with arclength $s \ll N^{1-1/k}$ and

$$\left| \frac{1}{N} \sum_{n=1}^{N} \chi_A(\mathbf{x}_{\pi(n)}) - \int_0^1 \chi_A(\mathbf{x}(t))\, dt \right| \le \varepsilon \tag{2.161}$$

for every measurable set $A \subseteq X$. If $X \subseteq \mathbf{R}^k$ or $X = \mathbf{R}^k/\mathbf{Z}^k$ we can use

$$\mathbf{x}(t) = \begin{cases} x_{\pi(i)} & \text{for } \frac{i-1}{N} \le t \le \frac{i-\varepsilon}{N}, 1 \le i < N, \\ x_{\pi(i)} + (x_{\pi(i+1)} - x_{\pi(i)}) \frac{t - \frac{i-\varepsilon}{N}}{\varepsilon/N} & \text{for } \frac{i-\varepsilon}{N} < t < \frac{i}{N}, 1 \le i < N, \\ x_{\pi(N)} & \text{for } 1 - \frac{1}{N} < t \le 1. \end{cases}$$

If $X = \mathbf{S}^k$ a similar construction works.

Finally it is clear that (2.161) implies (2.159). □

2.3.3 Metric Results

If (X, d) is a compact metric space and μ a positive normalized BOREL measure on X then there is a natural measure on the space of all sequences $(x_n)_{n\ge1}$, $x_n \in X$, namely the product measure μ_∞. It is well known (e.g. see KUIPERS and NIEDERREITER [983]) that μ_∞-almost all sequences are μ_∞-u.d. on X and μ_∞-almost no sequences are μ_∞-w.d. on X provided that μ is not concentrated on one point.

We want to prove similar theorems for continuous functions. In general there is no natural measure on the space of all continuous functions $\mathbf{x} : [0,\infty) \to X$. But on differentiable manifolds there is a diffusion process related to the heat equation or equivalently a BROWNian motion which can be used to introduce a measure on continuous functions.

Let X be a compact connected C^∞-RIEMANNian manifold without boundary and μ the normalized surface measure on X. Consider the heat equation on X

$$\frac{\partial u}{\partial t} = \Delta u, \qquad u = u(t,x),$$

where Δ denotes the LAPLACE-BELTRAMI operator on X. The heat equation describes a diffusion process (BROWNian motion), which can be interpreted as a MARKOV process on X. Applying a well-known theorem (see YOSIDA [1977]) the solution $u(t,x)$ of the initial value problem

$$\frac{\partial u}{\partial t} = \Delta u, \qquad u(0,x) = u_0(x)$$

is unique if $u_0(x)$ is a C^∞-function on X and the solution is given by

$$u(t,x) = \int_X p(t,x,y) u_0(y)\, d\mu(y), \tag{2.162}$$

where $p(t,x,y)$ are transition densities of a MARKOV process.

Let C_w denote the space of all continuous functions $\mathbf{x} : [0,\infty) \to X$ with $\mathbf{x}(0) = w \in X$. Then for fixed $0 < t_1 < t_2 < .... < t_n$ and a BOREL set $E \subseteq X^n$ the WIENER measure μ_w is defined by

$$\mu_w(\{\mathbf{x} \in C_w : (\mathbf{x}(t_1),\ldots,\mathbf{x}(t_n)) \in E\})$$
$$= \int_E p(t_n - t_{n-1}, x_{n-1}, x_n) \cdots p(t_2 - t_1, x_1, x_2) p(t_1, w, x_1)\, d\mu(x_1) \cdots d\mu(x_n)$$

and the σ-algebra on C_w is generated by the sets $\{\mathbf{x} \in C_w : (\mathbf{x}(t_1),\ldots,\mathbf{x}(t_n)) \in E\}$.

Obviously this concept covers the case of the torus $\mathbf{R}^k/\mathbf{Z}^k$ and that of the sphere $\mathbf{S}^k$. We will now prove the following theorem.

Theorem 2.98 *Let X be a compact connected* RIEMANN*ian manifold without boundary and $\mathcal{D}$ the system of all closed geodesic balls $\overline{B}(x,r) = \{y \in X : d(x,y) \le r\}$, $r > 0$. ($d(\cdot,\cdot)$ denotes the geodesic distance on X.) Then for μ_w-almost all functions $\mathbf{x}(t)$ with $\mathbf{x}(0) = w \in X$ we have*

$$\limsup_{T\to\infty} \frac{TD_T^{(\mathcal{D})}(\mathbf{x}(t))}{\sqrt{2T\log\log T}} = \alpha$$

with a certain positive constant α. Hence μ_w-almost all functions $\mathbf{x}(t)$ with $\mathbf{x}(0) = w \in X$ are μ-c.u.d.

Remark. It should be noted that a much more general theorem has been proved by BLÜMLINGER [234]. It covers more general diffusion processes on compact RIEMANNian manifolds. An explicit representation for the kernel $p(t,x,y)$ as provided by Lemma 2.100 is not needed. A further extension to fractal structures is provided by GRABNER and TICHY [675] (see also Section 2.1). Furthermore (and this will also be clear from the following proof) Theorem 2.98 holds for other discrepancy systems, too. For example, we can take the EUCLIDean systems $\mathcal{D}_s(A,\mathbf{R}^k/\mathbf{Z}^k)$ or $\mathcal{D}_h(A,\mathbf{R}^k/\mathbf{Z}^k)$. Furthermore it should be noted that the system of closed balls is indeed a discrepancy system (for details see Proposition 2.7 and BLÜMLINGER [233]).

Conversely if we restrict on μ-c.w.d. functions the situation changes drastically.

Theorem 2.99 *Let X be a compact connected* RIEMANN*ian manifold without boundary which is also a homogeneous space. Then μ_w-almost no functions $\mathbf{x}(t) \in C_W$ with $\mathbf{x}(0) = w \in X$ are μ-c.w.d.*

The proof of Theorem 2.98 is based on Theorem 1.192 by PHILIPP [1432]. In order to check the assumptions of Theorem 1.192 we use some properties of X and $p(t,x,y)$.

Lemma 2.100 *Let $(\varphi_i(x))_{i=0}^{\infty}$ be a complete system of orthonormal eigenfunctions of $-\Delta$ corresponding to the eigenvalues $0 = \lambda_0 < \lambda_1 \le \lambda_2 \le \cdots$; $\varphi_0(x) \equiv 1$. Then the probability densities $p(t,x,y)$ of the* BROWN*ian motion are given by*

$$p(t,x,y) = \sum_{i=0}^{\infty} \varphi_i(x)\varphi_i(y)e^{-\lambda_i t} \qquad (\text{for } t > 0). \tag{2.163}$$

Proof. As it is well-known (cf. BERGER et al. [185] or WARNER [1944]) the operator $-\Delta$ is selfadjoint and $\lim_{i\to\infty}\lambda_i = \infty$; thus the existence of a complete orthonormal system of eigenfunctions φ_i is evident. In the following we make use of the asymptotic relation (cf. BERARD [169], BERGER et al. [185], and CHAVEL [339])

$$\sum_{\lambda_i \le t} 1 \sim C_N t^{N/2}, \tag{2.164}$$

where C_N is a constant only depending on the dimension N of X. Let L be an arbitrary positive integer. Then for $t > 0$

$$\begin{aligned}
\left|\sum_{i=0}^{\infty} \lambda_i^L \varphi_i(x)\varphi_i(y)e^{-\lambda_i t}\right| &\le \sum_{i=0}^{\infty} \lambda_i^L \|\varphi_i\|_\infty^2 e^{-\lambda_i t} \\
&\le c(X)\sum_{i=0}^{\infty} \lambda_i^{L+K} e^{-\lambda_i t} \\
&\le c(X)\sum_{n=0}^{\infty}\sum_{n<\lambda_i\le n+1} \lambda_i^{L+K} e^{-\lambda_i t} \\
&\le c(X)\sum_{n=0}^{\infty}(n+1)^{L+K}e^{-nt}C_N(n+1)^{N/2} \\
&\le c(X)C_N\sum_{n=0}^{\infty}(n+1)^{L+K+N/2}e^{-nt} < \infty,
\end{aligned}$$

where we have used the inequality

$$\|\varphi_i\|_\infty \le c(X)\lambda_i^k \quad \text{for } i \ge 1$$

with constants $c(X)$ and $K = K(X)$ only depending the manifold X (cf. CHAVEL [339, p. 102, Thm. 8] or WARNER [1944, p. 256]). Thus we observe that the infinite series given in (2.163) and all its derivatives with respect to t are absolutely convergent.

By the orthonormality of the eigenfunctions φ_i we see that

$$\int_X \tilde{p}(t,x,y)\varphi_i(x)d\mu(x) = e^{-\lambda_i t}\varphi_i(y), \tag{2.165}$$

where $\tilde{p}$ denotes the infinite series on the right-hand side of (2.163). Thus the integral on the left is a solution of the heat equation with the initial condition $u(0,x) = \varphi_i(x)$. Obviously, (2.165) is also valid for finite linear combinations of the eigenfunctions $\varphi_i(x)$. (2.164) and the observation thereafter yields $\tilde{p}(t,x,.) \in C(X)$ for all $x \in X$ and $t > 0$. Thus $\tilde{p}(t,x,.)$ can be uniquely extended to a continuous linear map $C(X) \to C(X)$.

To see that the series (2.163) is actually the heat kernel $p(t,x,y)$ in (2.162) we note that $p(t,x,y)$ is a probability density. Hence

$$f \to \int_X p(t,x,y)f(x)\,d\mu(x)$$

is a continuous linear map which coincides on finite linear combinations of the eigenfunctions $\varphi_i(x)$ with the map

$$f \to \int_X \tilde{p}(t,x,y) f(x)\, d\mu(x).$$

Thus $p(t,x,y) \equiv \tilde{p}(t,x,y)$ and the proof of Lemma 2.100 is complete. □

Lemma 2.101 *Let X be a N-dimensional compact* RIEMANN*ian manifold and let $\overline{B}(x,r) = \{y \in X : d(x,y) \le r\}$ be the closed geodesic ball with center x and radius r; d denotes the usual geodesic metric and μ the surface measure on X. Then there exists a constant K only depending on X such that $(r_2 > r_1)$*

$$\mu(\overline{B}(\xi, r_2)) - \mu(\overline{B}(\xi, r_1)) \le K(r_2^N - r_1^N). \tag{2.166}$$

Proof. First we note that by the RINOW-HOPF-Theorem (cf. BISHOP and CRITTENDEN [224])

$$\overline{B}(x,r) = \exp \overline{B}_x(r),$$

where $\overline{B}_x(r)$ denotes the closed ball of radius r in the tangent space T_x and exp the exponential map, which defines a local diffeomorphism of $U \subseteq T_x$ onto some open set $V \subseteq X$. Furthermore, by the compactness of X there exists a sufficiently large radius R such that for all $\mathbf{x} \in X$ $\exp_x$ is surjective on X. Now we make use of the explicit formula for the JACOBIan determinant of $\exp_x$ (cf. [224, pp. 253]). Using the JACOBI equation (cf. [224, pp. 173]) we conclude from the compactness of X that the JACOBI determinant of the exponential map is bounded by some constant K_1 in every ball $\overline{B}_x(r)$ with $r \le R$. K_1 can be chosen independently of x.

Let $\overline{B}(x,r_2)$ and $\overline{B}(x,r_1)$ be two geodesic balls in X with $0 < r_1 < r_2$. Then we have

$$\mu(\overline{B}(x,r_2)) - \mu(\overline{B}(x,r_1)) \le K_1 \int_{\overline{B}_x(r_2)\setminus \overline{B}_x(r_1)} dx = K(r_2^N - r_1^N),$$

where dx denotes the EUCLIDean volume element in the tangent space T_x. Thus the proof of Lemma 2.101 is complete. □

Let Γ be the set of all quadruples $\gamma = (\xi_1, r_1, \xi_2, r_2)$ with $\xi_1, \xi_2 \in X$ and $r_1, r_2 > 0$. Then for every $\gamma \in \Gamma$ and for any positive integer n we define the random variables (δ a fixed positive and sufficiently small constant)

$$X_n(\gamma) = \delta \int_n^{n+1} \left(\chi_{\overline{B}(\xi_2,r_2)}(\mathbf{x}(t)) - \chi_{\overline{B}(\xi_1,r_1)}(\mathbf{x}(t))\right) dt, \qquad \mathbf{x}(t) \in C_w. \tag{2.167}$$

The next two lemmata will be essential to check the assumptions of Theorem 1.192. Recall that $\mathcal{M}_{S,T}$ denotes the σ-field generated by the variables X_n with $S \le n \le T$.

Lemma 2.102 *We have*

$$|\mu_w(A)\mu_w(B) - \mu_w(A \cap B)| \le c\mu_w(A)\mu_w(B)e^{-\lambda_1 M}$$

for $M > 1$, $c > 0$ a fixed constant and $A \in \mathcal{M}_{0,L}$, $B \in \mathcal{M}_{M+L,\infty}$.

Proof. First we note that $\mu_w(A\cap B) = \mu_w(A)\mu_w(B|A)$. By the MARKOV property of BROWNian motion (cf. ITO and MCKEAN [863]) we derive

$$\begin{aligned}\mu_w(B|A) &= \int\limits_X p(L,w,\xi|A)\mu_w(B|\mathbf{x}(L)=\xi)\,d\mu(\xi) \qquad (2.168)\\ &= \int\limits_X\int\limits_X p(L,w,\xi|A)\mu_w(B|\mathbf{x}(L+M)=\eta)p(M,\eta,\xi)\,d\mu(\xi)\,d\mu(\eta).\end{aligned}$$

By Lemma 2.100 we obtain

$$|1-p(M,\eta,\xi)| = \left|\sum_{i=1}^{\infty}\varphi_i(\eta)\varphi_i(\xi)e^{-M\lambda_i}\right| \le ce^{-\lambda_1 M} \qquad (2.169)$$

for some constant $c>0$. Hence by (2.168)

$$\begin{aligned}&|\mu_w(A\cap B)-\mu_w(A)\mu_w(B)|\\ &\le \mu_w(A)\int\limits_X\int\limits_X p(L,w,\xi|A)\mu_w(B|\mathbf{x}(L+M)=\eta)\\ &\quad\times\left|p(L+M,w,\eta)-p(M,\eta,\xi)\right|\,d\mu(\xi)\,d\mu(\eta) \le c'\mu_w(A)\mu_w(B)e^{-\lambda_1 M},\end{aligned}$$

for a suitable constant $c'>0$, which proves Lemma 2.102.□

Lemma 2.103 *Let $X_n(\gamma)$ be the random variables defined in (2.167) and $S_N^2(\gamma)$ as in Theorem 1.192. Then for $N\to\infty$*

$$S_N^2(\gamma) = \sigma^2(\gamma)N + \mathcal{O}(1)$$

with $0\le\sigma^2(\gamma)<1$, where the $\mathcal{O}$-constant is uniform in $\gamma\in\Gamma$.

Proof. Set

$$f_\gamma(\mathbf{x}(t)) = \delta(\chi_{\overline{B}(\xi_2,r_2)}(\mathbf{x}(t)) - \chi_{\overline{B}(\xi_1,r_1)}(\mathbf{x}(t)))$$

for $\gamma\in\Gamma$ and $\mathbf{x}(t)\in\mathcal{C}_w$. First we prove

$$\int\limits_1^{N+1}\mathbf{E}(f_\gamma(\mathbf{x}(t)))dt = \delta N(\mu_1-\mu_2)+\mathcal{O}(1), \qquad (2.170)$$

where $\mu_i = \mu(\overline{B}(\xi_i,r_i)) = \mu(\overline{B}_i)$ $(i=1,2)$ and $\mathbf{E}$ denotes the expectation with respect to the WIENER measure. We have

$$\int\limits_1^{N+1}\mathbf{E}(f_\gamma(\mathbf{x}(t)))dt = \delta\int\limits_1^{N+1}\int\limits_{\overline{B}_2}\sum_{n=0}^{\infty}\varphi_n(w)\varphi_n(y)e^{-\lambda_n t}\,d\mu(y)\,dt$$

$$
\begin{aligned}
&-\delta \int_1^{N+1} \int_{\overline{B}_1} \sum_{n=0}^{\infty} \varphi_n(w)\varphi_n(y) e^{-\lambda_n t}\, d\mu(y)\, dt \\
= \;& +\delta N(\mu_2-\mu_1) - \delta \int_{\overline{B}_2} \sum_{n=1}^{\infty} \varphi_n(w)\varphi_n(y) \frac{e^{-\lambda_n(N+1)} - e^{-\lambda_n}}{\lambda_n}\, d\mu(y) \\
& \delta \int_{\overline{B}_1} \sum_{n=1}^{\infty} \varphi_n(w)\varphi_n(y) \frac{e^{-\lambda_n(N+1)} - e^{-\lambda_n}}{\lambda_n}\, d\mu(y) + \mathcal{O}(1) \\
= \;& \delta N(\mu_2-\mu_1) + \mathcal{O}(1),
\end{aligned}
$$

which follows immediately from the estimates in the proof of Lemma 2.100. Next we have

$$
\begin{aligned}
S_N^2(\gamma) &= \mathbf{E}\left(\int_1^{N+1} f_\gamma(\mathbf{x}(t)) - \mathbf{E}(f_\gamma(\mathbf{x}(t)))\, dt \right)^2 \\
&= \mathbf{E}\left(\int_1^{N+1} f_\gamma(\mathbf{x}(t))\, dt - \delta N(\mu_2-\mu_1) + O(1) \right)^2 \qquad (2.171) \\
&= \mathbf{E}\left(\int_1^{N+1} (f_\gamma(\mathbf{x}(t)) - \delta(\mu_2-\mu_1))\, dt \right)^2 + \mathcal{O}(1).
\end{aligned}
$$

Now we obtain

$$
\begin{aligned}
&\mathbf{E}\left(\int_1^{N+1} (f_\gamma(\mathbf{x}(t)) - \delta(\mu_2-\mu_1))\, dt \right)^2 \\
&= \;2 \int_1^{N+1} \int_1^{t} \mathbf{E}\big((f_\gamma(\mathbf{x}(t)) - \delta(\mu_2-\mu_1))(f_\gamma(\mathbf{x}(s)) - \delta(\mu_2-\mu_1))\big)\, ds\, dt \\
&= 2 \quad \delta(\mu_1-\mu_2) \int_1^{N+1} \int_X f_\gamma(x)(t-1)p(t,w,x)\, d\mu(x)\, dt \qquad (2.172) \\
&\quad +2\delta(\mu_1-\mu_2) \int_1^{N+1} \int_1^{t} \int_X f_\gamma(x) p(s,w,x)\, d\mu(x)\, ds\, dt \\
&\quad +N^2\delta^2(\mu_2-\mu_1)^2 + 2 \int_1^{N+1} \int_1^{t} \mathbf{E}(f_\gamma(\mathbf{x}(t)) f_\gamma(\mathbf{x}(s)))\, ds\, dt.
\end{aligned}
$$

Using (2.169) we get

$$2\delta(\mu_1-\mu_2)\int_1^{N+1}\int_X f_\gamma(x)(t-1)p(t,w,x)\,d\mu(x)\,dt$$
$$+2\delta(\mu_1-\mu_2)\int_1^{N+1}\int_1^t\int_X f_\gamma(x)p(s,w,x)\,d\mu(x)\,ds\,dt$$
$$=-2\delta^2N^2(\mu_2-\mu_1)^2+2\delta N(\mu_1-\mu_2)\int_X\sum_{n=1}^\infty f_\gamma(x)\varphi_n(w)\varphi_n(x)\frac{e^{-\lambda_n}}{\lambda_n}\,d\mu(x)\,+\,\mathcal{O}(1).$$

In order to compute the third integral we split it into three parts:

$$\int_1^{N+1}\int_1^t \mathbf{E}\big(f_\gamma(\mathbf{x}(t))f_\gamma(\mathbf{x}(s))\big)\,ds\,dt=\int_1^2\int_1^t+\int_2^{N+1}\int_1^{t-1}+\int_2^{N+1}\int_{t-1}^t=I+II+III.$$

Obviously, $I=\mathcal{O}(1)$.

For the computation of II we set

$$A_{\gamma,n,m}=\int_X\int_X f_\gamma(z)f_\gamma(y)\varphi_n(y)\varphi_n(z)\varphi_m(w)\varphi_m(y)\,d\mu(y)\,d\mu(z)$$

and obtain

$$\begin{aligned}II&=\int_2^{N+1}\int_1^{t-1}\int_X\int_X f_\gamma(z)f_\gamma(y)p(t-s,z,y)p(s,w,z)\,d\mu(z)\,d\mu(y)\,ds\,dt\\&=\int_2^{N+1}\int_1^{t-1}\sum_{n=0}^\infty\sum_{m=0}^\infty A_{\gamma,n,m}e^{-\lambda_n(t-s)}e^{-\lambda_m s}\,ds\,dt\\&=\int_2^{N+1}\Bigg(\sum_{\substack{n,m\\ \lambda_m=\lambda_n}}A_{\gamma,n,m}e^{-\lambda_n t}(t-2)\\&\qquad+\sum_{\substack{n,m\\ \lambda_m\neq\lambda_n}}A_{\gamma,n,m}\frac{e^{(t-1)(\lambda_n-\lambda_m)}-e^{\lambda_n-\lambda_m}}{\lambda_n-\lambda_m}e^{-\lambda_n t}\Bigg)\,dt,\end{aligned}$$

the second sum being convergent by (2.165) and

$$\frac{e^{(t-1)u}-e^u}{u}<(t-1)e^{(t-1)u}-e^u.$$

Furthermore we have

$$\begin{aligned}
&A_{\gamma,0,0}\frac{1}{2}(N-1)^2+\sum_{\substack{1\le n,m\\ \lambda_n=\lambda_m}} A_{\gamma,n,m}\left(\frac{e^{-2\lambda_n}}{\lambda_n^2}-\frac{e^{-\lambda_n(N+1)}}{\lambda_n^2}(\lambda_n(N-1)+1)\right)\\
&+\sum_{1\le m}(A_{\gamma,0,m}+A_{\gamma,m,0})\left(\frac{N-1}{\lambda_m}e^{-\lambda_m}-\frac{e^{-N\lambda_m}-e^{-\lambda_m}}{\lambda_m^2}\right)\\
&+\sum_{\substack{1\le n,m\\ \lambda_n\neq\lambda_m}} A_{\gamma,n,m}\left(\frac{e^{-\lambda_m-N\lambda_n}-e^{-\lambda_m-\lambda_n}}{\lambda_n(\lambda_n-\lambda_m)}-\frac{e^{-\lambda_n-N\lambda_m}-e^{-\lambda_n-\lambda_m}}{\lambda_m(\lambda_n-\lambda_m)}\right)\\
&=A_{\gamma,0,0}\left(\frac{N^2}{2}-N\right)+N\sum_{1\le m}(A_{\gamma,0,m}+A_{\gamma,m,0})\frac{e^{-\lambda_m}}{\lambda_m}+\mathcal{O}(1),
\end{aligned}$$

where we used the mean-value theorem to obtain

$$\frac{\lambda_m e^{-\lambda_m-N\lambda_n}-\lambda_n e^{-\lambda_n-N\lambda_m}}{\lambda_n\lambda_m(\lambda_n-\lambda_m)}=\frac{e^{-\lambda_n-\lambda_m}}{\lambda_n\lambda_m}\tilde{\lambda}(N-1)e^{-(N-1)\tilde{\lambda}},$$

with $\lambda_1\le\tilde{\lambda}\in(\min(\lambda_n,\lambda_m),\max(\lambda_n,\lambda_m))$ and

$$\sum_{\substack{1\le n,m\\ \lambda_n\neq\lambda_m}}=\mathcal{O}(1).$$

For the last double integral we obtain using (2.169)

$$\begin{aligned}
III &= \int_2^{N+1}\int_{t-1}^{t}\int_X\int_X f_\gamma(z)f_\gamma(y)p(t-s,z,y)p(s,w,z)\,d\mu(z)\,d\mu(y)\,ds\,dt\\
&= \int_2^{N+1}\int_{t-1}^{t}\int_X\int_X f_\gamma(z)f_\gamma(y)p(t-s,z,y)\,d\mu(z)\,d\mu(y)\,ds\,dt+\mathcal{O}(1)\\
&= \int_2^{N+1}\int_0^{1}\int_X\int_X f_\gamma(z)f_\gamma(y)p(u,z,y)\,d\mu(z)\,d\mu(y)\,du\,dt+\mathcal{O}(1)\\
&= C_1N+\mathcal{O}(1).
\end{aligned}$$

Combining (2.171), (2.172) and the results for the three integrals I, II, III yields

$$S_N^2(\gamma)=O(1)+\sigma^2(\gamma)N \text{ with } \sigma^2(\gamma)=\mathcal{O}(\delta^2\mu(\text{supp } f_\gamma)), \tag{2.173}$$

where supp f_γ denotes the support of f_γ. Thus the desired limit $\sigma^2(\gamma)=\lim_{N\to\infty}S_N^2(\gamma)/N$ exists uniformly in γ. Choosing δ sufficiently small we immediately obtain

$$\sigma^2(\gamma)<1. \qquad \square$$

Proof of Theorem 2.98. As mentioned above the proof relies on an application of Theorem 1.192. We have to define a suitable subclass Γ_θ of Γ and a subfamily $\mathcal{A}_\theta \subseteq \mathcal{A} = \{X_n(\gamma) : n = 1, 2, ... \text{ with } \gamma \in \Gamma\}$.

Let ξ be a fixed point in X and let R be sufficiently large such that the exponential mapping $\exp_\xi : \overline{B}_\xi(R) \to X$ is surjective (cf. proof of Lemma 2.101). Since $\exp_\xi$ is a C^∞-mapping, we see that

$$d(\exp_\xi \eta_1, \exp_\xi \eta_2) \le K_o|\eta_1 - \eta_2| \text{ for all } \eta_1, \eta_2 \in \overline{B}_\xi(R), \tag{2.174}$$

where d denotes the geodesic distance on X and $|\cdot|$ the distance in the tangent space T_ξ. Now we choose a fixed basis of orthogonal vectors of length $2R$ in the tangent space T_ξ. For every non-negative integer θ we define the class Ξ_θ to be the class of all points η in T_ξ such that all coordinates of η have dyadic digit representation $\pm 0, a_1, \dots, a_\theta$ ($a_i \in \{0, 1\}$). Note that with our choice of the basis in T_ξ the coordinates of any point in $\overline{B}_\xi(R)$ have absolute value < 1. Let A_θ be the class of all dyadic rationals r of the form $r = 1$ or $r = 0, a_1 ... a_\theta$ and let $\rho = \sup_{x,y\in X} d(x, y)$ denote the diameter of X. Then Γ_θ is defined to be the class of all quadruples $\gamma = (\xi_1, r_1, \xi_2, r_2)$ with $\xi_1, \xi_2 \in \exp_\xi \Xi_\theta$ and $\frac{r_1}{\rho}, \frac{r_2}{\rho} \in A_\theta$. Furthermore we define $\mathcal{A}_\theta = \{X_n(\gamma) : n = 1, 2, ... \text{ with } \gamma \in \Gamma_\theta\}$ for all $\theta = 0, 1, 2, ...$ Then we obtain

$$|\mathcal{A}_\theta| = |\Gamma_\theta| \le 2^{N(\theta+1)} 2^\theta 2^{N(\theta+1)} 2^\theta = 2^{2\theta+2N\theta+2N}. \tag{2.175}$$

Thus, condition *1.* in Theorem 1.192 is satisfied. Obviously the mixing condition *2.* is guaranteed by Lemma 2.102.

In the following we define the approximations $\overline{Y}_n(s)$ and $Y_n(s)$ (cf. *3.*). For any $\eta \in X, \theta \in \mathbb{N}_0$ there exists a point $\eta_\theta \in \exp_\xi \Xi_\theta$ such that

$$d(\eta, \eta_\theta) \le K_0\sqrt{N}2^{-\theta} \tag{2.176}$$

(cf. (2.174)). Hence for arbitrary r with $0 < r < R$

$$\overline{B}(\eta, r) \supseteq \overline{B}(\eta_\theta, \max(0, r - K_o\sqrt{N}2^{-\theta})). \tag{2.177}$$

Set

$$r_\theta = \left[2^\theta \frac{r}{\rho}\right] \frac{\rho}{2^\theta},$$

i.e. $r_\theta/\rho \in A_\theta$ and $r - \rho 2^{-\theta} \le r_\theta \le r$. Then we have by Lemma 2.101, (2.176) and (2.177)

$$\begin{aligned} \mu(\overline{B}(\eta, r)) - \mu(\overline{B}(\eta_\theta, r_\theta)) &\le \mu(\overline{B}(\eta, r)) - \mu(\overline{B}(\eta, \max(0, r_\theta - d(\eta, \eta_\theta)))) \\ &\le KNr^N(r - r_\theta + d(\eta, \eta_\theta)) \le \\ &\le KNr^N(\rho 2^{-\theta} + K_o\sqrt{N}2^{-\theta}) \\ &\le K_1 2^{-\theta} \end{aligned} \tag{2.178}$$

with a constant K_1 only depending on the manifold X. By verbally the same arguments we see that for arbitrary $r \quad (0 < r < R)$ and $\eta \in X, \theta \in \mathbf{N}_0$ there exists a point $\overline{\eta}_\theta \in \exp_\xi \Xi_\theta$ and a radius $\overline{r}_\theta$ with $\overline{r}_\theta/\rho \in A_\theta$ such that

$$\mu(\overline{B}(\overline{\eta}_\theta, \overline{r}_\theta)) - \mu(\overline{B}(\eta, r)) \leq K_2 2^{-\theta}. \tag{2.179}$$

Now let $\gamma = (\xi_1, r_1, \xi_2, r_2)$ be a fixed quadruple in Γ. For any $\theta \in \mathbf{N}_0$ let $\gamma_\theta = (\overline{\xi}_{1,\theta}, \overline{r}_{1,\theta}, \xi_{2,\theta}, r_{2,\theta})$ and $\overline{\gamma}_\theta = (\xi_{1,\theta}, r_{1,\theta}, \overline{\xi}_{2,\theta}, \overline{r}_{2,\theta})$ where $\xi_{i,\theta}, r_{i,\theta} \quad (i = 1, 2)$ are the approximations of ξ_i, r_i as considered in (2.178); $\overline{\xi}_{i,\theta}, \overline{r}_{i,\theta}$ are the approximations as considered in (2.179). Then we define

$$\begin{aligned}
\overline{X}_n(2\theta) &= \int_n^{n+1} \chi_{\overline{B}(\xi_{2,\theta}, r_{2,\theta})}(\mathbf{x}(t)) - \chi_{\overline{B}(\xi_{2,\theta-1}, r_{2,\theta-1})}(\mathbf{x}(t))\, dt \\
\overline{X}_n(2\theta+1) &= \int_n^{n+1} \chi_{\overline{B}(\overline{\xi}_{1,\theta}, \overline{r}_{1,\theta})}(\mathbf{x}(t)) \chi_{\overline{B}(\xi_{1,\theta-1}, r_{1,\theta-1})}(\mathbf{x}(t))\, dt \\
X_n(2\theta) &= \int_n^{n+1} \chi_{\overline{B}(\overline{\xi}_{2,\theta}, \overline{r}_{2,\theta})}(\mathbf{x}(t)) - \chi_{\overline{B}(\overline{\xi}_{2,\theta-1}, \overline{r}_{2,\theta-1})}(\mathbf{x}(t))\, dt \qquad (2.180) \\
X_n(2\theta+1) &= \int_n^{n+1} \chi_{\overline{B}(\xi_{1,\theta}, r_{1,\theta})}(\mathbf{x}(t)) - \chi_{\overline{B}(\xi_{1,\theta-1}, r_{1,\theta-1})}(\mathbf{x}(t))\, dt.
\end{aligned}$$

Now we set

$$\overline{Y}_n(s) = \sum_{\theta=1}^{s} \overline{X}_n(\theta) \quad \text{and} \quad Y_n(s) = \sum_{\theta=1}^{s} X_n(\theta). \tag{2.181}$$

Obviously, $\overline{Y}_n(s)$ and $Y_n(s)$ are random variables of the class $\mathcal{A}$ and $\overline{X}_n(\theta), X_n(\theta)$ are contained in the class $\mathcal{A}_\theta$. Furthermore we have

$$\left|\mathbf{E}(Y_n(s) - \overline{Y}_n(s))\right| < \int_n^{n+1} \sup_{y \in X} p(t, w, y)\, dt \left(\mu(\overline{B}(\overline{\eta}_{s-1}, \overline{r}_{s-1})) - \mu(\overline{B}(\eta_{s-1}, r_{s-1}))\right).$$

Applying Lemma 2.100 and (2.178), (2.179) we obtain

$$\left|\mathbf{E}(Y_n(s) - \overline{Y}_n(s))\right| \leq e^{-C_{2}s},$$

thus proving *3.* From (2.173) we conclude that

$$S_N^2(\theta) = S_N^2(\gamma_\theta) = \mathcal{O}(1) + \sigma^2(\gamma_\theta) N \text{ with } \sigma^2(\gamma_\theta) = \mathcal{O}(2^{-\theta}).$$

Hence

$$S_N^2(\theta) = \mathcal{O}(\theta^{-6} N) \qquad \text{for } \theta \leq \log N, \tag{2.182}$$

proving *4.* in Theorem 1.192.

By Lemmata 2.102 and 2.103 conditions *5.* and *6.* are satisfied, too, since $\mathcal{A}_\theta$ is a subclass of $\mathcal{A}$. Hence the assertion of Theorem 1.192 yields

$$\limsup_{N\to\infty} \frac{N D_N^{(\mathcal{D})}(\mathbf{x}(t))}{\sqrt{2N \log\log N}} = \alpha \tag{2.183}$$

for μ_w-almost all trajectories $\mathbf{x}(t) \in C_w$ and some $\alpha \geq 0$.

To see that $\alpha > 0$ we estimate $\sigma^2(\gamma)$ for $\gamma \in \Gamma$ with $r_1 = r_2$ and $f_\gamma \not\equiv 0$. Since for such γ the coefficients $A_{\gamma,0,\mu}$ vanish, the computation leading to (2.173) shows that

$$\begin{aligned}
\sigma^2(\gamma) &= \sum_{1\leq m} A_{\gamma,m,o} \frac{e^{-\lambda_m}}{\lambda_m} + \int_0^1 \int_X \int_X f_\gamma(z) f_\gamma(y) p(t,y,z) d\mu(y)\, d\mu(z)\, dt \\
&= \int_0^\infty \int_X \int_X f_\gamma(y) f_\gamma(z) p(t,y,z) d\mu(y)\, d\mu(z)\, dt \\
&= \int_0^\infty \int_X \int_X \int_X f_\gamma(y) f_\gamma(z) p(t/2,y,x) p(t/2,x,z)\, d\mu(x)\, d\mu(y)\, d\mu(z)\, dt \\
&= \int_0^\infty \int_X \left(\int_X f_\gamma(y) p(t/2,y,x)\, d\mu(y) \right)^2 d\mu(x)\, dt > 0.
\end{aligned}$$

Here we have used the CHAPMAN-KOLMOGOROV equation

$$\int_X p(t,x,y) p(s,z,x)\, d\mu(x) = p(t+s,z,y),$$

which is an immediate consequence of (2.163) and the orthonormality of φ_i. From

$$S_N^2(X_n(\gamma_1) + X_n(\gamma_2)) \leq S_N^2(X_n(\gamma_1)) + S_N^2(X_1(\gamma_2))$$

we see by taking $\gamma_1 = (\xi_1, r_1, \xi_2, 0), \gamma_2 = (\xi_1, 0, \xi_2, r_2)$ that $X_n(\gamma_1) + X_n(\gamma_2) = X_n(\gamma)$. Hence $\lim_{N\to\infty} \frac{1}{N} S_N^2(X_n(\gamma_i)) > 0$ for some $i \in \{1,2\}$ i.e. there exists a geodesic ball $\overline{B}$ corresponding to γ_i with $\sigma^2(\gamma_i) > 0$. Since $\alpha = \frac{1}{\delta} \sup_{\overline{B}} \sigma(\overline{B})$ we have $\alpha > 0$. In order to complete the proof of Theorem 2.98 we note that

$$\int_0^T (\chi_{\overline{B}}(\mathbf{x}(t))dt - \mu(\overline{B}))dt = \left(\int_0^{[T]} + \int_{[T]}^T \right) (\chi_{\overline{B}}(\mathbf{x}(t)) - \mu(\overline{B}))dt.$$

Hence we have

$$\frac{T\, D_T(\mathbf{x}(t))}{\sqrt{2T \log\log T}} = \frac{[T] D_{[T]}(s(t))}{\sqrt{2[T] \log\log[T]}} + o(1), \qquad T \to \infty$$

and the proof of Theorem 2.98 is complete. □

For the proof of Theorem 2.99 we will use the following

Lemma 2.104 *Let X be a compact connected* RIEMANN*ian manifold without boundary which is also a homogeneous space and U a neighbourhood of $w \in X$. Then we have*

$$\mu_w \left(\{ \mathbf{x} \in C_w : \mathbf{x}(t) \in U \text{ for all } 0 \leq t \leq T \} \right) > 0. \tag{2.184}$$

Proof. X is assumed to be homogeneous, i.e. it may be identified with the set of cosets G/K, in which G is the group acting on it and K is the centralizer (see SCHEMPP and DRESELER [1618]). Since X is homogeneous it is sufficient to consider $w = eK$, where e is the unit element of G. Choose a neighbourhood U_1 of e in G such that $U_1^2 K \subseteq U$. By the continuity of the diffusion process there exists $t_0 > 0$ and $x_0 \in U_1$ such that

$$p(t_0, eK, Vx_0K) = \int_{Vx_0K} p(t_0, eK, y)\, d\mu(y) > 0$$

for all neighbourhoods V of e. We choose a positive integer n with $2nt_0 > T$. By continuity of the multiplication there exists a neighbourhood $V_1 = V_1^{-1}$ of e with $(x_0 V_1^2 x_0^{-1})^n \subseteq U_1$. Set $\beta = p(t_0, eK, x_0V_1K) > 0$. Since $p(t, x, y)$ is symmetric and X is homogeneous we also have $p(t_0, eK, (x_0V_1)^{-1}K) = \beta > 0$. Hence, we obtain for every $v_1 \in V_1$

$$p(t_0, x_0v_1K, x_0v_1V_1x_0^{-1}K) = \beta$$

and

$$p(2t_0, eK, x_0V_1^2x_0^{-1}K) \geq \beta^2.$$

By induction it follows that for every t with $2(i-1)t_0 \leq t \leq 2it_0$ $(1 \leq i \leq n)$

$$\mu_w\left(\{\mathbf{x} \in \mathcal{C}_w : \mathbf{x}(t) \in (x_0V_1^2x_0^{-1})^i U_1K\}\right) \geq \beta^{2i}.$$

Since $(x_0V_1^2x_0^{-1})^n \subseteq U_1$ and $U_1^2K \subseteq U$ we obtain

$$\mu_w\left(\{\mathbf{x} \in \mathcal{C}_w : \mathbf{x}(t) \in U \text{ for all } 0 \leq t \leq 2nt_0\}\right) \geq \beta^{2n} > 0, \tag{2.185}$$

which proves the Lemma. □

Proof of Theorem 2.99. Let U be an open neighbourhood of $w = eK$ and let f be a continuous real-valued function on X with $\int_X f\, d\mu = 1$ such that the support of f is contained in the complement of U^2. Since X is compact there exist finitely many $a_1, \ldots, a_n \in G$ such that $a_1U, \ldots, a_nU$ cover X. Suppose now that $\mathbf{x}(t)$ (with $\mathbf{x}(0) = w$) is well distributed on X. Then we obtain for a suitable integer $T_0 > 0$

$$\int_A^{A+T_0} f(a_i^{-1}\mathbf{x}(t))\, dt > \frac{1}{2} \tag{2.186}$$

for all $A \geq 0$ and for all $1 \leq i \leq n$. Consider the MARKOV times

$$\tau_0 = 0, \quad \tau_i = \inf\{t > \tau_{i-1} : \mathbf{x}(t) \in U_{\mathbf{x}(\tau_{i-1})}\} \quad (i \geq 1),$$

which surely exist, since $\mathbf{x}(t)$ is w.d. Furthermore we have by the MARKOV property and by Lemma 2.104

$$\mu_w\left(\bigcup_{T_0=1}^{\infty} \{\mathbf{x} \in \mathcal{C}_w : \tau_i - \tau_{i-1} \leq T_0 \text{ for all } i \geq 1\}\right) = 0.$$

Now observe that every w.d. function $\mathbf{x} \in C_w$ satisfies $\tau_i - \tau_{i-1} < T_0$ for all $i \geq 1$. Namely, if $\tau_i - \tau_{i-1} \geq T_0$ for some $i \geq 1$ we would obtain $f(a_i^{-1}\mathbf{x}(t)) = 0$ for $\tau_{i-1} \leq t \leq \tau_i$ which means that

$$\int_{\tau_{i-1}}^{\tau_{i-1}+T_0} f(a_i^{-1}\mathbf{x}(t))\, dt = 0.$$

This contradicts (2.186). Thus, μ_w-almost no functions $\mathbf{x} \in C_w$ are w.d. □

Notes

In his fundamental paper Weyl [1953] also discussed c.u.d. functions mod 1, he established the Weyl criterion, and he applied c.u.d. linear functions to a problem of statistical mechanics. In fact, most properties concerning u.d. sequences have continuous analogues, e.g. Van der Corput's difference theorem [969, 785] (Theorem 2.80). There is also a corresponding Erdős-Turán-Koksma inequality (Theorem 2.77) and a Koksma-Hlawka inequality (Theorem 2.76). Of course these two theorems need the notion of discrepancy which was defined by Hlawka [785]. Furthermore, well distributed functions were introduced by Kuipers [971] and c.u.d. mod 1 with respect to summation methods by Hlawka [785] and Holewijn [841]. Kuipers [969] considered two other conceps of c.u.d.: c^I-u.d. and c^{II}-u.d. (see also Kuipers [970] and Kuipers and Meulenbeld [979, 980, 981]). Later it turned out that c^I-u.d. and c^{II}-u.d. are equivalent concepts [979, 980]. (For more details and further extensions see Kuipers and Niederreiter [983].)

Weyl [1953] already observed that every non-constant polynomial is c.u.d. mod 1. Other examples can be found in Hlawka [785], Müller [1222], Drmota [475, 477] and in Drmota and Tichy [487]; see also Tichy [1829]. Discrepancy bounds for the linear function $x(t) = \underline{\alpha}t$, $\underline{\alpha} \in \mathbf{R}^2$, can be found in Drmota [482] and Larcher [1014]. Beck [142] stated that for almost all $\underline{\alpha} \in \mathbf{R}^k$ we have $T\, D_T(\underline{\alpha}t) = O((\log T)^{k-1}(\log\log T)^{1+\varepsilon})$ $(\varepsilon > 0)$. Another interesting property of linear functions (which is called 'returns of an integral') is discussed by Moshchevitin [1208].

The first lower bound for the discrepancy of continuous functions in terms of the arc length $s(T)$ which is valid for infinitely many T is due to Taschner [1786]. Later Drmota [480] found a sharper bound (Theorem 2.91) which is optimal in the one dimensional case. Taschner [1787] also provided lower bounds for the discrepancy $D_T(x(t))$ on the torus $\mathbf{R}^k/\mathbf{Z}^k$ in terms of the arclength $s(T)$ which are satisfied for every T with $s(T) \geq s_0$. In fact he adapted Roth's method and obtained a lower bound for the L^2-discrepancy. Drmota and Tichy's approach [488] provides the same lower bound (for the usual discrepancy $D_T(x(t)) \gg s(T)^{k/(k-1)}$) in a much more general setting. In the case of the torus this lower bound is (almost) optimal, cf. Drmota [480, 486, 482] (Theorems 2.93, 2.94, and 2.95). Tichy [1840] considered c.u.d. of fractal curves and Drmota [484] investigated completely uniformly distributed functions. Various special problems on c.u.d. functions are considered in Tichy [1829].

Drmota [479] used Schmidt's integral equation method in order to obtain lower bounds for the discrepancy (with respect to rectangles in arbitrary position or with respect to balls) of the form $D_T(x(t)) \geq s(T)^{-1/2-1/(k-1)-\varepsilon}$. In [482] he sharpened this bound to $\gg s(T)^{-1/2-1/(k-1)}$ and showed that it is (almost) optimal (Theorem 2.96). An application of c.u.d. to the theory of turbulence is due to Bass [109]. Lower bounds for the c-discrepancy of multivariate functions are due to Grandits [695]; see also Taschner [1788].

There are several metric theorems. Hlawka [789] proved that almost all continuous functions (in the sense of Wiener measure) are c.u.d. mod 1. Fleischer [608] generalized Hlawka's theorem to the multidimensional case. Stackelberg [1735] provided a law of the iterated logarithm, too (compare also with Fleischer [609]). Loynes [1109] discussed c.u.d. of stochastic processes. Aumayr [57] provided a metric hereditary theorem of c.u.d. functions mod 1, see also Müller and Taschner [1223]. Later Blümlinger, Drmota, and Tichy [237, 238, 239] and Blümlinger [234] considered c.u.d. on compact manifolds and metric theorems and a law of the iterated logarithm with respect to the Brownian motion measure (Theorem 2.98). Drmota [477, 478] also proved that almost all functions are weakly well distributed. (Every weakly well distributed function is c.u.d.) Conversely Drmota [477] and Blümlinger, Drmota, and Tichy [238] showed that almost no functions are c.w.d. (Theorem 2.99).

2.4 Uniform Distribution in Abstract and Discrete Spaces

2.4.1 Uniform Distribution in Compact and Locally Compact Spaces

Abstact theory of uniform distribution is not the main scope of this book. Nevertheless we want to provide the basic notions and to present the main results. For more details see KUIPERS and NIEDERREITER [983] or HELMBERG [773].

Compact Spaces

The notion of uniform distribution modulo 1 can be generalized to compact spaces X:

Definition 2.105 *Let X be a compact (*HAUSDORFF*) space and μ a positive regular normalized* BOREL *measure on X. A sequence $(x_n)_{n\ge1}$, $x_n \in X$, is called* uniformly distributed with respect to μ *(μ-u.d.) if*

$$\lim_{N\to\infty} \frac{1}{N} \sum_{n=1}^{N} f(x_n) = \int_X f \, d\mu \tag{2.187}$$

holds for all continuous functions $f : X \to \mathbf{R}$.

Note that this definition is just a reformulation of the property that the (weighted) counting measure

$$\frac{1}{N} \sum_{n=1}^{N} \delta_{x_n} \tag{2.188}$$

(in which δ_x denotes the DIRAC measure concentrated on x) converges weakly to μ.

Obviously (2.187) need not be checked for all continuous functions $f : X \to \mathbf{R}$. It is sufficient to assume that (2.187) is satisfied for a so-called *convergence determining class* S of functions, e.g. if the linear space generated by S is dense in $C(X)$ then S is a convergence determining class. If X satisfies the second axiom of countability (i.e. X has a countable base) then there exists a countable convergence determining class. Thus the indiviual ergodic theorem applied to the ergodic shift $T(x_1, x_2, \ldots) = (x_2, x_3, \ldots)$ and to the function $F((x_n)_{n\ge1}) = f(x_1)$ immediately implies the following theorem

Theorem 2.106 *Let X be a compact* HAUSSDORFF *space with a countable topological base and μ a positive regular normalized* BOREL *measure on X. Then μ_∞-almost all sequences $(x_n)_{n\ge1}$ in X are μ-u.d.*

Especially it follows that there exist μ-u.d. sequences.

A very prominent special case of compact spaces is that of a compact group G. In this context μ will always denote the HAAR measure. In this case there exists a natural convergence determining class of functions related to representations of G.

Theorem 2.107 *Let $\{\mathbf{D}^{(\lambda)} : \lambda \in \Lambda\}$ be a system of representations of G that is obtained by choosing exactly one representation from each equivalence class of irreducible unitary representations of G. Let $\mathbf{D}^{(0)}$ be the trivial representation. Then a sequence $(x_n)_{n\ge 1}$ is u.d. in G if and only if*

$$\lim_{N\to\infty} \frac{1}{N} \sum_{n=1}^{N} \mathbf{D}^{(\lambda)}_{x_n} = 0$$

holds for all $\lambda \in \Lambda \setminus \{0\}$.

If $X = \mathbf{R}^k/\mathbf{Z}^k$ Theorem 2.107 is exactly WEYL's criterion (Theorem 1.19). Therefore Theorem 2.107 is called WEYL's criterion, too.

Locally Compact Spaces

The generalization of the notion of uniform distribution modulo 1 to uniform distribution on compact spaces is more or less a direct one, especially if we consider compact groups and the normalized HAAR measure. If X is a locally compact space and μ a (positive regular) BOREL measure with $\mu(X) = 1$ then it would be natural to define a sequence $(x_n)_{n\ge 1}$, $x_n \in X$, to be μ-u.d. by the property that (2.188) converges weakly to μ. However, if we consider the one point compactification $X' = X \cup \{\infty\}$ (with the measure $\mu'(E \cup \{\infty\}) = \mu(E)$) then $(x_n)_{n\ge 1}$ is μ-u.d. in X if and only if $(x_n)_{n\ge 1}$ is μ'-u.d. in X'. (Note that weak convergence on locally compact spaces is checked by a property of the kind (2.188) with continuous "test"-functions f with compact support.) Thus, if we want to introduce an essentially new notion of uniform distribution then we have to consider situations where $\mu(X) = \infty$.

If G is a locally compact group then the natural measure on G is a (left invariant or a right invariant) HAAR measure μ on G. If G is not compact then $\mu(G) = \infty$. Thus we cannot use the above concept of weak convergence of measures. Nevertheless several concepts of uniform distribution (with respect to a HAAR measure) on locally compact groups have been introduced. Most of them are equivalent to the usual uniform distribution if G is compact. In the following we report on some of these notions without proofs. (For more details and other notions see KUIPERS and NIEDERREITER [983] and RINDLER [1551, 1550]. All necessary facts about representation theory can be found in HEWITT and ROSS [776].) Especially it turns out that the HARTMAN, the RUBEL, the unitary, and the $\mathrm{L}^1(G)$ uniform distribution coincide with the usual uniform distribution if G is compact.

In what follows we will always assume that G is a locally compact group and μ a left invariant HAAR measure.

Definition 2.108 *A sequence $(x_n)_{n\geq 1}$, $x_n \in G$, is* HARTMAN *uniformly distributed (H-u.d.) if*

$$\lim_{N\to\infty} \frac{1}{N} \sum_{n=1}^{N} D_{x_n} = 0$$

holds for every irreducible, unitary, finite dimensional, nontrivial, continuous representation D of G.

In the abelian case this definition is due to HARTMAN [754]. Obviously this notion is equivalent to the usual one if G is compact by the (general) WEYL criterion. The following existence theorem is due to BERG, RAJAGOPALAN, and RUBEL [183] and to BENZINGER [167, 168].

Theorem 2.109 *If G/N is separable, where N denotes the* VON NEUMANN *kernel, then there exist H-u.d. sequences in G.*

If G is abelian or almost connected then the converse is also true.

The general case is still an open problem.

The VON NEUMANN kernel is the intersection of all kernels of irreducible, unitary, finite dimensional, continuous representations of G. G is almost connected if G/G_e is compact, where G_e is the component of the unit element e.

A slightly different notion of u.d. in locally compact groups is the following one due to RUBEL [1573].

Definition 2.110 *A sequence $(x_n)_{n\geq 1}$, $x_n \in G$, is* RUBEL *uniformly distributed (R-u.d.) if*

$$\lim_{N\to\infty} \frac{1}{N} \sum_{n=1}^{N} D_{x_n} = 0$$

holds for every irreducible, unitary, finite dimensional, nontrivial, continuous, periodic representation D of G.

A representation D is periodic if $G/\ker(D)$ is compact, where $\ker(D)$ denotes the kernel of D.

Obviously, every H-u.d. sequence is R-u.d., too. The notion of R-u.d. is discussed in detail by KUIPERS and NIEDERREITER [983]. For expample, one obtains the following charcterization.

Theorem 2.111 *A sequence $(x_n)_{n\geq 1}$ is R-u.d. in G if and only if $(x_nH)_{n\geq 1}$ is u.d. in G/H for every closed normal subgroup H of G for which G/H is compact.*

In contrast to HARTMAN u.d. the problem of existence is completely solved in case of RUBEL u.d. by BENZINGER [167, 168].

Theorem 2.112 *There exist R-u.d. sequences in G if and only if G is K-separable.*

G is K-separable if there exists a countable set F such that FH is dense in G/H if G/H is compact.

By using a direct generalization of Theorem 5.9. in Chapter 4 of KUIPERS and NIEDERREITER [983] and results of BENZINGER [168] it is possible to characterize those groups for which the notions of H-u.d. and R-u.d. coincide.

Theorem 2.113 *The notions of* HARTMAN *and* RUBEL *uniform distrubtion are the same if and only if all irreducible, unitary, finite dimensional, continuous representations of G are periodic or if no R-u.d. sequence exists.*

For example, if G is connected then this is the case.

Note that H-u.d. and R-u.d. sequences are defined via finite dimensional representations of G. However, it is also possible to use more general representations; see RINDLER [1545].

Definition 2.114 *Let U be a continuous representation of G such that for every $x \in G$, U_x is a unitary operator on a* HILBERT *space H. Furthermore let M be the subspace of H such that $U_x|_M = \mathrm{id}|_M$ for every $x \in G$ and let P denote the orthogonal projection of H onto M.*

A sequence $(x_n)_{n\geq 1}$, $x_n \in G$, is U-uniformly distributed (U-u.d.) if

$$\lim_{N\to\infty} \frac{1}{N} \sum_{n=1}^{N} U_{x_n} = P$$

holds in the strong operator topology.

Definition 2.115 *A sequence $(x_n)_{n\geq 1}$, $x_n \in G$, is unitary uniformly distributed (unitary u.d.) if $(x_n)_{n\geq 1}$ is U-u.d. for every non-trivial irreducible, unitary, continuous representation U of G.*

Note that in this case $M = \{0\}$ and $P = 0$.

Clearly, every unitary u.d. sequence is H-u.d. Furthermore, if every irreducible, unitary, and continuous representation is finite dimenional (i.e. G is a MOORE group) then the converse is also true.

The next notion is due to RINDLER [1542].

Definition 2.116 *For $y \in G$ let L_y be the isometric operator on $\mathrm{L}^1(G)$ defined by $L_y f(x) = f(y^{-1}x)$. Furthermore let $L_0(G)$ be the subspace of all $g \in \mathrm{L}^1(G)$ with $\int_G g\,d\mu = 0$.*

A sequence $(x_n)_{n\geq 1}$, $x_n \in G$, is $L^1(G)$-uniformly distributed ($L^1(G)$-u.d.) if

$$\lim_{N\to\infty} \frac{1}{N} \sum_{n=1}^{N} L_{x_n} = 0$$

in $L_0(G)$ in the strong operator topology.

Again it is clear that every $\mathrm{L}^1(G)$-u.d. sequence is H-u.d. and thus R-u.d.

These notions are related in the following way.

Theorem 2.117 *If G is reflexive then every $L^1(G)$-u.d. sequence is U-u.d. for every unitary representation U of G.*

The existence problem of $L^1(G)$-u.d. sequences is partially solved by RINDLER [1550].

Theorem 2.118 *If G is compact or abelian then $L^1(G)$-u.d. sequences exist if and only if G is separable.*

If G has a countable base then $L^1(G)$-u.d. sequences exist if and only if G is amenable.

There is a close connection between $L^1(G)$-u.d. sequences and u.d. sequences of measures which have been introduced by KERSTAN and MATTHES [919].

Definition 2.119 *A sequence of probabilty measures μ_n on G is called uniformly distributed on G if*

$$\lim_{n\to\infty} \|L_x f * (d\mu_n) - f * (d\mu_n)\|_1 = 0$$

for every $f \in L^1(G)$.

As usual $f * (d\mu)$ denotes the convolution

$$f(x) * (d\mu) = \int_G f(xy^{-1})\, d\mu(y).$$

Theorem 2.120 *A sequence $(x_n)_{n\geq 1}$ is $L^1(G)$-u.d. in (a locally compact group) G if and only if the sequence of (normalized) counting measures*

$$\mu_n = \frac{1}{n}\sum_{k=1}^{n} \delta_{x_k}$$

is uniformly distributed on G.

For a proof see RINDLER [1550].

2.4.2 Distribution Problems in Finite Sets

TIJDEMAN-MEIJER's Theorem

Let X be a finite set of order q. Then X with the discrete topology is a compact (metric) space. Let μ be a measure on X. A sequence $(x_n)_{n\geq 1}$ is μ-u.d. on X if

$$\lim_{N\to\infty} \frac{1}{N}\sum_{n=1}^{N} c_{\{x\}}(x_n) = \mu(x)$$

holds for every $x \in X$. Obviously the system of singletons $\{x\}$, $x \in X$, is a discrepancy system. Therefore the discrepancy $D_N(\mu, x_n)$ of $(x_n)_{n\geq 1}$ with respect to μ is given by

$$D_N(\mu, x_n) = \max_{x\in X}\left|\frac{1}{N}\sum_{n=1}^{N} \chi_{\{x\}}(x_n) - \mu(x)\right|.$$

It is interesting that for every probability measure μ there exists a sequence $(x_n)_{n\geq 1}$ such that $D_N(\mu, x_n) \leq 1/N$. More precisely, the following theorem holds.

Theorem 2.121 *Let X be a finite set of order $q \geq 2$. Then*

$$\sup_{\mu} \inf_{(x_n)} \sup_{N \geq 1} N\, D_N(\mu, x_n) = 1 - \frac{1}{2(q-1)}, \tag{2.189}$$

where the supremum is taken over all probability measures μ on X and the infimum over all sequences $(x_n)_{n \geq 1}$, $x_n \in X$.

The proof of Theorem 2.121 is divided into two major parts. First we show that the left hand side of (2.189) is $\leq 1 - 1/(2(q-1))$.

Proposition 2.122 *Let X be a finite set of order $q \geq 2$ and μ_j, $j \geq 1$, probability measures on X. Then there exists a sequence $(x_n)_{n \geq 1}$, $x_n \in X$, such that*

$$\sup_{N \geq 1, x \in X} \left| \sum_{n=1}^{N} \left(\chi_{\{x\}}(x_n) - \mu_n(x) \right) \right| \leq 1 - \frac{1}{2(q-1)}. \tag{2.190}$$

The proof of Proposition 2.122 makes use of the marriage theorem of P. Hall [719].

Lemma 2.123 *Let $I = \{1, 2, \ldots, L\}$ be a finite set and for each $n \in I$ let S_n be a subset of a set S. If for every subset $J \subseteq I$*

$$\left| \bigcup_{j \in J} S_j \right| \geq |J| \tag{2.191}$$

then there exists an injective function $\alpha : I \to S$ with $\alpha(n) \in S_n$, $n \in I$.

(Note that the assumtion of Lemma 2.123 is also necessary.)

In order to apply Lemma 2.123 we need proper sets S_n. For this purpose we introduce some notations. Fix a positive integer N. For every $x \in X$ and $n \leq N$ set

$$\Lambda_n(x) = \sum_{j=1}^{n} \mu_j(x)$$

and let $\overline{\Lambda}_n(x) = -[-\Lambda_n(x)]$ denote the smallest integer $\geq \Lambda_n(x)$. Furthermore set

$$L = \sum_{x \in X} \overline{\Lambda}_N(x) \geq \sum_{x \in X} \Lambda_N(x) = N.$$

If $L > N$ then we additionally define measures μ_j, $N < j \leq L$, by

$$\mu_j(x) = \frac{\overline{\Lambda}_N(x) - \Lambda_N(x)}{L - N}.$$

Now we can define $\Lambda_n(x)$ and $\overline{\Lambda}_n(x)$ ever for $N < n \leq L$. Note that

$$\Lambda_L(x) = \Lambda_N(x) + (L - N)\frac{\overline{\Lambda}_N(x) - \Lambda_N(x)}{L - N} = \overline{\Lambda}_N(x)$$

if $L > N$. Let S be the set of pairs (x, j) with $x \in X$, $1 \le j \le \overline{\Lambda}_L(x)$ (S may be regarded as a multiset, in which $x \in X$ has multiplicity $\overline{\Lambda}_L(x)$) and $S_n \subseteq S$, $1 \le n \le L$, is defined by

$$S_n = \bigcup_{x \in X} S_n(x),$$

with

$$S_n(x) = \{(x, j) \in S : [\Lambda_{n-1}(x), \Lambda_n(x)] \cap [j - 1 + d, j - d] \ne \emptyset\},$$

where we use the abbreviation $d = 1/(2(q-1))$ and $\Lambda_0(x) = 0$. Note that $|S_n(x)| \in \{0, 1, 2\}$,

$$|S_n| = \sum_{x \in X} |S_n(x)|, \tag{2.192}$$

and

$$|S| = \sum_{x \in X} \overline{\Lambda}_L(x) = L.$$

With help of the following two lemmata we prove that these subsets S_n satisfy the condition of Lemma 2.123.

Lemma 2.124 *If $n_0 + 1, n_0 + 2, \ldots, n_0 + v$ are v consecutive integers contained in the set $\{1, 2, \ldots, L\}$, $v \ge 1$, then*

$$\left| \bigcup_{m=1}^{v} S_{n_0+m} \right| \ge v.$$

Proof. By definition $(x, j) \in \bigcup_{m=1}^{v} S_{n_0+m}(x)$ if and only if

$$[\Lambda_{n_0}(x), \Lambda_{n_0+v}(x)] \cap [j - 1 + d, j - d] \ne \emptyset. \tag{2.193}$$

Set

$$\delta(x, n) = \begin{cases} 1 & \text{if } \{-\Lambda_n(x)\} > 1 - d, \\ 0 & \text{otherwise,} \end{cases}$$

$$\varepsilon(x, n) = \begin{cases} 1 & \text{if } \{\Lambda_n(x)\} > 1 - d, \\ 0 & \text{otherwise} \end{cases}$$

Then

$$\left| \bigcup_{m=1}^{v} S_{n_0+m}(x) \right| = \Lambda_{n_0+v}(x) - \Lambda_{n_0}(x) + \{-\Lambda_{n_0+v}(x)\} - \delta(x, n_0+v) + \{\Lambda_{n_0+v}(x)\} - \varepsilon(x, n_0).$$

Since

$$\sum_{x \in X} (\Lambda_{n_0+v}(x) - \Lambda_{n_0}(x)) = \sum_{x \in X} \sum_{j=n_0+1}^{v} \mu_j(x) = v,$$

we obtain from (2.192)

$$\begin{aligned}\left|\bigcup_{m=1}^{v} S_{n_0+m}\right| &= \sum_{x\in X}\left|\bigcup_{m=1}^{v} S_{n_0+m}(x)\right| \\ &= v+\sum_{x\in X}(\{-\Lambda_{n_0+v}(x)\}-\delta(x,n_0+v)) \qquad (2.194)\\ &\quad +\sum_{x\in X}(\{\Lambda_{n_0+v}(x)\}-\varepsilon(x,n_0)).\end{aligned}$$

Since

$$\begin{aligned}\sum_{x\in X}\{-\Lambda_n(x)\} &= \sum_{x\in X}\overline{\Lambda}_n(x)-\sum_{x\in X}\Lambda_n(x)=\sum_{x\in X}\overline{\Lambda}_n(x)-n,\\ \sum_{x\in X}\{\Lambda_n(x)\} &= \sum_{x\in X}\Lambda_n(x)-\sum_{x\in X}[\Lambda_n(x)]=n-\sum_{x\in X}[\Lambda_n(x)],\end{aligned}$$

all sums appearing in (2.194) are integers. Since $\delta(x,n)=1$ implies $\{-\Lambda_n(x)\}>1-d$ we always have $\{-\Lambda_n(x)\}-\delta(x,n)>-d$. Hence

$$\sum_{x\in X}(\{-\Lambda_n(x)\}-\delta(x,n))>-qd=\frac{-q}{2(q-1)}\ge -1,$$

and so

$$\sum_{x\in X}(\{-\Lambda_n(x)\}-\delta(x,n))\ge 0. \qquad (2.195)$$

Similarly $\varepsilon(x,n)=1$ gives $\{\Lambda_n(x)\}>1-d$ and therefore $\{\Lambda_n(x)\}-\varepsilon(x,n)>-d$. Thus

$$\sum_{x\in X}(\{\Lambda_n(x)\}-\varepsilon(x,n))\ge 0. \qquad (2.196)$$

Now Lemma 2.124 follows from (2.194), (2.195), and (2.196). □

Lemma 2.125 *Let*

$$\{n_1+1,\ldots,n_1+v_1\},\{n_2+1,\ldots,n_2+v_2\},\ldots,\{n_t+1,\ldots,n_t+v_t\}$$

denote t sets of consecutive integers from the set $\{1,\ldots,L\}$, $t\ge 2$, $v_1,\ldots,v_t\ge 1$, $n_1+v_1<n_2,\ldots,n_{t-1}+v_{t-1}<n_t$. Then

$$\left|\bigcup_{s=1}^{t}\bigcup_{m=1}^{v_s} S_{n_s+m}\right|\ge\sum_{s=1}^{t}v_s.$$

Proof. By (2.193) we have for every $x\in X$

$$\left|\bigcup_{m=1}^{v_s} S_{n_s+m}(x)\cup\bigcup_{m=1}^{v_{s+1}} S_{n_{s+1}+m}\right| = \left|\bigcup_{m=1}^{v_s} S_{n_s+m}(x)\right|+\left|\bigcup_{m=1}^{v_{s+1}} S_{n_{s+1}+m}(x)\right|-\alpha(x,s),$$

where

$$\alpha(x,s) = \begin{cases} 1 & \text{if } j-1+d \le \Lambda_{n_s+v_s}(x) \le \Lambda_{n_s+1}(x) \le j-d \text{ for some } j \in \mathbf{Z}, \\ 0 & \text{otherwise.} \end{cases}$$

Hence, by using (2.194), (2.195), and (2.196)

$$\begin{aligned}
\left|\bigcup_{s=1}^{t}\bigcup_{m=1}^{v_s} S_{n_s+m}\right| &\ge \sum_{s=1}^{t} v_s + \sum_{s=1}^{t}\sum_{x\in X}(\{-\Lambda_{n_s+v_s}(x)\} - \delta(x, n_s+v_s)) \\
&\quad + \sum_{s=1}^{t}\sum_{x\in X}(\{\Lambda_{n_s+v_s}(x)\} - \varepsilon(x, n_s+v_s)) - \sum_{s=1}^{t-1}\sum_{x\in X}\alpha(x,s) \\
&= \sum_{s=1}^{t} v_s + \sum_{x\in X}(\{-\Lambda_{n_t+v_t}(x)\} - \delta(x, n_t+v_t)) \\
&\quad + \sum_{x\in X}(\{\Lambda_{n_t+v_t}(x)\} - \varepsilon(x, n_t+v_t)) \\
&\quad + \sum_{s=1}^{t-1}\sum_{x\in X}(\{-\Lambda_{n_s+v_s}(x)\} - \delta(x, n_s+v_s) \\
&\qquad\quad + \{\Lambda_{n_s+v_s}(x)\} - \varepsilon(x, n_s+v_s) - \alpha(x,s)) \\
&\ge \sum_{s=1}^{t} v_s + \sum_{s=1}^{t-1}\sum_{x\in X}(\{-\Lambda_{n_s+v_s}(x)\} - \delta(x, n_s+v_s) \\
&\qquad\quad + \{\Lambda_{n_s+v_s}(x)\} - \varepsilon(x, n_s+v_s) - \alpha(x,s)).
\end{aligned}$$

Therefore it suffices to prove

$$\begin{aligned}
T_s &= \sum_{x\in X}(\{-\Lambda_{n_s+v_s}(x)\} - \delta(x, n_s+v_s) \\
&\qquad + \{\Lambda_{n_s+v_s}(x)\} - \varepsilon(x, n_s+v_s) - \alpha(x,s)) \ge 0
\end{aligned}$$

for all $s = 1, 2, \ldots, t-1$. If $\alpha(x,s) = 0$ for all $x \in X$ then by (2.195) and (2.196) we get $T_s \ge 0$. Suppose now that $\alpha(\tilde{x}, s) = 1$ for at least one $\tilde{x} \in X$. By the definition of $\alpha(x,s)$ we have $\delta(\tilde{x}, n_s+v_s) = \varepsilon(\tilde{x}, n_s+v_s) = 0$ and

$$\{-\Lambda_{n_s+v_s}(x)\} + \{\Lambda_{n_s+1}(x)\} \ge 1.$$

Thus

$$(\{-\Lambda_{n_s+v_s}(\tilde{x})\} - \delta(x, n_s+v_s)\{-\Lambda_{n_s+v_s}(\tilde{x})\} - \varepsilon(x, n_s+v_s) - \alpha(\tilde{x}, s)) \ge 0.$$

Hence, if a term in T_s is negative then $\alpha(x,s) = 0$, and by assumption there are at most $q-1$ of such terms. Since we always have $\{-\Lambda_n(x)\} - \delta(x,n) > -d$ and $\{\Lambda_n(x)\} - \varepsilon(x,n) > -d$ we obtain

$$T_s > (q-1)2d = -1.$$

Since T_s has to be an integer it follows that $T_s \geq 0$ which completes the proof of Lemma 2.125. □

Proof of Proposition 2.122. By Lemma 2.125 and Lemma 2.123 there exists an injective function $\alpha : \{1, \dots, L\} \to S$ with $\alpha(m) = (x_m, j_m) \in S_m$. Since $|S| = L$ this function is also surjective. Observe that $(x_m, j_m) \in S_m$ means that

$$j_m - 1 - d \leq \Lambda_m(x_m) \quad \text{and} \quad j_m - d \geq \Lambda_{m-1}(x_m), \tag{2.197}$$

which implies

$$\overline{\Lambda}_{m-1}(x_m) \leq j_m \leq \overline{\Lambda}_m(x_m). \tag{2.198}$$

For a fixed $n \leq L$ consider the sequence $\alpha(1), \dots, \alpha(n)$. For $x \in X$ let $j < \overline{\Lambda}_j(x)$. Since α is surjective there exists m such that $\alpha(m) = (x_m, j_m) = (x, j)$. By (2.198) $\overline{\Lambda}_{m-1}(x) \leq j < \overline{\Lambda}_n(x)$ which implies $m - 1 < n$ or $m \leq n$. Hence all pairs (x, j) with $1 \leq j \leq \overline{\Lambda}_n(x)$ occur in $\alpha(1), \dots, \alpha(n)$.

On the other hand, let $j > \overline{\Lambda}_n(x)$. By (2.198) $\alpha(m) = (x, j)$ satisfies $\overline{\Lambda}_m(x) \geq j > \overline{\Lambda}_n(x)$ which gives $m > n$. This implies that pairs (x, j) with $j > \overline{\Lambda}_n(x)$ do not occur in $\alpha(1), \dots, \alpha(n)$. (The pair $(x, \overline{\Lambda}_n(x))$ may or may not occur in $\alpha(1), \dots, \alpha(n)$.)

Consider the sequence $(x_n^{(N)})_{n \geq 1}$, by $x_n^{(N)} = x_n$ for $n \leq L$ and by $x_n^{(N)} = x_1$ for $n > L$. This means for $n \leq N \leq L$ and $x \in X$ we have

$$\sum_{j=1}^{n} \chi_{\{x\}}(x_j^{(N)}) = \overline{\Lambda}_n(x) \quad \text{or} \quad \sum_{j=1}^{n} \chi_{\{x\}}(x_j^{(N)}) = \overline{\Lambda}_n(x) - 1.$$

Suppose first that the left hand side equals $\overline{\Lambda}_n(x)$. Then $(x, \overline{\Lambda}_n(x)) = \alpha(m)$ occurs in $\alpha(1), \dots, \alpha(n)$. By (2.197) this implies

$$\overline{\Lambda}_n(x) - 1 - d \leq \Lambda_m(x) \leq \Lambda_n(x) \leq \overline{\Lambda}_n(x),$$

and so

$$\left| \sum_{j=1}^{n} \chi_{\{x\}}(x_j^{(N)}) - \Lambda_n(x) \right| = \overline{\Lambda}_n(x) - \Lambda_n(x) \leq 1 - d.$$

In the second case $(x, \overline{\Lambda}_n(x)) = \alpha(m)$ does not occur in $\alpha(1), \dots, \alpha(n)$. Hence $m > n$ and by (2.197)

$$\overline{\Lambda}_n(x) - d \geq \Lambda_{m-1}(x) \geq \Lambda_n(x) > \overline{\Lambda}_n(x) - 1,$$

which provides

$$\left| \sum_{j=1}^{n} \chi_{\{x\}}(x_j^{(N)}) - \Lambda_n(x) \right| = \Lambda_n(x) - (\overline{\Lambda}_n(x) - 1) \leq 1 - d.$$

Thus, for every $N \geq 1$ there exists a sequence $(x_n^{(N)})_{n \geq 1}$ such that

$$\max_{n \leq N, x \in X} \left| \sum_{j=1}^{n} \left(\chi_{\{x\}}(x_j) - \mu_j(x) \right) \right| \leq 1 - \frac{1}{2(q-1)}. \tag{2.199}$$

With help of this property it is quite easy to finish the proof of Proposition 2.122. Since X is finite there exists an element $\tilde{x} \in X$ such that $x_1^{(N)} = \tilde{x}$ for infinitely many N. We define $x_1 = \tilde{x}$ and proceed by induction. Suppose that the first n elements $x_1, \ldots, x_n$ are defined in a way that there are infinitely many sequences $(x_n^{(N)})_{n \ge 1}$ beginning with $x_1, \ldots, x_n$. Since X is finite there is an element $\tilde{\tilde{x}} \in X$ such that infinitely many sequences $(x_n^{(N)})_{n \ge 1}$ start with $x_1, \ldots, x_n, \tilde{\tilde{x}}$. Then we define $x_{n+1} = \tilde{\tilde{x}}$. In this way $(x_n)_{n \ge 1}$ is defined inductively.

Since for any n the first terms $x_1, \ldots, x_n$ of $(x_n)_{n \ge 1}$ are also the first n terms of some sequence $(x_n^{(N)})_{n \ge 1}$ with $N \ge n$ we have

$$\sum_{j=1}^{n} \chi_{\{x\}}(x_j^{(N)}) = \sum_{j=1}^{n} \chi_{\{x\}}(x_j)$$

for all $x \in X$. Hence (2.199) implies (2.190). □

Proposition 2.126 *Let X be a finite set of order $q \ge 2$. Then for every ε, $0 < \varepsilon < 1/(2q)$ there exists a probability measure μ on X such that for all sequences $(x_n)_{n \ge 1}$, $x_n \in X$,*

$$\sup_{N \ge 1} N\, D_N(\mu, x_n) > 1 - \frac{2}{2(q-1)} - \varepsilon. \tag{2.200}$$

Proof. Choose an irrational number ϑ with $1 - \varepsilon < (q-1)\vartheta < 1$ and fix an element $y \in X$. Set $\mu(y) = 1 - (q-1)\vartheta$ and $\mu(x) = \vartheta$ for $x \ne y$. Since $(n\vartheta)$ is u.d. mod 1 there exists N such that

$$\frac{1}{2(q-1)} < \{N\vartheta\} < \frac{1+\varepsilon}{2(q-1)}. \tag{2.201}$$

Suppose that there exists a sequence $(x_n)_{n \ge 1}$ such that

$$\sup_{N \ge 1} N\, D_N(\mu, x_n) \le 1 - \varepsilon - \frac{1}{2(q-1)}. \tag{2.202}$$

Then it follows from (2.201) that for $x \ne y$

$$\sum_{n=1}^{N} \chi_{\{x\}}(x_n) = [N\vartheta].$$

Since $\vartheta < 1/(q-1)$ and $1 - \varepsilon < (q-1)\vartheta$ we have

$$1 - \varepsilon - \frac{1}{q-1} < (q-2)\theta < \frac{q-2}{q-1} = 1 - \frac{1}{q-1}.$$

Consequently (2.201) implies $[(N+q-2)\vartheta] = [N\vartheta]$ and $\{(N+q-2)\}\vartheta\} = \{N\vartheta\} + (q-2)\vartheta$ with

$$1 - \frac{1}{2(q-1)} - \varepsilon < \{(N+q-2)\vartheta\} < 1 - \frac{1-\varepsilon}{2(q-1)}.$$

Hence by (2.202) we have for $x \neq y$

$$\sum_{n=1}^{N+q-2} \chi_{\{x\}}(x_n) = [N\vartheta] + 1 = \sum_{n=1}^{N} \chi_{\{x\}}(x_n) + 1,$$

which leads to the contradiction

$$q - 2 = \sum_{x \in X} \sum_{n=N+1}^{N+q-2} \chi_{\{x\}}(x_n) \geq q - 1.$$

□

$k(N)$-Distribution

As above we will consider finite sets $X = \{y_1, y_2, \ldots, y_q\}$ of order q. A measure μ on X is given by $p_j = \mu(y_j)$, $1 \leq j \leq q$. We adjoin the product measure μ_∞ to the space of sequences $(x_n)_{n\geq 1}$, $x_n \in X$. By the law of large numbers it is clear that

$$\lim_{N\to\infty} D_N(\mu, x_n) = 0 \qquad \mu_\infty\text{-almost surely.}$$

However much more can be said. Since the shift operator S (Bernoulli shift) on X defined by $S(x_1, x_2, \ldots) = (x_2, x_3, \ldots)$ is an ergodic transformation on X (cf. e.g. WALTERS [1935]) it follows from the BIRKHOFF ergodic theorem that

$$\lim_{N\to\infty} \frac{|\{1 \leq n \leq N-k : x_n x_{n+1} \ldots x_{n+k} = a_1 \ldots a_k\}|}{N} = \mu_k(A)$$

holds in probability for all blocks $A = a_1 \ldots a_k$ of a given constant length k ($\mu_k(A)$ is the k-fold product measure generated by $\mu(y_j) = p_j$.)

It is now natural to ask how fast (depending on N) k could grow such that this relation persists. In order to answer this question we introduce a special notion of discrepancy:

$$D_N^{(k,\phi)}(\mu, x_n) = \max_{A\in X^k} \sqrt{\frac{\phi(k)}{\mu_k(A)}} \left| \frac{|\{1 \leq n \leq N-k : x_n x_{n+1} \ldots x_{n+k} = A\}|}{N} - \mu_k(A) \right|, \tag{2.203}$$

where ϕ is a increasing function. We will call a sequence $(x_n)_{n\geq 1}$, $x_n \in X$, $(k(N), \phi)$-distributed with respect to μ if

$$\lim_{N\to\infty} D_N^{(k(N),\phi)}(x_1, \ldots, x_N) = 0.$$

By KOLMOGOROV's 0-1-law or by the fact that the set

$$\left\{(x_n)_{n\geq 1} : \lim_{N\to\infty} D_N^{(k(N),\phi)}(\mu, x_n) = 0\right\}$$

is invariant under the (ergodic) shift S, it follows that the only possible values for the probability of the set of $(k(N), \phi)$-distributed sequences are 0 and 1. The next theorem shows under which conditions almost all sequences are $(k(N), \phi)$-distributed.

Theorem 2.127 *Let $k(N) = \mathcal{O}(\log N)$ be a non-decreasing sequence of positive integers and $\phi(k)$ a increasing function such that*

$$\lim_{N\to\infty} \frac{N}{\phi(k(N))\log N} = \infty. \tag{2.204}$$

Then almost all sequences $(x_n)_{n\geq 1}$, $x_n \in X$, are $(k(N),\phi)$-distributed with respect to μ.

Conversely, if (2.204) is not satisfied and if

$$\frac{3}{4} < \limsup_{N\to\infty} \frac{k(N)}{\operatorname{lq} N} < \infty,$$

with $\operatorname{lq} = \log_q$ denoting the logarithm to base q, then almost no sequences $(x_n)_{n\geq 1}$, $x_n \in X$, are $(k(N),\phi)$-distributed with respect to μ.

A special (but important) case appears if the digits $y_1,\ldots,y_q$ are equidistributed, i.e. $\mu(y_j) = 1/q$, $1 \leq j \leq q$, and every word A of length k has the same probability $\mu_k(A) = q^{-k}$. Furthermore, it is appropriate to use $\phi(k) = q^k$. Here we have

$$D_N^{(k)}(x_n) = \max_{A\in X^k} q^k \left| \frac{|\{1 \leq n \leq N-k : x_n x_{n+1}\ldots x_{n+k} = A\}|}{N} - q^{-k} \right|. \tag{2.205}$$

The most important additional property in this case is that $k' \leq k''$ implies $D_N^{(k'')}(x_n) \leq D_N^{(k')}(x_n)$. (We will have to use this fact in the proof of Theorem 2.131.) We call a sequence $(x_n)_{n\geq 1}$, $x \in X$, $k(N)$-distributed if

$$\lim_{N\to\infty} D_N^{(k(N))}(x_n) = 0.$$

Obviously we can apply Theorem 2.127 for this special notion. Condition (2.204) is necessary and sufficient in this case.

Theorem 2.128 *Let $k(N)$ be a non-decreasing sequence of positive integers. Then the following 0-1-law holds*

$$\mu_\infty\left(\left\{(x_n)_{n\geq 1} : \lim_{N\to\infty} D_N^{(k(N))}(x_n) = 0\right\}\right) = \begin{cases} 1 & \textit{if } \operatorname{lq} n - \operatorname{lq}\operatorname{lq} n - k(n) \to \infty \\ 0 & \textit{otherwise.} \end{cases}$$

Remark. It should be mentioned that GOLDSTERN [662] and WINKLER [1963] provided explicit constructions of $k(N)$-u.d. sequences with low discepancy, where $\operatorname{lq} n - \operatorname{lq}\operatorname{lq} n - k(n) \to \infty$.

The proof of Theorem 2.127 will use multivariate correlation polynomials, which are a generalization of GUIBAS and ODLYZKO's correlation polynomials in one variable (cf. [708]). Using these polynomials we are able to compute the probability generating functions of the events we are interested in.

We now fix a word $A = a_1 a_2 \dots a_k$, $a_i \in X$ of length k. We are interested in the cardinalities of the following subsets of the set $\mathcal{S}_{\mathbf{r}}$, $\mathbf{r} = (r_1, \dots, r_q)$, of words containing r_j digits y_j, $1 \le j \le q$:

$$\begin{aligned} f_A(\mathbf{r}) &= |\{B \in \mathcal{S}_{\mathbf{r}} : B \text{ contains } A \text{ only at the end}\}|, \\ g_A(\mathbf{r}) &= |\{B \in \mathcal{S}_{\mathbf{r}} : B \text{ contains } A \text{ only at the beginning and at the end}\}|, \\ h_A(\mathbf{r}) &= |\{B \in \mathcal{S}_{\mathbf{r}} : B \text{ does not contain } A\}|. \end{aligned} \tag{2.206}$$

In order to compute the generating functions of these quantities we introduce the multivariate autocorrelation polynomial $[AA](z_1, \dots, z_q)$:

$$[z_1^{s_1} \cdots z_q^{s_q}][AA](z_1, \dots, z_q) = \begin{cases} 1 & \text{if } a_1 a_2 \dots a_{k-s_1-\dots-s_q} \\ & = a_{s_1+\dots+s_q+1} a_{s_1+\dots+s_q+2} \dots a_k \\ & \text{and the string } a_1 a_2 \dots a_{s_1+\dots+s_q} \\ & \text{contains } s_j \text{ digits } y_j, 1 \le j \le q, \\ 0 & \text{otherwise,} \end{cases}$$

where $[z_1^{s_1} \cdots z_q^{s_q}]P(z_1, \dots, z_q)$ denotes as usual the coefficient of $z_1^{s_1} \cdots z_q^{s_q}$ in $P(z_1, \dots, z_q)$. We are now ready to formulate

Proposition 2.129 *The generating functions of the combinatorial expressions (2.206) are given by*

$$\begin{aligned} F_A(z_1, \dots, z_q) &= \sum_{\mathbf{r} \ge 0} f_A(\mathbf{r}) z_1^{r_1} \cdots z_q^{r_q} \\ &= \frac{z_1^{A_1} \cdots z_q^{A_q}}{z_1^{A_1} \cdots z_q^{A_q} + (1 - z_1 - \dots - z_q)[AA](z_1, \dots, z_q)}, \\ G_A(z_1, \dots, z_q) &= \sum_{\mathbf{r} \ge 0} g_A(\mathbf{r}) z_1^{r_1} \cdots z_q^{r_q} \\ &= z_1^{A_1} \cdots z_q^{A_q} + \frac{(z_1 + \dots + z_q - 1) z_1^{A_1} \cdots z_q^{A_q}}{z_1^{A_1} \cdots z_q^{A_q} + (1 - z_1 - \dots - z_q)[AA](z_1, \dots, z_q)}, \\ H_A(z_1, \dots, z_q) &= \sum_{\mathbf{r} \ge 0} h_A(\mathbf{r}) z_1^{r_1} \cdots z_q^{r_q} \\ &= \frac{[AA](z_1, \dots, z_q)}{z_1^{A_1} \cdots z_q^{A_q} + (1 - z_1 - \dots - z_q)[AA](z_1, \dots, z_q)}, \end{aligned}$$

where A_j, $1 \le j \le q$, denote the number of y_j's in A.

Proof. We will show that

$$(1 - z_1 - \dots - z_q) H_A(z_1, \dots, z_q) + F_A(z_1, \dots, z_q) = 1, \tag{2.207}$$

$$z_1^{A_1} \cdots z_q^{A_q} H_A(z_1, \dots, z_q) = [AA](z_1, \dots, z_q) F_A(z_1, \dots, z_q), \tag{2.208}$$

and that

$$(1 - z_1 - \cdots - z_q)F_A(z_1, \ldots, z_q) + G_A(z_1, \ldots, z_q) = z_1^{A_1} \cdots z_q^{A_q}. \qquad (2.209)$$

Clearly (2.207), (2.208), and (2.209) provide the proposed representaions for $F_A(z_1, \ldots, z_q))$, $G_A(z_1, \ldots, z_q))$, and $H_A(z_1, \ldots, z_q))$.

To prove (2.207), consider the $h_A(r_1 - 1, r_2, \ldots, r_q)$ words of $r_1 - 1$ digits y_1 and r_j digits y_j, $2 \le j \le q$ that do not contain A as a subword, and adjoin to it (at the right end) y_1. Similarly, consider $h_A(r_1, \ldots, r_j - 1, \ldots, r_q)$ words that do not contain A as a subword, and adjoin to it (at the right end) y_j. In this way we obtain

$$\sum_{j=1}^{q} h_A(r_1, \ldots, r_j - 1, \ldots, r_q)$$

distinct words of r_j digits y_j, $1 \le j \le q$. Any one of them either contains an appearance of A or not, but if it does, then that appearance has to be in the rightmost position. Therefore we obtain

$$\sum_{j=1}^{q} h_A(r_1, \ldots, r_j - 1, \ldots, r_q) = h_A(\mathbf{r}) + f_A(\mathbf{r}),$$

and this is valid for all $r_j \ge 1$, $1 \le j \le q$. Multiplying by $z_1^{r_1} \cdots z_q^{r_q}$ and summing over $r_j \ge 1$, we easily obtain (2.207). (Note that the initial conditions for $r_j = 0$ always fit to (2.207).)

To obtain (2.208) consider anyone of the $h_A(r_1, \ldots, r_q)$ strings of length $r = r_1 + \ldots + r_q$ that do not contain A, and append A at its right end. Suppose that the first (from the left) appearance of A in this string has its rightmost character in position $r + s$. Then $0 < s \le |A|$. Furthermore, since A also appears in the last $|A|$ positions, the prefix of A of length s (with s_j digits y_j, $1 \le j \le q$) has to be equal to the suffix of A of length s. Surely all possible stings $f_A(r_1 + s_1, \ldots, r_q + s_q)$ can appear in the first $r + s$ positions. Furthermore there is a unique way to extend them to a string of length $r + |A|$ in which the last $|A|$ characters equal A. Therefore for $r \ge 0$ we have

$$h_A(r_1, \ldots, r_q) = \sum_{[z_1^{s_1} \cdots z_q^{s_q}][AA](z_1, \ldots, z_q) = 1} f_A(r_1 + s_1, \ldots, r_q + s_q),$$

which implies (2.208).

Finally the proof of (2.209) is almost the same as that of (2.207). Consider the $f_A(r_1, \ldots, r_j - 1, \ldots, r_q)$ words with r_i digits y_i, $i \ne j$ and $r_j - 1$ digits y_j that contain A only at the end, and adjoin to it (at the left end) y_j. In this way we obtain

$$\sum_{j=1}^{q} f_A(r_1, \ldots, r_j - 1, \ldots, r_q)$$

distinct words of r_j digits y_j, $1 \le j \le q$. Any one of them either contains an appearance of A at the leftmost position or not. Therefore we obtain

$$\sum_{j=1}^{q} f_A(r_1, \ldots, r_j - 1, \ldots, r_q) = f_A(\mathbf{r}) + g_A(\mathbf{r}),$$

which leads to (2.209). □

We use these functions to compute the probability generating function (p.g.f.)

$$\Phi_A^{(r)}(z) = \sum_{N \ge 0} p_A^{(r)}(N) z^N$$

of the probabilities

$$p_A^{(r)}(N) = \mu_\infty\left(\{(x_n)_{n\ge1} : |\{0 \le n \le N-k : x_{n+1}\ldots x_{n+k} = a_1 \ldots a_k\}| = r\}\right)$$

of all strings containing the substring A exactly r times. Obviously we have

$$\begin{aligned} \Phi_A^{(r)}(z) &= \frac{z^{-kr}}{\mu_k(A)^r} F_A(p_1 z, \ldots, p_q z)^2 G_A(p_1 z, \ldots, p_q z)^{r-1} \quad \text{for } r \ge 1, \\ \Phi_A^{(0)}(z) &= H_A(p_1 z, \ldots, p_q z). \end{aligned}$$

Inserting the results of Proposition 2.129 and setting

$$P(z) = \frac{1}{\mu_k(A)} [AA](p_1 z, \ldots, p_q z) \tag{2.210}$$

yields

$$\begin{aligned} \Phi_A^{(r)}(z) &= \frac{z^k}{\mu_k(A)} \frac{\left((1-z)(P(z) - \frac{1}{\mu_k(A)}) + z^k\right)^{r-1}}{((1-z)P(z) + z^k)^{r+1}}, \\ \Phi_A^{(0)}(z) &= \frac{P(z)}{(1-z)P(z) + z^k}. \end{aligned}$$

We split the proof of Theorem 2.127 into two parts. First we show that almost all sequences are $(k(N), \phi)$-uniformly distributed if $\lim_{N\to\infty} \frac{N\phi(k(N))}{\log N} = \infty$. Using our p.g.f. results we can write

$$p_A^{(r)}(N) = [z^N]\Phi_A^{(r)}(z) = \frac{1}{2\pi i} \oint_C \Phi_A^{(r)}(z) \frac{dz}{z^{N+1}}. \tag{2.211}$$

In order to be able to estimate the integral we will need information on the the zeros of the polynomial $(1-z)P(z) + z^k$.

Lemma 2.130 *The zero of smallest modulus z_0 of $(1-z)P(z)+z^k$ is real and positive and satisfies the estimate*

$$z_0 \ge 1 + C\mu_k(A)$$

for a positive constant C only depending on $p_1, \ldots, p_q$.

Proof. As $F_A(p_1 z, \ldots, p_q z)$ is a p.g.f. and $(1-z)P(z)+z^k$ is the denominator of this rational function, the zero z_0 of smallest modulus has to be real, positive, and ≥ 1. Next observe that

$$[AA](p_1, \ldots, p_q) \leq \sum_{r=1}^{k} \mu_r(a_1 \cdots a_r) = \mathcal{O}(1)$$

uniformly for all $A \in X^k$ and all $k \geq 1$, where the $\mathcal{O}$-constant depends on $p_1, \ldots, p_q$. Hence there exists a constant $c > 0$ such that $[AA](p_1 z, \ldots, p_q z) \leq c z^k$ for all $k \geq 1$ and real $z \geq 1$. Thus, if $z_0 \geq 1$ satisfies $(1-z_0)P(z_0) + z_0^k = 0$ then

$$z_0 - 1 = \mu_k(A) \frac{z_0^k}{[AA](p_1 z_0, \ldots, p_q z_0)} \geq \mu_k(A) \frac{1}{c} = C\mu_k(A).$$

□

Proof of Theorem 2.127. Suppose that

$$\psi(N) = \frac{N}{\phi(k(N)) \log N} \to \infty$$

as $N \to \infty$. We need estimates for the probability that the number of occurrences $Z_N(A)$ of a block A deviates too far from the mean value. Set

$$\begin{aligned} L_N(\delta_A) &= \mu_\infty(Z_N(A) < N\mu_k(A)(1-\delta_A)) \quad \text{and} \\ U_N(\delta_A) &= \mu_\infty(Z_N(A) > N\mu_k(A)(1+\delta_A)). \end{aligned} \tag{2.212}$$

These probabilities are sums of the $p_A^{(r)}(N)$ defined in (2.211):

$$\begin{aligned} L_N(\delta_A) &= \sum_{r < N\mu_k(A)(1-\delta_A)} p_A^{(r)}(N) \quad \text{and} \\ U_N(\delta_A) &= \sum_{r > N\mu_k(A)(1+\delta_A)} p_A^{(r)}(N). \end{aligned} \tag{2.213}$$

We will use the integral representation (2.211) to estimate these quantities.

Using the notations

$$\begin{aligned} Q(z) &= (1-z)P(z) + z^k, \\ a(z) &= \frac{z^k}{Q(z)^2}, \\ b(z) &= 1 + \frac{z-1}{\mu_k(A)Q(z)}, \end{aligned} \tag{2.214}$$

we have

$$\Phi_A^{(r)}(z) = \frac{1}{\mu_k(A)} a(z) b(z)^{r-1}$$

for $r \geq 1$, and so

$$U_N(\delta_A) = [z^N] \frac{1}{\mu_k(A)} a(z) \frac{b^j(z)}{1-b(z)}$$

with $j = [N\mu_k(A)(1+\delta_A)]$. As all the power series involved have positive coefficients we surely have

$$U_N(\delta_A) z^N \leq \frac{1}{\mu_k(A)} a(z) \frac{b^j(z)}{1-b(z)}$$

for every positive real z. Especially if we set $z = 1-\varepsilon$, where $\varepsilon < C\mu_k(A)$ (compare with Lemma 2.130) we obtain

$$U_N(\delta_A) \leq \frac{1}{\mu_k(A)} a(1-\varepsilon) \frac{b^j(1-\varepsilon)}{1-b(1-\varepsilon)} (1-\varepsilon)^{-N}.$$

Using the approximations

$$\begin{aligned}
a(1\pm\varepsilon) &= 1 + \mathcal{O}\left(\frac{1}{\mu_k(A)}\varepsilon\right), \\
b(1\pm\varepsilon) &= 1 \pm \frac{\varepsilon}{\mu_k(A)} + \mathcal{O}\left(\frac{\varepsilon^2}{\mu_k(A)^2}\right), \\
b^j(1\pm\varepsilon) &= \exp\left(\pm\frac{\varepsilon j}{\mu_k(A)} + \mathcal{O}\left(\frac{\varepsilon^2 j}{\mu_k(A)^2}\right)\right), \\
(1\pm\varepsilon)^{-n} &= \exp\left(\mp n\varepsilon + \mathcal{O}(n\varepsilon^2)\right)
\end{aligned}$$

yields

$$U_N(\delta_A) \leq \frac{1}{\varepsilon}\left(1 + \mathcal{O}\left(\frac{\varepsilon}{\mu_k(A)}\right)\right) \exp\left(\left(N - \frac{j}{\mu_k(A)}\right)\varepsilon + \mathcal{O}\left(\frac{\varepsilon^2 j}{\mu_k(A)^2}\right) + \mathcal{O}(N\varepsilon^2)\right).$$

Inserting

$$\varepsilon = \left(\mu_k(A)\frac{\log N}{N}\right)^{1/2}$$

into the above relations gives

$$U_N(\delta_A) \leq \exp\left(-\delta_A \left(N\mu_k(A)\log N\right)^{\frac{1}{2}} + C_1 \log N\right). \tag{2.215}$$

In the same way we treat the lower tail. Let now $j = [N\mu_k(A)(1-\delta_A)]$. From

$$L_N(\delta_A) = [z^N]\left(\frac{P(z)}{Q(z)} + \frac{a(z)}{\mu_k(A)} \frac{1-b^j(z)}{1-b(z)}\right)$$

we obtain

$$L_N(\delta_A) \leq \frac{P(1+\varepsilon)}{Q(1+\varepsilon)}(1+\varepsilon)^{-N} + \frac{1}{\mu_k(A)} j b^j(1+\varepsilon) a(1+\varepsilon)(1+\varepsilon)^{-N},$$

and by using the same value for ε as above

$$L_N(\delta_A) \le \exp\left(-\delta_A \left(N\mu_k(A)\log N\right)^{\frac{1}{2}} + C_2 \log N\right). \tag{2.216}$$

Combining this with (2.215) yields

$$\mu_\infty\left(\left\{(x_n)_{n\ge 1} : \left|\frac{Z_N(A)}{N} - \mu_k(A)\right| > \delta_A \mu_k(A)\right\}\right)$$
$$\le \exp\left(-\delta_A \left(N\mu_k(A)\log N\right)^{\frac{1}{2}} + C_3 \log N\right). \tag{2.217}$$

With

$$\delta_A = \psi(N)^{-1/4}(\mu_k(A)\phi(k(N)))^{-1/2}$$

we immediately obtain

$$\mu_\infty\left(D_N^{(k(N),\phi)}(\mu, x_n) > \psi(N)^{-1/4}\right) \tag{2.218}$$
$$\le q^{k(N)} \exp\left(-\psi(N)^{-\frac{1}{4}}\left(\frac{N}{\phi(k(N))\log N}\right)^{\frac{1}{2}} \log N + C_3 \log N\right)$$
$$\le \exp\left(-\psi(N)^{\frac{1}{4}} \log N + C' \log N\right). \tag{2.219}$$

Since

$$\sum_{N=1}^{\infty} \exp\left(-\psi(N)^{\frac{1}{4}} \mathrm{lq}\, N + C' \log N\right) < \infty,$$

the BOREL-CANTELLI lemma provides the first part of our Theorem.

If (2.204) is not satisfied then there exists a strictly increasing sequence of positive integers $(N_i)_{i\ge 1}$ and a constant $C > 0$ such that

$$\frac{N_i}{\phi(k(N_i))\log N_i} \le C$$

for all $i \ge 1$. In the following we will always assume that N is an element of $(N_i)_{i\ge 1}$.

W.l.o.g. we may assume that $p_1 + p_2 \le 2/q$ which implies $p_1 p_2 \le q^{-2}$. Next we introduce a set $\mathcal{A}_k$ of strings $A = a_1 a_2 \cdots a_k \in X^k$ of length k such that

$$a_j = \begin{cases} y_1 & \text{for } 1 \le j \le [k/3] + 1, \\ y_2 & \text{for } j = [k/3] + 2, \\ y_1 & \text{for } j = [2k/3] - 2, \\ y_2 & \text{for } [2k/3] - 2 \le j \le k. \end{cases}$$

It is very easy to see that every $A \in \mathcal{A}_k$ has only trivial autocorrelation and any two of them do not overlap each other. Note that $m = |\mathcal{A}_k| = c_k q^{k/3}$, in which $c_{k+3} = c_k > 0$, and that

$$\mu_k(\mathcal{A}_k) \le (p_1 p_2)^{k/3} \le q^{-2k/3}.$$

We need the p.g.f. $\varphi(z)$ of all strings not containing an element of $\mathcal{A}_k$. This function satisfies the equations

$$\begin{aligned}
\varphi(z) + \varphi_{A_1}(z) + \ldots + \varphi_{A_m}(z) &= z\varphi(z) + 1, \\
\varphi_{A_1}(z) &= z^k \mu_k(A_1)\varphi(z), \\
&\vdots \\
\varphi_{A_m}(z) &= z^k \mu_k(A_m)\varphi(z),
\end{aligned}$$

where $A_1, \ldots, A_m$ are the elements of $\mathcal{A}$ and $\varphi_{A_l}(z)$ ($l = 1, \ldots, m$) is the p.g.f. of the blocks ending with A_l but containing no further occurrence of any element of $\mathcal{A}$. Solving these equations yields

$$\varphi(z) = \frac{1}{1 - z + \mu_k(\mathcal{A})z^k}. \tag{2.220}$$

Note that the simplicity of these equations comes from the trivial overlap structure of the elements of $\mathcal{A}$.

Because of this simple overlap structure it is easy to see that

$$\varphi_{j_1 \ldots j_m}(z) = \frac{(j_1 + \ldots + j_m)!}{j_1! \cdots j_m!} \mu_k(A_1)^{j_1} \cdots \mu_k(A_m)^{j_m} z^{k(j_1 + \cdots + j_m)} \varphi(z)^{j_1 + \cdots + j_m + 1} \tag{2.221}$$

is the p.g.f. of all blocks containing A_l exactly j_l times ($l = 1 \ldots m$).

In order to complete the proof of Theorem 2.127 we need informations on the zeros of the polynomial $1 - z + \mu_k(\mathcal{A})z^k$. As in the first part of the proof of Lemma 2.130 the zero of smallest modulus z_0 of $1 - z + \mu_k(\mathcal{A})z^k$ is real and ≥ 1. Hence $0 = 1 - z_0 + \mu_k(\mathcal{A})z_0^k \geq 1 - z_0 + \mu_k(\mathcal{A})$ provides

$$z_0 \geq 1 + \mu_k(\mathcal{A}).$$

Now set

$$\begin{aligned}
M_N(\delta) &= \mu_\infty\left(\{(x_n)_{n\geq 1} : |Z_N(A_l) - N\mu_k(A_l)| \leq N\mu_k(A_l)\delta_{A_l},\ l = 1 \ldots m\}\right) \\
&= \sum_{\substack{|j_l - Nq^{-k}| \leq N\mu_k(A_l)\delta_{A_l} \\ l=1,\ldots,m}} [z^N]\varphi_{j_1 \ldots j_m}(z),
\end{aligned} \tag{2.222}$$

where

$$\delta_{A_l} = \frac{\delta}{\sqrt{\mu_k(A_l)\phi(k)}}$$

and $\delta > 0$ will be fixed in the sequel. Observe now that

$$[z^{N-kJ}]\varphi(z)^{J+1} \leq \varphi(1+\varepsilon)^{J+1}(1+\varepsilon)^{kJ-N}$$

for $\varepsilon \leq \mu_k(\mathcal{A})$. Inserting $\varepsilon = \mu_k(\mathcal{A}) - J/N$ and performing similar calculations as in the first part of the proof yields

$$[z^{N-kJ}]\varphi(z)^{J+1} \leq \frac{1}{\mu_k(\mathcal{A})} \frac{N^J e^{-N\mu_k(\mathcal{A})}}{J^J e^{-J}} \exp\left(\mathcal{O}(k\mu_k(\mathcal{A})^2 N)\right). \tag{2.223}$$

By assumption there exists $\theta > 0$ with $N \le q^{(\frac{4}{3}-\theta)k}$. Hence

$$k\mu_k(\mathcal{A})^2 N \ll kq^{-4k/3} q^{(\frac{4}{3}-\theta)k} = o(1)$$

as $N \to \infty$. Thus the term $\exp(\mathcal{O}(k\mu_k(\mathcal{A})^2 N))$ has no influence.

Now use STIRLING's formula and insert (2.223) into (2.222) to obtain

$$M_N(\delta) \ll \frac{\sqrt{N}}{\mu_k(\mathcal{A})} \prod_{l=1}^{m} e^{-N\mu_k(A_l)} \sum_{|j_l - N\mu_k(A_l)| \le N\mu_k(A_l)\delta_{A_l}} \frac{(N\mu_k(A_l))^{j_l}}{j_l!}. \tag{2.224}$$

Obviously we have

$$e^{-x} \sum_{|j-x|\le \gamma x} \frac{x^j}{j!} \le e^{-x}\left(e^x - \frac{x^{(1+\gamma)x}}{((1+\gamma)x)!}\right) \le 1 - \frac{C'}{\sqrt{x}} e^{-(\gamma^2/2+\mathcal{O}(\gamma^3))x}$$

for sufficiently large $x > 0$ and sufficiently small $\gamma > 0$ such that $(1+x)\gamma$ is an integer. Hence

$$\begin{aligned} e^{-N\mu_k(A_l)} \sum_{|j_l - N\mu_k(A_l)| \le N\mu_k(A_l)\delta_{A_l}} \frac{(N\mu_k(A_l))^{j_l}}{j_l!} &\le 1 - \frac{C'}{\sqrt{N\mu_k(A_l)}} \exp\left(-\frac{\delta^2 N}{3\phi(k(N))}\right) \\ &\le 1 - \frac{C'}{\sqrt{N\mu_k(A_l)}} \exp\left(-\frac{C\delta^2}{3}\log N\right) \\ &= 1 - \frac{C'}{\sqrt{\mu_k(A_l)}} N^{C\delta^2/3-1/2}, \end{aligned}$$

which leads to

$$M_N(\delta) \ll \frac{\sqrt{N}}{\mu_k(\mathcal{A})} \exp\left(-C' N^{-C\delta^2/3-1/2} \sum_{l=1}^{m} \frac{1}{\sqrt{\mu_k(A_l)}}\right).$$

Clearly, we have

$$\sum_{l=1}^{m} \frac{1}{\sqrt{\mu_k(A_l)}} = \frac{1}{(p_1p_2)^{([k/3+2])/2}} \left(\frac{1}{\sqrt{p_1}} + \cdots + \frac{1}{\sqrt{p_q}}\right)^{k-2([k/3+2])} \gg q^{5k/6},$$

which gives

$$\sum_{l=1}^{m} \frac{1}{\sqrt{\mu_k(A_l)}} \ge N^{5/8}.$$

Furthermore, $k(N) = \mathcal{O}(\log N)$ implies

$$\mu_k(\mathcal{A}) \gg \left(\min_{1\le j\le q} p_j\right)^{2k/3} = c^k \gg N^{-C''}$$

for some constant $C'' > 0$. Thus

$$M_N(\delta) \ll N^{C''+\frac{1}{2}} \exp\left(-C'N^{-C\delta^2/3+1/8}\right).$$

By choosing $\delta > 0$ small enough it follows that

$$\lim_{i\to\infty} M_{N_i}(\delta) = 0.$$

Since

$$\mu_\infty\left(\{(x_n)_{n\geq 1} : D_N^{(k(N),\phi)}(\mu, x_n) \leq \delta\}\right) \leq M_N(\delta)$$

the proof of Theorem 2.127 is complete. □

GOLDSTERN's Theorem

Let $X = \{y_1, \ldots, y_q\}$ be a finite set of order q with the uniform measure $\mu(y_j) = 1/q$, $1 \leq j \leq q$. Theorem 2.128 says that for any fixed sequence $k(N)$ with

$$\operatorname{lq} N - \operatorname{lq}\operatorname{lq} N - k(N) \to \infty \tag{2.225}$$

almost all sequences $(x_n)_{n\geq 1} \in X^{\mathbf{N}}$ are $k(N)$-distributed. Hence almost all sequences are $k(N)$-distributed for a countable system of (different) sequences $k(N)$ satisfying (2.225). But it might be that this property does not remain true if the phrase "for a countable system of sequences $k(N)$ satisfying (2.225)" is replaced by "for all sequences $k(N)$ satisfying (2.225)." However, there is a remarkable theorem (due to GOLDSTERN [663]) which allows to intersect uncountable many sets of measure 1 and to obtain again a set of measure 1 (under proper assumptions). For example, we get the following sharpened version of Theorem 2.128.

Theorem 2.131 *Almost all seqences $(x_n)_{n\geq 1} \in X^{\mathbf{N}}$ are $k(N)$-distributed for all sequences $k(N)$ satisfying (2.225).*

Theorem 2.131 is an application of an abstact measure theoretic concept. We need some notations. Let $\mathbf{N}_0$ denote the set of non-negative integers and ${\mathbf{N}_0}^{\mathbf{N}_0}$ the space of all functions (sequences) $x : \mathbf{N}_0 \to \mathbf{N}_0$. ${\mathbf{N}_0}^{\mathbf{N}_0}$ inherits its topology from the discrete topology on $\mathbf{N}_0$. Since $\mathbf{N}_0$ with the discrete topology is also a metric space it follows that ${\mathbf{N}_0}^{\mathbf{N}_0}$ is a metric space, too. For example, one can use the metric $d(x_1, x_2) = 2^{-n_0}$, in which $n_0 = \min\{n \in \mathbf{N}_0, x_1(n) \neq x_2(n)\}$, or we identify $x \in {\mathbf{N}_0}^{\mathbf{N}_0}$ with the irrational mumber $x' \in (0,1)$ via the continued fraction expansion $x' = [0, x(0), x(1), \ldots]$ and set $d'(x_1, x_2) = |x_1' - x_2'|$. Furthermore we will use the following ordering on ${\mathbf{N}_0}^{\mathbf{N}_0}$: $x_1 \leq x_2$ if and only if $x_1(n) \leq x_2(n)$ for all $n \geq 0$. (Note that $x_1 \leq x_2$ is not equivalent to $x_1' \leq x_2'$. The subset $\mathbf{N}_0 \uparrow \mathbf{N}_0 \subseteq {\mathbf{N}_0}^{\mathbf{N}_0}$ denotes the set of all sequences which diverge to infinity.

Definition 2.132 *A* perfect Polish space *C is a complete separable metric space.*

A subset $A \subseteq C$ is called analytic (or Σ_1^1-set) if it is the projection of a Borel set in $C \times C'$ for some Polish space C'. A subset of C is coanalytic (or Π_1^1) if its complement is analytic.

Furthermore, the space ${N_0}^{N_0}$ and the space of infinite sequences $(x_n)_{n\geq 1}$, $x_n \in X$, $|X| = q$, with the infinite product measure μ_∞ may be considered as complete Polish spaces.

There are various equivalent definitions of analytic sets, e.g. $A \subseteq C$ is analytic if and only if it is the continuous image of some BOREL set of a Polish space. It is also possible to define them as projections of closed sets in $C \times C'$. In what follows we make use of the fact that projections of analytic sets are again analytic. Let $M \subseteq C$ be the projection of an analytic set $A \subseteq C \times C'$, i.e. $M = \{x \in C : \text{there exists } y \in C' \text{ with } (x,y) \in A\}$. By definition there exists a BOREL set $B \subseteq C \times C' \times C''$ for some Polish space C'' such that $A = \{(x,y) \in C \times C' : \text{there exists } z \in C'' \text{ with } (x,y,z) \in B\}$. Hence

$$M = \{x \in C : \text{there exist } (y,z) \in C' \times C'' \text{ with } (x,y,z) \in B\}$$

is the projection of a BOREL set, i.e. analytic.

We also want to note that all analytic sets are measurable (see MOSCHOVAKIS [1206, 2H.8]).

Definition 2.133 *Let C be a perfect Polish space and $B \subseteq C \times {N_0}^{N_0}$. We say that $f \subseteq B$ unifomizes B if f may be interpreted as a function $D \to {N_0}^{N_0}$, where*

$$D = \{t \in C : \textit{there exists } x \in {N_0}^{N_0} \textit{ such that } (t,x) \in B\}.$$

Analytic sets are related to this concept via the VON NEUMANN selection theorem (see [1206, 4E.9]).

Proposition 2.134 *Every Σ_1^1-set in $C \times {N_0}^{N_0}$ can be uniformized by a measurable function.*

Corollary 2.135 *Let C be a perfect Polish space and μ a complete* BOREL *probabilty measure (i.e. μ is σ-additive, all Borel sets are measurable, and all subsets $B \subseteq C \times {N_0}^{N_0}$ is a Σ_1^1-set) and assume that for every $t \in C$ there exists $x \in {N_0}^{N_0}$ such that $(t,x) \in B$.*

Then there exists a random variable $X : C \to {N_0}^{N_0}$ (i.e. a measurable function in the sense that preimages of BOREL *sets are measurable) such that*

$$\mu(\{t \in C : (t, X(t)) \in B\}) = 1.$$

Furthermore for every random variable $T : C \to C$ there exists a random variable $X : C \to {N_0}^{N_0}$ such that

$$\mu(\{t \in C : (T(t), X(t)) \in B\}) = 1.$$

Proof. By assumption

$$\{t \in C : \text{there exists } x \in {N_0}^{N_0} \text{ such that } (t,x) \in B\} = C.$$

Hence by Proposition 2.134 there exists a measurable function $f \subseteq B$ with domain C. So for all $t \in C$ we have $(t, f(t)) \in B$. Thus $X = f$ is a proper choice.

Moreover, if $T : C \to C$ is a random variable then we have for $X = f \circ T$: $(T(t), X(t)) = (T(t), f(T(t))) \in B$. □

We will also make use of the following property.

Lemma 2.136 *Let $X : C \to {\mathbf{N}_0}^{\mathbf{N}_0}$ be a random variable. Then there exists a family $(x_n)_{n\in\mathbf{N}_0}$ of functions in ${\mathbf{N}_0}^{\mathbf{N}_0}$ with*

$$\mu(\{t \in C : \text{there exists } n, \text{ such that } X(t) \le x_n\}) = 1.$$

Proof. Call a Borel set $A \subseteq C$ of positive measure *good* if there is a function $x_A \in {\mathbf{N}_0}^{\mathbf{N}_0}$ such that

$$A \subseteq \{t \in C : X(t) \le x_A\}.$$

If there are finitely many good sets covering C (up to a measure zero) then we are done. Otherwise, let A_n, $n \in \mathbf{N}_0$, be a maximal antichain of good sets (with respect to $\subseteq$). Such a sequence can be found using ZORN's lemma. Again, if $\bigcup_{n\in\mathbf{N}_0} A_n$ covers C up to a zero set we are done. Assume that $A = C \setminus \bigcup_{n\in\mathbf{N}_0} A_n$ has positive measure ε. Now define a sequence $(k_n)_{n\in\mathbf{N}_0}$ of non-negative integers such that for each n the set $\{t \in C\colon X(t)(n) > k_n\}$ has measure $< \varepsilon/2^{n+3}$. Thus the set

$$\{t \in C : \text{for all } n \in \mathbf{N}_0\colon X(t)(n) \le k_n\} \cap A$$

has measure $\ge \varepsilon - \varepsilon/4 > 0$. Hence it is good, which contradicts the maximality of the family A_n, $n \in \mathbf{N}_0$. □

The first main result is the following theorem.

Theorem 2.137 *Let C be a perfect Polish space and μ a complete* BOREL *probabilty measure on C. Assume that for each $x \in {\mathbf{N}_0}^{\mathbf{N}_0}$ we have a measurable set $W_x \subseteq C$ with the following properties:*

- $\{(x,t) : x \in {\mathbf{N}_0}^{\mathbf{N}_0}, t \in W_x\}$ *is a Π^1_1-set in ${\mathbf{N}_0}^{\mathbf{N}_0} \times C$.*
- $\mu(W_x) = 1$ *for all $x \in {\mathbf{N}_0}^{\mathbf{N}_0}$.*
- *If $x \le x'$ then $W_{x'} \subseteq W_x$.*

Then

$$\mu\left(\bigcap_{x\in{\mathbf{N}_0}^{\mathbf{N}_0}} W_x\right) = 1.$$

Proof. Let id: $C \to C$ denote the identity function. Clearly we have

$$\mu\left(\{t \in C : \mathrm{id}(t) \in W_x\}\right) = \mu(W_x) = 1.$$

We will now show that the same is true if we replace $x \in {\mathbf{N}_0}^{\mathbf{N}_0}$ by a random variable $X : C \to {\mathbf{N}_0}^{\mathbf{N}_0}$, i.e.

$$\mu\left(\{t \in C : t \in W_{X(t)}\}\right) = 1.$$

By Lemma 2.136 we can find a family of functions $x_n : C \to \mathbf{N_0}^{\mathbf{N_0}}$, $n \in \mathbf{N_0}$, such that

$$\mu(\{t \in C : \text{there exists } n \in \mathbf{N_0} \text{ with } X(t) \leq x_n(t)\}) = 1.$$

Using the anti-monotonicity of the family W_x, $x \in \mathbf{N_0}^{\mathbf{N_0}}$, we get

$$\mu\left(\{t \in C : \text{there exists } n \in \mathbf{N_0} \text{ with } W_{X(t)} \supseteq W_{x_n(t)}\}\right) = 1,$$

and so

$$\mu\left(\{t \in C : W_{X(t)} \supseteq \bigcap_n W_{x_n(t)}\}\right) = 1.$$

Hence

$$\begin{aligned} \mu(\{t \in C : t \in W_{X(t)}\}) &\geq \mu\left(\{t \in C : t \in \bigcap_n W_{x_n(t)}\}\right) \\ &= \mu\left(\bigcap_{x \in \mathbf{N_0}^{\mathbf{N_0}}} W_x\right) = 1 \end{aligned}$$

for all random variables $X : C \to \mathbf{N_0}^{\mathbf{N_0}}$. In other words, if we set

$$B = \{(x,t) : x \in \mathbf{N_0}^{\mathbf{N_0}}, t \notin W_x\}$$

then we have

$$\mu(\{t \in C : (T(t), X(t)) \in B\}) = 0$$

for all $X : C \to \mathbf{N_0}^{\mathbf{N_0}}$. By assumption B is a Σ_1^1-set. Hence, by Corollary 2.135 it is impossible that for every $t \in C$ there exists $x \in \mathbf{N_0}^{\mathbf{N_0}}$ with $t \notin W_x$. Therefore there exists $t \in C$ such that for every $x \in \mathbf{N_0}^{\mathbf{N_0}}$ we have $t \notin W_x$. The set $\bigcap_{x \in \mathbf{N_0}^{\mathbf{N_0}}} W_x$ is nonempty.

We now repeat the above arguments for $A \cap W_x$ instead of W_x, where A is any Borel set with positive measure $\mu(A) > 0$ and conclude that $A \cap \bigcap_{x \in \mathbf{N_0}^{\mathbf{N_0}}} W_x$ is nonempty. Hence $W = \bigcap_{x \in \mathbf{N_0}^{\mathbf{N_0}}} W_x$ has outer measure 1. If follows from the assumption that the complement of

$$W = \{t \in C : t \in W_x \text{ for every } x \in \mathbf{N_0}^{\mathbf{N_0}}\}$$

is a projection of an analytic set in $C \times \mathbf{N_0}^{\mathbf{N_0}}$. Hence W is a Π_1^1-set and therefore measurable. Thus $\mu(W) = 1$. □

In fact we will use a little bit different version.

Theorem 2.138 *Let C be a perfect Polish space and μ a complete* BOREL *probabilty measure on C. Assume that for each $y \in \mathbf{N_0} \uparrow \mathbf{N_0}$ we have a set $U_y \subseteq C$ with the following properties:*

- *$\{(y,t) : y \in \mathbf{N_0} \uparrow \mathbf{N_0},\, t \in U_y\}$ is a Π_1^1-set in $\mathbf{N_0} \uparrow \mathbf{N_0} \times C$.*

- $\mu(U_y) = 1$ *for all* $y \in \mathbf{N}_0 \uparrow \mathbf{N}_0$.
- *If* $y \le y'$ *then* $U_y \subseteq U_{y'}$.

Then

$$\mu\left(\bigcap_{y\in\mathbf{N}_0\uparrow\mathbf{N}_0} U_y\right) = 1.$$

Proof. We show that Theorem 2.137 implies Theorem 2.138. For every unbounded $x \in \mathbf{N}_0{}^{\mathbf{N}_0}$ we define its "inverse" function $x^* \in \mathbf{N}_0 \uparrow \mathbf{N}_0$ by $x^*(n) = \min\{m : x(m) \ge n\}$. (For bounded $x \in \mathbf{N}_0{}^{\mathbf{N}_0}$ we set $x^* = x_0$, where $x_0 \in \mathbf{N}_0 \uparrow \mathbf{N}_0$ is some fixed sequence.) For a given family $(U_y : y \in \mathbf{N}_0 \uparrow \mathbf{N}_0)$ set $W_x = U_{x^*}$ for all unbounded $x \in \mathbf{N}_0{}^{\mathbf{N}_0}$ and $W_x = C$ otherwise. We will show that

$$\bigcap_{y\in\mathbf{N}_0\uparrow\mathbf{N}_0} U_y = \bigcap_{x\in\mathbf{N}_0{}^{\mathbf{N}_0}} W_x.$$

The inclusion "$\subseteq$" is clear since $W_x = U_{x^*}$ or $W_x = C$. For the converse inclusion, it is enough to see that for every $y \in \mathbf{N}_0 \uparrow \mathbf{N}_0$ there exists $x \in \mathbf{N}_0{}^{\mathbf{N}_0}$ with $x^* \le y$. In fact, for every $y \in \mathbf{N}_0 \uparrow \mathbf{N}_0$ we can find a sequence x which increases so fast that for all n, $x(y(n)) \ge n$, e.g. $x(m) = \max\{k : y(k) \le m\}$. Hence $x^*(n) = \min\{m : x(m) \ge n\} \le y(n)$.

Next, we show that the function $f : \mathbf{N}_0{}^{\mathbf{N}_0} \to \mathbf{N}_0{}^{\mathbf{N}_0}$, $x \mapsto x^*$ is BOREL which means that preimages of open sets are BOREL sets. Let $\mathbf{N}_{0u}^{\mathbf{N}_0}$ denote the set of unbounded sequences $x \in \mathbf{N}_0{}^{\mathbf{N}_0}$ and f' the restriction of f to $\mathbf{N}_{0u}^{\mathbf{N}_0}$. It suffices to show that f' is BOREL, since either $f^{-1}(U) = (f')^{-1}(U)$ or $f^{-1}(U) = (f')^{-1}(U) \cup (\mathbf{N}_0{}^{\mathbf{N}_0} \setminus \mathbf{N}_{0u}^{\mathbf{N}_0})$ and $(\mathbf{N}_0{}^{\mathbf{N}_0} \setminus \mathbf{N}_{0u}^{\mathbf{N}_0})$ is a BOREL set. Obviously the sets $U_{kn} = \{y \in \mathbf{N}_0{}^{\mathbf{N}_0} : y(n) = k\}$ constitute a subbase of the topology of $\mathbf{N}_0{}^{\mathbf{N}_0}$. Since

$$\begin{aligned}
(f')^{-1}(U_{kn}) &= \{x \in \mathbf{N}_{0u}^{\mathbf{N}_0} : x^*(n) = k\} \\
&= \{x \in \mathbf{N}_{0u}^{\mathbf{N}_0} : x(n) \ge k\} \cap \bigcap_{m<k} \{x \in \mathbf{N}_{0u}^{\mathbf{N}_0} : x(m) < k\} \\
&= \mathbf{N}_{0u}^{\mathbf{N}_0} \cap \{x \in \mathbf{N}_0{}^{\mathbf{N}_0} : x(n) \ge k\} \cap \bigcap_{m<k} \{x \in \mathbf{N}_0{}^{\mathbf{N}_0} : x(m) < k\}
\end{aligned}$$

is surely a BOREL set we have proved that f' (and f) are BOREL functions.

Note that BOREL functions also have the property that preimages of BOREL sets are BOREL and that preimages of analytic resp. of coanalytic sets are analytic resp. coanalytic. First let U denote the system of all sets $U \subseteq \mathbf{N}_0{}^{\mathbf{N}_0}$ such that $f^{-1}(U)$ is BOREL. Clearly, U contains all open sets. Furthermore, if $U \in \mathcal{U}$ then $\mathbf{N}_0{}^{\mathbf{N}_0} \setminus U_1 \in \mathcal{U}$ and if $U_1, U_2, \ldots \in \mathcal{U}$ then $\bigcup_{n\ge1} U_n \in \mathcal{U}$. Hence U contains all BOREL sets. Secondly, we show that preimages of Π_1^1-sets are again Π_1^1-sets. (The proof for analytic sets is exactly the same.) Let $A \subseteq \mathbf{N}_0{}^{\mathbf{N}_0}$ be a Π_1^1 set. By definition there exists a BOREL set $B \subseteq \mathbf{N}_0{}^{\mathbf{N}_0} \times C'$ for some Polish space C' such that $A = \{x \in \mathbf{N}_0{}^{\mathbf{N}_0} : (x, y) \in B \text{ for all } y \in C'\}$. Furthermore let $f'' : \mathbf{N}_0{}^{\mathbf{N}_0} \times C' \to \mathbf{N}_0{}^{\mathbf{N}_0} \times C'$, $(x, y) \mapsto (x^*, y)$.

Then f'' is a BOREL function, too, which implies that $B' = g'^{-1}(B)$ is a BOREL set. Hence, if we set $A' = g^{-1}(A)$ then

$$\begin{aligned} A' &= \{x \in \mathbf{N_0}^{\mathbf{N_0}} \,:\, f(x) = x^* \in A\} \\ &= \{x \in \mathbf{N_0}^{\mathbf{N_0}} \,:\, (x^*, y) \in B \text{ for all } y \in \mathbf{N_0}^{\mathbf{N_0}}\} \\ &= \{x \in \mathbf{N_0}^{\mathbf{N_0}} \,:\, (x, y) \in B' \text{ for all } y \in \mathbf{N_0}^{\mathbf{N_0}}\}. \end{aligned}$$

Thus A' is a Π_1^1-set.

Now set $g : \mathbf{N_0}^{\mathbf{N_0}} \times C \to \mathbf{N_0}^{\mathbf{N_0}} \times C$, $(x,t) \mapsto (x^*, t)$. Since f is BOREL, g and its restriciton g' to $\mathbf{N_0}_u^{\mathbf{N_0}} \times C$ are BOREL, too. Especially we have

$$(g')^{-1}\left(\{(y,t) \,:\, y \in \mathbf{N_0} \uparrow \mathbf{N_0},\, t \in U_y\}\right) = \{(x,t) \,:\, x \in \mathbf{N_0}_u^{\mathbf{N_0}},\, t \in W_x\},$$

which implies that $\{(x,t) \,:\, x \in \mathbf{N_0}^{\mathbf{N_0}},\, t \in W_x\}$ is a Π_1^1-set.

Hence all assumptions of Theorem 2.137 are satisfied. Thus it follows that

$$\mu\left(\bigcap_{y \in \mathbf{N_0} \uparrow \mathbf{N_0}} U_y\right) = \mu\left(\bigcap_{x \in \mathbf{N_0}^{\mathbf{N_0}}} W_x\right) = 1.$$

□.

Proof of Theorem 2.131. Let $C = X^{\mathbf{N_0}}$ equipped with the product measure μ_∞. Let k range over all functions in $\mathbf{N_0}^{\mathbf{N_0}}$ such that $\varphi_k(N) = [\mathrm{lq}\, N - \mathrm{lq}\,\mathrm{lq}\, N - k(N)] \geq 0$ and $\varphi_k(N) \to \infty$ as $N \to \infty$. Let R_k be the set of $k(N)$-uniformly distributed sequences in $X^{\mathbf{N_0}}$, and let $R = \bigcap_k R_k$. By definition of $k(N)$-uniform distribution it follows that $k \leq k'$ implies $R_{k'} \subseteq R_k$. Furthermore it is easy to check that

$$\varphi_{\varphi_k}(N) = [\mathrm{lq}\, N - \mathrm{lq}\,\mathrm{lq}\, N - [\mathrm{lq}\, N - \mathrm{lq}\,\mathrm{lq}\, N - k(N)]] = k(N).$$

Now, for each function $y \in \mathbf{N_0} \uparrow \mathbf{N_0}$ let

$$U_y = \begin{cases} R_{\varphi_y} & \text{if } y \leq \mathrm{lq} - \mathrm{lq}\mathrm{lq}, \\ X^{\mathbf{N_0}} & \text{otherwise.} \end{cases}$$

By this definition we have the property that $y \leq y'$ implies $U_y \subseteq U_{y'}$. Since

$$\begin{aligned} &\{(y,t) \,:\, y \in \mathbf{N_0} \uparrow \mathbf{N_0},\, t \in U_y\} \\ &= \bigcap_{m \geq 1} \bigcup_{N \geq 1} \bigcap_{A \in X^{\varphi_y(N)}} \Bigg\{ (y,t) \in \mathbf{N_0} \uparrow \mathbf{N_0} \times X^{\mathbf{N_0}} \,: \\ &\quad q^{\varphi_y(N)} \left| \frac{|\{1 \leq n \leq N - \varphi_y(N) \,:\, t(n), \ldots, t(n + \varphi_y(N)) = A\}|}{N} - q^{-\varphi_y(N)} \right| \leq \frac{1}{m} \Bigg\}, \end{aligned}$$

this set is a BOREL set and hence Π_1^1. Furthermore, by Theorem 2.127 $\mu_\infty(U_y) = 1$. Thus, by Theorem 2.138,

$$\mu_\infty\left(\bigcap_{y \in \mathbf{N_0} \uparrow \mathbf{N_0}} U_y\right) = 1.$$

For $\varphi_k(n) \to \infty$ let $y = \varphi_k$. Since $k = \varphi_y$ we obtain $R_k = R_{\varphi_y} = U_y$ which implies that $R = \bigcap_k R_k$ has measure 1, too. □

2.4.3 Combinatorial Concepts of Discrepancy

Let us consider the following problem. There is a finite set X and a system $\mathcal{R}$ of subsets of X. We want to find a 2-coloring of X such that for every $R \in \mathcal{R}$ the difference between the number of points in R with color 1 and that of color 2 is as small as possible.

Definition 2.139 *Let X be a finite set and $\mathcal{R}$ a system of subsets of X. A mapping $c : X \to \{-1, 1\}$ is called* 2-coloring *of X. We define the* discrepancy of c on $\mathcal{R}$ *by*

$$\operatorname{disc}(\mathcal{R}, c) = \max_{R \in \mathcal{R}} \left| \sum_{x \in R} c(x) \right|$$

and the discrepancy of $\mathcal{R}$ *by*

$$\operatorname{disc}(\mathcal{R}) = \min_{c : X \to \{-1,1\}} \operatorname{disc}(\mathcal{R}, c).$$

A general upper bound for $\operatorname{disc}(\mathcal{R})$ can be obtainted by using probabilistic methods.

Theorem 2.140 *Let X be a finite set and $\mathcal{R}$ a system of subsets of X. Then*

$$\operatorname{disc}(\mathcal{R}) = \mathcal{O}\left(\sqrt{\max_{R \in \mathcal{R}} |R| \cdot \log |\mathcal{R}|}\right). \tag{2.226}$$

Proof. Let $Y \subseteq X$. Then $c(Y)$ may be regarded as a random variable with $\mathbf{E}\, c(Y) = 0$ and $\mathbf{V} c(Y) = |Y|$. Hence, by Lemma 2.15

$$\mathbf{P}\left[|c(Y)| > \lambda \sqrt{|Y|}\right] < 2e^{-\lambda^2/2}.$$

Setting $\lambda = \sqrt{2 \log(4|\mathcal{R}|)}$ we obtain

$$\mathbf{P}\left[|c(Y)| > \sqrt{2 \log(4|\mathcal{R}|) \cdot |Y|}\right] < \frac{1}{2|\mathcal{R}|}$$

and

$$\mathbf{P}\left[\operatorname{disc}(\mathcal{R}) > \sqrt{2 \log(4|\mathcal{R}|)} \cdot \max_{R \in \mathcal{R}} |R|\right] < \frac{1}{2}.$$

Hence there exists a coloring $c : X \to \{-1, 1\}$ with (2.226). □

Although this bound is (in general) best possible there are important cases where substantial improvements are possible.

Definition 2.141 *Let X be a finite set and $\mathcal{R}$ a system of subsets of X. The* (primal) shatter function $\pi_{\mathcal{R}}$ *is defined by*

$$\pi_{\mathcal{R}}(m) = \max_{A \subseteq X,\, |A| \le m} |\{R \cap A : R \in \mathcal{R}\}|.$$

For example, consider any finite subset $X \subseteq \mathbf{R}^k$ and let $\mathcal{R}$ be the system $X \cap B$, where B is a ball. Then $\pi_{\mathcal{R}}(m) = \mathcal{O}(m^k)$

Theorem 2.142 *Let X be a finite set and $\mathcal{R}$ a system of subsets of X. Suppose that there exist constants $k, C \geq 1$ such that $\pi_{\mathcal{R}}(m) \leq Cm^k$. Then*

$$\mathrm{disc}(\mathcal{R}) = \mathcal{O}\left(n^{\frac{1}{2}-\frac{1}{2k}}(\log n)^{1+\frac{1}{2k}}\right) \qquad \text{if } k > 1$$

and

$$\mathrm{disc}(\mathcal{R}) = \mathcal{O}\left((\log n)^{\frac{5}{2}}\right) \qquad \text{if } k = 1.$$

The proof of Theorem 2.142 requires some preliminaries.

Definition 2.143 *Let $\mathcal{R}$ be a system of subsets of X and for any subset $Y \subseteq X$ set*

$$\mathcal{R}|_Y = \{R \cap Y \,:\, R \in \mathcal{R}\}.$$

We say that a subset $Y \subseteq X$ is shattered by $\mathcal{R}$ if $\mathcal{R}|_Y = 2^Y$, i.e. $\mathcal{R}$ contains all subsets of Y. The VAPNIK-CHERVONENKIS *dimension (VC-dimension) of $(X, \mathcal{R})$ is the maximal size of a shattered subset of X. (If there are shattered subsets of any size then we say that the VC-dimension is infinite.)*

Definition 2.144 *Let $\mathcal{R}$ be a system of subsets of X. A subset $S \subseteq X$ is called ε-net for $(X, \mathcal{R})$ if $S \cap R \neq \emptyset$ for every set $R \in \mathcal{R}$ with $|R| > \varepsilon|X|$.*

Proposition 2.145 *Let $\mathcal{R}$ be a system of subsets of X and suppose that $(X, \mathcal{R})$ has finite VC-dimension. Then for every $r > 1$ there exists a $(1/r)$-net for $(X, \mathcal{R})$ of size $\mathcal{O}(r \log r)$.*

Lemma 2.146 *Let $\Phi_d(n)$ be defined by*

$$\Phi_d(n) = \sum_{k=0}^{d} \binom{n}{k} \qquad \text{if } d < n$$

and $\Phi_d(n) = 2^n$ otherwise. Furthermore, let X be a finite set of size n and $\mathcal{R}$ be a system of subsets of X. If $(X, \mathcal{R})$ has VC-dimension d then $|\mathcal{R}| \leq \Phi_d(n)$.

Proof. The assertion is trivially true for $d = 0$ and $n = 0$. Assume the assertion to be true for any finite space of VC-dimension at most $d-1$, and for any space of VC-dimension d with at most $n-1$ elements, for some $d \geq 1$ and $n \geq 1$.

Now, let $(X, \mathcal{R})$ be a space of VC-dimension d with $|X| = n$ and $x \in X$. Consider the spaces $\mathcal{S} - x = (X - \{x\}, \mathcal{R} - x)$, where $\mathcal{R} - x = \{R - \{x\} : R \in \mathcal{R}\}$ and $\mathcal{S}^{(x)} = (X - \{x\}, \mathcal{R}^{(x)})$, with $\mathcal{R}^{(x)} = \{R \in \mathcal{R} : x \notin R, R \cup \{x\} \in \mathcal{R}\}$. Obviously $\mathcal{S} - x$ is of VC-dimension at most d; hence, by assumption, $|\mathcal{R} - x| \leq \Phi_d(n-1)$. We show that $\mathcal{S}^{(x)}$ is of VC-dimension at most $d - 1$.

Let A be a subset of $X - \{x\}$ that can be shattered by $\mathcal{R}^{(x)}$. Then it is easy to see that $A \cup \{x\}$ can be shattered by $\mathcal{R}$. Namely, for $A' \subseteq A$ there is an $R \in \mathcal{R}^{(x)}$ with

$A' = A \cap R$. Since $x \notin R$, $A' = (A \cup \{x\}) \cap R$ and $A' \cup \{x\} = (A \cup \{x\}) \cap (R \cup \{x\})$, where both R and $R \cup \{x\}$ are in $\mathcal{R}$. Since $A \cup \{x\}$ can be shattered by $\mathcal{R}$, $|A \cup \{x\}| \leq d$, we have $|A| \leq d - 1$. Thus $\mathcal{S}^{(x)}$ is of VC-dimension at most $d - 1$.

Because $\mathcal{S}^{(x)}$ is of VC-dimension at most $d-1$, by assumption, $|\mathcal{R}^{(x)}| \leq \Phi_{d-1}(n-1)$. Observing that $|\mathcal{R}| = |\mathcal{R} - x| + |\mathcal{R}^{(x)}|$, this yields

$$|\mathcal{R}| \leq \Phi_d(n-1) + \Phi_{d-1}(n-1) = \Phi_d(n).$$

□

Corollary 2.147 *Let $Y \subseteq X$. Then $|\mathcal{R}|_Y| \leq \Phi_d(|Y|)$.*

As above, let $\mathcal{R}$ be a system of subsets of X. For $\varepsilon > 0$ set

$$\mathcal{R}_\varepsilon = \{R \in \mathcal{R} : |R| > \varepsilon |X|\}.$$

Furthermore, let Q_ε^m be the set of $\mathbf{x} = (x_1, \ldots, x_m) \in X^m$ such that there exists $R \in \mathcal{R}_\varepsilon$ with $x_i \notin R$, $1 \leq i \leq m$, and let J_ε^{2m} be the set of $(\mathbf{x}, \mathbf{y}) \in X^m \times X^m$, $\mathbf{y} = (y_1, \ldots, y_m)$ such that there exists $R \in \mathcal{R}_\varepsilon$ with $x_i \notin R$, $1 \leq i \leq m$, but $y_j \in R$ for at least by $\varepsilon m/2$ indices j, $1 \leq j \leq m$.

Lemma 2.148 *Let $\varepsilon > 0$ and $m \geq 8/\varepsilon$. Then*

$$|Q_\varepsilon^m| \leq 2|J_\varepsilon^{2m}| \cdot |X|^{-m}.$$

Proof. It is convenient to use the language of probability theory. $\mathbf{P}^m$ will denote the counting measure on X^m, i.e. $\mathbf{P}^m(Z) = |Z|/|X|^m$.

For $R \in \mathcal{R}_\varepsilon$ let

$$Z_R = \{\mathbf{y} \in X^m : y_i \in R \text{ for at least } \varepsilon m/2 \text{ indices } i, 1 \leq i \leq m\}.$$

We first claim that $\mathbf{P}^m(Z_R) > \frac{1}{2}$ for all $R \in \mathcal{R}_\varepsilon$. To establish this, we show that $\mathbf{P}^m(\bar{Z}_R) < \frac{1}{2}$, where $\bar{Z}_R = X^m - Z_R$. Since $\mathbf{P}^1(R) \geq \varepsilon$ for each $R \in \mathcal{R}_\varepsilon$ and $\mathbf{y}$ is in $\bar{Z}_R$ only if $y_i \in R$ for fewer than $\varepsilon m/2$ indices i, $\mathbf{P}^m(\bar{Z}_R)$ is maximized as $\mathbf{P}^1(R)$ approaches ε. In this case, for random $\mathbf{y} \in X^m$ the expected number of indices i such that $y_i \in R$ is εm and the variance is $\varepsilon(1-\varepsilon)m$. Thus for each $\mathbf{y} \in \bar{Z}_R$, the number of y_i's in R differs by at least $\varepsilon m/2$ from the expected value. Hence by CHEBYSHEV's inequality

$$\mathbf{P}^m(\bar{Z}_R) \leq \frac{\varepsilon(1-\varepsilon)m}{(\varepsilon m/2)^2} < \frac{4}{\varepsilon m} \leq \frac{1}{2}.$$

Now consider a fixed $\mathbf{x} \in Q_\varepsilon^m$. By definition, there exists $R_\mathbf{x} \in \mathcal{R}_\varepsilon$ such that $x_i \notin R_x, 1 \leq i \leq m$. From the above, it follows that $\mathbf{y} \in Z_{R_\mathbf{x}}$ for more than half of the $\mathbf{y} \in X^m$; hence $(\mathbf{x}, \mathbf{y}) \in J_\varepsilon^{2m}$ for more than half of the $\mathbf{y} \in X^m$. Thus $\mathbf{P}^{2m}(J_\varepsilon^{2m}) > \frac{1}{2}\mathbf{P}^m(Q_\varepsilon^m)$, which proves the lemma. □

Lemma 2.149 *Suppose that $(X, \mathcal{R})$ has finite VC-dimension d and let $\varepsilon > 0$. Then*

$$|J_\varepsilon^{2m}| \leq |X|^{2m} \Phi_d(2m) 2^{-\varepsilon m/2}.$$

Proof. For each j, $1 \le j \le (2m)!$, let π_j be a distinct permutation of the indices $1, \dots, 2m$. For each $\mathbf{x} \in X^{2m}$, let $\Theta(\mathbf{x}) = |\{j : \pi_j(\mathbf{x}) \in J_\varepsilon^{2m}\}|$. It is easily verified that

$$\mathbf{P}^{2m}(J_\varepsilon^{2m}) \le \max_{\mathbf{x} \in X^{2m}} (\Theta(\mathbf{x})/(2m)!).$$

Consider a fixed $\mathbf{x} \in X^{2m}$. Let E be the set of distinct elements of X that appear in $\mathbf{x}$. For each permutation $\pi_j(\mathbf{x})$ in J_ε^{2m} there is a subset T of E that is a witness to the fact that $\pi_j(\mathbf{x}) \in J_\varepsilon^{2m}$ in the sense that there exists $R \in \mathcal{R}_\varepsilon$ with $T = R \cap E$, all occurrences of members of T (and so of R) appear in the second half of $\pi_j(\mathbf{x})$, and there are at least $\varepsilon m/2$ such occurrences. However, a given T can be a witness for only a small proportion of all permutations of $\mathbf{x}$. In particular, if there are l occurrences of members of T in $\mathbf{x}$ and $l \ge \varepsilon m/2$, then T is a witness for at most

$$\frac{\binom{m}{l}}{\binom{2m}{l}} = \frac{m(m-1)\cdots(m-l+1)}{2m(2m-1)\cdots(2m-l+1)} \le 2^{-l} \le 2^{-\varepsilon m/2}$$

of all permutations of $\mathbf{x}$ (and if $l < \varepsilon m/2$, then T is a witness for no permutation of $\mathbf{x}$). Since $|E| \le 2m$ and $(X, \mathcal{R})$ is of VC-dimension d, by Corollary 2.147 there are at most $\Phi_d(2m)$ distinct subsets of E induced by intersections with $R \in \mathcal{R}_\varepsilon$. Hence, there are at most $\Phi_d(2m)$ distinct witnesses. It follows that

$$\frac{\Theta(\mathbf{x})}{(2m)!} \le \Phi_d(2m) 2^{-\varepsilon m/2},$$

which implies the lemma. □

Proof of Proposition 2.145. Suppose that $(X, \mathcal{R})$ has finite VC-dimension d, that $\varepsilon > 0$, and that $m \ge 8/\varepsilon$. Then by Lemma 2.149 the probability that a random subset Y of size m fails to be an ε-net is less that

$$2\Phi_d(2m)2^{-\varepsilon m/2} = \mathcal{O}\left((2m)^d 2^{-\varepsilon m/2}\right).$$

Hence, if $m \ge C\frac{1}{\varepsilon}\log\frac{1}{\varepsilon}$ we have

$$2\Phi_d(2m)2^{-\varepsilon m/2} \le \frac{1}{2}$$

for a sufficiently large constant C just depending on d. □

Lemma 2.150 *Let $\mathcal{R}$, $\mathcal{S}$ be systems of subsets of an n-point set X, $|\mathcal{S}| > 1$, $|S| \le s$ for every $S \in \mathcal{S}$, and*

$$\prod_{R \in \mathcal{R}} (|R| + 1) \le 2^{(n-1)/5}.$$

Then there exists a mapping $c : X \to \{-1, 0, +1\}$, such that the value of c is nonzero for at least $n/10$ elements of X, $c(R) = 0$ for every $R \in \mathcal{R}$, and $|c(S)| \le \sqrt{2s\ln(4|\mathcal{S}|)}$ for every $S \in \mathcal{S}$.

Proof. Let C_0 be the set of all colorings $c : X \to \{-1,+1\}$, and let C_1 be the subcollection of mappings c with $|c(S)| \leq \sqrt{2s\ln(4|S|)}$ for all $S \in \mathcal{S}$. We have seen in the proof of Theorem 2.140 that $|C_1| \geq \frac{1}{2}|C_0| = 2^{n-1}$.

Now let us define a mapping $\nu : C_1 \to \mathbf{Z}^{|\mathcal{R}|}$, assigning to a coloring c the $|\mathcal{R}|$-component integer vector $\nu(c) = (c(R); R \in \mathcal{R})$. Since $|c(R)| \leq |R|$ and $c(R) - |R|$ is even for every R, the image of ν contains at most

$$\prod_{R\in\mathcal{R}} (|R|+1) \leq 2^{(n-1)/5}$$

integer vectors. Hence there is a vector $\nu_0 = \nu(c_0)$ such that ν maps at least $2^{4(n-1)/5}$ elements of C_1 to ν_0. Let C_2 be the collection of all $c \in C_1$ with $\nu(c) = \nu_0$. Let us pick one $c_0 \in C_2$ and for every $c \in C_2$, we define a new mapping $c' : X \to \{-1,0,1\}$ by $c'(x) = (c(x) - c_0(x))/2$. Then $c'(R) = 0$ for every $R \in \mathcal{R}$, and also $c'(S) \leq \sqrt{2s\ln(4|S|)}$ for every $S \in \mathcal{S}$. Let C_2' be the collection of c' for all $c \in C_2$.

To prove the lemma, it remains to show that there is a mapping $c' \in C_2'$ whose value is nonzero in at least $n/10$ points of X. The number of mappings $X \to \{-1,0,+1\}$ with at most $n/10$ nonzero elements is bounded by

$$\sum_{q=0}^{\lfloor n/10\rfloor} \binom{n}{q} 2^q,$$

and standard estimates show that this number is smaller than $2^{4(n-1)/5} \leq |C_2'|$. Hence there exists a mapping $c' \in C_2'$ with at least $n/10$ nonzero values. □

Lemma 2.151 *Let $\mathcal{R}$ be a system of subsets of X such that $(X,\mathcal{R})$ has finite VC-dimension. If $\mathcal{R}' = \{R_1 \setminus R_2 : R_1, R_2 \in \mathcal{R}\}$ then $(X,\mathcal{R}')$ has finite VC-dimension, too.*

Proof. Since $(X,\mathcal{R})$ has finite VC-dimension there exists a finite set $Y \subseteq X$ such that $R \cap Y \neq Y$ for every $R \in \mathcal{R}$. Hence $(R_1 \setminus R_2) \cap Y \neq Y$ for every $R_1, R_2 \in \mathcal{R}$ and $(X,\mathcal{R}')$ has finite VC-dimension. □

Proof of Theorem 2.142. We first describe a construction of a partial coloring for a set system using the previous Lemma 2.150, which will then be applied iteratively.

Let $\mathcal{R}$ be a system of subsets of X with $\pi_{\mathcal{R}}(m) = \mathcal{O}(m^d)$. Let us define another system by $\mathcal{R}' = \{R_1 \setminus R_2 : R_1, R_2 \in \mathcal{R}\}$. Since $(X,\mathcal{R})$ has bounded VC-dimension it follows from Lemma 2.151 that $(X,\mathcal{R}')$ has a bounded VC-dimension, too. Let $N \subseteq X$ be a $(1/r)$-net for $(X,\mathcal{R}')$ of size $\mathcal{O}(r\log r)$, where r is a parameter to be fixed later (the existence of such N is guaranteed by Proposition 2.145).

Let us call two sets $R_1, R_2 \in \mathcal{R}$ equivalent if $R_1 \cap N = R_2 \cap N$. Since the sets of $\mathcal{R}$ have at most $\mathcal{O}((r\log r)^d)$ distinct intersections with N, this equivalence has at most $\mathcal{O}((r\log r)^d)$ classes. Let a collection $\mathcal{L}$ contain exactly one set of each equivalence class. For a set $R \in \mathcal{R}$, let L_R be the member of $\mathcal{L}$ equivalent to R.

Let us put

$$\mathcal{S} = \{R \backslash L_R; R \in \mathcal{R}\} \cup \{L_R \backslash R; R \in \mathcal{R}\}.$$

For every R, $L_R \backslash R$ and $R \backslash L_R$ contain no points of N, and thus by the $(1/r)$-net property of N, the cardinality of any set of $\mathcal{S}$ is at most n/r. Also we have $|\mathcal{S}| \leq 2|\mathcal{R}| = \mathcal{O}(n^d)$.

We want to apply Lemma 2.150 on the set systems $\mathcal{L}$ and $\mathcal{S}$, hence we need an estimate on $\prod_{L \in \mathcal{L}}(|L|+1)$. This is bounded by $(n+1)^{|\mathcal{L}|} \leq (n+1)^{(Kr \log r)^d}$ for some constant K. Thus if we set $r = cn^{1/d}/(\log n)^{1+1/d}$ for a sufficiently small positive constant c, we derive that the product is bounded by $2^{(n-1)/5}$ as required. Then the size of sets of $\mathcal{S}$ is bounded by $s = n/r = \mathcal{O}(n^{1-1/d}(\log n)^{1+1/d})$, and Lemma 2.150 guarantees the existence of a mapping $c : X \to \{-1, 0, 1\}$, such that $c(L) = 0$ for all $L \in \mathcal{L}$

$$|c(S)| \leq \sqrt{2s \ln(4|\mathcal{S}|)} = \mathcal{O}(n^{1/2-1/2d}(\log n)^{1+1/2d}),$$

and the set $Y_1 = \{x \in X; c(x) \neq 0\}$ has at least $n/10$ elements.

If R is a set of $\mathcal{R}$, we can write

$$R = (L_R \cup S_1) \backslash S_2,$$

where $S_1 = R \backslash L_R \in \mathcal{S}, S_2 = L_R \backslash R \in \mathcal{S}, S_1$ and L_R are disjoint, and S_2 is contained in L_R. Hence

$$|c(R)| = |c(R \cap Y_1)| \leq |c(L_R)| + |c(S_1)| + |c(S_2)| = \mathcal{O}(n^{1/2-1/2d}(\log n)^{1+1/2d}).$$

In order to finish the proof, we apply the construction described above inductively. We set $X_1 = X$, and we obtain a partial coloring c_1, nonzero on a set Y_1, as above. We set $X_2 = X_1 \backslash Y_1$, and we obtain a partial coloring c_2 of X_2 by applying the above construction on the set system $(X_2, \mathcal{R}|_{X_2})$, etc. We repeat this construction until the size of the set X_k becomes trivially small (e.g., smaller than a suitable constant). Then we define $Y_k = X_k$ and we let c_k be the constant mapping with value 1 on Y_k.

Let $R \in \mathcal{R}$. Then we have

$$|c(R)| \leq |c_1(R \cap Y_1)| + |c_2(R \cap Y_2)| + \ldots + |c_k(R \cap Y_k)|. \tag{2.227}$$

For every i, $|c_i(R)|$ is bounded by

$$\mathcal{O}\left(n_i^{1/2-1/2d}(\log n_i)^{1+1/2d}\right),$$

where $n_i = |X_i| \leq (9/10)^{i-1}n$. Thus, for $d > 1$ the summands on the right hand side of 2.227 decrease exponentially, and we obtain

$$\operatorname{disc}(\mathcal{R}) = \mathcal{O}\left(n^{1/2-1/2d}(\log n)^{1+1/2d}\right),$$

as claimed.

In a similar way we derive $\operatorname{disc}(\mathcal{R}) = \mathcal{O}(\log^{5/2} n)$ for $d = 1$. □

2.4.4 Linear Recurring Sequences

A sequence of integers $(x_n)_{n\geq 1}$ is said to be uniformly distributed modulo m (u.d. mod m) if every residue class appears with the same frequency. In other words $(x_n)_{n\geq 1}$ induces a u.d. sequence on the finite set $X = \mathbf{Z}/m\mathbf{Z}$. It is obvious how to generalize this concept to (commutative) rings R modulo an ideal I, where the factor ring R/I is finite, i.e. I has finite norm $N(I) = |R/I|$. We will call such an ideal I an admissible ideal of R.

Definition 2.152 *Let R be a commutative ring. A sequence $(x_n)_{n\geq 1}$ of elements of R is called uniformly distributed mod I if I is an ideal of finite norm $N(I) = |R/I|$ and*

$$\lim_{N\to\infty} \frac{|\{n \leq N : x_n \equiv x \bmod I\}|}{N} = \frac{1}{N(I)}$$

holds for every element $x \in R$.

It is quite natural to consider linear recurring sequences $(x_n)_{n\geq 1}$ and to ask for criteria for uniform distribution of sequences of this kind.

Definition 2.153 *Let R be a commutative ring with unit element. Let $(x_n)_{n\geq 1}$, $x_n \in R$ be a sequence satisfying*

$$\sum_{k=0}^{K} a_k x_{n+k} = 0$$

for all $n \geq 1$, where $a_k \in R$, $0 \leq k \leq K$. The polynomial

$$c(x) = \sum_{k=0}^{K} a_k x^k \in R[x]$$

is called a characteristic polynomial of $(x_n)_{n\geq 1}$. [Note that a monic characteristic polynomial of a sequence need not be unique, even if we require that it has minimal degree].

If $(x_n)_{n\geq 1}$ admits a monic characteristic polynomial $c(x)$ then $(x_n)_{n\geq 1}$ is called linear recurring sequence with characteristic polynomial $c(x)$.

The following two theorems due to TURNWALD [1880] provide an answer to the above question. Here and in the following P, P_1, P_2 will always denote admissible prime ideals. If R is a DEDEKIND domain then R/P is a finite field. Hence $N(P)$ is a power of a prime p. For the basic properties of DEDEKIND domains we refer to NARKIEWICZ [1262]. A DEDEKIND domain can be characterized as an integrally closed NOETHERIAN ring such that every non-zero prime ideal is maximal. We just want to mention that the most well-known examples of DEDEKIND domains are the maximal orders in algebraic number fields. We will frequently use the usual notation $P|I$ if P divides I and the crucial property $N(IJ) = N(I)N(J)$. From this it immediately follows that a sequence is u.d. mod J provided that it is u.d. mod I and $J \supseteq I$.

Theorem 2.154 *Let I be an admissible ideal of the* DEDEKIND *domain R and let $(x_n)_{n\geq 1}$ be a linear recurring sequence (of elements of R) with characteristic polynomial $c(x)$. Assume that for every prime divisor P of I, $c(x)$ splits into linear factors modulo P and that the factors incongruent to x appear with multiplicity at most two. Then $(x_n)_{n\geq 1}$ is u.d. mod I if and only if the following conditions hold:*

(i) *If $P|I$ then $(x_n)_{n\geq 1}$ is u.d. mod P.*

(ii) *If $P^2|I$ and $N(P)=2$ or $N(P)=3$ then $(x_n)_{n\geq 1}$ is u.d. mod P^2.*

(iii) *If $P^2|I$ and $N(P)=p\geq 5$ then $p\not\equiv 0 \bmod P^2$.*

(iv) *If $P^3|I$ and $N(P)=2$ or $N(P)=3$ then $N(P)\not\equiv 0 \bmod P^2$.*

(v) *If $P_1|I$ and $P_2|I$, $P_1\neq P_2$, then $N(P_1)\neq N(P_2)$.*

Observe that for $R=\mathbf{Z}$ only the first two conditions have to be checked. The others are trivially satisfied.

If the characteristic polynomial has degree 2 then we can be much more precise.

Theorem 2.155 *Let I be an admissible ideal of the* DEDEKIND *domain R and let $(x_n)_{n\geq 1}$ be a linear recurring sequence (of elements of R) with characteristic polynomial $c(x)=x^2-c_1x-c_0$. Then $(x_n)_{n\geq 1}$ is u.d. mod I if and only if the following conditions hold:*

(i) *If $P|I$ then $N(P)=p$, $c_1^2+4c_0\equiv 0\ (P)$, $c_0\not\equiv 0\ (P)$; $2x_2\not\equiv c_1x_1\ (P)$ for $p>2$, $x_2\not\equiv x_1\ (P)$ for $p=2$.*

(ii) *If $P^2|I$ and $p=2$, then $c_1\not\equiv 0\ (P^2)$, $c_0\not\equiv 1\ (P^2)$.*

(iii) *if $P^2|I$ and $p=3$, then $c_0+c_1^2\not\equiv 0\ (P^2)$.*

(iv) *If $P^2|I$ and $N(P)=p\geq 5$ then $p\not\equiv 0 \bmod P^2$.*

(v) *If $P^3|I$ and $N(P)=2$ of $N(P)=3$ then $N(P)\not\equiv 0 \bmod P^2$.*

(vi) *If $P_1|I$ and $P_2|I$, $P_1\neq P_2$, then $N(P_1)\neq N(P_2)$.*

Lemma 2.156 *Let I be an ideal of R and let (x_n) be a linear recurring sequence with characteristic polynomial $c(x)$. If $\sum a_kx^k\equiv 0\ (c(x),I)$ then $\sum a_kx_{n+k}\equiv 0\ (I)$ for all $n\geq 1$.*

Proof. By assumption there exists a polynomial $\sum b_kx^k$ with coefficients in I such that $\sum(a_k-b_k)x^k$ is a multiple of $c(x)$. Hence, $\sum(a_k-b_k)x_{n+k}=0$ and $\sum a_kx_{n+k}\equiv\sum(a_k-b_k)x_{n+k}\equiv 0\ (I)$. (Note that every multiple of a characteristic polynomial is again a characteristic polynomial.) □

For the following calculations it is useful to observe that the congruences $f(x)\equiv 0$ $(c(x),I)$ and $g(x)\equiv 0\ (c(x),J)$ imply $f(x)g(x)\equiv 0\ (c(x),IJ)$.

Lemma 2.157 *Assume that* $x^n(x^l - 1) \equiv 0\ (c(x), I)$. *Then*

$$x^{2n}(x^{kl} - 1) \equiv x^{2n} k(x^l - 1)\ (c(x), I^2)$$

and

$$x^{3n}(x^{kl} - 1) \equiv x^{3n} k(x^l - 1) + x^{3n}\binom{k}{2}(x^l - 1)^2\ (c(x), I^3)$$

for $k > 0$.

Proof. From

$$x^{kl} - 1 = \left(1 + (x^l - 1)\right)^k - 1 = \sum_{j=1}^{k}\binom{k}{j}(x^l - 1)^j$$

and $x^{2n}(x^{kl} - 1)^2 \equiv 0\ (c(x), I^2)$ we deduce

$$x^{2n}(x^{kl} - 1) \equiv x^{2n}\binom{k}{1}(x^l - 1)\ (c(x), I^2).$$

Analogously, the second assertion follows from $x^{3n}(x^{kl} - 1)^3 \equiv 0\ (c(x), I^3)$. □

Lemma 2.158 *Let* I *be an ideal of* R *and assume that* $a \equiv 0\ (I)$ *for a positive integer* a. *If* $x^{n_0}(x^l - 1) \equiv 0\ (c(x), I^{h_0})$ $(n_0 \geq 0; l, h_0 > 0)$ *then for every linear recurring sequence* (x_n) *with characteristic polynomial* $c(x)$ *we have*

(i) $x_{n+a^h l} \equiv x_n\ (I^{h_0+h})$ *for* $h \geq 0$, $n > 2^h n_0$,

(ii) $x_{n+kl} \equiv x_n + k(x_{n+l} - x_n)\ (I^{2h_0})$ *for* $n > 2n_0$,

(iii) $x_{n+ka^h l} \equiv x_n + ka^{h-1}(x_{n+al} - x_n)\ (I^{2h_0+h})$ *for* $h > 0$, $n > 3 \cdot 2^{h-1} n_0$,

(iv) $x_{n+ka^h l} \equiv x_n + ka^h(x_{n+l} - x_n)\ (I^{2h_0+h})$ *for* $h > 0$, $n > 3 \cdot 2^{h-1} n_0$, *a odd*.

Proof. By Lemma 2.156 it suffices to show that

(i') $x^{2^h n_0}(x^{a^h l} - 1) \equiv 0\ (c(x), I^{h_0+h})$,

(ii') $x^{2n_0}(x^{kl} - 1) \equiv x^{2n_0} k(x^l - 1)\ (c(x), I^{2h_0})$,

(iii') $x^{3\cdot 2^{h-1} n_0}(x^{ka^h l} - 1) \equiv x^{3\cdot 2^{h-1} n_0} ka^{h-1}(x^{al} - 1)\ (c(x), I^{2h_0+h})$,

(iv') $x^{3\cdot 2^{h-1} n_0}(x^{ka^h l} - 1) \equiv x^{3\cdot 2^{h-1} n_0} ka^h(x^l - 1)\ (c(x), I^{2h_0+h})$.

To simplify the notation we write n instead of n_0 in the sequel.

From $x^{2^h n}(x^{a^h l} - 1) \equiv 0\ (c(x), I^{h_0+h})$ we conclude that

$$x^{2^{h+1} n}\left(x^{a^{h+1} l} - 1\right) \equiv x^{2^{h+1} n} a\left(x^{a^h l} - 1\right)\ (c(x), I^{2h_0+2h})$$

by Lemma 2.157. Since $x^{2^h n}a(x^{a^h l}-1) \equiv 0\ (c(x), I^{h_0+h+1})$ and $h_0+h+1 \le 2h_0+2h$, this proves (i') by induction, the case $h=0$ being trivial. Lemma 2.157, again, proves (ii').

Since $2h_0+1 \le 3h_0$, Lemma 2.157 shows that

$$x^{3n}(x^{kal}-1) \equiv x^{3n}k(x^{al}-1) + x^{3n}\binom{k}{2}(x^{al}-1)^2\ (c(x), I^{2h_0+1}).$$

From $x^n(x^{al}-1) \equiv 0\ (c(x), I^{h_0})$ and $x^{2n}(x^{al}-1) \equiv 0\ (c(x), I^{h_0+1})$ (case $h=1$ of (i')) we conclude

$$x^{3n}(x^{al}-1)^2 \equiv 0\ (c(x), I^{2h_0+1}).$$

This gives case $h=1$ of (iii'). By (i') we have $x^{2^h n}(x^{ka^h l}-1) \equiv 0\ (c(x), I^{h_0+h})$ so that, by Lemma 2.157,

$$x^{3\cdot 2^h n}(x^{aka^h l}-1) \equiv x^{3\cdot 2^h n}a(x^{ka^h l}-1) + x^{3\cdot 2^h n}\binom{a}{2}(x^{ka^h l}-1)^2\ \left(c(x), I^{3(h_0+h)}\right).$$

Since $x^{2\cdot 2^h n}(x^{ka^h l}-1)^2 \equiv 0\ (c(x), I^{2(h_0+h)})$ and $2h_0+h+1 \le 2(h_0+h)$, the second term vanishes mod I^{2h_0+h+1}, and we obtain

$$x^{3\cdot 2^h n}(x^{ka^{h+1} l}-1) \equiv x^{3\cdot 2^h n}a(x^{ka^h l}-1)\ \left(c(x), I^{2h_0+h+1}\right).$$

The proof of (iii') now follows by induction since $a \equiv 0\ (I)$.

If a is odd, then $\binom{a}{2} = a(a-1)/2 \equiv 0\ (I)$. Hence

$$x^{3n}\binom{a}{2}(x^l-1)^2 \equiv 0\ (I^{2h_0+1}),$$

and Lemma 2.157 gives

$$x^{3n}(x^{al}-1) \equiv x^{3n}a(x^l-1) + x^{3n}\binom{a}{2}(x^l-1)^2 \equiv x^{3n}a(x^l-1)\ \left(c(x), I^{2h_0+1}\right).$$

Since $a^{h-1} \equiv 0\ (I^{h-1})$, this implies

$$x^{3\cdot 2^{h-1}n}a^{h-1}(x^{al}-1) \equiv x^{3\cdot 2^{h-1}n}a^h(x^l-1)\ \left(c(x), I^{2h_0+h}\right).$$

Now (iv') follows from (iii'). □

Lemma 2.159 **(i)** *If (x_n) is u.d. mod P, then $N(P)=p$, $x_{j+n(p-1)} \equiv x_j + n(x_{j+p-1}-x_j)\ (P)$, $x_{j+p-1}-x_j \not\equiv 0\ (P)$ for $j \ge j_0$, and $x_{j+p^h(p-1)} \equiv x_j\ (P^h)$ for $j \ge j_0(h)$ and $h \ge 1$.*

(ii) *If $N(P)=p$ and $p>2$, then $x_{j+np^h(p-1)} \equiv x_j + np^{h-1}(x_{j+p(p-1)}-x_j)\ (P^{h+1})$ for $j \ge j_0(h)$ and $h \ge 1$.*

(iii) *If $N(P) = p$ and $p \geq 5$, then $x_{j+p(p-1)} \equiv x_j + p(x_{j+p-1} - x_j)$ (P^2) for $j \geq j_0$.*

Proof. By definition of u.d. mod P, $N(P)$ is finite. Let n_0 be the multiplicity of x in the factorization of $c(x)$ modulo P. Then setting $q = N(P)$ we have

$$x^{n_0}(x^{q-1} - 1)^2 \equiv 0\,(c(x), P)\,,$$

since every linear factor incongruent to x is a divisor of $x^{q-1} - 1$ and the multiplicity is assumed to be at most 2. Hence,

$$x^{n_0}(x^{p(q-1)} - 1) \equiv x^{n_0}(x^{q-1} - 1)^p \equiv 0\,(c(x), P)\,.$$

By Lemma 2.156 this implies that (x_n) has period (not necessarily minimal) $p(q-1)$ modulo P. If (x_n) is u.d. mod P every period length must be divisible by $q = N(P)$; hence, we conclude $q = p$. From

$$x^{n_0}(x^{n(p-1)} - 1) = x^{n_0}\left((1 + (x^{p-1} - 1))^n - 1\right) = x^{n_0}\sum_{j=1}^{n}\binom{n}{j}(x^{p-1} - 1)^j$$

and

$$x^{n_0}(x^{p-1} - 1)^2 \equiv 0\,(c(x), P)\,,$$

we deduce

$$x^{n_0}(x^{n(p-1)} - 1) \equiv x^{n_0} n(x^{p-1} - 1)\,(c(x), P)\,.$$

Again, by Lemma 2.156 this implies $x_{j+n(p-1)} - x_j \equiv n(x_{j+p-1} - x_j)$ (P) for $j \geq j_0 = n_0 + 1$. We now apply Lemma 2.158 (with $I = P$, $a = p$, $h_0 = 1$, $l = p(p-1)$) to obtain (after a change of notation)

$$x_{j+p^h(p-1)} \equiv x_j \;\left(P^h\right) \text{ for } h \geq 1,\ j \geq j_0(h)$$

and

$$x_{j+np^h(p-1)} \equiv x_j + np^{h-1}(x_{j+p(p-1)} - x_j)\;\left(P^{h+1}\right) \text{ for } h \geq 1,\ j \geq j_0(h),\ p > 2.$$

If $x_{j+p-1} - x_j \equiv 0$ (P) for some $j \geq j_0$, from $x_{j+n(p-1)} - x_j \equiv n(x_{j+p-1} - x_j)$ (P) we see that the residue x_j mod P appears at least p times (for $n = 0, \ldots, p-1$) in a period of length $p(p-1)$; but if (x_n) is u.d. mod P every residue must appear $p-1$ times, since the number of residues is $p = N(P)$. This concludes the proof of (i) and (ii). To prove (iii) we first remark that

$$\begin{aligned} x^{2n_0}(x^{p(p-1)} - 1) &= x^{2n_0}\sum_{j=1}^{p}\binom{p}{j}(x^{p-1} - 1)^j \\ &\equiv x^{2n_0}\left(p(x^{p-1} - 1) + (x^{p-1} - 1)^p\right)\left(c(x), P^2\right), \end{aligned}$$

since $\binom{p}{j} \equiv 0\ (P)$ for $1 \le j \le p-1$ and $x^{n_0}(x^{p-1}-1)^2 \equiv 0\ (c(x), P)$. For $p \ge 5$ we have $x^{2n_0}(x^{p-1}-1)^p \equiv 0\ (c(x), P^2)$; hence,

$$x^{2n_0}(x^{p(p-1)}-1) \equiv x^{2n_0}p(x^{p-1}-1)\ (c(x), P^2).$$

which implies $x_{j+p(p-1)} - x_j \equiv p(x_{j+p-1} - x_j)\ (P^2)$ for $j \ge 2n_0 + 1$. □

Remark. If $c(0) \not\equiv 0\ (P)$ then we may take $n_0 = 0$. Hence the conditions given in the lemma hold for $j \ge 1$.

Lemma 2.160 *Assume that $p = 2$ and (x_n) is u.d. mod P^2. If $p \equiv 0\ (P^2)$, then (x_n) is not u.d. mod P^3; if $p \not\equiv 0\ (P^2)$, (x_n) is u.d. mod P^h for all $h \ge 1$.*

Proof. As in the proof of the preceding lemma, we conclude that $x^{n_0}(x-1)^2 \equiv 0\ (c(x), P)$, since (x_n) is u.d. mod P. We define $r(x) = \sum r_k x^k$ to be the residue of $x^{n_0}(x^2-1)$ modulo $c(x)$. From $x^{n_0}(x^2-1) \equiv x^{n_0}(x-1)^2 \equiv 0\ (c(x), P)$ we see that $r(x) \equiv 0\ (c(x), P)$. Hence, $r(x)$ is divisible by $c(x)$ mod P, which implies $r_k \equiv 0\ (P)$ for all k, since the degree of $r(x)$ mod P is smaller than the degree of of $c(x)$ mod P ($=\deg(c(x))$ since the leading coefficient is 1). Observing that $x^{n_0}(x^{2k}-1) \equiv 0\ (c(x), P)$, we see that this implies

$$\begin{aligned} x^{n_0}r(x)^2 &\equiv x^{n_0}\sum r_k^2 x^{2k} \equiv \sum r_k^2\left(x^{n_0}(x^{2k}-1) + x^{n_0}\right) \\ &\equiv x^{n_0}\sum r_k^2 \equiv x^{n_0}\left(\sum r_k\right)^2 (c(x), P^3). \end{aligned}$$

From $x^{n_0}(x^2-1) \equiv r(x) \bmod (c(x))$ we deduce

$$x^{2n_0}(x^2-1)^2 \equiv r(x)^2 \bmod (c(x)).$$

Hence,

$$\begin{aligned} x^{3n_0}(x^4-1) &= 2x^{3n_0}(x^2-1) + x^{3n_0}(x^2-1)^2 \equiv 2x^{3n_0}(x^2-1) + x^{n_0}r(x)^2 \\ &\equiv 2x^{3n_0}(x^2-1) + x^{n_0}\left(\sum r_k\right)^2 (c(x), P^3), \end{aligned}$$

and, by Lemma 2.156,

$$x_{j+3n_0+4} - x_{j+3n_0} \equiv 2(x_{j+3n_0+2} - x_{j+3n_0}) + x_{j+n_0}\left(\sum r_k\right)^2 (P^3).$$

From $x^{n_0}(x^2-1) \equiv 0\ (c(x), P)$ and

$$x^{2n_0}(x^4-1) \equiv 2x^{2n_0}(x^2-1) + x^{2n_0}(x^2-1)^2 \equiv 0\ (c(x), P^2),$$

we see that $x_{j+2} \equiv x_j\ (P)$ and $x_{j+4} \equiv x_j\ (P^2)$ for $j \ge j_0 = 2n_0 + 1$. Since (x_n) is u.d. mod P^2, each of the four residues mod P^2 must appear once in a period, which implies $x_{j+2} \not\equiv x_j\ (P^2)$ for $j \ge j_0$, i.e., $x_{j+2} - x_j$ lies in the unique residue class mod

P^2 that belongs to P but not to P^2. We conclude that $x_{j+3} - x_{j+1} \equiv x_{j+2} - x_j \ (P^2)$ for $j \geq j_0$. Since $x_{j+1} - x_j \equiv 1 \ (P)$, we finally obtain

$$\sum r_k \equiv \sum r_k(x_{j+1+k} - x_{j+k}) \equiv (x_{j+n_0+3} - x_{j+n_0+1}) - (x_{j+n_0+2} - x_{j+n_0}) \equiv 0 \ (P^2)$$

(taking into account that $x^{n_0}(x^2-1) \equiv \sum r_k x^k \bmod (c(x))$ implies $x_{j+n_0+2} - x_{j+n_0} \equiv \sum r_k x_{j+k}$). Hence, $(\sum r_k)^2 \equiv 0 \ (P^3)$ and

$$x_{j+3n_0+4} - x_{j+3n_0} \equiv 2(x_{j+3n_0+2} - x_{j+3n_0}) \ (P^3) \,.$$

If $2 \equiv 0 \ (P^2)$, this means that (x_n) has period 4 mod P^3; since 4 is not divisible by $N(P^3) = 2^3$, (x_n) is not u.d. mod P^3. Now let us assume $2 \not\equiv 0 \ (P^2)$; then the above relation yields $x_{j+4} - x_j \not\equiv 0 \ (P^3)$ for sufficiently large j. Lemma 2.158 now gives ($I = P$, $a = 2$, $h_0 = 1$, $l = 2$, $k = 1$)

$$x_{j+2^{h+1}} \equiv x_j + 2^{h-1}(x_{j+4} - x_j) \ (P^{h+2}) \ \text{ for } h > 0,\ j \geq 3 \cdot 2^{h-1} n_0 .$$

Hence,

$$x_{j+2^{h+1}} \equiv x_j \ (P^{h+1}) \ \text{ and } x_{j+2^{h+1}} \not\equiv x_j \ (P^{h+2}) \ \text{ for } j \geq j_0(h), h > 0.$$

By assumption, (x_n) is u.d. mod P^2, i.e., every residue appears once in a period of length 4. Since $x_{j+4} \not\equiv x_j \ (P^3)$, this implies that every residue mod P^3 appears once in a period of length 8. Inductively we conclude the analogous statement modulo P^h for all h, i.e., (x_n) is u.d. mod P^h for all h. □

Proof of Theorem 2.154. We assume first that (x_n) is u.d. mod I. Then (x_n) is u.d. mod every divisor of I. This proves (i) and (ii). For $P^2|I$ and $p \geq 5$ we deduce

$$x_{j+p(p-1)} \equiv x_j + p(x_{j+p-1} - x_j) \ (P^2)$$

from Lemma 2.159(iii). Assume $p \equiv 0 \ (P^2)$; then $x_{j+p(p-1)} \equiv x_j \ (P^2)$ (for sufficiently large j), which is impossible since the period must be divisible by $N(P^2) = p^2$. If $P^3|I$ and $p = 2$, the preceding lemma implies $p \not\equiv 0 \ (P^2)$. For the remaining case $p = 3$ we use Lemma 2.159(ii) to obtain

$$x_{j+p^2(p-1)} \equiv x_j + p(x_{j+p(p-1)} - x_j) \ (P^3)$$

so that

$$x_{j+p^2(p-1)} \equiv x_j \ (P^3) \quad \text{if } p \equiv 0 \ (P^2) \,.$$

Since $p^2(p-1)$ is not divisible by $N(P^3) = p^3$, (x_n) cannot be u.d. mod P^3; hence P^3 cannot divide I. This concludes the proof of (iii) and (iv).

Assume $P_i|I$ ($i = 1,2$) and $P_1 \neq P_2$; then P_1P_2 divides I. Hence, the period of (x_n) modulo P_1P_2 must be divisible by $N(P_1P_2)$. By Lemma 2.159(i) we have $N(P_i) = p_i$ and (x_n) has period $p_i(p_i - 1)$ mod P_i. From $N(P_1) = N(P_2)$ we obtain

$p_1 = p_2 = p$, so that (x_n) has period $p(p-1)$ mod $P_1P_2 = P_1 \cap P_2$. Since $p(p-1)$ is not divisible by $N(P_1P_2) = p^2$, we arrive at a contradiction. Hence, $N(P_1) \neq N(P_2)$.

Now we assume (i)–(v). If $P|I$ then, by (i), (x_n) is u.d. mod P, so that Lemma 2.159(i) shows $N(P) = p$ and $x_{j+p^h(p-1)} \equiv x_j$ (P^h) for $h \geq 1$, $j \geq j_0(h)$. By (v), a rational prime p belongs only to one prime ideal. Hence, every divisor of I may be written in the form $\prod P_i^{h_i} \cdot P^k$, $p_1 < p_2 < \ldots < p$; $\prod P_i^{h_i}$ may be the empty product. In order to prove that (x_n) is u.d. mod I, we may therefore proceed inductively and show that (x_n) is u.d. mod $\prod P_i^{h_i} \cdot P^{k+1}$ provided (x_n) is u.d. mod $\prod P_i^{h_i} \cdot P^k$, $P^{k+1}|I$, and $p_1 < p_2 < \ldots < p$. The first step is given by (i). If $p = 2$, the assertion follows from (ii, (iv), and Lemma 2.160. In the following we assume $p > 2$. Define $l = \prod p_i^{h_i} \cdot (p_i - 1)$; then (x_n) has period l modulo $\bigcap P_i^{h_i} = \prod P_i^{h_i}$, so that $x_{j+nl} - x_j \equiv n(x_{j+l} - x_j)$ $(\prod P_i^{h_i})$ for sufficiently large j. By Lemma 2.159 we have

$$x_{j+np^k(p-1)} - x_j \equiv n(x_{j+p^k(p-1)} - x_j)\,(P^{k+1}) \quad \text{for } k \geq 0, j \geq j_0(k).$$

Hence,

$$x_{j+nlp^k(p-1)} - x_j \equiv n(x_{j+lp^k(p-1)} - x_j)\left(\prod P_i^{h_i} \cdot P^{k+1}\right).$$

If we can prove $x_{j+lp^k(p-1)} - x_j \not\equiv 0$ (P^{k+1}), then the last congruence means that $x_{j+nlp^k(p-1)}$ $(n = 0, \ldots, p-1)$ runs through the p residues mod $\prod P_i^{h_i} \cdot P^{k+1}$ belonging to the residue x_j mod $\prod P_i^{h_i} \cdot P^k$. Since (x_n) has period $lp^k(p-1)$ mod $\prod P_i^{h_i} \cdot P^k$, this implies that (x_n) is u.d. mod $\prod P_i^{h_i} \cdot P^{k+1}$ provided (x_n) is u.d. mod $\prod P_i^{h_i} \cdot P^k$. From $x_{j+lp^k(p-1)} - x_j \equiv l(x_{j+p^k(p-1)} - x_j)$ (P^{k+1}) and $(l, p) = 1$ (since $p_i < p$ for all i), we see that it remains to prove $x_{j+p^k(p-1)} - x_j \not\equiv 0$ (P^{k+1}). If $k = 0$, this follows from Lemma 2.159(i). If $k \geq 1$ and $p \geq 5$ we have, by Lemma 2.159

$$x_{j+p^k(p-1)} - x_j \equiv p^{k-1}(x_{j+p(p-1)} - x_j) \equiv p^k(x_{j+p-1} - x_j)\,(P^{k+1}).$$

Since, by (iii), $p \not\equiv 0$ (P^2) and $x_{j+p-1} - x_j \not\equiv 0$ (P), this proves the assertion in this case. Now suppose $p = 3$. By (ii), (x_n) is u.d. mod P^2. If $x_{j+p(p-1)} - x_j \equiv 0$ (P^2) for some $j \geq j_0$, then $x_{j+np(p-1)} - x_j \equiv n(x_{j+p(p-1)} - x_j)$ (P^2) implies that the residue x_j appears p times (for $n = 0, \ldots, p-1$) in a period of length $p^2(p-1)$. Since there are p^2 residues mod P^2, each of them must apear $(p-1)$ times. Hence, $x_{j+p(p-1)} - x_j \not\equiv 0$ (P^2) for sufficiently large j. Consequently, the assertion is equivalent to $p^{k-1} \not\equiv 0$ (P^k), i.e., $p \not\equiv 0$ (P^2) if $k > 1$. To conclude the proof we remark that $k > 1$ and $P^{k+1}|I$ imply $P^3|I$; hence, $p \not\equiv 0$ (P^2) follows from (iv). □

Proof of Theorem 2.155. We will first show that for every linear recurring sequence (x_n) with characteristic polynomial $c(x)$ and for every admissible prime ideal P, $c(x)$ must have a non-trivial multiple factor mod P if (x_n) is u.d. mod P. Assume that $c(x)$ has no multiple factors mod P except possibly the factor x whose multiplicity we denote by n_0. Let d denote the degree of the splitting field of $R|P$. Then $c(x)$ divides $x^{n_0}(x^{N(P)^d-1} - 1)$ mod P, i.e. $x^{n_0}(x^{N(P)^d-1} - 1) \equiv 0$ $(c(x), P)$. By Lemma 2.156 (x_n) has period $N(P)^d - 1$. Since this number is not divisible by $N(P)$, (x_n) cannot be u.d. mod P.

Hence we only have to show that (i)–(iii) is equivalent to condition (i) and (ii) of Theorem 2.154. Let P be a prime ideal. If (x_n) is u.d. mod P then $N(P) = p$ by Lemma 2.159(i). The conditions $c_0 \not\equiv 0\ (P)$ and $c_1^2 + 4c_0 \equiv 0\ (P)$ are surely satisfied since they are equivalent to $c(x) \equiv (x-a)^2\ (P)$ for some $a \not\equiv 0\ (P)$. From $c(x) \equiv (x-a)^2$ we conclude

$$ax_n \equiv ((x_2 - ax_1)(n-1) + ax_1)\, a^{n-1}\ (P).$$

Since $a^{p-1} = a^{N(P)-1} \equiv 1\ (P)$, we obtain

$$a(x_{j+n(p-1)} - x_j) \equiv n(x_2 - ax_1)(p-1)a^{j-1} \equiv -n(x_2 - ax_1)a^{j-1}\ (P).$$

If $x_2 - ax_1 \equiv 0\ (P)$, this means that (x_n) has period $p-1$ mod P; hence (x_n) is not u.d. mod P. If $x_2 - ax_1 \not\equiv 0\ (P)$, then the subsequences $(x_{j+n(p-1)})$ $(j = 0, \ldots, p-2)$ are u.d. mod P, being nontrivial arithmetic sequences mod P. Since (x_n) is the union of these subsequences, we finally obtain that (x_n) is u.d. mod P if and only if $x_2 - ax_1 \not\equiv 0\ (P)$. For $p = 2$ this means $x_2 - x_1 \not\equiv 0\ (P)$; for $p > 2$ the condition is equivalent to $2x_2 \not\equiv c_1x_1\ (P)$ since $2 \not\equiv 0\ (P)$ and $2a \equiv c_1\ (P)$.

Let (x_n) be u.d. mod P. Assume $p = 2$ first. We have to show that (x_n) is u.d. mod P^2 if and only if $c_0 \not\equiv 1\ (P^2)$ and $c_1 \not\equiv 0\ (P^2)$. Since (x_n) is u.d. mod P, we have $c_0 \equiv 1\ (P)$, $c_1 \equiv 0\ (P)$, and $x_{j+1} \equiv x_j + 1\ (P)$ for $j \geq 1$. By Lemma 2.159(i) and the following remark, $x_{j+4} \equiv x_j\ (P^2)$ for $j \geq 1$. Hence, (x_n) is u.d. mod P^2 if and only if $x_1 \not\equiv x_3\ (P^2)$ and $x_2 \not\equiv x_4\ (P^2)$. The second condition may be replaced by $x_4 - x_2 \equiv x_3 - x_1\ (P^2)$. Since

$$\begin{aligned}(x_4 - x_2) - (x_3 - x_1) &= (c_1x_3 + (c_0 - 1)x_2) - (c_1x_2 + (c_0 - 1)x_1)\\ &\equiv c_1(x_3 - x_2) + (c_0 - 1)(x_2 - x_1) \equiv c_1 + (c_0 - 1)\ (P^2)\end{aligned}$$

and

$$\begin{aligned}x_3 - x_1 &= c_1x_2 + (c_0 - 1)x_1 \equiv c_1(x_1 + 1) + (c_0 - 1)x_1\\ &\equiv (c_1 + c_0 - 1)x_1 + c_1\ (P^2),\end{aligned}$$

we obtain the conditions $c_1 + (c_0 - 1) \equiv 0\ (P^2)$ and $c_1 \not\equiv 0\ (P^2)$, which are equivalent to $c_0 - 1 \not\equiv 0\ (P^2)$ and $c_1 \not\equiv 0\ (P^2)$.

Now assume $p = 3$. We prove that (x_n) is u.d. mod P^2 if and only if $c_0 + c_1^2 \not\equiv 0$ (P^2). By Lemma 2.159 we have $x_{j+6} \equiv x_j\ (P)$ and $x_{j+6n} \equiv x_j + n(x_{j+6} - x_j)\ (P^2)$ for $j \geq 1$. If, for some j, $x_{j+6} - x_j \equiv 0\ (P^2)$, then x_j appears three times (for $n = 0, 1, 2$) in a period of length 18; hence, (x_n) is not u.d. mod P^2 in this case. If $x_{j+6} - x_j \not\equiv 0\ (P^2)$, then x_{j+6n} $(n = 0, 1, 2)$ runs through the three residues mod P^2 belonging to the residue class of x_j mod P. Since (x_n) is u.d. mod P and $x_{j+6} \equiv x_j$ (P), this implies that (x_n) is u.d. mod P^2 provided $x_{j+6} \not\equiv x_j\ (P^2)$ for all j. As in the first part of the proof we have $c(x) \equiv (x-a)^2\ (P)$, $a \not\equiv 0\ (P)$; hence $c_0 \equiv -a^2$ (P), $c_1 \equiv 2a\ (P)$. From $c(x) = c(a) + (2a - c_1)(x-a) + (x-a)^2$ we conclude that $c(a)(x-a) + (x-a)^3 \equiv 0\ (c(x), P^2)$, since $2a - c_1 \equiv 0\ (P)$, and $(x-a)^2 \equiv 0\ (c(x), P)$. Observing that

$$x^3 - a^3 = (x-a)^3 + 3ax(x-a) \quad \text{and} \quad x(x-a) \equiv a(x-a)\ (c(x), P),$$

we obtain

$$x^3 - a^3 \equiv -c(a)(x-a) + 3a^2(x-a) \equiv \left(3a^2 - c(a)\right)(x-a) \ (c(x), P^2).$$

Since $x^3 - a^3 \equiv (x-a)^3 \equiv 0 \ (c(x), P)$ and $x^3 + a^3 \equiv x^3 - a^3 + 2a^3 \equiv 2a^3 \ (c(x), P)$, this yields

$$x^6 - a^6 = (x^3 - a^3)(x^3 + a^3) \equiv 2a^3 \left(3a^2 - c(a)\right)(x-a) \ (c(x), P^2).$$

From $a^2 = a^{N(P)-1} \equiv 1 \ (P)$ we obtain

$$a^6 = 1 + 3(a^2-1) + 3(a^2-1)^2 + (a^2-1)^3 \equiv 1 \ (P^2).$$

Hence (by Lemma 2.156),

$$x_{j+6} - x_j \equiv 2a^3 \left(3a^2 - c(a)\right)(x_{j+1} - ax_j) \ (P^2).$$

Since $x_{j+1} - ax_j \equiv a(x_j - ax_{j-1}) \equiv \ldots \equiv a^{j-1}(x_2 - ax_1) \ (P)$ and $x_2 - ax_1 \not\equiv 0 \ (P)$, $x_{j+6} - x_j \not\equiv 0 \ (P^2)$ is seen to be equivalent to $3a^2 - c(a) \not\equiv 0 \ (P^2)$. Taking $a = 2c_1$ (a was only subject to the condition $c_1 \equiv 2a \ (P)$), we finally conclude that (x_n) is u.d. mod P^2 if and only if

$$c_1^2 + c_0 \equiv 3 \cdot 4c_1^2 - (4c_1^2 - 2c_1^2 - c_0) \not\equiv 0 \ (P^2). \square$$

We will now discuss linear recurring sequences of first order in more detail, namely inhomogeneous recurring sequences of the form

$$x_{n+1} = ax_n + b.$$

It is easy to show that they also admit the following representations:

$$(a-1)x_n = a^{n-1}((a-1)x_1 + b) - b, \quad \text{and} \tag{2.228}$$

$$x_{n+k} - x_n = a^{n-1}(a^{k-1} + \cdots + a + 1)((a-1)x_1 + b). \tag{2.229}$$

Since sequences of this kind satisfy the second order linear recurrence

$$x_{n+2} - (a+1)x_{n+1} + ax_n = 0$$

we can apply Theorem 2.155 to obtain the following property.

Theorem 2.161 *Let I be an admissible ideal of the* DEDEKIND *domain R and let $(x_n)_{n\geq 1}$, $x_n \in R$, be a sequence satisfying $x_{n+1} = ax_n + b$ with $a, b \in R$. Then (x_n) is u.d. mod I if and only if the following conditions hold:*

(i) *If $P|I$ then $N(P) = p$, $a \equiv 1 \ (P)$, and $b \not\equiv 0 \ (P)$.*

(ii) *If $P^2|I$ and $p = 2$, then $a \not\equiv -1 \ (p^2)$.*

(iii) *If $P^2|I$ and $p > 2$, then $p \not\equiv 0$ (P^2).*

(iv) *If $P^3|I$ and $p = 2$, then $p \not\equiv 0$ (P^2).*

(v) *If $P_i|I$ $(i = 1,2)$ and $P_1 \neq P_2$, then $p_1 \neq p_2$.*

You only have to use $c_1 = a + 1$, $c_0 = -a$, and completely elementary calculations to see that these conditions for a and b are equivalent to those for c_1 and c_0 in Theorem 2.155.

Therefore the inhomogeneous linear case $x_{n+1} = ax_n + b$ is completely solved, too. However, we can do even more.

Definition 2.162 *Let I be an admissible ideal of a* DEDEKIND *domain R and let $\varphi(I) = |(R/I)^*|$ be the number of invertible residue classes mod I. A sequence $(x_n)_{n\geq 1}$, $x_n \in R$, is called weakly uniformly distributed mod I (w.u.d. mod I) if x_n is invertible mod I for infinitely many n and*

$$\lim_{N\to\infty} \frac{|\{n \leq N : x_n \equiv x \bmod I\}|}{|\{n \leq N : x_n I \in (R/I)^*\}|} = \frac{1}{\varphi(I)}$$

for every $x \in R$ which is invertible mod I.

Note that every u.d. sequence mod I is w.u.d. mod I, too. Furthermore, recall that if the ideal I has the prime ideal factorization $I = \prod P_j^{r_j}$ $(r_j > 0)$ then

$$\varphi(I) = \prod N(P_j)^{r_j - 1}(N(P_j) - 1).$$

Later we will implicitly use that (x_n) is w.u.d. mod J provided that it is w.u.d. mod I and $J|I$.

Theorem 2.163 *(*TICHY *and* TURNWALD *[1849]) Let I be an admissible ideal of a* DEDEKIND *domain R. Then a sequence $(x_n)_{n\geq 1}$, $x_n \in R$, satisfying $x_{n+1} = ax_n + b$ with $a, b \in R$ is w.u.d. mod I if and only if the following conditions hold:*

(i) *If $P^h|I$ then (x_n) is w.u.d. mod P^h.*

(ii) *If $a \not\equiv 1$ (P) for some $P|I$ with $2 \not\equiv 0$ (P), then $Q^2 \not| I$ and $ab \equiv 0$ (Q) for every $Q|I$ with $N(Q) = 2$.*

(iii) *For $P^h|I$, $P^{h+1} \not| I$, we put $l(P) = p^h$ if $a \equiv 1$ (P), and $l(P) = p^{h-1}(N(P) - 1)$ if $a \not\equiv 1$ (P). Then $l(P_1)$ and $l(P_2)$ are coprime if $P_1, P_2|I$, $P_1 \neq P_2$, and $N(P_i) > 2$ $(i = 1,2)$.*

(iv) *If $ab \equiv 1$ (P) for some $P|I$ with $N(P) = 2$, then $ab \equiv 1$ (Q) for every Q with $Q^2|I$ and $N(Q) = 2$. If $ab \equiv 1$ (P_i) for some $P_i|I$ with $N(P_i) = 2$ $(i = 1,2)$, then $x_1 \equiv 1$ (P_1) implies $x_1 \equiv 1$ (P_2).*

(v) *There is at most one P with $N(P) = 2$ and $P^2|I$.*

It will be convenient to introduce the following notion.

Definition 2.164 *Let I be an admissible ideal of a* DEDEKIND *domain R. A linear recurring sequence $(x_n)_{n\geq 1}$, $x_n \in R$, covers I if every invertible residue class mod I occurs in a period of (x_n) mod I. (Since $N(I) = |R/I|$ is finite any linear recurring sequence (x_n) is periodic mod I.)*

Obviously, every w.u.d. sequence mod I covers I. However, it is not obvious that the converse statement is also true. In fact, we will prove that if (x_n) with $x_{n+1} = ax_n + b$ covers I then (i)–(v) of Theorem 2.163 are satisfied. Conversely if (i)–(v) hold then (x_n) is w.u.d. mod I. The following observation will be an essential step in the proof:

If (x_n) is w.u.d mod J with period l, $x_{n+kl} \equiv u_n + k(x_{n+l} - x_n)$ (JP), and $x_{n+l} \not\equiv x_n$ (JP) for all $n \geq n_0$ then (x_n) is w.u.d. mod JP provided that $N(P) = p$.

In the following lemmas and corollarys we consider always sequences (x_n) with $x_{n+1} = ax_n + b$.

Lemma 2.165 *If $a \equiv 1$ (P) and $N(P) > 2$, then the following conditions are equivalent (and (x_n) has period p mod P):*

(i) *(x_n) covers P.*

(ii) *$N(P) = p$, $b \not\equiv 0$ (P).*

(iii) *(x_n) is w.u.d. mod P.*

Proof. From $x_n \equiv x_1 + (n-1)b$ (P) we conclude that (x_n) has period p mod P. Hence, if (i) holds then $p \geq N(P) - 1$, i.e. $N(P) = p$. If $b \equiv 0$ (P), then (x_n) is constant mod P. Since $N(P) - 1 \geq 2$, we see that (i) implies (ii). If (ii) holds, then every residue occurs precisely once in a period of length p, which implies (iii). As noted above, (iii) implies (i). □

Lemma 2.166 *If $a \not\equiv 1$ (P) and $N(P) > 2$, then the following conditions are equivalent (and (x_n) has period $N(P) - 1$ mod P):*

(i) *(x_n) covers P.*

(ii) *a is a primitive root mod P, $b \equiv 0$ (P), $x_1 \not\equiv 0$ (P).*

(iii) *(x_n) is w.u.d. mod P.*

Proof. From (2.228) we see that (x_n) has period $N(P) - 1$ mod P. If a is not a primitive root, then there are less than $N(P) - 1$ different residues in a period of length $N(P) - 1$. Assume that (x_n) covers P. Then a is a primitive root and $(a-1)x_1 + b \not\equiv 0$ (P). Then for all n we have $(a-1)x_n \not\equiv -b$ (P), which implies $b \equiv 0$ (P) and $x_1 \not\equiv 0$ (P). Thus (i) implies (ii). From this, Lemma 2.166 follows immediately. □

Lemma 2.167 *For $N(P) = 2$ the following conditions are equivalent (and (x_n) has period 2 mod P):*

(i) *(x_n) covers P.*

(ii) *$x_n \equiv 1$ (P) for some $n > 0$.*

(iii) *$b \equiv 1$ (P) or $a \equiv 1$ (P), $b \equiv 0$ (P), $x_1 \equiv 1$ (P).*

(iv) *(x_n) is w.u.d. mod P.*

Proof. Assume that $x_n \equiv 1$ (P) for some $n > 1$ and $b \equiv 0$ (P). Then from $x_n \equiv a^{n-1}x_1$ (P) we conclude $a \equiv 1$ (P) and $x_1 \equiv 1$ (P). Hence (ii) implies (iii). If $a \equiv 0$ (P), then (x_n) has period 1 mod P; if $a \equiv 1$ (P), then $x_{n+2} \equiv x_n$ (P). Thus, in both cases, (x_n) has period 2 mod P. □

Corollary 2.168 *If (x_n) is w.u.d. mod I then (x_n) is w.u.d. mod P for $P|I$.*

Proof. If (x_n) is w.u.d. mod I then it covers I and, consequently, covers P. Thus the assertion follows from the previous lemmas. □

Lemma 2.169 *If (x_n) covers P^2, then $N(P) = p$.*

Proof. Assume $N(P) > p$. Then, by Lemma 2.165 and Lemma 2.166, a is a primitive root mod P. Hence putting $l = N(P) - 1$ we have $\frac{a^l-1}{a-1} \equiv 0$ (P). Since

$$\frac{a^{pl}-1}{a^l-1} = \sum_{j=1}^{p} \binom{p}{j} (a^l - 1)^{j-1} \equiv 0 \ (P),$$

this yields $\frac{a^{pl}-1}{a^l-1} \equiv 0$ (P^2) and by (2.229) we conclude $x_{n+pl} - x_n \equiv 0$ (P^2). Since pl is smaller than the number $N(P)(N(P) - 1)$ of invertible residues mod P^2, we se that (x_n) cannot cover P^2. □

Lemma 2.170 *Assume that $2 \equiv 0$ (P) and that (x_n) is w.u.d. mod P with $b \equiv 0$ (P). Then (x_n) has period 2^{h-1} mod P^h if it covers P^h for some $h \geq 2$ and:*

(i) *If (x_n) covers P^2, then $N(P) = 2$ and $(a-1)x_1 + b \not\equiv 0$ (P^2); if these conditions hold, then (x_n) is w.u.d. mod P^2 and (x_n) has period 1 mod P.*

(ii) *If (x_n) covers P^3, then $a + 1 \not\equiv 0$ (P^2); if (x_n) covers P^2 and $a + 1 \not\equiv 0$ (P^2), then (x_n) is w.u.d. mod P^3.*

(iii) *If (x_n) covers P^4, then $2 \not\equiv 0$ (P^2); if (x_n) covers P^3 and $2 \not\equiv 0$ (P^2), then (x_n) is w.u.d. mod P^h for all h (≥ 1).*

Proof. If (x_n) covers P^2, then $N(P) = 2$ and $a \equiv x_1 \equiv 1$ (P) (hence $x_n \equiv 1$ (P) for all n) by Lemma 2.169 and Lemma 2.167. Then from (2.229) we obtain $x_{n+2} - x_n \equiv 0$ (P^2). Hence $x_1 \not\equiv x_2$ (P^2), i.e. $(a-1)x_1 + b \not\equiv 0$ (P^2). Conversely, if $N(P) = 2$, then $a \equiv x_1 \equiv 1$ (P) by Lemma 2.167. Hence $x_{n+2} \equiv x_n$ (P^2) and $x_1 \not\equiv x_2$ (P^2) if $(a-1)x_1 + b \not\equiv 0$ (P^2). This proves (i). Assume that (x_n) covers P^2. If $a + 1 \equiv 0$ (P^2), then, by (2.229), (x_n) has period 2 mod P^3. Since there are 4 invertible residues mod P^3, (x_n) cannot cover P^3. If, however, $a+1 \not\equiv 0$ (P^2) then (by (2.229) again) $x_{n+2}-x_n \not\equiv 0$ (P^3), since $(a-1)x_1+b \not\equiv 0$ (P^2) by (i). Thus in this case (x_n) is w.u.d. mod P^3 with period 4. This proves (ii). Since $a^3+a^2+a+1 = (a^2+1)(a+1) \equiv 2(a+1)$ (P^3), (2.229) yields $x_{n+4} - x_n \equiv 0$ (P^4) if $a \equiv 1$ (P) and $2 \equiv 0$ (P^2). Thus (x_n) cannot cover P^4 if $2 \equiv 0$ (P^2). Assume that $2 \not\equiv 0$ (P^2) and (x_n) covers P^3. From $a^2 \equiv 1$ (P^2) we see that $a^{2^{h-1}}+1 \equiv 2 \not\equiv 0$ (P^2) for $h \geq 2$; hence $a+1 \not\equiv 0$ (P^2) yields by induction $a^{2^h-1}+\ldots+a+1 = (a^{2^{h-1}}+1)(a^{2^{h-1}-1}+\ldots+a+1) \not\equiv 0$ (P^{h+1}) for all $h \geq 1$. thus (2.229) gives $x_{n+2^h} - x_n \not\equiv 0$ (P^{h+2}); note that $(a-1)x_1+b \not\equiv 0$ (P^2). Hence (x_n) is w.u.d. mod P^{h+2} with period 2^{h+1} if this holds for h replaced by $h-1$. This concludes the proof of Lemma 2.170. □

Lemma 2.171 *Assume that $2 \equiv 0$ (P) and that (x_n) is w.u.d. mod P with $b \not\equiv 0$ (P). Then (x_n) has minimal period 2^h mod P^h if it covers P^h for some $h \geq 2$, and:*

(i) *If (x_n) covers P^2, then $N(P) = 2$ and $a \equiv 1$ (P) and $a+1 \not\equiv 0$ (P^2); if these conditions hold, then (x_n) is w.u.d. mod P^2 and (x_n) has minimal period 2 mod P.*

(ii) *If (x_n) covers P^3, then $2 \not\equiv 0$ (P^2); if (x_n) covers P^2 and $2 \not\equiv 0$ (P^2), then (x_n) is w.u.d. mod P^h for all h (≥ 1).*

Proof. If (x_n) covers P^2, then $N(P) = 2$ holds by Lemma 2.169; from $x_{n+1} - x_n = a^{n-1}((a-1)x_1+b)$ we conclude $a \not\equiv 0$ (P), i.e. $a \equiv 1$ (P). Then $x_{n+1} \equiv x_n + 1$ (P) and only every second x_n is invertible and (x_n) must have period at least 4 mod P^2. Hence $x_{n+2}-x_n = a^{n-1}(a+1)((a-1)x_1+b)$ implies $a+1 \not\equiv 0$ (P^2) if (x_n) covers P^2. Conversely, if $a \equiv 1$ (P) and $a+1 \not\equiv 0$ (P^2), then $x_{n+2}-x_n \not\equiv 0$ (P^2) for all n. Thus (x_n) is w.u.d. mod P^2 with minimal period 4. Assume that (x_n) covers P^2. If $2 \equiv 0$ (P^2) then $a^2 \equiv 1$ (P^2) implies $x_{n+4} - x_n = a^{n-1}(a+1)(a^2+1)((a-1)x_1+b) \equiv 0$ (P^3). Hence (x_n) cannot cover P^3, since only every second x_n is invertible. But if $2 \not\equiv 0$ (P^2) then from $a+1 \not\equiv 0$ (P^2) we deduce $a^{2^h-1}+\ldots+a+1 \not\equiv 0$ (P^{h+1}) for $h \geq 1$ (cf. the proof of Lemma 2.170). Thus (2.229) yields $x_{n+2^h} - x_n \not\equiv 0$ (P^{h+1}) for $h \geq 1$, since $(a-1)x_1+b \not\equiv 0$ (P). Hence (x_n) is w.u.d. mod P^{h+1} with minimal period 2^{h+1} if this holds with h replaced by $h-1$. The assertion follows now by induction. □

Lemma 2.172 *Assume $a \equiv 1$ (P) and $N(P) > 2$. If (x_n) covers P^2, then $N(P) = p$, $b \not\equiv 0$ (P) and $p \not\equiv 0$ (P^2); if these conditions hold, then (x_n) is w.u.d. mod P^h for all h (≥ 1) and (x_n) has period p^h mod P^h.*

Proof. If (x_n) covers P, then $N(P) = p$ and $b \not\equiv 0$ (P), by Lemma 2.165. Hence $p > 2$ and $a^{p-1}+\ldots+a+1 = \sum_{j=1}^{p}\binom{p}{j}(a-1)^{j-1} \equiv p$ (P^2). If $p \equiv 0$ (P^2), then (2.229) implies $x_{n+p} - x_n \equiv 0$ (P^2) and (x_n) cannot cover P^2. Now assume that $N(P) = p$, $b \not\equiv 0$ (P), and $p \not\equiv 0$ (P^2). Then $a^{p-1}+\ldots+a+1 \not\equiv 0$ (P^2) and, by (2.229), $x_{n+p} - x_n \not\equiv 0$ (P^2) for all n. Since (x_n) is a second order linear recurrence with characteristic polynomial $c(x) = (x-a)(x-1)$ and $x^p - 1 \equiv (x-1)^p \equiv 0$ $(c(x), P)$, Lemma 2.158 implies $x_{n+kp^{h+1}} - x_n \equiv kp^h(x_{n+p} - x_n)$ (P^{h+2}) for all $h \geq 0$,

$n > 0$. Thus (x_n) is w.u.d. mod P^{h+2} with period p^{h+2} if this holds with h replaced by $h-1$. Hence, using Lemma 2.165, the assertion follows by induction. □

Lemma 2.173 *Assume $a \not\equiv 1$ (P) and $N(P) > 2$. If (x_n) covers P^2, then $N(P) = p$ and $a^{p-1} \not\equiv 1$ (P^2); if these conditions hold, then (x_n) is w.u.d. mod P^2 with period $p(p-1)$ if it is w.u.d. mod P. If (x_n) covers P^3, then $p \not\equiv 0$ (P^2). If (x_n) covers P^2 and $p \not\equiv 0$ (P^2) then (x_n) is w.u.d. mod P^h with period $p^{h-1}(p-1)$ mod P^h for all $h \geq 1$.*

Proof. Assume that (x_n) covers P. Then, by Lemma 2.166, $a \not\equiv 0$ (P), $b \equiv 0$ (P), and $x_1 \not\equiv 0$ (P). If (x_n) covers P^2, then $N(P) = p$ (by Lemma 2.169) and $a^{p-1} \not\equiv 1$ (P^2), since otherwise $\frac{a^{p-1}-1}{a-1} \equiv 0$ (P^2) yields $x_{n+p-1} - x_n \equiv 0$ (P^2), by (2.229). If, however, $a^{p-1} \not\equiv 1$ (P^2) then (2.229) gives $x_{n+p-1} - x_n \not\equiv 0$ (P^2) for all n. Since (x_n) is a second order linear recurrence with characteristic polynomial $c(x) = (x-a)(x-1)$ and $x^{p-1} - 1 \equiv 0$ $(c(x), P)$, Lemma 2.158 implies $x_{n+kp^h(p-1)} - x_n \equiv kp^h(x_{n+p-1} - x_n)$ (P^{h+2}) for all $h \geq 0$, $n \geq 0$ (note that $p > 2$). If $p \equiv 0$ (P^2), then for $h = 1$ we obtain $x_{n+p(p-1)} - x_n \equiv 0$ (P^3) and (x_n) cannot cover P^3. If, however, $p \not\equiv 0$ (P^2), $N(P) = p$, and $a^{p-1} \not\equiv 1$ (P^2), then (x_n) is w.u.d. mod P^{h+2} with period $p^{h+1}(p-1)$, if this holds with h replaced by $h-1$; for $h = 0$ we need not require $p \not\equiv 0$ (P^2). Using Lemma 2.166, the assertion follows inductively. □

Corollary 2.174 *(x_n) covers P^h if and only if it is w.u.d. mod P^h.*

Proof. This follows immediately from the preceding lemmas. □

Lemma 2.175 *Let (x_n) be an arbitrary sequence in R. Assume that (x_n) is w.u.d. mod J_i with period l_i $(i = 1, 2)$ where J_1, J_2 are coprime non-zero ideals and l_1, l_2 are coprime positive integers. Then (x_n) is w.u.d. mod J_1J_2 with period l_1l_2.*

Proof. Note that l_1l_2 is a period of (x_n) mod $J_1 \cap J_2 = J_1J_2$. Hence we have to show that the number of indices n in a period of length l_1l_2 with $x_n \equiv y$ (J_1J_2) is the same for all y that are invertible mod J_1J_2. This congruence is equivalent to $x_n \equiv y$ (J_i) for $i = 1, 2$, hence this means that n belongs to a certain number (independent of y) of residue classes mod l_i $(i = 1, 2)$. Since l_1, l_2 are coprime, the number of corresponding residue classes mod l_1l_2 is equal to the product of those numbers, hence independent of y. □

Proof of Theorem 2.163. Assume first that (x_n) covers I. Then (i) follows from Corollary 2.174. If $a \not\equiv 1$ (P) and $2 \not\equiv 0$ (P) then, by Lemma 2.166, (x_n) has period $N(P) - 1$ mod P. Hence (x_n) has period $N(P) - 1$ mod PQ if $N(Q) = 2$ (since (x_n) has period 2 mod Q by Lemma 2.167). Consequently, all x_n in a period must be invertible mod PQ if (x_n) covers PQ. But if $ab \not\equiv 0$ (Q) then $x_{n+1} \equiv x_n + 1$ (Q) and only one half of the x_n is invertible; hence $ab \equiv 0$ (Q). Then (2.229) implies $x_{n+2} - x_n \equiv 0$ (Q^2) for $n \geq 2$. Thus (x_n) has period $N(P) - 1$ mod PQ^2 and it cannot cover PQ^2. This proves (ii).

Assume $P_i | I$, $P_1 \neq P_2$, $N(P_i) > 2$ $(i = 1, 2)$, and $d = (l(P_1), l(P_2)) > 1$. Let h_i be the multiplicity of P_i in I. If $a \equiv 1$ (P_i) then from Lemma 2.165 and Lemma 2.172 we see that $p_i > 2$ and (x_n) has period $l(P_i) = \frac{p_i}{p_i - 1}\varphi(P_i^{h_i})$ mod $P_i^{h_i}$; if $a \not\equiv 1$ (P_i) then from Lemma 2.166 and Lemma 2.173 we see that (x_n) has period $l(P_i) = \varphi(P_i^{h_i})$ mod $P_i^{h_i}$. Hence (x_n) has period $\frac{1}{d}l(P_1)l(P_2)$ mod $P_1^{h_1}P_2^{h_2}$, which implies $\frac{1}{d}l(P_1)l(P_2) \geq \varphi(P_1^{h_1})\varphi(P_2^{h_2})$. Now note that $l(P_i) \leq \frac{5}{4}\varphi(P_i^{h_i})$ if $p_i \neq 3$ and $l(P_i) \leq \frac{3}{2}\varphi(P_i^{h_i})$ if $p_i = 3$. Since $\frac{1}{2} \cdot \frac{3}{2} \cdot \frac{5}{4} < 1$ and $\frac{1}{3} \cdot \frac{3}{2} \cdot \frac{3}{2} < 1$, we must have $d = 2, p_1 = p_2 = 3$. But then from the definition of $l(P_i)$ we immediately obtain $a \not\equiv 1$ (P_i), in which case $l(P_i) = \varphi(P_i^{h_i})$ so that the above inequality fails again. This contradiction proves (iii).

Assume $ab \equiv 1$ (P) for some $P|I$ with $N(P) = 2$, and $ab \equiv 0$ (Q) for some Q with $Q^2|I$ and $N(Q) = 2$. Since (2.229) implies $x_{n+2} - x_n \equiv 0$ (Q^2) for $n \geq 2$, (x_n) has period 2 mod PQ^2. Hence every x_n in a period must be invertible mod PQ^2. But only one half of the x_n is invertible mod P, since $x_{n+1} \equiv x_n + 1$ (P). This contradiction proves the first part of (iv). Assume $ab \equiv 1$ (P_i), $N(P_i) = 2$, $x_1 \equiv 1$ (P_1), and $x_1 \equiv 0$ (P_2). Then from $x_{n+1} \equiv x_n + 1$ (P_i) we see that no x_n is invertible mod P_1P_2. This completes the proof of (iv).

Let $P_i^2|I$ and $N(P_i) = 2$ $(i = 1, 2)$; $P_1 \neq P_2$. Then (x_n) has period 4 mod P_i^2 by Lemma 2.170 and Lemma 2.171. Hence (x_n) has period 4 mod $P_1^2P_2^2$, which implies that all x_n in a period are invertible mod P_i $(i = 1, 2)$. Hence $ab \equiv 0$ (P_i) and (by (2.229)) (x_n) has period 2 mod P_i^2. But then (x_n) has period 2 mod $P_1^2P_2^2$, which is impossible. Thus (v) holds.

Now, in order to prove the second part of the Theorem, we assume that (i), (ii), (iii), (iv), (v) hold. We shall use the following obvious observation.

Suppose (x_n) to be w.u.d. mod P, $N(P) = 2$, and $ab \equiv 0$ (P). Then (x_n) is w.u.d. mod JP, if (x_n) is w.u.d. mod J and $P \nmid J$. (Note that the assumption implies $x_n \equiv 1$ (P) for $n \geq 1$.)

From Lemmas 2.165, 2.166, 2.172, 2.173 we see that (i) and (iii) imply that (x_n) has the coprime periods $l(P_1)$, $l(P_2)$ mod $P_1^{h_1}$, $P_2^{h_2}$, if $P_1 \neq P_2$, $P_i^{h_i}|I$, and $N(P_i) > 2$ $(i = 1, 2)$. Hence, by (i) and Lemma 2.175, (x_n) is w.u.d. modulo the the product of the divisors $P_i^{h_i}$ with $N(P_i) > 2$. Hence, by (ii) and the above observation, (x_n) is w.u.d. mod I if $a \not\equiv 1$ (P) for some $P|I$ with $2 \not\equiv 0$ (P). Now assume $a \equiv 1$ (P) for all $P|I$ with $2 \not\equiv 0\,(P)$. Then, by (iii) and Lemma 2.175, it is sufficient to prove that (x_n) is w.u.d. modulo the product of the divisors of I that are powers of primes of norm 2, since (by Lemmas 2.167, 2.170, 2.171) the period of (x_n) modulo powers of these primes is a power of 2. Note that $l(P)$ is odd if $p \neq 2$, $a \equiv 1$ (P) or if $p = 2$, $N(P) > 2$ (since then Lemma 2.165 and Lemma 2.169 imply $l(P) = N(P) - 1$). Assume first that $ab \equiv 0$ (P) for all $P|I$ with $N(P) = 2$. Then from (i), (v), and the above observation it is clear that (x_n) is w.u.d. modulo the product of the corresponding prime powers. So it remains to assume $ab \equiv 1$ (P) for some $P|I$ with $N(P) = 2$. Now note that the above observation also holds if $ab \equiv 1$ (P), provided that x_n is invertible mod P if it is invertible mod J. Hence our assertion follows from (iv) and (v). (Note that the second part of (iv) means that x_n is invertible mod P_1 if and only if it is invertible mod P_2.) Thus the proof of the Theorem is complete. □

Corollary 2.176 *Let $a, b, x_1, m \in \mathbf{Z}$, $m > 1$. Then a sequence $(x_n)_{n\geq 1}$ satisfying $x_{n+1} = ax_n + b$ with $a, b \in \mathbf{Z}$ is w.u.d. mod m if and only if the following conditions* *hold: (p, q prime)*

(i') *If $p|m$, $p > 2$, and $a \equiv 1$ (p), then $b \not\equiv 0$ (p).*
If $p|m$, $p > 2$, and $a \not\equiv 1$ (p), then a is a primitive root mod p, $b \equiv 0$ (p) and $x_1 \not\equiv 0$ (p); if $p^2|m$, then $a^{p-1} \not\equiv 1$ (p^2).
If $2|m$, then $b \equiv 1$ (2) or $a \equiv 1$ (2), $b \equiv 0$ (2), $x_1 \equiv 1$ (2); if $4|m$ and $b \equiv 1$ (2), then $a \equiv 1$ (4); if $4|m$ and $b \equiv 0$ (2), then $(a-1)x_1 + b \not\equiv 0$ (4); if $8|m$ and $b \equiv 0$ (2), then $a \equiv 1$ (4).

(ii') *If $a \not\equiv 1$ (p) for some $p|m$ with $p > 2$, then $4 \nmid m$ and $2|m$ implies $ab \equiv 0$ (2).*

(iii') *There is at most one $p|m$ with $p > 2$ and $a \not\equiv 1$ (p). If it exists, then $p \not\equiv 1$ (q) for all $q|m$, $q > 2$.*

Proof. The sequence (x_n) is w.u.d. mod I if and only if conditions (i), (ii), (iii), (iv), (v) of 2.163 hold. (iv) and (v) are satisfied trivially in the case $R = \mathbf{Z}$. From Lemmas 2.165, 2.166, 2.167, 2.170, 2.171, 2.172, 2.173 it follows, that in this case (i) and (i') are equivalent; (ii') and (iii') are obvious reformulations of (ii) and (iii), respectively. □

Notes

Uniformly distributed sequences in locally compact groups (in the sense of Rubel) were studied by Benzinger [167, 168]; see also Veech [1903, 1904]. Rindler [1541, 1543] studies the class $C(G)$ of all sequences (c_n) (in a compact metrizable group G) such that $(c_n x_n)$ is u.d. for all u.d. sequences (x_n); also well distributed sequences are considered. In the case of the torus group the class $C(G)$ ("almost constant sequences" was characterized by Rauzy [1501, 1503, 1506]. Rindler [1555] gave a new proof and an extension to compact metrizable groups; for quantitative aspects and generalizations (in the classical case) to weighted means see Tichy and Turnwald [1847]; see also Coquet [407]. Almost constant sequences in locally compact abelian groups were considered by Rindler [1553] and Losert and Rindler [1107]; see also Gröchenig [700]. Rindler [1542] considered L^1-u.d. sequences on locally compact separable groups; for amenable groups the existence of such sequences is shown, in the case of $G = \mathbf{R}$ it is proved that the sequences (n^a), $1/2 \leq a \leq 1$ are $L^1(\mathbf{R})$-u.d.. Of course, for compact groups L^1-u.d. is equivalent to the usual concept of u.d.; see also Rindler [1549, 1547]. A generalization to sequences of transformations is due to Losert [1097]. Rindler [1545] extended a result of Schoißengeier [1646, 1647] on well-distribution in locally compact groups and considers also u.d. with respect to infinite dimensional representations. Rindler [1552] investigated u.d. in quotient groups, and in [1548] the concept of almost u.d. in locally compact abelian groups. Schmitt [1641, 1642, 1643] and Rindler[1544, 1556] studied u.d. sequences in linear spaces ("linear equidistribution"); in $\mathbf{R}^k$ this concept coincides with the relative equidistribution in the sense of Gerl [651, 652]; see also Gerl [655]. Rindler [1551] gave a short proof of the existence of u.d. sequences in separable locally compact groups. Gröchenig, Losert and Rindler [702, 703] studied relations between Hartman and unitary u.d. in solvable groups. Hlawka [801] introduced u.d. chains; further results are due to Rindler [1546]. Schneider [1644] computed corresponding homology groups and Tichy [1810, 1814] proved discrepancy bounds. Rindler and Schoißengeier [1557] gave a characterization of differentiable functions using u.d. sequences. Further results on u.d. in locally compact groups are due to Hansel and Troallic [732]. By the same authors in [731] the construction of Baayen and Hedrlin [58] for well distributed sequences in compact spaces is extended to the locally compact case. Special results on u.d. in compact abelian groups and on u.d. sequences with repetitions are due to Topuzoglu [1870, 1869]. For further results on u.d. in locally compact groups we refer to Rindler [1550], Maxones, Muthsam and Rindler [1154], Maxones and Rindler [1156, 1157, 1158]. Mauclaire [1143, 1142, 1144] investigated the distribution of additive arithmetic functions and of generalized

Rudin-Shapiro sequences with values in compact abelian groups. For more details on Rudin-Shapiro sequences we refer to the Notes of Section 3.1.

Dupain and Lesca [506] and Dupain [500] studied density properties of subsequences of u.d. sequences; see also Thomas and Dupain [509]. Losert and Rindler [1106] continued these investigations (from a general point of view) and obtained very satisfactory results, which are based on the concept of "essential index sequences", i.e. sequences of positive integers such that the corresponding subsequences of an arbitrary u.d. sequence is again u.d.. A characterization of essential index sequences is due to Rauzy [1509]. Gröchenig [701] considered essential index sequences for well distribution. Losert and Tichy [1108] proved that almost all subsequences of a u.d. multi-sequence $(x_{n_1,\dots,n_s})$ are again u.d., provided that the summation is performed in such a way that all indices tend to ∞ simultaneously. For "independent summation" counterexamples are given, also weighted means are considered; see also Losert [1102] and Tichy [1809, 1812, 1816].

Losert and Rindler [1106] considered well-distributed sequences on topological semigroups (using representations). Answering a question of Veech [1906] they constructed a sequence (r_n) of positive integers having the following property: let (a_n) be any sequence generating a dense subsemigroup of an arbitrary topological semigroup G, then the sequence $a_{r_1}, a_{r_1}a_{r_2}, a_{r_1}a_{r_2}a_{r_3}, \dots, a_{r_1}\cdots a_{r_n}, \dots$ is well distributed. For relations between topological dynamics and u.d. of sequences we refer to Veech [1906, 1907]. Hewitt and Katznelson [775] discussed several generalizations of u.d.. Saint-André [1590] gave a a survey on u.d. and harmonic analysis. Sun [1763] gave a probabilistic extension of the well-known result that almost no sequence is well distributed.

Uniform distribution of sequences of positive real numbers which are given as additive semigroup generated by an increasing sequence was investigated by Borel [245, 247, 248, 250] and Losert [1101, 1103]; see also Müller [1219]. In a series of papers Losert [1099, 1098, 1100] investigated uniform distribution in compact, separable, non-metrizable spaces and groups. The existence of u.d. (and even well-distributed) sequences is proved in dyadic spaces, in particular for compact groups. On the other hand, several well-known properties of u.d. sequences in metrizable spaces cannot be extended to the non-metrizable case, e.g. the rearrangement theorem of von Neumann, which says that any dense sequence can be rearranged into a u.d. sequence (c.f. Kuipers and Niederreiter [983]). In [1100] it is shown (by examples) that the class of all u.d. sequences can be empty, a non-empty null set, non-measurable (of interior measure 0 and exterior measure 1) or a set of measure 1. Furthermore, the existence of a u.d. sequence, in general, does not imply the existence of a well-distributed sequence. Mercourakis [1188] continued the investigations of Losert on the existence of well-distributed sequences.

A Borel set E of the unit circle is called a W-set if there exists an increasing sequence (n_k) of positive integers such that for every $x \in E$ the sequence $(n_k x)$ has an asymptotic distribution different from the uniform distribution. Lyons [1114] proved that for a finite complex Borel measure μ on the unit circle the Fourier coefficients $\hat{\mu}(n)$ tend to 0 (as $|n| \to \infty$), if and only if every W-set is a μ-null set; see also Lyons [1115]. Blümlinger [231, 235] could solve this characterization problem in the general case of compact and even locally compact abelian groups.

Gillard [657] established a nice result on p-adic extensions and Iwasawa theory; he made essentially use of p-adic uniform distribution. For u.d. in p-adic numbers we refer to Beer [154] and Decomps-Guilloux [445, 444]; for a p-adic analogon of Koksma-Hlawka's theroem see Bertrandias [211]. Anashin [37] studied u.d. sequences of p-adic integers. Grandet-Hugot [688] investigated u.d. in the group of adeles of $\mathbb{Q}$, see also [690, 692, 689, 691]. In [693, 694] several notions of u.d. in locally compact groups were discussed; see also Schreiber [1661]. Vaaler [1886] established p-adic analogues of metric results of the Gál-Koksma type. Equidistribution of rational functions mod p was investigated by Odoni and Spain [1389].

Horbowicz [845] established a new criterion for u.d. of sequences modulo 1. Tichy [1831] extended it to compact spaces; see also Salát [1594]. Hwang [858] proved the following result: Assume either that every $d_j - d_1$ is rational, or that the set of minimum and maximum points of f on the open interval $(0,1)$ is finite. Then there are rationals r and r' such that $f(r) \le f(r + d_j)$ and $f(r') \ge f(r' + d_j)$ for all j; see also Tichy [1833] for a generalizations to compact spaces. Fleischer [610, 611] introduced the concept of discrepancy operator on compact spaces; for more recent results on such operators and the connections to the classical discrepancy see Amstler [36]. For ergodic properties of averaging operators and u.d. on compact groups we refer to Rosenblatt [1561]. Laszló and Salát [1045] proved that the set of all u.d. sequences of real numbers is a dense $F_{\sigma\delta}$-set of first category in the space of all sequences with Frechet metric. Connections between u.d. and concepts of differential geometry can be found in Ermine [572, 573]. General rearrangement theorems are due to Niederreiter [1286, 1305]. Larcher [1019] proved the following theorem for compact metric spaces without isolated points: For any dense sequences (x_n) , (y_n) in the metric space and any sequence (ε_n) of positive numbers, there exist permutations σ, τ of the set of positive integers such that the distance of $x_{\sigma(n)}$ and $y_{\tau(n)}$ is $\leq \varepsilon_n$ for all n. Talayan [1779] investigated (using the continuum hypothesis) rearrangement properties of sequences of functions $f_n(t)$ defined on an interval J.

The study of transformations of uniformly distributed sequences was initiated in two papers by Hlawka and Mück [839, 840]. More recently several authors considered the problem which functions f map arbitrary uniformly distributed sequences on such sequences, see Bosch [269], Porubsky, Šalát and Strauch [1452], Porubsky and Strauch [1454], and Tichy and Winkler [1851]. For related topological results we refer to Schmeling and Winkler [1621], and for a very recent contribution to Winkler [1967]. Binder [220, 221] gave a characterization of Riemann integrable functions on compact metric spaces via u.d. sequences extending a well-known result of De Bruijn and Post [436]. For related results concerning partitions we refer to Chersi and Volcic [348]. Transformation of sequences mod 1 by matrices are considered by Myerson and Pollington [1230]. In [792] Hlawka presents a metric analogon of a theorem by Veech [1905]. Uniform distribution on compact Riemannian manifolds was systematically investigated by Blümlinger [233]; for a special example see Davis [435]. Bauer [111] developed a discrepancy theory in separable metric spaces with respect to a given measure μ. The discrepancy of a sequence of probability measures μ_n is defined as the Prohorov metric $\rho((1/N)\sum_{n=1}^{N}\mu_n, \mu)$. In the case of the k-dimensional torus this notion of discrepancy for a sequence of point measures $\mu_n = \delta_{x_n}$ coincides with the classical discrepancy $D_N(x_n)$. Mück and Philipp [1214] proved inequalities relating different kinds of discrepancy concepts, such as isotropic discrepancy, Prohorov metric, etc.

Obata [1386, 1385] and Blümlinger and Obata [240] investigated permutations preserving arithmetic means and u.d. of sequences; for related investigations see Rindler [1554].

In a series of papers [116, 115, 117, 118, 121, 127, 131, 132, 135, 140] Beck investigated combinatorial discrepancy problems, especially 2-coloring problems; see also Beck and Fiala [152]. For instance, in [135] he proved an upper bound for balanced 2-coloring in the cube. Further results in combinatorial discrepancy theory can be found in Beck and Spencer [151] and Wagner [1922]. For a survey see Beck and Sós [150]

Matousek and Spencer [1136] proved that there exists a two-coloring of the first n integers for which all arithmetic progressions have discrepancy $O(n^{1/4})$, improving an earlier result of Beck [117]. Conversly, Roth [1568] has shown the existence of a constant C such that, for every two-coloring, the discrepancy is greater than $Cn^{1/4}$. Thus Roth's lower bound is optimal. However, the proof of [1136] is not constructive. Matousek, Welzl and Wernisch [1137] considered the discrepancy of a two-coloring of a finite set with respect to a given set system. They proveed upper bounds which imply some of Beck's bounds for geometric discrepancies (see Sections 1.3 and 2.1). Matousek [1135] obtained precise bounds for the discrepancy with respect to half-spaces. Spencer [1729] established an upper bound for a certain "probabilistic" discrepancy.

Razborov, Szemeredi and Wigderson [1522] constructed for any positive integer N a subset of residues mod N of size $(\log N)^{O(1)}$, which is nearly uniformly distributed in every arithmetic progression mod N. Alon and Mansour [32] consider special "ε-discrepancy sets" in the space of n-dimensional integral vectors mod p. Alon and Peres [33] established quantitative versions of the fact that every sufficiently large set X in [0,1] has a dilation nX mod 1 with small maximal gap and even small discrepancy. The proof is based on a second-moment argument which reduces the problem to an estimate on the number of edges in a certain graph; see also Berend and Peres [182].

Kirschenhofer and Tichy [925] investigated double sequences over finite sets (and more general on compact spaces). They introduced a concept of uniform distribution which is based on counting submatrices. This work was continued in [926, 927, 929, 928], where especially the results of Niederreiter [1273], Meijer [1161] and Tijdeman [1853] on u.d. in finite sets were extended to the k-block discrepancy. Niederreiter [1288] applied the combinatorial discrepancy results of Meijer and Tijdeman to a distribution problem with respect to convex sets. Tijdeman [1855] established an algorithm for finding a sequence with optimal discrepancy. Winkler [1962] proved an inequality involving the "finite set" discrepancy and the usual discrepancy of a corresponding digital sequence. Flajolet, Kirschenhofer and Tichy [603, 604] studied $k(N)$-uniformly distributed 0, 1-sequences. They obtained essentially best possible metric results which were extended to general discrete spaces by Grabner [678]. Goldstern [662] provided a construction of such optimal distributed sequences. Drmota and Winkler [492] extended these metric results to $k(N)$-u.d. sequences modulo 1. The results are somehow weaker than in the discrete case; also a construction of such sequences is provided.

Arnoux and Rauzy [53] investigated sequences over some finite alphabet by counting the number $P(n)$ of subwords of length n of the given sequence ("subword complexity"). They constructed a sequence of complexity $2n+1$. Rote [1564] found a sequence of complexity $P(n) = 2n$. If $P(n) \leq n$ for some n, then the word is ultimately periodic, and $P(n)$ is in fact bounded. The lowest possible complexity for an interesting infinite word is thus $P(n) = n+1$. Sequences with complexity $P(n) \leq n+1$ are called Sturmian sequences, see Coven and Hedlund [421], Rauzy [1517, 1518] and the recent contributions by Mignosi and Seebold [1191], Berstel and Seebold [195] and Tijdeman [1858] for a characterization of complementary triples of Sturmian bisequences. Tapsoba [1780] described automata calculating the complexity of automatic sequences. A recent contribution on the complexity of sequences is due to Arnoux and Mauduit [50]. As a general reference on automatic sequences and their arithmetic description we mention here the fundamental paper of Christol, Kamae, Mendès France and Rauzy [364]; see also Mendès France [1176]. For connections with the so called folding sequences we refer to Dekking, Mendès France and van der Poorten [448]; see also Allouche and Bousquet-Mélou [22], Allouche [18], Allouche and Bacher [21] and Mendès France [1183, 1185]. A recent contribution on folded continued fractions is due to Allouche, Lubiw, Mendès France, van der Poorten and Shallit [26]. An important example is the Thue-Morse sequence 0110100110010110 ... generated by the substitutions $0 \to 01$ and $1 \to 10$ starting with 0; for other definitions and porperties of this sequence we refer to Mendès France [1175]. A survey on connections between number theory dynamical systems and substitutions is due to Rauzy [1520]. Allouche and Mendès France [27] considered applications of the Rudin-Shapiro sequence to the Ising-model; see also Mendès France [1179, 1184]; for automata related to the Ising-model we refer to Mendès France [1186]. Note that the Rudin-Shapiro sequence is defined as $(-1)^{s(n)}$, where $s(n)$ counts the number of 11-blocks in the binary expansion of n. Allouche and Liardet [25] considerd generalizations and studied several interesting properties concerned with correlation functions and spectral measure. Allouche and Shallit [31] consider the complexity of generalized Rudin-Shapiro sequences. For further results on generalized Rudin-Shapiro sequences and extensions of the Thue-Morse sequence as well as on automatic and regular sequences (in general) we refer to Allouche et al. [20], Allouche and Shallit [30], Allouche [16, 17, 19], Allouche and Bousquet-Mélou [23] and Allouche, Morton and Shallit [28], Borel and Laubie [266] and Mendès France [1181]. For connections to random walks we mention Dekking [446]. For connections between ultimately periodic sequences and recognizibility we refer to Bruyère et al. [307, 306]. Recently, also various kinds of discrete billard sequences have been studied from the point of view of complexity, see for instance Shiokawa and Tamura [1673] (for billards in the cube), Arnoux, Mauduit, Shiokawa and Tamura [52, 51], and the thesis of Hubert [855].

Kiss [931] studied the distribution of $[R_{n+1}/R_n]$ mod m, where R_n is a linear recurring sequence. Kopetzky [951] considered the sequence $([\alpha n], [\beta n])$ in $\mathbf{Z}^2$. A discrepancy bound for the sequence $([n^c])$, $0 < c < 1$, is due to Goutziers [672].

In a series of papers Kuipers and Shiue investigated special sequences in integers and in more general rings. In [988] it is proved that the sequence of Lucas numbers is not u.d. mod m for any integer $m \geq 2$. For other results in this direction we refer to [985, 972, 986, 987], Velez [1908], and Erlebach and Velez [571]. Niederreiter [1272] proved that the sequence of Fibonacci numbers is u.d. mod m if m is a power of 5; see also Bundschuh [310]. Niederreiter [1299] extended a theorem of Rieger [1540] on the distribution mod m of linear recurring sequences; see also Rieger [1540] and for more simple proofs Turnwald [1881]. Niederreiter and Shiue [1355, 1358] characterized the u.d. of linear recurring sequences up to order 4 over a finite field. Conditions for u.d. mod m of second order linear recurring sequences in integers were established by Bundschuh [311], Bundschuh and Shiue [313], Shiue and Hu [1678], Shiue [1676], and Webb and Long [1950]. A complete characterization of u.d. mod m of such sequences is due to Bumby [309]; see also Nathanson [1265]. A somewhat different approach, an extension to second order linear recurring sequences in Dedekind domains and some general results for recurrences of higher order were given by Turnwald [1880]. Conditions for the u.d. of third order linear recurring sequences in integers are due to Knight and Webb [939]. A complete characterization over Dedekind domains (including various results for higher order recurrences) can be found in Turnwald [1879]; Tichy and Turnwald [1845] quoted the result for recurrences of order 3 in rational integers. Weak uniform distribution of special second order linear recurring sequences was investigated by Tichy and Turnwald [1849] (for inhomogeneous recurrences of first order see Theorem 2.163) and Turnwald [1882] (where a complete solution for the Fibonacci sequence is given). For further contributions on u.d. of recurring sequences in integers we refer to Cavior [329, 330], Jacobson and Velez [868], Nagasaka [1235, 1236], Nagasaka and Ando [1237], Nagasaka and Shiue [1242], Tichy [1835], Turnwald [1881], and Webb [1951]. In a series of papers Somer [1718, 1719, 1720, 1721, 1722, 1723] investigated the structure of the periods of linear recurring sequences in integers mod m; see also Banks and Somer [103], Niederreiter, Schinzel and Somer [1353] and Carroll, Jacobson and Somer [327]. Shparlinskij [1680] studied the distribution of blocks in integer valued recurrences.

Niederreiter and Lo [1348] systematically investigated u.d. sequences in algebraic integers; they extended the classical concept of Banach-Buck density to algebraic integers. For further contributions to u.d. and densities we refer to Pastéka [1409, 1410, 1411, 1412, 1414, 1413]. For various results on density measures and related topics see Corwin and Pfeffer [415], Pastéka and Salát [1417], Pastéka [1415], Salát and Tijdeman [1596, 1597], Grekos [696], Grekos and Volkmann [697], and Powell and Salát [1458]. Niederreiter and Shiue [1357, 1356] studied u.d. sequences in rings of integral matrices and weakly u.d. sequences in finite fields; for u.d. in finite fields see also Myerson [1224]. A detailed survey on u.d. of integer sequences is due to Narkiewicz [1260]; see also Narkiewicz [1259, 1258, 1261] and Narkiewicz and Rayner [1263]. For results on permutation polynomials we refer to the book by Lidl, Mullen and Turnwald [1090]. Radoux [1495] studied the distribution mod p (prime) of the sequence of Bell numbers; see also Wagstaff [1934]. Shparlinskij [1686] considered the distribution of values of recurring sequences and the Bell numbers in finite fields. For results on the distribution of fractional parts of recurrent sequences see Shparlinskij [1684]. A general property of weakly u.d. sequences mod m was shown by Uchiyama [1883]. Deshouillers, Haji-Diab [452] investigated u.d. of polynomial sequences $[p(n)]$ mod m. General distribution properties of sequences of integers and relations to Novoselov's polyadic numbers are due to Pastéka and Porubsky [1416]; see also Pastéka and Tichy [1418]. A collection of some results concerning u.d. of integer sequences is due to Werbinski [1952].

Sequences of polynomials over finite fields were considered in Kuipers [973, 974, 975]. An application of the notion of independence to u.d. sequences mod m is given in Kuipers and Shiue [990]; see also Burke [317], Kuipers [977] and Kuipers and Niederreiter [982]. A detailed study of (statistical) independence of sequences in compact spaces is due to Niederreiter [1283], where the author also extends a theorem of Dupain [500] on subsequences; see also Dupain and Lesca [506]. For statistical independence we refer to the Notes of Section 1.6; and to Rauzy [1506].

Houndonougho [850] studied the sequence (θ^n) in formal power series; see also Long and Webb [1094] and Webb [1948]. Rhin [1533] provided discrepancy bounds for sequences of formal power series over finite fields; see also Deshouillers [454, 455] and Allouche and Deshouillers [24]. A recent contribution to the u.d. in formal power series over finite fields is due to Car [324].

Uniform distribution in the ring of Gaussian integers is discussed by various authors such as Kuipers, Niederreiter, Shiue; see [984, 992] and Burke and Kuipers [318]. This work was continued in the case of quaternions in [993]. Meijer and Shiue [1163] considered special sequences in $\mathbf{Z}_{g_1}\times\cdots\times\mathbf{Z}_{g_k}$ with coprime g_j. Kuipers and Shiue [991] established an elementary criterion for u.d. mod m of rational integers. Kuipers [976, 978] and Kuipers and Shiue [989] investigated the u.d. of the logarithms of a linear recurring sequences. Thus Benford's law is established for a wide class of linear recurring sequences; see also Kanemitsu, Nagaska, Rauzy, and Shiue [894], Katz and Cohen [913], Jager and Liardet [870]. (More on Benford's law can be found in the Notes of Section 2.2.) Recently, Tichy [1808] extended these investigations by providing discrepancy bounds. A discrepancy problem with applications to linear recurrences was discussed by Kiss and Tichy [935].

A famous problem on sequences in discrete spaces is the so called $3n+1$-problem, or Syracuse Problem. Consider the following operator on the set of integers:

$$T(n) := \begin{cases} \frac{1}{2}n & \text{if } n \text{ is even,} \\ \frac{1}{2}(3n+1) & \text{if } n \text{ is odd} \end{cases}$$

Now choose a starting number $x \in \mathbf{N}$, and look at its $3n+1$ trajectory $\{T^k(x) : k \geq 0\}$, where $T^k = T \circ \ldots \circ T$ denotes the k-fold iterate of T for $k \geq 1$, and $T^0(x) = x$. The famous and unsolved $3n+1$ conjecture says that any $3n+1$ trajectory eventually hits 1, for any starting number $x \in \mathbf{N}$.

For an extensive survey on the literature on this conjecture we refer to Lagarias [1004] and Müller [1220, 1221]; recent contributions to this problem are due to Applegate and Lagarias [45, 46], Clark [370], Eliahou [556], Korec [953, 954], Sander [1599] and Wirsching [1968]. For extensions of the $3n+1$-problem see Franco and Pomerance [619] and Mignosi [1190].

Chapter 3

Applications

3.1 Numerical Integration, Approximation, and Mathematical Finance

3.1.1 Integration and Approximation of Continuous Functions

Numerical integration was an early application of uniformly distributed sequences. Independently, KOROBOV [957] and HLAWKA [787, 788] developed the method of good lattice points (or optimal coefficients). Afterwards various authors continued these investigations and applied this approach to different problems such as integral equations, global optimization, approximations etc. The basic idea of crude Monte Carlo integration, say of functions f on $I^k = [0,1]^k$, is to choose N integration points $\mathbf{x}_1, \ldots \mathbf{x}_N$ randomly in I^k and to approximate the integral $I(f) = \int_{I^k} f(\mathbf{x})d\mathbf{x}$ by the arithmetic mean $I_N(f) = \frac{1}{N}\sum_{n=1}^{N} f(\mathbf{x}_n)$. In practice one chooses deterministically constructed Quasi-Monte Carlo points $\mathbf{x}_n$ instead of "actual" random points. A good choice for the points is the initial segment of a sequence with small discrepancy, since by the KOKSMA-HLAWKA inequality

$$|I_N(f) - I(f)| \leq V(f) D_N(\mathbf{x}_n), \tag{3.1}$$

where $V(f)$ is the total variation of f (in the sense of HARDY and KRAUSE, see Section 1.1).

Various kinds of such sequences are known. Most of them are based on the one-dimensional VAN DER CORPUT sequence. We recall its definition (see also Section 1.3):

Let $b \geq 2$ be an integer and

$$n = \sum_{j=0}^{\infty} d_j(n)\, b^j, \qquad d_j \in \{0, 1, \ldots, b-1\}$$

the digit expansion of the integer $n \geq 1$ in base $b \geq 2$. Then the VAN DER CORPUT sequence $(\gamma_b(n)_{n\geq 1})$ in base b is defined by

$$\gamma_b(n) = \sum_{j=0}^{\infty} d_j(n)\, b^{-j-1}.$$

Definition 3.1 (HALTON **sequence**, [729]) *For given dimension k the k-dimensional* HALTON *sequence $(\mathbf{x}_n)$ in I^k is defined by $\mathbf{x}_n = (\gamma_{b_1}(n), \dots, \gamma_{b_k}(n))$, where $b_1, \dots, b_k$ are given coprime positive integers.*

Remark. As it was shown in [729], the HALTON **sequence** has a discrepancy of order $\mathcal{O}((\log N)^k/N)$.

Definition 3.2 (HAMMERSLEY **sequence**, [730]) *For given k, N the k-dimensional* HAMMERSLEY *sequence $(\mathbf{x}_n)$ of size N is defined by $\mathbf{x}_n = (\gamma_{b_1}(n), \dots, \gamma_{b_{k-1}}(n), \frac{n}{N})$, where $b_1, \dots, b_{k-1}$ are given coprime positive integers.*

Remark. As it was shown in [730], the HAMMERSLEY **sequence** has a discrepancy of order $\mathcal{O}((\log N)^{k-1}/N)$.

Note that the HAMMERSLEY sequence is a point set of size N which cannot be extended to an infinite sequence. In the following we will discuss briefly good lattice point sequences and later net sequences. The net sequences are very suitable for various applications; we will just mention the most recent ones, for instance some specific problems in mathematical finance.

Remark. Usually, sequences the discrepancy of which is bounded from above by $\mathcal{O}((\log N)^k/N)$, are called **low discrepancy sequences**. It is a well known conjecture that this is the optimal order of magnitude for infinite sequences in the k-dimensional unit cube. This conjecture is open for $k \geq 2$. (See Section 1.3.)

We will also describe recent algorithms for computing the discrepancy. We will focus on the L^2-discrepancy, since it is the natural quantity measuring the average-case complexity of numerical integration.

A very important class of integration points are given by good lattice points (g.l.p.). A g.l.p. is a point $\mathbf{g} \in \mathbf{Z}^k$ such that the point sequence $\mathbf{x}_n = \{\frac{n}{N}\mathbf{g}\}$, $n = 0, \dots, N-1$ has small discrepancy. One can show the existence of points $\mathbf{g} \in \mathbf{Z}$ such that $N D_N(\mathbf{x}_n) = \mathcal{O}\left((\log N)^k\right)$; i.e. this g.l.p. sequence is a low discrepancy sequence. In Section 3.3 we will apply g.l.p. sequences for the approximate solution of certain partial differential equations.

The method of good lattice points

In Section 1.4 we discussed the uniform distribution of the fractional parts $\mathbf{x}_n = \{n\underline{\alpha}\} \in I^k$, $n = 0, 1, \dots$, where $\underline{\alpha} = (\alpha_1, \dots, \alpha_k) \in \mathbf{R}^k$ is such that $1, \alpha_1, \dots, \alpha_k$ are linearly independent over the rationals. Discrete versions of these sequences are obtained if we consider points α with rational coordinates. If $\underline{\alpha} \in \mathbf{Q}^k$ is such a point and the positive integer N is a common denominator of its coordinates, then $\underline{\alpha} = N^{-1}\mathbf{g}$ with $\mathbf{g} \in \mathbf{Z}^k$. We are thus led to the point set

$$\mathbf{x}_n = \left\{\frac{n}{N}\mathbf{g}\right\} \in I^k \qquad \text{for } n = 0, 1, \ldots, N-1. \tag{3.2}$$

Note that it is not necessary to consider the points $\mathbf{x}_n$ for $n \geq N$, since they just replicate the points in (3.2). With the points in (3.2), we get the Quasi-Monte Carlo approximation

$$\int_{I^k} f(\mathbf{u})d\mathbf{u} \approx \frac{1}{N}\sum_{n=0}^{N-1} f\left(\left\{\frac{n}{N}\mathbf{g}\right\}\right). \tag{3.3}$$

This approximation is particularly suited for periodic integrands. Let f be a periodic function of $\mathbf{R}^k$ with period interval I^k (or, equivalently, with period 1 in each of its k variables). Then, first of all, we can drop the fractional parts in (3.3) to get the simpler form

$$\int_{I^k} f(\mathbf{u})d\mathbf{u} \approx \frac{1}{N}\sum_{n=0}^{N-1} f\left(\frac{n}{N}\mathbf{g}\right). \tag{3.4}$$

Furthermore, suppose that f is represented by the absolutely convergent FOURIER series

$$f(\mathbf{u}) = \sum_{\mathbf{h}\in\mathbf{Z}^k} \hat{f}(\mathbf{h})\, e(\mathbf{h}\cdot\mathbf{u}) \qquad \text{for } \mathbf{u} \in \mathbf{R}^k$$

with FOURIER coefficients

$$\hat{f}(\mathbf{h}) = \int_{I^k} f(\mathbf{u})\, e(-\mathbf{h}\cdot\mathbf{u})\, d\mathbf{u} \qquad \text{for } \mathbf{h} \in \mathbf{Z}^k.$$

Then, since the exact value of the integral in (3.4) is given by $\hat{f}(0)$, we obtain

$$\begin{aligned}
\frac{1}{N}\sum_{n=0}^{N-1} f\left(\frac{n}{N}\mathbf{g}\right) - \int_{I^k} f(\mathbf{u})d\mathbf{u} &= \frac{1}{N}\sum_{n=0}^{N-1}\sum_{\mathbf{h}\in\mathbf{Z}^k} \hat{f}(\mathbf{h}) e\left(\frac{n}{N}\mathbf{h}\cdot\mathbf{g}\right) - \hat{f}(0) \\
&= \frac{1}{N}\sum_{\mathbf{h}\in\mathbf{Z}^k} \hat{f}(\mathbf{h}) \sum_{n=0}^{N-1} e\left(\frac{n}{N}\mathbf{h}\cdot\mathbf{g}\right) - \hat{f}(0) \\
&= \frac{1}{N}\sum_{\substack{\mathbf{h}\in\mathbf{Z}^k \\ \mathbf{h}\neq\mathbf{0}}} \hat{f}(\mathbf{h}) \sum_{n=0}^{N-1} e\left(\frac{n}{N}\mathbf{h}\cdot\mathbf{g}\right).
\end{aligned}$$

Now the last inner sum is equal to 0 if $\mathbf{h}\cdot\mathbf{g} \not\equiv 0 \bmod N$ and equal to N if $\mathbf{h}\cdot\mathbf{g} \equiv 0 \bmod N$, and so

$$\frac{1}{N}\sum_{n=0}^{N-1} f\left(\frac{n}{N}\mathbf{g}\right) - \int_{I^k} f(\mathbf{u})d\mathbf{u} = \sum_{\substack{\mathbf{h}\neq\mathbf{0} \\ \mathbf{h}\cdot\mathbf{g}\equiv 0 \bmod N}} \hat{f}(\mathbf{h}). \tag{3.5}$$

Thus the integration error in (3.4) can be expressed as a sum of certain FOURIER coefficients of f.

According to the k-dimensional RIEMANN-LEBESGUE lemma, $\hat{f}(\mathbf{h})$ tends to zero as $\|\mathbf{h}\|_\infty$ tends to infinity. The rate of convergence of $\hat{f}(\mathbf{h})$ toward zero serves as a regularity condition on f. In our context it is useful to consider $\mathbf{E}_\alpha^k(C)$ (see Definition 1.34), which is defined to be the class of all continuous periodic functions f on $\mathbf{R}^k$ with period interval I^k such that

$$|\hat{f}(\mathbf{h})| \leq Cr(\mathbf{h})^{-\alpha} \qquad \text{for all nonzero } \mathbf{h} \in \mathbf{Z}^k,$$

where $\alpha > 1$ and $C > 0$ are real constants. Furthermore, $\mathbf{E}_\alpha^k$ will denote the class of all f with $f \in \mathbf{E}_\alpha^k(C)$ for some $C > 0$.

It is easily seen that if $f \in \mathbf{E}_\alpha^k$, then its FOURIER series is absolutely convergent and represents f. An important sufficient condition for the membership of a periodic function f on $\mathbf{R}^k$ with period interval I^k in the regularity class $\mathbf{E}_\alpha^k$ is the following. Let $\alpha > 1$ be an integer and suppose that all partial derivatives

$$\frac{\partial^{m_1+\cdots+m_k} f}{\partial u_1^{m_1} \cdots \partial u_k^{m_k}} \qquad \text{with } 0 \leq m_i \leq \alpha - 1 \text{ for } 1 \leq i \leq s$$

exist and are of bounded variation on I^k in the sense of HARDY and KRAUSE; then $f \in \mathbf{E}_\alpha^k(C)$ with a value of C which can be given explicitly (see ZAREMBA [1984]). A more restrictive sufficient condition is the following: If $\alpha > 1$ is an integer and all partial derivatives

$$\frac{\partial^{m_1+\cdots+m_k} f}{\partial u_1^{m_1} \cdots \partial u_k^{m_k}} \qquad \text{with } 0 \leq m_i \leq \alpha \text{ for } 1 \leq i \leq s$$

exist and are continuous on $\mathbf{R}^k$, then $f \in \mathbf{E}_\alpha^k(C)$ with an explicit value of C.

Definition 3.3 *For a real number $\alpha > 1$, for $\mathbf{g} \in \mathbf{Z}^k$, and for an integer $N \geq 1$, we put*

$$P_\alpha(\mathbf{g}, N) = \sum_{\substack{\mathbf{h} \neq 0 \\ \mathbf{h}\cdot\mathbf{g} \equiv 0 \bmod N}} r(\mathbf{h})^{-\alpha}.$$

Theorem 3.4 *For any real numbers $\alpha > 1$ and $C > 0$, for any $\mathbf{g} \in \mathbf{Z}^k$ and any integer $N \geq 1$, we have*

$$\max_{f \in \mathbf{E}_\alpha^k(C)} \left| \frac{1}{N} \sum_{n=0}^{N-1} f\left(\frac{n}{N}\mathbf{g}\right) - \int_{I^k} f(\mathbf{u}) d\mathbf{u} \right| = C\, P_\alpha(\mathbf{g}, N).$$

Proof. For $f \in \mathbf{E}_\alpha^k(C)$, the bound

$$\left| \frac{1}{N} \sum_{n=0}^{N-1} f\left(\frac{n}{N}\mathbf{g}\right) - \int_{I^k} f(\mathbf{u}) d\mathbf{u} \right| \leq C\, P_\alpha(\mathbf{g}, N).$$

follows immediately from (3.5) and Definitions 1.34 and 3.3. Now let f_0 be the special function

$$f_0(\mathbf{u}) = C \sum_{\mathbf{h}\in\mathbf{Z}^k} r(\mathbf{h})^{-\alpha} e(\mathbf{hu}) \qquad \text{for } \mathbf{u} \in \mathbf{R}^k.$$

Then $f_0 \in \mathbf{E}_\alpha^k(C)$, and

$$\frac{1}{N}\sum_{n=0}^{N-1} f_0\left(\frac{n}{N}\mathbf{g}\right) - \int_{I^k} f_0(\mathbf{u})d\mathbf{u} = CP_\alpha(\mathbf{g}, N)$$

by (3.5) and Definition 3.3. □

Theorem 3.4 shows that, for given α and N, the lattice point $\mathbf{g}$ should be chosen in such a way that $P_\alpha(\mathbf{g}, N)$ is small. Via an averaging argument it can be shown that there are such sequences with discrepancy of order $\mathcal{O}((\log N)^k/N)$; for a proof see for instance KUIPERS and NIEDERREITER [983] or NIEDERREITER [1336]. For a detailed discussion of g.l.p.'s and the more general lattice rules for numerical integration we refer to DAVIS and RABINOWITZ [434] and NIEDERREITER [1336]. For the more recent literature on lattice rules see the Notes.

Computation of the Discrepancy

There are recent efforts to design efficient algorithms for computing the discrepancy (see DOBKIN ET AL. [467, 468]). The aim here is to follow HEINRICH [761] and to give an algorithm of worst case complexity $\mathcal{O}(N(\log N)^k)$ which computes the L^2-discrepancy.

Let $A = ((\mathbf{x}_1, p_1), \ldots, (\mathbf{x}_N, p_N))$ be an array with $\mathbf{x}_n = (x_n^{(1)}, \ldots, x_n^{(k)}) \in I^k$, $p_n \in \mathbf{R}$ $(1 \le n \le N)$, and let the quadrature be given by

$$Q\,f = \sum_{n=1}^{N} p_n f(\mathbf{x}_n)$$

for any continuous function $f \in C(I^k)$. For a $\mathbf{t} = (t_1, \ldots, t_k) \in I^k$, we let

$$\Delta(\mathbf{t}) = \sum_{n=1}^{N} p_n \chi_{[\mathbf{0},\mathbf{t})}(\mathbf{x}_n) - \int_{I^k} \chi_{[\mathbf{0},\mathbf{t})}(\mathbf{x})\, d\mathbf{x},$$

where $[\mathbf{0}, \mathbf{t}) = \prod_{j=1}^{k}[0, t_j]$. If $p_n = 1/N$ for all n, then $\Delta(\mathbf{t})$ measures, as usual, the local deviation of the empirical distribution of the point set $\{\mathbf{x}_n : n = 1, \ldots, N\}$ from the uniform distribution. In the present notation the L^2-discrepancy of A has the form

$$D_N^{(2)}(A) = \left(\int_{I^k} \Delta(\mathbf{t})^2 dt\right)^{1/2}.$$

The order of the smallest possible L^2-discrepancy is given by

$$\inf_{|A|=N} D_N^{(2)}(A) = \mathcal{O}\left(N^{-1}(\log N)^{(k-1)/2}\right);$$

see ROTH [1567], DAVENPORT [429] and Sections 1.3 and 2.1. For more special results we refer to TEMLYAKOV [1796], FROLOV [627] and BYKOVSKII [320].

As a starting point for further investigations the following explicit formula may be used:

$$\begin{aligned}(D_N^{(2)}(A))^2 &= \int_{I^k}\left(\prod_{l=1}^k t_l\right)^2 dt - 2\sum_{n=1}^N p_n \int_{I^k}\prod_{l=1}^k t_l \chi_{[0,t_j)}(x_n^{(l)})\,dt \\ &\quad + \sum_{n,j=1}^N p_n p_j \int_{I^k}\prod_{l=1}^k \chi_{[0,t_l)}(x_n^{(l)})\chi_{[0,t_l)}(x_j^{(l)})\,dt \\ &= 3^{-k} - 2^{1-k}\sum_{n=1}^N p_n \prod_{l=1}^k (1-(x_n^{(l)})^2) + \\ &\quad + \sum_{n,j=1}^N p_n p_j \prod_{l=1}^k (1-\max(x_n^{(l)}, x_j^{(l)})). \end{aligned} \tag{3.6}$$

This formula was first pointed out and used for the numerical investigation of various low discrepancy sets by WARNOCK [1945] and many experimental investigations were based on it.

The second term of (3.6) is computed in $\mathcal{O}(N)$ operations, the straightforward computation of the third term requires $\mathcal{O}(N^2)$ operations (arithmetic operations or comparisons). This makes the computation of $D_N^{(2)}(A)$ for large A a highly complex task.

The following algorithm D is defined recursively and will accomplish a slightly more general task in $\mathcal{O}(N(\log N)^k)$ operations. Given another array $B = ((\mathbf{y}_1, q_1), \ldots, (\mathbf{y}_M, q_N))$ with $\mathbf{y}_m = (y_m^{(1)}, \ldots, y_m^{(k)}) \in I^k$, $q_m \in \mathbf{R}$, $1 \le m \le M$, the algorithm will compute

$$D(A,B,k) = \sum_{n=1}^N \sum_{m=1}^M p_n q_m \prod_{l=1}^k (1-\max(x_n^{(l)}, y_m^{(l)})). \tag{3.7}$$

The main idea of HEINRICH is to suppose that we know that the first coordinates of A are all not greater than those of B, i.e.

$$x_n^{(1)} \le y_m^{(1)} \qquad (1 \le n \le N,\ 1 \le m \le M).$$

Then (3.7) simplifies to

$$D(A,B,k) = \sum_{n=1}^N \sum_{m=1}^M p_n' q_m' \prod_{l=2}^k (1-\max(x_n^{(l)}, y_m^{(l)})),$$

where $p_n' = p_n$ and $q_m' = (1-y_m^{(1)})q_m$. Hence we have reduced the dimension of the problem by 1. Suppose now that $k = 1$. Then after reduction we are left with the double sum

$$\sum_{n=1}^{N}\sum_{m=1}^{M} p'_n q'_m = \left(\sum_{n=1}^{N} p'_n\right)\left(\sum_{m=1}^{M} q'_m\right)$$

which now can be computed in $\mathcal{O}(N+M)$ operations. The algorithm is recursive and applies the divide-and-conquer strategy to reduce the dimension.

We assume now that A is sorted in such a way that

$$x_1^{(1)} \leq x_2^{(1)} \leq \ldots \leq x_N^{(1)}. \tag{3.8}$$

This can be achieved by an initial sorting in $\mathcal{O}(N \log N)$ time. (This initial sorting is not a part of the algorithm D, but in the recursion the algorithm will take care of such an ordering by itself). We formally also include the case $B = \emptyset$ and the case $k = 0$. In the latter we suppose $A = ((\mathbf{x}_1, p_1), \ldots, (\mathbf{x}_N, p_N))$ and $B = ((\mathbf{y}_1, q_1), \ldots, (\mathbf{y}_M, q_M))$.

Algorithm D

INPUT: A, B, k as above, A satisfying $A \neq \emptyset$ and (3.8)

OUTPUT: $D(A, B, k)$

CASE 1: $M = 0$ (i.e. $B = \emptyset$)

$$D(A, B, k) = 0$$

CASE 2: $k = 0$, $M \geq 1$

$$D(A, B, 0) = \left(\sum_{n=1}^{N} p_n\right)\left(\sum_{m=1}^{M} q_m\right)$$

CASE 3: $N = 1$, $k \geq 1$, $M \geq 1$

$$D(A, B, k) = p_1 \sum_{m=1}^{M} q_m \prod_{l=1}^{k} (1 - \max(x_{1l}, y_m^{(l)}))$$

CASE 4: $N > 1$, $k \geq 1$, $M \geq 1$

Set $p = [\frac{N}{2}]$, $\xi = x_p^{(1)}$. Form new arrays A_1, A_2, B_1, B_2 as follows:

$$A_1 = ((\mathbf{x}_1, p_1), \ldots, (\mathbf{x}_p, p_p)) \tag{3.9}$$

$$A_2 = ((\mathbf{x}_{p+1}, p_{p+1}), \ldots, (\mathbf{x}_N, p_N)) \tag{3.10}$$

To define the arrays B_1, B_2, we treat the elements of B consecutively. All elements whose first coordinate is not greater than ξ go into B_1, the rest goes into B_2. Precisely, we put

$$B_1 = ((\mathbf{y}_{m_1}, q_{m_1}), \ldots, (\mathbf{y}_{m_q}, q_{m_q})) \tag{3.11}$$

$$B_2 = ((\mathbf{y}_{m_{q+1}}, q_{m_{q+1}}), \ldots, (\mathbf{y}_{m_n}, q_{m_n})), \tag{3.12}$$

where q and the m_j are defined through the relations

$$\begin{aligned} y_{m_j,1} &\leq \xi \qquad (j = 1, \ldots, q) \\ y_{m_j,1} &> \xi \qquad (j = q+1, \ldots, M) \end{aligned}$$

and

$$\begin{aligned} m_1 < m_2 &< \ldots < m_q, \\ m_{q+1} < m_{q+2} &< \ldots < m_M. \end{aligned}$$

Let P' be the projection of $\mathbf{R}^k$ onto $\mathbf{R}^{k-1}$ given by omitting the first coordinate. Put

$$\begin{aligned} \mathbf{x}'_n &= P'\mathbf{x}_n & (n = 1, \ldots, N) \\ \mathbf{y}'_m &= P'\mathbf{y}_m & (m = 1, \ldots, M) \\ p'_n &= p_n & (n = 1, \ldots, p) \\ p'_n &= p_n(1 - x_n^{(1)}) & (n = p+1, \ldots, N) \\ q'_{m_j} &= q_{M_j} & (j = 1, \ldots, q) \\ q'_{m_j} &= q_{m_j}(1 - y_{m_j}^{(1)}) & (j = q+1, \ldots, n) \end{aligned}$$

Form the sets A'_1, A'_2, B'_1, B'_2 defined by literally putting primes to the symbols in (3.9)-(3.12). Obtain A''_1, A''_2 from A'_1, A'_2 by sorting with respect to the (new) first coordinate, so that (3.8) holds for these new arrays. In the case $k = 1$ the definitions above have to be interpreted in the appropriate way: the primed arrays consist only of the p'_n and q'_m and the sorting step is omitted.

Finally, we set

$$D(A, B, k) = D(A_1, B_1, k) + D(A_2, B_2, k) + D(A''_1, B'_2, k-1) + D(A''_2, B'_1, k-1)$$

This recursion completes case 4 and the algorithm.

It is easy to see that

$$\begin{aligned} D(A''_1, B'_2, k-1) &= D(A_1, B_2, k) \\ D(A''_2, B'_1, k-1) &= D(A_2, B_1, k) \end{aligned}$$

and hence the algorithm indeed computes the desired quantity (3.7).

Now we estimate the maximal number of operations $L(N, M, k)$ over all possible inputs of size N, M, and dimension k. Let us assume that we use a sorting algorithm with (worst-case) number of operations of at most $c_{sort} M \log M$ (the logarithm to the base 2), with some constant $c_{sort} > 0$.

Theorem 3.5 *For each $k \geq 0$ there exists a constant $c_k > 0$ such that for all $N \geq 1$, $M \geq 0$:*

$$L(N, M, k) \leq c_k (N + M)(\log N + 1)^k. \tag{3.13}$$

Proof. For $M = 0$ we have $L(N, 0, k) = 0$ and (3.13) holds trivially. For $k = 0$, $M \geq 1$ we are in case 2 and have

$$L(N, M, 0) = N + M - 1,$$

so we put $c_0 = 1$. For $k \geq 1$ we define

$$c_k = \max(3k + 1, (\log 1.5)^{-1}(c_{k-1} + c_{sort} + 3)). \tag{3.14}$$

By induction on N we shall prove that (3.13) holds for all $k \geq 1$, $M \geq 1$. For $N = 1$, which is case 3, it follows that

$$L(1, M, k) = (3k + 1)M \leq c_k(M + 1). \tag{3.15}$$

Now we fix $N > 1$, put $p = [\frac{N}{2}]$ and $\sigma(k) = 1$ if $k > 1$, $\sigma(k) = 0$ if $k = 1$. From case 4 we deduce

$$\begin{aligned} L(N, M, k) \leq \max_{0 \leq q \leq M} \{ & n + 2(N - p + M - q) + \sigma(k) c_{sort}(p \log p + (N - p)\log(N - p)) \\ & + L(p, q, k) + L(N - p, M - q, k) \\ & + L(p, M - q, k - 1) + L(N - p, q, k - 1) + 3 \}. \end{aligned} \tag{3.16}$$

For $N > 1$ we have

$$\max(p, N - p) = \left\lceil \frac{N + 1}{2} \right\rceil \leq 2N/3.$$

From this and the induction hypothesis, we obtain

$$\begin{aligned} L(N, M, k) & \leq \max_{0 \leq q \leq M} \{ 3(N + M) + \sigma(k) c_{sort} N (\log N + 1) \\ & \quad + c_k(p + q)(\log p + 1)^d + c_k(N - p + M - q)(\log(N - p) + 1)^k \\ & \quad + c_{k-1}(p + M - q)(\log p + 1)^{k-1} + c_{k-1}(N - p + q)(\log(N - p) + 1)^{k-1} \} \\ & \leq (3 + c_{sort})(N + M)(\log N + 1)^{k-1} + c_k(N + M)(\log(2N/3) + 1)^k \\ & \quad + c_{k-1}(N + M)(\log N + 1)^{k-1}. \end{aligned}$$

From

$$\begin{aligned} (\log(2N/3) + 1)^k & = (\log N + 1 - \log 1.5)^k \\ & \leq (\log N + 1 - \log 1.5)(\log N + 1)^{k-1} \\ & = (\log N + 1)^k - \log 1.5(\log N + 1)^{k-1}, \end{aligned} \tag{3.17}$$

we conclude

$$\begin{aligned} L(N, M, k) & \leq (3 + c_{sort} + c_{k-1})(N + M)(\log N + 1)^{k-1} \\ & \quad + c_k(N + M)(\log N + 1)^k - c_k \log 1.5(N + M)(\log N + 1)^{k-1} \\ & \leq c_k(N + M)(\log N + 1)^k. \end{aligned}$$

Having algorithm D, it is clear how to compute $D_N^{(2)}(A)$: We determine the first two terms of (3.6), then we sort A, so that it has nondecreasing first coordinates, and finally we apply $D(A, A)$. Clearly, this takes not more than $\mathcal{O}(N(\log N)^k)$ operations. □

The presented algorithm can be improved in the average case sense. The complexity of the improved algorithm is of the same order; for details we refer to HEINRICH [761].

The computation of the star-discrepancy is much more difficult. Explicit formulas were established by BUNDSCHUH and ZHU; see [314] and the Notes of this section. A general theorem is due to LIN ACHAN [2], which we state without proof. It gives an algorithm for computing the star-discrepancy of point sequences in $[0,1)^k$ with respect to weights $p_1, \dots, p_N$ with $\sum_{n=1}^N p_n = 1$:

$$D_N^*(\mathbf{x}_n, p_n) = \sup_J \left| \sum_{n=1}^N p_n \chi_J(\mathbf{x}_n) - \lambda_k(J) \right|.$$

Theorem 3.6 *Let $\mathbf{x}_n = (x_n^{(1)}, x_n^{(2)}, \dots, x_n^{(k)})$ for $1 \le n \le N$ be a sequence of k-dimensional points in I^k. Assuming that $x_{\sigma_i(1)}^{(i)} \le x_{\sigma_i(2)}^{(i)} \le \cdots \le x_{\sigma_i(N)}^{(i)}$ and $\sigma_1(1) = 1, \sigma_1(2) = 2, \dots, \sigma_1(N) = N$ where $\sigma_i(1), \sigma_i(2), \dots, \sigma_i(N)$ for $1 \le i \le N$ is a proper permutation of $1, 2, \dots, N$. Then its star-discrepancy $D_N^*(\mathbf{x}_n, p_n)$ with respect to arbitrary positive weights $p_1, p_2, \dots, p_N$ with $\sum_{n=1}^N p_n = 1$ can be computed by*

$$D_N^*(\mathbf{x}_n, p_n) = \max_{0 \le m_1, m_2, \dots, m_k \le N} \Delta_N(m_1, m_2, \dots, m_k)$$

where $\Delta_N(m_1, m_2, \dots, m_k)$ is defined by

$$\begin{aligned}\Delta_N(m_1, \dots, m_k) \quad = \quad & \max\{x_{\sigma_1(m_1+1)}^{(1)} x_{\sigma_2(m_2+1)}^{(2)} \cdots x_{\sigma_k(m_k+1)}^{(k)} \\ & \quad -(p_{\sigma_l(1)} + p_{\sigma_l(2)} + \cdots + p_{\sigma_l(m)}), \\ & (p_{\sigma_l(1)} + p_{\sigma_l(2)} + \cdots + p_{\sigma_l(m)}) \\ & \quad -x_{\sigma_1(m_1)}^{(1)} x_{\sigma_2(m_2)}^{(2)} \cdots x_{\sigma_k(m_k)}^{(k)}\}\end{aligned}$$

and $\{\sigma_l(1), \sigma_l(2), \dots, \sigma_l(m)\}$ is given by

$$\{\sigma_l(1), \sigma_l(2), \dots, \sigma_l(m)\} = \bigcap_{i=1}^N \{\sigma_i(1), \sigma_i(2), \dots, \sigma_i(m_i)\}.$$

Approximation of Functions

In the following we will discuss the approximation of differentiable functions applying low discrepancy sequences (see PROINOV [1473]). Suppose we have a finite sequence

$x_1, \ldots, x_N$ in the unit interval $[0,1]$ and a finite sequence $p_1, \ldots, p_N$ of nonnegative numbers such that

$$\sum_{n=1}^{N} p_n = 1.$$

The numbers $p_1, \ldots, p_N$ are considered as weights of the numbers $x_1, \ldots, x_N$, respectively. Further let

$$P_0 = 0 \quad \text{and} \quad P_n = \sum_{\nu=1}^{n} p_\nu, \quad 1 \le n \le N,$$

(see also Section 2.2.) We define the functions g and h on $[0,1]$ by

$$g(x) = \sum_{n=1}^{N} p_n \chi_{[0,x)}(x_n) - x$$

and

$$h(x) = |x - x_k| \qquad \text{if} \quad x \in [P_{k-1}, P_k), \quad 1 \le k \le N.$$

Then

$$D_N^*(x_n, p_n) = \|g\|_\infty = \sup_{0 \le x \le 1} |g(x)|$$

is the discrepancy of $x_1, \ldots, x_N$ with respect to the weights $p_1, \ldots, p_N$. Furthermore

$$D_N^{(p)}(x_n, p_n) = \|g\|_p = \left(\int_0^1 |g(x)|^p dx \right)^{1/p}, \qquad 0 < p < \infty,$$

is the L^p-discrepancy of $x_1, \ldots, x_N$ with respect to the weights $p_1, \ldots, p_N$.

As a measure of distribution of $x_1, \ldots, x_N$ with respect to the weights $p_1, \ldots, p_N$, except for the discrepancies D_N^* and $D_N^{(p)}$, we will also use the number

$$\Delta_N^{(p)} = \left(\int_0^1 |g(x)|^{p+1} dx \right) \Big/ \left(\int_0^1 |g(x)|^p dx \right), \qquad 0 \le p < \infty.$$

Obviously, $\Delta_N^{(0)} = D_N^{(1)}$. Furthermore let $\omega_f(c)$ be the modulus of continuity of f as defined in Section 1.1.

The following lemma is due to PROINOV [1466].

Lemma 3.7 *Let $x_1 \le \ldots \le x_N$. Then for $0 < p < \infty$, we have*

$$D_N^{(p)}(x_n, p_n) = \|h\|_p$$

and

$$D_N^*(x_n, p_n) = \|h\|_\infty = \max_{1 \le k \le N} \max\{|x_k - P_{k-1}|, |x_k - P_k|\}.$$

We will now investigate the approximation of r-times differentiable functions on $[0,1]$ by means of functions $L_N^{(r)}(f;x)$ which are defined on $[0,1]$ by

$$L_N^{(r)}(f;x) = \sum_{j=0}^{r} \frac{f^{(j)}(x_k)}{j!}(x-x_k)^j \qquad \text{if} \quad x \in [P_{k-1}, P_k), \quad 1 \leq k \leq N.$$

Let BW^r be the set of all functions for which $f^{(r)}$ is bounded. We shall assume in what follows that $[a,b) = [a,b]$ if $b = 1$. Further, we suppose that

$$x_1 \leq x_2 \leq \ldots \leq x_N.$$

Theorem 3.8 *Let r be a positive integer and $0 < p < \infty$. Then for every function $f \in BW^r$, we have*

$$\|f - L_N^{(r)}(f)\|_p \leq \frac{(D_N^{(pr)})^r}{(r-1)!} \int_0^1 (1-x)^{r-1} \omega_{f^{(r)}}(x D_N^*) dx, \tag{3.18}$$

where $D_N^{(pr)}$ and D_N^* are the $L^{(pr)}$-discrepancy and the star-discrepancy of the sequence (x_n) with respect to the weights p_n.

Proof. Using TAYLOR's formula with the remainder in integral form it follows easily (see [1485]) that for $x \in [P_{k-1}, P_k)$, $1 \leq k \leq N$, we have

$$\begin{aligned} & f(x) - L_N^{(r)}(f;x) \\ = \; & \frac{(x-x_k)^r}{(r-1)!} \int_0^1 (1-t)^{r-1} [f^{(r)}(x_k + (x-x_k)t) - f^{(r)}(x_k)] dt. \end{aligned}$$

Therefore,

$$\begin{aligned} & \|f - L_N^{(r)}(f)\|_p \\ = \; & \left(\sum_{k=1}^{N} \int_{P_{k-1}}^{P_k} |f(x) - L_N^{(r)}(f;x)|^p dx \right)^{1/p} \\ = \; & \frac{1}{(r-1)!} \left(\sum_{k=1}^{N} \int_{P_{k-1}}^{P_k} |x-x_k|^{pr} \right. \\ & \times \left. \left| \int_0^1 (1-t)^{r-1} [f^{(r)}(x_k + (x-x_k)t) - f^{(r)}(x_k)] dt \right|^p dx \right)^{1/p}. \end{aligned} \tag{3.19}$$

From (3.19), we obtain the estimate

$$\begin{aligned} & \|f - L_N^{(r)}(f)\|_p \\ \leq \; & \frac{1}{(r-1)!} \left(\sum_{k=1}^{N} \int_{P_{k-1}}^{P_k} |x-x_k|^{pr} \right. \end{aligned}$$

$$\times \left(\int_0^1 (1-t)^{r-1} \omega_{f^{(r)}}(|x - x_k| t)\, dt \right)^p dx \Bigg)^{1/p}$$

$$= \frac{1}{(r-1)!} \left(\int_0^1 h(x)^{pr} \left(\int_0^1 (1-t)^{r-1} \omega_{f^{(r)}}(t\, h(x))\, dt \right)^p dx \right)^{1/p}. \quad (3.20)$$

From Lemma 3.7, it follows that for all $x, t \in [0,1]$,

$$\omega_{f^{(r)}}(t\, h(x)) \le \omega_{f^{(r)}}(t\|h\|_\infty) = \omega_{f^{(r)}}(tD_N^*).$$

Hence we get from (3.20)

$$\|f - L_N^{(r)}(f)\|_p \le \frac{1}{(r-1)!} \left(\int_0^1 h(x)^{pr} dx \right)^{1/p} \int_0^1 (1-t)^{r-1} \omega_{f^{(r)}}(tD_N^*)\, dt.$$

From this and Lemma 3.7, we obtain (3.18) and the theorem is proved. □

Integration of Singular Functions

In the following we want to discuss briefly Quasi-Monte Carlo methods for the numerical integration of singular functions. A first class of functions was investigated by Sobol' [1702]. A certain type of point singularities was considered by Klinger [937]. In this case explicit error bounds can be derived. For any point sequence $\mathbf{x}_1, \mathbf{x}_2, \ldots$ and function f on I^k we will use the following notation

$$I_N(f) := \frac{1}{N} \sum_{n=1}^{N} f(\mathbf{x}_n) \qquad \text{and} \qquad I(f) := \int_{I^k} f(\mathbf{x})\, d\mathbf{x}.$$

Theorem 3.9 *Let $(\mathbf{x}_n)_{n \ge 1}$ be the k-dimensional* Halton *sequence and let*

$$f(\mathbf{x}) = \frac{1}{|\mathbf{x}^{(1)}|^{\beta_1} + \cdots + |\mathbf{x}^{(k)}|^{\beta_k}} \quad \textit{with} \quad \frac{1}{\beta_1} + \cdots + \frac{1}{\beta_k} > 1.$$

Then we have

$$|I_N(f) - I(f)| = \mathcal{O}\left(N^{-1 + \left(\sum_{m=1}^{k} \frac{1}{\beta_m}\right)^{-1}} (\log N)^k \right).$$

Sketch Proof. We introduce a function g which is equal to f except for some rectangular neighbourhood of the origin

$$R_N = \prod_{m=1}^{k} \left[0, c N^{-\frac{1}{\beta_m} \left(\sum_{j=1}^{k} \frac{1}{\beta_j}\right)^{-1}} \right]$$

and define

$$g_{|R_N} = \max_{\mathbf{x} \in I^k - R_N} f(\mathbf{x}).$$

Now, the approximation error $E_N(f) := |I_N(f) - I(f)|$ may be estimated as follows:

$$E_N(f) \leq |I_N(f) - I_N(g)| + |I_N(g) - I(g)| + |I(g) - I(f)|. \tag{3.21}$$

It therefore suffices to find bounds for the three terms on the righthand side. The first term can actually be proved to be equal zero if we select c, which is independent of N, small enough. It can be shown by using only some ideas of elementary number theory that then no point $\mathbf{x}_1, \ldots, \mathbf{x}_N$ of the HALTON sequence falls into R_N.

Furthermore g is of bounded variation and $V(g) \leq c' \max_{\mathbf{x} \in I^k - R_N} f(\mathbf{x})$. Hence we can apply the KOKSMA-HLAWKA inequality (Theorem 1.14) and since the discrepancy of the HALTON sequence satisfies

$$D_N(\mathbf{x}_1, \ldots, \mathbf{x}_N) = \mathcal{O}\left(\frac{(\log N)^k}{N}\right),$$

we obtain

$$|I_N(g) - I(g)| \leq D_N(\mathbf{x}_1, \ldots, \mathbf{x}_N) \cdot V(f) = \mathcal{O}\left(N^{-1+\left(\sum_{m=1}^k \frac{1}{\beta_m}\right)^{-1}} (\log N)^k\right).$$

Finally, we get for the third term in (3.21)

$$|I(g) - I(f)| \leq c'' \int_{R_N} f(\mathbf{x})\, d\mathbf{x} = \mathcal{O}\left(N^{-1+\left(\sum_{m=1}^k \frac{1}{\beta_m}\right)^{-1}}\right).$$

The last equality can be proved e.g. by induction on the dimension k. □

Remark. Similar results also hold for net sequences (see the following subsection) or for singularities in the interior of the unit cube. In both cases, however, we have to skip a finite number of the points $\mathbf{x}_n$ in the summation for I_N. This is in fact necessary as it is possible to find for any uniformly distributed sequence $\mathbf{x}_1, \mathbf{x}_2, \ldots$ a function f with a singularity in $(0,1)^k$ such that

$$\limsup_{N\to\infty} I_N(f) = \infty.$$

3.1.2 Nets and *(t,s)*-sequences

The general theory of *(t,m,s)*-nets and *(t,s)*-sequences was developed by NIEDERREITER [1317, 1319]), although the first special cases of such sequences were given by SOBOL [1701] and later by FAURE [582]. For a detailed definition and discussion of these two important examples of net sequences we also refer to BECK and CHEN [143]. The *(t,s)*-sequences have, concerning discrepancy, among all known low discrepancy sequences the best theoretical properties. Note that in this subsection the dimension always is denoted by s instead of k since the notation (t,s)-sequences is now standard. The parameter t describes the quality of the distribution of the sequence. First, we want to motivate the introduction of these sequences by the following observation.

Let $b \geq 2$ be an integer and

$$n = \sum_{j=0}^{\infty} d_j(n)\, b^j, \qquad d_j \in \{0, 1, \ldots, b-1\}$$

the digit expansion of the integer $n \geq 0$ in the base b. Now consider the integers $n = kb^m, \ldots, (k+1)b^m - 1$, where $k \geq 0$ is an arbitrary integer. It is clear that the digits $d_j(n)$ with $0 \leq j \leq m-1$ run through all possible values of $\{0, 1, \ldots, b-1\}^m$ whereas all other digits remain constant. Reflecting these digital expansions with respect to the decimal point (i.e. constructing the points $x_n = \gamma_b(n)$ of the VAN DER CORPUT sequence) we obtain numbers in $[0, 1)$ whose m leading digits attain all possible values from $\{0, 1, \ldots, b-1\}^m$, while the remaining digits stay fixed. In other words, each interval of the form

$$[ab^{-m}, (a+1)b^{-m}), \qquad a = 0, 1, \ldots, b^m - 1,$$

contains exactly one point x_n with $n \in \{kb^m, \ldots, (k+1)b^m - 1\}$. This leads us to the following three definitions, which are of basic importance for further considerations on the *(t,s)*-sequences.

Definition 3.10 *Let $b \geq 2$ be some chosen base, and $d_i \geq 0$ and $0 \leq a_i < b^{d_i}$ for $1 \leq i \leq s$ integers. We define an s-dimensional*[1] elementary interval *as an interval of the form*

$$E = \prod_{i=1}^{s} [a_i b^{-d_i}, (a_i + 1) b^{-d_i}).$$

We see easily that $\lambda_s(E) = b^{-d}$ with $d = \sum_{i=1}^{s} d_i$.

Definition 3.11 *Let t and m be integers satisfying $0 \leq t \leq m$. We define a (t,m,s)-net in base b as a point set $P \in I^s$ with $|P| = b^m$, such that*

$$A(E, b^m, \mathbf{x}_n) = b^t$$

for all elementary intervals E with $\lambda_s(E) = b^{t-m}$.

This definition immediately implies that every *(t,m,s)*-net in base b satisfies $\Delta(E; b^m) = 0$, for every elementary interval E, where we write

$$\Delta(J; N) = \frac{1}{N} A(J, N, \mathbf{x}_n) - \lambda_s(J)$$

for the local discrepancy over J.

Definition 3.12 *Let $t \geq 0$ be an integer. A sequence $(\mathbf{x}_n)_{n \geq 1}$ in I^s is called a (t,s)-sequence in b if for all $k \geq 0$ and $m > t$ the point set*

$$P = \{\mathbf{x}_n : kb^m < n \leq (k+1)b^m\}$$

is a (t,m,s)-net in base b.

[1] In this context it is usual to use the notation s for the dimension.

There are three questions that immediately arise: first, if there are *(t,s)*-sequences in b for all possible values of t, s and b; secondly, if these sequences really satisfy the relation $D_N^* = \mathcal{O}(\log^s N/N)$; and finally, how one could construct these sequences efficiently.

We note that each *(t,m,s)*-net in b is for all $t \leq u \leq m$ also a (u, m, s)-net in b. This follows from the fact that each elementary interval E with $\lambda_s(E) = b^{u-m}$ can be represented as a union of b^{u-t} disjoint elementary intervals E_j with $\lambda_s(E_j) = b^{t-m}$, $j = 1, \ldots, b^{u-t}$, by counting the elements in E. The same result is obtained for the sequences. This means that smaller values of the parameter t imply better uniform distribution. However, we are going to see that it would not be possible to choose $t = 0$, i.e. the smallest allowed value for t, for each dimension s.

In the following theorem we state the discrepancy bounds for the *(t,m,s)*-nets and *(t,s)*-sequences as given in NIEDERREITER [1317]:

Theorem 3.13 *The star discrepancy D_N^* of a (t,m,s)-net $\mathbf{x}_1, \ldots, \mathbf{x}_N$ in b with $m > 0$ satisfies*

$$D_N^*(\mathbf{x}_1, \ldots \mathbf{x}_N) \leq K_s(b)\, b^t \frac{(\log N)^{s-1}}{N} + \mathcal{O}\left(b^t \frac{(\log N)^{s-2}}{N}\right),$$

with

$$K_s(b) = \left(\frac{b-1}{2 \log b}\right)^{s-1}$$

if $s = 2$ or $b = 2$, $s = 3, 4$; otherwise

$$K_s(b) = \frac{1}{(s-1)!} \left(\frac{\lfloor b/2 \rfloor}{\log b}\right)^{s-1}.$$

The star discrepancy D_N^* of a (t,s)-sequence $(\mathbf{x}_n)_{n \geq 1}$ in base b with $m > 0$ satisfies

$$D_N^*(\mathbf{x}_1, \ldots \mathbf{x}_N) \leq N_s(b)\, b^t \frac{(\log N)^s}{N} + \mathcal{O}\left(b^t \frac{(\log N)^{s-1}}{N}\right),$$

where

$$N_s(b) = \frac{1}{s} \left(\frac{b-1}{2 \log b}\right)^s$$

if $s = 2$ or $b = 2$, $s = 3, 4$; otherwise

$$N_s(b) = \frac{1}{s!} \frac{b-1}{2 \lfloor b/2 \rfloor} \left(\frac{\lfloor b/2 \rfloor}{\log b}\right)^s.$$

For $N_s(b)$ we have

$$\lim_{s \to \infty} N_s(b) = 0,$$

which does not apply to the constant of the corresponding relation for the HALTON sequence, whose limit tends to infinity.

The proof of the above theorem only uses those properties of nets and *(t,s)*-sequences, which have been defined above. We can conclude that, presumed such sequences exist, they really attain the order of magnitude of the discrepancy which characterizes low-discrepancy sequences. We may observe again that the discrepancy bound for the low values of t decreases, i.e. the sequence is better uniformly distributed for lower values of t. Recently NIEDERREITER and XING [1368, 1369, 1370] proposed a construction, through which very small values of t may be obtained.

One general construction principle for *(t,s)*-sequences is given in NIEDERREITER [1319]. We give here a form where the bijections ψ_r and $\eta_{i,j}$ from NIEDERREITER [1319] are set equal to identity. This choice is relatively practical for implementation purposes. Moreover, the whole theory is based on some fixed finite field $\mathbf{F}_b$, where b is a prime power. For computer implementations it is more practical to assume that b is a prime number. So, in order to construct a *(t,s)*-sequence in base b we choose a finite field $\mathbf{F}_b$ with $|\mathbf{F}_b| = b$ and elements $c_{jr}^{(i)} \in \mathbf{F}_b$ for $1 \leq i \leq s$, $j \geq 1$, $r \geq 0$, where for all sufficiently large j and fixed i and r we have $c_{jr}^{(i)} = 0$. For $n = 0, 1, \ldots$ let

$$n = \sum_{r=0}^{\infty} a_r(n) b^r, \qquad a_r(n) \in \mathbf{F}_b,$$

be the digit expansion of n in base b. We set

$$x_n^{(i)} = \sum_{j=1}^{\infty} x_{nj}^{(i)} b^{-j} \quad \text{for} \quad 1 \leq i \leq s \quad \text{and} \quad n \geq 1,$$

where

$$x_{nj}^{(i)} = \sum_{r=0}^{\infty} c_{jr}^{(i)} a_r(n) \quad \in \quad \mathbf{F}_b$$

for $1 \leq i \leq s$, $j \geq 1$, $n \geq 1$. Addition and multiplication are operations in $\mathbf{F}_b$ and all sums are actually finite. We define a sequence of s-dimensional points by

$$\mathbf{x}_n = (x_n^{(1)}, \ldots, x_n^{(s)}) \in I^s \quad \text{for} \quad n = 1, 2, \ldots. \tag{3.22}$$

The right choice of the elements $c_{jr}^{(i)}$ for getting a *(t,s)*-sequence is specified in the following theorem.

Theorem 3.14 *Let b be an arbitrary prime power and let $p_1, \ldots, p_s \in \mathbf{F}_b[x]$ be pairwise coprime, where $s \geq 1$ is arbitrary and deg(p_i) $= e_i \geq 1$ for $1 \leq i \leq s$. For $1 \leq i \leq s$ and $j \geq 1$ let $g_{ij} \in \mathbf{F}_b[x]$ with $\gcd(g_{ij}, p_i) = 1$ and*

$$\lim_{j \to \infty} (je_i - \deg(g_{ij})) = \infty$$

for $1 \leq i \leq s$. For $0 \leq k < e_i, 1 \leq i \leq s$, and $j \geq 1$ let

$$\frac{x^k g_{ij}(x)}{p_i(x)^j} = \sum_{r=w}^{\infty} a^{(i)}(j, k, r) x^{-r-1}, \tag{3.23}$$

be a series expansion defining the elements $a^{(i)}(j,k,r) \in \mathbf{F}_b$. *Furthermore set*

$$c_{jr}^{(i)} = a^{(i)}(q+1,u,r) \in \mathbf{F}_b \qquad (3.24)$$

for $1 \le i \le s$, $j \ge 1$, $r \ge 0$ *and* $j-1 = qe_i + u$ *where* $0 \le u < e_i$.

Under these presumptions the sequence from (3.22) with $c_{jr}^{(i)}$ *from (3.24) is a (t,s)-sequence in base b with*

$$t = \sum_{i=1}^{s}(e_i - 1).$$

The proof, using formal LAURENT series over $\mathbf{F}_b$, may be found in NIEDERREITER [1319]. The construction itself may be realized by applying linear reccurring sequences. For numerical purposes one usually chooses $g_{ij}(x) \equiv 1$. There are suggestions for computer implementations of these sequence, which include a "super-fast" algorithm for $b = 2$. For more details see BRATLEY, FOX and NIEDERREITER [294].

It is important to choose the parameter t as small as possible. Therefore, if we fix some b, we take for $p_1, \ldots, p_s$ the first s irreducible polynomials in $\mathbf{F}_b$, (i.e. from $i \le h$ follows $\deg(p_i) \le \deg(p_k)$ for $1 \le i \le h \le s$). With this choice and with $e_i = \deg(p_i)$, $1 \le i \le s$, we may define

$$T_b(s) = \sum_{i=1}^{s}(e_i - 1)$$

as the minimal value of t. For the asymptotic behaviour it follows from NIEDERREITER [1319, Theorem 2] that

$$T_b(s) = \mathcal{O}(s \log s).$$

The elements $c_{jr}^{(i)}$ are not uniquely determined. In any case we have shown the existence of a $(T_b(s), s)$-sequence in base b for (all prime powers b and) all dimensions $s \ge 1$. This implies also that a $(0, s)$-sequence exists for the small values of s only.

One special case of the *(t,s)*-sequences are the FAURE sequences. They are specified as $(0, s)$-sequences in base q, where q is the smallest prime number with $q \ge s$. In this case we can take for p_i the polynomials

$$p_i(x) = x + a, \qquad \text{with} \qquad a = 0, 1, \ldots, s-1.$$

They are irreducible, pairwise coprime, and $\deg(p_i) = 1$. Hence $t = 0$.

The discrepancy of the FAURE sequences satisfies the relation

$$D_N^* \le F_s(q)\frac{(\log N)^s}{N} + \mathcal{O}\left(\frac{(\log N)^{s-1}}{N}\right)$$

with

$$F_s(q) = \frac{1}{s!}\left(\frac{q-1}{2\log q}\right)^s.$$

For this constant we also have

$$\lim_{s\to\infty} F_s(q) = 0.$$

Moreover, in general it is greater than $N_s(q)$. In NIEDERREITER [1319] there is a formula for the minimal $N_s(q)$ (with given s); the value of q is declared to be the least odd prime power with $q \geq s$. The following asymptotical relation is valid for both sequences:

$$\lim_{s\to\infty} \frac{\log N_s}{s \log\log s} = -1.$$

The construction of the FAURE sequence is theoretically not very complicated – in principal one has the main fraim as above for the *(t,s)*-sequences, just the elements $c_{jr}^{(i)}$ are not defined via a linear reccurence, but in terms of binomial coefficients. Nevertheless, its implementation on a computer turns out to work quite slowly.

The third and oldest *(t,s)*-sequence is the so called SOBOL sequence. It has been called by SOBOL [1701] a LP_τ-sequence and corresponds to a (τ, s)-sequence in base 2. Its discrepancy satisfies also the relation $D_N^* \leq S_s(\log N)^s/N + \mathcal{O}((\log N)^{s+1}/N)$ with

$$S_s = \frac{2^{t(s)}}{s!(\log 2)^s},$$

and

$$K\frac{s \log s}{\log\log s} \leq t(s) \leq \frac{s\log s}{\log 2} + \mathcal{O}(s \log\log s)$$

for $K > 0$. As the corresponding HALTON bound, this constant grows superexponentially with s, although it is not nearly as large as the HALTON constant. We have

$$\log S_s = \mathcal{O}(s \log\log s)$$

and

$$\lim_{s\to\infty} S_s = \infty.$$

The construction of these sequences uses linear recurrences in the field $\mathbf{F}_2$ as well, but the characteristic polynomials are primitive ones.The theory uses monocyclic operators, matrices and determinants and one can show that this approach is equivalent to a construction involving formal LAURENT series.

In new constructions of *(t,s)*-sequences some new methods for gaining good elements $c_{jr}^{(i)}$ have been proposed. The parameter which was tried to be optimized here is t. This construction uses elliptic function fields over small finite fields as opposed to the one proposed in NIEDERREITER [1317] where a rational function field was used.

For instance, NIEDERREITER and XING [1368] established a construction yielding a (t, s)-sequence in base 2 satisfying the optimal bound

$$T_2(s) = \mathcal{O}(s).$$

For very recent contributions in this direction see NIEDERREITER and XING [1369].

Numerical results concerning star discrepancy and L_2-discrepancy, as well as some critical remarks on discrepancy as an efficient error measure for practical purposes was given in MOROKOFF and CAFLISCH [1205]; see also SOBOL and SHUKHMAN [1713].

3.1.3 Mathematical Finance

We will discuss briefly two examples from mathematical finance, a recent field of application of Quasi-Monte Carlo methods, where high-dimensional integrals occur.

First we consider a risk model introduced by GERBER [650] with a process of claims

$$S_t = \sum_{i=0}^{N(t)} X_i$$

of an insurance company. The claim-number-process $N(t) \sim Poisson(\lambda t)$ is modeled by a POISSON-process with parameter λt and $N(0) = 0$; the single claims X_i are independently identically distributed with cumulative distribution function $F_X(x)$ and finite mean μ. In order to compensate the claims X_i, the insurance company is being paid a premium of $c > \lambda\mu$ per unit time. The insurance wants to pay a dividend to its shareholders or its clients, and decides to do this in the following way: Dividends are paid whenever the free reserve of the company reaches a given barrier, so that the free reserve stays on this barrier until the next claim occurs.

We assume now that we have a linear barrier $b+at$, and that our initial free reserve is given by x. We are interested in the expectation of the amount of dividends that are paid, preceding possible ruin. Ruin occurs in the event that the free reserve drops below zero.

The expectation of the dividends $g(x,b)$ can then be determined by the fixed point theorem of BANACH as shown in GERBER [650] and TICHY [1826], using the following operator A given by

$$\begin{aligned} Ag(x,b) &= \int_0^{\frac{b-x}{c-a}} \lambda e^{-(\lambda+\delta)t} \int_0^{x+ct} g(x+ct-y, b+at)dF(y)dt + \\ &+ \int_{\frac{b-x}{c-a}}^{\infty} \lambda e^{-(\lambda+\delta)t} \int_0^{b+at} g(b+at-y, b+at)dF(y)dt + \\ &+ \frac{c-a}{\lambda+\delta} e^{-(\lambda+\delta)\frac{b-x}{c-a}}. \end{aligned}$$

We iterate this formula to get an approximate solution for g (see SIEGL and TICHY [1688]). For n iterations we have to compute $2n$-dimensional integrals. In order to do this, we use Quasi-Monte Carlo methods rather than standard quadrature formulas which would need excessive computation time.

Now we compute the integral by taking N points of dimension $2i_{max}$. For $k = 1,\ldots,N$ we compute the value of g at level i by

$$g_k^i(x,b) \tag{3.25}$$

$$= e^{-\delta t_k^i} f(y_k^i) \begin{cases} g_k^{i-1}(x+ct_k^i - y_k^i, b+at_k^i)e^y(1-e^{-x-ct_k^i}) & , \text{ if } t_k^i \le u \\ g_k^{i-1}(b+at_k^i - y_k^i, b+at_k^i)e^y(1-e^{-b-at_k^i}) & , \text{ otherwise} \end{cases}$$

$$+ \frac{c-a}{\lambda+\delta} e^{-(\lambda+\delta)\frac{b-x}{c-a}} \tag{3.26}$$

for $i > 1$ and $g^0 \equiv \frac{1}{2}$. Then the approximation of g in (x, b) is given by $g^{i_{max}} = \frac{1}{N}\sum_{k=1}^{N} g_k^{i_{max}}$. The t_k^i are exponentially distributed pseudo-random variates with parameter λ. Because of this, we can eliminate the multiplication by $\lambda e^{-\lambda t}$ due to

$$\int f(t)h(t)dt = \int f(t)dH(t) = \mathbf{E}(f(T)). \tag{3.27}$$

The variate y_k^i is exponentially distributed cut off at $\min(x+ct_k^i, b+at_k^i)$. The integral equation is therefore multiplied by $e^y(1-e^{-h})$ to balance this effect.

The integral equation above is closely related to the corresponding simulation procedure. To compute the expectation of the dividend payments, we define claim-times t_i with $t_0 = 0$, where $t_i = t_{i-1} + \varepsilon$ and ε is an exponentially distributed random variate. For all claims i with $x_{i-1} + c\varepsilon \geq b + at_i$ we set $\sigma_i = \sigma_{i-1} + \frac{c-a}{\delta}\left(e^{-\delta t_i} - e^{-\delta\left(t_{i-1} + \frac{b_{i-1}-x_{i-1}}{c-a}\right)}\right)$ and $\sigma_0 = 0$. We terminate execution of the procedure when $t_i > t_{max}$ or $x_i < 0$. Here t_{max} is an approximation of the time where the discounted dividend payments get small enough to be negligible. The final value of σ_i is designated by σ. Then we have for the expectation

$$\mathbf{E}[g(x,b)] \approx \frac{1}{N}\sum_{j=1}^{N}\sigma(i). \tag{3.28}$$

The results can be compared with exact solutions obtained in SIEGL and TICHY [1688]. The advantage of the integral equation is the fact that the corresponding error-estimate works on the supremum norm, whereas the simulation procedure relies on the fact that errors average out for large N, thereby giving better mean error and worse maximal error.

For the integral equation we can use Quasi-Monte Carlo methods as well. To get good approximations we need to compute some high dimensional integrals. The use of the HALTON sequence is not desireable because we need a very high number of evaluations to overcome the structural deficits of this method. However, for a high number of evaluations other sequences will result in better discrepancy. It turns out that this implementation of the recursion method is closely related to the simulation procedure. The results support this close connection, because the pseudo random generator gives comparable results to the Quasi-Monte Carlo evaluation. Net sequences seem to be more suitable for applications of this kind.

Recent work on a second specific field of application has generated a big interest of financial institutions in Quasi-Monte Carlo methods. The paper of PASKOV and TRAUB [1406] introduces Quasi-Monte Carlo methods to evaluate **collateralized mortage based securities (CMOs).** CMOs were first considered in 1983 in response to investor demands for mortage-backed securities that are targeted at different segments of the yield curve. Since then, the market has evolved considerably and the structures of CMOs have become far more complex.

The prepayment option of the underlying mortages makes the CMOs difficult to analyse. CMOs solve this problem by dividing the cash flow from a pool of mortages into tranches. This makes the analysis of the performance of a CMO easier. By selecting different pools of mortages different risks and yields are generated. For more details on CMOs, we refer to FABOZZI [576].

PASKOV and TRAUB study a CMO that consists of ten tranches. The underlying pool of mortages has a 30 year maturity and cash flows are obtained monthly. This implies 360 cash flows. The actual rules are rather complex and they are given in the prospectus describing the financial product. We want to estimate the expected value of the sum of present values of future cash flows for each of the tranches.

For $j = 1, \ldots, 360$, let C be the monthly payment on the underlying pool of mortages, and let i_j be the appropriate interest in month j. By w_j we denote the percentage prepaying in month j and with $a_{360-j+1}$ we denote the remaining annuity after month j.

Recall that the remaining annuity a_j is given by

$$a_j = 1 + v_0 + \cdots + v_0^{j-1}$$

with discount rates $v_j = 1/(1+m_0)$ and i_0 the current monthly interest rate. In this notation C and a_j are constants and i_j and w_j are stochastic variables to be determined below. We now describe the interest rate model. Assume that the interest rate i_j is of the form

$$i_j = K_0 e^{\xi_j} i_{j-1} = K_0^j i_0 e^{\xi_1 + \cdots + \xi_k},$$

where $\{\xi_j\}_{j=1}^{360}$ are independent normally distributed random variates with mean 0 and variance σ^2, and K_0 is a given constant. In our case $\sigma^2 = 0.0004$ is chosen. Suppose that the prepayment model w_j, as a function of i_j, is computed as

$$\begin{aligned} w_j &= w_j(\xi_1, \ldots, \xi_j) = K_1 + K_2 \arctan(K_3 i_j + K_4) \\ &= K_1 + K_2 \arctan(K_3 K_0^j i_0 e^{\xi_1 + \cdots + \xi_j} + K_4), \end{aligned}$$

where K_1, K_2, K_3, and K_4 are given constants. The cash flow in month j, $j = 1, 2, \ldots, 360$ is

$$\begin{aligned} M_j &= M_j(\xi_1, \ldots, \xi_j) \\ &= C \prod_{l=1}^{j-1} (1 - w_l(\xi_1, \ldots, \xi_l)) + (1 - w_l(\xi_1, \ldots, \xi_l) + w_l(\xi_1, \ldots, \xi_l) a_{360-k+1}). \end{aligned}$$

The cash flow is distributed to the tranches according to the rules of the CMO under consideration. Let $G_{j;T}(\xi_1, \ldots, \xi_j)$ be the portion of the cash flow M_j for month j directed to tranche T. The form of this function is very complex. It suffices to observe that it is a continuous function. To find the present value of the tranche T for month j, we have to multiply by the discount factor, that is the product of the yearly discounts

$$u_j(\xi_1, \ldots, \xi_{k-1}) = v_0 v_1(\xi_1) \cdots v_{j-1}(\xi_1, \ldots, \xi_{j-1}).$$

Summing up over all periods j for tranch T gives the present Value PV_T,

$$PV_T(\xi_1,\ldots,\xi_{360}) = \sum_{j=1}^{360} G_{j;T}(\xi_1,\ldots,\xi_j)u_j(\xi_1,\ldots,\xi_j).$$

We want to compute the expected value $\mathbf{E}[PV_T]$. By change of variables, it is easy to see that:

$$\mathbf{E}[PV_T]= \int_{[0,1]^{360}} PV_T(y_1(x_1),\ldots,y_{360}(x_{360}))dx_1\cdots dx_{360},$$

where $y_j = y_j(x_j)$ is implicitly given by

$$x_i = \frac{1}{\sqrt{2\pi\sigma}} \int_{-\infty}^{y_i} e^{-t^2/(2\sigma)}dt.$$

Therefore, our problem is reduced to a problem of computing ten multivariate integrals over the 360-dimensional unit cube. We stress that after generating a point $x_1,\ldots,x_{360}$ the point $y_1,\ldots,y_{360}$ has to be computed by finding the value of the inverse normal cumulative distribution function at each x_j, $j = 1,2,\ldots,360$. Due to the complex form of PV_T it is very difficult to find a good bound for the variation of PV_T. It is widely believed that low discrepancy sequences are effective in general only for dimension < 30. Although the CMO problem is apparently of higher dimension some tranches are of lower dimension. The effect of the lower dimension is higher than that of the higher dimensions.

The SOBOL sequence is particularly suited for such integrands. This sequence is a special dyadic sequence of the net type; see SOBOL [1701]. PASKOV and TRAUB find that the Quasi-Monte Carlo algorithms converge significantly faster than Monte Carlo algorithms, and the convergence is smoother.

Using the method of PASKOV and TRAUB, the integration terminates 2 to 5 times faster with the SOBOL sequence than the Monte Carlo algorithm. The risk-model introduced before will yield smaller absolute errors with Quasi-Monte Carlo methods than with classical simulation. This indicates the superiority of these methods in practical applications.

The use of Quasi-Monte Carlo methods and the advance of high-performance computers makes it possible to model many insurance related problems which were until now considered unaccessible. Most notable among these problems is the use of stochastic interest rates, realistic claimsize and claimnumber distributions and dynamic market shares. At the TU Graz our group is working on a project that deals with these topics.

3.1.4 Average Case Analysis

In contrast to the worst case analysis of an algorithms the average case analysis requires a probability distribution on the space of all possibel input data. Hence, specific parameters of the algorithms can be considered as random variables. Usually one is interested in the expected value and the variance in order to get a first insight of the "typical" behaviour of the algorithm.

In the following we want to analyze the average case behaviour of numerical integration of continuous functions. For this purpose we equip the class of continuous functions on the unit cube with the classical WIENER measure (c.f. ITO and MC KEAN [863]); see also Section 2.3. We will present a result of WOŹNIAKOWSKI [1972] which yields an optimal construction for the integration points. The construction is essentially based on the L^2-discrepancy $D_N^{(2)}(\mathbf{x}_n)$ of a point sequence $\mathbf{x}_n$. We will use the following point sequence $(\mathbf{z}_n^*)$ defined by

$$\{\mathbf{z}_1^*, \ldots, \mathbf{z}_N^*\} = \left\{ \left(\frac{n+t^*}{N}, u_n^{(1)}, \ldots, u_n^{(k-1)} \right) : 0 \le n + t^* < N \right\},$$

where $u_n^{(j)}$ denote the components of the $(k-1)$-dimensional HALTON sequence. For $t^* = 0$ this is just the HAMMERSLEY sequence. Let us remark here that this sequence is an example of a sequence with optimal L^2-discrepancy, cf. ROTH [1569, 1570].

We now recall the basic properties of WIENER measure. Let $\mathcal{F}$ be the class of all real-valued continuous functions on the k-dimensional unit cube $I_k = [0,1]^k$. The WIENER measure μ_w is a probability measure on $\mathcal{F}$, (for functions in one variable see also Section 2.3). It is a Gaussian measure with mean value 0 and covariance kernel

$$R(\mathbf{x}, \mathbf{y}) = \int_{\mathcal{F}} f(\mathbf{x}) f(\mathbf{y}) \, d\mu_w(f) = \min(\mathbf{x}, \mathbf{y}) = \prod_{j=1}^{k} \min(x^{(j)}, y^{(j)}),$$

for all points $\mathbf{x} = (x^{(1)}, \ldots, x^{(k)}), \mathbf{y} = (y^{(1)}, \ldots, y^{(k)}) \in [0,1]^k$. (Note that in Section 2.3 functions in one variable with values on a Riemannian manifold were considered.)

Furthermore, set $\mathbf{x}_n^* = \mathbf{1} - \mathbf{z}_n^*$. We approximate the integral of a continuous function f by the arithmetic mean extended over the integration points $\mathbf{x}_n^*$,

$$I(f) = \int_{I^k} f(\mathbf{x}) \, d\mathbf{x} \approx \frac{1}{N} \sum_{n=1}^{N} f(\mathbf{x}_n^*) = I_N^*(f). \tag{3.29}$$

In the following we will show that $I_N^*(f)$ is an optimal algorithm in the sence of average case complexity with respect to WIENER measure.

Usually one wants to establish an algorithms with small costs. Costs depend on computation time and on the approximation error. In case of approximating the integral $I(f)$ by the arithmetic mean $I_N(f)$, cost $I_N(f)$ is proportional to N since there are cN function evaluations. (For every evaluation we assume constant costs c) and N operations ($N-1$ additions and 1 division.) The average error of numerical integration with respect to WIENER measure is given by

$$E^{avg}(I_N) = \left(\int_{\mathcal{F}} (I(f) - I_N(f))^2 \, d\mu_W(f) \right)^{\frac{1}{2}}.$$

The average case ϵ-complexity is then given as the minimal cost of all algorithms I_N with average error $\leq \epsilon$:

$$comp^{avg}(\epsilon, \mathcal{F}) = \inf\{\text{cost}\, I_N : E^{avg}(I_N) \leq \epsilon\}$$

Lemma 3.15 *Let f be a continuous function and $I(f) = \int_{I^k} f(\mathbf{x})\, d\mathbf{x}$ and. Then*

$$\int_{\mathcal{F}} I(f)^2 \, d\mu_w(f) = 3^{-k}$$

and

$$\int_{\mathcal{F}} I(f) f(\mathbf{y}) \, d\mu_w(f) = \prod_{j=1}^{k} y^{(j)} \left(1 - \frac{y^{(j)}}{2}\right).$$

Proof. First we consider the case $k = 1$.

$$\begin{aligned}
\int_{\mathcal{F}} I(f)^2 \, d\mu_w(f) &= \int_0^1 \int_0^1 \int_{\mathcal{F}} f(x) f(y) \, d\mu_w(f) \, dx \, dy \\
&= \int_0^1 \int_0^1 \min(x, y) \, dx \, dy \\
&= \int_0^1 \int_0^y \min(x, y) \, dx \, dy + \int_0^1 \int_y^1 \min(x, y) \, dx \, dy \\
&= \int_0^1 \int_0^y x \, dx \, dy + \int_0^1 \int_y^1 y \, dx \, dy \\
&= \int_0^1 \frac{y^2}{2} dy + \int_0^1 y\,(1 - y) \, dy \\
&= \int_0^1 \frac{y^2}{2} + y\,(1 - y) \, dy = \frac{1}{3}
\end{aligned}$$

For $k > 1$ we obtain

$$\begin{aligned}
\int_{\mathcal{F}} I(f)^2 \, d\mu_w(f) &= \int_{I^k} \int_{I^k} \int_{\mathcal{F}} f(\mathbf{x}) f(\mathbf{y}) \, d\mu_w(f) \, d\mathbf{x} \, d\mathbf{y} \\
&= \int_{I^k} \int_{I^k} \prod_{j=1}^{k} \min\{x^{(j)}, y^{(j)}\} \, d\mathbf{x} \, d\mathbf{y} \\
&= \prod_{j=1}^{k} \left(\int_0^1 \int_0^1 \min(x^{(j)}, y^{(j)}) \, dx^{(j)} \, dy^{(j)} \right) = 3^{-k}
\end{aligned}$$

The second identity can be proved in a similar way. □

Lemma 3.16 *Let $I_N(f) = \frac{1}{N}\sum_{n=1}^{N} f(\mathbf{x}_n)$ for an arbitrary given sequence $(\mathbf{x}_n)$ and let $I(f)$ be as above. Then the L^2-discrepancy of $(\mathbf{z}_n) = (1-x_n^{(1)}, \ldots, 1-x_n^{(k)})$ can be computed by*

$$\left(D_N^{(2)}(\mathbf{z}_n)\right)^2 = \int_{\mathcal{F}} (I(f) - I_N(f))^2\, d\mu_w(f).$$

Proof. We apply Lemma 3.15 and proceed as in (3.6):

$$\begin{aligned}
&\int_{\mathcal{F}} (I(f) - I_N(f))^2\, d\mu_w(f) \\
&= \int_{\mathcal{F}} I^2(f) - 2\, I_N(f) I(f) + I_N^2(f)\, d\mu_w(f) \\
&= 3^{-k} - 2\int_{\mathcal{F}} I(f)\frac{1}{N}\sum_{n=1}^{N} f(\mathbf{x}_n)\, d\mu_w(f) + \int_{\mathcal{F}} \left(\frac{1}{N}\sum_{n=1}^{N} f(\mathbf{x}_n)\right)^2 d\mu_w(f) \\
&= 3^{-k} + \frac{2}{N}\sum_{n=1}^{N}\int_{\mathcal{F}} I(f) f(\mathbf{x}_n)\, d\mu_w(f) + \frac{1}{N^2}\sum_{n,m=1}^{N}\int_{\mathcal{F}} f(\mathbf{x}_n) f(\mathbf{x}_m)\, d\mu_w(f) \\
&= 3^{-k} + \frac{2}{N}\sum_{n=1}^{N}\prod_{j=1}^{k} x_n^{(j)}\left(1 - \frac{x_n^{(j)}}{2}\right) + \frac{1}{N^2}\sum_{n,m=1}^{N}\prod_{j=1}^{k}\min(x_n^{(k)}, x_m^{(k)}).
\end{aligned}$$

For the left hand side we derive

$$\begin{aligned}
\left(D_N^{(2)}(\mathbf{z}_n)\right)^2 &= \int_{I^k}\left(t^{(1)}\cdots t^{(k)} - \frac{1}{N}\sum_{n=1}^{N}\chi_{[0,t)}(\mathbf{z}_n)\right)^2 dt \\
&= \int_{I^k}\left(t^{(1)}\cdots t^{(k)}\right)^2 dt - \frac{2}{N}\sum_{n=1}^{N}\int_{I^k} t^{(1)}\cdots t^{(k)}\chi_{[0,t)}(\mathbf{z}_n)\, dt \\
&\quad + \frac{1}{N^2}\sum_{n,m=1}^{N}\int_{I^k}\chi_{[0,t)}(\mathbf{z}_n)\chi_{[0,t)}(\mathbf{z}_m)\, dt \\
&= \prod_{j=1}^{k}\int_0^1\left(t^{(j)}\right)^2 dt^{(j)} - \frac{2}{N}\sum_{n=1}^{N}\prod_{j=1}^{k}\int_0^1 t^{(j)}\chi_{[0,t^{(j)})}(z_n^{(j)})\, dt^{(j)} \\
&\quad + \frac{1}{N^2}\sum_{n,m=1}^{N}\prod_{j=1}^{k}\int_0^1\chi_{[0,t^{(j)})}(z_n^{(j)})\chi_{[0,t^{(j)})}(z_m^{(j)})\, dt^{(j)} \\
&= 3^{-k} - \frac{2}{N}\sum_{n=1}^{N}\prod_{j=1}^{k}\int_{z^{(j)}}^1 t^{(j)}\, dt^{(j)} \\
&\quad + \frac{1}{N^2}\sum_{n,m=1}^{N}\prod_{j=1}^{k}\int_{\max(z_n^{(j)}, z_m^{(j)})}^1 dt^{(j)}
\end{aligned}$$

$$= 3^{-k} - \frac{2}{N}\sum_{n=1}^{N}\prod_{j=1}^{k}\frac{1-(z_n^{(j)})^2}{2} + \frac{1}{N^2}\sum_{n,m=1}^{N}\prod_{j=1}^{k}\left(1-\max(z_n^{(j)}z_m^{(j)})\right).$$

Comparing with (3.6) we derive the desired result. □

The following lemma is a slight generalization of the main lemma in ROTH [1567], where a lower discrepancy bound was obtained. (As above we use the notation $\mathbf{1} = (1,\dots,1)$.)

Lemma 3.17 *There exists a positive number γ_k such that for all reals c_n and x_n*

$$\int_{I^k}\left(\sum_{n=1}^{N} c_n\chi_{[0,\mathbf{t})}(\mathbf{1}-\mathbf{x}_n) - t_1\cdots t_k\right)^2 dt_1\dots dt_k \geq \gamma_k\frac{1}{N^2}(\log N)^{k-1}. \tag{3.30}$$

Proof. Similarly to the proof of Proposition 2.19 we set

$$D(\mathbf{t}) = \sum_{n=1}^{N} c_n\chi_{[0,\mathbf{t})}(\mathbf{1}-\mathbf{x}_n) - t_1\cdots t_k$$

and

$$F(\mathbf{t}) = \sum_{\|\mathbf{r}\|_1=n} f_{\mathbf{r}}(\varrho),$$

where n is defined by $2N \leq 2^n < 4N$ and on every $\mathbf{r}$-box $I = I_1\times\cdots\times I_k$ (i.e. I_j are of the form $I_j = [m_j2^{-r_j},(m_j+1)2^{-r_j})$, where $0\leq m_j < 2^{r_j}$, see Section 1.3) we have

$$f_{\mathbf{r}}(\mathbf{t}) = \pm R_{\mathbf{r}}(\mathbf{t}),$$

where the sign is chosen such that (for every $\mathbf{r}$-box I)

$$\int_I f_{\mathbf{r}}(\mathbf{t})D(\mathbf{t})\,d\mathbf{t} \geq 0. \tag{3.31}$$

Now consider an $\mathbf{r}$-box $I = I_1\times\cdots I_k$, $I_j = [m_j2^{-r_j},(m_j+1)2^{-r_j})$, not containing a point $\mathbf{x}_1,\dots,\mathbf{x}_N$. As in the proof of Proposition 2.19 we obtain

$$\begin{aligned}\int_I R_{\mathbf{r}}(\mathbf{t})D(\mathbf{t})\,d\mathbf{t} &= (-1)^k\int_{I'}(-\lambda_k(B(\mathbf{t}))\,d\mathbf{t} \\ &= (-1)^{k+1}2^{-\|\mathbf{r}\|_1-2k},\end{aligned} \tag{3.32}$$

where $I' = I_1'\times\cdots\times I_k'$, $I_j' = [m_j2^{-r_j},(m_j+1/2)2^{-r_j})$, and $B(\mathbf{t}) = \prod_{j=1}^{k}[t_j,t_j+2^{-r_j-1}]$. Note that for every $\mathbf{r}$ with $\|\mathbf{r}\|_1 = n$ there are exactly 2^n $\mathbf{r}$-boxes. Since $N\leq\frac{1}{2}2^n$ (3.32) implies

$$\int_{[0,1)^k} f_{\mathbf{r}}(\mathbf{t})D(\mathbf{t})\,d\mathbf{t} \geq 2^{-n-2k-1}.$$

Hence we obtain

$$\int_{[0,1)^k} F(\mathbf{t})D(\mathbf{t})\,d\mathbf{t} \geq \binom{n+k-1}{k-1} 2^{-n-2k-1} \gg \frac{(\log N)^{k-1}}{N}.$$

Finally, since

$$\int_{[0,1)^k} F(\mathbf{t})^2\,d\mathbf{t} = \binom{n+k-1}{k-1} \ll (\log N)^{k-1},$$

a direct application of the CAUCHY-SCHWARZ inequality yields (3.30). □

Theorem 3.18 *The average case ϵ-complexity can be estimated by*

$$comp^{avg}(\epsilon, \mathcal{F}) = \mathcal{O}\left(\epsilon^{-1}(\log \epsilon^{-1})^{\frac{k-1}{2}}\right)$$

and the algorithm $I_N^*(f)$ defined by (3.29) is optimal.

Proof. Let

$$N = c\epsilon^{-1}\left(\log \epsilon^{-1}\right)^{(k-1)/2}.$$

Thus we obtain for $T_N = D_N^{(2)}(\mathbf{z}_n^*)$

$$\begin{aligned} T_N &\leq c'\frac{1}{N}(\log N)^{\frac{k-1}{2}} \\ &= \frac{c'}{c}\epsilon\frac{\left(\log\left(\epsilon^{-1}(\log \epsilon^{-1})^{\frac{k-1}{2}}\right)\right)^{\frac{k-1}{2}}}{(\log \epsilon^{-1})^{\frac{k-1}{2}}} \\ &= \frac{c'}{c}\epsilon\frac{\left(\log \epsilon^{-1} + \frac{k-1}{2}\log\log \epsilon^{-1}\right)^{\frac{k-1}{2}}}{(\log \epsilon^{-1})^{\frac{k-1}{2}}} \\ &= \frac{c'}{c}\epsilon\left(1 + \frac{k-1}{2}\frac{\log\log \epsilon^{-1}}{\log \epsilon^{-1}}\right)^{\frac{k-1}{2}}. \end{aligned}$$

Since

$$\lim_{x\to\infty}\frac{\log x}{x} = 0,$$

we can choose c sufficiently large such that T_N can be made arbitrary small. Because of Lemma 3.16 we conclude that the error of $I_N^*(f)$ does not exceed ϵ. Thus the upper bound for the complexity is proved.

In order to obtain the lower bound we proceed as follows. First we consider the case where the information is non adaptive. Since numerical integration is a linear problem we know from the general theory as described in TRAUB, WASILKOWSKI and WOŹNIAKOWSKI [1873] that also the optimal solution algorithm $\hat{I}_N^*(f)$ is linear. Thus this algorithm is given by

$$\hat{I}_N^*(f) = \sum_{n=1}^{N} c_n f(\mathbf{x}_n)$$

with reals c_n. As in Lemma 3.16 we obtain:

$$\int_{\mathcal{F}} (I(f) - \hat{I}_N^*(f))^2 d\mu_w(f) = \int_{I^k} \left(\sum_{n=1}^{N} c_n \chi_{[0,\mathbf{t})}(\mathbf{1} - \mathbf{x}_n) - t^{(1)} \cdots t^{(k)} \right)^2 dt.$$

By Lemma 3.17 this expression is bounded from below by $\gamma_k N^{-2}(\log N)^{k-1}$. Thus we have shown the lower bound.

Now we consider the problem if the use of adaptive information can improve the complexity. However, since $N^{-2}(\log N)^{k-1}$ is a convex function in N, the square mean of the minimal error can be bounded by a convex function. Thus by the above result we obtain that adaptive information cannot improve the complexity of the algorithm. □

Notes

In a series of papers Hlawka systematically investigated several applications of uniformly distributed sequences. He built up a theory ("Number-theoretic Analysis") applying uniformly distributed sequences to different kinds of problems of numerical analysis and applied mathematics. In [796] he studied the numerical analytic continuation of multivariate complex functions on polycylinders. This work was continued by Taschner [1782] and Larcher [1015, 1016]. The analytic continuation of special Dirichlet series by means of Kronecker sequences is discussed in Hlawka [808]; see also [810]. In [803] he studied applications of uniform distribution to the approximation theorem of Weierstraß; the error is estimated in terms of the discrepancy. Haslinger [755] established results on Newton interpolation polynomials and u.d. A discrepancy estimate for polynomials of best approximation is due to Blatt and Mhaskar [228]. Applications of u.d. sequences to the Müntz approximation theorem are due to Hlawka [818]. For applications of this kind it is important to use discrepancies defined via polynomials (instead of characteristic functions), see also Hlawka [798, 799, 797], Schmidt [1640] and Tichy [1822]. In a recent contribution Klinger and Tichy [938] could extend the concept of polynomial discrepancy to spherical sequences.

For the average case analysis of numerical integration we refer to the monograph of Woźniakowski and Traub [1873], to Novak [1372, 1371] and Woźniakowski [1972]. Morokoff and Caflisch [1205] analyzed several types of discrepancies and gave a simplified proof of Woźniakowski's result. Recently, Wasilkowski and Woźniakowski [1947] applied the average case analysis approach to obtain an upper bound 1.4778842 for the smallest p for which there exists a $D > 0$ such that for all k and all $\varepsilon \leq 1$ there exist $D\varepsilon^{-p}$ points with k-dimensional L^2-discrepancy at most ε.

Proinov [1461, 1466] considered integration schemes involving u.d. double sequences; see also Proinov and Kirov [1485, 1486]. For an application of discrepancy to the summation of multiple sums see Proinov and Peeva [1487]. Diamond and Vaaler [459] proved metric results on the approximation error for numerical integration using the $(n\alpha)$-sequence. Trendafilov [1874, 1875] established error bounds for the numerical integration of vector-valued functions. For numerical integration of singular functions we refer to Sobol [1702], Klinger [937], Driver, Lubinsky, Petruska and Sarnak [474] and Myerson [1227]; for related reults see also Bouleau [275] and Oskolkov [1397].

Shparlinskij [1687] gave an extensive survey of different kinds of measures for the uniform distribution of point sequences. Furthermore he discussed applications of these quantities to the solution of linear equations, to Chebyshev and Hermite interpolation problems, and to special codes; see also [1681, 1683, 1685] and Fischer and Riesler [600]. Babaev [59] estimated the number of operations needed to compute an integral of a multivariate continuous function. Tichy [1826] established approximative formulas for the numerical evaluation of Fourier coefficients with respect to various kinds of orthogonal systems such as multivariate trigonometric, Hermite, and Laguerre functions; see also Hlawka [795] for the classical Fourier coefficients and Blümlinger and Tichy [241]. The main point in these investigations is a general inequality for the total variation of the product of functions. An extension is due to Reitgruber [1525]. Blümlinger and Tichy [242] and Blümlinger [232] used this inequality to study a special Banach algebra of functions; for multivariate functions of bounded variations see also Reitgruber [1526]. Blaga [225] presented a Monte Carlo approximation of an integral extended over a simplex. For numerical experience on integration with quasirandom sequences we refer to Radović, Sobol and Tichy [1497] and Sobol and Shukhman [1714, 1713, 1712, 1715]. Sobol and Tutunnikov [1716] studied a variance reducing multiplier for Monte Carlo integrations. An important contribution on pseudorandom numbers and optimal coefficients is due to Niederreiter [1293].

For numerical integration and approximation of smooth functions using u.d. sequences and L^p-discrepancy with respect to general weights we refer to Proinov [1475, 1468, 1473] and Proinov and Kirov [1485, 1486]; for related results see also Proinov [1464, 1461], Stegbuchner [1739], and Zinterhof [1998]. Sobol [1705, 1706] considered numerical integration and the determination of the extremal values of multivariate Lipschitz functions, see also Sobol and Bakin [1711]. Levitan, Markovich, Rozin and Sobol [1079] and Sobol [1709, 1708] investigated various types of quasirandom sequences and applications. Bruneau [304, 305] studied continuity problems of lacunary Fourier series. For further contributions to numerical integration using Quasi-Monte Carlo methods and applications of diophantine approximations we refer to Niederreiter [1270, 1313, 1278]. Zinterhof [1999] introduced the notion of gratis lattice points for multidimensional integration. A survey on numerical integration via good lattice points is due to Haber [712]. Niederreiter [1320] and Wang [1940] gave surveys on Quasi-Monte Carlo methods for numerical integration. Sobol [1704, 1707] gave popular introductions on u.d. and various applications. For an introduction into Monte Carlo methods see Kalos and Whitlock [888] and Sobol [1710].

Glinkin [659] established quadrature formulas for convex functions. For further results and on quadrature formulas see also Hua and Wang [851, 852, 853], Israilov [862], Kuzyutin [997], and De Clerck [437]. Sarkar and Prasad [1600] studied integral equations by means of Quasi-Monte Carlo methods using Halton and Faure sequences and worked out a detailed comparison of the sequences. Fredholm integral equations concerned with a problem in radiosity were investigated by Keller [917].

Niederreiter and Sloan [1359] established upper and lower bounds for the discrepancy of nodes in lattice rules for multi-dimensional numerical integration, they also extend the applicability of lattice rules to non-periodic functions; see also Niederreiter [1330, 1337], Squire [1731], and Niederreiter and Sloan [1361] for Fibonacci lattice rules in two dimensions. Masry and Cambanis [1129] investigated trapezoidal Monte-Carlo integration and gave estimates for the mean square approximation; Mikhajlov [1193, 1194] studied weighted Monte-Carlo methods. A variant of the Koksma-Hlawka inequality for vertex-modified integration rules is due to Niederreiter and Sloan [1362]. For further results on lattice rules we refer to Korobov [962], Niederreiter [1331, 1332, 1339] and Niederreiter and Sloan [1360]. A survey is due to Sloan and Joe [1697].

The proof of existence of good lattice points with respect to general moduli is due to Niederreiter [1295], for applications of good lattice points to optimization see Niederreiter [1312]. Larcher [1021] showed that the existence theorem of Niederreiter on good lattice points is essentially best possible. Existence theorems on good lattice points in three dimensions are due to Larcher and Niederreiter [1036]. Borosh and Niederreiter [267] established a systematic search on optimal multipliers. Further results on good lattice points can be found in Larcher [1017, 1018] and Zaremba [1987, 1985]. Wang, Xu and Zhang [1939] established tables for good lattice points mod m in k dimensions for $m \leq 5.5 \cdot 10^7$ and $k \leq 18$. Sugiura [1762] published tables for g.l.p. formulae in 3, 4, 5, and 6 dimensions. Wang [1937, 1938, 1936] proved discrepancy bounds for a lattice-type point set with applications to numerical integration. For further results in that direction see Zhu [1994, 1995].

Applications of u.d. to Pythagorean triples can be found in Hlawka [805]. In [807] the same author studied the distribution of the angles in the construction of the square root snail; see also Teufel [1798].

Drmota [481] considered applications of Quasi-Monte Carlo methods to search processes with applications in electrical engeneering; for applications of c.u.d. functions in signal processing see also Drmota and Tichy [490]. Hlawka [832] considered applications of u.d. sequences to Buffon's problem; see also [831]. In forthcoming papers [779] the same author considers various applications to probability theory and statistics.

Applications of uniformly distributed sequences to mathematical linguistics can be found in Hlawka [806]. This work was continued by Tichy [1815], Kirschenhofer and Tichy [924] and Pechlaner [1423]. Moriguti [1202] and later Tichy [1836] studied distribution properties of "computer rational" numbers.

In Hlawka [813] linear difference equations in several variables are studied by means of uniformly distributed sequences. In the two papers [816] and [817] Hlawka replaced the lemma of Du Bois-Reymond of variational calculus by a quantitative version yielding an estimate for the involved function in terms of the discrepancy; see also Tichy [1824]. Further investigations on variational calculus and uniform distribution can be found in Hlawka [819]. In [790] and [794] the same author investigated several applications of uniformly distributed sequences in kinetic gas theory. Hlawka [833] discussed uniform distribution and entropy. Mendès France [1177, 1178] considered applications of u.d. to the entropy of planar curves. Hlawka [828] studied various geometric and physical applications of uniform distribution, such as potentials and convex billiards. For a survey on several applications of u.d. sequences we refer to Hlawka [838].

Hlawka [795] studied first applications of u.d. sequences to compute extremal values of functions. Later many authors analyzed problems of global optimization by means of the dispersion of sequences (see Notes of Section 1.1). Lambert [1008, 1009] investigated how to implement low-dispersion sequences for applications in optimization and gave a survey on such sequences.

Methods for the computation of the discrepancy can be found in Niederreiter [1271], Zaremba [1988], and De Clerck [439]. Bundschuh and Zhu [315] proved an explicit formula in two and three dimensions which was extended by Zhu [1996] to arbitrary dimensions. Theorem 3.6 is due to Lin Achan [2]. For explicit formulas to compute the discrepancy we also refer to Niederreiter [1271, 1274].

Sequences of the net type were first introduced by Sobol [1701] in the dyadic case; see also Srinivasan [1732]. Recently, the discrepancy of this kind of sequences in two dimensions was analyzed precisely by Faure [593]. Faure [582] obtained a construction of such low discrepancy sequences with respect to an arbitrary prime base b and arbitrary dimension. For results of this kind see also Faure [589, 591, 592] and the Notes of Section 1.4. A systematic approach to net sequences is due to Niederreiter [1317, 1319, 1328, 1336, 1333]. A detailed discussion of constructions (based on finite rings) of net sequences is due to Larcher, Niederreiter and Schmid [1039]. For using non-archimedian techniques and for metric results see Larcher and Niederreiter [1038] and the earlier papers by Larcher [1034] and Niederreiter [1343]. In Larcher and Niederreiter [1038] the authors define diophantine approximation constants for formal Laurent series over finite fields and show connections with distribution properties of Kronecker-type sequences, focussing on probabilistic results on the distribution of pseudo-random sequences constructed by the digital method. Recently Niederreiter and Xing [1368, 1975] constructed net-sequences obtained from algebraic function fields over finite fields. A special example is discussed which originates in the elliptic curve $y^2 + y = x^3 + x + 1$ over the prime field of characteristic 2. In the very recent contribution [1369] Niederreiter and Xing construct digital (t, s)-sequences in prime power base q for which the quality parameter t has optimal order of magnitude. The construction is based on algebraic function fields of order q which contain many places of degree 1 relative to the genus.

Larcher [1030], Larcher, Lauß, Niederreiter and Schmid [1035], Larcher and Schmid [1040], Larcher, Schmid and Wolf [1041] and Larcher and Traunfellner [1044] studied applications of net sequences to the numerical integration of high-dimensional Walsh series. For more combinatorial aspects of net sequences and relations to orthogonal arrays we refer to Niederreiter [1335]; see also Mullen and Whittle [1217] and Hansen, Mullen and Niederreiter [733]. Niederreiter [1315] gives various estimates for the discrepancy of different kinds of point sets; see also [1313]. Proinov and Tonchev [1488] gave an upper bound for three-dimensional net-sequences, Xiao [1974] suggested a new quantity for measuring the distribution of nets. Tables of net-sequences are due to Mullen, Mahalanabis and Niederreiter [1216]. For further estimates concerning low discrepancy sequences we refer to Shi [1669, 1671]. For randomly permuted net sequences we refer to Owen [1402].

Niederreiter and Osgood [1349] considered the approximation of an integral arising from a discretization method for solving retarded ordinary differential equations. Halton [728] proposed methods for the simulation of particle streams by means of Monte Carlo methods. Caflisch and Moskowitz [321] studied applications of Quasi-Monte Carlo methods to Feynman-Kac integrals and Spanier [1728] considered applications of Quasi-Monte Carlo methods to particle transport problems. For applications of Quasi-Monte Carlo methods to the numerical evaluation of moments of multivariate distributions on special geometric domains we refer to Wang and Fang [1941, 1942].

3.2 Spherical Problems

3.2.1 Spherical Designs and CHEBYSHEV Quadrature

By a k-dimensional spherical t-design we mean a pointset $\mathbf{x}_1, \dots, \mathbf{x}_N$ on the unit sphere $\mathbf{S}^{k-1}$ such that

$$\frac{1}{N}\sum_{n=1}^{N} p(\mathbf{x}_n) = \int_{\mathbf{S}^{k-1}} p(\mathbf{x})d\sigma^*(\mathbf{x}),$$

where σ^* denotes the normalized surface measure on $\mathbf{S}^{k-1}$, holds for all polynomials p of degree $\leq t$.

The existence of such designs for arbitrary t and k was shown bei SEYMOUR and ZASLAVSKY [1666] and WAGNER was the first to prove an upper bound for the number of points N. In [1926] he showed that

$$N \ll c_k t^{8k^2}.$$

As we will see later on, the main idea is to reduce this question to the problem of finding a bound for the number of points in the CHEBYSHEV quadrature formula. This is a very interesting question for itself and the first to consider this problem was BERNSTEIN in 1937.

A CHEBYSHEV type quadrature formula is a formula of the type

$$\int_{-1}^{1} f(t)\omega(t)dt = \frac{1}{N}\sum_{i=1}^{N} f(x_i) + R(f). \tag{3.33}$$

For the case $\omega(t) = \frac{1}{2}$ BERNSTEIN [194] obtained the widely known result that if formula (3.33) has degree $2n-1$, then the number of points N must satisfy $N \gg n^2$. In another paper [193], which is only available in Russian, BERNSTEIN also proved the converse; i.e. there exists a CHEBYSHEV type quadrature formula of degree $2n-1$ where N satisfies

$$N \leq 4\sqrt{2}(n+1)(n+4).$$

Recently KUIJLAARS [967, 968] could extend BERNSTEIN's method to get bounds for the case of the ultraspherical weight function

$$\omega_\alpha(t) = C_\alpha(1-t^2)^\alpha, \ \alpha \geq 0$$

and we will present the result below.

Let $P_n(t)$ be the orthogonal polynomial of degree n with respect to some weight function $\omega(t)$ and $1 > \xi_{1,n}, > \cdots > \xi_{n,n} > -1$ be the zeros of $P_n(t)$. Furthermore let $Q_{n-1}(t)$ be the orthogonal polynomial of degree $n-1$ with respect to the weight function $(1-t^2)\omega(t)$ and analogously $1 = \eta_{0,n} > \eta_{1,n} > \cdots > \eta_{n-1,n} > \eta_{n,n} = -1$ be the zeros of Q_{n-1}. We assume that $P_n(1) > 0$ and $Q_{n-1}(1) > 0$. Also consider for every real a

$$P_n(t,a) = P_n(t) - a(1 - \operatorname{sgn}(a)\, t)Q_{n-1}(t),$$

where $\operatorname{sgn}(a)$ denotes the sign of a. Let $1 > \xi_{1,n}(a) > \cdots > \xi_{n,n}(a) > -1$ be the zeros of $P_n(t,a)$. We have $\xi_{i,n} < \xi_{i,n}(a) < \eta_{i-1,n}$ for positive a and $\eta_{i,n} < \xi_{i,n}(a) < \xi_{i,n}$ if $a < 0$.

Now the LOBATTO quadrature formula is defined by

$$\int_{-1}^{1} f(t)\omega(t)dt = \sum_{i=1}^{n} \lambda_n(\eta_{i,n})f(\eta_{i,n}) + R(f) \tag{3.34}$$

The numbers $\lambda_n(x)$ can be characterized by a certain extremal property, see KARLIN and STUDDEN [905]. That is

$$\lambda_n(x) = \min \int_{-1}^{1} f(t)\omega(t)dt,$$

where the minimum is taken over all polynomials $f(t)$ of degree $\leq 2n-1$ which satisfy

$$f(t) \geq 0 \text{ on } [-1,1] \qquad \text{and} \qquad f(x) = 1.$$

Next we need to introduce the function

$$\pi_n(x) = \sum_{j=1}^{i} \lambda_n(\xi_j(a)),$$

which may be represented in a similar way as λ_n by

$$\pi_n(x) = \min \int_{-1}^{1} f(t)\omega(t)dt,$$

where the minimum is taken over all polynomials $f(t)$ of degree $\leq 2n-1$ which satisfy

$$f(t) \geq 0 \text{ on } [-1,1], \qquad \text{and} \qquad f(x) \geq 1 \text{ on } [x,1].$$

Now we formulate the main theorem.

Theorem 3.19 *Let* $\infty = a_1 > a_2 > \cdots > a_{l+1} \geq 0 \geq b_1 > \cdots > b_l \geq -\infty$, *and let* N *be an even number such that*

$$\pi_n(\xi_{i,n}(a_{i+1})) - \pi_n(\xi_{i,n}(a_i)) \geq \frac{1}{N}, \qquad i = 1,\ldots,l,$$

$$\pi_n(\xi_{i,n}(b_{i+1})) - \pi_n(\xi_{i,n}(b_i)) \geq \frac{1}{N}, \qquad i = 1,\ldots,l-1,$$

Then there exist $1 > x_1 > \cdots > x_n > 0$ *and positive integers* $A_1,\ldots,A_n$, *such that the quadrature formula*

$$\int_{-1}^{1} f(t)\omega(t)dt = \frac{1}{N}\sum_{i=1}^{n} A_i(f(x_i) + f(-x_i)) + R(f)$$

has degree $2n-1$.

Remark. Note that this formula is a CHEBYSHEV type quadrature formula in which many codes coincide; the node x_i has multiplicity A_i.

We will apply the above result to deduce some bound for the number of nodes of a CHEBYSHEV type quadrature formula for the weight function

$$\omega_\alpha = C_\alpha(1-t^2)^\alpha, \quad \text{with } \alpha \geq 0.$$

Lemma 3.20 *Let n be a positive integer. Then the function*

$$F(x) := \frac{\pi_n(x) - \int_x^1 \omega_\alpha(t)\,dt}{(1+x)^{1+\alpha}}$$

is decreasing on $[-1,1]$.

Proof. Let $-1 < x_0 < y_0 < 1$ and $f(t)$ be the polynomial of degree $\leq 2n-1$ which satisfies

$$f(t) \geq 1,\ t \in [x_0, 1], \quad f(t) \geq 0,\ t \in [-1,1], \quad \pi(x_0) = \int_{-1}^{1} f(t)\omega_\alpha(t)\,dt.$$

Then $g(s) := f(-1 + \frac{1+x_0}{1+y_0}(1+s))$ is a polynomial of degree $\leq 2n-1$ with

$$g(s) \geq 1,\ s \in [y_0, 1], \quad g(s) \geq 0,\ s \in [-1,1],$$

and therefore

$$\begin{aligned}
\pi_n(y_0) &\leq \int_{-1}^{1} g(s)\omega_\alpha(s)\,ds. \\
&= \frac{1+y_0}{1+x_0}\int_{-1}^{-1+2\frac{1+x_0}{1+y_0}} f(t)\omega_\alpha\left(-1 + \frac{1+y_0}{1+x_0}(1+t)\right)dt.
\end{aligned}$$

Using

$$\omega_\alpha\left(-1 + \frac{1+y_0}{1+x_0}(1+t)\right) \leq \left(\frac{1+y_0}{1+x_0}\right)^\alpha \omega_\alpha(t)$$

and the fact that $f(t) \geq 1$ on $[x_0, 1]$, we obtain the estimate

$$\pi_n(y_0) \leq \left(\frac{1+y_0}{1+x_0}\right)^{1+\alpha}\left(\pi_n(x_0) - \int_{x_0}^{1}\omega_\alpha(t)\,dt\right) + \int_{y_0}^{1}\omega_\alpha(s)\,ds.$$

□

Lemma 3.21 *Let $\alpha \geq 0$. There exists a constant $B_1 > 0$ and n_0 such that for $n \geq n_0$ and $0 \leq x \leq y \leq 1$,*

$$\pi_n(x) - \pi_n(y) \geq (y-x)\frac{B_1}{n^{2\alpha}}.$$

In the following proofs the numbers $K_1, K_2, \ldots$ denote positive constants which do not depend on x and n.

Proof. It suffices to prove that whenever $\pi_n(x)$ is differentiable at $x \in [0,1]$ then

$$-\pi_n'(x) \geq \frac{B_1}{n^{2\alpha}}.$$

We first assume $x \in [0, \xi_{1,n}]$. Computing $F'(x)$, we get

$$|\pi_n'(x)| \geq \omega_\alpha(x) - (1+\alpha)\,\frac{\pi_n(x) - \int_x^1 \omega_\alpha(t)\,dt}{1+x}.$$

Let $x^- = \xi_{i,n}(a)$ and put $x^+ = \xi_{i+1,n}(a)$, such that $x^- \geq x > x^+$. Then by the MARKOV-STIELTJES inequality, c.f. FREUD [624], and since $\pi(x)$ is decreasing, we have

$$\pi_n(x) - \int_x^1 \omega_\alpha(t)\,dt \leq \int_{x^+}^1 \omega_\alpha(t)\,dt - \int_x^1 \omega_\alpha(t)\,dt \leq \int_{x^+}^{x^-} \omega_\alpha(t)\,dt.$$

Using the well known asymptotic properties for the zeros of the JACOBI polynomials, see e.g. SZEGŐ [1769], we obtain

$$x^- - x^+ \leq \frac{K_1}{n} \text{ and } \omega_\alpha(x^+) \leq K_2\omega_\alpha(x).$$

This yields, if n is large enough,

$$(1+\alpha)\,\frac{\pi_n(x) - \int_x^1 \omega_\alpha(t)\,dt}{1+x} \leq \frac{1+\alpha}{1+x}\,\frac{K_1\,K_2}{n}\omega_\alpha(x) \leq \frac{1}{2}\omega_\alpha(x).$$

Since

$$\omega_\alpha(x) \geq \omega_\alpha(\xi_{1,n}) \geq \frac{K_3}{n^{2\alpha}},$$

we have finished the proof for the first case. Secondly assume $x \in [\xi_{1,n}, 1]$. Suppose $x = \xi_{1,n}(a), a \geq 0$ and let

$$\pi_n(x) = \lambda_n(x) = \int_{-1}^1 f(t)\omega_\alpha(t)\,dt,$$

where

$$f(t) = C_1(t - \xi_{2,n}(a))^2 \cdots (t - \xi_{n,n}(a))^2 \cdot (t+1)$$

and C_1 is chosen such that $f(x) = 1$. Obviously the polynomial $f(t)$ is increasing and convex on $[\xi_{2,n}(a), 1]$. Therefore, if $y > x$

$$f(y) > \frac{y - \xi_{2,n}(a)}{x - \xi_{2,n}(a)}$$

and on the other hand we have $\pi_n(y) \leq \pi_n(x)/f(y)$. It follows that

$$\pi_n(x) - \pi_n(y) \geq \pi_n(x)\left(1 - \frac{1}{f(y)}\right) \geq (y-x)\,\frac{\pi_n(x)}{y - \xi_{2,n}(a)} \geq (y-x)\frac{\pi_n(1)}{1 - \xi_{2,n}}.$$

Since $(1-\xi_{2,n}) \le \frac{K_4}{n^2}$ the only fact that remains to prove is

$$\pi_n(1) \ge \frac{K_5}{n^{2+2\alpha}}.$$

The zeros of $(1-t^2)P_{n-1}^{(\alpha+1,\alpha+1)}(t)$ are the nodes for the LOBATTO quadrature formula (3.34) and so we deduce

$$\pi_n(1) = \lambda_n(1) = \frac{1}{2\, P_{n-1}^{(\alpha+1,\alpha+1)}(1)} \int_{-1}^{1} (1+t) P_{n-1}^{(\alpha+1,\alpha+1)}(t) \omega_\alpha(t)\, dt.$$

Using RODRIGUES' formula and integration by parts, we can evaluate the above integral explicitely and therefore estimate

$$\pi_n(1) \ge \frac{1}{2} \frac{\Gamma(2\alpha+3)}{(n+2\alpha+1)^{2+2\alpha}},$$

which completes the proof. □

For the orthogonal polynomials $Q_{n-1}(t)$ with respcect to the weight function $(1-t^2)\omega_\alpha(t)$ we now take $\frac{d}{dt}P_n^{(\alpha,\alpha)}(t)$. Thus

$$P_n(t,a) = P_n^{(\alpha,\alpha)}(t) - a(1-\operatorname{sgn}(a)\, t) \frac{d}{dt} P_n^{(\alpha,\alpha)}(t).$$

Lemma 3.22 *Let* $a_i = \frac{1}{i-1}, i = 1,\ldots,l+1$ *and* $b_i = -\frac{i-1}{n^2}, i = 1\ldots l$. *Then for certain positive constants* B_2, B_3, *we have*

$$\xi_{i,n}(a_i) - \xi_{i,n}(a_{i+1}) \ge \frac{B_2}{n^2}, \quad i = 1,\ldots,l \tag{3.35}$$

$$\xi_{i,n}(b_i) - \xi_{i,n}(b_{i+1}) \ge \frac{B_3}{n^2}, \quad i = 1,\ldots,l-1 \tag{3.36}$$

Proof. First we prove (3.35) for $i \ge 2$. For $x \in (\xi_{i,n}, \eta_{i-1,n})$ we have

$$A(x) := \frac{P_n^{(\alpha,\alpha)}(x)}{(1-x)P_n^{(\alpha,\alpha)'}(x)} > 0$$

and using the differential equation for the ultraspherical polynomials

$$A'(x) = \frac{1-\alpha A(x)}{1-x} + \frac{(1+\alpha)A(x) + n(n+2\alpha+1)A^2(x)}{1+x}.$$

Note that $A(\xi_{i,n}) = a$ and we obtain for the inverse function $\xi_{i,n}(a)$

$$\begin{aligned} \xi_{i,n}'(a) &= \frac{1}{A'(\xi_{i,n}(a))} \\ &\ge \left[\frac{1}{1-\eta_{i-1,n}} + (1+\alpha)a + n(n+2\alpha+1)a^2\right]^{-1}. \end{aligned}$$

It follows immediatly that

$$\frac{\xi_{i,n}(a_i)-\xi_{i,n}(a_{i+1})}{a_i-a_{i+1}} \geq \left[\frac{1}{1-\eta_{i-1,n}}+(1+\alpha)a_i+n(n+2\alpha+1)a_i^2\right]^{-1}.$$

Since

$$1-\eta_{i-1,n} \geq K_1\frac{(i-1)^2}{(n+\alpha+1)^2}$$

(cf. SZEGŐ [1769, Section 8.9]) an easy computation yields

$$\xi_{i,n}(a_i)-\xi_{i,n}(a_{i+1}) \geq \frac{K_1}{2(1+K_1)(n+\alpha+1)^2}$$

and we have proved (3.35) for $i \geq 2$.

For the case $i=1$ we deduce from the definition of $P_n(t,a)$

$$1-\xi_{1,n}(1)=\frac{P_n^{(\alpha,\alpha)}(\xi_{1,n}(1))}{P_n^{(\alpha,\alpha)'}(\xi_{1,n}(1))}=\frac{(\xi_{1,n}(1)-\xi_{1,n})P_n^{(\alpha,\alpha)'}(\xi_{1,n}(\tau))}{P_n^{(\alpha,\alpha)'}(\xi_{1,n}(1))},$$

where $\tau \in (\xi_{1,n},\xi_{1,n}(1))$. Since $P_n^{(\alpha,\alpha)'}(x)$ is increasing on $(\xi_{1,n},1)$, cf. (4.7.14) in SZEGŐ [1769], we get

$$1-\xi_{1,n}(1) \geq \frac{(\xi_{1,n}(1)-\xi_{1,n})P_n^{(\alpha,\alpha)'}(\xi_{1,n})}{P_n^{(\alpha,\alpha)'}(1)}.$$

From Theorem 8.9.1 in SZEGŐ [1769] we find

$$P_n^{(\alpha,\alpha)'}(\xi_{1,n}) \approx n^{\alpha+2}$$

and from the differential equation

$$P_n^{(\alpha,\alpha)'}(1) \approx n^{\alpha+2}.$$

Hence

$$1-\xi_{1,n}(1) \geq \frac{\xi_{1,n}(1)-\xi_{1,n}}{K_2}$$

and therefore

$$1-\xi_{1,n}(1) \geq \frac{1-\xi_{1,n}}{K_2+1} \geq \frac{K_3}{n^2}.$$

The estimate (3.36) can be proved in a very similar way. □

Now we are ready to prove the main result for ultraspherical weight functions.

Theorem 3.23 *For any n there exists a* CHEBYSHEV *type quadrature formula of degree n*

$$\int_{-1}^{1} f(t)\omega_\alpha(t)dt = \frac{1}{N}\sum_{i=1}^{N}(f(x_i)+f(-x_i))+R(f)$$

with the number of nodes satisfying

$$N \le K n^{2+2\alpha},$$

where K only depends on α.

Proof. Using the numbers $a_1, \ldots, a_{l+1}$ and $b_1, \ldots, b_l$ from Lemma 3.22 and combining Lemma 3.21 with (3.35),(3.36) we obtain

$$\begin{aligned} \pi_n(\xi_{i,n}(a_{i+1})) - \pi_n(\xi_{i,n}(a_i)) &\ge \frac{B_1 B_2}{n^{2+2\alpha}}, \; i = 1, \ldots, l \\ \pi_n(\xi_{i,n}(b_{i+1})) - \pi_n(\xi_{i,n}(b_i)) &\ge \frac{B_1 B_3}{n^{2+2\alpha}}, \; i = 1, \ldots, l-1 \end{aligned}$$

Now the result follows immediatly from Theorem 3.19. □

From now on we consider a $k-1$-dimensional sphere. The above result can be applied to the construction of k-dimensional spherical designs and we proceed as in WAGNER [1926].

Consider the polynomial $p(x_1, \ldots, x_k) = x_1^{m_1} \ldots x_k^{m_k}$, where the non negative integers m_i satisfy $m_1 + m_2 + \cdots + m_k \le t$. If we introduce spherical coordinates, we obtain

$$\begin{aligned} I &:= \int_{\mathbf{S}^{k-1}} p(x_1, \ldots, x_k)\, d\sigma(x_1, \ldots, x_k) \\ &= \int_0^{2\pi} (\cos\phi)^{m_{k-1}} (\sin\phi)^{m_k}\, d\phi \\ &\quad \cdot \prod_{j=1}^{k-2} \int_{-1}^{1} (1-u_j^2)^{\frac{1}{2}(m_k + \cdots + m_{j+1})} u_j^{m_j} (1-u_j^2)^{\frac{k-j-2}{2}}\, du_j =: I_{k-1} \prod_{j=1}^{k-2} I_j. \end{aligned}$$

For each coordinate $u_j, 1 \le j \le k-2$, we choose a set A_j according to Theorem 3.19 with $\alpha = \frac{k-3}{2}, \ldots, \frac{1}{2}, 0$ and $n = t$. It is clear, by writing the integrand of the first integral in terms of the exponential function, that the pointset $\{0, \frac{2\pi}{n}, \ldots, \frac{2\pi}{n}(n-1)\}$ integrates I_{k-1} exactly for any $n > t$. Now consider for a moment that all m_j, $1 \le j \le k-2$ are even. Then the first factor in the integrand of I_j is a polynomial in u_j and by the the result above A_j integrates I_j exactly. Now if one m_j is odd then the integral I_j and therefore I equals zero. This is why it suffices to take $A_j^* = A_j \cup -A_j$ to integrate I_j. Let $A = \{(\theta_1, \ldots, \theta_{k-2}, \phi)\}$ be the set of points on $\mathbf{S}^{k-1}$ obtained by letting $\cos\theta_1, \ldots, \cos\theta_{k-2}$ and ϕ run independently through the averaging sets $A_1^*, \ldots, A_{k-2}^*, A_{k-1}$. Then A is a spherical t-design and its size N can be easily estimated to satisfy

$$N \le c_k\, t \cdot t^{2+\frac{k-3}{2}} \cdots t^{2+0} = c_k\, t^{1+2(k-2)+\frac{k-3}{2}(k-2)} = c_k\, t^{1+(k-2)\frac{k+1}{2}}.$$

Hence we obtain the following corollary.

Corollary 3.24 *For any positive integer t and $k \ge 3$ there exists a k-dimensional spherical t-design of size N, where N satisfies*

$$N \ll t^{1+(k-2)\frac{k+1}{2}}.$$

3.2.2 Slice Dispersion and Polygonal Approximation of Curves

In the following we describe an application of the dispersion of spherical point sequences to a problem of computational geometry. We follow two papers by ROTE and TICHY [1566, 1565].

Definition 3.25 *The dispersion of a point set $A = \{\mathbf{x}_1, \ldots, \mathbf{x}_N\} \subseteq \mathbf{S}^k$ with respect to a given family $\mathcal{R}$ of subsets of $\mathbf{S}^k$, called ranges, is defined by*

$$d^{\mathcal{R}}(A) = \sup\{\sigma^*(R) : R \in \mathcal{R}, A \cap R = \emptyset\}, \tag{3.37}$$

where σ^* is the normalized surface measure on $\mathbf{S}^k$.

Note that this notion of dispersion is slightly different to that treated in Section 1.1 which measures the radius of the largest ball not intersecting $\{\mathbf{x}_1, \ldots, \mathbf{x}_N\}$. Originally the disperio was introduced by HLAWKA [800] and later investigated in more general form in NIEDERREITER [1302]. The dispersion can be applied for analyzing optimization algorithms, see NIEDERREITER [1302], cf. also NIEDERREITER and PEART [1350] and Chapter 6 in the monograph by NIEDERREITER [1336]. In TICHY [1839] optimization problems on the sphere are considered. For further references see the Notes of Section 1.1.

The range spaces that are usually considered on the sphere are the range spaces $\mathcal{C}$ of spherical caps and the system $\mathcal{S}$ of shperical slices. (A slice is the intersection of two half-spheres.) Note that there exist lower bounds for the discrepancy with respect to caps and for the discrepancy with respect to slices which are of the same order of magnitude (see Theorem 2.22).

In the following we will consider the *dispersion* with respect to $\mathcal{S}$, and we will explain an application of spherical slice dispersions to the piecewise linear approximation of curves in threedimensional space.

Obviously, for any point set A on the sphere $\mathbf{S}^k$, the following elementary relation holds between cap and slice dispersion:

$$d^{\mathcal{S}}(A) \le c_k \cdot d^{\mathcal{C}}(A)^{1/k}, \tag{3.38}$$

for some positive constant c_k. This is obvious from the fact that any slice contains a cap of corresponding volume.

We also have the trivial lower bound $d^{\mathcal{S}}(A) \ge 1/N$ for every N-point set A. The following proposition states that this bound can be achieved, up to a constant factor.

Proposition 3.26 *For every N there is a point set $A = \{\mathbf{x}_1, \ldots, \mathbf{x}_N\}$ with N points on the sphere $\mathbf{S}^k$ which has slice dispersion $d^{\mathcal{S}}(A) = \mathcal{O}(1/N)$.*

Sketch Proof. For the construction of such a sequence w. l. o. g. we may assume that $\mathbf{S}^k$ is the unit sphere in $(k+1)$-dimensional space. Then we evenly distribute the N points over the $\binom{k+1}{2}$ two-dimensional coordinate hyperplanes (on the sphere): On each of the "coordinate circles" C_i, $i = 1, \ldots, \binom{k+1}{2}$, which are the intersections of the coordinate planes with the sphere, we place the corresponding points equidistantly.

The detailed poof that this example has the desired order of dispersion is based on geometric arguments and can be found in ROTE and TICHY [1565]. □

In the following we will show how the slice dispersion on the sphere in three-dimensional space arises in a problem of piecewise linear approximation of curves in space.

For instance, in *robotics* it is an important problem to approximate a "general" curve by simple curves like straight lines, circles etc., because the arm of the robot can only run along such simple curves. The most important case is the approximation by a polygonal line.

Let us first consider a twice continuously differentiable spatial curve $\mathbf{x}(s)$ parameterized by its arc length s. We want to construct a sequence of points $\mathbf{x}(s_0)$, $\mathbf{x}(s_1), \ldots, \mathbf{x}(s_M)$ such that the polygonal line with vertices at these points is an ε-approximation of the curve segment C: $\{\mathbf{x}(s) : s_0 \leq s \leq s_M\}$, where ε is an arbitrarily given (small) positive number. Let us recall here the definition of an ε-approximation: A (closed) point set A is an ε-approximation of the (closed) set B if their HAUSDORFF-distance

$$\delta(A, B) = \max\left\{\max_{x\in A} d(x, B), \max_{y\in B} d(y, A)\right\}$$

is not greater than ε (d denoting the Euclidean distance). We consider a line segment L_k with endpoints $\mathbf{x}(s_k), \mathbf{x}(s_{k+1})$, and denote the tangent vector by $\dot{\mathbf{x}}(s_k)$. Then $\delta(C, L_k) < \varepsilon$ is guaranteed provided that

$$|\mathbf{x}(s_{k+1}) - \mathbf{x}(s_k) - \dot{\mathbf{x}}(s_k)(s_{k+1} - s_k)| \leq \varepsilon.$$

Applying TAYLOR's formula yields

$$|\ddot{\mathbf{x}}(s_k)|(s_{k+1} - s_k)^2 \leq \frac{\varepsilon}{2}$$

as an approximate sufficient condition that the polygonal line is an ε-approximation. Thus we obtain the following iteration procedure for computing the vertices of the polygonal approximation:

$$s_{k+1} - s_k = \sqrt{\frac{\varepsilon}{2\kappa(s_k)}},$$

where $\kappa(s_k)$ denotes the curvature in the point $\mathbf{x}(s_k)$. This method, of course, has one disadvantage: one has to know arc length and curvature in advance. In the following we describe a different method which makes use of low-dispersed spherical point sequences.

Let us first consider a plane curve C: $\{\mathbf{x}(t) : 0 \leq t \leq \sigma\}$, which we want to approximate piecewise linearly. We successively construct the vertices $\mathbf{x}(t_k)$ for $0 = t_0 < t_1 < \cdots < t_M = \sigma$ as follows:

Suppose that t_k has already been constructed. Let u_k be the largest value ($t_k < u_k \leq \sigma$) for which there is a direction (i.e., a unit vector) $\mathbf{w}_k$ such that for any $t \in [t_k, u_k]$ there is a scalar λ with $|\mathbf{x}(t) - \mathbf{x}(t_k) - \lambda\mathbf{w}_k| \leq \varepsilon$. Then we set

$$t_{k+1} = \max\{t \leq u_k : \mathbf{x}(t) = \mathbf{x}(t_k) + \lambda\mathbf{w}_k \text{ for some } \lambda\}. \tag{3.39}$$

In other words, the curve C between t_k and t_{k+1} is contained in the infinitely long strip of width 2ε centered at the line through $\mathbf{x}(t_k)$ and $\mathbf{x}(t_{k+1})$, and t_{k+1} is the largest possible value with this property. For more details and references on the geometric background we refer to ROTE and TICHY [1565].

However, since the curve is only guaranteed to be contained in the infinitely long ε-strip, there are "pathological" examples where this procedure does not necessarily lead to an ε-approximation. For situations arising in practice the procedure works. For example, requiring that the curve is smooth and has an upper bound less than $1/\varepsilon$ on the curvature suffices to ensure that the algorithm yields an ε-approximation. This would however exclude polygonal curves, which are especially important in practical applications, where one wants to approximate one such curve by another polygonal curve with fewer vertices.

What we need is a condition on the local "growth rate" of the arc length,

$$d(\mathbf{x}(s), \mathbf{x}(s + \Delta s)) \geq \rho \cdot \Delta s, \text{for all } \Delta s \leq \lambda \text{ and } 0 \leq s < s + \Delta s \leq \sigma. \tag{3.40}$$

Here $\rho \leq 1$ is the parameter determining the growth rate and the parameter λ makes the condition local. We will say that a curve satisfying (3.40) has *λ-local minimum growth rate* at least ρ.

Note that this local condition does not prevent the curve from crossing itself after making a "big" loop. If the curve should have cusps it must be subdivided at these points before applying the algorithm.

The following lemma states that this condition is sufficient for the correctness of our algorithm. Since we will need the spatial case we already formulate the lemma in arbitrary dimensions.

Lemma 3.27 *Let C be a (continuous) curve from A to B whose $(2\varepsilon/\rho)$-local minimum growth rate is bigger than ρ, for some $\rho > 0$. If C is contained in the cylinder with axis AB and radius ε then the segment AB is an $(\varepsilon\sqrt{1 + 1/\rho^2})$-approximation of C.*

Proof. Let us assume w. l. o. g. that $A = \mathbf{x}(0)$ is the origin and $B = \mathbf{x}(\sigma)$ lies on the positive x-axis. It is sufficient to show that the x-coordinate never goes below $-\varepsilon/\rho$ and never exceeds the x-coordinate of B by more than ε/ρ. By symmetry, we just have to prove the first statement. Let $\mathbf{x}(s_0)$ with $s_0 \in [0, \sigma]$ be a point on the curve with negative x-coordinate. Consider the two nearest points $\mathbf{x}(s_1)$ and $\mathbf{x}(s_2)$ on the curve with x-coordinates equal to 0: $0 \leq s_1 < s_0 < s_2 \leq \sigma$, $x(s_1) = x(s_2) = 0$, and $x(s) < 0$ for $s_1 < s < s_2$. We show that $s_2 - s_1 \leq 2\varepsilon/\rho$, from which $x(s_0) \geq -\varepsilon/\rho$ follows. Otherwise, assume that $s_2 - s_1 > 2\varepsilon/\rho =: \Delta s$. By continuity, we can find two intermediate points $\mathbf{x}(s')$ and $\mathbf{x}(s' + \Delta s)$ with $s_1 \leq s' < s' + \Delta s \leq s_2$ which have the same x-coordinate: If we let s' vary from s_1 to $s_2 - \Delta s$ we initially have $0 = x(s_1) > x(s_1 + \Delta s)$, and at the end we have $x(s_2 - \Delta s) < x((s_2 - \Delta) + \Delta s) = 0$; thus, there must be a crossover point s'. By (3.40), $d(\mathbf{x}(s'), \mathbf{x}(s' + \Delta s)) \geq \rho' \cdot \Delta s = \rho'(2\varepsilon/\rho) > 2\varepsilon$, where $\rho' > \rho$ is the local minimum growth rate. But then $\mathbf{x}(s')$ and $\mathbf{x}(s' + \Delta)$ cannot both be contained in the cylinder with radius ε around the x-axis, a contradiction. $\square$

Remark. One way to ensure (3.40) is to require that the curve has *locally increasing chords*:

$$d(\mathbf{x}(s_1), \mathbf{x}(s_4)) \geq d(\mathbf{x}(s_2), \mathbf{x}(s_3)), \tag{3.41}$$

for all $0 \leq s_1 \leq s_2 \leq s_3 \leq s_4 \leq \sigma$ and $s_4 - s_1 \leq \lambda$.

A stronger condition than (3.41) is to require that the curve locally has *no angles sharper than* $\pi/2$, i. e., for any three consecutive points $\mathbf{x}(s_1)$, $\mathbf{x}(s_2)$, and $\mathbf{x}(s_3)$ with $0 \leq s_1 \leq s_2 \leq s_3 \leq \sigma$ and $s_3 - s_1 \leq \lambda$, the angle at $\mathbf{x}(s_2)$ in the triangle $\mathbf{x}(s_1)\mathbf{x}(s_2)\mathbf{x}(s_3)$ is at least $\pi/2$. Clearly, this implies the local increasing chords property.

A direct generalization of the above procedure to the three-dimensional case is not possible for computational reasons. Thus we use projections to reduce the three-dimensional case to the two-dimensional situation.

Let $\mathbf{x}^i(t)$, $(i = 1, \ldots, n)$ be orthogonal projections $P^i(\mathbf{x})$ of the spatial curve C: $\mathbf{x}(t)$, $0 \leq t \leq 1$, onto M suitably chosen planes $E_1, \ldots, E_M$. We compute the interpolation points $\mathbf{x}(t_k), k = 1, \ldots,$ recursively and suppose that $t_0 = 0, t_1, \ldots, t_k$ are known. For each projection $i = 1, \ldots, M$, let S^i be the set of parameter values $t' > t_k$ such that there exists a scalar λ between 0 and 1 satisfying $|\mathbf{x}^i(t) - \mathbf{x}^i(t_k) - \lambda(\mathbf{x}^i(t') - \mathbf{x}^i(t_k))| \leq \varepsilon$ for any $t \in (t_k, t']$. S^i is the set of possible parameter values for the next interpolation point t_{k+1} when seen in the i-th projection direction. Each set S^i is a finite union of intervals, and we define t_{k+1} as the largest value in the intersection of these M sets.

Now we connect the points $\mathbf{x}(t_k)$ by line segments. In each projection i, the curve between $\mathbf{x}^i(t_k)$ and $\mathbf{x}^i(t_{k+1})$ lies in the infinite 2ε-strip centered at the line through $\mathbf{x}^i(t_k)$ and $\mathbf{x}^i(t_{k+1})$. Note that it is not sufficient to determine just the maximum possible parameter value t_{k+1} by (3.39) in each projection and take the minimum of these values, because this value might not be contained in each S^i.

Although the 2ε-strips cover the curve in each projection, the spatial curve between $\mathbf{x}(t_k)$ and $\mathbf{x}(t_{k+1})$ does not necessarily lie in the infinite cylinder with radius ε centered at the line through $\mathbf{x}(t_k)$ and $\mathbf{x}(t_{k+1})$.

We will now discuss how to choose the n projection planes to ensure that this holds for a cylinder with radius $A\varepsilon$, for a constant $A > 1$ which we want to be as small as possible.

Denoting the direction of the projection P^i by p_i we set for any straight line g

$$\mathcal{Z}(p_1, \ldots, p_M; g) = \bigcap_{i=1}^{M} \{ x : \delta(P^i(x), P^i(g)) \leq 1 \}.$$

This set is an intersection of M parallel slabs. The intersection of $\mathcal{Z}(p_1, \ldots, p_M; g)$ with an orthogonal plane of g is a convex symmetric polygon Z, whose edges are parallel to the projections of p_i onto Z. The distance of the edges to the center of Z is 1. Setting A equal to the maximal distance of a vertex of Z to the center, the constant A fulfills the desired property if we take the line through the points $\mathbf{x}(t_k)$ and $\mathbf{x}(t_{k+1})$ as g. Clearly,

$$A = \frac{1}{\cos \alpha/2},$$

where α is the maximal angle between two adjacent edges. If we consider the projection directions $p_1, \ldots, p_M$ and the line g as points on the sphere S^2, α is the opening angle of the largest empty slice with corners at the two points corresponding to g. Since we want α to be small for all directions g, we have to choose these N points exactly in such a way that the slice-dispersion is minimal. This can be achieved by taking the point set in Proposition 3.26 for the two-dimensional sphere S^2.

Assuming a local smoothness property like (3.40), we can conclude by Lemma 3.27 that the polygonal curve is an $\mathcal{O}(\varepsilon)$-approximation of C.

Notes

In the fundamental paper [811] Hlawka starts a systematic investigation of uniformly distributed sequences on products of spheres; see also a reprint of this paper in the Selecta Volume [829]. Similar investigations were independently done by Freeden, see [622, 623] and the very recent work of Cui and Freeden [425]. In a series of papers Hlawka [822, 823, 824, 820, 821, 827] developed his approach of Green function discrepancy on Riemannian manifolds.. The Euclidean, the non-Euclidean and the projective space are discussed in detail. This concept is based on the so-called Lipschitz discrepancy which is also known in probability theory under the name Vaserstein metric; see also Blümlinger [233]. In [826, 835, 834] Hlawka provides a detailed investigation of the Radon transform by means of u.d. sequences. Binder, Hlawka, and Schoißengeier [222] present many further applications, e.g. an application to a mechanical problem.

Different kinds of distribution problems on the sphere are considered in Stolarsky [1743], Harman [737], Beck [125], and Wagner [1924, 1923, 1925, 1928, 1929]. E.g. in [1925] (essentially) optimal lower bounds for sums $\sum |x_j - x_k|^\alpha$, $x_1, \ldots, x_N$ denoting a finite point set on the sphere, were proved. Also products of such differences were considered; the proofs heavily depend on tools from spherical harmonic analysis. Wagner's investigation [1926, 1930] on averaging sets was extended to multisets by Durner [510]. For general results on averaging sets we refer to Seymour and Zaslavsky [1666].

For results on irregularities of distribution on the sphere see Section 2.1. Gerl [653, 654] investigates u.d. of sequences on the sphere in connection with orthogonal projections of the sequence. Guralnik, Zemach, and Warnock [709] established a new algorithm for random sampling in the interior or on the surface of a sphere.

Lubotzky, Philipps and Sarnak [1110, 1111] established an important construction of a low discrepancy sequence on the 2-sphere S^2. The method is based on choosing the points to be the orbit under the action of a free subgroup (generated by three rotations around the coordinate axes in $\mathbf{R}^3$ with angle $\arccos(-3/5)$) of $SO(3)$ on S^2. Of particular importance are Hecke operators and Deligne's theorem concerning Ramanujan's conjecture on the τ-function. For the L^2-discrepancy with respect to spherical caps it is shown that $1/\sqrt{N}$ is the correct order of magnitude (apart from a possible logarithmic factor in the numerator); see also Colin de Verdiere [375]. Applications of this kind of sequences were given by Tichy [1837, 1838].

3.3 Partial Differential Equations

3.3.1 The Heat Equation

In the following we are concerned with the one-dimensional, periodic heat equation

$$\begin{aligned} \frac{\partial u}{\partial t} &= \frac{\partial^2 u}{\partial x^2} \\ u(x,0) &= u_0(x) \\ u(x+1,t) &= u(x,t). \end{aligned} \tag{3.42}$$

This equation could be numerically solved using usual methods of finite differences. If we choose M equidistant spatial grid points, write Δt for the time step and u_i^n for the approximation to $u(i\Delta x, n\Delta t)$, then we get the difference equations of the form

$$\begin{aligned} u_i^{n+1} &= (1-2\lambda)u_i^n + \lambda(u_{i+1}^n + u_{i-1}^n) \\ u_i^0 &= u_0(i\Delta x) \\ u_M^n &= u_0^n \\ u_{M-1}^n &= u_{-1}^n \\ u_{M+1}^n &= u_1^n. \end{aligned} \tag{3.43}$$

We have here $\lambda = \Delta t/(\Delta x)^2$ and by the classical COURANT-FRIEDRICHS-LEVY condition $\lambda \leq 1/2$.

The simulation model approximates the solution of these difference equations, and the convergence in the error analysis is also based on the exact solution of the difference equations, not on the heat equation itself. Here we are going to describe the Quasi-Monte Carlo method for simulation, i.e. some sort of random walk using Quasi-random numbers. The main idea of this method is reordering the particles according to their spatial position. This happens in order to break correlations between the sequence elements. Moreover, one single sequence can be used for all time steps. If we write $x(j)$ for the spatial position of the jth particle and $x_i = i\Delta x = i/M$, then the reordering means relabeling the particles so that

$$x_1 \leq x(1) \leq x(2) \leq \cdots \leq x(N) \leq x_M. \tag{3.44}$$

The above model was described and the convergence of the method was proved by MOROKOFF and CAFLISCH [1204]. The sequence used for the proof was a two-dimensional HAMMERSLEY-type sequence of length N which can be constructed by pairing N terms of some one-dimensional low discrepancy sequence with the sequence

$$\{\frac{1}{2N}, \frac{3}{2N}, \ldots, \frac{(2N-1)}{2N}\}. \tag{3.45}$$

Before stating Theorem 3.28 (that provides an estimate for the weak measure of the error) we have to introduce the following notation. In order to get the weak formulation we have to sum the difference equation against a test function over the grid points. We write X_j^n for the position of the jth particle at time $t^n = n\Delta t$ and $\phi(x)$ for a piecewise continuous, periodic function.

The weak formulation has the form

$$\begin{aligned} &\sum_{i=1}^{M} \phi(x_i)u(x_i, t^{n+1}) \\ &= (1-2\lambda)\sum_{i=1}^{M} \phi(x_i)u(x_i, t^n) + \lambda \sum_{i=1}^{M} u(x_i, t^n)(\phi(x_{i+1}) + \phi(x_{i-1})). \end{aligned}$$

The approximation of this formulation may be written as

$$\sum_{i=1}^{M} \phi(x_i)u(x_i,t^n) \approx \frac{1}{N}\sum_{j=1}^{N} \phi(X_j^n), \tag{3.46}$$

where $\phi(x)$ is some test function. Now we are able to define the weak measure of error as

$$E_N^n = \sup_{\{(s,r):0\le s\le r\le 1\}} |d_N^n(s,r)|,$$

where

$$d_N^n(s,r) = \frac{1}{N}\sum_{j=1}^{N} \phi_{s,r}(X_j^n) - \sum_{i=1}^{M} \phi_{s,r}(x_i)u(x_i,t^n)$$

and $\phi_{s,r}(x)$ is the characteristic function

$$\phi_{s,r}(x) = \chi_{[s,r]}(x), \qquad 0 \le s \le r \le 1.$$

Theorem 3.28 *Let $(b_n)_{n\ge1}$ be a one dimensional sequence in $I = [0,1]$, such that the two dimensional* HAMMERSLEY*-type sequence formed from any contiguous subsequence of $(b_n)_{n\ge1}$ of length N has discrepancy bounded by D_N. If $(b_n)_{n\ge1}$ is used in the simulation of the discrete periodic one dimensional heat equation with particle reordering, then the weak measure of error at time step n, E_N^n, satisfies*

$$E_N^n \le 5nD_N + E_N^0. \tag{3.47}$$

where E_N^0 is the error in representing the initial data by the particle approximation.

In what follows we want to give a sketch proof of this theorem.

Sketch Proof. For computing the error for one time step we define an another characteristic function

$$\Psi_j(c) = \chi_{(\frac{j-1}{N},\frac{j}{N}]}.$$

We define also the function $\Theta(X_j^n, b)$ by

$$\Theta(X_j^n,b) = \begin{cases} X_j^n - \Delta x, & 0 \le b \le \lambda \\ X_j^n + \Delta x, & \lambda < b \le 2\lambda \\ X_j^n, & 2\lambda < b \le 1 \end{cases}.$$

Furthermore, in order to establish a relationship between $\phi(X_j^n)$ and $\phi(X_j^{n+1})$, we define the operator K^{n+1} as

$$K^{n+1}\phi(b,c) = \sum_{j=1}^{N} \phi(\Theta(X_j^n,b))\Psi_j(c).$$

From the integration of K^{n+1} over I^2 we get

$$\int_{I^2} K^{n+1}\phi(b,c)\,db\,dc = \frac{1-2\lambda}{N}\sum_{j=1}^{N}\phi(X_j^n) + \frac{\lambda}{N}\sum_{j=1}^{N}\left(\phi(X_j^n+\Delta x) + \phi(X_j^n - \Delta x)\right),$$

which is exactly the approximation of the weak solution. We compute this integral with the Quasi-Monte Carlo method using the HAMMERSLEY-type sequence of the length N. Here $(c_j)_{j=1}^N$ is the sequenece defined in (3.45) and $(b_j)_{j=1}^N$ is the given one dimensional sequence. The elements of $(b_j)_{j=1}^N$ decide whether a particle moves left or right. The elements of $(c_j)_{j=1}^N$ help us to determine through $\Psi_j(c)$ which particles moves next.

The error of the Quasi-Monte Carlo integration is defined as

$$\delta_N^n(s,r) = \frac{1}{N}\sum_{j=1}^{N} K^{n+1}\phi_{s,r}(b_j,c_j) - \int_{I^2} K^{n+1}\phi_{s,r}(b,c)\,db\,dc.$$

It is easily shown that the relation

$$d_N^{n+1} = \delta_N^n(s,r) + (1-2\lambda)d_N^n(s,r) + e_N^n(s,r), \tag{3.48}$$

with

$$e_N^n(s,r) = \lambda(d_N^n(s+\Delta x, r+\Delta x) + d_N^n(s-\Delta x, r-\Delta x))$$

holds. Thus

$$|e_N^n(s,r)| \le 2\lambda E_N^n. \tag{3.49}$$

Now we write the term $\delta_N^n(s,r)$ in the form

$$\delta_N^n(s,r) = \frac{1}{N}\sum_{k=1}^{N}\sum_{j=1}^{N}\phi_{s,r}(\Theta(X_j^n,b_k))\Psi_j(c_k) - \int_{I^2}\sum_{j=1}^{N}\phi_{s,r}(\Theta(X_j^n,b))\Psi_j(c)\,db\,dc.$$

If we regard that the elements X_j^n are reordered, then we can show by an easy geometrical consideration that the function $\sum_{j=1}^N \phi_{s,r}(\Theta(X_j^n,b))\Psi_j(c)$ is a union of maximally 5 disjoint, axis parallel rectangles in I^2. Here the reordering is crucial. In this way we get

$$|\delta_N^n(s,r)| \le 5D_N.$$

Hence with (3.49) we have

$$E_N^{n+1} \le 5D_N + (1-2\lambda)E_N^n + 2\lambda E_N^n,$$

or

$$E_N^{n+1} \le 5D_N + E_N^n.$$

The final result follows from repeated substitution. □

MOROKOFF and CAFLISCH [1204] applied the described method for the numerical simulation and came to the conclusion that the error in the Quasi-Monte Carlo simulation is significantly less than the corresponding error for a standard Monte Carlo, i.e. random or pseudorandom, simulation. They observed that the best possible convergence rate is $\mathcal{O}(1/N)$, which can be attained by using low discrepancy sequences to determine the particle motion. On the other hand, they compared this computation with the one done with a pseudorandom sequence. In this case the expected convergence of the order $N^{-1/2}$ was achieved.

The generalisation of this method for higher dimensions turns out to be very complicated. The main tool of the method, namely the reordering of the particles at the suitable place, may not be done so easily as in the one dimensional case.

3.3.2 The BOLTZMANN Equation

In the following we describe a method for computing an approximate solution of the BOLTZMANN equation (due to LÉCOT [1048, 1049]). We are concerned with the crude simplified case of an infinite spatially homogeneous and isotropic gas (the velocity distribution is radially symmetric) whose differential cross section $\sigma(g)$ equals k/g where g is the relative speed and k is some constant. The discretisation scheme is organized as follows:
The initial velocity distribution (in $\mathbf{R}^3$) f^0 is approximated by a sum of N DIRAC measures, the time is discretized by steps of length Δt and the approximation f^n of the velocity distribution at time $t^n = n\Delta t$ is obtained from f^{n-1} in two phases. First an intermediary measure g^n is defined by a direct one-step method. It is a linear combination of N DIRAC masses and also of N^2 surface measures on spheres $\mathbf{S}^2$. The coefficients may be interpreted as collision probabilities, the points where the DIRAC masses are concentrated as velocities of uncollided molecules and the spheres as sets of new velocities of colliding molecules. The intermediate step which leads from g^n to f^n is presented as a numerical quadrature of some function on $[0,1)^4$, with low discrepancy points as nodes. These numbers sample new discrete velocities according to the preceding collision probabilities and probability measures.

Let $k > 0$ and f_0 be a positive real function such that

$$\int_{\mathbf{R}_+} v^2 f_0(v)\, dv = 1.$$

Let $T > 0$ and f be a regular function which satisfies the simplified BOLTZMANN equation formulated as follows:

$$\frac{\partial f}{\partial t}(|\mathbf{v}'|, t) = \frac{k}{\pi} \int_{\mathbf{R}^3 \times S_+^2} (f(|\mathbf{v}'|, t) f(|\mathbf{w}'|, t) - f(|\mathbf{v}|, t) f(|\mathbf{w}|, t)) \frac{\mathbf{n}(\mathbf{v}-\mathbf{w})}{|\mathbf{v}-\mathbf{w}|}\, d\mathbf{w}\, d\mathbf{n},$$

for all $\mathbf{v} \in \mathbf{R}^3$ and all $t \in [0, T]$ where

$$\begin{aligned} S_+^2 &= \{\mathbf{n} \in \mathbf{S}^2 : \mathbf{n}(\mathbf{v}-\mathbf{w}) > 0\}, \\ \mathbf{v}\prime &= \mathbf{v} - (\mathbf{n}(\mathbf{v}-\mathbf{w}))\mathbf{n}, \quad \mathbf{w}' = \mathbf{w} + (\mathbf{n}(\mathbf{v}-\mathbf{w}))\mathbf{n} \end{aligned}$$

and for all $\mathbf{v} \in \mathbf{R}^3$ $f(|\mathbf{v}|, 0) = f_0(|\mathbf{v}|)$.

From this equation which describes the evolution of the gas, a weak formulation which is suited to a numerical treatment has been derived. Let $B(\mathbf{R}_+)$ be the space of all bounded measurable functions everywhere defined on $\mathbf{R}_+$. Let us denote $I = [0, 1)$. Then for all $\phi \in B(\mathbf{R}_+)$ ([1048]):

$$\frac{d}{dt} \int_{\mathbf{R}_+} \phi(v) v^2 f(v,t)\, dv =$$
$$k\pi \int_{\mathbf{R}_+^2 \times I^2} (\phi(v') - \phi(v))\, v^2 f(v,t) w^2 f(w,t)\, dv\, dw\, da\, db,$$

where $v' = (a, b; w, v)$, with

$$(a, b; w, v) = 2^{-1/2} \cdot \left(v^2 + w^2 + ((v^2 + w^2)^2 - 4(2b-1)^2 v^2 w^2)^{1/2} (2a - 1)\right)^{1/2}.$$

We numerically solve this equation by choosing two nonnegative integers M and N and defining

(i) a time step $\Delta t = T/M$, assuming $q = 4k\pi\Delta t < 1$;

(ii) a sequence $V^0 = \{v_i^0 : 1 \le i \le N\}$, such that from $i \le j$ follows $v_i^0 \le v_j^0$;

(iii) $f^0 = \frac{1}{N} \sum_{1 \le i \le N} \delta_{v_i^0}$,

where $\delta_{v_i^0}$ is the DIRAC measure at v_i^0 and f^0 approximates the initial velocity density $v^2 f_0(v)$.

For $1 \le n \le M$ we generate $V^n = \{v_i^n : 1 \le i \le N\} \subset \mathbf{R}_+$ and $f^n = \frac{1}{N} \sum_{1 \le i \le N} \delta_{v_i^n}$ as follows:

(i) A RADON measure g^n on $\mathbf{R}_+$ is defined for all $\phi \in B(\mathbf{R}_+)$ by

$$\int_{\mathbf{R}_+} \phi(v) g^n(dv) = \int_{\mathbf{R}_+} \phi(v) f^{n-1}(dv)$$
$$+ \; q \int_{\mathbf{R}_+^2 \times I^2} (\phi(v') - \phi(v)) f^{n-1}(dv) f^{n-1}(dw)\, da\, db.$$

(ii) For $1 \le i \le N$, $1 \le j \le N$ let $\chi_{j,i}$ be the characteristic function of $[\frac{q}{N}(j-1), \frac{q}{N}j) \times [\frac{i-1}{N}, \frac{i}{N})$
and χ_i be the characteristic function of $[q, 1) \times [\frac{i-1}{N}, \frac{i}{N})$.
If $K^n\phi$ is defined on I^4 by

$$\begin{aligned} K^n\phi(a,b,c,d) &= \sum_{1 \le i \le N} \phi(v_i^{n-1}) \chi_i(c,d) \\ &+ \sum_{1 \le i \le N} \sum_{1 \le j \le N} \phi(a, b; v_j^{n-1}, v_i^{n-1}) \chi_{j,i}(c,d), \end{aligned}$$

then

$$\int_{\mathbf{R}_+} \phi(v) g^n(dv) = \int_{I^4} K^n \phi(z)\, dz.$$

Now let us choose a sequence in I^4

$$Z^n = \{z_{(n-1)N+\ell} = (a_{(n-1)N+\ell}, b_{(n-1)N+\ell}, c_{(n-1)N+\ell}, d_{(n-1)N+\ell}) : 1 \le \ell \le N\}$$

such that any subset $I^3 \times [\frac{i-1}{N}, \frac{i}{N})$, $1 \le i \le N$ contains one term of Z^n. We call this property (P). Now we define for all $\phi \in B(\mathbf{R}_+)$ f^n as follows:

$$\int_{\mathbf{R}_+} \phi(v) f^n(dv) = \frac{1}{N} \sum_{1 \le \ell \le N} K^n \phi(z_{(n-1)N+\ell}).$$

According to the property (P), the possibility of a collision for a given molecule i within the time interval $(t^{n-1}, t^n]$ is considered once.
If $\chi_{j,i}(c_{(n-1)N+\ell}, d_{(n-1)N+\ell}) = 1$, then molecule i collides with molecule j and its new velocity v_i^n equals $[a_{(n-1)N+\ell}, b_{(n-1)N+\ell}; v_j^{n-1}, v_i^{n-1}]$.
If $\chi_i(c_{(n-1)N+\ell}, d_{(n-1)N+\ell}) = 1$, then molecule i does not collide and $v_i^n = v_i^{n-1}$.

The crutial tool of LÉCOT [1048] is sorting the sequence V^n after each step n with respect to the order on $\mathbf{R}$. This additional requirement permits us to estimate the errors of the described method by means of the discrepancies of Z^n.

In order to analyse the convergence properties of the described schemes, we proceed as follows. We write $t^n = n\Delta t$ and $\phi_r = \chi_{[0,r)}$ for $r > 0$. Apart from the discrepancy D_N and the star discrepancy D_N^* of a point set we need the star discrepancy $D_N^*(V^n, f)$ of the sequence V^n relative to the density $v^2 f(v, t^n)$. This discrepancy defines the error of the method at time t^n by

$$d_N^n(r) = \frac{1}{N} \sum_{1 \le i \le N} \phi_r(v_i^n) - \int_{\mathbf{R}_+} \phi_r(v) v^2 f(v,t)\, dv \tag{3.50}$$

and

$$D_N^*(V^n, f) = \sup_{r>0} |d_N^n(r)|. \tag{3.51}$$

For computing this error we have to regard the relation (see [1048]):

$$d_N^n(r) = d_N^{n-1}(r) + q e_N^{n-1}(r) + \varepsilon^{n-1}(r) + \delta_N^{n-1}(r),$$

whose parts are defined as follows:

$$\begin{aligned} e_N^n(r) &= \int_{\mathbf{R}_+^2 \times I^2} (\phi_r(v') - \phi_r(v)) f^n(dv) f^n(dw)\, da\, db \\ &\quad - \int_{\mathbf{R}_+^2 \times I^2} (\phi_r(v') - \phi_r(v)) v^2 f(v, t^n) w^2 f(w, t^n)\, dv\, dw\, da\, db \end{aligned}$$

where $v' = (a, b; w, v)$,

$$\varepsilon^n(r) = \int_{[t^n, t^{n+1}] \times \mathbf{R}_+} \phi_r(v) \left(\frac{\partial f}{\partial t}(v, t^n) - \frac{\partial f}{\partial t}(v, t) \right) v^2 \, dv \, dt,$$

$$\delta_N^n(r) = \frac{1}{N} \sum_{1 \le \ell \le N} K^{n+1} \phi_r(z_{nN+\ell}) - \int_{I^4} K^{n+1} \phi_r(z) \, dz.$$

Now we state the bounds for this terms. The error of the method itself is then estimated by means of the discrepancies of the sequences in $[0, 1)^4$ which perform the quadratures. This ensures the convergence when uniformly distributed sequences are used.

The bounds in the first two theorems are achieved by combining a classical error analysis of a one-step method with an estimation of a quadrature error.

Theorem 3.29 *The error term $e_N^n(r)$ can be estimated by*

$$|e_N^n(r)| \le 3 D_N^*(V^n, f).$$

Theorem 3.30 *If f is twice continuously differentiable with respect to t, the error term $\varepsilon^n(r)$ is estimated through*

$$|\varepsilon^n(r)| \le \Delta t \int_{[t^n, t^{n+1}] \times \mathbf{R}_+} \left| \frac{\partial^2 f}{\partial t^2}(\nu, t) \right| \nu^2 \, d\nu \, dt.$$

In order to estimate $\delta_N^n(r)$ we need to bound the difference between the exact measure of a set and its Quasi-Monte Carlo approximation: this is achieved by an application of NIEDERREITER and WILLS [1367] (Theorem 1.18) which we reformulate here in a suitable way.

Lemma 3.31 *Let E be a measurable subset of I^k. For $\varepsilon > 0$ let*

$$E_\varepsilon = \{\mathbf{z} \in I^k : \exists \mathbf{z}\prime \in E, |\mathbf{z} - \mathbf{z}\prime| < \varepsilon\},$$

$$E_{-\varepsilon} = \{\mathbf{z} \in I^k : \forall \mathbf{z}\prime \in I^k \setminus E, |\mathbf{z} - \mathbf{z}\prime| \ge \varepsilon\},$$

(where $|\cdot|$ is the Euclidean norm in $\mathbf{R}^k$). Then for every sequence $\mathbf{x}_1, \ldots, \mathbf{x}_N \in I^k$, if there exists $C > 1$ such that for all $\varepsilon > 0$

$$\max\{\lambda_k(E_\varepsilon \setminus E), \lambda_k(E \setminus E_{-\varepsilon})\} \le C\varepsilon$$

holds, then we have

$$\left| \frac{1}{N} \sum_{n=1}^{N} \chi_E(\mathbf{x}_n) - \lambda_k(E) \right| \le (4Ck^{1/2} + 2K + 1) \, D_N(\mathbf{x}_n)^{1/k}.$$

In our case we may construct convenient sets E_ε which satisfy $C = 5$. This leads to:

Theorem 3.32 *We have*

$$|\delta_N^n(r)| \le 52 D_N(Z^{n+1})^{1/4}.$$

An estimation of the error of the method is then obtained by combining inequalities stated in the last three theorems. As an additional regularity requirement on f, let us assumc that it is twice continously differentiable with respect to t and that

$$v^2 \frac{\partial^2 f}{\partial t^2}(v,t) \in \mathrm{L}^1(\mathbf{R}_+ \times (0,T)).$$

Theorem 3.33 *If the discrepancies of all Z^n are bounded by some D_N, then for all $1 \le n \le M$*

$$\begin{aligned} D_N^*(V^n, f) &\le e^{12k\pi t^n} D_N^*(V^0, f) \\ &+ \Delta t \int_{[t^n, t^{n+1}] \times \mathbf{R}_+} e^{12k\pi(t^n - t)} \left| \frac{\partial^2 f}{\partial t^2}(\nu, t) \right| \nu^2 \, d\nu \, dt \\ &+ \frac{13}{3k\pi} e^{12k\pi t^n} \frac{D_N^{1/4}}{\Delta t}. \end{aligned}$$

The third term on the right-hand side of the last formula has an unpleasant Δt in the denominator. But the computational experiments show that, for fixed N, $D_N^*(V^M, f)$ does not grow to infinity when Δt tends towards 0. The following theorem provides another estimation.

Theorem 3.34 *If all the sequences $Z^n t$ equal*

$$Z^* = \{(a_\ell, b_\ell, c_\ell, d_\ell) : 1 \le \ell \le N\} \subset (0,1)^4 \cup \{0\}$$

then, for $\Delta t \le \frac{1}{4k\pi} \min_{1 \le \ell \le N}\{c_\ell, c_\ell > 0\}$,

$$\begin{aligned} D_N^*(V^n, f) &\le e^{12k\pi t^n} D_N^*(V^0, f) \\ &+ \Delta t \int_{[t^n, t^{n+1}] \times \mathbf{R}_+} e^{12k\pi(t^n - t)} \left| \frac{\partial^2 f}{\partial t^2}(\nu, t) \right| \nu^2 \, d\nu \, dt \\ &+ \frac{1}{3} e^{12k\pi t^n}. \end{aligned}$$

The accuracy of the above estimations have been assesed in LÉCOT [1048] by computation of effective errors in an example where an exact solution was known. It has been shown that using low discrepancy sequences, especially the HAMMERSLEY set which posesses the property (P), was more reliable in this context than doing computations with classical pseudo-random numbers. However, the error increased as Δt decreased.

The improvement of the convergence properties of the described method with respect to the time step was attempted in LÉCOT [1049] and the pseudo-random scheme is compared with the low discrepancy ones associated to the HALTON sequence and the FAURE sequence in base 5. Both of these sequences satisfy the property (P). Morover, the low discrepancy sequences superceded the pseudo-random ones significantly. In contrast to the method using HAMMERSLEY set, the error did not grow when Δt decreased. The difference in the performance of the two low discrepancy sequences was small, with a slight advantage for the HALTON sequence.

3.3.3 The FOURIER Approach

In HUA and WANG [851] several kinds of partial differential equations were solved via a FOURIER series expansion combined with an application of the good lattice point method. In the following we describe this method considering a general class of initial value problems for periodic functions:

$$\frac{\partial^2}{\partial t^2}u(t,\mathbf{x}) = -(-D)^p u(t,\mathbf{x}),$$
$$u(0,\mathbf{x}) = f(\mathbf{x}), \quad \frac{\partial}{\partial t}u(t,\mathbf{x})\Big|_{t=0} = g(\mathbf{x}), \tag{3.52}$$

where p is a positive integer and $D = \sum_{i=1}^{k}\sum_{j=1}^{k} a_{ij}\partial^2/(\partial x_i \partial x_j)$ a second-order differential operator with positive definite coefficient matrix $A = (a_{ij})$; f, g are supposed to be in $\mathbf{E}_k^\alpha(C)$ with $\alpha - 2p > 1$; see Section 3.1. The most important cases in applications are $p = 1$ and $p = 2$, where D is the Laplacian operator. The solution of (3.52) can be found by FOURIER analysis setting

$$u(t,\mathbf{x}) = \sum_{\mathbf{h}\in\mathbf{Z}^k} c(t,\mathbf{h})e(\mathbf{h}\cdot\mathbf{x}). \tag{3.53}$$

Inserting into (3.52) we obtain

$$\sum_{\mathbf{h}\in\mathbf{Z}^k} \frac{\partial^2}{\partial t^2}c(t,\mathbf{h})e(\mathbf{h}\cdot\mathbf{x}) = \sum_{\mathbf{h}\in\mathbf{Z}^k} c(t,\mathbf{h})(4\pi^2[\mathbf{h},\mathbf{h}])^p e(\mathbf{h}\cdot\mathbf{x}), \tag{3.54}$$

where $[\mathbf{h},\mathbf{h}]$ is the inner product defined by $\mathbf{h}A\mathbf{h}^*$ ($\mathbf{h}^*$ denotes the transposed of $\mathbf{h}$). By comparing coefficients of $e(\mathbf{h}\cdot\mathbf{x})$, the differential equation

$$\frac{\partial^2}{\partial t^2}c(t,\mathbf{h}) = -c(t,\mathbf{h})(4\pi^2[\mathbf{h},\mathbf{h}])^p \tag{3.55}$$

follows immediately. Denoting the FOURIER coefficients of the functions f and g by $F(\mathbf{h})$ and $G(\mathbf{h})$ we derive initial conditions

$$c(0,\mathbf{h}) = F(\mathbf{h}), \quad \frac{\partial}{\partial t}c(t,\mathbf{h})\Big|_{t=0} = G(\mathbf{h}). \tag{3.56}$$

From this the FOURIER series expansion for the solution of (3.52) follows immediately:

$$u(t,\mathbf{x}) = F(0) + \sum_{\mathbf{h}\neq 0}\{A(\mathbf{h})\exp(i(4\pi^2)^p[\mathbf{h},\mathbf{h}]^p t)$$
$$+B(\mathbf{h})\exp(-i(4\pi^2)^p[\mathbf{h},\mathbf{h}]^p t)\}e(\mathbf{h}\cdot\mathbf{x}), \tag{3.57}$$

where

$$A(\mathbf{h}) = \frac{1}{2}\{F(\mathbf{h}) - i(4\pi^2)^{-p}[\mathbf{h},\mathbf{h}]^{-p}G(\mathbf{h})\},$$
$$B(\mathbf{h}) = \frac{1}{2}\{F(\mathbf{h}) - i(4\pi^2)^{-p}[\mathbf{h},\mathbf{h}]^{-p}G(\mathbf{h})\}. \quad (3.58)$$

(Since, $\alpha - 2p > 1$, the infinite series are absolutely convergent). Next we apply the method of good lattice points in order to approximate the FOURIER coefficients $F(\mathbf{h})$ and $G(\mathbf{h})$; see Section 3.1. By means of a well-known result from HUA and WANG [851], we have the error estimate

$$\sup_{f\in \mathbf{E}_k^\alpha(C)} \sum_{\|\mathbf{h}\|<[M^{\alpha/(2\alpha-1)}]} \left| F(\mathbf{h}) - \frac{1}{m}\sum_{n=1}^{m} f(\omega_n)e(-\omega_n\mathbf{h}) \right| \leq \gamma M^{-\alpha(\alpha-1)/(2\alpha-1)+\epsilon} \quad (3.59)$$

with an explicit constant γ and arbitrary $\epsilon > 0$. This yields the following approximation of: $u(t,\mathbf{x})$

$$\begin{aligned} \tilde{u}(t,\mathbf{x}) \;=\; & \sum_{\|\mathbf{h}\|<[M^{\alpha/(2\alpha-1)}]} \left\{ \frac{1}{m}\sum_{n=0}^{m-1} f(\omega_n)\exp\left(-2\pi i n\frac{a\cdot\mathbf{h}}{m}\right)\cos\left((4\pi^2)^p[\mathbf{h},\mathbf{h}]^p t\right) \right. \\ & + (4\pi^2)^{-p}[\mathbf{h},\mathbf{h}]^{-p}\frac{1}{m}\sum_{n=0}^{m-1} g(\omega_n) \\ & \left. \times \exp\left(-2\pi i n\frac{a\cdot\mathbf{h}}{m}\right)\sin\left((4\pi^2)^p[\mathbf{h},\mathbf{h}]^p t\right)\right\}. \end{aligned} \quad (3.60)$$

From (3.59) we derive the error bound ($\epsilon > 0$):

$$|u(t,\mathbf{x}) - \tilde{u}(t,\mathbf{x})| \leq \tilde{\gamma}(C,\alpha,k\epsilon)M^{-\alpha(\alpha-1)/(2\alpha-1)+\epsilon}. \quad (3.61)$$

In the following we consider the spherical analogue of the above initial-value problem;

for a more detailed presentation see TICHY [1838] and TICHY and TOMANTSCHGER [1844]. Let $\mathbf{F}_\alpha^k(C)$ denote the class of all functions f defined on the k-dimensional sphere $\mathbf{S}^k$:

$$f(x) = \sum_{n=0}^{\infty}\sum_{j=1}^{Z(k,n)} c(j,n)H_{n,j}, \quad (3.62)$$

with

$$|c(j,n)| \leq \frac{C}{n^\alpha}, \quad \alpha > 1, C > 0, \quad (3.63)$$

where $H_{n,j}$, $j = 1,\ldots,Z(k,n)$ denotes a basis of the system of spherical harmonics of order n. $Z(k,n) = (2n+k-1)(n+j-2)!/(n!(k-1)!)$ is the dimension of the space of all spherical harmonics of order n. (For a detailed discussion of spherical harmonics we refer to MÜLLER [1218].) The following initial-value problem is considered:

$$\frac{\partial^2}{\partial t^2}u(t,x) = -(-\Delta)^p u(t,x),$$

$$u(t,x)|_{t=0} = f(x), \quad \left.\frac{\partial u}{\partial t}\right|_{t=0} = g(x), \qquad f,g \in \mathbf{F}_\alpha^k(C), \tag{3.64}$$

where Δ denotes the spherical Laplacian. In order to find the solution of (3.64) we set

$$u(t,x) = \sum_{n=0}^{\infty} \sum_{j=1}^{Z(k,n)} c(t,j,n) H_{n,j}(x). \tag{3.65}$$

Using

$$\Delta H_{n,j} = -n(n+k-1)H_{n,j}$$

and inserting into the differential equation yields

$$\frac{\partial^2}{\partial t^2}c(t,j,n) = c(t,j,n)(-n^p(n+k-1)^p).$$

Hence

$$c(t,j,n) = A_{j,n}\cos\left(-n^{p/2}(n+k-1)^{p/2}t\right) + B_{j,n}\sin\left(-n^{p/2}(n+k-1)^{p/2}t\right) \tag{3.66}$$

and

$$u(t,x) = \sum_{n=0}^{\infty} \sum_{j=1}^{Z(k,n)} \left\{ A_{n,j}\cos\left(n^{p/2}(n+k-1)^{p/2}t\right) + B_{n,j}\sin\left(-n^{p/2}(n+k-1)^{p/2}t\right) \right\} H_{n,j}. \tag{3.67}$$

$A_{n,j}$ and $B_{n,j}$ can be determined by FOURIER analysis of the functions f and g. Let $F(n,j)$ and $G(n,j)$ denote the spherical FOURIER coefficients of f and g. Then

$$A_{n,j} = F(n,j), \qquad B_{n,j} = \frac{-G(n,j)}{n^{p/2}(n+k-1)^{p/2}}. \tag{3.68}$$

In order to find an approximative solution we have to compute the FOURIER coefficients numerically. This can be done by low discrepancy sequences on the sphere. In the following we will apply HLAWKA's concept of GREEN function discrepancy; see HLAWKA [822, 823, 824, 820, 821].

The above type of initial value problems can be generalized directly to arbitrary compact Riemannian manifolds X of dimension m with metrical tensor

$$ds^2 = \sum_{i,j=1}^{m} g_{ij}\, dx_i\, dx_j.$$

Let $\Delta = \frac{\Delta}{2\sqrt{g}} \sum_{i=1}^{m} \frac{\partial}{\partial x_i} \left(\sqrt{g} \sum_{j=1}^{m} g^{ij} \frac{\partial}{\partial x_j} \right)$ denote the LAPLACE-BELTRAMI operator on X (in local coordinates $x_i, i = 1, \ldots m$) and $g = \det(g_{ij})$ and (g^{ij}) the inverse of (g_{ij}). Then we consider the problem

$$\frac{\partial^2}{\partial f^2} u(t,x) = -(-\Delta)^p u(t,x) u(0,x) = f(x),\ u_t(0,x) = g(x) \tag{3.69}$$

with twice continuously differentiable initial conditions f and g. The cases $p = 1$ (wave equation) and $p = 2$ (plate equation) are of special interest in mathematical physics, see for instance COURANT and HILBERT [417]. We note that every twice continuously differentiable function on X can be expanded with respect to the eigenfunctions Φ_n of $-\Delta$. The corresponding eigenvalues are nonnegative numbers and we order them with respect to their magnitude: $0 = \lambda_{\leq} \lambda_1 \leq \lambda_2 \ldots \to \infty$. Furthermore $(\Phi_n), n = 0, 1 \ldots$ can be taken as an orthonormal system with respect to the normalized surface measure μ on X. The eigenfunctions of $-\Delta$ satisfy the integral equation

$$\Phi_j(x) + \lambda_j \int_X G(x,y) \Phi_j(y) d\mu(y) = 0, \tag{3.70}$$

where $G(x,y)$ is the GREEN function on x. Now we proceed as in the spherical case. This leads to the explicit formula

$$u(t,x) = \sum_{j=0}^{\infty} \left(F_j \cos(\lambda_j^{p/2} t) + G_j \lambda_j^{-p/2} \sin(\lambda_j^{p/2} t) \right) \Phi_j(x) \tag{3.71}$$

for the solution of (3.69). F_j, G_j denote the FOURIER coefficients of f, g, respectively:

$$F_j = \int_X f(x) \Phi_j(x) d\mu(x) \quad , \quad G_j = \int_X g(x) \Phi_j(x) \mu(x). \tag{3.72}$$

In order to compute such surface integrals we will use HLAWKA's approach. We consider a print distribution $\mathbf{x}_1, \ldots, \mathbf{x}_N$ of N points on the manifold X. Then we take the iterated kernels $G^{(r)}(x,y)$, $r = 1, 2, \ldots$, defined by

$$G^{(r)}(x,y) = \int_X G^{(r-1)}(x,z) G(z,y)\, d\mu(z)$$

and by $G^{(1)}(x,y) = G(x,y)$. It can be shown that for sufficiently large r these kernels have a uniformly convergent FOURIER expansion

$$G^{(r)}(x,y) = \sum_{j=1}^{\infty} \frac{\Phi_j(x) \Phi_j(y)}{\lambda_j^r}.$$

The crucial point of HLAWKA's method is to estimate

$$E_N(f) = |I_N(f) - I(f)|$$

in terms of the so-called GREEN function discrepancy. Here again we use the notation

$$I_N(f) = \frac{1}{N}\sum_{n=1}^{N} f(\mathbf{x}_n) \quad \text{and} \quad I_(f) = \int_X f(x)\,d\mu(x).$$

This GREEN function discrepancy $D_N^{(r)}(\mathbf{x}_n)$ (of order r) is defined by

$$D_N^{(r)}(\mathbf{x}_n) = \sup_{y\in X}\left|\frac{1}{N}\sum_{n=1}^{N} G^{(r)}(\mathbf{x}_n, y)\right|. \tag{3.73}$$

Extending the classical KOKSMA-HLAWKA inequality HLAWKA [811] proved the error bound

$$E_N(f) \le V^{(r)}(f) D_N^{(r)}(\mathbf{x}_n), \tag{3.74}$$

where

$$V^{(r)}(f) = \int_X \left|\Delta^{(r)} f(y)\right| d\mu(y)$$

with $\Delta^{(r)}$ denoting the r-times iterated LAPLACE-BELTRAMI operator.

For a fixed positive integer M we consider the following approximation $\hat{u}_M(t,x)$ of the solution $u(t,x)$ of our initial value problem.

$$\begin{aligned}\hat{u}_M(t,x) &= \sum_{j=0}^{M}\frac{1}{N}\sum_{n=1}^{N} f(\mathbf{x}_n)\Phi_j(\mathbf{x}_n)\cos(\lambda_j^{p/2}t)\Phi_j(x) \\ &\quad + \sum_{j=0}^{M}\lambda_j^{-p/2}\frac{1}{N}\sum_{n=1}^{N} g(\mathbf{x}_n)\Phi_j(\mathbf{x}_n)\sin(\lambda_j^{p/2}t)\Phi_j(x). \end{aligned}\tag{3.75}$$

Now we can approximate the FOURIER coefficients $I(f\Phi_j)$, $I(g\Phi_j)$ by the sums $I_N(f\Phi_j)$, $I_N(g\Phi_j)$ and the errors can be estimated in term of $D_N^{(r)}(\mathbf{x}_n)$ and $V^{(r)}$. This yields the following approximative solution of the differential equation (3.69):

$$\begin{aligned}|u(t,x) - \hat{u}_M(t,x)| &\le D^{(r)}(\omega_N)\left[\sum_{j=0}^{M} V^{(r)}(f\Phi_j) + \sum_{j=0}^{M}\lambda_j^{-p/2} V^{(r)}(g\Phi_j)\right] \\ &\quad + \mathcal{O}(M^{1-2r}). \end{aligned}\tag{3.76}$$

The parameters r and M have to be chosen sufficiently large.

Notes

In [814] Hlawka gives error bounds for the numerical solution of various partial differential equations of mathematical physics, including the wave equation, the telegraph-, the Klein-Gordon- and the Helmholtz equation. In Hlawka [825] systems of second order ordinary and partial differential equations are studied by diophantine properties of Kronecker sequences. Tichy [1837] considered a special class of partial differential equations using the Fourier approach. Reitgruber and Tichy [1527] apply good lattice points to the numerical solution of a class of partial differential equations.

Taschner [1789, 1790] used u.d. sequences for the numerical solution of systems of equations and for solving certain initial value problems.

Various papers are devoted to the numerical solution of the Boltzmann equation by Quasi-Monte Carlo methods. This field of applications was initiated by Neunzert and Wick [1267]. Nanbu [1257] presented a direct simulation scheme for the Boltzmann equation. Babovsky and Illmer [60] showed that the discrete measures given by the Nanbu simulation method converge with respect to the weak topology to a solution of the Boltzmann equation. Lécot [1048, 1049, 1047, 1050] used net sequences and derived error bounds. In [1047] an algorithm for generating low-discrepancy sequences on vector computers is presented.

3.4 Random Number Generators

3.4.1 Basic Concepts of Randomness

The main problem in random number generation is the following: Given a distribution function F on $\mathbf{R}$, generate a sequence of real numbers that simulates a sequence of independent and identically distributed random variables with distribution function F. Applications of random numbers in Monte Carlo methods have been discussed above: in numerical integration, in optimization, in computational geometry in solving partial differential equations, integral equations etc. They are also important in computational statistics, in the implementation of probabilistic algorithms, and in cryptology. For a detailed introduction to random number generators we refer to NIEDERREITER [1336] and to LAGARIAS [1005] for applications in cryptology.

It is common to make a distinction between uniform and non uniform random numbers on $[0, 1]$. Non-uniform random numbers have a distribution function different from $F(x) = x$. Usually non-uniform random numbers are produced by transforming uniform ones. Many transformation methods have been developed, such as the inversion method, the rejection method, the composition method and others. For a detailed discussion of non uniform random number generation we refer to DEVROYE [457].

There are three different methods used in the analysis of random numbers: structural, complexity-theoretic, and statistical. Structural requirements refer to aspects such as period length, lattice structure and so on. The complexity-theoretic approach to randomness is important for instance in applications to cryptology. Such an approach to a definition of randomness for finite strings of digits is due to KOLMOGROV [943], CHAITIN [331, 332] and MARTIN-LÖF [1126]; for a recent contribution in that direction see CALUDE and JÜRGENSEN [322]. If $b \geq 2$ is an integer and σ_N is a string of length N consisting of elements of $\mathbf{Z}_b$, then its KOLMOGROV complexity is defined as the shortest length of a program that generates σ_N on a TURING machine. The string σ_N is designated as random if it has the maximum KOLMOGROV complexity among all strings of length N over $\mathbf{Z}_b$. Recently, the existence of secure private-key crypto systems was shown to be equivalent to the existence of special class of random bit generators (so called uniform polynomial generators, see HASTAD [756]).

In the following we will discuss the statistical requirements imposed on random numbers in more detail. The statistical requirements are based on distribution and independence properties. Usually these properties are investigated by well-known statistical tests, such as the uniformity test, the serial test, the gap test, the run test, the spectral test etc. KNUTH [940] proposes a hierarchy of definitions for a sequence (x_n) of uniform random numbers. On the lowest level of this hierarchy is his definition $R1$ which just means that (x_n) is completely uniformly distributed. A much stronger version of randomness is described by his definition $R4$. It proposes to call a sequence (x_n) in $[0,1]$ random if for every effective algorithm that specifies a sequence (b_n) of distinct positive integers, the sequence (x_{b_n}) is completely uniformly distributed. Since there are only countably many such effective algorithms, it follows that almost all sequences satisfy KNUTH's definition $R4$. Furthermore, Theorem 1.183 says that (x^n) satisfies KNUTH's definition $R4$ for almost all $x > 1$.

A random or pseudo random sequence certainly is supposed to have several properties which are typical in a probabilistic sense. But of course we cannot require simultaneously all properties which define a set of probability one, since

$$\bigcap\{M : \mathbf{P}(M) = 1\} = \emptyset.$$

But even KNUTH's definition $R4$ is not appropriate for computational matters: Although the set of sequences fulfilling it has probability 1 no pseudo random number from an effective algorithm can, by definition, be random in the sense of $R4$.

If one tries to apply KOLMOGOROV's complexity concept to sequences which can be generated by a deterministic program, one gets aware of a similar dilemma: The better the algorithm used for the generation of the sequence, the worse, by definition, is the randomness behaviour. This shows that for applications another approach has to be found. KOLMOGOROV's notion of randomness is too restrictive.

Since a pseudo random sequence produced by a computer program is, by definition, a deterministic function, we have to single out a (small) number of certain properties which are extremely relevant to what we intend to call random. The first idea is to require that pseudo random numbers (for short PRN) are uniformly distributed. This means that the discrepancy tends to 0. Thus the convergence behaviour of the discrepancy might be regarded a test for randomness.

But of course this is not enough. There are many other properties one likes that PRN should have or, in other words, there are many other tests they should pass. Hence it is of principal interest to analyze in a systematic manner which and how many tests can reasonably be required. This has been done by WINKLER [1965]. Several approaches are investigated there. The concepts are motivated by probabilistic $0-1$-laws and an analysis of different classes of tests, applied to PRN-sequences. The common idea of all these concepts is to consider sequences of test functions t_n, where each t_n depends only on the first n members of the sequence. Such a test sequence defines a set of sequences (x_n) such that $t_n((x_n)) \to 0$ (for instance). Sets which can be defined in this way by tests are called testable. In varying this concept one gets a hierarchy of notions of testability. GOLDSTERN [664] points out that this hierarchy is essentially equivalent to the BOREL hierarchy of open, closed, F_σ, G_δ-sets and so on. By means of this equivalence and descriptive set theory he elucidates the situation in a satisfactory way and shows that, roughly spoken, the first glance BOREL complexity of testable sets like that of u.d. sequences is the true one.

The main results in WINKLER [1965] are of the following type: Let $\mathcal{T}$ be the system of all one-sets which are testable in some specified sense and have certain invariance properties. Then the set

$$R = \bigcap_{M \in \mathcal{T}} M$$

of all sequences that fulfill all tests of the corresponding type either is empty or contains sequences which are definitely not random in any useful sense.

The consequence is that there is still no better way than to use very special statistical tests for describing the random behaviour of a PRN. Indeed, two of the most familiar ones are discrepancy tests:

I. Uniformity test. This is a test for the empirical distribution of an initial segment $x_1, x_2, \ldots, x_N$ of the sequence. The test is performed by calculating (or bounding) the discrepancy or the star discrepancy of the sequence.

II. Serial test. This is a multidimensional version of the uniformity test; it severely tests the independence of PRN. Let (x_n) be a given sequence of PRN in $[0,1)$ and set $\mathbf{x}_n^{(k)} = (x_n, x_{n+1}, \ldots, x_{n+k-1}) \in [0,1)^k$. Then the test-quantity is just the k-discrecpancy $D_N^{(k)}(x_n)$ given by

$$D_N^{(k)}(x_n) = D_N\left(\mathbf{x}_1^{(k)}, \ldots, \mathbf{x}_N^{(k)}\right),$$

or the corresponding star-discrepancy; see Section 1.4.

For practical applications of Monte Carlo methods users need deterministic algorithms for producing PRN-sequences that behave like random numbers. Of course, the PRN-sequences must pass statistical tests for randomness to be suitable for simulation purposes. Naturally, different kinds of applications need different kinds of PRN. Thus it is very important to have a list of generators ready for practical applications. In a recent research project guided by HELLEKALEK (University of Salzburg) a data base of PRN generators will be established. We hope that in a few years this data base can be reached via Internet.

3.4.2 Examples of Number Theoretic Generators

Following mainly LAGARIAS [1005] we will discuss briefly several examples of number-theoretic random number generators; see also LAGARIAS [1006].

EXAMPLE 1 (MULTIPLICATIVE CONGRUENTIAL GENERATOR)

$$y_{n+1} = ay_n + c \mod M,$$

where $0 \leq y_n \leq M-1$. Generators of this form are widely used in practice in Monte Carlo methods; see KNUTH [940] and NIEDERREITER [1336]. The parameter M (modulus), is a large integer, a (multiplier) an integer with $1 \leq a < M$ and $\gcd(a, M) = 1$, and $c \in \mathbf{Z}_M$. Taking the initial value y_0 in $\mathbf{Z}_M$ we get a periodic sequence with period length $\leq M$. The period is equal to M if and only if $\gcd(c, M) =$

1, $a \equiv 1 \bmod p$ for every prime p dividing M, and $a \equiv 1 \bmod 4$ if 4 divides M. The following three cases are the standard ones for practical implementations:

i) M prime, a primitive root mod M, $c = 0, y_0 \neq 0$.
ii) $M = 2^t, a \equiv 5 \bmod 8$, $c \equiv 1 \bmod 2$
iii) $M = 2^t, a \equiv 5 \bmod 8$, $c = 0$, $y_0 \equiv 1 \bmod 2$.

From the sequence (y_n) in $\mathbf{Z}_M$ we derive the linear congruential PRN $x_n \in [0, 1)$ by

$$x_n = \frac{y_n}{M} \quad , \quad n = 0, 1, \ldots \tag{3.77}$$

The performance of linear PRN under the uniformity test is well known in all three cases i), ii), iii):

$$D_N^*(x_1, \ldots, x_{N-1}) = \mathcal{O}\left(N^{-1} M^{1/2} (\log M)^2\right), \tag{3.78}$$

for all N less or equal to the period length; the implied $\mathcal{O}$-constant is an absolute one; see NIEDERREITER [1336]. The important question of the performance of a linear congruential PRN sequence (x_n) under the k-dimensional serial test was explored much later. As a typical result we mention the bound ($k \geq 2$):

$$D_{M-1}^{(k)}(x_n) = \mathcal{O}\left(M^{-1} (\log M)^k \log\log(M+1)\right); \tag{3.79}$$

see NIEDERREITER [1296, 1336]. It follows also easily that for PRN of type i) the points $\mathbf{x}_n^{(k)} = (x_{n+1}, \ldots, x_{n+k-1}), n = 0, 1 \ldots$ form a regular lattice pattern in $[0, 1]^k$.

The lattice structure of linear conguential PRN provides a basis for so-called lattice tests that can be applied to these PRN, see also KNUTH [940]. For instance, one may determine the minimum number of parallel hyperplanes on which all points $\mathbf{x}_n^{(k)}$ lie. This number should be as large as possible. Another possibility is to determine the maximum distance between adjacent hyperplanes, taken over all families of parallel hyperplanes which contain all points $\mathbf{x}_n^{(k)}$. This distance should be as small as possible. In the following we state the result of a geometric analysis of PRN due to AFFLERBACH [5] without proof.
Let us consider the case of a multiplicative congruential generator

$$x_i \equiv b \cdot x_{i-1} \bmod 2^e, \; 0 < x_i < 2^e, \; e > 3, \tag{3.80}$$

with $b \equiv 5 \bmod 8$ and x_0 odd. As it is well known in this case the maximal period length 2^{e-2} is obtained.

Theorem 3.35 *Let $d = \gcd(n, 2^{e-2})$ be greater than 1. For all odd initial values x_0 the periodic continuation (with period 2^e) of the set of all non-overlapping vectors generated by the recurrence (3.80) with $b \equiv 5$ modulo 8 is a grid, which is described by the shift vector*

$$\mathbf{g}_0(x_0) = x_0(1, b, \ldots, b^{n-1})$$

and the lattice basis

$$\begin{aligned} \mathbf{g}_1 &= 4d(1, b, \ldots, b^{n-1}), \\ \mathbf{g}_2 &= (0, 2^e, 0 \ldots, 0), \\ &\vdots \\ \mathbf{g}_n &= (0, 0, \ldots, 0, 2^e). \end{aligned}$$

EXAMPLE 2 (POWER GENERATOR)

$$y_{n+1} = y_n^d \mod M$$

There are two important special cases of the power generator both occuring when $M = p_1 \cdot p_2$ is a product of two odd primes. The first case occurs when $\gcd(d, \varphi(M)) = 1$, where $\varphi(M) = (p_1 - 1)(p_2 - 1)$ is EULER's totient function. Then the map $x \mapsto s^d \mod M$ is bijective on the multiplicative residue group mod M, and this operation is the encryption operation of the RSA public key cryptosystem (see RIVEST, SHAMIR, and ADLEMAN [1558]), where d, M are publicly known. This generator is called RSA generator.

The second case occurs when $d = 2, M = p_1 p_2$ with $p_1, p_2 \equiv 3 \mod 4$. This generator is called the SQUARE GENERATOR, see BLUM, BLUM and SHUB [230].

EXAMPLE 3 (QUADRATIC GENERATOR)

$$y_{n+1} = a y_n^2 + b y_n + c \mod M$$

Quadratic generators are special cases of non-linear first order congruential methods

$$y_{n+1} \equiv f(y_n) \mod M \qquad (\text{for } n = 0, 1, \ldots). \tag{3.81}$$

We derive uniform PRN x_n by $x_n = \frac{1}{M} y_n$ in $[0, 1)$. Clearly, the period of (x_n) and (y_n) is $\leq M$. If f is a polynomial of degree d with integer coefficients and $M = p$ is a prime, then the following discrepancy bound can be shown

$$D_p^{(k)}(x_n) \leq 1 - \left(1 - \frac{1}{p}\right)^k + (d-1) p^{-1/2} \left(\frac{4}{\pi^2} \log p + 1,72\right)^k \tag{3.82}$$

for $2 \leq k \leq d$, see NIEDERREITER [1336]. The bound is based on the WEIL-STEPANOV bound for character sums [1740].

The quadratic congruential method was proposed by KNUTH; see [940]. Here we use the modulus $M = 2^\alpha, \alpha \geq 2$. For the corresponding sequence $x_0, x_1, \ldots$ of quadratic congruential pseudrandom numbers, we have $\text{per}(x_n) = M$ (maximal period) if and only if a is even, $b \equiv a + 1 \mod 4$, and c is odd.

The lattice structure of quadratic congruential PRN was analyzed by EICHENAUER and LEHN [517]. For a given dimension $k \geq 2$, the set of all nonoverlapping k-tuples $\mathbf{x}_{nk}^{(k)} = (x_{nk}, x_{nk+1}, \ldots, x_{nk+k-1}) \in I^k, n = 0, 1, \ldots, M-1$, is the same as the intersection of I^k with a union of $2^{\omega-\eta}$ grids with explicitly known shift vectors and lattice bases. Here η and ω are determined by $\gcd(k, M) = 2^\eta$, $\gcd(a, M) = 2^\beta$, and $\omega = \max\left(\lfloor\frac{1}{2}(\alpha - \beta + 1)\rfloor, \eta\right)$.

EICHENAUER-HERMANN and NIEDERREITER [555] investigated the performance of quadratic congruential PRN under the two-dimensional serial test. Let $D_M^{(2)}$ be the discrepancy $D_M^{(k)}(x_n)$ for $k = 2$ and let β be as above; we always assume full period. Then, for $k = 2$, we have

$$D_M^{(2)} < \frac{2}{M} + \frac{4}{7}(2\sqrt{2}+1)2^{\beta/2}M^{-1/2}\left(\frac{2}{\pi}\log M + \frac{2}{5}\right)^2 .$$

This is the best possible bound in the sense that, if $\beta \le \alpha - 2$ and $b \equiv 1 \bmod 2^{\beta+1}$, then

$$D_M^{(k)} \ge \frac{1}{(\pi+2)\sqrt{2}} 2^{\beta/2} M^{-1/2} \qquad \text{for all } k \ge 2.$$

The upper bound for $D_M^{(2)}$ suggests that it is reasonable to choose the parameter a in such a way that $\beta = 1$, i.e. that $a \equiv 2 \bmod 4$. Then the criterion for per $(x_n) = M$ implies that $b \equiv 3 \bmod 4$, and, in this case, we have the lower bound

$$D_M^{(k)} \ge \frac{1}{3(\pi+2)} M^{-1/2} \qquad \text{for all } k \ge 2.$$

For further literature on quadratic generators we refer to the Notes.

EXAMPLE 4 (DISCRETE EXPONENTIAL GENERATOR)

$$y_{n+1} = g^{y_n} \mod M$$

A special case of importance occurs when M is an odd prime and g is a primitive root mod M. The problem of recovering y_n given y_{n+1}, g, M is the ***discrete logarithm problem***, see ODLYZKO [1387].

EXAMPLE 5 (INVERSIVE CONGRUENTIAL GENERATOR)

$$y_{n+1} = a\bar{y}_n + b \mod M,$$

where $M = p \ge 5$ is a prime modulus and $\bar{c}$ is defined by $c \cdot \bar{c} \equiv 1 \bmod p$ for $c \ne 0$ and $\bar{0} = 0$. The parameters a, b are coprime numbers of $\mathbf{Z}_p$ with $a \ne 0$. The PRN $x_n = y_n/p$ are called inversive congruential pseudorandom numbers of modulus p. These PRN were introduced by EICHENAUER and LEHN [516]. Later they were investigated in a series of papers by these authors, NIEDERREITER and others; see the Notes. The period length is always $\le p$ and it can be shown that

$$D_p^{(k)}(x_n) \le 1 - \left(1 - \frac{1}{p}\right)^k + \left(\frac{2k-2}{\sqrt{p}} + \frac{k-1}{p}\right)\left(\frac{4}{\pi^2}\log p + 1,72\right)^k \qquad (3.83)$$

for $k \ge 2$; c.f. NIEDERREITER [1336]. The proof depends on the WEIL-STEPANOV bound for exponential sums over $\mathbf{Z}_p$.

EXAMPLE 6 (FEEDBACK SHIFT REGISTER SEQUENCES)

$$y_{n+d} = f(y_{n+d-1}, \ldots, y_n),$$

for a fixed function f. The most important case is the one where f is a linear function in d variables over $\mathbf{Z}$:

$$y_{n+d} \equiv \sum_{j=0}^{d-1} a_j\, y_{n+j} \mod p \quad (\text{for } n = 0, 1, \ldots),$$

p being a prime. By the digital multistep method due to TAUSWORTHE [1794] the sequence (y_n) is transformed into a sequence (x_n) of uniform PRN in the following way: choose an integer with $2 \leq m \leq d$ and put

$$x_n = \sum_{j=1}^{m} y_{mn+j-1} p^{-j} \qquad (\text{for } n = 0, 1 \ldots).$$

EXAMPLE 7 (THE MATRIX METHOD)

$$\mathbf{y}_{n+1} \equiv \mathbf{A} \cdot \mathbf{y}_n \mod M, \qquad (\text{for } n = 0, 1, \ldots)$$

where $\mathbf{A}$ is a non-singular $r \times r$-matrix mod $M, r \geq 2$. The sequence $\mathbf{y}_0 \neq 0, \mathbf{y}_1, \ldots$ is a sequence of pseudorandom vectors mod M, which can be transformed into uniform pseudorandom vectors $\mathbf{x}_n$ in $[0, 1)^r$ via the usual tansformation

$$\mathbf{x}_n = \frac{1}{M}\mathbf{y}_n \qquad (u = 0, 1, \ldots).$$

For recent references on matrix generators we refer to the Notes and the survey article of NIEDERREITER [1345].

EXAMPLE 8 (CELLULAR ATOMATA)

A one-dimensional cellular automaton consists of s sites $(a_1, \ldots, a_s)$, which may take integer values between 0 and $d - 1$, for some fixed d. These values are updated in at discrete times according to a fixed rule ϕ of the form

$$a_i^{(n+1)} = \phi(a_{i-r}^{(n)}, a_{i-r+1}^{(n)}, \ldots, a_{i+r}^{(n)}),$$

where the subscripts are interpreted mod s. The quantity r is called the width of the cellular automaton. WOLFRAM [1970] has proposed two binary ($d = 2$) cellular automata of width 1 for use in pseudrandom number generation. These have the rules

$$a_i^{(n+1)} \equiv a_{i-1}^{(n)} + a_i^{(n)} + a_{i+1}^{(n)} + a_i^{(n)} a_{i+1}^{(n)} \mod 2$$

and

$$a_i^{(n+1)} \equiv 1 + a_{i-1}^{(n)} + a_{i+1}^{(n)} + a_i^{(n)} a_{i+1}^{(n)} \mod 2,$$

respectively. These generators have the advantage that they can be efficiently implemented on some parallel processors. For further relations between pseudorandom numbers and cellular automata we refer to TEZUKA and FUSHIMI [1803].

Notes

There exists a vaste literature on lattice tests of pseudorandom sequences, see for instance Knuth [940], Niederreiter [1336], Dieter [463, 464, 465] and Mendès France [1169]. Afflerbach and Grothe [6] presented an algorithm for the computation of Minkowski-reduced lattice bases; for the lattice structure of linear congruential generators see [4, 5]. Afflerbach and Grothe [7] proposed the so called Beyer quotients q_k (defined as the maximal dilatation of a fundamental k-corner with k-shortest edges of the lattice embedded in $\mathbf{R}^k$) as estimators for the distribution quality of matrix generators. Altman [34] discussed several applications of Marsaglia's lattice test. For further results on the lattice structure of linear generators we refer to Couture and L'Ecuyer [419, 418], Eichenauer-Hermann [533], de Matteis, Eichenauer-Hermann and Grothe [440], and Marsaglia [1123], who proposed a well-known lattice test.

Afflerbach and Weilbächer [8] proposed an algorithm for the computation of the two-dimensional discrepancy of linear generators. Eichenauer and Lehn [515] presented an algorithm for the period length of generalized Fibonacci generators; see also Brent [295] and Mascagni, Cuccaro, Pryor, and Robinson [1128]. Fishman [601] provided an exhaustive analysis of multiplicative congruential generators with modulus 2^{32}. Parker [1404] investigated the period of the Fibonacci generator. Denzer and Ecker [451] considered optimal multipliers with prime moduli for linear congruential generators. A discrepancy bound for linear congruential generators is due to Levin [1073, 1074]. For further results in this direction see also Bhavsar, Gujar, Horton and Lambrou [218].

Linear congruential multi-step generators (i.e. k-th order integer-valued linear recurring generators) were investigated in a series of papers by Niederreiter [1276, 1282, 1291, 1290, 1292, 1293, 1297, 1298, 1301, 1300, 1303, 1309, 1306, 1307, 1310, 1314, 1318, 1322]. In these papers the generators are mainly tested by estimating the discrepancy of the sequence $(y_n/M, \cdots, y_{n+k-1}/M)$ in $[0,1]^k$, where M is a sufficiently large and suitably chosen module. Also digital constructions of random points are used. The discrepancy is estimated via exponential sums. Fushimi [638] established further results on the Tausworthe method. Fushimi and Tezuka [640] considered the k-distribution of generalized feedback shift register pseudorandom numbers. L'Ecuyer, Blouin, and Couture [1055] reported on an extensive computer search for good multiple recursive generators.

The combination of several construction principles for pseudorandom numbers can lead to an increased period length, to an improved statistical performance, and to a "better" lattice structure. The original proposals of Wichman and Hill [1956, 1957] were based on the combination of linear generators; see also L'Ecuyer [1051], L'Ecuyer and Tezuka [1057], Levitan and Sobol [1080], and Härtel [753]. For more recent contributions on combined generators we refer to Couture and L'Ecuyer [419], L'Ecuyer [1054] and Tezuka and L'Ecuyer [1804]. Couture, L'Ecuyer and Tezuka [420] investigated the lattice structure of combined generators in spaces of formal Laurent series; see also L'Ecuyer, Blouin and Couture [1056], and Tezuka [1800]. Sezgin [1667] studied a special composite random number generator.

Recently, pseudorandom matrix generators became of great interest because they are suitable for parallel computing. We refer here to Aluru, Prabhu and Gustafson [35], Anderson [38], Eddy [511], Grothe [704, 705], Niederreiter [1327, 1329, 1341, 1344, 1346] and Zhang and Zhou [1991]. The period length of pseudorandom matrix generators was studied by Eichenauer, Grothe and Lehn [548]. For long-range correlations in multiplicative generators we refer to Eichenauer-Hermann and Grothe [547] and De Matteis and Pagnutti [441]. For a detailed investigation on pseudorandom generation using parallel computing we refer to Burton and Page [319]. For an implementation of a distirbuted generator see Chen and Whitlock [341].

An excellent survey on the development (until 1978) of Quasi-Monte Carlo methods and (linear) random number generators given by Niederreiter [1296]; the more recent development can be found in [1336]. Niederreiter [1345] provides an up-to-date survey, especially on matrix generators and on compound generators, including a comprehensive list of references. Further important surveys on random number generation are due to Deák [442], Dieter [465, 466] and L'Ecuyer [1053]; see also L'Ecuyer [1052], Niederreiter [1326, 1334, 1340, 1338] and Tezuka [1802].

Eichenauer and Niederreiter [519], Eichenauer, Grothe and Lehn [514] discussed Marsaglia's lattice test in connection with nonlinear pseudorandom number generators. Niedereiter [1323] continued these investigations by studying statistical independence properties involving discrepancy bounds; see also Niederreiter [1321]. Fushimi [639] discussed generators whose subsequences are k-distributed. Quadratic generators were considered by Eichenauer and Lehn [517], Eichenauer-

Hermann [529, 536, 545, 544, 540], Eichenauer-Hermann and Niederreiter [520, 555]. Further contributions on quadratic and nonlinear generators can be found in Cusick [427], Eichenauer-Hermann [530, 536, 542] and Eichenauer-Hermann and Niederreiter [554].

In a series of papers Eichenauer and Lehn, Eichenauer, Lehn and Topuzoglu [518], [516], Eichenauer-Hermann, and Topuzoglu [521], Eichenauer-Hermann and Grothe [527], Eichenauer-Hermann and Ickstadt [550] and Eichenauer-Hermann [523, 522, 528, 529, 525, 524, 532, 535, 533, 534, 530, 531, 537, 538, 541, 544] systematically investigated inversive congruential generators. Further results on inversive generators can be found in Huber [854], Emmerich [560], Niederreiter [1324, 1342, 1325, 1341], Flahive and Niederreiter [602] and Eichenauer-Hermann and Niederreiter [526, 549, 553]. The last two authors especially applied bounds for Kloosterman sums for estimating corresponding exponential sums. Chou [361] characterized inversive maximal period polynomials over finite fields; see also [362]. In [363] he extended his results to inversive generators with respect to general moduli; Chou [362] discussed the period length of inversive vector generators. Eichenauer-Hermann [539] investigated a compound version of the nonlinear congruential method, which is more suitable from the point of view of implementation. In general such compound generators are constructed via $\mathbf{x}_n \equiv \mathbf{x}_n^{(1)} + \cdots \mathbf{x}_n^{(r)} \pmod 1$, where $\mathbf{x}_n^{(i)}$ are pseudo random points generated by given congruential generators modulo (distinct) prime numbers p_i. Further work on the analysis of compound generators is due to Eichenauer-Hermann [543], Eichenauer-Hermann and Emmerich [546], and Eichenauer-Hermann and Larcher [551, 552].

Antipov [40, 41] established results on the complete u.d. of multiplicative generators; see also Szczuka [1767]. Antipov [42, 43] investigated period properties of congruential generators; here we also mention Tezuka [1801]. A doubly multiplicative congruential method was studied by Miyazaki [1198]. Blum, Blum and Shub [230] studied a special unpredictable generator. Different kinds of generators were studied by Rizzi [1559] (based on primitive polynomials), by Chassaing [338], by Goldreich, Krawczyk and Luby [660] by Moriyama, Nakamura and Yamuda [1203] and by Sugita [1761] (based on irrational rotations). Magliveras, Oberg and Surkan [1116] described a random number generator based on permutation groups and Marsaglia, Zaman, Tsang and Wai [1124] proposed a "universal" generator which performs well under several tests of independence. Edwards [512] investigated random bit sequences by means of next bit tests for special random generators. Shparlinski [1682] analyzed the generator $x_{n+1} = [ax_n]$ (proving discrepancy bounds).

Gut, Egorov and Il'in [710, 711] investigated normal pseudorandom numbers by means of the Walsh transform of u.d. pseudorandom numbers. Jerum, Valiant and Vazirani [873] and Yuen [1980] investigated combinatorial aspects of random number generators. Tezuka [1799] proposed the spectral test based on Walsh functions for testing pseudorandom sequences. For more on statistical tests on pseudorandom numbers see Dobris [469]. Here we also refer to a detailed investigation of Hellekalek and his group (Salzburg) who work out a systematic study of various tests on uniformly distributed pseudorandom numbers; see for instance Auer and Hellekalek [55, 56]. The generation of non-uniformly distributed pseudorandom sequences is covered by the excellent monograph of Devroy [457]. Special references in this context are Boswell, Gore, Patil, and Toillic [274], and Hörmann [848]. Szczuka and Zielinski [1768] give a survey on the generation of gamma-distributed random variables.

Niederreiter [1333] established a digital construction over finite fields for producing an analogon of good lattice point sequences. This constructon was simplyfied by Larcher [1031]. Niederreiter [1316] considered relashionships between continued fractions for formal power series and applications to pseudorandom sequences. Cugiani [423] and Rotondi [1571] applied p-adic methods for describing pseudorandom sequences; see also Cugiani and Rotondi [424] for vector generation.

Wikramanatra [1960] discussed empirical tests of additive congruential generators and Neumann [1266] presented implementations of congruential generators. Rubinstein [1574] established an algorithm for generating random points in simple geometric regions. Schulze [1662] discussed a deterministic simulation of random variables via Quasi-Monte Carlo methods. For various generators related to cryptology we refer to Meier and Staffelbach [1160], Pararin [1419], and Sadeghiyan and Pieprzyk [1586].

Bibliography

[1] A. G. Abercrombie. Beatty sequences and multiplicative number theory. *Acta Arith.*, 70:195–207, 1995.

[2] L. Achan. Discrepancy in $[0,1]^s$. *preprint.*

[3] R. Adler, M. Keane, and M. Smorodinsky. A construction of a normal number for the continued fraction transformation. *J. Number Theory*, 13:95–105, 1981.

[4] L. Afflerbach. *Lineare Kongruenz-Generatoren zur Erzeugung von Pseudo-Zufalls-zahlen und ihre Gitterstruktur.* Fachbereich Mathematik der Techn. Hochschule Darmstadt, 1983.

[5] L. Afflerbach. The sub-lattice structure of linear congruential random number generators. *Manuscr. Math*, 55:455–465, 1986.

[6] L. Afflerbach and H. Grothe. Calculation of Minkowski-reduced lattice basis. *Computing*, 35:269–276, 1985.

[7] L. Afflerbach and H. Grothe. The lattice structure of pseudo-random vectors generated by matrix generators. *J. Comput Appl.Math.*, 23:127–131, 1988.

[8] L. Afflerbach and R. Weilbächer. The exact determination of rectangle discrepancy for linear congruential pseudorandom numbers. *Math. Comput.*, 53:343–354, 1989.

[9] M. Ajtai, I. Havas, and J. Komlos. Every group admits a bad topology. *Studies in Pure Mathematiacs (In the memory of P. Turán, Budapest)*, 21–34, 1983.

[10] M. Akita, K. Goto, and T. Kano. A problem of Orlicz in the Scottish Book. *Proc. Japan Acad. A*, 62:267–269, 1986.

[11] M. Akita, S. Iseka, and T. Kano. On the convergence of $\sum_{n=1}^{\infty} n^{-\alpha} \exp(2\pi i\, n^{\beta}\, \theta)$. *Number theory and combinatorics, Proc. Conf., Tokyo/Jap., Okayama/Jap. and Kyoto/Jap. 1984*, 1–20, 1985.

[12] R. Alexander. Geometric methods in the study of irregularities of distribution. *Combinatorica*, 10:115–136, 1990.

[13] R. Alexander. Principles of a new method in the study of irregularities of distribution. *Invent. Math.*, 103:279–296, 1991.

[14] K. Alladi. New concepts in arithmetic functions. *Matscience Report*, 83, 1975.

[15] I.A. Allakov. On the distribution of the fractional part of the sequence αp^2 with the prime numbers of an arithmetic progession. *Izv. Akad. Nauk SSSR, Ser. Mat.*, 6:7–11, 1990.

[16] J.-P. Allouche. Tours de Hanoi, automates finis et fractions continues. *Sémin. Théor. Nombres Bordeaux*, Exp. 19:8 p., 1987.

[17] J.-P. Allouche. Des nouvelles des automates. *Publ. Math. Orsay*, 1–4, 1992.

[18] J.-P. Allouche. The number of factors in paperfolding sequence. *Bull. Austral. Math. Soc.*, 46:23–31, 1992.

[19] J.-P. Allouche. Sur la complexité des suites infinies. *Bull. Belg. Math. Soc. - Simon Stevin*, 1:133–143, 1994.

[20] J.-P. Allouche, A. Arnold, J. Berstel, S. Brlek, W. Jockusch, S. Plouffe, and B. Sagan. A relative of the Thue-Morse sequence. *Discrete Math.*, 139:455–461, 1995.

[21] J.-P. Allouche and R. Bacher. Toeplitz sequences, paperfolding, towers of hanoi and progression-free sequences of integers. *Enseign. Math. II. Ser. 38*, 38:315–327, 1992.

[22] J.-P. Allouche and M. Bosquet-Mélou. Canonical positions for the factors in paperfolding sequences. *Theor. Comput. Sci.*, 129:263–278, 1994.

[23] J.-P. Allouche and M. Bosquet-Mélou. Facteurs des suites de Rudin-Shapiro generalisees. *Bull. Belg. Math. Soc. - Simon Stevin*, 1:145–164, 1994.

[24] J.-P. Allouche and J.-M. Deshouillers. Repartition de la suite des puissances d'une serie formelle algebrique. *Publ. Math. Orsay*, 37–47, 1988.

[25] J.-P. Allouche and P. Liardet. Generalized Rudin-Shapiro sequences. *Acta Arith.*, 60:1–27, 1991.

[26] J.-P. Allouche, A. Lubiw, M. Mendès France, A.J. van der Poorten, and J. Shallit. Convergents of folded continued fractions. *Acta Arith.*, 77:77–96, 1996.

[27] J.-P. Allouche and M. Mendès France. Suite de Rudin-Shapiro et modèle d'Ising. *Bull. Soc. Math. Fr.*, 113:273–283, 1985.

[28] J.-P. Allouche, P. Morton, and J. Shallit. Pattern spectra, substring enumeration, and automatic sequences. *Theor. Comput. Sci.*, 94:161–174, 1992.

[29] J.-P. Allouche and O. Salon. Sous-suites polynomiales de certaines suites automatiques. *J. Theor. Nombres Bordx.*, 5:111–121, 1993.

[30] J.-P. Allouche and J. Shallit. The ring of k-regular sequences. *Theor. Comput. Sci.*, 198:163–197, 1992.

[31] J.-P. Allouche and J.O. Shallit. Complexité des suites de Rudin-Shapiro généralisés. *J. Theor. Nombres Bordx.*, 5:283–302, 1993.

[32] N. Alon and Y. Mansour. ε-discrepancy sets and their application for interpolation of sparse polynomials. *Inform. Processing Letters*, 54:337–342, 1995.

[33] N. Alon and Y. Peres. Uniform dilations. *Geometric and Functional Analysis*, 2:1–28, 1992.

[34] N.S. Altman. Bit-wise behavior of random number generators. *SIAM J. Sci. Stat. Comput.*, 9:941–949, 1988.

[35] S. Aluru, G.M. Prabhu, and J. Gustafson. A random number generator for parallel computers. *Parallel Comput.*, 18:839–847, 1992.

[36] C. Amstler. Discrepancy operators and numerical integration on compact groups. *Monatsh. Math.*, 119:177–186, 1995.

[37] V.S. Anashin. Uniformly distributed sequences of p-adic integers. *Math. Notes*, 55:109–133, 1994.

[38] S.L. Anderson. Random number generators on vector supercomputers and other advanced architectures. *SIAM Rev.*, 32:221–251, 1990.

[39] V.A. Andreeva. A generalization of a theorem of Koksma on uniform distribution. *C.R. Acad. Bulg. Sci.*, 40:9–12, 1987.

[40] M.V. Antipov. Uniform distribution and correlation of multiplicative sequences. *All-Union Conf.*, Pap. VI, Part 1:64–72, 1979.

[41] M.V. Antipov. Estimates of the error of integration in the practice of the Monte Carlo method. *Mathematical and imitation models of systems, Collect. Sci. Works, Novisibirsk*, 10–16, 1983.

[42] M.V. Antipov. On recurrent sequences of pseudo-random numbers. *Theory and appl. of statistical modelling, Collect. Sci. Works, Novosibirsk*, 3–9, 1988.

[43] M.V. Antipov. Restrictions in numerical Monte Carlo methods in case of using pseudorandom numbers. *Theory and appl. of statistiacal modeling. Collect. Sci. Works, Novosibirsk*, 3–9, 1989.

[44] R.G. Antonini. On the notion of uniform distribution mod 1. *Fibonacci Q.*, 29:230–234, 1991.

[45] D. Applegate and J. Lagarias. Density bounds for the $3x+1$ problem. I. Tree-search method. *Math. Comput.*, 64:411–426, 1995.

[46] D. Applegate and J. Lagarias. Density bounds for the $3x+1$ problem. II. Krasikov inequalities. *Math. Comput.*, 64:427–438, 1995.

[47] G.I. Arkhipov, A.A. Karatsuba, and V.N. Chubarikov. Multiple trigonometric sums. *Tr. Math. Inst. Steklova*, 151, 1980.

[48] G.I. Arkhipov, A.A. Karatsuba, and V.N. Chubarikov. Multiple trigonometric sums and their applications. *Izv. Akad. Nauk SSSR, Ser. Mat.*, 44:723–781, 1980.

[49] G.I. Arkhipov, A.A. Karatsuba, and V.N. Chubarikov. Multiple trigonometric sums (English). *Proc. Steklov Inst. Math.*, 151, 1982.

[50] P. Arnoux and C. Mauduit. Complexité de suites engendrées par des récurrences unipotentes. *Acta Arith.*, 76:85–97, 1996.

[51] P. Arnoux, C. Mauduit, I. Shiokawa, and J. Tamura. Rauzy's conjecture on billiards in the cube. *Tokyo J. Math.*, 17:211–218, 1994.

[52] P. Arnoux, C. Mauduit, I. Shiokawa, and J.I. Tamura. Complexity of sequences defined by billiard in the cube. *Bull. Soc. Math. Fr.*, 122:1–12, 1994.

[53] P. Arnoux and G. Rauzy. Représentations géométrique des suites de complexité $2n+1$. *Bull. Soc. Math. Fr.*, 119:199–215, 1991.

[54] E.Y. Atanassov. Note on the discrepancy of the van der Corput generalized sequences. *C.R. Acad. Bulg. Sci.*, 42:41–44, 1989.

[55] T. Auer and P. Hellekalek. Independence of uniform pseudorandom numbers, part I: the theoretical background. *Proc. Internat. Workshop Parallel Numerics '94*, 44–58, 1994.

[56] T. Auer and P. Hellekalek. Independence of uniform pseudorandom numbers, part II: empirical results. *Proc. Internat. Workshop Parallel Numerics '94*, 59–73, 1994.

[57] W.J. Aumayr. Metrische Sätze über die Vererbbarkeit der Gleichverteilung von aus Funktionen gewonnenen Folgen. *Österr. Akad. Wiss. SB II*, 194:309–321, 1985.

[58] P.C. Baayen and Z. Hedrlin. The existence of well distributed sequences in compact spaces. *Indag. Math.*, 27:221–228, 1965.

[59] A. Babaev. An estimate of the labor consumption of a certain integration method by the Monte Carlo method. *Izv. Akad. Nauk SSSR, Ser. Mat.*, 4:9–11, 1981.

[60] H. Babovsky and R. Illner. A convergence proof for Nanbu's simulation method for the full Boltzmann equation. *SIAM J. Numer. Anal.*, 26:45–65, 1989.

[61] L. Baggett. On functions that are trivial cocycles for a set of irrationals. *Proc. Am. Math. Soc.*, 104:1212–1215, 1988.

[62] A. Baker. *Transcendental number theory*. Cambridge Univ. Press, Cambridge, 1975.

[63] A. Baker and J. Coates. Fractional parts of powers of rationals. *Math. Proc. Cambridge Phil. Soc*, 77:269–279, 1975.

[64] R.C. Baker. A diophantine problem on groups. I. *Trans. Am. Math. Soc.*, 150:499–506, 1970.

[65] R.C. Baker. A diophantine problem on groups. II. *Proc. London Math. Soc., Ser. 3*, 21:757–768, 1970.

[66] R.C. Baker. A diophantine problem on groups. III. *Math. Proc. Cambridge Phil. Soc.*, 70:31–47, 1971.

[67] R.C. Baker. On a theorem of Erdős and Taylor. *Bull. London Math. Soc.*, 4:373–374, 1972.

[68] R.C. Baker. A diophantine problem on groups.IV. *Illinois J. Math.*, 18:552–564, 1974.

[69] R.C. Baker. Khinchin's conjecture and Marstrand's theorem. *Mathematika*, 21:248–260, 1975.

[70] R.C. Baker. On a metrical theorem of Weyl. *Mathematika*, 22:29–33, 1975.

[71] R.C. Baker. Some metrical theorems in strong uniform distribution. *J. London Math. Soc.*, 9:467–477, 1975.

[72] R.C. Baker. Dyadic methods in the measure theory of numbers. *Trans. Am. Math. Soc.*, 221:419–432, 1976.

[73] R.C. Baker. On the fractional parts of certain additive forms. *Math. Proc. Cambridge Phil. Soc.*, 79:463–467, 1976.

[74] R.C. Baker. Riemann sums and Lebesgue integrals. *Q. J. Math., Oxford II. Ser.*, 27:191–198, 1976.

[75] R.C. Baker. Fractional parts of several polynomials. *Q. J. Math., Oxford II. Ser.*, 28:453–471, 1977.

[76] R.C. Baker. Singular n-tuples and Hausdorff dimension. *Math. Proc. Cambridge Phil. Soc.*, 81:377–385, 1977.

[77] R.C. Baker. Fractional parts of several polynomials. II. *Mathematika*, 25:76–93, 1978.

[78] R.C. Baker. Exceptional sets in uniforms distribution. *Proc. Edinburgh Math. Soc.*, 22:145–160, 1979.

[79] R.C. Baker. Metric number theory and the large sieve. *J. London Math. Soc.*, 24:34–40, 1981.

[80] R.C. Baker. On the distribution modulo 1 of the sequence $\alpha n^3 + \beta n^2 + \gamma n$. *Acta Arith.*, 39:399–405, 1981.

[81] R.C. Baker. On the fractional parts of $\alpha n^2 + \beta n$. *Glasgow Math. J.*, 22:181–183, 1981.

[82] R.C Baker. On the fractional parts of $\alpha n^3 + \beta n^2 + \gamma n$. *Journées arithmétiques, Exeter 1980, London Math. Soc. Lect. Note Ser.*, 56:226–231, 1982.

[83] R.C. Baker. Small fractional parts of quadratic forms. *Proc. Edinburgh Math. Soc.*, 25:269–277, 1982.

[84] R.C. Baker. Small fractional parts of the sequence. *Mich. Math. J.*, 28:223–228, 1982.

[85] R.C. Baker. Weyl sums and diophantine approximation. *J. London Math. Soc.*, 25:25–34, 1982.
[86] R.C. Baker. Entire functions and uniform distribution modulo one. *Proc. London Math. Soc.*, 49:87–110, 1984.
[87] R.C. Baker. *Diophantine inequalities.* Clarendon Press, Oxford, 1986.
[88] R.C. Baker. Entire functions and discrepancy. *Monatsh. Math.*, 102:179–182, 1986.
[89] R.C. Baker. On the values of entire functions at the positive integers. *Publ. Math. Orsac*, 1–5, 1986.
[90] R.C. Baker. Correction to: Weyl sums and diophantine approximation. *J. London Math. Soc.*, 46:202–204, 1992.
[91] R.C. Baker and J. Brüdern. Pairs of quadratic forms modulo one. *Glasgow Math. J.*, 35:51–61, 1993.
[92] R.C. Baker, J. Brüdern, and G. Harman. The fractional part of αn^k for square-free n. *Q. J. Math., Oxford II. Ser.*, 42:421–431, 1991.
[93] R.C. Baker and G. Harman. Small fractional parts of quadratic and additive forms. *Math. Proc. Cambridge Phil. Soc.*, 90:5–12, 1981.
[94] R.C. Baker and G. Harman. Sequences with bounded logarithmic discrepancy. *Math. Proc. Cambridge Phil. Soc.*, 107:213–225, 1990.
[95] R.C. Baker and G. Harman. On the distribution of αp^k modulo one. *Mathematika*, 38:170–184, 1991.
[96] R.C. Baker and G. Kolesnik. On the distribution of p^α modulo one. *J. Reine Angew. Math.*, 356:174–193, 1985.
[97] R.C. Baker and S. Schäffer. Pairs of additive quadratic forms modulo one. *Acta Arith.*, 62:45–59, 1992.
[98] A. Balog. On the fractional part of p^α. *Arch. Math.*, 40:434–440, 1983.
[99] A. Balog. On the distribution of p^θ mod. 1. *Acta Math. Hung.*, 45:179–199, 1985.
[100] A. Balog. A remark on the distribution of αp modulo one. *Publ. Math. Orsay*, 6–24, 1986.
[101] A. Balog and J. Friedlander. Simultaneous diophantine approximation using primes. *Bull. London Math. Soc.*, 20:289–292, 1988.
[102] A. Balog and A. Perelli. Diophantine approximation by square-free numbers. *Ann. Sc. Norm. Super. Pisa Cl. Sci.*, 11:353–359, 1984.
[103] D. Banks and L. Somer. Period patterns of certain second-order linear recurrences modulo a prime. *Applications of Fibonacci Numbers*, 4:37–40, 1991.
[104] G. Barat. Sur un procédé universel d'extraction. *J. Theor. Nombres Bordx.*, 7:435–445, 1995.
[105] G. Barat and P.J. Grabner. Distribution properties of g-additive functions. *J. Number Theory*, 60:103–123, 1996.
[106] J. Bass. Fonctions pseudo-aléatoires et fonctions de Wiener. *C.R. Acad. Sci., Paris*, 247:1163–1165, 1958.
[107] J. Bass. Suites uniformément denses, moyennes trigonométriques, fonctions pseudo-aléatoires. *Bull. Soc. Math. Fr.*, 87:1–64, 1959.
[108] J. Bass. Construction d'une suite complètement èquirèpartie modulo 1. *Semin. Delange-Pisot-Poitou, 14e Annee 1972/73, Thèorie des Nombres*, Exp. 2:6 p., 1973.
[109] J. Bass. Stationary functions and their applications to the theory of turbulence. *J.Math. Anal. Appl.*, 47:354–399, 1974.
[110] N.L. Bassily. Distribution of q-ary digits in some sequences of integers. *Ann. Univ. Sci. Budap. Rolando Eötvös, Sect. Comput.*, 14:13–22, 1994.
[111] W. Bauer. Diskrepanz in separablen metrischen Räumen. *Monatsh. Math.*, 78:289–296, 1974.
[112] C. Baxa. On the discrepancy of the sequences $(n\alpha)$. *J. Number Theory*, 55:94–107, 1995.
[113] C. Baxa and J. Schoißengeier. Minimum and maximum order of magnitude of the discrepancy of $(n\alpha)$. *Acta Arith.*, 68:281–290, 1994.
[114] C. Baxa and J. Schoißengeier. On the discrepancy of the sequence $\alpha\sqrt{n}$. *J. London Math. Soc.*, to appear.
[115] J. Beck. Balanced two-colorings of finite sets in the square. *Combinatorica*, 1:327–335, 1981.
[116] J. Beck. Balancing families of integer sequences. *Combinatorica*, 1:209–216, 1981.
[117] J. Beck. Roth's estimate of the discrepancy of integer sequences is nearly sharp. *Combinatorica*, 1:319–325, 1981.

[118] J. Beck. Irregularities of two-colourings of the $n \times n$ square lattice. *Combinatorica*, 2:111–123, 1982.
[119] J. Beck. On an imbalance problem of G. Wagner concerning polynomials. *Stud. Sci. Math. Hung.*, 17:417–424, 1982.
[120] J. Beck. On a problem of K.F. Roth concerning irregularities of point distribution. *Invent. Math.*, 74:477–487, 1983.
[121] J. Beck. Cube-lattice with good distribution behaviour. *Stud. Sci. Math. Hung.*, 19:21–27, 1984.
[122] J. Beck. New results in the theory of irregularities of point distribution. *Lect. Notes Math.*, 1068:1–16, 1984.
[123] J. Beck. Some results and problems in "combinatorial discrepancy theory". *Topics in classical number theory, Colloq. Budapest 1981, Vol. I, Colloq. Math. Soc. Janos Bolyai*, 34:203–218, 1984.
[124] J. Beck. Some upper bounds in the theory of irregularities of distribution. *Acta Arith.*, 43:115–130, 1984.
[125] J. Beck. Sums of distances between points on a sphere - an application of the theory of irregularities of distribution to discrete geometry. *Mathematika*, 31:33–41, 1984.
[126] J. Beck. Irregularities of distribution and combinatorics. *London Math. Soc. Lect. Note Ser.*, 103:25–46, 1985.
[127] J. Beck. On irregularities of $+/-1$ sequences. *Österr. Akad. Wiss. SB II*, 195:13–23, 1986.
[128] J. Beck. Irregularities of distribution. *Acta Math.*, 159:1–49, 1987.
[129] J. Beck. On a problem of Erdős in the theory of irregularities of distribution. *Math. Ann.*, 277:233–247, 1987.
[130] J. Beck. Irregularities of distribution. *Proc. London Math. Soc.*, 56:1–50, 1988.
[131] J. Beck. On a lattice point problem of L. Moser. I. *Combinatorica*, 8:21–47, 1988.
[132] J. Beck. On a lattice point problem of L. Moser. II. *Combinatorica*, 8:159–176, 1988.
[133] J. Beck. On irregularities of point sets in the unit square. *Combinatorics*, 52:63–74, 1988.
[134] J. Beck. On the discrepancy of convex plane sets. *Monatsh. Math.*, 105:91–106, 1988.
[135] J. Beck. Balanced two-colorings of finite sets in the cube. *Discrete Math.*, 73:13–25, 1989.
[136] J. Beck. On a lattice-point problem of H. Steinhaus. *Stud. Sci. Math. Hung.*, 24:263–268, 1989.
[137] J. Beck. A two-dimensional van Aardenne-Ehrenfest theorem in irregularities of distribution. *Compos. Math.*, 72:269–339, 1989.
[138] J. Beck. Irregularities of distribution and category theorem. *Stud. Sci. Math. Hung.*, 26:81–86, 1991.
[139] J. Beck. The modulus of polynomials with zeros on the unit circle: A problem of Erdős. *Ann. of Math.*, 134:609–651, 1991.
[140] J. Beck. Quasi-random 2-colorings of point sets. *Random Struct. Algorithms*, 2:289–302, 1991.
[141] J. Beck. Randomness of $n\sqrt{2}$ mod 1 and a Ramsey property of the hyperbola. *Sets, graphs and numbers*, 60:23–66, 1992.
[142] J. Beck. Probabilistic diophantine approximation, I. Kronecker sequences. *Ann. Math.*, 140:451–502, 1994.
[143] J. Beck and W. W.L. Chen. *Irregularities of distribution.* Cambridge University Press, Cambridge, 1987.
[144] J. Beck and W.W.L. Chen. Irregularities of point distribution relative to convex polygons. III. *preprint.*
[145] J. Beck and W.W.L. Chen. Note on irregularities of distribution. *Mathematika*, 33:148–163, 1986.
[146] J. Beck and W.W.L. Chen. Irregularities of point distribution relative to convex polygons. *(Irregularities of partitions). Algorithms Comb.*, 8:1–22, 1989.
[147] J. Beck and W.W.L. Chen. Note on irregularities of distribution. II. *Proc. London Math. Soc.*, 61:251–272, 1990.
[148] J. Beck and W.W.L. Chen. Irregularities of point distribution relative to convex polygons. II. *Mathematika*, 40:127–136, 1993.
[149] J. Beck and W.W.L. Chen. Irregularities of point distribution relative to half planes. I. *Mathematika*, 40:102–126, 1993.

[150] J. Beck and V.T. Sós. Discrepancy theory. *Handbook of Combinatorics, Elsevier*, 1405–1446, 1995.
[151] J. Beck and J. Spencer. Well-distributed 2-colorings of integers relative to long arithmetic progressions. *Acta Arith.*, 43:287–294, 1984.
[152] Jòzsef Beck and T. Fiala. "Integer-making" theorems. *Discrete Appl. Math.*, 3:1–8, 1981.
[153] J.-F. Bedin and C. Deutsch. Distribution uniforme modulo 1 d'une fonction arithmétique rèele additive relativement à un support fondamental. *C.R. Acad. Sci., Paris*, 277:149–151, 1973.
[154] S. Beer. Die Diskrepanz von Differenzenfolgen im p-adischen. *Monatsh. Math.*, 76:289–294, 1972.
[155] H. Behnke. Über die Verteilung von Irrationalitäten mod 1. *Abh. Math. Semin. Univ. Hamburg*, 1:252–267, 1922.
[156] H. Behnke. Zur Theorie der diophantischen Approximationen I. *Abh. Math. Semin. Univ. Hamburg*, 3:261–318, 1924.
[157] R. Béjian. Sur certaines suites présentant une faible discrépance à l'origine. *C.R. Acad. Sci., Paris*, 286:135–138, 1978.
[158] R. Béjian and H. Faure. Discrèpance de la suite de van der Corput. *C.R. Acad. Sci., Paris*, 285:313–316, 1977.
[159] R. Béjian and H. Faure. Discrépance de la suite de van der Corput. *Semin. Delange-Pisot-Poitou, 19e Annee 1977/78, Theorie des Nombres*, Exp.13:14 p., 1978.
[160] A. Bellow. On "bad universal" sequences in ergodic theory II. *Lect. Notes Math.*, 1033, 1983.
[161] A. Bellow and V. Losert. On sequences of density zero in ergodic theory. *Contemp. Math.*, 26:49–60, 1984.
[162] A. Bellow and V. Losert. The weighted ergodic theorem and the individual ergodic theorem along subsequences. *Trans. Am. Math. Soc.*, 288:307–345, 1985.
[163] V.K. Bel'nov. Some theorems on the distribution of fractional parts. *Sib. Mat. Zh.*, 18:512–521, 1977.
[164] F. Benford. The law of anomalous numbers. *Proc. Am. Phi. Soc.*, 78:551–572, 1938.
[165] M. Bennett. Fractional parts of powers of rational numbers. *Proc. Chambridge Phil. Soc.*, 114:191–201, 1993.
[166] M.A. Bennett. Effective lower bounds for the fractional parts of powers of a dense set of rationals. *C. R. Math. Acad. Sci., Soc. R. Can.*, 15:201–206, 1993.
[167] L. Benzinger. Uniformly distributed sequences in locally compact groups. I. *Trans. Am. Math. Soc.*, 188:149–165, 1974.
[168] L. Benzinger. Uniformly distributed sequences in locally compact groups. II. *Trans. Am. Math. Soc.*, 188:167–178, 1974.
[169] P.H. Berard. *Spectral Geometry: Direct and invers problems.* Lect. Notes Math. 1202, Springer, Berlin, 1986.
[170] D. Berend. Multi-invariant sets on tori. *Trans. Am. Math. Soc.*, 280:509–532, 1983.
[171] D. Berend. Multi-invariant sets on compact abelian groups. *Trans. Am. Math. Soc.*, 286:505–535, 1984.
[172] D. Berend. Actions of sets of integers on irrationals. *Acta Arith.*, 48:275–306, 1987.
[173] D. Berend. Dense (mod 1) dilated semigroups of algebraic numbers. *J. Number Theory*, 26:246–256, 1987.
[174] D. Berend. A recurrence property of smooth functions. *Isr. J. Math.*, 62:32–36, 1988.
[175] D. Berend. Density modulo 1 in local fields. *Acta Arith.*, 52:267–282, 1989.
[176] D. Berend. Irrational dilations of approximate multiplicate semigroups of integers. *Theorie des nombres, C. R. Conf. Int., Quebec/Can. 1987*, pages 29–40, 1989.
[177] D. Berend. IP-sets on the circle. *Can. J. Math.*, 42:575–589, 1990.
[178] D. Berend and M. D. Boshernitzan. On a result of Mahler on the decimal expansions of $(n\alpha)$. *Acta Arith.*, 66:315–322, 1994.
[179] D. Berend and M.D. Boshernitzan. Numbers with complicated decimal expansions. *Acta Math. Hung.*, 66:113–126, 1995.
[180] D. Berend and C. Frougny. Computability by finite automata and Pisot bases. *Math. Syst. Theory*, 27:275–282, 1994.
[181] D. Berend and G. Kolesnik. Distribution modulo 1 of some oscillating sequences. *Isr. J. Math.*, 71:161–179, 1990.

[182] D. Berend and Y. Peres. Asymptotically dense dilations of sets on the circle. *J. London Math. Soc.*, 47:1–17, 1993.

[183] D. Berg, M. Rajagopalan, and L.A. Rubel. Uniform distribution in locally compact groups. *Trans. Am. Math. Soc.*, 133:435–446, 1968.

[184] V. Bergelson. Weakly mixing PET. *Ergodic Theory Dyn. Syst.*, 7:337–349, 1987.

[185] M. Berger, P. Gauduchon, and E. Mazet. *Le spectre d'une Variété Riemannienne*. Lect. Notes Math. 194, Springer, Berlin, 1971.

[186] I. Berkes. Critical LIL behavior of the trigonometric system. *Trans. Am. Math. Soc.*, 347:515–530, 1995.

[187] I. Berkes and W. Philipp. The size of trigonometric and Walsh series and uniform distribution mod.1. *J. London Math. Soc*, 50:454–464, 1994.

[188] V. I. Bernik. Ergodische Eigenschaft linear unabhängiger Polynome. *Ser. fiz.-mat. Nauk*, 6:47–53, 1972.

[189] V. I. Bernik. Erzeugte Extremalflächen. *Mat. Sbornik*, Ser. 103:480–489, 1977.

[190] V. I. Bernik. An application of the Hausdorff dimension to the theory of diophantine approximations. *Acta Arith.*, 42:219–253, 1983.

[191] V.I. Bernik and E.I. Kovalevskaya. Distribution of the points of Schmidt's curve for almost all values of the argument. *Dokl. Akad. Nauk Belarusi*, 36:777–781, 1992.

[192] V.I. Bernik and N.A. Pereverseva. The method of trigonometric sums and lower estimates of Hausdorff dimension. *New Trends Probab. Stat.*, 2:75–81, 1991.

[193] S. N. Bernstein. On quadrature formulas with positive coefficients. *Izv. Akad. Nauk SSSR, Ser. Mat.*, 4:479–503, 1937. In Russian.

[194] S. N. Bernstein. Sur les formules de quadrature de Cotes et Tchebycheff. *C. R. Acad. Sci. URSS*, 14:323–326, 1937.

[195] J. Berstel and P. Séébold. A characterization of Sturmian morphisms. *Lect. Notes Comput. Sci.*, 711:281–290, 1993.

[196] A. Bertand-Mathis. Nombres normaux dans diverses bases. *Ann. Inst. Fourier*, 45:1205–1222, 1995.

[197] M.-J. Bertin and D. W. Boyd. A characterization of two related classes of Salem numbers. *J. Number Theory*, 50:309–317, 1995.

[198] M.J. Bertin. Generalisation des suites de Pisot et de Boyd. *Acta Arith.*, 57:211–223, 1991.

[199] M.J. Bertin. Generalisation des suites de Pisot et de Boyd II. *Acta Arith.*, 59:215–219, 1991.

[200] M.J. Bertin. k-nombres de Pisot et de Salem. *Advances in number theory (Proceedings of the third conference of the Canadian Number Theory Association 1991)*, 391–397, 1993.

[201] M.J. Bertin. k-nombres de Pisot et de Salem. *Acta Arith.*, 68:113–131, 1994.

[202] M.J. Bertin, A. Decomps-Guilloux, M. Grandet-Hugot, M. Pathiaux-Delefosse, and J.P. Schreiber. *Pisot and Salem numbers*. Birkhaeuser Verlag, Basel, 1992.

[203] M.J. Bertin and M. Pathiaux-Delefosse. Conjecture de Lehmer et petits nombres de Salem. *Queen's Papers in Pure and Appl. Math.*, 81:144, 1989.

[204] A. Bertrand. Développements en base de Pisot et répartition modulo 1. *C.R. Acad. Sci., Paris*, 285:419–421, 1977.

[205] A. Bertrand. Répartition modulo 1 et développement en base 0. *C.R. Acad. Sci., Paris*, 289:1–4, 1979.

[206] A. Bertrand-Mathis. Développement en base θ, répartition modulo un de la suite (θ^n). *Bull. Soc. Math. Fr.*, 114:271–323, 1986.

[207] A. Bertrand-Mathis. Ensembles intersectifs et récurrence de Poincaré. *Isr. J. Math.*, 55:184–198, 1986.

[208] A. Bertrand-Mathis. Ensembles intersectifs et recurrence de Poincaré. *Publ. Math. Orsay*, 55–72, 1988.

[209] A. Bertrand-Mathis and B. Volkmann. On (ε, k)-normal words in connecting dynamical systems. *Monatsh. Math.*, 107:267–279, 1989.

[210] F. Bertrandias. Colloque sur la répartition asymptotique modulo 1. *Breukelen Lectures*, 1962.

[211] F. Bertrandias. Théorème de Koksma en p-adique. *Sém. Delange-Pisot*, 6:16, 1967.

[212] J. Bésineau. Indépendance statistique d'ensembles liés à la fonction "somme des chiffres". *Sémin. Théor. Nombres Bordeaux*, Exp. 21:20 p., 1971.

[213] J. Bésineau. Indépendence statistique d'ensembles liés à la fonction 'somme des chiffres'. *Semin. Delange-Pisot-Poitou, 13e Annee 1971/72, Theorie des Nombres*, Exp. 23:8 p., 1973.

[214] J. Bésineau. Presque périodicité et sous-suites. *C.R. Acad. Sci., Paris*, 278:203–206, 1974.
[215] J. Bésineau. Ensembles d'entiers à caractères presque périodique et équirépartition. *Lect. Notes Math.*, 475:1–12, 1975.
[216] F. Beukers. Fractional parts of powers of rationals. *Math. Proc. Comb. Philos Soc.*, 90:13–20, 1981.
[217] F. Beukers. Fractional parts of powers of 3/2. *Prog. Math.*, 22:13–18, 1982.
[218] V.C. Bhavsar, U.G. Gujar, J.D. Horton, and L.A. Lambrou. Evaluation of the discrepancy of the linear congruential pseudo-random number sequences. *BIT*, 30:258–267, 1990.
[219] Ch. Biester, P.J. Grabner, G. Larcher, and R.F. Tichy. Adaptive search in Quasi-Monte-Carlo optimization. *Math. Comput.*, 64:807–818, 1995.
[220] C. Binder. Über einen Satz von de Bruijn und Post. *Österr. Akad. Wiss. SB II*, 179:233–251, 1971.
[221] C. Binder. Une remarque sur la charactérisation des fonction (r, pi)-integrables. *Lect. Notes Math.*, 475:13–17, 1975.
[222] C. Binder, E. Hlawka, and J. Schoißengeier. Über einige Beispiele für Anwendungen der Theorie der Gleichverteilung. *Math. Slovaca*, 43:427–446, 1993.
[223] A. Bisbas. A note on the distribution of digits in dyadic expansions. *C.R. Acad. Sci., Paris*, 318:105–109, 1994.
[224] R.L. Bishop and R.J. Crittenden. *Geometry of Manifolds*. Academic Press, New York, 1964.
[225] P. Blaga. Monte Carlo integration on a simplex. *Stud. Univ. Babes-Bolyai*, 33:19–26, 1988.
[226] J. M. Blanchard. Non literal transducers and some problems of normality. *J. Theor. Nombres Bordx.*, 5:303–321, 1993.
[227] J. M. Blanchard, J. M. Dumont, and A. Thomas. Generic sequences, transducers and multiplication of normal numbers. *Isr. J. Math.*, 80:257–287, 1992.
[228] H.-P. Blatt and H. N. Mhaskar. A discrepancy theorem concerning polynomials of best approximation in $L^p_w[-1,1]$. *Monatsh. Math.*, 120:91–103, 1995.
[229] R. Blecksmith, M. Filaseta, and C. Nicol. A result on the digits of a^n. *Acta Arith.*, 64:331–339, 1993.
[230] L. Blum, M. Blum, and M. Shub. A simple unpredictable pseudo-random number generator. *SIAM J. Comput.*, 15:364–383, 1986.
[231] M. Blümlinger. Rajchman measures on compact groups. *Math. Ann.*, 284:55–62, 1989.
[232] M. Blümlinger. Topological algebras of functions of bounded variation. II. *Manuscr. Math.*, 65:377–384, 1989.
[233] M. Blümlinger. Asymptotic distribution and weak convergence on compact Riemannian manifolds. *Monatsh. Math.*, 110:277–188, 1990.
[234] M. Blümlinger. Sample path properties of diffusion processes on compact manifolds. *Lect. Notes Math.*, 1452:6–19, 1990.
[235] M. Blümlinger. Characterization of measures in the group C^*-algebra of a locally compact group. *Math. Ann.*, 289:393–402, 1991.
[236] M. Blümlinger. Slice discrepancy and irregularities of distribution on the sphere. *Mathematika*, 38:105–116, 1991.
[237] M. Blümlinger, M. Drmota, and R.F. Tichy. Metrische Sätze der C-Gleichverteilung auf der Sphäre. *Lect. Notes Math.*, 1262:14–21, 1987.
[238] M. Blümlinger, M. Drmota, and R.F. Tichy. Asymptotic distribution of functions on compact homogeneous spaces. *Ann. Mat. Pura Appl.*, 152:79–93, 1988.
[239] M. Blümlinger, M. Drmota, and R.F. Tichy. A uniform law of interated logarithm for Brownian motion on compact Riemannian manifolds. *Math. Z.*, 201:495–507, 1989.
[240] M. Blümlinger and N. Obata. Permutation preserving Cesàro mean, densities of natural numbers and uniform distibution of sequences. *Ann. Inst. Fourier*, 41:665–678, 1991.
[241] M. Blümlinger and R.F. Tichy. Bemerkungen zu einigen Anwendungen gleichverteilter Folgen. *Österr. Akad. Wiss. SB II*, 195:253–265, 1986.
[242] M. Blümlinger and R.F. Tichy. Topological algebras of functions of bounded variation. *Manuscr. Math.*, 65:245–255, 1989.
[243] R.P. Boas. *Entire Functions*. Academic Press, New York, 1954.
[244] V.P. Bobrovskij. w-uniform distributions. *Collect. Sci. Works*, 18–24, 1985.
[245] J.-P. Borel. Equirépartition modulo 1 et semi-groupes additifs. *C.R. Acad. Sci., Paris*, 287:743–745, 1978.

[246] J.-P. Borel. Nombres premiers géneralisés et équirépartition modulo 1. *Semin. Delange-Pisot-Poitou, 19e Annee 1977/78, Theor. des Nombres*, Exp. 7:7 p., 1978.
[247] J.-P. Borel. Equirépartition modulo 1 de la suite $\{\alpha x_n\}$ où x_n décrit un semi-groups additif de nombres réels. *Sémin. Théor. Nombres 1978-1979*, Exp. 7:13 p., 1979.
[248] J.-P. Borel. Equirépartition modulo 1 de semi-groups additifs de nombres réels. *Semin. Delange-Pisot-Poitou, 20e Annee 1978/79, Theories des nombres*, Exp. 31:9 p., 1980.
[249] J.-P. Borel. Quelques résultats d'équirépartition liés aux nombres premiers généralisés de Beurling. *Acta Arith.*, 38:255–272, 1980.
[250] J.-P. Borel. Un problème d'equirépartition modulo 1 lié aux partitions. *Bull. Soc. Math. Fr.*, 108:229–250, 1980.
[251] J.-P. Borel. Repartition modulo 1 suivant la measure de Dirac à l'origine. *C.R. Acad. Sci., Paris*, 294:5–8, 1982.
[252] J.-P. Borel. Sous-groupes de R liés à la répartition modulo 1 des suites. *Ann. Fac. Sci. Toulouse, Math.*, 5:217–235, 1983.
[253] J.-P. Borel. Discrépances de suites homothétiques. *Monatsh. Math.*, 104:175–189, 1987.
[254] J.-P. Borel. Parties d'ensembles b-normaux. *Manuscr. Math.*, 62:317–335, 1988.
[255] J.-P. Borel. Suites es mesures auto-similaires. *Sémin. Théor. Nombres Bordeaux*, Exp. 2:16 p., 1988.
[256] J.-P. Borel. Suites et mesures ayant des propriétés d'auto-similarité. *Publ. Math. Orsay*, 15–38, 1988.
[257] J.-P. Borel. Ensembles normaux et suites bornees. *Publ. Math. Orsay*, 13–29, 1989.
[258] J.-P. Borel. Self-similar measures and sequences. *J. Number Theory*, 31:208–241, 1989.
[259] J.-P. Borel. Suites de longueur minimale associées à un ensemble normal donné. *Isr. J. Math.*, 64:229–250, 1989.
[260] J.-P. Borel. Sur certains ensembles normaux. *J. Theor. Nombres Bordx.*, 1:67–79, 1989.
[261] J.-P. Borel. Polynomes a coefficients positifs multiples d'un polynome donne. *Lect. Notes Math.*, 1415:97–115, 1990.
[262] J.-P. Borel. Suites dont la discrepance est comparable a un logarithme. *Monatsh. Math.*, 110:207–216, 1990.
[263] J.-P. Borel. Sur certains sous-groupes de **R** lies a la suite des factorielles. *Colloq. Math.*, 62:21–30, 1991.
[264] J.-P. Borel. Sur une suite de fonctions liées à un procédé itératif-II. *J. Theor. Nombres Bordx.*, 5:235–261, 1993.
[265] J.-P. Borel. Symbolic representation of piecewise linear functions on the unit interval and application to discrepancy. *Theor. Comput. Sci.*, 123:61–87, 1994.
[266] J.-P. Borel and F. Laubie. Construction de mots de Christoffel. *C.R. Acad. Sci., Paris*, 313:483–485, 1991.
[267] I. Borosh and H. Niederreiter. Optimal multipliers for pseudo-random number generation by the linear congruential method. *BIT*, 23:65–74, 1983.
[268] D. Borwein and W. Gawronski. On certain sequences of plus and minus ones. *Canadian J. Math.*, 30:170–179, 1978.
[269] W. Bosch. Functions that preserve the uniform distribution of sequences. *Trans. Am. Math. Soc.*, 307:143–152, 1988.
[270] M.D. Boshernitzan. Homogeneously distributed sequences and Poincaré sequences of integers of sublacunary growth. *Monatsh. Math.*, 96:173–181, 1983.
[271] M.D. Boshernitzan. Density modulo 1 of dilations of sublacunary sequences. *Adv. Math.*, 108:104–117, 1994.
[272] M.D. Boshernitzan. Elementary proof of Furstenberg's diophantine result. *Proc. Am. Math. Soc.*, 122:67–70, 1994.
[273] M.D. Boshernitzan. Uniform distribution and Hardy fields. *J. Anal. Math.*, 62:225–240, 1994.
[274] M.T. Boswell, S.D. Gore, G.P. Patil, and C. Taillie. The art of computer generation of random variables. *Rao, C.R. (ed.) Computational statistics. Amsterdam*, 9:661–721, 1993.
[275] N. Bouleau. A remark on random and equidistributed sequences. *Potential Anal.*, 1:379–384, 1992.
[276] J. Bourgain. Théorèmes ergodiques ponctuels pour certains ensembles arithmétiques. *C.R. Acad. Sci., Paris*, 305:397–402, 1987.
[277] J. Bourgain. Almost sure convergende and boundes entropy. *Isr. J. Math.*, 63:79–97, 1988.

[278] J. Bourgain. On the maximal ergodic theorem for certain subsets of the integers. *Isr. J. Math.*, 61:39–72, 1988.

[279] J. Bourgain. On the pointwise ergodic theorem on L^p for arithmetic sets. *Isr. J. Math.*, 61:73–84, 1988.

[280] J. Bourgain. On $a(p)$-subsets of squares. *Isr. J. Math.*, 67:291–311, 1989.

[281] J. Bourgain. Pointwise ergodic theorems for arithmetic sets. *Publ. Math. IHES*, 69:5–45, 1989.

[282] J. Bourgain. Double recurrence and almost sure convergence. *J. Reine Angew. Math.*, 404:140–161, 1990.

[283] J. Bourgain. Problems of almost everywhere convergence related to harmonic analysis and number theory. *Isr. J. Math.*, 71:97–127, 1990.

[284] D. Bowman. Approximation of $\lfloor n\alpha + s \rfloor$ and the zero of $\{n\alpha + s\}$. *J. Number Theory*, 50:128–144, 1995.

[285] D. W. Boyd. Pisot sequences which satisfy no linear recurrence. *Acta Arith.*, 48:191–195, 1987.

[286] D. W. Boyd. Best approximation of real numbers by Pisot numbers. *Lect. Notes Pure Appl. Math.*, 147:9–16, 1993.

[287] D. W. Boyd. Linear recurrence relations for some generalized Pisot sequences. *Advances in number theory (Proceedings of the third conference of the Canadian Number Theory Association 1991)*, 333–340, 1993.

[288] D. W. Boyd and W. Parry. Limit points of the Salem numbers. *Number theory, Proc. 1st Conf. Can. Number Theory Assoc.*, pages 27–35, 1990.

[289] D. W. Boyd and J. M. Steele. Monotone subsequences in the sequence of fractional parts of multiples of an irrational. *J. Reine Angew. Math.*, 306:49–59, 1979.

[290] D.W. Boyd. Some integer sequences related to Pisot sequences. *Acta Arith.*, 34:295–305, 1992.

[291] D.W. Boyd. On beta expansions for Pisot numbers. *Math. Comput.*, 65:841–860, 1996.

[292] D.W. Boyd. On the beta expansion for Salem numbers of degree 6. *Math. Comput.*, 65:861–875, 1996.

[293] Jeff Boyle. An application of Fourier series to the most significant digit problem. *Am. Math. Monthly*, 101:879–886, 1994.

[294] P. Bratley, B.L. Fox, and H. Niederreiter. Implementation and tests of low-discrepancy sequences. *ACM Trans. Model. Comput. Simulation*, 2:195–213, 1992.

[295] R.P. Brent. On the periods of generalized Fibonacci recurrences. *Math. Comput.*, 63:389–401, 1994.

[296] A. Broglio and P. Liardet. Predictions with automata. *Symbolic dynamics and its appl., Proc. Am. Math. Soc. Conf. in honor of R.L. Adler,/New Haven/CT (USA) 1991, Contemp. Math.*, 135:111–124, 1992.

[297] G. Brown. Some new applications of Riesz products. *Proc. Cent. Math. Appl. Aust. Natl. Univ.*, 29:1–13, 1991.

[298] G. Brown and W. Moran. Schmidt's conjecture on normality for commuting matrices. *Invent. Math.*, 111:449–463, 1993.

[299] G. Brown, W. Moran, and Ch. Pearce. Riesz products and normal numbers. *J. London Math. Soc.*, 32:12–18, 1985.

[300] G. Brown, W. Moran, and D. Pollington. Normality to non-integer bases. *C.R. Acad. Sci., Paris*, 316:1241–1244, 1993.

[301] J. L. jun. Brown and R. L. Duncan. Weak convergence of monotone functions and Weyl's criterion. *Portugaliae Math.*, 33:69–75, 1974.

[302] T.C. Brown. A characterization of the quadratic irrationals. *Canad. Math. Bull.*, 34:36–41, 1991.

[303] T.C. Brown and P.J.-S. Shiue. Sums of fractional parts of integer multiples of an irrational. *J. Number Theory*, 50:181–192, 1995.

[304] M. Bruneau. Comportement local des fonctions et approximation sur le tore. *Semin. Delange-Pisot-Poitou, 16e annee 1974/75, Theorie des Nombres*, Exp. 5:10 p., 1975.

[305] M. Bruneau. Comportement local des fonctions à série de Fourier lacunaire. *Acta Arith.*, 30:297–305, 1976.

[306] V. Bruyère, G. Hansel, C. Michaux, and R. Villemaire. Correction to: "Logic and p-recognizable sets of integers". *Bull. Belg. Math. Soc. Simon Stevin*, 1:577, 1994.
[307] V. Bruyère, G. Hansel, C. Michaux, and R. Villemaire. Logic and p-recognizable sets of integers. *Bull. Belg. Math. Soc. Simon Stevin*, 1:191–238, 1994.
[308] Z. Bukovska and T. Salát. Topological results of sequences $\{n_k x\}_{k=1}^{\infty}$ and their applications in the theory of trigonometric series. *Czech. Math. J.*, 43:115–123, 1993.
[309] R.T. Bumby. A distribution property for linear recurrences of the second order. *Proc. Am. Math. Soc.*, 50:101–106, 1975.
[310] P. Bundschuh. On the distribution of Fibonacci numbers. *Tamkang J. Math.*, 5:75–79, 1974.
[311] P. Bundschuh. On a problem on uniform distribution. *Top. Number Theory*, 13:15–17, 1976.
[312] P. Bundschuh. Konvergenz unendlicher Reihen und Gleichverteilung mod 1. *Arch. Math.*, 29:518–523, 1977.
[313] P. Bundschuh and J.-S. Shiue. Solution of a problem on the uniform distribution of integers. *Atti Accad. Naz. Lincei*, 55:172–177, 1974.
[314] P. Bundschuh and Y. Zhu. The formulas of exact calculation of the discrepancy of low-dimensional finite point sets. I. *Chin. Sci. Bull.*, 38:1318–1319, 1993.
[315] P. Bundschuh and Y. Zhu. A method for exact calculation of the discrepancy of low-dimensional finite point sets. I. *Abh. Math. Semin. Univ. Hamburg*, 63:115–133, 1993.
[316] T. Burg, M. Drmota, and R.F. Tichy. Some new results in summability theory. *Lect. Notes Math.*, 1452:20–30, 1990.
[317] J. R. Burke. A general notation of independence of sequences of integers. *Internat. J. Math. Math. Sci.*, 16:515–518, 1993.
[318] J. R. Burke and L. Kuipers. Asymptotic distribution and independence of sequences of Gaussian integers. *Simon Stevin*, 50:3–21, 1976.
[319] F.W. Burton and R.L. Page. Distributed random number generation. *J. Funct. Program.*, 2:203–212, 1992.
[320] V.A. Bykovskii. *On the correct order of the error of optimal cubature formulas in spaces with dominant derivative, and on quadratic deviations of grids.* Computing Center Far-Eastern Scientific Center, Acad. Sci USSR, Vladivostok, 1985. (Russian) R. Zh. Mat. 1986, 7B 1663.
[321] R. E. Caflisch and B. Moskowitz. Modified Monte Carlo Methods Using Quasi-Random Sequences. *Lect. Notes Stat.*, 106:1–15, 1995.
[322] C. Calude and H. Jürgensen. Randomness as an invariant for number representations. *Lect. Notes Comput. Sci.*, 812, 1994.
[323] D. G. Cantor. On families of Pisot E-sequences. *Ann. Sci. École Norm. Supér.*, 9:283–308, 1976.
[324] M. Car. Répartition modulo 1 dans un corps de séries formelles sur un corps fini. *Acta Arith.*, 69:229–242, 1995.
[325] F. Cardin. On the geometrical Cauchy problem for the Hamilton-Jacobi equation. *Buovo Cimento B*, 104:525–544, 1989.
[326] D. Carlson. Good sequences of integers. *J. Number Theory*, 7:91–104, 1975.
[327] D. Carroll, E. Jacobson, and L. Somer. Distribution of two-term recurrence sequences mod p^e. *Fibonacci Q.*, 32:260–265, 1994.
[328] F. S. Cater and R. B. Crittenden. The distribution of sequences modulo one. *Acta Arith.*, 28:429–432, 1976.
[329] St. R. Cavior. Uniform distribution for prescribed moduli. *Fibonacci Q.*, 15:209–210, 1977.
[330] St. R. Cavior. Uniform distribution (mod m) of recurrent sequences. *Fibonacci Q.*, 15:265–267, 1977.
[331] G.J. Chaitin. On the length of programs for computing finite binary sequences. *J. Assoc. Comput. Mach.*, 13:547–569, 1966.
[332] G.J. Chaitin. *Information, Randomness & Incompleteness. Papers on Algorithmic Information Theory.* World Scientific Publishers, Singapore, 1987.
[333] H. Chaix and H. Faure. Discrépance et diaphonie des suites de van der Corput généralisées. *C.R. Acad. Sci., Paris*, 310:315–320, 1990.
[334] H. Chaix and H. Faure. Discrépance et diaphonie des suites de van der Corput généralisées (II). *C.R. Acad. Sci., Paris*, 311:65–68, 1990.
[335] H. Chaix and H. Faure. Discrépance et diaphonie des suites de van der Corput généralisées (III). *C.R. Acad. Sci., Paris*, 312:755–758, 1991.

[336] H. Chaix and H. Faure. Discrépance et diaphonie en dimension un. *Acta Arith.*, 63:103–141, 1993.

[337] D.G. Champernowne. The construction of decimals normal in the scale of ten. *London Math. Soc.*, 8:254–260, 1933.

[338] Ph. Chassaing. An optimal random number generator on Z. *Stat. Probab. Lett.*, 7:307–309, 1989.

[339] I. Chavel. *Eigenvalues in Riemannian Geometry*. Academic Press, Orlando, 1984.

[340] G. Chen. A note on almost uniform distributions. *Bull. Inst. Math., Acad. Sin.*, 1:229–238, 1973.

[341] J. Chen and P. Whitlock. Implementation of a distributed pseudorandom number generator. *Lect. Notes Stat.*, 106:168–185, 1995.

[342] W.W.L. Chen. On irregularities of distribution. *Mathematika*, 27:153–170, 1980.

[343] W.W.L. Chen. Irregularities of point distribution in unit cubes. *Journées arithmétiques, Exeter 1980, London Math. Soc. Lect. Note Ser.*, 56:232–236, 1982.

[344] W.W.L. Chen. On irregularities of distribution. *Q. J. Math., Oxford II. Ser.*, 34:257–279, 1983.

[345] W.W.L. Chen. On irregularities of distribution and approximate evaluation of certain functions. *Q. J. Math., Oxford II. Ser.*, 36:173–182, 1985.

[346] W.W.L. Chen. On irregularities of distribution and approximate evaluation of certain functions. *Proc. Math.*, 70:75–86, 1987.

[347] W.W.L. Chen. On irregularities of distribution iii. *J. Austral. Math. Soc.*, 60:228–244, 1996.

[348] F. Chersi and A. Volcic. λ-equidistributed sequences of partitions and a theorem of the De Bruijn-Post type. *Ann. Mat. Pura Appl.*, 162:23–32, 1992.

[349] G.H. Choe. Ergodicity and irrational rotations. *Proc. R. Ir. Acad. Sect A*, 93:193–202, 1993.

[350] G.H. Choe. Spectral types of uniform distribution. *Proc. Am. Math. Soc.*, 120:715–722, 1994.

[351] G.H. Choe. Weighted normal numbers. *Bull. Austral. Math. Soc.*, 52:177–181, 1995.

[352] K.-K. Chong and M.-C. Liu. The fractional parts of a sum of polynomials. *Monatsh. Math.*, 81:195–202, 1976.

[353] G. Choquet. Algorithmes adaptés aux suites $(k\theta^n)$ et aux chaines associées. *C.R. Acad. Sci., Paris*, 290:719–724, 1980.

[354] G. Choquet. Construction effective de suites $(k(3/2)^n)$. Etude des mesures (3/2)-stables. *C.R. Acad. Sci., Paris*, 291:69–74, 1980.

[355] G. Choquet. Les fermés (3/2)-stables de π and structure des fermés dénombrables. *C.R. Acad. Sci., Paris*, 291:239–244, 1980.

[356] G. Choquet. Répartition des nombres $k(3/2)^n$. *C.R. Acad. Sci., Paris*, 290:575–580, 1980.

[357] G. Choquet. θ-jeux récursifs et application aux suites $(k\theta^n)$. *C.R. Acad. Sci., Paris*, 290:863–868, 1980.

[358] G. Choquet. θ-fermés and θ-chaines et θ-cycles, (pour $\theta = 3/2$). *C.R. Acad. Sci., Paris*, 292:5–10, 1981.

[359] G. Choquet. Nombres 3/2-constructibles. *Publ. Math. Univ. Pierre Marie Curie*, 78:3, 1985.

[360] G. Choquet. Peut-on renouveler ses champs d'intére? *Atti Semin. Mat. Fis. Univ. Modena*, 1:241–255, 1990.

[361] W.-S. Chou. On inversive maximal period polynomials over finite fields. *Appl. Algebra Engrg. Comm. Comput.*, 6:245–250, 1995.

[362] W.-S. Chou. The period lengths of inversive pseudorandom vector generations. *Finite Fields Appl.*, 1:126–132, 1995.

[363] W.-S. Chou. The period lengths of invrsive congruential recursions. *Acta Arith.*, 73:325–341, 1995.

[364] G. Christol, T. Kamae, M. Mendès France, and G. Rauzy. Suites algébriques, automates et substitutions. *Bull. Soc. Math. Fr.*, 108:401–419, 1980.

[365] V.N. Chubarikov. Estimates of multiple trigonometric sums over prime numbers. *Izv. Akad. Nauk. SSSR, Ser. Mat.*, 49:1031–1067, 1985.

[366] F.R.K. Chung and R.L. Graham. On the set of distance determined by the union of arithmetic progression. *Ars combinat.*, 1:57–76, 1976.

[367] F.R.K. Chung and R.L. Graham. On irregularities of distribution of real sequences. *Proc. Natl. Acad. Sci.*, 78:4001, 1981.

[368] F.R.K. Chung and R.L. Graham. On irregularities of distribution. *Topics in classical number theory, Colloq. Budapest 1981, Vol. I, Colloq. Math. Soc. Janos Bolyai*, 37:181–222, 1984.

[369] J. Cigler and G. Helmberg. Neuere Entwicklungen der Theorie der Gleichverteilung. *Jber. Deutsch. Math.-Verein.*, 64:1–50, 1961.

[370] D. Clark. Second order difference equations related to the Collatz $3n+1$ conjecture. *J. Differ. Equations Appl.*, 1:73–85, 1995.

[371] T. Cochrane. Trigonometric approximation and uniform distribution modulo one. *Proc. Am. Math. Soc.*, 103:695–702, 1988.

[372] P. Codecá. A certain property of local discrepancy of Farey sequences. *Atti. Accad. Sci*, XIII, Ser. 8:163–173, 1981.

[373] P. Codecá and A. Perelli. On the uniform distribution (mod 1) of the Farey fractions and spaces. *Math. Ann.*, 279:413–422, 1988.

[374] J. Coffey. Uniformly distributed d-sequences. *Measure and measurable dynamics, Proc. Conf. Rochester, Contemp. Math.*, 94:97–112, 1989.

[375] Y. Colin de Verdiére. Distribution de points sur une sphére. *Sémin. Bourbaki*, 41, No. 703:177–178, 1989.

[376] R.J. Cook. Small fractional parts of quadratic forms in many variables. *Mathematika*, 27:25–29, 1980.

[377] R.J. Cook. Small fractional parts of quadratic and cubic polynomials in many variables. *Topics in classical number theory, Colloq. Budapest 1981, Vol. I, Colloq. Math. Soc. Janos Bolyai*, 34:281–303, 1984.

[378] R.J. Cook. Small values of indefinite quadratic forms and polynomials in many variables. *Stud. Sci. Math. Hung.*, 19:265–272, 1984.

[379] J. Coquet. Sur les fonctions q-multiplicatives presque-périodiques. *C.R. Acad. Sci., Paris*, 281:63–65, 1975.

[380] J. Coquet. Sur les fonctions q-multiplicatives pseudo-aléatoires. *C.R. Acad. Sci., Paris*, 282:175–178, 1976.

[381] J. Coquet. Remarques sur les nombres de Pisot-Vijayaroghavan. *Acta Arith.*, 32:79–87, 1977.

[382] J. Coquet. Contribution à l'étude harmonique de suites arithmétiques. *Prépublications*, IV:111, 1978.

[383] J. Coquet. Fonctions q-multiplicatives. Application aux nombres de Pisot- Vijayaraghavan. *Sémin. Théor. Nombres Bordeaux*, Exp. 17:15 p., 1978.

[384] J. Coquet. On the uniform distribution modulo one of some subsequences of polynomial sequences. *J. Number Theory*, 10:291–296, 1978.

[385] J. Coquet. Sur certaines suites pseudo-aléatoires. *Acta Sci. Math.*, 40:228–235, 1978.

[386] J. Coquet. Répartition modulo 1 des suites q-additives. *Commentat. Math.*, 21:23–42, 1979.

[387] J. Coquet. Sur certaines suites uniformément équiréparties modulo un (II). *Bull. Soc. R. Sci. Liége*, 48:426–431, 1979.

[388] J. Coquet. Sur la mesure spectrale des suites multiplicatives. *Ann. Inst. Fourier*, 29:163–170, 1979.

[389] J. Coquet. Corrélation de suites arithmétiques. *Sémin. Delange-Pisot-Poitou, 21e Annee 1979/80, Theorie des Nombres*, Exp. 12:12 p., 1980.

[390] J. Coquet. On the uniform distribution modulo one of subsequences of polynomial sequences. *J. Number Theory*, 12:244–250, 1980.

[391] J. Coquet. Sur certaines suites pseudo-aléatoires. *Monatsh. Math.*, 90:27–35, 1980.

[392] J. Coquet. Sur certaines suites pseudo-aléatoires. II. *Acta Math. Hung.*, 36:139–146, 1980.

[393] J. Coquet. Sur certaines suites uniformément équiréparties modulo 1. *Acta Arith.*, 36:157–162, 1980.

[394] J. Coquet. Sur l'équirépartition des suites à croissance lente et des suites non décroissantes. *Bull. Soc. Math. Fr.*, 108:251–258, 1980.

[395] J. Coquet. Types de répartition complète des suites. *Ann. Fac. Sci. Toulouse, Math.*, 2:137–155, 1980.

[396] J. Coquet. Une remarque sur les suites équiréparties a croissance lente. *Acta Arith.*, 36:143–146, 1980.

[397] J. Coquet. Ensembles normaux et équirépartition complète. *Acta Arith.*, 39:49–52, 1981.

[398] J. Coquet. Représentation des entiers naturels et indépendance. *Ann. Inst. Fourier*, 31:1–25, 1981.

[399] J. Coquet. Harmonic properties of some arithmetical sequences. *Manuscr. Math.*, 39:233–243, 1982.

[400] J. Coquet. Répartition de la somme des chiffres associée à une fraction continue. *Bull. Soc. R. Sci. Liège*, 51:161–165, 1982.

[401] J. Coquet. Représentations des entiers naturels of suites uniformément equiréparties. *Ann. Inst. Fourier*, 32:1–5, 1982.

[402] J. Coquet. Représentations lacunaires des entiers naturels. *Arch. Math.*, 38:184–188, 1982.

[403] J. Coquet. Graphes connexes représentation des entiers et équirépartition. *J. Number Theory*, 16:363–375, 1983.

[404] J. Coquet. Répartition de suites polynomials. *Monatsh. Math.*, 95:111–116, 1983.

[405] J. Coquet. Représentation lacunaires des entiers naturels. *Arch. Math.*, 41:238–242, 1983.

[406] J. Coquet. Sur la représentation des multiples d'un entier dans une base. *Journées arithmétiques, Orsay 1982, Publ. Math. Orsay*, 83:20–37, 1983.

[407] J. Coquet. Permutations des entiers et répartition des suites. *Publ. Math. Orsay*, 25–39, 1986.

[408] J. Coquet, T. Kamae, and M. Mendès France. Sur la mesure spectrale de certaines suites arithmétiques. *Bull. Soc. Math. Fr.*, 105:369–384, 1978.

[409] J. Coquet and P. Liardet. Répartitions uniformes des suites et indépendance statistique. *Compos. Math.*, 51:215–236, 1984.

[410] J. Coquet and P. Liardet. A metric study involving independent sequences. *J. Anal. Math.*, 49:15–53, 1987.

[411] J. Coquet and M. Mendès France. Suites à spectre vide et suites pseudo-aléatoires. *Acta Arith.*, 32:99–106, 1977.

[412] J. Coquet and G. Rhin. Fourier-Bohr spectrum of sequence related to continued fraction. *J. Number Theory*, 17:327–336, 1983.

[413] J. Coquet, G. Rhin, and Ph. Toffin. Représentations des entiers naturels et indépendance statistique. *Ann. Inst. Fourier*, 31:1–15, 1981.

[414] J. Coquet and Ph. Toffin. Représentations des entiers naturels et indépendance statistique. *Bull. Sci. Math.*, 105:289–298, 1981.

[415] L. Corwin and C. Pfeffer. On the density of sets in $(A/\mathbf{Q})^n$ defined by polynomials. *Colloq. Math.*, 68:1–5, 1995.

[416] J. Couot. Théorie ergodique de l'équirépartition. *Lect. Notes Math.*, 475:26–88, 1975.

[417] R. Courant and D. Hilbert. *Methoden der Mathematischen Physik I,II.* Springer, Berlin, 1968.

[418] R. Couture and P. L'Ecuyer. On the lattice structure of certain linear congruential sequences related to AWC/SWB generators. *Math. Comput.*, 62:799–808, 1994.

[419] R. Couture and P. L'Ecuyer. Orbits and lattices for linear random number generators with composite moduli. *Math. Comput.*, to appear.

[420] R. Couture, P. L'Ecuyer, and S. Tezuka. On the distribution of k-dimensional vectors for simple and combined Tausworthe sequences. *Math. Comput.*, 60:749–761, S11–S16, 1993.

[421] E. M. Coven and A. Hedlund. Sequences with minimal block growth. *Math. Systems Theory*, 7:138–153, 1973.

[422] D. Crisp, W. Moran, A. Pollington, and P.J.-S. Shiue. Substitution invariant cutting sequences. *J. Theor. Nombres Bordx.*, 5:123–137, 1993.

[423] M. Cugiani. p-adic numbers and pseudo-random sequences. *Rend., Sci. Mat. Appl.*, A 117:237–252, 1983.

[424] M. Cugiani and R. Rotondi. Generation of equidistributed vectors in a p-adic domain. *Rend., Sci. Mat. Appl., A*, 123:21–39, 1989.

[425] J. Cui and W. Freeden. Equidistribution on the sphere. preprint.

[426] T.W. Cusick. Zaremba's conjecture and sums of the divisor function. *Math. Comput.*, 61:171–176, 1993.

[427] T.W. Cusick. Properties of the x^2-mod N pseudorandom number generator. *IEEE Trans. Inform. Theory*, 41:1155–1159, 1995.

[428] H. Daboussi and M. Mendès France. Spectrum, almost-periodicity and equidistribution modulo 1. *Stud. Sci. Math. Hung.*, 9:173–180, 1975.

[429] H. Davenport. Note on irregularities of distribution. *Mathematika*, 3:131–135, 1956.

[430] H. Davenport and P. Erdős. Note on normal numbers. *Canad. J. Math.*, 4:58–63, 1953.

[431] H. Davenport and P. Erdős. A theorem on uniform distribution. *Magyar Tud. Akad. Mat. Kutató Int. Közl.*, 8:3–11, 1963.

[432] H. Davenport, P. Erdős, and W.J. LeVeque. On Weyl's criteron for uniform distribution. *Michigan Math. J.*, 10:311–314, 1963.

[433] H. Davenport and W.J. LeVeque. Uniform distribution relative to a fixed sequence. *Michigan Math. J.*, 10:315–319, 1963.

[434] P.J. Davis and P. Rabinowitz. *Methods of numerical integration.* Academic Press Inc., Orlando, 1984.

[435] R. Davis. Measures not approximable or not specifiable by means of balls. *Mathematika*, 18:157–160, 1971.

[436] N.G. De Bruijn and K.A. Post. A remark on uniformly distributed sequences and Riemann integrability. *Indag. Math.*, 30:149–150, 1986.

[437] L. De Clerck. A proof of Niederreiter's conjecture concerning error bounds for quasi-Monte Carlo integration. *Adv. Appl. Math.*, 2:1–6, 1981.

[438] L. De Clerck. *De exacte berekening van de sterdiscrepantie van de rijen von Hammersley in 2 dimensies.* Katholieke Universiteit Leuven, Leuven, Belgium, 1984.

[439] L. De Clerck. A method for exact calculation of the stardiscrepancy of plane sets applied to the sequences of Hammersley. *Monatsh. Math.*, 101:261–278, 1986.

[440] A. De Matteis, J. Eichenauer-Herrmann, and H. Grothe. Computation of critical distances within multiplicative congruential pseudorandom number sequences. *J. Comput. Appl. Math.*, 39:49–55, 1992.

[441] A. De Matteis and S. Pagnutti. Parallelization of random number generators and long-range correlations. *Numer. Math.*, 53:595–608, 1988.

[442] I. Deák. *Random Number Generators and Simulation.* Akademiai Kiado, Budapest, 1989.

[443] M. A.B. Deakin. Another derivation of Benford's law. *Austral. Math. Soc. Gaz.*, 20:162–163, 1993.

[444] A. Decomps-Guilloux. Répartition de dans les adéles. *Semin. Delange-Pisot-Poitou, 7e Annee 1965/66, Theorie des Nombres*, Exp. 5:14 p., 1967.

[445] A. Decomps-Guilloux. Théorèmes d'approximation dans les adèles. *Algèbre Théor. Nombres*, 19, No. 8:14 p., 1967.

[446] F.M. Dekking. Marches automatiques. *J. Theor. Nombres Bordx.*, 5:93–100, 1993.

[447] F.M. Dekking and M. Mendès France. Uniform distribution modulo one: a geometrical viewpoint. *J. Reine Angew. Math.*, 329:143–153, 1981.

[448] F.M. Dekking, M. Mendès France, and A. van der Poorten. Folds! *Math. Intelligencer*, 4:130–138, 173–181, 190–195; see also L. Auteurs, Corrigendum, Math. Intelligencer 5 (1983), 1982.

[449] H. Delange. On some arithmetic functions. *Illinois J. Math.*, 2:81–87, 1958.

[450] M. Denker, C. Grillenberger, and K. Sigmund. *Ergodic Theory on Compact Spaces.* Lect. Notes Math. 527, Springer, Berlin, 1976.

[451] V. Denzer and A. Ecker. Optimal multipliers for linear congruential pseudorandom number generators with prime moduli. *BIT*, 28:803–808, 1988.

[452] J.-M. Deshouillers and A. Hajj-Diab. Sur la distribution modulo M des suites polynomials $[P(n)]$. *J. Reine Angew. Math.*, 323:193–199, 1981.

[453] J.-M. Deshouillers. Sur la fonction de répartition des certaines fonctions arithmétiques définies sur l'ensemble des nombres premiers moins un. *Sémin. Théor. Nombres Bordeaux*, Exp. 20:16 p., 1970.

[454] J.-M. Deshouillers. La répartition modulo 1 des puissances de rationnels dans l'anneau des séries formelles sur un corps fini. *Sémin. Théor. Nombres 1979-1980*, Exp. 5:22 p., 1980.

[455] J.-M. Deshouillers. La répartition modulo 1 des puissance d'un élément dans. *Symp. Durham*, 2:69–72, 1981.

[456] J.-M. Deshouillers. Geometric aspects of Weyl sums. *Banach Cent. Publ.*, 17:75–82, 1985.

[457] L. Devroye. *Non-uniform random variate generation.* Springer-Verlag, New York, 1986.

[458] P. Diaconis. The distribution of leading digits and uniform distribution mod 1. *Ann. Probab.*, 5:72–81, 1976.

[459] H. G. Diamond and J. D. Vaaler. Metric theorems on the estimation of integrals by sums. *Sémin. Théor. Nombres Bordaux*, 41:7, 1982.

[460] H. Dickinson. The Hausdorff dimension of systems of simultaneously small linear forms. *Mathematika.*, 40:367–374, 1993.

[461] H. Dickinson. The Hausdorff dimension of sets arising in metric diophantine approximation. *Acta Arith.*, 68:133–140, 1994.

[462] U. Dieter. Beziehungen zwischen Dedekindschen Summen. *Abh. Math. Semin. Univ. Hamburg*, 21:109–125, 1957.

[463] U. Dieter. Pseudo-random numbers: The exact distribution of pairs. *Math. Comput.*, 25:855–883, 1971.

[464] U. Dieter. How to calculate shortest vectors in a lattice. *Math Comput.*, 29:827–833, 1975.

[465] U. Dieter. Probleme bei der Erzeugung gleichverteilter Zufallszahlen, in Zufallszahlen und Simulationen. Teubner, Stuttgart:7–20, 1986.

[466] U. Dieter. Erzeugung von gleichverteilten Zufallszahlen. Jahrbuch Überblicke Mathematik 1993:25–44, 1993.

[467] D. Dobkin and D. Eppstein. Computing the discrepancy. *Proceedings of the Ninth Annual Symposium on Computational Geometry*, 47–52, 1993.

[468] D. Dobkin and D. Gunopulos. Computing the rectangle discrepancy. Preprint, Princeton University, 1994.

[469] G.V. Dobris. Generation and statistical tests of sequences of n-distributed pseudo-random numbers. *Program. Comput. Software*, 12:94–99, 1986.

[470] N.M. Dobrovol'skij. An effective proof of a theorem of Roth on quadratic dispersion. *Russ. Math. Surv.*, 39:117–118, 1984.

[471] A.F. Dowidar. Summability methods and distribution of sequences of integers. *J. Natur. Sci. Math.*, 12:337–341, 1972.

[472] A.F. Dowidar. On the generalized uniform distribution (mod 1). *Publ. Math. Debrecen*, 19:35–38, 1973.

[473] A.F. Dowidar. On the generalized uniform distribution (mod 1). *Proc. Math. phys. Soc.*, 36:97–100, 1975.

[474] K.A. Driver, D.S. Lubinsky, G. Petruska, and P. Sarnak. Irregular distribution of $\{n\beta\}$, $n = 1, 2, 3, \ldots$, quadrature of singular integrands, and curious basic hypergeometric series. *Indag. Math.*, 2:469–481, 1991.

[475] M. Drmota. *Gleichverteilte Funktionen auf Mannigfaltigkeiten*. Technische Universität Wien, 1986.

[476] M. Drmota. Irregularities of distributions with respect to polytopes. *Mathematika*, 43:108–119, 1996.

[477] M. Drmota. Gleichmäßig gleichverteilte und schwach gleichmäßig gleichverteilte Funktionen modulo 1. *Lect. Notes Math.*, 1262:22–36, 1987.

[478] M. Drmota. Metrische Sätze über schwach gleichmäßig gleichverteilte Funktionen. *Anz. Österr. Akad. Wiss.*, 113–118, 1987.

[479] M. Drmota. Untere Schranken für die C-Diskrepanz. *Österr. Akad. Wiss. SB II*, 196:107–117, 1987.

[480] M. Drmota. An optimal lower bound for the discrepancy of c-uniformly distributed functions modulo 1. *Indag. Math.*, 50:21–28, 1988.

[481] M. Drmota. Such- und Prüfprozesse mit praktischen Gitterpunkten. *Anz. Österr. Akad. Wiss.*, 125:23–28, 1988.

[482] M. Drmota. Irregularities of continuous distributions. *Ann. Inst. Fourier*, 39:501–527, 1989.

[483] M. Drmota. On irregularities of distribution on the hyperbolic plane. *Lect. Notes Math.*, 1452:31–42, 1990.

[484] M. Drmota. On iterated weighted means of bounded sequences and uniform distribution. *Tsukuba J. Math.*, 15:249–260, 1991.

[485] M. Drmota. Irregularities of distribution and convex sets. *Grazer Math. Ber.*, 318:9–16, 1992.

[486] M. Drmota. On linear diophantine equations and Fibonacci numbers. *J. Number Theory*, 44:315–327, 1993.

[487] M. Drmota and R.F. Tichy. C-uniform distribution of entire functions. *Rend. Semin. Mat. Univ. Padova*, 79:49–58, 1988.

[488] M. Drmota and R.F. Tichy. C-uniform distribution on compact metric spaces. *J. Math. Analyis. Appl.*, 129:284–292, 1988.

[489] M. Drmota and R.F. Tichy. A note on Riesz means and well distribution. *Math. Nachr.*, 139:267–279, 1988.

[490] M. Drmota and R.F. Tichy. Deterministische Approximation stoachastisch-ergodischer Signale. *Messen-Steuern-Regeln*, 32:109–114, 1989.

[491] M. Drmota, R.F. Tichy, and R. Winkler. Completely uniformly distributed sequences of matrices. *Lect. Notes Math.*, 1452:43–57, 1990.

[492] M. Drmota and R. Winkler. $s(n)$-uniform distribution modulo 1. *J. Number Theory*, 50:213–225, 1995.

[493] V. Drobot. On dispersion and Markov constants. *Acta Math. Hung.*, 47:89–93, 1986.

[494] V. Drobot. Gaps in the sequence $n^2\vartheta \,(mod\, 1)$. *Int. J. Math. Math. Sci.*, 10:131–134, 1987.

[495] R.J. Duffin and A.C. Schaeffer. Khintchine's problem in metric diophantine approximation. *Duke Math. J.*, 8:243–255, 1941.

[496] J.-M. Dumont, T. Kamae, and S. Takahashi. Minimal cocycles with the scaling property and substitutions. preprint.

[497] J.-M. Dumont and A. Thomas. Une modification multiplicative des nombres g normaux. *Ann. Fac. Sci. Toulouse, Math.*, 8:367–373, 1987.

[498] J.-M. Dumont and A. Thomas. Digital sum moments and substitutions. *Acta Arith.*, 64:205–225, 1993.

[499] J.-M. Dumont and A. Thomas. Modifications de nombres normaux par des transducteurs. *Acta Arith.*, 68:153–170, 1994.

[500] Y. Dupain. Répartition des sous-suites d'une suite donnée. *Bull. Soc. Math. Fr.*, 37:59–62, 1974.

[501] Y. Dupain. Intervalles à restes majorés pour la suite. *Acta Math. Hung.*, 29:289–303, 1977.

[502] Y. Dupain. Intervalles à restes majorés pour la suite. *Astérisque*, 41–42:193–197, 1977.

[503] Y. Dupain. Intervalles à restes majorés pour la suite. *Bull. Soc. Math. Fr.*, 106:153–160, 1978.

[504] Y. Dupain. Discrépance de la suite. *Ann. Inst. Fourier*, 29:81–106, 1979.

[505] Y. Dupain, R.R. Hall, and G. Tenenbaum. Sur l'équirépartition modulo 1 de certaines fonctions de diviseurs. *J. London Math. Soc.*, 26:397–411, 1982.

[506] Y. Dupain and J. Lesca. Répartition des sous-suites d'une suite donnée. *Acta Arith.*, 23:307–314, 1973.

[507] Y. Dupain and V. T. Sós. On the one-sided boundedness of discrepancy-function of the sequence. *Acta Arith.*, 37:363–374, 1980.

[508] Y. Dupain and V. T. Sós. On the discrepancy of $(n\alpha)$ sequences. *Topics in classical number theory, Colloq. Budapest 1981, Vol. I, Colloq. Math. Soc. Janos Bolyai*, 34:355–387, 1984.

[509] Y. Dupain and A. Thomas. Sous-suites équiréparties d'une suite donnée. *Acta Arith.*, 26:285–292, 1975.

[510] A. Durner. On averaging multisets. *Monatsh. Math.*, 121:41–53, 1996.

[511] W.F. Eddy. Random number generators for parallel processors. *J. Comput. Appl. Math.*, 31:63–71, 1990.

[512] K.J. Edwards. Perfect pseudorandom numbers. *Bull., Inst. Math.*, 24:80–81, 1988.

[513] R.B. Eggleton and R.J. Simpson. Beatty sequences and Langford sequences. *Discrete Math.*, 111:165–178, 1993.

[514] J. Eichenauer and H. Grothe. Marsaglia's lattice test and non-linear congruential pseudorandom number generators. *Metrika*, 35:241–250, 1988.

[515] J. Eichenauer and J. Lehn. Eine Bemerkung zur Periodenlängenbestimmung bei einem verallgemeinerten Fibonacci-Generator. *Elem. Math.*, 39:81–84, 1984.

[516] J. Eichenauer and J. Lehn. A non-linear congruential pseudo-random number generator. *Stat. Hefte*, 27:315–326, 1986.

[517] J. Eichenauer and J. Lehn. On the structure of quadratic congruential sequences. *Manuscr. Math.*, 58:129–140, 1987.

[518] J. Eichenauer, J. Lehn, and A. Topuzoglu. A nonlinear congruential pseudorandom number generator with power of two modulus. *Math. Comput.*, 51:757–759, 1988.

[519] J. Eichenauer and H. Niederreiter. On Marsaglia's lattice test for pseudorandom numbers. *Manuscr. Math.*, 62:245–248, 1988.

[520] J. Eichenauer and H. Niederreiter. On the discrepancy of quadratic congruential pseudorandom numbers. *J. Comput. Appl. Math.*, 34:243–249, 1991.

[521] J. Eichenauer and A. Topuzoglu. On the period length of congruential pseudorandom number sequences generated by inversions. *J. Comput. Appl. Math.*, 31:87–96, 1990.

[522] J. Eichenauer-Hermann. Inversive congruential pseudorandom numbers avoid the planes. *Math. Comput.*, 56:297–301, 1991.

[523] J. Eichenauer-Hermann. On the discrepancy of inversive congruential pseudorandom numbers with prime power modulus. *Manuscr. Math.*, 71:153–161, 1991.

[524] J. Eichenauer-Hermann. Construction of inversive congruential pseudorandom numbers generators with maximal period length. *J. Comput. Appl. Math.*, 40:345–349, 1992.

[525] J. Eichenauer-Hermann. Inversive congruential pseudorandom numbers: a tutorial. *Internat. Statist. Rev.*, 60:167–176, 1992.

[526] J. Eichenauer-Hermann. Lower bounds for the discrepancy of inversive congruential pseudorandom numbers with power of two modulus. *Math. Comput.*, 58:775–779, 1992.

[527] J. Eichenauer-Hermann. A new inversive congruential pseudorandom number generator with power of two modulus. *ACM Trans. Model. Comput. Simulation*, 2:1–11, 1992.

[528] J. Eichenauer-Hermann. On the autocorrelation structure of inversive congruential pseudorandom number sequences. *Statist. Papers*, 33:261–268, 1992.

[529] J. Eichenauer-Hermann. A remark on the discrepancy of quadratic congruential pseudorandom numbers. *J. Comput. Appl. Math.*, 43:383–387, 1992.

[530] J. Eichenauer-Hermann. Equidistribution properties of nonlinear congruential pseudorandom numbers. *Metrika*, 40:333–338, 1993.

[531] J. Eichenauer-Hermann. Explicit inversive congruential pseudorandom numbers: the compound approach. *Computing*, 51:175–182, 1993.

[532] J. Eichenauer-Hermann. Inversive congruential pseudorandom numbers. *Z. Angew. Math. Mech.*, 73:T644–T647, 1993.

[533] J. Eichenauer-Hermann. The lattice structure of nonlinear congruential pseudorandom numbers. *Metrika*, 40:115–120, 1993.

[534] J. Eichenauer-Hermann. On the discrepancy of inversive congruential pseudorandom numbers with prime power modulus, II. *Manuscripta Math.*, 79:239–246, 1993.

[535] J. Eichenauer-Hermann. Statistical independence of a new class of inversive congruential pseudorandom numbers. *Math. Comput.*, 60:375–384, 1993.

[536] J. Eichenauer-Hermann. Compound nonlinear congruential pseudorandom numbers. *Monatsh. Math.*, 117:213–222, 1994.

[537] J. Eichenauer-Hermann. Improved lower bounds for the discrepancy of inversive congruential pseudorandom numbers. *Math. Comput.*, 62:783–786, 1994.

[538] J. Eichenauer-Hermann. On generalized inversive congruential pseudorandom numbers. *Math. Comput.*, 63:293–299, 1994.

[539] J. Eichenauer-Hermann. On the discrepancy of quadratic congruential pseudorandom numbers with power of two modulus. *J. Comput. Appl. Math.*, 53:371–376, 1994.

[540] J. Eichenauer-Hermann. Discrepancy bounds for nonoverlapping pairs of quadratic congruential pseudorandom numbers. *Arch. Math.*, 65:362–368, 1995.

[541] J. Eichenauer-Hermann. Nonoverlapping pairs of explicit inversive congruential pseudorandom numbers. *Monatsh. Math.*, 119:49–61, 1995.

[542] J. Eichenauer-Hermann. Pseudorandom number generation by nonlinear methods. *Internat. Stat. Rev.*, 63:247–255, 1995.

[543] J. Eichenauer-Hermann. A unified approach to the analysis of compound pseudorandom numbers. *Finite Fields Appl.*, 1:102–114, 1995.

[544] J. Eichenauer-Hermann. Modified explicit inversive congruential pseudorandom numbers with power of two modulus. *Statistics Comput.*, to appear.

[545] J. Eichenauer-Hermann. Quadratic congruential pseudorandom numbers: distribution of triples. *J. Comput. Appl. Math.*, to appear.

[546] J. Eichenauer-Hermann and F. Emmerich. Compound inversive congruential pseudorandom numbers: an average-case analysis. *Math. Comput.*, 65:215–225, 1996.

[547] J. Eichenauer-Hermann and H. Grothe. A remark on long-range correlationsin multiplicative pseudo-random number generators. *Numer. Math.*, 56:609–611, 1989.

[548] J. Eichenauer-Hermann, H. Grothe, and J. Lehn. On the period length of pseudorandom vector sequences generated by matrix generators. *Math. Comput.*, 52:145–148, 1989.

[549] J. Eichenauer-Hermann and Niederreiter H. Kloosterman-type sums and the discrepancy of nonoverlapping pairs of inversive congruential pseudorandom numbers. *Acta Arith.*, 65:185–194, 1993.

[550] J. Eichenauer-Hermann and K. Ickstadt. Explicit inversive congruential pseudorandom numbers with power of two modulus. *Math. Comput.*, 62:787–797, 1994.

[551] J. Eichenauer-Hermann and G. Larcher. Average behaviour of compound nonlinear congruential pseudrorandom numbers. *Finite Fields Appl.*, 2:111–123, 1996.

[552] J. Eichenauer-Hermann and G. Larcher. Average equidistribution properties of compound nonlinear congruential pseudorandom numbers. *Math. Comput.*, to appear, 1996.

[553] J. Eichenauer-Hermann and H. Niederreiter. Bounds for exponential sums and their applications to pseudorandom numbers. *Acta Arith.*, 67:269–281, 1994.

[554] J. Eichenauer-Hermann and H. Niederreiter. On the statistical independence of nonlinear congruential pseudorandom numbers. *ACM Trans. Model. Comput. Simulation*, 4:89–95, 1994.

[555] J. Eichenauer-Hermann and H. Niederreiter. An improved upper bound for the discrepancy of quadratic congruential pseudorandom numbers. *Acta Arith.*, 69:193–198, 1995.

[556] Shalom Eliahou. The $3x+1$ problem: new lower bounds on nontrivial cycle lengths. *Discrte Math.*, 118:45–56, 1993.

[557] P.D.T.A. Elliott. On distribution functions (mod 1): quantitative Fourier inversion. *J. Number Theory*, 4:509–522, 1972.

[558] P.D.T.A. Elliott. *Probabilistic Number Theory. 1. Mean Value Theorems.* Springer, Berlin, 1979.

[559] P.D.T.A. Elliott. *Probabilistic Number Theory. 2. Central Limit Theorems.* Springer, New York, 1980.

[560] F. Emmerich. Pseudorandom vector generation by the compound inversive method. *Math. Comput.*, to appear.

[561] P. Erdős. On the distribution function of additive functions. *Ann. of Math.*, 47:1–20, 1946.

[562] P. Erdős. Problems and results on diophantine approximations. *Compos. Math.*, 16:52–69, 1964.

[563] P. Erdős and I.S. Gál. On the law of the iterated logarithm. I,II. *Indag. Math.*, 17:65–76, 77–84, 1955.

[564] P. Erdős, M. Joo, and I. Joo. On a problem of Tamas Varga. *Bull. Soc. Math. Fr.*, 120:507–521, 1992.

[565] P. Erdős and J.F. Koksma. On the uniform distribution modulo 1 of lacunary sequences. *Indag. Math.*, 11:79–88, 1949.

[566] P. Erdős and J.F. Koksma. On the uniform distribution modulo 1 of sequences $\{f(n,\delta)\}$. *Indag. Math.*, 11:299–302, 1949.

[567] P. Erdős and T.K. Sheng. Distribution of rational points on the real line. *J. Austral. Math. Soc. A*, 20:124–128, 1975.

[568] P. Erdős and S.J. Taylor. On the set of points of convergence of a lacunary trigonometric series and the equidistribution properties of related sequences. *Proc. London Math. Soc.*, 7:598–615, 1957.

[569] P. Erdős and P. Turán. On a problem in the theory of uniform distribution. I. *Indag. Math.*, 10:370–378, 1948.

[570] P. Erdős and P. Turán. On a problem in the theory of uniform distribution. II. *Indag. Math.*, 10:406–413, 1948.

[571] L. Erlebach and W. Yslas Vélez. Equiprobability in the Fibonacci sequence. *Fibonacci Q.*, 21:189–191, 1983.

[572] J.-L. Ermine. Courants ergodiques et répartition géométrique. *Sémin. Théor. Nombres 1977-1978*, Exp. 19:19 p., 1978.

[573] J.-L. Ermine. Courants ergodiques et répartition geométrique. *Lect. Notes Math.*, 807:180–198, 1980.

[574] G.R. Everest. Corrigenda to: Uniform distribution and lattice point counting. *J. Austral. Math. Soc. A*, 56:144, 1994.

[575] G.R. Everest and T.M. Gagen. A decomposition for the inverse regulator of a number field. *Ark. Mat.*, to appear.

[576] F. J. Fabozzi. *Handbook of Mortage Backed Securities.* Probus Publishing Co., 1992.

[577] A.S. Fainleib. A generalization of Esseen's inequality and its application in prohabilistic number theory. *Izv. Akad. Nauk SSSR, Ser. Mat.*, 32:859–879, 1968.

[578] K.J. Falconer. *The Geometry of Fractal Sets.* Cambridge University Press, Cambridge, 1985.

[579] A.H. Fan. Lacunarité à la Hadamard et équirépartition. *Colloq. Math.*, 66:151–163, 1993.

[580] H. Faure. Discrépance de suites associées à un système de numération. *C.R. Acad. Sci., Paris*, 286:293–296, 1978.

[581] H. Faure. Discrépances de suites associées à un système de numération. *Bull. Soc. Math. Fr.*, 109:143–182, 1981.

[582] H. Faure. Discrépance de suites associées à un système de numération (en dimension s). *Acta Arith.*, 41:337–351, 1982.

[583] H. Faure. Suites à faible discrépance en dimension s. *Journées arithmétiques, Exeter 1980, London Math. Soc. Lect. Note Ser.*, 56:284–290, 1982.

[584] H. Faure. Etude des restes pour les suites de van der Corput généralisées. *J. Number Theory*, 16:376–394, 1983.

[585] H. Faure. Lemme de Bohl pour les suites de Van der Corput généralisées. *J. Number Theory*, 22:4–20, 1986.

[586] H. Faure. On the star-discrepancy of generalized Hammersley sequences in two dimensions. *Monatsh. Math.*, 101:291–300, 1986.

[587] H. Faure. Discrépance quadratique de suites infinies en dimension un. (Quadratic discrepancy of infinte sequences in dimension one). *Théorie des nombres (C. R. Conf. Int. Quebec 1987)*, pages 207–212, 1989.

[588] H. Faure. Discrépance quadratique de la suite de van der Corput et de sa symétrique. *Acta Arith.*, 55:333–350, 1990.

[589] H. Faure. Using permutations to reduce discrepancy. *J. Comput. Appl. Math.*, 31:97–103, 1990.

[590] H. Faure. Good permutations for extreme discrepancy. *J. Number Theory.*, 42:47–56, 1992.

[591] H. Faure. Suggestions for quasi-Monte-Carlo users. *Lect. Notes Pure Appl. Math.*, 141:269–278, 1993.

[592] H. Faure. Methodes quasi-Monte-Carlo multidimennelles. *Theor. Comput. Sci.*, 123:131–137, 1994.

[593] H. Faure. Discrepancy lower bound in two dimensions. *Lect. Notes Stat.*, 106:198–204, 1995.

[594] H. Faure and H. Chaix. Minoration de discrépance en dimension 2. *C.R. Acad. Sci., Paris*, 319:1–4, 1994.

[595] H. Faure and H. Chaix. Minoration de discrépance en dimension deux. *Acta Arith.*, 76:149–164, 1996.

[596] S. Ferenczi, J. Kwiatkowski, and C. Mauduit. A density theorem for (multiplicity, rank) pairs. *J. Anal. Math.*, 65:45–75, 1995.

[597] P. Filipponi and R. Menicocci. Some probabilistic aspects of the terminal digits of Fibonacci. *Fibonacci Q.*, 33:325–331, 1995.

[598] G. Fiorito. On properties of periodically monotone sequences. *Appl. Math. Comput.*, 72:259–275, 1995.

[599] G. Fiorito, R. Musmeci, and M. Strano. Uniform distribution and applications to a class of recurring series. *Matematiche (Catania)*, 48:123–133, 1993.

[600] B. Fischer and L. Riesler. Newton interpolation in Fejer and Chebyshev points. *Math. Comput.*, 53:265–278, 1989.

[601] G.S. Fishman. Multiplicative congruential random number generators with modulus 2^β: An exhaustive analysis for $\beta = 32$ and a partial analysis for $\beta = 48$. *Math. Comput.*, 54:331–344, 1990.

[602] M. Flahive and H. Niederreiter. On inversive congruential generators for pseudorandom numbers. *Finite Fields, Coding Theory, and Advances in Comm. and Comp., (G. Mullen and P.J.-S. Shiue eds.)*, 75–80, 1992.

[603] P. Flajolet, P. Kirschenhofer, and R. Tichy. Deviations from uniformity in random strings. *Probab. Theory Relat. Fields*, 80:139–150, 1988.

[604] P. Flajolet, P. Kirschenhofer, and R.F. Tichy. Discrepancy of sequences in discrete spaces. *(Irregularities of partitions). Algorithms Comb.*, 8:61–70, 1989.

[605] L. Flatto. z-numbers and β-transformations. *Symbolic dynamic and its applications*, 181–201, 1991.

[606] L. Flatto, J.C. Lagarias, and A.D. Pollington. On the range of fractional parts $\{\xi(p/q)^n\}$. *Acta Arith.*, 70:125–147, 1995.
[607] B.J. Flehinger. On the probability that a random integer has initial digit A. *Am. Math. Monthly*, 73:1056–1061, 1966.
[608] W. Fleischer. Das Wienersche Maß einer gewissen Menge von Vektorfunktionen. *Monatsh. Math.*, 75:193–197, 1971.
[609] W. Fleischer. Eine Diskrepanzabschätzung für stetige Funktionen. *Z. Wahrscheinlichkeitstheorie verw. Gebiete*, 23:18–21, 1972.
[610] W. Fleischer. Diskrepanzbegriff für kompakte Räume. *Anz. Österr. Akad. Wiss.*, 127–131, 1981.
[611] W. Fleischer. Ein Diskrepanzbegriff für kompakte Räume (a notion of discrepancy for compact spaces). *Lect. Notes Math.*, 1114:14–15, 1985.
[612] W. Fleischer and H. Stegbuchner. Über eine Ungleichung in der Theorie der Gleichverteilung mod. 1. *Ber. Math. Inst. Univ. Salzburg*, 4:45–54, 1981.
[613] W. Fleischer and H. Stegbuchner. Über eine Ungleichung in der Theorie der Gleichverteilung mod. 1. *Österr. Akad. Wiss. SB II*, 191:133–139, 1982.
[614] O.M. Fomenko and E.P. Golubeva. On the distribution of the sequence $bp^{\frac{3}{2}}$ modulo 1. *Zap. Nauchn. Semin. Leningr. Otd. Mat. Inst. Stekl.*, 91:31–39, 1979.
[615] A.H. Forrest. The limit points of certain classical exponential series and other continuous cocycles. preprint.
[616] A.S. Fraenkel. Complementary systems of integers. *Am. Math. Monthly*, 84:114–115, 1977.
[617] A.S. Fraenkel. Iterated floor function, algebraic numbers, discrete chaos, Beatty subsequences, simigroups. *Trans. Am. Math. Soc.*, 341:639–664, 1994.
[618] A.S. Fraenkel and R. Holzman. Gap problems for integer part and fractional part sequences. *J. Number Theory*, 50:66–86, 1995.
[619] Z. Franco and C. Pomerance. On a conjecture of crandall conserning the $qx + 1$ problem. *Math. Comput.*, 64:1333–1336, 1995.
[620] J.N. Franklin. Deterministic simulation of random processes. *Math. Comput.*, 17:28–59, 1963.
[621] J.N. Franklin. Equidistribution of matrix-power residues modulo one. *Math. Comput.*, 18:560–568, 1964.
[622] W. Freeden. Über eine Klasse von Integralformeln der mathematischen Geodäsie. *Geodätisches Institut Aachen*, 1979.
[623] W. Freeden. On integral formulas of the (unit) sphere and their applications to numerical computations of integrals. *Computing*, 25:131–146, 1980.
[624] G. Freud. *Orthogonale Polynome*. Birkhäuser Verlag, Basel, 1969.
[625] E. Fried and V.T. Sós. A generalization of the three-distance theorem for groups. *Algebra Univers.*, 29:136–149, 1992.
[626] J.B. Friedlander and H. Ivaniec. On the distribution of the sequence $n^2\theta$ (mod 1). *Can. J. Math.*, 39:338–344, 1987.
[627] K.K. Frolov. An upper estimate of the discrepancy in the L^p-metric. *Sov. Math.*, 21:840–842, 1980.
[628] C. Frougny and B. Solomyak. Finite beta-expansions. *Ergodic Theory Dynamical Systems*, 12:1333–1336, 1992.
[629] A. Fuchs and G. Letta. Le problème du premier chiffre décimal pour les nombres premiers. *Rend. Accad. Naz. Sci. Detta XL*, 18:81–87, 1994.
[630] A. Fujii. On the uniformity of the distribution of the zeros of the Riemann zeta function. *Comment. Math. Univ. St. Pauli*, 31:99–113, 1982.
[631] A. Fujii. Diophantine approximation, Kronecker's limit formula and the Riemann hypothesis. *Théorie des nombres (C. R. Conf. Int. Quebec 1987)*, pages 240–250, 1989.
[632] A. Fujii. Some problems of diophantine approximation in the theory of the Riemann zeta function IV. *Comment. Math. Univ. St. Pauli*, 43:217–244, 1994.
[633] H. Furstenberg. Disjointness in ergodic theory, minimal sets, and a problem in diophantine approximation. *Math. Systems Theory*, 1:1–49, 1967.
[634] H. Furstenberg. *Recurrence in Ergodic Theory and Combinatorial Number Theory*. Princeton University Press, Princeton, 1981.
[635] H. Furstenberg. Nonconventional ergodic averages. *Proc. Symp. Pure Math.*, 50:43–56, 1990.

[636] H. Furstenberg, H. Keynes, and L. Shapiro. Prime flows in topological dynamics. *Isr. J. Math.*, 14:26–38, 1973.

[637] H. Furstenberg and B. Weiss. Simultaneous diophantine approximation and IP-sets. *Acta Arith.*, 49:413–426, 1988.

[638] M. Fushimi. Increasing the orders of equidistribution of the leading bits of the Tausworthe sequence. *Inf. Process. Lett.*, 16:189–192, 1983.

[639] M. Fushimi. Designing a uniform random number generator whose subsequences are k-distributed. *SIAM J. Comput.*, 17:79–99, 1988.

[640] M. Fushimi and S. Tezuka. The k-distribution of generalized feedback shift register pseudorandom numbers. *Commun. ACM*, 26:516–523, 1983.

[641] P. Gabriel, M. Lemanczyk, and P. Liardet. Ensemble d'invariants pour les produits croisés de anzai. *Suppl. au Bull. Soc. Math. Fr.*, 119:1–102, 1991.

[642] I.S. Gál and L. Gál. The discrepancy of the sequence $\{(2^n x)\}$. *Indag. Math.*, 26:129–143, 1964.

[643] I.S. Gál and J.F. Koksma. Sur l'ordre de grandeur des fonctions sommables. *C.R. Acad. Sci., Paris*, 227:1321–1323, 1948.

[644] I.S. Gál and J.F. Koksma. Sur l'ordre de grandeur des fonctions sommables. *Indag. Math.*, 12:192–207, 1950.

[645] J. Galambos. Uniformly distributed sequences mod 1 and Cantor's series representation. *Czech. Math. J.*, 26:636–641, 1976.

[646] J. Galambos. Correction to my paper: Uniformly distributed sequences mod 1 and Cantor's series representation. *Czech. Math. J.*, 27:672, 1977.

[647] T.H. Ganelius. *Tauberian Remainder Theorems.* Lect. Notes Math. 232, Springer, Berlin, 1971.

[648] R.L. Garifullina. Distribution of the fractional parts of the matrix exponential function. *J. Sov. Math. transl. from Issled Prikl. Mat.*, 41:1396–1400, 1975.

[649] J.F. Geelen and R.J. Simpson. A two-dimensional Steinhaus theorem. *Australas. J. Comb.*, 8:169–197, 1993.

[650] H. U. Gerber. On the probability of ruin in the presence of a linear dividend barrier. *Scand. Actuarial J.*, 105–115, 1981.

[651] P. Gerl. Relative Gleichverteilung in kompakten Räumen. *Math. Z.*, 121:24–50, 1971.

[652] P. Gerl. Relative Gleichverteilung in kompakten Räumen II. *Monatsh. Math.*, 75:410–422, 1971.

[653] P. Gerl. Gleichverteilung auf der Kugel. *Arch Math.*, 24:203–207, 1973.

[654] P. Gerl. Gleichverteilung auf Kugeln im R^n. *Österr. Akad. Wiss. SB II*, 182:21–30, 1974.

[655] P. Gerl. Quelques généralisations de l'équirépartition. *Lect. Notes Math.*, 475:100–103, 1975.

[656] A. Ghosh. The distribution of αp^2 modulo 1. *Proc. London Math. Soc.*, 42:252–269, 1981.

[657] R. Gillard. Extensions abéliennes et répartition modulo 1. *Astérique*, 61:83–93, 1979.

[658] N.M. Glazunov. On equidistribution of the values of Kloosterman sums. *Dokl. Akad. Nauk Ukrain SSR, Ser. A*, 9–12, 1983.

[659] I.A. Glinkin. On a best quadrature formula on the class of convex functions. *Mat. Zametki*, 35:697–707, 1984.

[660] O. Goldreich, H. Krawczyk, and M. Luby. On the existence of pseudorandom generators. *SIAM J. Comput.*, 22:1163–1175, 1993.

[661] M. Goldstern. Eine Klasse vollständig gleichverteilter Folgen. *Lect. Notes Math.*, 1262:37–45, 1987.

[662] M. Goldstern. Vollständige Gleichverteilung in diskreten Räumen. *Lect. Notes Math.*, 1262:46–49, 1987.

[663] M. Goldstern. An application of Shoenfield's absoluteness theorem to the theory of uniform distribution. *Monatsh. Math.*, 116:237–243, 1993.

[664] M. Goldstern. The complexity of uniform distribution. *Math. Slovaca*, 44:491–500, 1994.

[665] M. Goldstern, R.F. Tichy, and G. Turnwald. Distribution of the ratios of the terms of a linear recurrence. *Monatsh. Math.*, 107:35–55, 1989.

[666] K. Goto. On the Wiener-Schoenberg theorem for asymptotic distribution functions. *Proc. Jap. Acad.*, Ser. A 57:420–423, 1981.

[667] K. Goto. Some examples of Benford sequences. *Math. J. Okayama Univ.*, 34:225–232, 1992.

[668] K. Goto and T. Kano. Uniform distribution of some special sequences. *Proc. Japan Acad. A*, 61:83–86, 1985.

[669] K. Goto and T. Kano. Some necessary conditions for (M, λ_n) uniform distribution mod 1. *Comment. Math. Univ. St. Pauli*, 35:85–91, 1986.

[670] K. Goto and T. Kano. Remarks to our former paper: "Uniform distribution of some special sequences". *Proc. Japan Acad. A*, 68:348–350, 1992.

[671] K. Goto and T. Kano. Discrepancy inequalities of Erdős-Turán and of Le Veque. *Surikaisekikenkyusho Kokyuroku*, 387:35–47, 1993.

[672] C.J. Goutziers. The discrepancy of the sequence $([n^c])_{n=1}^{\infty}$, $0 < c < 1$. *Delft Progress Rep.*, 2:219–224, 1977.

[673] P. Grabner. *Harmonische Analyse, Gleichverteilung und Ziffernentwicklungen.* TU Vienna, 1989.

[674] P. Grabner. Erdős-Turán type discrepancy bounds. *Monatsh. Math.*, 111:127–135, 1991.

[675] P. Grabner and R.F. Tichy. Equidistribution and Brownian motion on the Sierpiński gasket. *Monatsh. Math.*, to appear, 1997.

[676] P.J. Grabner. On digit expansions with respect to second order linear recurring sequences. *Lect. Notes Math.*, 1452:58–64, 1990.

[677] P.J. Grabner. Ziffernentwicklungen bezüglich linearer Rekursionen. *Österr. Akad. Wiss. SB II*, 199:1–21, 1991.

[678] P.J. Grabner. Block distribution in random strings. *Ann. Inst. Fourier*, 43:539–549, 1993.

[679] P.J. Grabner, P. Kiss, and R.F. Tichy. Diophantine approximation in terms of linear recurrent sequences. *CMS Conf. Proc. (Proceedings of the fourth conference of the Canadian Number Theory Association 1994)*, 15:187–195, 1995.

[680] P.J. Grabner, P. Liardet, and R.F. Tichy. Odometers and systems of numeration. *Acta Arith.*, 70:103–123, 1995.

[681] P.J. Grabner, O. Strauch, and Tichy R.F. Maldistribution in higher dimensions. *Mathematica Pannonica*, to appear.

[682] P.J. Grabner and R.F. Tichy. Contributions to digit expansions with respect to linear recurrences. *J. Number Theory*, 36:160–169, 1990.

[683] P.J. Grabner and R.F. Tichy. Remark on an inequality of Erdős-Turán-Koksma. *Anz. Österr. Akad. Wiss.*, 127:15–22, 1990.

[684] P.J. Grabner and R.F. Tichy. α-expansions, linear recurrences and the sum-of-digits function. *Manuscr. Math.*, 70:311–324, 1991.

[685] P.J. Grabner and R.F. Tichy. Remarks on statistical independence. *Math. Slovaca*, 44:91–94, 1994.

[686] P.J. Grabner, R.F. Tichy, and R. Winkler. On the stability of the quotients of successive terms of linear recurring sequences. *Number Theoretic and Algebraic Methods in Computer Science*, 185–192, 1993.

[687] S.W. Graham and G. Kolesnik. *Van der Corput's Method of Exponential Sums.* Cambridge University Press 126, Cambridge, 1991.

[688] M. Grandet-Hugot. Equirépartition dans les adèles. *C.R. Acad. Sci., Paris*, 280:873–876, 1975.

[689] M. Grandet-Hugot. Equirépartition dans les adèles. *Sémin. Delange-Pisot-Poitou, 16e Annee 1974/75, Theorie des Nombres*, Exp. 22:7 p., 1975.

[690] M. Grandet-Hugot. Quelque résultats concernant l'équirépartition dans l'anneau des adéles d'un corps de nombres algébriques. *Bull. Sci. Math.*, 99:91–111, 1975.

[691] M. Grandet-Hugot. Etude de différents types d'équirépartition dans un anneau d'adèles. *Bull. Sci. Math.*, 100:3–16, 1976.

[692] M. Grandet-Hugot. Quelques résultats concenrant l'équirépartition dans l'anneau des adèles d'un corps de nombres algébriques rectificatif. *Bull. Sci. Math.*, 99:243–247, 1976.

[693] M. Grandet-Hugot. Introduction aux exposés sur l'équirépartition. *Sémin. Delange-Pisot-Poitou, 18e Annee 1976/77, Theorie des Nombres*, Exp. 61:3 p., 1977.

[694] M. Grandet-Hugot. Quelques notions sur l'équirépartition dans les groupes abéliens compacts et localement compacts. *Sémin. Delange-Pisot-Poitou, 18e annee 1976/77, Theorie des Nombres*, Exp. 21:8 p., 1977.

[695] P. Grandits. C-Diskrepanz von Flächen im Raum. *Österr. Akad. Wiss. SB II*, 59–81, 1990.

[696] G. Grekos. Répartition des densités des sous-suites d'une suite d'entiers. *J. Number Theory*, 10:177–191, 1978.

[697] G. Grekos and B. Volkmann. On densities and gaps. *J. Number Theory*, 26:129–148, 1987.

[698] G. Griso. Equirépartition et zéros de la fonction zeta (d'après E. Hlawka). *Semin. Delange-Pisot-Poitou, 19e Annee 1977/78, Theor. des Nombres*, Exp. 23:7 p., 1978.

[699] S.A. Gritsenko. On a problem of I.M. Vinogradov. *Math. Notes*, 39:341–349, 1986.

[700] K. Gröchenig. Almost constant sequences and well-distribution in compact groups. *Monatsh. Math.*, 100:171–182, 1985.

[701] K. Gröchenig. Wesentliche Indexfolgen und gleichmäßige Gleichverteilung. *Österr. Akad. Wiss. SB II*, 195:225–238, 1986.

[702] K. Gröchenig, V. Losert, and H. Rindler. Separabilität, Gleichverteilung und Fastperiodizität. *Anz. Österr. Akad. Wiss.*, 117–119, 1984.

[703] K. Gröchenig, V. Losert, and H. Rindler. Uniform distribution in solvable groups. *Lect. Notes Math.*, 1210:97–107, 1986.

[704] H. Grothe. Matrix generators for pseudo-random vector generation. *Stat. Hefte*, 28:233–238, 1987.

[705] H. Grothe. *Matrixgeneratoren zur Erzeugung gleichverteilter Pseudozufallsvektoren*. Techn. Hochschule Darmstadt, 1988.

[706] V.S. Grozdanov. On the diophany of one class one-dimensional sequences. *Int. J. Math. Math. Sci.*, 19:115–124, 1996.

[707] P.M. Gruber and C.G. Lekkerkerker. *Geometry of Numbers*. North-Holland, Amsterdam, 1987.

[708] L.J. Guibas and A.M. Odlyzko. String overlaps, pattern matching, and nontransitive games. *J. Comb. Theory, Ser. A*, 30:183–208, 1981.

[709] G. Guralnik, Ch. Zemach, and T. Warnock. An algorithm for uniform random sampling of points in and on a hypersphere. *Inf. Process Lett.*, 21:17–21, 1985.

[710] R.E. Gut, V.V. Egorov, and V.N. Il'in. A method for generating normal pseudorandom numbers. *Zh. Vychisl. Mat. Mat. Fiz.*, 26:1254–1256, 1986.

[711] R.E. Gut, V.V. Egorov, and V.N. Il'in. A method of generating normal pseudorandom numbers. *U.S.S.R. Comput. Math. Math. Phys.*, 26:192–193, 1986.

[712] S. Haber. A number-theoretic problem in numerical approximation of integrals. *Approximation theory III*, 473–480, 1980.

[713] J.A. Haight. On multiples of certain real sequences. *Acta Arith.*, 49:303–306, 1988.

[714] I.J. Haland. Uniform distribution of generalized polynomials. *J. Number Theory*, 45:327–366, 1993.

[715] I.J. Haland. Uniform distribution of generalized polynomials of the product type. *Acta Arith.*, 67:13–27, 1994.

[716] I.J. Haland and D. Knuth. Polynomials involving the floor function. *Math. Scand.*, 76:194–200, 1995.

[717] G. Halász. Remarks on the ramainder in Birkhoff's ergodic theorem. *Acta Math. Hung.*, 28:389–395, 1976.

[718] G. Halász. On Roth's method in the theory of irregularities of point distribution. *Recent progress in analytic number theory*, 2:79–94, 1981.

[719] P. Hall. On representatives of subsets. *J. London Math. Soc.*, 10:26–30, 1935.

[720] R.R. Hall. The divisors of integers. *Acta Arith.*, 26:41–46, 1974.

[721] R.R. Hall. The divisors of integers. II. *Acta Arith.*, 28:129–135, 1975.

[722] R.R. Hall. Sums of imaginary powers of the divisors of integers. *J. London Math. Soc.*, 9:571–580, 1975.

[723] R.R. Hall. The distribution of $f(d)$ (mod 1). *Acta Arith.*, 31:91–97, 1976.

[724] R.R. Hall. A new definition of the density of an integer sequence. *J. Austral. Math. Soc. A*, 26:487–500, 1978.

[725] R.R. Hall. The divisor density of integer sequences. *J. London Math. Soc.*, 24:41–53, 1981.

[726] R.R. Hall and G. Tenenbaum. Les ensembles de multiples et la densité divisorielle. *J. Number Theory*, 22:308–333, 1986.

[727] P.R. Halmos. *Measure Theory*. Springer, New York, 1974.

[728] J. H. Halton. Pseudo-random trees: Multiple independent sequence generators for parallel and branching computations. *J. Comput. Phys.*, 84:1–56, 1989.

[729] J.H. Halton. On the efficiency of certain quasi-random sequences of points in evaluating multi-dimensional integrals. *Numer. Math.*, 2:84–90, 1960.
[730] J.M. Hammersley. Monte Carlo methods for solving multiple problems. *Ann. New York Acad. Sci.*, 86:844–874, 1960,.
[731] G. Hansel and J.-P. Troallic. Suites uniformément distribuées et suites bien distribuées and une approche combinatoire. *Quad. Ric.Sci.*, 109:101–110, 1981.
[732] G. Hansel and J.P. Troallic. Suites uniformément distribuées et fonctions faiblement presque-périodiques. *Bull. Soc. Math. Fr.*, 108:207–212, 1980.
[733] T. Hansen, G.L. Mullen, and H. Niederreiter. Good parameters for a class of node sets in quasi-Monte Carlo integration. *Math. Comput.*, 61:225–234, 1993.
[734] G.H. Hardy and J.E. Littlewood. Some problems of Diophantine approximation. III: The fractional part of $n^k\vartheta$. *Acta Math.*, 37:155–191, 1914.
[735] G.H. Hardy and J.E. Littlewood. Some problems of diophantine approximation: The lattice points of a right-angled triangle II. *Abh. Math. Semin. Univ. Hamburg*, 1:212–249, 1922.
[736] G. Harman. Trigonometric sums over primes. *Mathematika*, 28:249–254, 1981.
[737] G. Harman. Sums of distances between points of a sphere. *Int. J. Math. Math. Sci.*, 5:707–714, 1982.
[738] G. Harman. On the distribution of αp modulo one. *Mathematika*, 30:104–116, 1983.
[739] G. Harman. On the distribution of αp modulo one. *J. London Math. Soc.*, 27:9–18, 1983.
[740] G. Harman. Trigonometric sums over primes. *Glasgow Math. J.*, 24:23–37, 1983.
[741] G. Harman. Metric Diophantine approximation with two restricted variables, III. *J. Number Theory*, 29:364–375, 1988.
[742] G. Harman. Some theorems in the metric theory of Diophantine approximation. *Math. Proc. Cambridge Phil. Soc.*, 99:385–394, 1988.
[743] G. Harman. Metrical theorems on fractional parts of real sequences. *J. Reine Angew. Math.*, 396:192–211, 1989.
[744] G. Harman. Some cases of the Duffin and Schaeffer conjecture. *Q. J. Math., Oxford II. Ser.*, 41:395–404, 1990.
[745] G. Harman. Diophantine approximation by prime numbers. *J. London Math. Soc.*, 44:218–226, 1991.
[746] G. Harman. Fractional and integral parts of p^λ. *Acta Arith.*, 58:141–152, 1991.
[747] G. Harman. Metrical theorems on fractional parts of real sequences, II. *J. Number Theory*, 44:47–57, 1993.
[748] G. Harman. Small fractional parts of additive forms. *Philos. Trans. Roy. Sci. London Ser. A*, 345:327–338, 1993.
[749] G. Harman. Numbers badly approximable by fractions with prime denominator. *Math. Proc. Cambridge. Philos. Soc.*, 118:1–5, 1995.
[750] G. Harman. Small fractional parts of additive forms in prime variables. *Q. J. Math., Oxford II. Ser.*, 46:321–332, 1995.
[751] G. Harman. Towards an arithmetical analysis of the continuum. *London Math. Soc. Lect. Note Ser.*, 215:127–138, 1995.
[752] G. Harman and K. Matsumoto. Discrepancy estimates for the value-distribution of the Riemann zeta-function IV. *J. London Math. Soc.*, 50:17–24, 1994.
[753] F. Härtel. A random number generator for parallel processes. *Operations research proceedings 1994, Berlin: Springer Verlag*, 239–243, 1995.
[754] S. Hartman. Remarks on equidistributions on noncompact groups. *Comput. Math.*, 16:66–71, 1971.
[755] F. Haslinger. Newton'sche Interpolationspolynome und Gleichverteilung. *Lect. Notes Math.*, 1114:16–18, 1985.
[756] J. Hastad. Pseudorandom generators under uniform assumptions. *Proc. 22nd ACM Symp. on Theory of Computing*, 395–404, 1990.
[757] D.R. Heath-Brown. The fractional part of αn^k. *Mathematika*, 35:28–37, 1988.
[758] D.R. Heath-Brown. Weyl's inequality and Hua's inequality. *Lect. Notes Math.*, 1380:87–92, 1989.
[759] E. Hecke. Über analytische Funktionen und die Verteilung von Zahlen mod. Eins. *Abh. Math. Semin. Univ. Hamburg*, 1:54–76, 1922.

[760] H.A. Heilbronn. On the distribution of the sequence $n^2\theta$ (mod 1). *Q. J. Math, Oxford II Ser.*, 19:249–256, 1948.
[761] S. Heinrich. Efficient algorithms for computing the L_2-discrepancy. *Math. Comput.*, to appear, 1996.
[762] P. Hellekalek. On the rate of convergence of the L_n-discrepancy to the extreme discrepancy. *Monatsh. Math.*, 88:1–6, 1979.
[763] P. Hellekalek. On regularities of the distribution of special sequences. *Monatsh. Math.*, 90:291–295, 1980.
[764] P. Hellekalek. Regularities in the distribution of special sequences. *J. Number Theory*, 18:41–55, 1984.
[765] P. Hellekalek. Ergodicity of a class of cylinder flows related to irregularities of distribution. *Compos. Math.*, 61:129–136, 1987.
[766] P. Hellekalek. On the boundedness of Weyl sums. *Monatsh. Math.*, 114:199–208, 1992.
[767] P. Hellekalek. Weyl sums over irrational rotations. *Grazer Math. Ber.*, 318:29–44, 1992.
[768] P. Hellekalek. General discsrepancy estimates II: The Haar function system. *Acta Arith.*, 67:313–322, 1994.
[769] P. Hellekalek. General discsrepancy estimates: The Walsh function system. *Acta Arith.*, 67:209–218, 1994.
[770] P. Hellekalek. General discsrepancy estimates III: The Erdős-Turán-Koksma Inequality for the Haar function system. *Monatsh. Math.*, 120:25–45, 1995.
[771] P. Hellekalek and G. Larcher. On functions with bounded remainder. *Ann. Inst. Fourier*, 39:17–26, 1989.
[772] P. Hellekalek and G. Larcher. On Weyl sums and skew products over irrational rotations. *Theor. Comput. Sci.*, 65:189–196, 1989.
[773] G. Helmberg. *Theory of Uniform Distribution.* Manuscipt, Österr. Akad. Wiss.
[774] G. Helmberg. On subsequences of normal sequences. *Teubner-Texte*, 94:70–71, 1987.
[775] E. Hewitt and Y. Katznelson. Diophantine approximation in certain solenoids. *Bull. Greek Math. Soc.*, 18:157–175, 1977.
[776] E Hewitt and K.A. Ross. *Abstract Harmonic Analysis.* Springer, Berlin, 1963.
[777] T.P. Hill. Base-invariance implies Benford's law. *Proc. Am. Math. Soc.*, 1994.
[778] T.P. Hill. The significant-digit phenomenon. *Am. Math. Monthly*, 102:322–327, 1995.
[779] E. Hlawka. Statistik und Gleichverteilung. *preprint.*
[780] E. Hlawka. Integrale auf konvexen Körpern I. *Monatsh. Math.*, 54:1–36, 1950.
[781] E. Hlawka. Integrale auf konvexen Körpern II. *Monatsh. Math.*, 54:81–99, 1950.
[782] E. Hlawka. Zur formalen Theorie der Gleichverteilung in kompakten Gruppen. *Rend. Circ. Mat. Palermo*, 4:33–47, 1955.
[783] E. Hlawka. Folgen auf kompakten Räumen. *Abh. Math. Semin. Univ. Hamburg*, 20:223–241, 1956.
[784] E. Hlawka. Folgen auf kompakten Räumen. II. *Math. Nachr.*, 18:188–202, 1958.
[785] E. Hlawka. Über C-Gleichverteilung. *Ann. Mat. Pura. Appl.*, 49:311–326, 1960.
[786] E. Hlawka. Funktionen von beschränkter Variation in der Theorie der Gleichverteilung. *Ann. Mat. Pura Appl.*, 54:325–333, 1961.
[787] E. Hlawka. Zur angenäherten Berechnung mehrfacher Integrale. *Monatsh. Math.*, 66:140–151, 1962.
[788] E. Hlawka. Uniform distribution modulo 1 and numerical analysis. *Compos. Math.*, 16:92–105, 1964.
[789] E. Hlawka. Ein metrischer Satz in der Theorie der C-Gleichverteilung. *Monatsh. Math.*, 74:436–447, 1970.
[790] E. Hlawka. Mathematische Modelle der kinetischen Gastheorie. *Sympos. Math.*, 4:81–97, 1970.
[791] E. Hlawka. Discrepancy and Riemann integration. *Studies in Pure Mathematics*, Academic Press, New York:121–129, 1971.
[792] E. Hlawka. Ein metrisches Gegenstück zu einem Satz von W.A. Veech. *Monatsh. Math.*, 76:436–447, 1972.
[793] E. Hlawka. Über eine Methode von E. Hecke in der Theorie der Gleichverteilung. *Acta Arith.*, 24:11–31, 1973.

[794] E. Hlawka. Mathematische Modelle der kinetischen Gastheorie. *Rheinisch-Westfäl. Akad. Wiss., Vorträge*, 240:19, 1974.

[795] E. Hlawka. Anwendung zahlentheoretischer Methoden auf Probleme der numerischen Mathematik. *Österr. Akad. Wiss. SB II*, 184:217–225, 1975.

[796] E. Hlawka. Numerische analytische Fortsetzung in Polyzylindern. *Österr. Akad. Wiss. SB II*, 184:307–331, 1975.

[797] E. Hlawka. Zur quantitativen Theorie der Gleichverteilungen. *Österr. Akad. Wiss. SB II*, 184:355–365, 1975.

[798] E. Hlawka. Zur Theorie der Gleichverteilung. *Anz. Österr. Akad. Wiss.*, 13–14, 1975.

[799] E. Hlawka. Zur Theorie der Gleichverteilung. *Anz. Österr. Akad. Wiss.*, 23–24, 1975.

[800] E. Hlawka. Abschätzung von trigonometrischen Summen mittels diophantischer Approximationen. *Österr. Akad. Wiss. SB II*, 185:43–50, 1976.

[801] E. Hlawka. On some concepts, theorems and problems in the theory of uniform distribution. *Top. Number Theory*, 13:97–109, 1976.

[802] E. Hlawka. Über die Gleichverteilung gewisser Folgen, welche mit den Nullstellen der Zetafunktion zusammenhängen. *Österr. Akad. Wiss. SB II*, 184:459–471, 1976.

[803] E. Hlawka. Der Approximationssatz von Weierstraß und die Theorie der gleichverteilten Folgen. *Monatsh. Math.*, 88:137–170, 1979.

[804] E. Hlawka. *Theorie der Gleichverteilung*. Bibliographisches Inst., Mannheim, Wien, 1979.

[805] E. Hlawka. Approximation von Irrationalzahlen und pythagoräische Tripel. *Bonn. Math. Schr.*, 121:1–32, 1980.

[806] E. Hlawka. Gleichverteilung und mathematische Linguistik. *Österr. Akad. Wiss. SB II*, 189:209–248, 1980.

[807] E. Hlawka. Gleichverteilung und Quadratwurzelschnecke. *Monatsh. Math.*, 89:19–44, 1980.

[808] E. Hlawka. Über einige Reihen, welche mit den Vielfachen von Irrationalzahlen zusammenhängen. *Acta Arith.*, 37:285–306, 1980.

[809] E. Hlawka. Über einige Sätze, Begriffe und Probleme in der Theorie der Gleichverteilung. *Österr. Akad. Wiss. SB II*, 189:437–490, 1980.

[810] E. Hlawka. Über einige Reihen, die mit den Vielfachen von Irrationalzahlen zusammenhängen. *Österr. Akad. Wiss. SB II*, 190:33–61, 1981.

[811] E. Hlawka. Gleichverteilung auf Produkten von Sphären. *J. Reine Angew. Math.*, 330:1–43, 1982.

[812] E. Hlawka. Gleichverteilung und das Konvergenzverhalten von Potenzreihen am Rande des Konvergenzkreises. *Manuscr. Math.*, 44:231–263, 1983.

[813] E. Hlawka. Lineare Differenzengleichungen in mehrern Variablen. *J. Reine. Angew. Math.*, 339:166–178, 1983.

[814] E. Hlawka. Näherungslösungen der Wellengleichung und verwandter Gleichungen durch zahlentheoretische Methoden. *Österr. Akad. Wiss. SB II*, 193:359–442, 1984.

[815] E. Hlawka. *The theory of uniform distribution*. Academic Publishers, Berkhamsted, Herts.; England, 1984.

[816] E. Hlawka. Bemerkung zum Lemma von Du Bois-Reymond I. *Lect. Notes Math.*, 1114:26–29, 1985.

[817] E. Hlawka. Bemerkung zum Lemma von Du Bois-Reymond II. *Lect. Notes Math.*, 1114:30–39, 1985.

[818] E. Hlawka. Gleichverteilung und ein Satz von Müntz. *J. Number Theory*, 24:35–46, 1986.

[819] E. Hlawka. Über die direkten Methoden der Variationsrechnung und Gleichverteilung (on direct methods of the calculus of variations and uniform distribution). *Lect. Notes Math.*, 1262:50–85, 1987.

[820] E. Hlawka. Beiträge zur Theorie der Gleichverteilung und ihre Anwendungen IV: Der sphärische Fall, die einpunktig kompaktifizierten $\mathbf{R}^n$ und $\mathbf{C}^n$, der reelle und komplexe projektive Raum. *Österr. Akad. Wiss. SB II*, 197:209–259, 1988.

[821] E. Hlawka. Beiträge zur Theorie der Gleichverteilung und ihre Anwendungen V: Der Fall der unitären Gruppe. *Österr. Akad. Wiss. SB II*, 197:261–289, 1988.

[822] E. Hlawka. Beiträge zur Theorie der Gleichverteilung und ihren Anwendungen I. Einleitung. *Österr. Akad. Wiss. SB II*, 197:1–94, 1988.

[823] E. Hlawka. Beiträge zur Theorie der Gleichverteilung und ihren Anwendungen II. Der euklidische Fall. *Österr. Akad. Wiss. SB II*, 197:95–120, 1988.

[824] E. Hlawka. Beiträge zur Theorie der Gleichverteilung und ihren Anwendungen III. Der nichteuklidische Fall. *Österr. Akad. Wiss. SB II*, 197:121–154, 1988.
[825] E. Hlawka. Eine Anwendung diophantischer Approximationen auf die Theorie von Differentialgleichungen (an application of diophantine approximation to the theory of differential equations). *Aequationes Math.*, 35:232–253, 1988.
[826] E. Hlawka. Zur Radontransformation. *Österr. Akad. Wiss. SB II.*, 198:331–379, 1989.
[827] E. Hlawka. Korrekturen und Bemerkungen zur Theorie der Gleichverteilung und ihren Anwendungen. I-V. *Österr. Akad. Wiss. SB II*, 199:193, 1990.
[828] E. Hlawka. Näherungsformeln zur Berechnung von mehrfachen Integralen mit Anwendungen auf die Berechnungen von Potentialen, Induktionskoeffizienten und Lösungen von Gleichungssystemen. *Lect. Notes Math.*, 1452:65–111, 1990.
[829] E. Hlawka. *Selecta.* Springer-Verlag, Berlin, 1990.
[830] E. Hlawka. Über eine Klasse von gleichverteilten Folgen. *Acta Arith.*, 53:389–402, 1990.
[831] E. Hlawka. Buffons Nadelproblem und verwandte Probleme behandelt mit der Theorie der Gleichverteilung II. *Geom. Dedicata*, 44:105–110, 1992.
[832] E. Hlawka. Buffons Nadelproblem und verwandte Probleme behandelt mit der Theorie der Gleichverteilung. *J. Number Theory*, 43:93–108, 1993.
[833] E. Hlawka. Gleichverteilung — Entropie. *Expo. Math.*, 11:3–46, 1993.
[834] E. Hlawka. Radontransformation III. *Österr. Akad. Wiss. SB II.*, 202:95–106, 1993.
[835] E. Hlawka. Zur Radontransformation II. *Österr. Akad. Wiss. SB II.*, 202:59–88, 1993.
[836] E. Hlawka. Die Entwicklung der Theorie der Gleichverteilung von Hermann Weyl. *Versl. Gewone Vergad. Afd. Natuurkd.*, 103:85–92, 1994.
[837] E. Hlawka and C. Binder. Über die Entwicklung der Theorie der Gleichverteilung in den Jahren 1909 bis 1916. *Arch. Hist. Exact. Sci.*, 36:197–240, 1986.
[838] E. Hlawka, F.J. Firneis, and P. Zinterhof. Zahlentheoretische Methoden in der numerischen Mathematik - Probleme und Methoden. *Schriftenr. Österr. Comput. Ges.*, 12:9–126, 1981.
[839] E. Hlawka and R. Mück. A transformation of equidistributed sequences. *Appl. Number Theory*, 371–388, 1972.
[840] E. Hlawka and R. Mück. Über eine Transformation von gleichverteilten Folgen. *Computing*, 9:127–138, 1972.
[841] P.J. Holewijn. *Contirbutions to the theory of asymptotic distribution mod 1.* 1965.
[842] J.J. Holt. On a form of the Erdős-Turán inequality. *Acta Arith.*, 74:61–66, 1996.
[843] J.J. Holt and J.D. Vaaler. The Beurling-Selberg extremal functions for a ball in Euclidean space. *Atti Accad. Naz. Lincei*, to appear.
[844] J. Horbowicz. An asymptotic relation between the extreme discrepancy and the L_p-discrepancy. *Monatsh. Math.*, 90:297–301, 1980.
[845] J. Horbowicz. Criteria for uniform distribution. *Indag. Math.*, 43:301–307, 1981.
[846] J. Horbowicz. On well distribution modulo 1 and systems of numeration. *Acta Arith.*, 48:53–62, 1987.
[847] J. Horbowicz and H. Niederreiter. Weighted exponential sums and discrepancy. *Acta Math. Hung.*, 54:89–97, 1989.
[848] W. Hörmann. A note on the quality of Random Variates generated by the ratio of uniforms method. *ACM Trans. Model. Comput. Simulation*, 4:96–106, 1994.
[849] B. Host. Nombres normaux, entropie, translations. *Isr. J. Math.*, 91:419–428, 1995.
[850] V. Houndonougbo. Mesure de répartition d'une suite dans un corps de séries formelles sur un corps fini. *C.R. Acad. Sci., Paris*, 288:997–999, 1979.
[851] L.-K. Hua and Y. Wang. On uniform distribution and numerical analysis. *Sci. Sinica*, 16:184–198, 1975.
[852] L.-K. Hua and Y. Wang. *Applications of number theory to numerical analysis.* Springer-Verlag, Berlin, 1981.
[853] L.-K. Hua and Y. Wang. Applications of number theory to numerical analysis. *Recent progress in analytic number theory*, 2:111–118, 1981.
[854] K. Huber. On the period length of generalized inversive pseudorandom generators. *Appl. Algebra Eng. Comm. Comput.*, 5:255–260, 1994.
[855] P. Hubert. Dynamique symbolique des billards polygonaux rationnels. *These, Univ. Marseille II*, 1995.
[856] M.N. Huxley. The fractional parts of smooth sequence. *Mathematika*, 35:292–296, 1988.

[857] H.-K. Hwang. *Théorèmes limites pour les structures combinatoire et les functions arithmétiques*. L'École Polytechnique, 1994.
[858] J.S. Hwang. A problem on continuous and periodic functions. *Pac. J. Math.*, 117:143–147, 1985.
[859] K.-H. Indlekofer, A. Jarai, and I. Katai. On some properties of attractors generated by iterated function systems. *Acta Sci. Math.*, 60:411–427, 1995.
[860] J. Isbell and St. Schanuel. On the fractional parts of n/j, $j = o(n)$. *Proc. Am. Math. Soc.*, 60:65–67, 1977.
[861] S.A. Ismatullaev. Bemerkung zu einem Satz von A.G. Postnikov. *Izv. Akad. Nauk*, 19:74–75, 1975.
[862] V.I. Israilov. On applications of number-theoretical methods in the theory of cubature formulae. *Vopr. Vychisl. Prikl. Mat.*, 65:135–148, 1981.
[863] K. Ito and H.P. McKean. *Diffusion Processes and Their Sample Paths*. Springer, New York, 1975.
[864] S. Ito. Some skew product transformations associated with continued fractions and their invariant measure. *Tokyo J. Math.*, 9:115–133, 1986.
[865] S. Ito and I. Shiokawa. A construction of β-normal sequences. *J. Math. Soc.*, 27:20–23, 1975.
[866] S. Ito and S. Yasumoti. On continued fractions, substitutions and characteristic sequences $[nx+y]-[(n-1)x+y]$. *Japan J. Math.*, 16:287–306, 1990.
[867] K. Jacobs. Gleichverteilung mod 1. *Selecta Math.*, 4:57–93, 1972.
[868] E. Jacobson and W.Y. Velez. Uniform and f-uniform distribution of recurrence sequences over dedekind domains. *J. Sichuan Univ., Nat. Sci. Ed.*, 26:98–103, 1989.
[869] H. Jager and J. De Jonge. The circular dispersion spectrum. *J. Number Theory*, 49:360–384, 1994.
[870] H. Jager and P. Liardet. Distribution arithmétiques des dénominateurs de convergents de fractions continues. *Indag. Math.*, 50:181–197, 1988.
[871] M. Jäger. The Hausdorff-dimension of the boundary of a unit-interval of a number system. *Ann. Univ. Sci. Budap. Rolando Eötvös, Sect. Comput.*, 14:79–90, 1994.
[872] T. Jech. The logarithmic distribution of leading digits and finitely additive measures. *Discrete Math.*, 108:53–57, 1992.
[873] M. R. Jerrum, L. G. Valiant, and V. V. Vazirani. Random generation of combinatorial structures from a uniform distribution. *Theor. Comput. Sci.*, 43:169–188, 1986.
[874] G. Ji and H. Lu. On dispersion and Markov Constants. *Monatsh. Math.*, 121:69–77, 1996.
[875] C.H. Jia. On the distribution of αp modulo one. *J. Number Theory*, 45:241–253, 1993.
[876] Ch. R. Johnson and M. Newman. Triangles generated by powers of triplets on the unit circle. *J. Res. nat. Bur. Standards*, 77:137–141, 1973.
[877] R. Jones, J. Olsen, and M. Wierdl. Subsequence ergodic theorems for l^p-concentrations. *Trans. Am. Math. Soc.*, 331:837–850, 1992.
[878] A.A. Judin. Über das Maß großer Absolutbeträge einer trigonometrischen Summe. *Teoretikocisl. Issled, Spektru Markov*, 163–171, 1973.
[879] A. Del Junco and J. M. Steele. Growth rates for monotone subsequences. *Proc. Am. Math. Soc.*, 71:179–182, 1978.
[880] J. Kaczorowski. The k-functions in multiplicative number theory, I. On complex explicit formulæ. *Acta Arith.*, 56:195–211, 1990.
[881] J. Kaczorowski. The k-functions in multiplicative number theory, II. Uniform distribution of zeta zeros. *Acta Arith.*, 56:213–224, 1990.
[882] J. Kaczorowski. The k-functions in multiplicative number theory, III. Uniform distribution of zeta zeros; discrepancy. *Acta Arith.*, 57:199–210, 1991.
[883] J. Kaczorowski. The k-functions in multiplicative number theory, IV. On a method of A. E. Ingham. *Acta Arith.*, 57:231–244, 1991.
[884] J.-P. Kahane. Sur les coefficients de Fourier-Bohr. *Studia Math.*, 21:103–106, 1961.
[885] J.-P. Kahane. Sur une note de Charles Pisot. *Acta Arith.*, 34:267–272, 1979.
[886] S. Kakutani. Ergodic theory of shift transformations. *Proceedings of the 5th Berkely symposium on mathematical statistics*, 2:405–414, 1967.
[887] S. Kakutani. Strictly ergodic dynamical systems. *Proceedings of the 6th Berkely symposium on mathematical statistics*, 2:319–326, 1972.
[888] M.H. Kalos and P. A. Whitlock. *Monte Carlo Methods*. Wiley-Interscience, New York, 1986.

[889] T. Kamae. Subsequences of normal sequences. *Isr. J. Math.*, 16:121–149, 1973.

[890] T. Kamae. Normal numbers and ergodic theory. *Lect. Notes Math.*, 550:253–269, 1976.

[891] T. Kamae. Sum of digits to different bases and mutual singularity of their spectral measures. *Osaka J. Math.*, 15:569–574, 1978.

[892] T. Kamae and M. Mendès France. Van der Corput's difference theorem. *Isr. J. Math.*, 31:335–342, 1978.

[893] T. Kamae and B. Weiss. Normal numbers and selection rules. *Isr. J. Math.*, 21:101–110, 1975.

[894] S. Kanemitsu, K. Nagasaka, G. Rauzy, and J.-S. Shiue. On Benford's law: the first digit problem. *Lect. Notes Math.*, 1299:158–169, 1988.

[895] H. Kano. A remark on Wagner's ring of normal numbers. *Arch. Math.*, 60:46–50, 1993.

[896] H. Kano. General constructions of normal numbers of Korobov type. *Osaka J. Math.*, 30:909–919, 1993.

[897] H. Kano and I. Shiokawa. Rings of normal and nonnormal numbers. *Isr. J. Math.*, 84:403–416, 1993.

[898] T. Kano. Sur la théorie de l'équirépartition mod 1. *Actes Symp. Math.*, 3:37–54, 1981.

[899] T. Kano. A counterexample to a conjecture of Erdős in the theory of uniform distribution. *Surikaisekikenkyusho Kokyuroku*, 86:187–195, 1994.

[900] A.A.. Karatsuba. On a diophatine inequatlity (Russian). *Acta Arith.*, 53:309–324, 1989.

[901] A.A. Karatsuba. Distribution of inverse values in a residue ring modulo a given number. *Dokl. Akad. Nauk. SSSR*, 333:138–139, 1993.

[902] A.A. Karatsuba. The distribution of inverses in a residue ring modulo a given modulus. *Russ. Acad. Sci. Dokl. Math.*, 48:452–454, 1994.

[903] A.A. Karatsuba. Analogues of Kloosterman sums. *Izv. Ross. Akad. Nauk Ser. Mat.*, 59:93–102, 1995.

[904] A.A. Karatsuba and I.M. Vinogradov. The method of trigonogmetric sums in number theory. *Proc. Steklov Inst. Math.*, 168:3–30, 1984.

[905] S. Karlin and W. J. Studden. *Tchebycheff systems: with applications in analysis and statistics.* Interscience, New York, 1966.

[906] G. Károlyi. Geometric discrepancy theorems in higher dimensions. *Stud. Sci. Math. Hung.*, 30:59–94, 1995.

[907] G. Károlyi. Irregularities of point distributions relative to homothetic convex bodies. *Monatsh. Math.*, 120:247–279, 1995.

[908] I. Kátai. Distribution mod 1 of additive functions on the set of divisors. *Acta Arith.*, 30:209–212, 1976.

[909] I. Kátai. A remark on a theorem of H. Daboussi. *Acta Math. Hung.*, 47:223–225, 1986.

[910] I. Kátai. Uniform distribution of sequences connected with arithmetical functions. *Acta Math. Hung.*, 51:401–408, 1988.

[911] I. Kátai. Distribution of q-additive function. *Probability theory and appl.*, 309–318, 1992.

[912] N.M. Katz. *Gauss sums, Kloosterman sums and monodrony groups.* Ann. Math. Stud., 116, Princeton, 1988.

[913] T. M. Katz and D. I.A. Cohen. The first digit property for exponential sequences is independent of the underlying distribution. *Fibonacci Q.*, 24:2–7, 1986.

[914] R.M. Kaufman. On the distribution of $\{\sqrt{p}\}$. *Mat. Zametki*, 26:497–504, 1979.

[915] M. Keane and G. Rauzy. Stricte ergodicité des échanges d'intervalles. *Math. Z.*, 62:203–212, 1980.

[916] M. Keane, M. Smorodinsky, and B. Solomyak. On the morphology of γ-expansions with deleted digits. *Tans. Am. Math. Soc.*, 347:955–966, 1995.

[917] A. Keller. A quasi-Monte-Carlo algorithm for the global illumination problem in the radiosity setting. *Lect. Notes Stat.*, 106:239–251, 1995.

[918] J.H.B. Kemperman. Distributions modulo 1 of slowly changing sequences. *Nieuw. Arch. Wiskunde*, 21:138–163, 1973.

[919] J. Kerstan and K. Matthes. Gleichverteilungseigenschaften von Faltungen von Verteilungsgesetzen auf lokalkompakten abelschen Gruppen. *Math. Nachr.*, 37:267–312, 1968.

[920] H. Kesten. On a conjecture or Erdős and Szüsz related to uniform distribution mod 1. *Acta Arith.*, 12:193–212, 1966/67.

[921] A. Khintchine. Ein Satz über Kettenbrüche mit arithmetischen Anwendungen . *Math. Z.*, 18:289–306, 1923.

[922] A. Khintchine. Einige Sätze über Kettenbrüche, mit Anwendungen auf die Theorie der Diophantischen Approximationen. *Math. Ann.*, 92:115–125, 1924.

[923] H. Ki and T. Linton. Normal numbers and subsets of n with given densities. *Fund. Math.*, 144:163–179, 1994.

[924] P. Kirschenhofer and R.F. Tichy. Gleichverteilung und formale Sprachen. *Österr. Akad. Wiss. SB II*, 189:291–319, 1980.

[925] P. Kirschenhofer and R.F. Tichy. On uniform distribution of double sequences. *Manuscr. Math.*, 35:195–207, 1981.

[926] P. Kirschenhofer and R.F. Tichy. Über Eigenschaften und Anwendungen der s-Diskrepanz. *Österr. Akad. Wiss. SB II*, 190:195–205, 1981.

[927] P. Kirschenhofer and R.F. Tichy. Gleichverteilung in diskreten Räumen (uniform distribution on discrete spaces). *Lect. Notes Math.*, 1114:66–76, 1985.

[928] P. Kirschenhofer and R.F. Tichy. Some distribution properties of $0,1$-sequences. *Manuscr. Math.*, 54:205–219, 1985.

[929] P. Kirschenhofer and R.F. Tichy. Zur Diskrepanz von 0,1-Folgen (discrepancy of 0,1-sequences). *J. Number Theory*, 21:156–175, 1985.

[930] P. Kiss. Note on distribution of the sequences $n\theta$ modulo a linear recurrences. *Discuss. Math.*, 7:135–139, 1985.

[931] P. Kiss. A distribution property of second-order linear recurrences. *Fibonacci Numbers and Their Applications*, Math. Appl., D. Reidel Publ. Co., 28:121–130, 1986.

[932] P. Kiss. Results on the ratios of the terms of second order linear recurrences. *Math. Slovaca*, 41:257–260, 1991.

[933] P. Kiss and Zs. Sinka. On the ratios of the terms of second order linear recurrences. *Period. Math. Hung.*, 23:139–143, 1991.

[934] P. Kiss and R.F. Tichy. Distribution of the ratios of the terms of a second order linear recurrences. *Indag. Math.*, 48:79–86, 1986.

[935] P. Kiss and R.F. Tichy. A discrepancy problem with applications to linear recurrences. *Proc. Japan Acad. A*, 65:135–138, 1989.

[936] P. Kiss and R.F. Tichy. On asymptotic distribution modulo a subdivision. *Publ. Math. Debrecen*, 37:186–191, 1990.

[937] B. Klinger. Numerical integration of singular integrands using low-discrepancy sequences. *preprint*, 1996.

[938] B. Klinger and R.F. Tichy. Polynomial discrepancy of point sequences. *preprint*, 1996.

[939] M.J. Knight and W. A. Webb. Uniform distribution of third order linear recurrence sequences. *Acta Arith.*, 36:1980, 1980.

[940] D.E. Knuth. *The Art of Computer Programming, Vol. 2, Seminumerical Algorithms, 2nd ed.* Addison Wesley, Reading, 1981.

[941] J.F. Koksma. *Diophantische Approximationen*. Springer, Berlin, 1936.

[942] J.F. Koksma. Some theoremson diophantine inequalities. *Math. Centrum Amsterdam*, Scriptum no. 5, 1950.

[943] A.N. Kolmogorov. Three approaches to the definition of the concept "quantity of information". *Prob. Peredachi Inf.*, 1:3–11, 1965.

[944] T. Komatsu. The fractional part of $(n\theta+\phi)$ and Beatty sequences. *J. Theor. Nombres Bordx.*, 7:387–406, 1995.

[945] T. Komatsu. On the characteristic word of the inhomogeneous Beatty sequences. *Bull. Austral. Math. Soc.*, 51:337–351, 1995.

[946] T. Komatsu. A certain power series associated with a Beatty sequence. *Acta Arith.*, 76:109–129, 1996.

[947] A.G. Konheim. Mantissa distribution. *Math. Comput.*, 19:143–144, 1965.

[948] R.J. Kooman. *Convergence properties of recurrence sequences*. Univ. Leiden, 1989.

[949] R.J. Kooman and R. Tijdeman. Convergence properties of linear recurrence sequences. *Nieuw Arch. Wiskd. IV*, 8:13–25, 1990.

[950] N. Kopecek, G. Larcher, R.F. Tichy, and G. Turnwald. On the discrepancy of sequences associated with the sum-of-digits function. *Ann. Inst. Fourier*, 37:1–17, 1987.

[951] H.G. Kopetzky. Über natürliche Zahlen n mit der Eigenschaft $([\alpha n],[\beta n]) = 1$. *Acta Arith.*, 35:345–352, 1979.

[952] H.G. Kopetzky and F.J. Schnitzer. On the dispersion spectrum. *Monatsh. Math.*, 112:115–124, 1991.

[953] I. Korec. The $3x+1$ problem, generalized Pascal triangles and cellular automata. *Math. Slovaca*, 42:547–563, 1992.

[954] I. Korec. A density estimate for the $3x+1$ problem. *Math. Slovaca*, 44:85–89, 1994.

[955] I. Kőrnyei. On a theorem of Pisot. *Publ. Math. Debrecen*, 34:169–179, 1987.

[956] A.N. Korobov. Continued fractions of some normal numbers. *Mat. Zametiki*, 47:28–33, 1990.

[957] N. M Korobov. The approximate computation of multiple integrals. *Dokl. Akad. Nauk SSSR*, 124:1207–1210, 1959. (in Russian).

[958] N.M. Korobov. *Number-theoretical methods in approximate analyis (Russian)*. Fizmatgiz, Moscow, 1963.

[959] N.M. Korobov. Multiple trigonometric sums. *Vestn. Mosk. Univ., Ser. I*, 22–29, 1981.

[960] N.M. Korobov. Estimates of trigonometric sums and character sums. *Diophantine approximations*, 1:42–47, 1985.

[961] N.M. Korobov. *Trigonometric sums and their applications*. Nauka, Moskva, 1989.

[962] N.M. Korobov. Quadrature formulas with combined grids. *Math. Notes*, 55:159–164, 1994.

[963] N.M Korobov and D.A. Mit'kin. Lower bounds of complete trigonometrical sums. *Vestn. Mosk. Univ., Ser I*, 54–57, 1977.

[964] E.I. Kovalevskaja. Metric theorems on the approximation of zero by a linear combination of polynomials with integral coefficients. *Acta Arith.*, 25:93–104, 1973.

[965] A.H. Kruse. Estimates of $\sum_{k=1}^{k} k^{-s}\langle kx\rangle^{-t}$. *Acta Arith.*, 12:229–261, 1967.

[966] J. Kubilius. Probabilistic methods in the theory of distribution of arithmetic functions. *Aktual'nye probl. analit. Teor. Cisel*, 81–118, 1974.

[967] A. Kuijlaars. The minimal number of nodes in Chebyshev type quadrature formulas. Math. preprint series 14, Dep. of Math, Univ. of Amsterdam, 1992.

[968] A. Kuijlaars. Chebychev-type quadrature for analytic weights on the circle and the interval. *Indag. Math.*, 6:419–432, 1995.

[969] L. Kuipers. *De asymptotische verdeling modulo 1 van de waarden van meetbare functies*. Vrije Univ. Amsterdam, 1947.

[970] L. Kuipers. On real periodic functions and functions with periodic derivatives. *Indag. Math.*, 12:34–40, 1950.

[971] L. Kuipers. Continuous distribution modulo 1. *Nieuw Arch. voor Wisk.*, 10:87–82, 1962.

[972] L. Kuipers. Uniform distribution of sequences of integers. *Math. Colloqu. Univ. Cape Town*, 6:15–19, 1971.

[973] L. Kuipers. Uniform distribution of sequences of polynomials in $GF[p^r, x]$. *Math. Colloqu. Univ. Cape Town*, 6:20–25, 1971.

[974] L. Kuipers. A remark on a theorem of L. Carlitz. *Mat. Vesnik*, 9:113–116, 1972.

[975] L. Kuipers. A remark on asymptotic distribution in $GF[p^r, x]$. *Rev. Roum. Math. Pures Appl.*, 18:1217–1221, 1973.

[976] L. Kuipers. Letter to the editor. Generalized Fibonacci numbers and uniform distribution mod 1. *Fibonacci Q.*, 14:214, 253, 276, 281, 1976.

[977] L. Kuipers. Einige Bemerkungen zu einer Arbeit von G.J. Rieger. *Elem. Math.*, 34:32–34, 1979.

[978] L. Kuipers. A property of the Fibonacci sequence (F_m). *Fibonacci Q.*, 20:112–113, 1982.

[979] L. Kuipers and B. Meulenbeld. Asymptotic C-distribution. I. *Indag. Math.*, 11:425–431, 1949.

[980] L. Kuipers and B. Meulenbeld. Asymptotic C-distribution. II. *Indag. Math.*, 11:432–437, 1949.

[981] L. Kuipers and B. Meulenbeld. Some theorems in the theory of uniform distribution. *Indag. Math.*, 12:53–56, 1949.

[982] L. Kuipers and H. Niederreiter. Asymptotic distribution mod m and independence of sequences of integers. *Proc. Japan Acad. A*, 50:256–260, 261–265, 1974.

[983] L. Kuipers and H. Niederreiter. *Uniform distribution of sequences*. Wiley-Interscience Publ., 1974.

[984] L. Kuipers, H. Niederreiter, and J. Shiue. Uniform distribution of sequences in the ring of Gaussian integers. *Bull. Inst. Math., Acad. Sin.*, 3:311–325, 1975.

[985] L. Kuipers and J.-S. Shiue. On the distribution modulo m of sequences of generalized Fibonacci numbers. *Tamkang J. Math.*, 2:181–186, 1971.
[986] L. Kuipers and J.-S. Shiue. A distribution property of a linear recurrence of the second order. *Atti Accad. Naz. Lincei*, 52:6–10, 1972.
[987] L. Kuipers and J.-S. Shiue. A distribution property of the sequence of Fibonacci numbers. *Fibonacci Q.*, 10:375–376, 1972.
[988] L. Kuipers and J.-S. Shiue. A distribution property of the sequence of Lucas numbers. *Elem. Math.*, 27:10–11, 1972.
[989] L. Kuipers and J.-S. Shiue. Remark on a paper by Duncan and Brown on the sequence of logarithms of certain recursive sequences. *Fibonacci Q.*, 11:292–294, 1973.
[990] L. Kuipers and J.-S. Shiue. Asymptotic distribution modulo m of sequences of integers and the notion of independence. *Atti Accad. Naz. Lincei*, VIII, Ser. 11:63–90, 1974.
[991] L. Kuipers and J.-S. Shiue. Ein Gleichverteilungskriterium. *Elem. Math.*, 31:137–139, 1976.
[992] L. Kuipers and J.S. Shiue. On a criterion for uniform distribution of a sequence in the ring of Gaussian integers. *Rev. Roum. Math. Pures Appl.*, 25:1059–1063, 1980.
[993] L. Kuipers and J.S. Shiue. Uniform distribution of sequences in rings of integral quaternions. *Glasgow Math. J.*, 23:21–29, 1982.
[994] L. Kuipers and J.S. Shiue. On the L^p-discrepancy of certain sequences. *Fibonacci Q.*, 26:157–162, 1988.
[995] M.F. Kulikova. Numerical treatment of the problem of construction of Bernoulli-normal sequences of characters. *Stud. in Number Theory*, 6:100–108, 1975.
[996] S. Kunoff. $N!$ has the first digit property. *Fibonacci Q.*, 25:365–367, 1987.
[997] V.F. Kuzyutin. On errors of approximate integration formulas on some classes of periodic functions. *Primen Funkts. Anal. Teor. Priblizh.*, 113–117, 1982.
[998] Y. Lacroix. Metric properties of generalized Cantor products. *Acta Arith.*, 63:61–77, 1993.
[999] Y. Lacroix and A. Thomas. Number systems and repartition. *J. Number Theory*, 49:308–318, 1994.
[1000] M. Laczkovich. Equidecomposibility and discrepancy; a solution of Tarski's cirle-squaring problem. *J. Reine Angew. Math.*, 404:77–117, 1990.
[1001] M. Laczkovich. Decomposition of sets with small boundary. *J. London Math. Soc.*, 46:58–64, 1992.
[1002] M. Laczkovich. Uniformly spread discrete sets in $\mathbf{R}^d$. *J. London Math. Soc.*, 46:39–57, 1992.
[1003] M. Laczkovich. Discrepancy estimates for sets with small boundary. *Stud. Sci. Math. Hung.*, 30:105–109, 1995.
[1004] J. C. Lagarias. The $3x + 1$ problem and its generalizations. *Am. Math. Monthly*, 92:3–21, 1985.
[1005] J.C. Lagarias. Pseudorandom number generators in cryptography and number theory. *Proc. of Symposia in Appl. Math.*, 42:115–143, 1990.
[1006] J.C. Lagarias. Pseudorandom numbers. *Probability and Algorithms*, 65–85, 1992.
[1007] J.P. Lambert. Quasi-Monte Carlo, low discrepancy sequences, and ergodic transformations. *J. Comput. Appl. Math.*, 12/13:419–423, 1985.
[1008] J.P. Lambert. Quasi-random sequences for optimization and numerical integration. *Numerical integration. Recent developments, software and applications*, Ser. C 203:193–203, 1987.
[1009] J.P. Lambert. Quasi-random sequences in numerical practice. *Numer. Math.*, 86:273–284, 1988.
[1010] J.P. Lambert. A sequence well dispersed in the unit square. *Proc. Am. Math. Soc.*, 103:383–388, 1988.
[1011] B. Lapeyre and G. Pagès. Familles de suites à discrépance faible obtenues par itération de transformations de $[0,1]$. *C.R. Acad. Sci., Paris*, 308:507–509, 1989.
[1012] B. Lapeyre, G. Pages, and K. Sab. Sequences with low discrepancy generalisation and application to Robbins-Monro algorithm. *Statistics*, 21:251–272, 1990.
[1013] G. Larcher. Metric results on the approximation of zero by linear combinations of independent and of dependent rationals. *preprint*.
[1014] G. Larcher. The continuous irrational rotation: Irregularities of distribution. *preprint*.
[1015] G. Larcher. *Numerische analytische Fortsetzung von Funktionen*. Univ. Salzburg, 1982.
[1016] G. Larcher. *Numerische analytische Fortsetzung von Funktionen. Arbeitsberichte des Instituts für Mathematik 4/82*. Univ. Salzburg, 1982.

[1017] G. Larcher. Optimale Koeffizienten bezüglich zusammengesetzter Zahlen. *Monatsh. Math.*, 100:127–135, 1985.

[1018] G. Larcher. On the distribution of sequences connected with good lattice points. *Monatsh. Math.*, 101:135–150, 1986.

[1019] G. Larcher. Quantitative rearrangement theorems. *Compos. Math.*, 60:251–259, 1986.

[1020] G. Larcher. Über die isotrope Diskrepanz von Folgen. *Arch. Math.*, 46:240–249, 1986.

[1021] G. Larcher. A best lower bound for good lattice points. *Monatsh. Math.*, 104:45–51, 1987.

[1022] G. Larcher. A new extremal property of the Fibonacci ratio. *Fibonacci Q.*, 26:247–255, 1988.

[1023] G. Larcher. On the distribution of s-dimensional Kronecker-sequences. *Acta Arith.*, 51:335–347, 1988.

[1024] G. Larcher. On the distribution of sequences connected with digit-representation. *Manuscr. Math.*, 61:33–42, 1988.

[1025] G. Larcher. On the distribution of the multiples of an s-tupel of real numbers. *J. Number Theory*, 31:367–372, 1989.

[1026] G. Larcher. An inequality with applications in diophantine approximation. *Lect. Notes Math.*, 1452:132–138, 1990.

[1027] G. Larcher. Bemerkung zur Diskrepanz einer Klasse von Folgen. *Anz. Österr. Akad. Wiss.*, 1:1–5, 1991.

[1028] G. Larcher. Corrigendum to the paper "On the distribution of s-dimensional Kronecker sequences". *Acta Arith.*, 60:93–95, 1991.

[1029] G. Larcher. On the cube-discrepancy of Kronecker-sequences. *Arch. Math.*, 57:362–369, 1991.

[1030] G. Larcher. A class of low-discrepancy point-sets and its application to numerical integration by number-theoretical methods. *Grazer Math. Ber.*, 318:69–80, 1993.

[1031] G. Larcher. Nets obtained from rational functions over finite fields. *Acta Arith.*, 63:1–13, 1993.

[1032] G. Larcher. Zur Diskrepanz verallgemeinerter Ziffernsummenfolgen. *Österr. Akad. Wiss. SB II*, 202:179–185, 1993.

[1033] G. Larcher. On the two-dimensional Kronecker-sequence and a class of ergodic skew-products. *Arch. Math.*, 63:231–237, 1994.

[1034] G. Larcher. On the distribution of an analog to classical Kronecker sequences. *J. Number Theory*, 52:198–215, 1995.

[1035] G. Larcher, A. Lauß, H. Niederreiter, and W.Ch. Schmid. Optimal polynomials for (t, m, s)-nets and numerical integration of multivariate Walsh series. *SIAM J. Numer. Anal.*, to appear.

[1036] G. Larcher and H. Niederreiter. Optimal coefficients modulo prime powers in the three-dimensional case. *Ann. Mat. Pura Appl.*, 155:299–315, 1989.

[1037] G. Larcher and H. Niederreiter. Kronecker-type sequences and nonarchimedean diophantine approximations. *Acta Arith.*, 63:379–396, 1993.

[1038] G. Larcher and H. Niederreiter. Generalized (t, s)-sequences, Kronecker-type sequences, and diophantine approximations of formal Laurent series. *Trans. Am. Math. Soc.*, 347:2051–2073, 1995.

[1039] G. Larcher, H. Niederreiter, and W.Ch. Schmid. Digital Nets and Sequences Constructed over Finite Rings and their Application to Quasi-Monte Carlo Integration. *Monatsh. Math.*, 121:231–253, 1996.

[1040] G. Larcher and W.Ch. Schmid. On the numerical integration of high-dimensional Walsh-series by quasi-Monte Carlo methods. *Math. Comput. Simul.*, 38:127–134, 1995.

[1041] G. Larcher, W.Ch. Schmid, and R. Wolf. Representation of functions as Walsh series to different bases and an application to the numerical integration of high-dimensional Walsh series. *Math. Comput.*, 63:701–716, 1994.

[1042] G. Larcher and R.F. Tichy. Arithmetical properties of the standard Gray-code. *Österr. Akad. Wiss. SB II*, 197:449–461, 1988.

[1043] G. Larcher and R.F. Tichy. Some number-theoretical properties of generalized sum-of-digit functions. *Acta Arith.*, 52:183–196, 1989.

[1044] G. Larcher and C. Traunfellner. On the numerical integration of Walsh series by number-theoretic methods. *Math. Comput.*, 63:277–291, 1994.

[1045] V. Lašzló and T. Salát. The structure of some sequence spaces, and uniform distribution (mod 1). *Period. Math. Hung.*, 10:89–98, 1979.

[1046] A. Laurincikas. On the distribution of complex-valued multiplicative functions. *J. Theor. Nombres Bordx.*, 8:183–203, 1996.

[1047] C. Lécot. An algorithm for generating low discrepancy sequences on vector computers. *Parallel Comput.*, 11:113–116, 1989.

[1048] C. Lécot. A direct simulation Monte Carlo scheme and uniformly distributed sequences for solving the Boltzmann equation. *Computing*, 41:41–57, 1989.

[1049] C. Lécot. Low discrepancy sequences for solving the Boltzmann equation. *J. Comput. Appl. Math.*, 25:237–249, 1989.

[1050] C. Lecot. A quasi-Monte Carlo method for the Boltzmann equation. *Math. Comput.*, 56:621–644, 1991.

[1051] P. L'Ecuyer. Efficient and portable combined random number generators. *Commun. ACM*, 31:742–749,774, 1988.

[1052] P. L'Ecuyer. Random numbers for simulation. *Commun. ACM*, 33:85–97, 1990.

[1053] P. L'Ecuyer. Uniform random number generation. *Ann. Oper. Res.*, 53:77–120, 1994.

[1054] P. L'Ecuyer. Combined multiple recursive random number generators. *Oper. Res.*, to appear.

[1055] P. L'Ecuyer, F. Blouin, and R. Couture. A search for good multiple recursive random number generators. *ACM Trans. Model. Comput. Simulation*, 3:87–98, 1993.

[1056] P. L'Ecuyer, F. Blouin, and R. Couture. A search for good multiple recursive random number generators. *ACM Trans. Model. Comput. Simulation*, 3:87–98, 1993.

[1057] P. L'Ecuyer and S. Tezuka. Structural properties for two classes of combined random number generators. *Math. Comput.*, 57:735–746, 1991.

[1058] D. Leitmann. On the uniform distribution of some sequences. *J. London Math. Soc.*, 14:430–432, 1976.

[1059] M. Lemanczyk and C. Mauduit. Ergodicity of a class of cocycles over irrational rotations. *J. London Math. Soc.*, 49:124–132, 1994.

[1060] J. Lesca. Sur la répartition modulo 1 de la suite $n\alpha$. *Acta Arith.*, 20:345–352, 1972.

[1061] E. Lesigne, C. Mauduit, and B. Mosse. Le theorem ergodique le long d'une suite q-multiplicative. *Compos. Math.*, 93:49–79, 1994.

[1062] V.F. Lev. Diaphony and square discrepancy of multidimensional cells. *Math. Notes*, 47:556–564, 1990.

[1063] W.J. LeVeque. Note on a theorem of Koksma. *Proc. Am. Math. Soc.*, 1:380–383, 1950.

[1064] W.J. LeVeque. The distribution modulo 1 of trigonometric sequences. *Duke Math. J.*, 20:367–374, 1953.

[1065] W.J. LeVeque. On uniform distribution modulo a subdivision. *Pac. J. Math*, 3:757–771, 1953.

[1066] W.J. LeVeque. An inequality connected with Weyl's criterion for uniform distribution. *Proc. Symp. Pure Math.* Vol VIII, Am. Math. Soc., Providence, R. I., 31:22–30, 1965.

[1067] M.B. Levin. Über die Gleichverteilung der Folge λ^n. *Mat. Sbornik*, 98:207–222, 1975.

[1068] M.B. Levin. On the distribution of fractional parts of the exponential function. *Izvestija vyss. ucebn. Zaved*, 11:50–57, 1977.

[1069] M.B. Levin. Sur les nombres absolument normaux. *Vestn. Mosk. Univ., Ser I*, 31–37, 1979.

[1070] M.B. Levin. On the completely uniform distribution of fractional parts of the exponential function. *Tr. Semin. Im. I. G. Petrovskogo*, 7:245–256, 1981.

[1071] M.B. Levin. Completely uniform distribution of fractional parts of the exponential function. *J. Sov. Math.*, 31:3247–3256, 1985.

[1072] M.B. Levin. Uniform distribution of the matrix exponential function. *Studies in number theory. Analytical number theory, Interuniv. Sci. Collect., Saratov*, 10:46–62, 1988.

[1073] M.B. Levin. On the choice of parameters in pseudorandom number generators. *Sov. Math., Dokl.*, 40:101–105, 1989.

[1074] M.B. Levin. Effectivization of a theorem of Koksma. *Mat. Zametki*, 47:163–166, 1990.

[1075] M.B. Levin. Jointly absolutely normal numbers. *Math. Notes*, 48:1213–1220, 1990.

[1076] M.B. Levin. On simultaneously absolutely normal numbers. *Mat. Zametki*, 48:61–71, 1990.

[1077] M.B. Levin. On the discrepancy of Markov-normal sequences. preprint.

[1078] M.B. Levin and I.E. Shparlinkij. Über die Gleichverteilung der Bruchteile rekurrenter Folgen. *Usp. Mat. Nauk*, 34:203–204, 1979.

[1079] Yu.L. Levitan, N.I. Markovich, S.G. Rozin, and I.M. Sobol. On quasirandom sequences for numerical computations. *Zh. Vychisl. Mat. Mat. Fiz.*, 28:755–759, 1988.

[1080] Yu.L. Levitan and I.M. Sobol. On a pseudo-random number generators for personal computers. *Matem. Modelirovanie*, 2:119–126, 1990.
[1081] H. Z. Li. On the distribution of αn^k modulo 1. *Acta Math. Sin.*, 37:122–128, 1994.
[1082] P. Liardet. Répartition et ergodicité. *Semin. Delange-Pisot-Poitou, 19e Annee 1977/78, Theor. des Nombres*, Exp. 10:12 p., 1978.
[1083] P. Liardet. Discrépance sur le cercle. *Primaths. I, Univ. Marseille*, 1979.
[1084] P. Liardet. Propriétes génériques de processur croisés. *Isr. J. Math.*, 39:303–325, 1981.
[1085] P. Liardet. Regularities of distribution. *Compos. Math.*, 61:267–293, 1987.
[1086] P. Liardet. Proprietes harmoniques de la numeration suivant Jean Coquet. *Publ. Math. Orsay*, 1–35, 1988.
[1087] P. Liardet. Some metric properties of subsequences. *Acta Arith.*, 55:119–135, 1990.
[1088] P. Liardet and D. Volný. Constructions of smooth and analytic cocycles over irrational circle rotations. *preprint.*
[1089] P. Liardet and D. Volný. Sums of continuous and differentiable functions in dynamical systems. *preprint.*
[1090] R. Lidl, G.L. Mullen, and G. Turnwald. *Dickson polynomials.* Longman Scientific & Technical, Harlow, 1993.
[1091] M.Ch. Liu and K.M. Tsang. On the distribution (mod 1) of polynomials of a prime variable. *Nagoya Math.*, 85:241–249, 1982.
[1092] M. Loève. *Probability Theory, 3rd ed.* Van Nostrand, Princeton, 1963.
[1093] G. Lohöfer and D.H. Mayer. On a theorem by Florek and Slater on recurrence properties of circle maps. *Publ. Res. Inst. Math. Sci.*, 26:335–357, 1990.
[1094] C.T. Long and W.A. Webb. Normality in $GF(q,x)$. *Atti Accad. Naz. Lincei*, 54:848–853, 1974.
[1095] N. Loraud. β-shift, sysémes de numération et automates. *J. Theor. Nombres Bordx.*, 7:473–498, 1995.
[1096] G.G. Lorentz. A contribution to the theory of divergent sequences. *Acta Math.*, 80:167–190, 1948.
[1097] V. Losert. Almost constant sequences of transformation. *Monatsh. Math.*, 85:105–113, 1978.
[1098] V. Losert. On the existence of uniformly, distributed sequences in compact topological spaces. *Trans. Am. Math. Soc.*, 246:463–471, 1978.
[1099] V. Losert. Uniformly distributed sequences on compact, separable, non-metrizable groups. *Acta Sci. Math.*, 40:107–110, 1978.
[1100] V. Losert. On the existence of uniformly distributed sequences in compact topological spaces. *Monatsh. Math.*, 87:247–260, 1979.
[1101] V. Losert. Equirépartition des suites définies par des semi-groups additifs. *C.R. Acad. Sci., Paris*, 292:573–575, 1981.
[1102] V. Losert. Gleichverteilte Folgen und Folgen, für die fast alle Teilfolgen gleichverteilt sind. *Lect. Notes Math.*, 1114:84–97, 1985.
[1103] V. Losert. Gleichverteilung von Folgen, die durch additive Halbgruppen definiert sind. *Lect. Notes Math.*, 1114:77–83, 1985.
[1104] V. Losert. The Borel property for simple Riesz means. *Monatsh. Math.*, 102:217–226, 1986.
[1105] V. Losert, W.G. Nowak, and R.F. Tichy. On the asymptotic distribution of the powers of $s \times s$-matrices. *Compos. Math.*, 45:273–291, 1982.
[1106] V. Losert and H. Rindler. Uniform distribution and the mean ergodic theorem. *Invent. Math.*, 50:65–74, 1978.
[1107] V. Losert and H. Rindler. Almost constant sequences. *Astérsique*, 61:133–143, 1979.
[1108] V. Losert and R.F. Tichy. On uniform distribution of subsequences. *Probab. Theory Relat. Fields*, 72:517–528, 1986.
[1109] R.M. Loynes. Some results in the probilistic theory of asymptotic uniform distribution modulo 1. *Z. Wahrscheinlichkeitsth. verw. Gebiete*, 26:33–41, 1973.
[1110] A. Lubotzky, R. Phillips, and P. Sarnak. Hecke operators and distributing points on the sphere. *Commun. Pure Appl. Math.*, 39:149–186, 1986.
[1111] A. Lubotzky, R. Phillips, and P. Sarnak. Hecke operators and distributing points on S^2. *Commun. Pure Appl. Math.*, 40:401–420, 1987.
[1112] J. van de Lune. On the distribution of a specific number-theoretical sequence. *Afd. zuivere Wisk*, 004:8, 1969.

[1113] W.A.J. Luxemburg and J. Korevaar. Entire functions and Müntz-Szász type approximation. *Trans. Am. Math. Soc.*, 157:23–37, 1971.

[1114] R. Lyons. Fourier-Stieltjes coefficients and asymptotic distribution modulo 1. *Ann. of Math.*, 122:155–170, 1985.

[1115] R. Lyons. Mixing and asymptotic distribution modulo 1. *Ergodic Theory Dyn. Syst.*, 8:597–619, 1988.

[1116] S.S. Magliveras, B.A. Oberg, and A.J. Surkan. A new random number generator from permutation groups. *Rend. Semin. Mat. Fis. Milano*, 54:203–223, 1984.

[1117] K. Mahler. On the translation properties of a simple class of arithmetical functions. *J. Math. and Phys.*, 6:158–163, 1927.

[1118] K. Mahler. On the fractional parts of the powers of a rational number. II. *Mathematika*, 4:122–124, 1957.

[1119] K. Mahler. Arithmetical properties of the digits of the multiples of an irrational number. *Bull. Austral. Math. Soc.*, 8:191–203, 1973.

[1120] E. Manstavicius. Über die Verteilung additiver arithmetischer Funktionen (mod 1). *Litov. Mat. Sb.*, 13:101–108, 1973.

[1121] E. Manstavicius. Natural divisors and the Brownian motion. *J. Theor. Nombres Bordx.*, 8:159–171, 1996.

[1122] E. Manstavicius and R. Skrabutenas. Summation of values of multiplicative functions on semigroups. *Lith. Math. J.*, 33:255–264, 1993.

[1123] G. Marsaglia. The mathematics of random number generators. *The Unreasonable Effectiveness of Number Theory, Proc. Symp. Applied Math.*, 46:73–90, 1992.

[1124] G. Marsaglia, A. Zaman, and W.W. Tsang. Toward a universal random number generator. *Stat. Probab. Lett.*, 9:35–39, 1990.

[1125] J. M. Marstrand. On Khichin's conjecture about strong uniform distribution. *Proc. London Math. Soc.*, 21:540–556, 1970.

[1126] P. Martin-Löf. The definition of random sequences. *Inform. Control*, 9:602–619, 1966.

[1127] F.J. Martinelli. Construction of generalized normal numbers. *Pac. J. Math.*, 76:117–122, 1978.

[1128] M. Mascagni, S.A. Cuccaro, D.V. Pryor, and M.L. Robinson. A fast, high quality, and reproducible parallel lagged-Fibonacci pseudorandom number generator. *J. Comput. Phys.*, 119:211–219, 1995.

[1129] E. Masry and St. Cambanis. Trapezoidal Monte Carlo integration. *SIAM J. Numer. Anal.*, 27:225–246, 1990.

[1130] B. de Mathan. Sur les suites eutaxiques. *Sémin. Théor. Nombres Bordeaux*, Exp. 29:10 p., 1971.

[1131] B. de Mathan. Un critère de non-eutaxie. *C.R. Acad. Sci., Paris*, 273:433–436, 1971.

[1132] B. de Mathan. Sur un probléme de densité modulo 1. *C.R. Acad. Sci., Paris*, 287:277–279, 1978.

[1133] B. de Mathan. Un ensemble exceptionel pour un probléme de répartition modulo 1. *Sémin. Théor. Nombres 1977-1978*, Exp. 25:9 p., 1978.

[1134] B. de Mathan. Numbers contravening a condition in density modulo 1. *Acta Math. Hung.*, 36:237–241, 1980.

[1135] J. Matousek. Tight upper bounds for the discrepancy of half-spaces. *Discrete Comput. Geom.*, 13:593–601, 1995.

[1136] J. Matousek and J. Spencer. Discrepancy in arithmetic progressions. *J. Am. Math. Soc.*, 9:195–204, 1996.

[1137] J. Matousek, E. Welzl, and L. Wernisch. Discrepancy and approximations for bounded vc-dimension. *Combinatorica*, 13:455–466, 1993.

[1138] K. Matsumoto. Discrepancy estimates for the value-distribution of the Riemann zeta-function. II. *Number Theory and Combinatorics*, World Scientific Singapore:265–278, 1985.

[1139] K. Matsumoto. Discrepancy estimates for the value-distribution of the Riemann zeta-function. I. *Acta Arith.*, 48:167–190, 1987.

[1140] K. Matsumoto. Discrepancy estimates for the value-distribution of the Riemann zeta-function. III. *Acta Arith.*, 50:315–337, 1988.

[1141] J.-L. Mauclaire. Sur la répartition des fonctions q-additives I. *J. Theor. Nombres Bordx.*, 5:79–91, 1993.

[1142] J.-L. Mauclaire. A characterization of generalized rudin-shapiro sequences with values in a locally compact abelian group. *Acta Arith.*, 68:213–217, 1994.

[1143] J.-L. Mauclaire. On the distribution of the values of an additive arithmetical function with values in a locally compact abelian group. *Acta Arith.*, 68:201–212, 1994.

[1144] J.-L. Mauclaire. On the regularity of additive arithmetical functions with values in a locally compact group. *Ann. Univ. Sci. Budap. Rolando Eötvös, Sect. Comput.*, 14:135–144, 1994.

[1145] C. Mauduit. Automates finis et équirépartition modulo 1. *C.R. Acad. Sci., Paris*, 299:121–123, 1984.

[1146] C. Mauduit. Automates finis et ensembles normaux. *Ann. Inst. Fourier*, 36:1–25, 1986.

[1147] C. Mauduit. Substitutions et equirepartition modulo 1. *Publ. Math. Orsay*, 85–89, 1988.

[1148] C. Mauduit. Sur l'ensemble normal des substitutions de longueur quelconque. *J. Number Theory*, 29:235–250, 1988.

[1149] C. Mauduit. Caractérisation des ensembles normaux substitutifs. *Invent. Math.*, 95:133–147, 1989.

[1150] C. Mauduit. Substitutive normal sets. *Number theory, Elementary and analytic, Proc. Conf. Budapest, Colloq. Math. Soc. Janos Bolyai*, 51:1317-323, 1990.

[1151] C. Mauduit. Proprietes arithemtiques des substitutions. *Prog. Math.*, 102:177–190, 1992.

[1152] C. Mauduit and Rivat J. Répartition des fonctions q-multiplicatives dans la suite $([n^c])_{n\in\mathbf{N}}, c > 1$. *Acta Arith.*, 71:171–179, 1995.

[1153] C. Mauduit and B. Mosse. Suites de G_q orbite finie. *Acta Arith.*, 57:69–82, 1991.

[1154] W. Maxones, H. Muthsam, and H. Rindler. Bemerkung zu einem Satz von E. Hlawka. *Anz. Österr. Akad. Wiss.*, 82–83, 1976.

[1155] W. Maxones, I. Richards, H. Rindler, and J. Schoißengeier. Ein metrisches Ergebnis über die Diskrepanz von Folgen. *Österr. Akad. Wiss. SB II*, 185:405–409, 1977.

[1156] W. Maxones and H. Rindler. Bemerkungen zu einer Arbeit von P. Gerl "Gleichverteilung auf lokalkompakten Gruppen". *Math. Nachr.*, 79:193–199, 1977.

[1157] W. Maxones and H. Rindler. Einige Resultate über unitär gleichverteilte Maßfolgen. *Anz. Österr. Akad. Wiss.*, 11–13, 1977.

[1158] W. Maxones and H. Rindler. Asymptotisch gleichverteilte Netze von Wahrscheinlichkeitsmaßen auf lokalkompakten Gruppen. *Colloq. Math.*, 40:131–145, 1978.

[1159] D. H. Mayer. On the distribution of recurrence times in nonlinear system. *Lett. Math. Phys.*, 16:139–143, 1988.

[1160] W. Meier and O. Staffelbach. The self-shrinking generator. *Kluwer Int. Ser. Eng. Comput. Sci.*, 276:287–295, 1994.

[1161] H.G. Meijer. On a distribution problem in finite sets. *Proc. Nederl. Akad. Wet.*, A 76:9–17, 1973.

[1162] H.G. Meijer and H. Niederreiter. Equirépartition et théorie des nombres premiers. *Lect. Notes Math.*, 475:104–112, 1975.

[1163] H.G. Meijer and J.S. Shiue. Uniform distribution in $\mathbf{Z}_g$ and $\mathbf{Z}_{g_1} \times \cdots \times \mathbf{Z}_{g_t}$. *Proc. Nederl. Akad. Wet.*, A 79:200–212, 1976.

[1164] M. Mendès France. Ensembles normaux. *Sémin. Théor. Nombres Bordeaux*, Exp. 6:6 p., 1969.

[1165] M. Mendès France. Les suites à spectre vide et la répartition modulo 1. *Sémin. Théor. Nombres Bordeaux*, Exp. 13:19 p., 1971.

[1166] M. Mendès France. Les suites à spectre vide et la répartition modulo 1. *J. Number Theory*, 5:1–15, 1973.

[1167] M. Mendès France. Les suites additives et leur répartition (mod 1). *Sémin. Théor. Nombres Bordeaux*, Exp. 8:6 p., 1974.

[1168] M. Mendès France. Les ensembles de Bésineau. *Semin. Delange-Pisot-Poitou, 15e annee 1973/74, Theorie des Nombres*, Exp. 7:6 p., 1975.

[1169] M. Mendès France. Suites de nombres au hasard (d'après Knuth). *Sémin. Théor. Nombres Bordeaux*, Exp. 6:11 p., 1975.

[1170] M. Mendès France. A characterization of Pisot numbers. *Mathematika*, 23:32–34, 1976.

[1171] M. Mendès France. Indépendance statistique et nombres de Pisot. *Sémin. Théor. Nombres Bordeaux*, Exp. 13:6 p., 1976.

[1172] M. Mendès France. Indépendance statistique, somme des chiffres, mesure spectrale. *Semin. Delange-Pisot-Poitou, 18e Annee 1976/77, Theor. des Nombres*, Exp. G 9:4 p., 1977.

[1173] M. Mendès France. Le théorème de van der Corput. *Sémin. Théor. Nombres 1977-1978*, Exp. 4:6 p., 1978.

[1174] M. Mendès France. Les ensembles de van der Corput. *Semin. Delange-Pisot-Poitou, 19e Annee 1977/78, Theor. des Nombres*, Exp. 12:5 p., 1978.

[1175] M. Mendès France. A propos de la suite de Morse. *Sémin. Théor. Nombres 1978-1979*, Exp. 13:11 p., 1979.

[1176] M. Mendès France. Sur les décimales des nombres algébriques réels. *Sémin. Théor. Nombres 1979-1980*, Exp. 28:7 p., 1980.

[1177] M. Mendès France. Entropie, dimension et thermodynamique des courbes planes. *Theorie des nombres, Semin. Delange-Pisot-Poitou, Paris 1981/82, Prog. Math.*, 38:153–177, 1983.

[1178] M. Mendès France. Entropy of curves and uniform distribution. *Topics in classical number theory, Colloq. Budapest 1981, Colloq. Math. Soc. Janos Bolyai*, 34:1051–1067, 1984.

[1179] M. Mendès France. La chaine d'Ising imaginaire ou la suite de Rudin-Shapiro. *Groupe Etude Théor. Anal. Nombres*, 35:7, 1985.

[1180] M. Mendès France. A diophantine inequality. *Österr. Akad. Wiss. SB II*, 195:105–108, 1986.

[1181] M. Mendès France. Suites de Rudin-Shapiro, theme et variation. *Sémin. Théor. Nombres Bordeaux*, Exp. 5:11 p., 1987.

[1182] M. Mendès France. A diophantine problem. *(Irregularities of Partitions). Algorithms Comb.*, 8:129–135, 1989.

[1183] M. Mendès France. The inhomogeneous Ising chain and paperfolding. *Springer Proc. Phys.*, 47:195–202, 1990.

[1184] M. Mendès France. The Ising transducer. *Ann. Inst. Henri Poincaré, Phys. Théor.*, 52:259–265, 1990.

[1185] M. Mendès France. The Rudin-Shapiro sequence, Ising chain, and paperfolding. *Prog. Math.*, 85:367–382, 1990.

[1186] M. Mendès France. Opacity of an automaton. application to the inhomogeneous Ising chain. *Commun. Math. Phys.*, 1–12, 1991.

[1187] A.McD. Mercer. A note on some irrational decimal fractions. *Am. Math. Monthly*, 101:567–568, 1994.

[1188] S. Mercourakis. Some remarks on countably determined measures and uniform distribution of sequences. *Monatsh. Math.*, 121:79–111, 1996.

[1189] K.D. Merrill. Comhomology of step functions under irrational rotations. *Isr. J. Math.*, 52:320–340, 1985.

[1190] F. Mignosi. On a generalization of the $3x+1$ problem. *J. Number Theory*, 55:28–45, 1995.

[1191] F. Mignosi and P. Séébold. Morphismes sturmiens et règles de Rauzy. *J. Theor. Nombres Bordx.*, 5:221–233, 1993.

[1192] M. Mignotte. A characterization of integers. *Am. Math. Monthly*, 84:278–281, 1977.

[1193] G.A. Mikhajlov. A criterion for uniform optimality of weighted Monte Carlo methods, and its applications. *Sov. Math.*, 30:816–820, 1984.

[1194] G.A. Mikhajlov. Optimization of weighted Monte Carlo methods. *Nauka*, R. 2.00:240, 1987.

[1195] A. Miklavc. Elementary proofs of two theorems on the distribution of numbers $n\theta$ (mod 1). *Proc. Am. Math. Soc.*, 39:279–280, 1973.

[1196] G. Miles and R.K. Thomas. On the polynomial uniformity of translation of the n-torus. *Adv. Math.*, 2:219–229, 1978.

[1197] R.A. Mitchell. Error estimates arising from certain pseudorandom sequences in a quasi-random search method. *Math. Comput.*, 55:289–297, 1990.

[1198] H. Miyazaki. Pseudorandom numbers generated by the doubly multiplicative congruential method. *Sel. Pap. 3rd Int. Meet. Stat.*, 319–325, 1987.

[1199] H. L. Montgomery and H. Niederreiter. Estimation optimale de sommes exponentielles. *Bull. Greek Math. Soc.*, 18:291–301, 1977.

[1200] W. Moran and C.E.M. Pearce. Discrepancy results for normal numbers. *Harmonic analysis and operator algebras*, 16:203–210, 1988.

[1201] W. Moran and A.D. Pollington. Metrical results on normality to distinct bases. *J. Number Theory*, 54:180–189, 1995.

[1202] Sigeiti Moriguti. A theory of computer rational numbers. *Japan J. Appl. Math.*, 1:253–271, 1984.

[1203] Y. Moriyama, T. Nakamura, and M. Yamada. New random number generator based on binary and ternary m-sequences. *Kushiro Natl. Coll. Technol.*, 23:51–54, 1989.
[1204] W.J. Morokoff and R. E. Caflisch. A quasi-Monte Carlo approach to particle simulation of the heath equation. *SIAM J. Numer. Anal.*, 30:1558–1573, 1993.
[1205] W.J. Morokoff and R. E. Caflisch. Quasi-random sequences and their discrepancies. *SIAM J. Comput.*, 15:1251–1279, 1994.
[1206] Y.N. Moschovakis. *Descriptive Set Theory*. North-Holland, Amsterdam, 1980.
[1207] N.G. Moshchevitin. On the distribution of fractional parts of a system of linear functions. *Vestn. Mosk. Univ., Ser. I*, 26–31, 1990.
[1208] N.G. Moshchevitin. Recent results on asymptotic behavior of integrals of quasiperiodic functions. *Am. Math. Soc. Transl.*, 168:201–209, 1995.
[1209] D.A. Moskvin. Über die Trajektorien der ergodischen Endomorphismen eines zwei- dimensionalen Torus, welche auf einer glatten Kurve beginnen. *Aktual'nye Probl. analit. Teor.*, 138–167, 1974.
[1210] D.A. Moskvin. Metric theory of automorphisms of the two-dimensional torus. *Izv. Akad. Nauk*, 45:69–100, 1981.
[1211] D.A. Moskvin. On the metric theory of automorphisms of the two-dimensional torus. *Math. USSR.*, 18:61–88, 1982.
[1212] D.A. Moskvin. Metric properties of ergodic endomorphisms of a multi-dimensional torus. *J. Sov. Math.*, 50:1854–1876, 1987.
[1213] C.J. Mozzochi. *On the pointwise convergence of Fourier series*. Lect. Notes Math. 199, Springer, Berlin, 1971.
[1214] R. Mück and W. Philipp. Distance of probability measures and uniform distribution mod 1. *Math. Z.*, 142:195–202, 1975.
[1215] R.H. Muhutdinov. Kriterium für die ungleichmäßige Verteilung gebrochener Teile der Exponentialfunktion. *Izv. Akad. Nauk*, 19:23–26, 1975.
[1216] G.L. Mullen, A. Mahalanabis, and H. Niederreiter. Tables of (t, m, s)-net and (t, s)-sequence parameters. *Lect. Notes Stat.*, 106:58–86, 1995.
[1217] G.L. Mullen and G. Whittle. Point sets with uniformity properties and orthogonal hypercubes. *Monatsh. Math.*, 113:265–273, 1992.
[1218] C. Müller. *Spherical Harmonics*. Lect. Notes Math. 17, Springer, Berlin, 1966.
[1219] H. Müller. Eine Note zur Gleichverteilung additiv erzeugter Folgen. *Arch. Math.*, 44:255–258, 1985.
[1220] H. Müller. Das $3n + 1$ Problem. *Mitt. Math. Ges. Hamburg*, 12:231–251, 1991.
[1221] H. Müller. Über eine Klasse 2-adischer Funktionen im Zusammenhang mit dem "$3x + 1$"-Problem. *Abh. Math. Semin. Univ. Hamburg*, 64:293–302, 1994.
[1222] W. Müller. *C-Gleichverteilung Modulo 1*. Technische Universiät Wien, 1983.
[1223] W. Müller and R.J. Taschner. Ein metrischer Satz der C-Gleichverteilung. *Monatsh. Math.*, 97:207–212, 1984.
[1224] G. Myerson. A combinatorial problem in finite fields. *Q.J. Math., Oxford II. Ser.*, 31:219–231, 1980.
[1225] G. Myerson. How small can a sum of roots of unity be? *Am. Math. Monthly*, 93:457–459, 1986.
[1226] G. Myerson. Dedekind sums and uniform distribution. *J. Number Theory*, 28:233–239, 1988.
[1227] G. Myerson. On ignoring the singularity. *SIAM J. Numer. Anal.*, 28:1803–1807, 1991.
[1228] G. Myerson. Discrepancy and distance between sets. *Indag. Math.*, 3:193–201, 1992.
[1229] G. Myerson. A sampler of recent developments in the distribution of sequences. *Lect. Notes Pure Appl. Math.*, 147:163–190, 1993.
[1230] G. Myerson and A.D. Pollington. Notes on uniform distribution. *London Math. Soc. Lect. Note Ser.*, 154:211–212, 1990.
[1231] G. Myerson and A.D. Pollington. Notes on uniform distribution modulo one. *J. Austral. Math. Soc. A*, 49:264–272, 1990.
[1232] K. Nagasaka. On Hausdorff dimension of non-normal sets. *Ann. Inst. Stat. Math.*, 23:515–521, 1971.
[1233] K. Nagasaka. The theory of Hausdorff dimension and its applications. *Proc. Symp. RIMS*, 124–144, 1977.

[1234] K. Nagasaka. La dimension de Hausdorff de certains ensembles dans [0, 1]. *Proc. Japan Acad. A*, 54:109–112, 1978.

[1235] K. Nagasaka. Distribution property of recursive sequences defined by (mod m). *Fibonacci Q.*, 22:76–81, 1984.

[1236] K. Nagasaka. Weakly uniform distribution mod m for certain recursive sequences and for monomial sequences. *Tsukuba J. Math.*, 9:159–166, 1985.

[1237] K. Nagasaka and S. Ando. Symmetric recursive sequences mod m. *Applications of Fibonacci numbers*, 3:17–28, 1988.

[1238] K. Nagasaka and Ch. Batut. Note sur les nombres normaux. *Nanta Math.*, 13:57–68, 1980.

[1239] K. Nagasaka, S. Kanemitsu, and J.S. Shiue. Benford's law: The logarithmic law of first digit. *J. Number Theory*, 51:361–391, 1990.

[1240] K. Nagasaka and J.-S. Shiue. On a theorem of Koksma on discrepancy. *Algebraic Structures and Number Theory*, World Scientific Singapore:208–224, 1990.

[1241] K. Nagasaka, J.-S. Shiue, and X. Yu. On multiplicative functions with regularity properties. *New Trends Probab. Stat.*, 2:155–163, 1992.

[1242] K. Nagasaka and J.S. Shiue. Asymptotic distribution and independence of sequences of g-adic integers. *Prospects of Math. Science*, 157–171, 1988.

[1243] M. Nair and A. Perelli. On the distribution of $p^{1/2}$ modulo one. *Number theory, Elementary and analytic, Proc. Conf. Budapest, Colloq. Math. Soc. Janos Bolyai*, 51:393–435, 1990.

[1244] R. Nair. On LeVeque's theorem about the uniform distribution (mod 1) of $(a_j \cos a_j x)$. *Isr. J. Math.*, 65:96–112, 1989.

[1245] R. Nair. On strong uniform distribution. *Acta Arith.*, 56:183–193, 1990.

[1246] R. Nair. Some theorems on metric uniform distribution using L^2-methods. *J. Number Theory*, 35:18–52, 1990.

[1247] R. Nair. On polynomials in primes and J. Bourgain's circle method approach to ergodic theorems. *Ergodic Th. Dynamical Syst.*, 11:485–499, 1991.

[1248] R. Nair. On the metrical theory of continued fractions. *Proc. Am. Math. Soc.*, 120:1041–1046, 1994.

[1249] R. Nair. On Riemann Sums and Lebesgue Integrals. *Monatsh. Math.*, 120:49–54, 1995.

[1250] H. Nakada. Geodesic flows and diophantine approximations. *Stability theory and related topics in dynamical systems*, 6:125–127, 1989.

[1251] H. Nakada and G. Wagner. Duffin-Schaeffer theorem of diophantine approximation for complex numbers. *Journees arithmetiques, Exp. Congr., Luminy/Fr. 1989, Asterisque*, 198-200:259–263, 1991.

[1252] Y. N. Nakai and I. Shiokawa. A class of normal numbers. *Japan J. Math*, 16:17–29, 1990.

[1253] Y. N. Nakai and I. Shiokawa. A class of normal numbers II. *London Math. Soc. Lect. Note*, 154:204–210, 1990.

[1254] Y. N. Nakai and I. Shiokawa. Discrepancy estimates for a class of normal numbers. *Acta Arith.*, 62:271–284, 1992.

[1255] Y. N. Nakai and I. Shiokawa. Normality of numbers generated by the values of polynomials at primes. preprint.

[1256] M. Nakajima and Y. Ohkubo. Weyl's type criterion for general distribution mod 1. *Proc. Japan Acad. A*, 65:315–317, 1989.

[1257] K. Nanbu. Direct simulation scheme derived from the Boltzmann equation. I. Monocomponent ases. *J. Phys. Soc. Japan*, 49:2042–2049, 1980.

[1258] W. Narkiewicz. On a kind of uniform distribution for systems of multiplicative functions. *Lith. Math. J.*, 22:127–137, 1982.

[1259] W. Narkiewicz. Uniform distribution of sequences of integers. *Journées arithmétiques, Exeter 1980, London Math. Soc. Lect. Note Ser.*, 56:202–210, 1982.

[1260] W. Narkiewicz. *Uniform distribution of sequences of integers in residue classes*, volume 1087. Lect. Notes Math., Berlin, 1984.

[1261] W. Narkiewicz. Correction to the paper: On a kind of uniform distribution of values of multiplicative functions in residue classes. *Acta Arith.*, 46:301–304, 1986.

[1262] W. Narkiewicz. *Elementary and Analytic Theory of Algebraic Numbers.* Springer, Berlin, 1990.

[1263] W. Narkiewicz and F. Rayner. Distribution of values of $\sigma_2(n)$ in residue classes. *Monatsh. Math.*, 94:133–141, 1982.

[1264] I.P. Natanson. *Theorie der Funktionen einer reellen Veränderlichen.* Adademie-Verlag, Berlin, 1961.
[1265] M.B. Nathanson. Linear recurrences and uniform distribution. *Proc. Am. Math. Soc.*, 48:289–291, 1975.
[1266] P. Neumann. Einige Gedanken zur Programmierung und zum Test von Zufallszahlengeneratoren. *Wiss. Z. Techn. Univ. Dres.*, 37:181–182, 1988.
[1267] H. Neunzert and J. Wick. Die Theorie der asymptotischen Verteilung und die numerische Lösung von Integrodifferentialgleichungen. *Numer. Math.*, 21:234–243, 1973.
[1268] S. Newcomb. Note on the frequency of use of the different digits in natural numbers. *Am. J. Math.*, 4:34–40, 1881.
[1269] H. Niederreiter and J. Schoißengeier. Almost periodic functions and uniform distribution mod 1. *J. Reine Angew. Math.*, 291:189–203, 1977.
[1270] H. Niederreiter. On a number-theoretical integration method. *Aequationes Math.*, 8:304–311, 1971.
[1271] H. Niederreiter. Discrepancy and convex programming. *Ann. Math. Pura Appl.*, 93:89–97, 1972.
[1272] H. Niederreiter. Distribution of Fibonacci numbers mod 5^k. *Fibonacci Q.*, 10:373–374, 1972.
[1273] H. Niederreiter. A distribution problem in finite sets. *Appl. Number Theory*, 237–248, 1972.
[1274] H. Niederreiter. Methods for estimating discrepancy. *Appl. Number Theory*, 203–236, 1972.
[1275] H. Niederreiter. On a class of sequnces of lattice points. *J. Number Theory*, 4:477–502, 1972.
[1276] H. Niederreiter. On the distribution of pseudo-random numbers generated by the linear congruential method. *Math. Comput.*, 26:793–795, 1972.
[1277] H. Niederreiter. Uniform distribution of lattice points. *Proc. Number Theory, Univ. Colorado, Boulder 1972*, pages 162–166, 1972.
[1278] H. Niederreiter. Application of diophantine approximation to numerical integration. *Diophantine Approx. Appl.*, 129–199, 1973.
[1279] H. Niederreiter. The distribution of Farey points. *Math. Ann.*, 201:341–345, 1973.
[1280] H. Niederreiter. Metric theorems on the distribution of sequences. *Analytic Number Theory*, 24:195–212, 1973.
[1281] H. Niederreiter. Zur quantitativen Theorie der Gleichverteilung. *Monatsh. Math.*, 77:55–62, 1973.
[1282] H. Niederreiter. On the distribution of pseudo-random numbers generated by the linear congruential method. *Math. Comput.*, 28:1117–1132, 1974.
[1283] H. Niederreiter. Indépendance de suites. *Lect. Notes Math.*, 475:120–131, 1975.
[1284] H. Niederreiter. On a paper of Blum, Eisenberg, and Hahn concerning ergodic theory and the distribution of sequences in the Bohr group. *Acta Sci. Math.*, 37:103–108, 1975.
[1285] H. Niederreiter. Quantitative versions of a result of Hecke in the theory of uniform distribution mod 1. *Acta Arith.*, 28:321–339, 1975.
[1286] H. Niederreiter. Rearrangement theorems for sequences. *Astérisque*, 24–25:243–261, 1975.
[1287] H. Niederreiter. Résultats nouveaux dans la théorie quantitative de l'équirépartition. *Lect. Notes Math.*, 475:132–154, 1975.
[1288] H. Niederreiter. Well-distributed sequences with respect to system of convex sets. *Proc. Am. Math. Soc.*, 47:305–310, 1975.
[1289] H. Niederreiter. An application of the Hilbert-Montgomery-Vaughan inequality to the metric theory of uniform distribution mod 1. *J. London Math. Soc.*, 13:497–506, 1976.
[1290] H. Niederreiter. On the distribution of pseudo-random numbers generated by the linear congruential method. *Math. Comput.*, 30:571–597, 1976.
[1291] H. Niederreiter. Some new exponential sums with applications to pseudo-random numbers. *Top. Number Theory*, 13:209–232, 1976.
[1292] H. Niederreiter. Statistical independence of linear congruential pseudo-random numbers. *Bull. Am. Math. Soc.*, 82:927–929, 1976.
[1293] H. Niederreiter. Pseudo-random numbers and optimal coefficients. *Advances Math.*, 26:99–181, 1977.
[1294] H. Niederreiter. Ergodic sequences of measures and a problem in additive number theory. *Arch. Math.*, 31:21–32, 1978.
[1295] H. Niederreiter. Existence of good lattice points in the sense of Hlawka. *Monatsh. Math.*, 86:203–219, 1978.

[1296] H. Niederreiter. Quasi-Monte Carlo methods and pseudo-random numbers. *Bull. Am. Math. Soc.*, 84:957–1041, 1978.
[1297] H. Niederreiter. The serial test for linear congruential pseudo-random numbers. *Bull. Am. Math. Soc.*, 84:273–274, 1978.
[1298] H. Niederreiter. Nombres pseudo-aleatoires et équirepartition. *Astérique*, 61:155–164, 1979.
[1299] H. Niederreiter. Verteilung von Resten rekursiver Folgen. *Arch. Math.*, 34:526–533, 1980.
[1300] H. Niederreiter. Statistical independence properties of Tausworthe pseudo-random numbers. *Comb. and Computing*, 163–168, 1981.
[1301] H. Niederreiter. Statistical tests for tausworthe pseudo-random numbers. *Proc. 2nd Pannonian Symp.*, 265–274, 1982.
[1302] H. Niederreiter. A quasi-Monte Carlo method for the approxmate computation. *Studies in pure mathematics*, 523–529, 1983.
[1303] H. Niederreiter. Applications des corps finis aux nombres pseudo-aléatoires. *Sémin. Théor. Nombres Bordeaux*, Exp. 38:9 p., 1983.
[1304] H. Niederreiter. Distribution mod 1 of monotone sequences. *Indag. Math.*, 46:315–327, 1984.
[1305] H. Niederreiter. A general rearrangement theorem for sequences. *Arch. Math.*, 43:530–534, 1984.
[1306] H. Niederreiter. Optimal multipliers for linear congruential pseudo-random numbers: The decimal case. *Statistics and probability*, 3:255–269, 1984.
[1307] H. Niederreiter. The performance of k-step pseudorandom number generators under the uniformity test. *SIAM J. Sci. Stat. Comput*, 5:798–810, 1984.
[1308] H. Niederreiter. Quasi-Monte Carlo methods for global optimization. *Math. Statistics and Applications*, B:251–267, 1985.
[1309] H. Niederreiter. The seriel test for pseudo-random numbers generated by the linear congruential method. *Numer. Math.*, 46:51–68, 1985.
[1310] H. Niederreiter. Distribution properties of feedback shift register sequences. *Probl. Control Inf. Theory*, 15:19–34, 1986.
[1311] H. Niederreiter. Dyadic fractions with small partial quotients. *Monatsh. Math.*, 101:309–315, 1986.
[1312] H. Niederreiter. Good lattice points for quasirandom search methods. *Lect. Notes Control Inf. Sci.*, 84:647–654, 1986.
[1313] H. Niederreiter. Low-discrepancy point sets. *Monath. Math.*, 102:155–167, 1986.
[1314] H. Niederreiter. A pseudorandom vector generator based on finite field arithmetic. *Math. Jap.*, 31:759–774, 1986.
[1315] H. Niederreiter. Pseudozufallszahlen und die Theorie der Gleichverteilung. *Österr. Akad. Wiss. SB II*, 109–138, 1986.
[1316] H. Niederreiter. Continued fractions for formal power series, pseudorandom numbers, and linear complexity of sequences. *Contributions to General Algebra*, 5:221–233, 1987.
[1317] H. Niederreiter. Point sets and sequences with small discrepancy. *Monatsh. Math.*, 104:273–337, 1987.
[1318] H. Niederreiter. A statistical analysis of generalized feedback shift register pseudorandom number generators. *SIAM J. Sci. Stat. Comput.*, 8:1035–1051, 1987.
[1319] H. Niederreiter. Low discrepancy and low-dispersion sequences. *J. Number Theory*, 30:51–70, 1988.
[1320] H. Niederreiter. Quasi-Monte Carlo methods for multi-dimensional numerical integration. *Numerical integration*, 85:157–171, 1988.
[1321] H. Niederreiter. Remarks on nonlinear congruential pseudorandom numbers. *Metrika*, 35:321–328, 1988.
[1322] H. Niederreiter. The serial test for digital k-step pseudorandom numbers. *Math. J. Okayama Univ.*, 30:93–119, 1988.
[1323] H. Niederreiter. Statistical independence of nonlinear congruential pseudorandom numbers. *Monatsh. Math.*, 106:149–159, 1988.
[1324] H. Niederreiter. The serial test for congruential pseudorandom numbers generated by inversions. *Math. Comput.*, 52:135–144, 1989.
[1325] H. Niederreiter. Lower bounds for the discrepancy of inversive congruential pseudorandom numbers. *Math. Comput.*, 55:277–287, 1990.

[1326] H. Niederreiter. Pseudorandom numbers generated from shift register sequences. *Lect. Notes in Math.*, 1452:165–177, 1990.

[1327] H. Niederreiter. Statistical independence properties of pseudroandom vectors produced by matrix generators. *J. Comput. Appl. Math.*, 31:139–151, 1990.

[1328] H. Niederreiter. Constructions of low-discrepancy point sets and sequences. *Sets, Graphs, and Numbers, Conf. Proc., Colloq. Math. Soc. Janos Bolyai*, 60:529–559, 1991.

[1329] H. Niederreiter. Recent trends in random number and random vector generation. *Ann. Oper. Res.*, 31:323–345, 1991.

[1330] H. Niederreiter. The existence of efficient lattice rules for multidimensional numerical integration. *Math. Comput.*, 58:305–314, 1992.

[1331] H. Niederreiter. Existence theorems for efficient lattice rules. *NATO ASI Ser., Ser. C*, 357:71–80, 1992.

[1332] H. Niederreiter. Lattice rules for multiple integration. *Lect. Notes Econ. Math. Syst.*, 379:15–26, 1992.

[1333] H. Niederreiter. Low-discrepancy point sets obtained by digital constructions over finite fields. *Czech. Math. J.*, 42:143–166, 1992.

[1334] H. Niederreiter. New methods for pseudorandom number and pseudorandom vector generation. *Proc. 1992 Winter Simulation Conf.*, 264–269, 1992.

[1335] H. Niederreiter. Orthogonal arrays and other combinatorial aspects in the theory of uniform point distributions in unit cubes. *Discrete Math.*, 106/107:361–367, 1992.

[1336] H. Niederreiter. *Random Number Generation and Quasi-Monte Carlo Methods.* SIAM Conf. Ser. Appl. Math. Vol. 63, Philadelphia, 1992.

[1337] H. Niederreiter. Supplement to: The existence of efficient lattice rules for multidimensional numerical integration. *Math. Comput.*, 58:S7–S16, 1992.

[1338] H. Niederreiter. Finite fields, pseudorandom numbers, and quasirandom points. *Finite Fields, Coding Theory, and Advances in Comm. and Comp. (G.L. Mullen and P.J.-S. Shiue, eds.)*, 375–394, 1993.

[1339] H. Niederreiter. Improved error bounds for lattice rules. *J. Complexity*, 9:60–75, 1993.

[1340] H. Niederreiter. Pseudorandom numbers and quasirandom points. *Z. Angew. Math. Mech.*, 73:T648–T652, 1993.

[1341] H. Niederreiter. On a new class of pseudorandom numbers for simulation methods. *J. Comput. Appl. Math.*, 56:159–167, 1994.

[1342] H. Niederreiter. Pseudorandom vector generation by the inversive method. *ACM Trans. Model. Comput. Simulation*, 4:191–212, 1994.

[1343] H. Niederreiter. Low-discrepancy sequences and non-archimedean diophantine approximations. *Stud. Sci. Math. Hung.*, 30:111–122, 1995.

[1344] H. Niederreiter. The multiple-recursive matrix method for pseudorandom number generation. *Finite Fields Appl.*, 1:3–30, 1995.

[1345] H. Niederreiter. New developments in uniform pseudorandom number and vector generation. *Lect. Notes Stat.*, 106:87–120, 1995.

[1346] H. Niederreiter. Pseudorandom vector generation by the multiple-recursive matrix method. *Math. Comput.*, 64:279–294, 1995.

[1347] H. Niederreiter and J. Horbowicz. Optimal bounds for exponential sums in terms of discrepancy. *Colloq. Math.*, 52:355–366, 1988.

[1348] H. Niederreiter and S.K. Lo. Uniform distribution of sequences of algebraic integers. *Math. J. Okayama*, 18:13–29, 1975.

[1349] H. Niederreiter and Ch.F. Osgood. A uniform distribution question related to numerical analysis. *Math. Comput.*, 30:366–370, 1976.

[1350] H. Niederreiter and P. Peart. Localization of search in quasi-Monte Carlo methods for global optimization. *SIAM J. Sci. Stat. Comput.*, 7:660–664, 1986.

[1351] H. Niederreiter and W. Philipp. On a theorem of Erdős and Turán on uniform distribution. *Proc. Number Theory, Univ. Colorado, Boulder 1972*, pages 180–182, 1972.

[1352] H. Niederreiter and W. Philipp. Berry-Esseen bounds and a theorem of Erdős and Turán on uniform distribution mod 1. *Duke Math. J.*, 40:633–649, 1973.

[1353] H. Niederreiter, A. Schinzel, and L. Somer. Maximal frequencies of elements in second-order linear recurring sequences over a finite field. *Elem. Math.*, 46:139–143, 1991.

[1354] H. Niederreiter and C.P. Schnorr. The distribution of values of Kloosterman sums. *Arch. Math.*, 56:270–277, 1991.

[1355] H. Niederreiter and J.-S. Shiue. Equidistribution of linear recurring sequences in finite fields. *Indag. Math.*, 39:397–405, 1977.

[1356] H. Niederreiter and J.-S. Shiue. Weak equidistribution of sequences in finite fields. *Contrib. Gen. Algebra*, 6:203–212, 1988.

[1357] H. Niederreiter and J.S. Shiue. Uniform distribution of sequences in rings of integral matrices. *Galsgow Math. J.*, 20:169–178, 1979.

[1358] H. Niederreiter and J.S. Shiue. Equidistribution of linear recurring sequences in finite fields. *Acta Arith.*, 38:197–207, 1980.

[1359] H. Niederreiter and I.H. Sloan. Lattice rules for multiple integration and discrepancy. *Math. Comput.*, 54:303–312, 1990.

[1360] H. Niederreiter and I.H. Sloan. Quasi-Monte Carlo methods with modified vertex weights. *ISNM, Int. Ser. Numer. Math.*, 112:253–265, 1993.

[1361] H. Niederreiter and I.H. Sloan. Integration of nonperiodic functions of two variables by Fibonacci lattice rules. *J. Comput. Appl. Math.*, 51:57–70, 1994.

[1362] H. Niederreiter and I.H. Sloan. Variants of the Koksma-Hlawka Inequality for Vertex-Modified Quasi-Monte Carlo Integration Rules. *Math. Comput. Modelling*, 23:69–77, 1996.

[1363] H. Niederreiter and R.F. Tichy. Beiträge zur Diskrepanz bezüglich gewichteter Mittel. *Manuscr. Math.*, 42:85–99, 1983.

[1364] H. Niederreiter and R.F. Tichy. Solution of a problem of Knuth on complete uniform distribution of sequences. *Mathematika*, 32:26–32, 1985.

[1365] H. Niederreiter and R.F. Tichy. Metric theorems on uniform distribution and approximation theory. *Asterisque*, 147/148:319–323, 1987.

[1366] H. Niederreiter, R.F. Tichy, and G. Turnwald. An inequaltiy for difference of distribution functions. *Arch. Math.*, 54:166–172, 1990.

[1367] H. Niederreiter and J.M. Wills. Diskrepanz und Distanz von Maßen bezüglich konvexer und Jordanscher Mengen. *Math. Z.*, 144:125–134, 1975.

[1368] H. Niederreiter and C. Xing. Low-discrepancy sequences obtained from algebraic function fields over finite fields. *Acta Arith.*, 72:281–298, 1995.

[1369] H. Niederreiter and C. Xing. Explicit global function fields over the binary field with many rational places. *Acta Arith.*, 75:383–396, 1996.

[1370] H. Niederreiter and C. Xing. Low-discrepancy sequences and global function fields with many rational places. *Finite Fields Appl.*, 2:241–273, 1996.

[1371] E. Novak. *Deterministic and stochastic error bounds in numerical analysis.* Lect. Notes Math. 1349, Springer, Berlin, 1988.

[1372] E. Novak. Stochastic properties of quadrature formulas. *Numer. Math.*, 53:609–620, 1988.

[1373] W.G. Nowak. Die Diskrepanz der Doppelfolgen und einige Verallgemeinerungen. *Österr. Akad. Wiss. SB II*, 383–409, 1978.

[1374] W.G. Nowak. Optimale Diskrepanzabschätzungen bei speziellen Dreiecksdoppelfolgen. *Anz. Österr. Akad. Wiss.*, 179–182, 1978.

[1375] W.G. Nowak. Einige spezielle metrische Resultate zur quantitativen Theorie der Gleichverteilung modulo 1. *Anz. Österr. Akad. Wiss.*, 12–16, 1980.

[1376] W.G. Nowak. Über die Diskrepanz von trigonometrischen Folgen und von Potenzen komplexer Zahlen. *Österr. Akad. Wiss. SB II*, 189:249–276, 1980.

[1377] W.G. Nowak. Zur Gleichverteilung mod 1 der Potenzen von Quaternionen mit Norm größer als 1. *Arch. Math.*, 34:243–248, 1980.

[1378] W.G. Nowak. Ein kurzer Beweis eines Satzes von Sierpiński. *Sect. Math.*, 24:153–156, 1981.

[1379] W.G. Nowak. Zur asymptotischen Verteilung der Potenzen komplexer Zahlen. *Arch. Math.*, 40:69–72, 1983.

[1380] W.G. Nowak and R.F. Tichy. Einige weitere Resultate in Analogie zu einem Gleichverteilungssatz von Koksma. *Monatsh. Math.*, 92:203–220, 1981.

[1381] W.G. Nowak and R.F. Tichy. Über die Verteilung mod 1 der Potenzen reeller (2×2)-Matrizen. *Indag. Math.*, 43:219–230, 1981.

[1382] W.G. Nowak and R.F. Tichy. Zur Verteilung mod 1 der Potenzen komplexer Zahlen und gewisser Verallgemeinerungen im $\mathbf{R}^s$. *Arch. Math.*, 38:236–242, 1982.

[1383] W.G. Nowak and R.F. Tichy. An improved estimate on the distribution mod 1 of powers of real matrices. *Compos. Math.*, 49:283–289, 1983.

[1384] O.V. Nuzhdin and I.M. Sobol'. A new measure of irregularity of distribution. *J. Number Theory*, 39:367–373, 1991.

[1385] N. Obata. Density of natural numbers and the Lévy group. *J. Number Theory*, 30:288–297, 1988.

[1386] N. Obata. A note on certain permutation groups in the infinite dimensional rotation group. *Nagoya Math. J.*, 109:91–107, 1988.

[1387] A.M. Odlyzko. Discrete logrithms in finite fields and their cryptographic significance. *Lect. Notes Comput. Sci.*, 209:244–316, 1985.

[1388] A.M. Odlyzko and L. B. Richmond. On the unimodality of high convolutions of discrete distributions. *Ann. Probab.*, 13:299–306, 1985.

[1389] R.W.K. Odoni and P.G. Spain. Equidistribution of values of rational functions (mod p). *Proc. Edinburgh Math. Soc.*, 125:911–929, 1995.

[1390] Y. Ohkubo. On the weighted uniform distribution of sequences $(a_n x)$. *Comment. Math. Univ. St. Pauli*, 32:61–76, 1983.

[1391] Y. Ohkubo. Discrepancy with respect to weighted means of some sequences. *Proc. Japan Acad. A*, 62:201–204, 1986.

[1392] Y. Ohkubo. The weighted discrepancies of some slowly increasing sequences. *Math. Nachr.*, 174:239–251, 1995.

[1393] M. Olivier. Répartition des valeurs de la fonction "sommes des chiffres". *Sémin. Théor. Nombres Bordeaux*, Exp. 16:7 p., 1971.

[1394] M. Olivier. Répartition des valeurs de la fonction "somme des chiffres". *Semin. Delange-Pisot-Poitou, 12e Annee 1970/71, Theor. des Nombres*, Exp. 15:5 p., 1972.

[1395] I. Oren. Ergodicity of cylinder flows arising from irregularities of distribution. *Isr. J. Math.*, 127–138:44, 1983.

[1396] V.A. Oskolkov. Hardy-Littlewood problems on uniform distribution of arithmetic progressions. *Izv. Akad. Nauk*, 54:159–172, 1990.

[1397] V.A. Oskolkov. The Hardy-Littlewood problem for regular and uniformly distributed number sequences. *Izv. Ross. Akad. Nauk Ser. Mat.*, 58:153–166, 1994.

[1398] A. Ostrowski. Bemerkungen zur Theorie der Diophantischen Approximationen. *Abh. Math. Semin. Univ. Hamburg*, 1:77–98, 1922.

[1399] A.M. Ostrowski. On rational approximation to an irrational number. *Rend. Semin. Mat. Fis. Milano*, 47:241–256, 1977.

[1400] A.M. Ostrowski. On the distribution function of certain sequences (mod 1). *Acta Arith.*, 37:85–104, 1980.

[1401] A.M. Ostrowski. On the error term in multidimensional diophantine approximation. *Acta Arith.*, 41:163–183, 1982.

[1402] A.B. Owen. Randomly permuted (t, m, s)-nets and (t, s)-sequences. *Lect. Notes Stat.*, 106:299–317, 1995.

[1403] G. Pagès. van der Corput sequences, Kakutani transforms and one-dimensional numerical integration. *J. Comput. Appl. Math.*, 44:21–39, 1992.

[1404] J.D. Parker. The period of the Fibonacci random number generator. *Discrete Appl. Math.*, 20:145–164, 1988.

[1405] W. Parry. On the β-expansion of real numbers. *Acta Math. Hung.*, 11:401–416, 1960.

[1406] S. H. Paskov and J.F. Traub. Computing high dimensional integrals with applications to finance. *Journal of portfolio management*, 1995. to appear.

[1407] M. Pastéka. On distribution functions of sequences. *Acta Math. Univ. Comen.*, 50/51:227–235, 1987.

[1408] M. Pastéka. Solution of one problem from the theory of uniform distribution. *C.R. Acad. Bulg. Sci.*, 41:29–31, 1988.

[1409] M. Pastéka. Convergence of series and submeasures of the set of positive integers. *Math. Slovaca*, 40:273–278, 1990.

[1410] M. Pastéka. Covering densities. *Math. Slovaca*, 42:593–614, 1992.

[1411] M. Pastéka. Some properties of Buck's measure density. *Math. Slovaca*, 42:15–32, 1992.

[1412] M. Pastéka. Remarks on Buck's measure density. *Tatra Mountains Math. Publ.*, 3:191–200, 1993.

[1413] M. Pastéka. Densities. *Atti Sem. Mat. Fis. Univ. Modena*, 42:601–624, 1994.

[1414] M. Pastéka. Measure density of some sets. *Math. Slovaca*, 44:515–524, 1994.

[1415] M. Pastéka. A note about the submeasures and Fermat last theorem. *Ricerche Mat.*, 43:79–90, 1994.

[1416] M. Pastéka and S. Porubský. On distribution of sequences of integers. *Math. Slovaca*, 43:521–539, 1993.

[1417] M. Pastéka and T. Salát. Buck's measure density and sets of positive integers containing arithmetic progression. *Math. Slovaca*, 41:283–293, 1991.

[1418] M. Pastéka and R.F. Tichy. Distribution problems in Dedekind domains and submeasures. *Ann. Univ. Ferrara, Nuova Ser., Sez. VII*, 40:191–206, 1994.

[1419] J. Patarin. How to construct pseudorandom and super pseudorandom permutations from one single pseudorandom function. *Lect. Notes Comput. Sci.*, 658:256–266, 1993.

[1420] A.I. Pavlov. On the distribution mod one and Benford's law. *Izv. Akad. Nauk*, 45:760–774, 1981.

[1421] C.E.M. Pearce and M.S. Keane. On normal numbers. *J. Austral. Math. Soc. A*, 32:79–87, 1982.

[1422] P.B. Peart. The dispersion of the Hammersley sequence in the unit square. *Monatsh. Math.*, 94:249–262, 1982.

[1423] W. Pechlaner. Numerische Analysen zur Diskrepanz von Sprachen. *Österr. Akad. Wiss. SB II*, 190:245–251, 1981.

[1424] A. Perelli and U. Zannier. An Omega result in uniform distribution theory. *Q. J. Math., Oxford II. Ser.*, 38:95–102, 1987.

[1425] Y. Peres. Application of Banach limits to the study of sets of integers. *Isr. J. Math.*, 62:17–31, 1988.

[1426] Y.F. Pétermann. On Golomb's self describing sequence. *J. Number Theory*, 53:13–24, 1995.

[1427] G.M. Petersen. Sequences with the strong Weyl property. *J. Natl. Acad. Math.*, 2:107–110, 1984.

[1428] K. Petersen. On a series of cosecants related to a problem of ergodic theory. *Compos. Math.*, 26:313–317, 1973.

[1429] A. Pethő. Perfect powers in second order linear recurrences. *J. Number Theory*, 15:5–13, 1982.

[1430] A. Pethő and R.F. Tichy. S-unit equations, linear recurrences and digit expansions. *Publ. Math. Debrecen*, 42:145–154, 1993.

[1431] V. Petrov. *Sums of Independent Random Variables*. Springer, Berlin, 1975.

[1432] W. Philipp. Mixing sequences of random variables and probabilistic number theory. *Memoirs Am. Math. Soc.*, 114, 1971.

[1433] W. Philipp. Empirical distribution functions and uniform distribution mod 1. *Diophantine Approx. Appl.*, 211–234, 1973.

[1434] W. Philipp. Limit theormes for lacunary series and uniform distribution mod 1. *Acta Arith.*, 26:241–251, 1975.

[1435] W. Philipp. A functional law of the iterated logarithm for empirical distribution functions of weakly dependent random variables. *Ann. Probab.*, 5:319–350, 1977.

[1436] W. Philipp. Empirical distribution functions and strong approximation theorems for dependent random variables. A problem of Baker in probabilistic number theory. *Trans. Am. Math. Soc.*, 345:705–727, 1994.

[1437] W. Philipp and W. Stout. Invariance principles for martingales and sums of independent random variables. *Math. Z.*, 192:253–264, 1986.

[1438] R.G.E. Pinch. A sequence well distributed in the square. *Math. Proc. Cambridge Phil. Soc.*, 99:19–22, 1986.

[1439] R.S. Pinkham. On the distribution of first significant digits. *Ann. Math. Stat.*, 32:1223–1230, 1961.

[1440] M. Pollicott. A note on unfiform distribution for primes and closed orbits. *Isr. J. Math.*, 55:199–212, 1986.

[1441] M. Pollicott and K. Simon. The Hausdorff dimension of λ-expansions with deleted digits. *Trans. Am. Math. Soc.*, 347:967–983, 1994.

[1442] A.D. Pollington. The Hausdorff dimension of a set of non-normal well approximate numbers. *Lect. Notes Math.*, 751:256–264, 1979.

[1443] A.D. Pollington. On the density of sequences $\{n_k\xi\}$. *Illinois J. Math.*, 23:511–515, 1979.

[1444] A.D. Pollington. The Hausdorff dimension of certain sets related to sequences which are not dense mod 1. *Q. J. Math., Oxford II. Ser.*, 31:351–361, 1980.

[1445] A.D. Pollington. Progression arithmétiques généralisées et le problème des $(3/2)^n$. *C.R. Acad. Sci., Paris*, 292:383–384, 1981.

[1446] A.D. Pollington. On generalised arithmetic and geometric progression. *Acta Arith.*, 40:255–262, 1982.

[1447] A.D. Pollington. Sur les suites $(k\theta^n)$. *C.R. Acad. Sci., Paris*, 296:941–943, 1983.

[1448] A.D. Pollington and R.C. Vaughan. The k-dimensional Duffin and Schaffer conjecture. *J. Theor. Nombres Bordx.*, 1:81–88, 1989.

[1449] A.D. Pollington and R.C. Vaughan. The k-dimensional Duffin and Schaffer conjecture. *Mathematika*, 37:190–200, 1990.

[1450] G. Pólya and G. Szegö. *Aufgaben und Lehrsätze aus der Analysis I,II.* Springer, Berlin-Heidelberg-New York (3. Aufl.), 1964.

[1451] H. Porta and K.B. Stolarsky. Wythoff pairs as semigroup invariants. *Adv. Math.*, 85:69–82, 1991.

[1452] S. Porubsky, T. Salát, and O. Strauch. Transformation that preserve uniform distribution. *Acta Arith.*, 49:459–479, 1988.

[1453] S. Porubsky, T. Salát, and O. Strauch. On a class of uniformly distributed sequences. *Math. Slovaca*, 40:143–170, 1990.

[1454] S. Porubsky and O. Strauch. Transformations that preserve uniform distribution II. *Grazer Math. Ber.*, 318:173–182, 1992.

[1455] L.P. Postnikov. Geometry of numbers and diophantine approximation. *Inst. Fiziki i Mathematiki Akademii Nauk*, 241, 1975.

[1456] L. P. Postnikova. Quantitative Form des Borelschen Problems. *Acta Arith.*, 21:235–250, 1972.

[1457] L.P. Postnikova. Ein konstruktives Problem über die gebrochenen Anteile der Exponentialfunktion. *Saratov*, 4:80–88, 1972.

[1458] B.J. Powell and T. Salát. Convergence of subseries of the harmonic series and asymptotic densities of sets of positive integers. *Publ. Inst. Math., Nouv. Sér.*, 50, 64:60–70, 1991.

[1459] P.D. Proinov. Square discrepancy of symmetrical lattices. *Moscow. Univ. Math. Bull.*, 29:57–65, 1975.

[1460] P.D. Proinov. Über die quadratische Abweichung symmetrischer Gitter. *Vestn. Mosk. Univ., Ser. I*, 30:41–47, 1975.

[1461] P.D. Proinov. An error estimation from above of a general quadrature process with positive weights. *C.R. Acad. Bulg. Sci.*, 35:605–608, 1982.

[1462] P.D. Proinov. Points on constant type and estimates from above for the quadratic deviation of a certain class of infinite sequences. *C.R. Acad. Bulg. Sci.*, 35:753–755, 1982.

[1463] P.D. Proinov. Estimation of L^2-discrepancy of a class of infinite sequences. *C.R. Acad. Bulg. Sci.*, 36:37–40, 1983.

[1464] P.D. Proinov. Note on the convergence of the general quadrature process with positive weights,in "Constructive Function Theory 77". *Proc. Internat. Conf. Blagoevgrad*, 121–125, 1984.

[1465] P.D. Proinov. On the extreme and L^2-discrepancies of symmetric finite sequences. *Serdica*, 10:376–383, 1984.

[1466] P.D. Proinov. Uniformly distributed matrices and numerical integration. *Constr. theory of functions*, 704–709, 1984.

[1467] P.D. Proinov. Generalization of two results of the theory of uniform distribution. *Proc. Am. Math. Soc.*, 95:527–532, 1985.

[1468] P.D. Proinov. Numerical integration and approximation of differentiable functions. *C.R. Acad. Bulg. Sci.*, 38:187–190, 1985.

[1469] P.D. Proinov. On an inequality in the theory of uniform distribution. *C.R. Acad. Bulg. Sci.*, 38:1465–1468, 1985.

[1470] P.D. Proinov. On the L^2-discrepancy of some infinite sequences. *Serdica*, 11:3–12, 1985.

[1471] P.D. Proinov. Exact lower bounds for the p-discrepancy. *C.R. Acad. Bulg. Sci.*, 39:39–41, 1986.

[1472] P.D. Proinov. On irregulaities of distribution. *C.R. Acad. Bulg. Sci.*, 39:31–34, 1986.

[1473] P.D. Proinov. Numerical integration and approximation of differentiable functions, II. *J. Approximation Theory*, 50:373–393, 1987.

[1474] P.D. Proinov. Discrepancy and integration of continuous functions. *J. Approximation Theory*, 52:121–131, 1988.

[1475] P.D. Proinov. Integration of smooth functions and φ-discrepancy. *J. Approximation Theory*, 52:284–292, 1988.

[1476] P.D. Proinov. On the Erdős-Turán inequality on uniform distribution. I. *Proc. Japan Acad. A*, 64:27–28, 1988.

[1477] P.D. Proinov. On the Erdős-Turán inequality on uniform distribution. II. *Proc. Japan Acad. A*, 64:49–52, 1988.

[1478] P.D. Proinov. On the inequalities of Erdős-Turán and Berry-Esseen I. *Proc. Japan Acad. A*, 64:381–384, 1988.

[1479] P.D. Proinov. Symmetrization of the van der Corput generalized sequences. *Proc. Japan Acad. A*, 64:159–162, 1988.

[1480] P.D. Proinov. On the inequalities of Erdős-Turán and Berry-Esseen II. *Proc. Japan Acad. A*, 65:17-20, 1989.

[1481] P.D. Proinov and V.A. Andreeva. Note on a theorem of Koksma on uniform distribution. *C.R. Acad. Bulg. Sci.*, 39:41–44, 1986.

[1482] P.D. Proinov and E.Y. Atanassov. On the distribution of the van der Corput generalized sequences. *C.R. Acad. Sci., Paris*, 307:895–900, 1988.

[1483] P.D. Proinov and V.S. Grozdanov. Symmetrization of the van der Corput-Halton sequence. *C.R. Acad. Bulg. Sci.*, 40:5–8, 1987.

[1484] P.D. Proinov and V.S. Grozdanov. On the diaphony of the van der Corput-Halton sequence. *J. Number Theory*, 30:94–104, 1988.

[1485] P.D. Proinov and G.H. Kirov. On a quadrature formula for numerical integragtion in the class $C'[0,1]$. *C. R. Acad. Bulg. Sci.*, 36:1027–1030, 1983.

[1486] P.D. Proinov and G.H. Kirov. Application of uniformly distributed matrices for approximation of functions and numerical integration. *C. R. Acad. Bulg. Sci.*, 37:1625–1628, 1984.

[1487] P.D. Proinov and D.R. Peeva. Numerical summation of multiple sums. *Mathematics and education in Math., Proc. 14th Spring Conf., Sunny Beach/Bulg.*, 269–274, 1985.

[1488] P.D. Proinov and P.T. Tonchev. On the discrepancy of the three-dimensional nets of Sobol. *Mathematics and education in mathematics, Proc. 14th Spring Conf., Sunny Beach/Bulg.*, 275–279, 1985.

[1489] O. Prunner. *Verallgemeinerte Folgenspektren*. Univ. Wien, 1979.

[1490] L.D. Pustyl'nikov. On new estimates for the Weyl sums and the remainder term in the distribution of the fractional parts of a polynomial. *Russ. Math. Surv.*, 36:195–196, 1981.

[1491] L.D. Pustyl'nikov. On new estimates of Weyl sums and the remainder term in the law of distribution of fractional parts of a polynomial. *Usp. Mat. Nauk*, 36:203–204, 1981.

[1492] L.D. Pustyl'nikov. Generalized continued fractions and estimates for Weyl sums and the remainder term in the law of the distribution of the fractional parts of the values of a polynomial. *Mat. Zametki*, 56:144–148, 1994.

[1493] M. Queffelec. Application de la théorie des algèbres de mesures à l'étude des mesures spectrales. *Sémin. Théor. Nombres 1978-1979*, Exp. 20:9 p., 1979.

[1494] M. Queffelec. Mesures spectrales associées a certaines suites arithmétiques. *Bull. Soc. Math. Fr.*, 107:385–421, 1979.

[1495] Ch. Radoux. Propriétés de distribution de la suite des nombres de Bell réduite modulo p premier. *C.R. Acad. Sci., Paris*, 285:653–655, 1977.

[1496] Ch. Radoux. Suites à croissance presque géométrique et répartition modulo 1. *Bull. Soc. Math. Belg. Ser. A*, 42:659–671, 1990.

[1497] I. Radovic, I.M. Sobol', and R.F. Tichy. Quasi-Monte Carlo methods for numerical integration - a comparison of different low-discrepancy sequences. *Monte Carlo Methods and Applications*, 2:1–14, 1996.

[1498] R.A. Raimi. On the distribution of first significant digits. *Am. Math. Monthly*, 76:342–348, 1969.

[1499] L. Ramshaw. On the gap structure of sequences of points on a circle. *Indag. Math.*, 40:527–541, 1978.

[1500] L. Ramshaw. On the discrepancy of the sequence formed by the multiples of an irrational number. *J. Number Theory*, 13:138–175, 1981.
[1501] G. Rauzy. Croissance et répartition modulo 1. *Sémin. Théor. Nombres Bordeaux*, Exp. 28:9 p., 1972.
[1502] G. Rauzy. Fonctions entieres et répartition modulo 1. *Bull. Soc. Math. Fr.*, 100:409–415, 1972.
[1503] G. Rauzy. Etude de quelques ensembles de fonctions définis par des propriétés de moyenne. *Sémin. Théor. Nombres Bordeaux*, Exp. 20:18 p., 1973.
[1504] G. Rauzy. Fonctions entières et répartition modulo un. *Bull. Soc. Math. Fr.*, 101:185–192, 1973.
[1505] G. Rauzy. Fonctions entières et répartition modulo 1. *Bull. Soc. Math. Fr.*, 37:137–138, 1974.
[1506] G. Rauzy. Equirépartition et entropie. *Lect. Notes Math.*, 475:155–175, 1975.
[1507] G. Rauzy. Nombres normaux et processus déterministes. *Astérisque*, 24–25:263–265, 1975.
[1508] G. Rauzy. Nombres normaux et processus déterministes. *Acta Arith.*, 29:211–225, 1976.
[1509] G. Rauzy. Propriétés statistiques de suites arithmétiques. *Collection SUP. Le Math.*, 15:133, 1976.
[1510] G. Rauzy. Sur une suite liée à la discrépance de la suite $(n\alpha)$. *C.R. Acad. Sci., Paris*, 282:1323–1325, 1976.
[1511] G. Rauzy. Répartition de suites et équations fonctionnelles associées. *Monatsh. Math.*, 83:315–329, 1977.
[1512] G. Rauzy. Répartition modulo 1. *Astérisque*, 41–42:81–101, 1977.
[1513] G. Rauzy. Les zéros entiers des fonctions entièresde type exponentiel. *Sémin. Théor. Nombres Bordeaux*, Exp. 6:10 p., 1978.
[1514] G. Rauzy. Echanges d'intervalles et transformations induites. *Acta Arith.*, 34:315–328, 1979.
[1515] G. Rauzy. Discrépance d'une suite complètement équirépartie. *Ann. Fac. Sci. Toulouse, Math.*, 3:105–112, 1981.
[1516] G. Rauzy. Itération des endomorphismes d'un intervalle. *Theorie des nombres, Semin. Delange-Pisot-Poitou, Paris 1979-80, Prog. Math.*, 12:255–261, 1981.
[1517] G. Rauzy. Suites à termes dans un alphabet fini. *Sémin. Théor. Nombres Bordeaux*, Exp. 25:16 p., 1983.
[1518] G. Rauzy. Des mots en arithmétique. *Pub. Dép. Math.*, 6/B:103–113, 1984.
[1519] G. Rauzy. Ensembles à restes bornés. *Sémin. Théor. Nombres Bordeaux*, Exp. 24:12 p., 1984.
[1520] G. Rauzy. Rotations sur les groupes, nombres algebriques, et substitutions. *Sémin. Théor. Nombres Bordeaux*, Exp. 21:12 p., 1988.
[1521] T.V. Ravenstein. On the discrepancy of the sequence formed from multiples of an irrational number. *Bull. Austral. Math. Soc.*, 31:329–338, 1985.
[1522] A. Razborov, E. Szemeredi, and A. Wigderson. Construction of small sets that are uniform in arithmetic progressions. *Comb. Prob. and Comput.*, 2:513–518, 1993.
[1523] A. Reich. Eulerprodukte und diskrete Mittelwerte. *Monatsh. Math.*, 90:303–309, 1980.
[1524] A. Reich. Dirichletreihen und gleichverteilte Folge. *Analysis*, 1:303–312, 1981.
[1525] W. Reitgruber. Funktionen von beschränkter gewichteter Schwankung. *Österr. Akad. Wiss. SB II*, 196:463–494, 1987.
[1526] W. Reitgruber. *Numerische Analysis und Funktionen von beschränkter Schwankung.* Technische Universität Wien, 1987.
[1527] W. Reitgruber and R.F. Tichy. Numerische Lösung gewisser partieller Differentialgleichungen mittels guter Gitterpunkten. *Z. Angew. Math. Mech.*, 69:109–111, 1989.
[1528] A. Renyi. Representations for reals numbers and the ergodic properties. *Acta Math. Hung.*, 8:477–493, 1957.
[1529] M. Reversat. Un critére d'eutaxie. *C.R. Acad. Sci., Paris*, 277:405–407, 1973.
[1530] M. Reversat. Sur les approximations diophantiennes asymptotiques. *C.R. Acad. Sci., Paris*, 282:1395–1397, 1976.
[1531] M. Reversat. Sur les approximations diophantiennes simultanées asymptotiques. *Acta Arith.*, 34:329–348, 1979.
[1532] S.G. Revesz and I.Z. Ruzsa. On approximating Lebesgue inegrals by Riemmann sums. *Glasgow Math. J.*, 33:129–134, 1992.
[1533] G. Rhin. Répartition modulo 1 dans un corps de séries formelles sur un corps fini. *Dissertationes Math.*, 95:75, 1972.

[1534] G. Rhin. Equirépartition modulo 1 des suites lacunaires. *Sémin. Théor. Nombres Bordeaux*, Exp. 19:5 p., 1973.

[1535] G. Rhin. Répartition modulo 1 de $f(p_n)$ quand f est une fonction entière. *Semin. Delange-Pisot-Poitou, 14e annee 1972/73, Theorie des Nombres*, Exp. 20:3 p., 1973.

[1536] G. Rhin. Sur la répartition modulo 1 des suites $f(p)$. *Acta Arith.*, 23:217–248, 1973.

[1537] G. Rhin. Répartition modulo 1 de $f(p)$ quand f est une fonction entière. *C.R. Acad. Sci., Paris*, 280:1259–1261, 1975.

[1538] G. Rhin. Répartition modulo 1 de $f(p)$ quand f est une série entière. *Lect. Notes Math.*, 475:176–244, 1975.

[1539] G. Rhin. Sur la répartition modulo 1 de quelques suites arithmétiques. *Semin. Delange-Pisot-Poitou, 18e Annee, 1976/77, Theor. des Nombres,*, Exp. G3:4 p., 1977.

[1540] G.J. Rieger. Bemerkungen über gewisse nichtlineare Kongruenzen. *Elem. Math.*, 32:113–115, 1977.

[1541] H. Rindler. Ein Problem aus der Theorie der Gleichverteilung. *Math. Z.*, 135:73–92, 1973.

[1542] H. Rindler. Ein Gleichverteilungsbegriff für mittelbare Gruppen. *Österr. Akad. Wiss. SB II*, 182:107–119, 1974.

[1543] H. Rindler. Ein Problem aus der Theorie der Gleichverteilung. *Monatsh. Math.*, 78:51–67, 1974.

[1544] H. Rindler. Gleichverteilte Folge in Banachräumen. *Anz. Österr. Akad. Wiss.*, 93–95, 1974.

[1545] H. Rindler. Gleichverteilte Folgen von Operatoren. *Compos. Math.*, 29:201–211, 1974.

[1546] H. Rindler. Gleichverteilte Ketten. *Anz. Österr. Akad. Wiss.*, 162–166, 1974.

[1547] H. Rindler. Teilweise gleichverteilte Folgen. *Anz. Österr. Akad. Wiss.*, 213–217, 1974.

[1548] H. Rindler. A remark on Weyl's criterion on uniformly distributed sequences. *Proc. Nederl. Akad. Wet.*, A 78:93–95, 1975.

[1549] H. Rindler. Zur L^1-Gleichverteilung auf abelschen und kompakten Gruppen. *Arch. Math.*, 26:209–213, 1975.

[1550] H. Rindler. Gleichverteilte Folgen in lokalkompakten Gruppen. *Monatsh. Math.*, 82:207–235, 1976.

[1551] H. Rindler. Uniform distribution on locally compact groups. *Proc. Am. Math. Soc.*, 54:417–422, 1976.

[1552] H. Rindler. Uniformly distributed sequences in quotient groups. *Acta Sci. Math.*, 38:153–156, 1976.

[1553] H. Rindler. Fast konstante Folgen.II. *Anz. Österr. Akad. Wiss.*, 4–6, 1978.

[1554] H. Rindler. Eine Charakterisierung gleichverteilter Folgen. *Arch. Math.*, 32:185–199, 1979.

[1555] H. Rindler. Fast konstante Folgen. *Acta Arith.*, 35:189–193, 1979.

[1556] H. Rindler. Lineare Gleichverteilung. *Manuscr. Math.*, 30:103–106, 1979.

[1557] H. Rindler and J. Schoißengeier. Gleichverteilte Folgen und differenzierbare Funktionen. *Monatsh. Math.*, 84:125–131, 1977.

[1558] R. Rivest, A. Shamir, and L. Adleman. On digital signatures and public-key cryptosystems,. *Commun. ACM*, 21:120–126, 1978.

[1559] A. Rizzi. Generation of pseudo-random binary sequences by primitive polynomials. *Statistica*, 42:193–207, 1982.

[1560] A.D. Rogers. A functional from geometry with applications to discrepancy estimtes and the radon transform. *Trans. Am. Math. Soc.*, 341:275–313, 1994.

[1561] J. Rosenblatt. Uniform distribution in compact groups. *Mathematika*, 23:198–207, 1976.

[1562] J. Rosenblatt. Universally bad sequences in ergodic theory, almost everywhere convergence (II). *A. Bellow, R. Jones, eds., Academic Press*, 227–245, 1991.

[1563] J.M. Rosenblatt and M. Wierdl. Pointwise ergodic theorems via harmonic analysis. *London Math. Soc. Lect. Note Ser.*, 205:3–151, 1995.

[1564] G. Rote. Sequences with subword complexity $2n$. *J. Number Theory*, 46:196–213, 1994.

[1565] G. Rote and R.F. Tichy. Spherical dispersion with an application to polygonal approximation of curves. *Anz. Österr. Akad. Wiss*, 132:3–10, 1995.

[1566] G. Rote and R.F. Tichy. Quasi-Monte-Carlo methods and the dispersion of point sequences. *Math. Comput. Modelling*, 23:9–23, 1996.

[1567] K. F. Roth. On irregularities of distribution. *Mathematika*, 1:73–79, 1954.

[1568] K.F. Roth. Remark concerning integer sequences. *Acta Arith.*, 9:257–260, 1964.

[1569] K.F. Roth. On irregularities of distribution. *Acta Arith.*, 35:373–384, 1979.

[1570] K.F. Roth. On irregularities of distribution.IV. *Acta Arith.*, 37:67–75, 1980.

[1571] R. Rotondi. Programming and statistical control of p-adic pseudo-random sequences. *Rend., Sci. Math. Appl., A*, 121:3–19, 1987.

[1572] J. Rousseau-Egele. Sur un théorème d'équidistribution forte. *C.R. Acad. Sci., Paris*, 286:567–569, 1978.

[1573] L.A. Rubel. Uniform distribution in locally compact groups. *Comment. Math. Helv.*, 39:253–258, 1965.

[1574] R.Y. Rubinstein. Generating random vectors uniformly distributed inside and on the surface of different regions. *Eur. J. Oper. Res.*, 10:205–209, 1982.

[1575] W. Rudin. *Real and Complex Analysis*. McGraw-Hill, New York, 1974.

[1576] I.Z. Ruzsa. On difference-sequences. *Acta Arith.*, 25:151–157, 1974.

[1577] I.Z. Ruzsa. On difference sets. *Stud. Sci. Math. Hung.*, 13:319–326, 1978.

[1578] I.Z. Ruzsa. Uniform distribution, positive trigonometric polynomials and difference sets. *Sémin. Théor. Nombres Bordeaux*, Exp. 18:18 p., 1982.

[1579] I.Z. Ruzsa. On the uniform and almost uniform distribution of $(a_n x)$ mod 1. *Sémin. Théor. Nombres Bordeaux*, Exp. 20:21 p., 1983.

[1580] I.Z. Ruzsa. Connections between the uniform distribution of a sequence and its difference. *Topics in classical number theory, Colloq. Budapest 1981, Colloq. Math. Soc. Janos Bolyai*, 34:1419–1443, 1984.

[1581] I.Z. Ruzsa. The discrepancy of rectangles and squares. *Grazer Math. Ber.*, 318:135–140, 1992.

[1582] I.Z. Ruzsa. Generalizations of Kubilius' class of additive functions I. *New Trends Probab. Stat.*, 2:269–283, 1992.

[1583] I.Z. Ruzsa. On an inequality of Erdős and Turán concerning uniform distribution modulo one. I. *Sets, Graphs, and Numbers, Conf. Proc., Colloq. Math. Soc. Janos Bolyai*, 60:621–630, 1993.

[1584] I.Z. Ruzsa. On an inequality of Erdős and Turán concerning uniform distribution modulo one. II. *J. Number Theory*, 49:84–88, 1994.

[1585] B.P. Rynne. The Hausdorff dimension of sets of points whose simultaneous rational approximation by sequences of integer vectors have errors with small product. *J. Number Theory*, 48:75–79, 1994.

[1586] B. Sadeghiyan and J. Pieprzyk. A construction of super pseudorandom permutations from a single pseudorandom function. *Lect. Notes Comput. Sci.*, 658:267–284, 1993.

[1587] B. Saffari and R.C. Vaughan. On the fractional parts of x/n and related sequences. I. *Ann. Inst. Fourier*, 26:115–131, 1977.

[1588] B. Saffari and R.C. Vaughan. On the fractional parts of x/n and related sequences. II. *Ann. Inst. Fourier*, 27:1–30, 1977.

[1589] B. Saffari and R.C. Vaughan. On the fractional parts of x/n and related sequences. III. *Ann. Inst. Fourier*, 27:31–36, 1977.

[1590] M. Saint-André. *Equirépartition et analyse harmonique commutative*. Universite de Sherbrooke, Sherbrooke, Québec, 1972.

[1591] T. Salát. Zu einigen Fragen der Gleichverteilung (mod 1). *Czech. Math. J.*, 18:476–488, 1968.

[1592] T. Salát. Bemerkung über die Verteilung von Ziffern in Cantorschen Reihen. *Czech. Math. J.*, 23:497–499, 1973.

[1593] T. Salát. On exponents of convergence of subsequences. *Czech. Math. J.*, 34:362–370, 1984.

[1594] T. Salát. Criterion for uniform distribution of sequences and a class of Riemann integrable functions. *Math. Slovaca*, 37:199–203, 1987.

[1595] T. Salát. On good sequences of positive integers. preprint.

[1596] T. Salát and R. Tijdeman. On density measures of sets of positive integers. *Topics in classical number theory, Colloq. Budapest 1981, Vol. I, Colloq. Math. Soc. Janos Bolyai*, 34:1445–1457, 1981.

[1597] T. Salát and R. Tijdeman. Asymptotic densities of sets of positive integers. *Math. Slovaca*, 33:199–207, 1983.

[1598] J.W. Sander. On a conjecture of Zaremba. *Monatsh. Math.*, 104:133–137, 1987.

[1599] J.W. Sander. On the $(3n+1)$-conjecture. *Acta Arith.*, 55:241–248, 1990.

[1600] P.K. Sarkar and M.A. Prasad. A comparative study of pseudo and quasi random sequences for the solution of integral equations. *J. Comput. Phys.*, 68:66–88, 1987.

[1601] A. Sárközy. On the distribution of residues of products of integers. *Acta Math. Hung.*, 49:397–401, 1987.

[1602] P. Schatte. On random variables with logarithmic mantissa distribution relative to several bases. *Elektr. Inform. Kybernetik*, 17:293–295, 1981.

[1603] P. Schatte. On the asymptotic uniform distribution of sums reduced mod 1. *Math. Nachr.*, 115:257–281, 1983.

[1604] P. Schatte. On H_∞-summability and the uniform distribution of sequences. *Math. Nachr.*, 113:237–243, 1984.

[1605] P. Schatte. On the asymptotic logarithmic distribution of the floating-point mantissas of sums. *Math. Nachr.*, 127, 1984.

[1606] P. Schatte. Some estimates of the H_∞-uniform distribution. *Monatsh. Math.*, 103:233–240, 1987.

[1607] P. Schatte. Note on the discrepancy of well-distributed sequences. *Math. Nachr.*, 139:109–113, 1988.

[1608] P. Schatte. On mantissa distributions in computing and Benford's law. *J. Inf. Process. Cybern.*, 24:443–455, 1988.

[1609] P. Schatte. On the discrepancy of well-distributed sequences. *Math. Nachr.*, 136:345–358, 1988.

[1610] P. Schatte. On the uniform distribution of certain sequences and Benford's law. *Math. Nachr.*, 136:271–273, 1988.

[1611] P. Schatte. On well-distribution with respect to weighted means. *Math. Nachr.*, 138:125–130, 1988.

[1612] P. Schatte. On measures of uniformly distributed sequences and Benford's law. *Monatsh. Math.*, 107:245–256, 1989.

[1613] P. Schatte. On Benford's law for continued fractions. *Math. Nachr.*, 148:137–144, 1990.

[1614] P. Schatte. On a uniform law of the iterated logarithm for sums mod 1 and Benford's law. *Lith. Math. J.*, 31:133–142, 1991.

[1615] P. Schatte. On transformations of distribution functions on the unit interval – a generalization of the Gauss-Kuzmin-Levy theorem. *Z. Anal. Anwend.*, 12:273–283, 1993.

[1616] P. Schatte. On Benford's law to variable base. preprint.

[1617] P. Schatte and K. Nagasaka. A note on Benford's law for second order linear recurrences with perodical coefficients. *Z. Anal. Anwend.*, 10:251–254, 1991.

[1618] W. Schempp and B. Dreseler. *Einführung in die harmonische Analyse*. Teubner, Stuttgart, 1980.

[1619] J. Schiffer. Discrepancy of normal numbers. *Acta Arith.*, 47:175–186, 1986.

[1620] A. Schinzel, H.P. Schlickewei, and W.M. Schmidt. Small solutions of quadratic congruence and small fractional parts of quadratic forms. *Acta Arith.*, 37:241–248, 1980.

[1621] J. Schmeling and R. Winkler. Typical dimension of the graph of certain functions. *Monatsh. Math.*, 119:303–320, 1995.

[1622] J. Schmid. The joint distribution of the binary digits of integer multiples. *Acta Arith.*, 43:391–415, 1984.

[1623] W.M. Schmidt. Normalität bezüglich Matrizen. *J. Reine Angew. Math.*, 214/215:227–260, 1964.

[1624] W.M. Schmidt. Irregularities of distribution. *Q. J. Math., Oxford II. Ser.*, 19:181–191, 1968.

[1625] W.M. Schmidt. Irregularities of distribution. II. *Trans. Am. Math. Soc.*, 136:347–360, 1969.

[1626] W.M. Schmidt. Irregularities of distribution. III. *Pac. J. Math.*, 29:225–234, 1969.

[1627] W.M. Schmidt. Irregularities of distribution. IV. *Invent. Math.*, 7:55–82, 1969.

[1628] W.M. Schmidt. Irregularities of distribution. V. *Proc. Am. Math. Soc.*, 25:608–614, 1970.

[1629] W.M. Schmidt. Irregularities of distribution.VI. *Compos. Math.*, 24:63–74, 1972.

[1630] W.M. Schmidt. Irregularities of distribution.VII. *Acta Arith.*, 21:45–50, 1972.

[1631] W.M. Schmidt. Irregularities of distribution. VIII. *Trans. Am. Math. Soc.*, 198:1–22, 1974.

[1632] W.M. Schmidt. Irregularities of distribution.IX. *Acta Arith.*, 27:385–396, 1975.

[1633] W.M. Schmidt. Irregularities of distribution. X. *Number Theory and Algebra (Academic Press)*, 311–329, 1977.

[1634] W.M. Schmidt. *Lectures on irregularities of distribution*. Tata Inst. Lect. Notes on Mathematics, Bombay, 1977.

[1635] W.M. Schmidt. On the distribution modulo 1 of the sequence $\alpha n^2 + \beta n$. *Canadian J. Math.*, 29:819–826, 1977.
[1636] W.M. Schmidt. *Small fractional parts of polynomials.* Am. Math. Soc., Illinois, 1977.
[1637] W.M. Schmidt. *Diophantine Approximation.* Lect. Notes Math. 785, Springer, Berlin, 1980.
[1638] W.M. Schmidt. On cubic polynomials. I: Hua's estimate of exponential sums. *Monatsh. Math.*, 93:63–74, 1982.
[1639] W.M. Schmidt. Bounds for exponential sums. *Acta Arith.*, 44:281–297, 1984.
[1640] W.M. Schmidt. Bemerkungen zur Polynomdiskrepanz. *Österr. Akad. Wiss. SB II*, 202:173–177, 1995.
[1641] P. Schmitt. Lineare Gleichverteilung. *Manuscr. Math.*, 12:271–283, 1974.
[1642] P. Schmitt. Linear uniform distribution. *Lect. Notes Math.*, 475:245–252, 1975.
[1643] P. Schmitt. Lineare Gleichverteilung im $\mathbf{R}^s$. *Anz. Österr. Akad. Wiss.*, 139–142, 1975.
[1644] R. Schneider. Berechnung von Homologiegruppen auf gleichverteilten Ketten. *Österr. Akad. Wiss. SB II*, 194:215–228, 1985.
[1645] R. Schneider. Capacity of sets and uniform distribution of sequences. *Monatsh. Math.*, 104:67–81, 1987.
[1646] J. Schoißengeier. Gleichmäßige Gleichverteilung in lokalkompakten Räumen. *Anz. Österr. Akad. Wiss.*, 95–101, 1974.
[1647] J. Schoißengeier. Gleichmäßige Gleichverteilung in lokalkompakten Räumen. *Österr. Akad. Wiss. SB II*, 184:171–202, 1975.
[1648] J. Schoißengeier. Der Zusammenhang zwischen den Nullstellen der ζ-Funktion und Folgen der Form $g(p)$ mod 1, p-prim. *Österr. Akad. Wiss. SB II*, 187:183–195, 1978.
[1649] J. Schoißengeier. Eine neue Diskrepanz für gewisse Primzahlfolgen. *Österr. Akad. Wiss. SB II*, 187:219–224, 1978.
[1650] J. Schoißengeier. Über die Diskrepanz gewisser Folgen mod 1 mit Hilfe Diophantischer Approximation. *Math. Z.*, 159:169–173, 1978.
[1651] J. Schoißengeier. Über die Diskrepanz von Folgen (αn). *Österr. Akad. Wiss. SB II*, 187:225–236, 1978.
[1652] J. Schoißengeier. The connection between the zeros of the ζ-function and sequences $(g(p))$, p prime, mod 1. *Monatsh. Math.*, 87:21–52, 1979.
[1653] J. Schoißengeier. Zur Metrik der Diskrepanz. *Monatsh. Math.*, 87:313–316, 1979.
[1654] J. Schoißengeier. On the discrepancy of sequences (αn). *Acta Math. Hung.*, 38:29–43, 1981.
[1655] J. Schoißengeier. On the discrepancy of $(n\alpha)$. *Acta Arith.*, 44:241–279, 1984.
[1656] J. Schoißengeier. Über die Diskrepanz der Folgen $(n\alpha)$. *Lect. Notes Math.*, 1114:148–153, 1985.
[1657] J. Schoißengeier. Abschätzungen für $\sum B_1(n\alpha)$. *Monatsh. Math.*, 102:59–77, 1986.
[1658] J. Schoißengeier. On the discrepancy of $(n\alpha)$. II. *J. Number Theory*, 24:54–64, 1986.
[1659] J. Schoißengeier. The discrepancy of $(n\alpha)_{n\geq 1}$. *Math. Ann.*, 296:529–545, 1993.
[1660] J. Schoißengeier. On the longest gaps in the sequence $(n\alpha)$ mod 1. *Grazer Math. Ber.*, 318:155–166, 1993.
[1661] J.P. Schreiber. Approximations diophantiennes et problèmes additivs dans les groups abéliens localement compacts. *Bull. Soc. Math. Fr.*, 101:297–332, 1974.
[1662] R. Schulze. Determinierte Simulation von Zufallsvektoren. *Wiss. Z. Humboldt-Univ.*, 30:449–453, 1981.
[1663] F. Schweiger. *Ergodic theory of fibred systems and metric number theory.* Clarendon Press, Oxford, 1995.
[1664] F. Schweiger. The hidden entropy. *Anz. Österr. Akad. Wiss.*, 131:31–38, 1995.
[1665] W.A. Sentance. A further analysis of Benford's law. *Fibonacci Q.*, 11:490–494, 1973.
[1666] P. D. Seymour and T. Zaslavsky. Averaging sets. *Adv. Math.*, 52:213–240, 1984.
[1667] F. Sezgin. Some remarks on a new composite random number generator. *Comput. Math. Appl.*, 30:125–130, 1995.
[1668] L. Shapiro. Regularities of distribution. *Stud. in probability and ergodic theory*, 2:135–154, 1978.
[1669] S. Shi. Optimal uniform distributions generated by m-sequences. *Acta Math. Sin.*, 22:123–128, 1979.
[1670] S.Z. Shi. Estimate of error for quadrature of a multidimensional continuous function (Chinese). *Math. Numer. Sinica*, 3:360–364, 1982.

[1671] S. Shih. Une généralisation des "bons treillis". *C.R. Acad. Sci., Paris*, 290:527–530, 1980.
[1672] I. Shiokawa. Asymptotic distributions of digits in integers. *Number theroy, Elementary and analytic, Proc. Conf. Budapest, Colloq. Math. Soc. Janos Bolyai*, 51:505–525, 1990.
[1673] I. Shiokawa and J.-I. Tamura. Description of sequences defined by billiards in the cube. *Proc. Japan Acad. A*, 68:207–211, 1992.
[1674] I. Shiokawa and S. Uchiyama. On some properties of the dyadic Champernowne numbers. *Acta Math. Hung.*, 26:9–27, 1975.
[1675] B.M. Shirokov. Distribution of values of arithmetic functions in residue classes. *J. Sov. Math.*, 29:1356–1363, 1985.
[1676] J.-S. Shiue. A remark of a paper by Bundschuh. *Tamkang J. Math.*, 4:129–130, 1973.
[1677] J.-S. Shiue. On a paper by Johnson and Newman. *Soochow J. Math.*, 1:17–20, 1975.
[1678] J.-S. Shiue and Ming-H. Hu. Some remarks on the uniform distribution of a linear recurrence of the second order. *Tamkang J. Math.*, 4:101–103, 1973.
[1679] I.E. Shparlinskij. On complete uniform distribution. *Zh. Vychisl. Mat. Mat. Fiz.*, 19:1330–1333, 1979.
[1680] I.E. Shparlinskij. Distribution of fractional parts of recurrent sequences. *Zh. Vychisl. Mat. Mat. Fiz.*, 21:1588–1591, 1981.
[1681] I.E. Shparlinskij. On a property of a sequence of real numbers and the rate of convergence of some interpolation processes. *Proc. 2nd All-Union Conf. on Methods of Comput. Math. Krasnojarsk*, 47–48, 1982.
[1682] I.E. Shparlinskij. On a certain property of a multiplicative generator of pseudo-random numbers. *Zh. Vychisl. Mat. Mat. Fiz.*, 23:222–223, 1983.
[1683] I.E. Shparlinskij. On the rate of convergency of the Newton interpolation process and the size of some codes. *Uspechi Matem. Nauk*, 39:205–206, 1984.
[1684] I.E. Shparlinskij. Distribution of fractional parts of recurrent sequences. *Studies in number theory. Analytical number theory, Interuniv. Sci. Collect., Saratov*, 10:111–120, 1988.
[1685] I.E. Shparlinskij. On some generalizations of Chebyshev polynomials. *Sib. Mat. Zh.*, 31:217–218, 1990.
[1686] I.E. Shparlinskij. On the distribution of values of recurring sequences and the Bell numbers in finite fields. *Eur. J. Comb.*, 12:81–87, 1991.
[1687] I.E. Shparlinskij. On some characteristics of uniformity of distribution and their applications. *Comput. Algebra and Number Theory*, Kluwer, Dordrecht, *Math. Appl., Dordr.*, 325:227–241, 1995.
[1688] T. Siegl and R.F. Tichy. Lösungsverfahren eines Risikomodells bei exponentiell fallender Schadensverteilung. *Mitt., Schweizer Ver. Versicherungsmathematiker*, pages 95–118, 1996.
[1689] W. Sierpiński. Sur une courbe dont tout point est un point de ramification. *C.R. Acad. Sci., Paris*, 160:302–305, 1915.
[1690] K. Sigmund. Nombres normaux et théorie ergodique. *Lect. Notes Math.*, 532:202–215, 1976.
[1691] J.St.C.L. Sinnadurai. Note on a paper by A. Topuzoglu. *Indag. Math.*, 44:353–354, 1982.
[1692] M.M. Skriganov. The geometry of numbers and uniform distributions. *Sov. Math., Dokl.*, 43:878–881, 1991.
[1693] M.M. Skriganov. Constructions of uniform distributions in terms of geometry of numbers. *Algebra Anal.*, 6:200–230, 1994.
[1694] B.F. Skubenko. Cyclic sets of numbers and lattices. *Zap. Nauchn. Semin. Leningr. Otd. Mat. Inst. Stekl.*, 160:151–158, 1987.
[1695] N. B. Slater. Gaps and steps for the sequence $n\theta$ mod 1. *Math. Proc. Cambridge Phil. Soc.*, 63:1115–1123, 1967.
[1696] J. Slivka and N.C. Severo. Measures of sets decomposing the simply normal numbers in the unit interval. *Mathematika*, 41:164–172, 1994.
[1697] I.H. Sloan and S. Joe. *Lattaice methods for multiple integration.* Oxford Science Publication, Oxford, 1994.
[1698] B.G. Sloss and W.F. Blyth. Walsh functions and uniform distribution mod 1. *Tohoku Math. J.*, 45:555–563, 1993.
[1699] C.J. Smyth. Inequalities relating different definitions of discrepancy. *J. Austral. Math. Soc. A*, 17:81–87, 1974.
[1700] I.M. Sobol'. An accurate error estimate for multidimensional quadrature formulae for the functions of the class L_p (Russian). *Dokl. Akad. Nauk. SSSR*, 132:1041–1044, 1960.

[1701] I.M. Sobol'. The distribution of points in a cube and the approximate evaluation of integrals. *Zh. Vycisl. Mat. Mat. Fiz.*, 7:784–802, 1967.

[1702] I.M. Sobol'. Calculation of improper integrals using uniformly distributed sequences. *Sov. Math., Dokl.*, 14:734–738, 1973.

[1703] I.M. Sobol'. Uniformly distributed sequences with an additional uniform property. *U.S.S.R. Comput. Math. Math. Phys.*, 16:236–242, 1977.

[1704] I.M. Sobol'. Points, uniformly filing a multidimensional cube. *Mathematika, Kybernetika*, 2:32, 1985.

[1705] I.M. Sobol'. Determination of the extremal values of a function of several variables satisfying a general Lipschitz Condition. *U.S.S.R. Comput. Maths. Math. Phys.*, 28:112–118, 1988.

[1706] I.M. Sobol'. Quadrature formulae for functions of several variables satisfying a general Lipschitz Condition. *U.S.S.R. Comput. Maths. Math. Phys.*, 29:201–206, 1989.

[1707] I.M. Sobol'. *Die Monte-Carlo-Methode.* Deutscher Verlag der Wissenschaften, Berlin, 1991.

[1708] I.M. Sobol'. An efficient approach to multicriteria optimum design problems. *Surv. Math. Ind.*, 1:259–281, 1992.

[1709] I.M. Sobol'. A global search for multicriterial problems. *Multiple criteria decision making, Proc. 9th Int. Conf. Theory Appl. Bus. Ind. Gov., Fairfax/VA USA 1990*, 401–412, 1992.

[1710] I.M. Sobol. *A Primer for the Monte Carlo Method.* CRC Press, Boca Raton, 1994.

[1711] I.M. Sobol' and S.G. Bakin. On the crude multidimensional search. *J. Comput. Appl. Math.*, 56:283–293, 1994.

[1712] I.M. Sobol' and B. V. Shukhman. Random and quasirandom sequences: Numerical estimates of uniformity of distribution. *Österr. Akad. Wiss., SB II*, 201:161–167, 1992.

[1713] I.M. Sobol' and Boris V. Shukhman. On computational experiments in uniform distribution. *Österr. Akad. Wiss. SB II*, 201:161–167, 1992.

[1714] I.M. Sobol' and B.V. Shukhman. Integration with quasirandom sequences numerical experience. *Int. J. of Modern Physics C*, 6:263–275, 1995.

[1715] I.M. Sobol and B.V. Shukman. Random and quasirandom sequences: Numerical estimates of uniformity of distribution. *Math. Comput. Modelling*, 18:39–45, 1993.

[1716] I.M. Sobol' and A.V. Tutunnikov. A variante reducing multiplier for Monte Carlo integrations. *Math. Comput. Modelling*, 23:87–96, 1996.

[1717] B. Solomyak. Conjugates of beta-numbers and the zero-free domain for a class of analytic functions. *Proc. London Math. Soc.*, 68:477–498, 1994.

[1718] L. Somer. Distribution of residues of certain second-order linear recurrences modulo p. *Applications of Fibonacci Numbers*, 3:311–324, 1990.

[1719] L. Somer. Distribution of residues of certain second-order linear recurrences modulo p-II. *Fibonacci Q.*, 29:72–78, 1991.

[1720] L. Somer. Possible restricted periods of certain Lucas sequences modulo p. *Applications of Fibonacci Numbers*, 4:289–298, 1991.

[1721] L. Somer. Periodicity properties of kth-order linear recurrences with irreducible characteristic polynomial over a finite field. *Finite Fields, Coding Theory and Advances in Comm. and Comp. (G.L. Mullen and P.J.-S. Shiue, eds.)*, 195–207, 1993.

[1722] L. Somer. Upper bounds for frequencies of elements in second-order recurrences over a finite field. *Applications of Fibonacci Numbers*, 5:527–546, 1993.

[1723] L. Somer. Distribution of residues of certain second-order linear recurrences modulo p-III. *Applications of Fibonacci Numbers*, 6, 1996.

[1724] V.T. Sós. On the theory of diophantine approximations. I. *Acta Math. Hung.*, 8:461–472, 1957.

[1725] V.T. Sós. On the discrepancy of the sequence $(n\alpha)$. *Top. Number Theory*, 13:359–367, 1976.

[1726] V.T. Sós. On strong irregularities of the distribution of $(n\alpha)$ sequences. *Studies in pure Math.*, 685–700, 1983.

[1727] V.T. Sós and S.K. Zaremba. The mean-square discrepancies of some two-dimensional lattices. *Stud. Sci. Math. Hung.*, 14:255–271, 1979.

[1728] J. Spanier. Quasi-Monte Carlo Methods for Particle Transport Problems. *Lect. Notes Stat.*, 106:121–148, 1995.

[1729] J. Spencer. Sequences with small discrepancy relative to n events. *Compos. Math.*, 47:365–392, 1982.

[1730] V. Sprindzhuk. *Metric theory of Diophantine approximations*. Wiley (Translation from Russian), New-York, 1979.
[1731] W. Squire. Fibonacci cubature. *Fibonacci Q.*, 19:313–314, 1981.
[1732] S. Srinivasan. On two-dimensional Hammersley's sequences. *J. Number Theory*, 10:421–429, 1978.
[1733] S. Srinivasan and R.F. Tichy. Uniform distribution of prime power sequences. *Anz. Österr. Akad. Wiss.*, 33–36, 1993.
[1734] O. P. Stackelberg. Metric theorems related to the Kronecker-Weyl theorem mod m. *Monatsh. Math.*, 82:57–69, 1976.
[1735] O.P. Stackelberg. A uniform law of the iterated logarithm for functions C-uniformly distributed mod 1. *Indiana Univ. Math. J.*, 21:515–528, 1971.
[1736] J. M. Steele. Shortest paths through pseudo-random points in the d-cube. *Proc. Am. Math. Soc.*, 80:130–134, 1980.
[1737] H. Stegbuchner. Eine mehrdimensionale Version der Ungleichung von LeVeque. *Monatsh. Math.*, 87:167–169, 1979.
[1738] H. Stegbuchner. Zur quantitativen Theorie der Gleichverteilung mod 1. *Ber. Math. Inst. Univ. Salzburg*, 3:9–58, 1980.
[1739] H. Stegbuchner. Numerische Quadratur glatter Funktionen mit Gleichverteilungsmethoden. *Rend. Mat. Appl.*, 2:593–599, 1982.
[1740] S.A. Stepanov. On estimating rational trigonometric sums with prime denominator. *Tr. Mat. Inst. Steklova*, 112:346–371, 1971.
[1741] C. L. Stewart and R. Tijdeman. On infinite difference sets. *Canad. J. Math.*, 31:897–910, 1979.
[1742] M. Stewart. Irregularities of uniform distribution. *Acta Math. Hung.*, 37:185–221, 1981.
[1743] K.B. Stolarsky. Sums of distances between points on a sphere. *Proc. Am. Math. Soc.*, 41:575–582, 1973.
[1744] R.G. Stoneham. On absolute (j,ε)-normality in the rational fractions with applications to normal numbers. *Acta Arith.*, 22:277–286, 1973.
[1745] R.G. Stoneham. On the uniform ε-distribution of residues within the periods of rational fractions with applications to normal numbers. *Acta Arith.*, 22:371–389, 1973.
[1746] R.G. Stoneham. Some further results concerning (j,ε)-normality in the rationals. *Acta Arith.*, 26:83–96, 1974.
[1747] R.G. Stoneham. Normal recurring decimals, normal periodic systems, (j,ε)-normality, and normal numbers. *Acta Arith.*, 28:349–361, 1976.
[1748] O. Strauch. Duffin-Schaeffer conjecture and some new types of real sequences. *Acta Math. Univ. Comen.*, 40-41:233–265, 1982.
[1749] O. Strauch. Two properities of the sequence $n\alpha$ (mod1). *Acta Math. Univ. Comen.*, 44/45:67–73, 1984.
[1750] O. Strauch. Some applications of Franel's integral I. *Acta Math. Univ. Comen.*, 50/51:237–246, 1987.
[1751] O. Strauch. Some applications of Franel-Kluyver's integral.II. *Math. Slovaca*, 39:127–140, 1989.
[1752] O. Strauch. On the L^2-discrepancy of distances of points from a finite sequence. *Math. Slovaca*, 40:245–259, 1990.
[1753] O. Strauch. An improvement of an inequality of Koksma. *Indag. Math.*, 3:113–118, 1992.
[1754] O. Strauch. L^2-Discrepancy. *Math. Slovaca*, 44:601–632, 1994.
[1755] O. Strauch. A new moment problem of distribution functions in the unit interval. *Math. Slovaca*, 44:171–211, 1994.
[1756] O. Strauch. Uniformly maldistributed sequences in a strict sense. *Monatsh. Math.*, 120:153–164, 1995.
[1757] O. Strauch. A new type of criterions for limit laws of sequences. *Math. Slovaca*, to appear.
[1758] E. Strzelecki. On sequences $\{\xi t_n (\text{mod } 1)\}$. *Canadian Math. Bull.*, 18:727–738, 1975.
[1759] W. Stute. Convergence rates for the isotropic discrepancy. *Ann. Probab.*, 5:707–723, 1977.
[1760] I. Stux. On the uniform distribution of prime powers. *Commun. pure appl. Math.*, 27:729–740, 1974.
[1761] H. Sugita. Pseudo-random number generator by means of irrational rotation. *Monte Carlo Methods Appl.*, 1:35–57, 1995.

[1762] H. Sugiura. $3, 4, 5, 6$ dimensional good lattice points formulae. *Lect. Notes Numer. Appl. Anal.*, 14:181–195, 1995.

[1763] Y. Sun. On metric theorems in the theory of uniform distribution. *Compos. Math.*, 86:15–21, 1993.

[1764] Y. Sun. Some properties of uniformly distributed sequences. *J. Number Theory*, 44:273–280, 1993.

[1765] B. Sury. Fractional parts of log p and a digit function. *Expo. Math.*, 11:381–384, 1993.

[1766] S. Swierczkowski. On successive settings of an arc on the circumference of a circle. *Fund. Math.*, 46:187–189, 1958.

[1767] A. Szczuka. Completely uniformly distributed sequences as pseudorandom number generators. *Rocz. Pol. Tow. Mat., Ser. III, Mat. Stosow.*, 35:57–86, 1992.

[1768] A. Szczuka and R. Zielinski. Generators of pseudorandom numbers and gamma distributions. *Rocz. Pol. Tow. Mat., Ser. III, Mat. Stosow.*, 36:99–114, 1993.

[1769] G. Szegő. *Orthogonal Polynomials.* Am. Math. Soc. Colloq. Publ. 23. Am. Math. Soc., New York, 2nd edition, 1959.

[1770] P. Szüsz and B. Volkmann. On numbers with given digit distribution. *Arch. Math.*, 52:237–244, 1989.

[1771] P. Szüsz and B. Volkmann. A combinatorial method for constructing normal numbers. *Forum Math.*, 6:399–414, 1994.

[1772] S. Takahashi. An asmyptotic property of a gap sequence. *Proc. Japan Acad. A*, 38:101–104, 1962.

[1773] S. Takahashi. On lacunary trigonometric series. *Proc. Japan Acad. A*, 41:503–506, 1965.

[1774] S. Takahashi. On the lacunary Fourier series. *Tohoku Math. J.*, 19:79–85, 1967.

[1775] S. Takahashi. On lacunary trigonometric series. II. *Proc. Japan Acad. A*, 44:766–770, 1968.

[1776] S. Takahashi. On the central limit theorem for lacunary trigonometric series. *Tohoku Math. J.*, 20:289–295, 1968.

[1777] S. Takahashi. Lacunary trigonometric sum and probability. *Tohoku Math. J.*, 22:502–510, 1970.

[1778] S. Takahashi. On the law of the iterated logarithm for lacunary trigonometric series. *Tohoku Math. J.*, 24:319–329, 1972.

[1779] F.A. Talalyan. Uniformly distributed sequences and rearrangements. *Dokl., Akad. Nauk Arm. SSR*, 84:164–168, 1987.

[1780] T. Tapsoba. Automates calculant la complexité des suites automatiques. *J. Theor. Nombres Bordx.*, 6:127–134, 1994.

[1781] A. Tarski. Problème 38. *Fund. Math.*, 7:381, 1925.

[1782] R.J. Taschner. Probleme der analytischen Fortsetzung. *Österr. Akad. Wiss. SB II*, 185:459–484, 1977.

[1783] R.J. Taschner. Der Differenzensatz von van der Corput und gleichverteilte Funktionen. *Anz. Österr. Akad. Wiss.*, 115–118, 1978.

[1784] R.J. Taschner. Kriterien in der Theorie der Gleichverteilung. *Monatsh. Math.*, 86:221–237, 1978.

[1785] R.J. Taschner. Der Differenzensatz von van der Corput und gleichverteilte Funktionen. *J. Reine Angew. Math.*, 307/308:325–340, 1979.

[1786] R.J. Taschner. Gleichverteilte Doppelfolgen und eine Abschätzung der C-Diskrepanz. *Monatsh. Math.*, 88:321–330, 1979.

[1787] R.J. Taschner. The discrepancy of C-uniformly distributed multidimensional functions. *J. Math. Anal. Appl.*, 78:400–404, 1980.

[1788] R.J. Taschner. Eine untere Abschätzung der Diskrepanzen mehrdimensionaler Flüsse. *Anz. Österr. Akad. Wiss.*, 162–167, 1980.

[1789] R.J. Taschner. Eine zahlentheoretische Methode zur Bestimmung von Näherungslösungen einer Anfangswertaufgabe. *Österr. Akad. Wiss. SB II*, 189:277–284, 1980.

[1790] R.J. Taschner. Eine zahlentheoretische Methode zur Bestimmung von Näherungslösungen nichtlinearer Gleichungssysteme. *Österr. Akad. Wiss. SB II*, 189:285–289, 1980.

[1791] R.J. Taschner. Eine Ungleichung von van der Corput und Kemperman. *Monatsh. Math.*, 91:139–152, 1981.

[1792] R.J. Taschner. Metrische Sätze über die Vererbbarkeit der Gleichverteilung. *Österr. Akad. Wiss. SB II*, 193:193–200, 1985.

[1793] R.J. Taschner. A general vision of van der Corput's difference theorem. *Pac. J. Math.*, 104:231–239, 1987.
[1794] R.C. Tausworthe. Random numbers generated by linear recurrence modulo two. *Math. Comput.*, 19:201–209, 1965.
[1795] N. Temirgaliev. Upper estimate of the discrepancy of an "algebraic" lattice. *Vestn. Akad. Nauk Kaz. SSR*, 11:60–64, 1990.
[1796] V. N. Temlyakov. On a way of obtaining lower estimates for the errors of quadrature formulas. *Math. USSR Sbornik*, 71:247–257, 1992. (English transl.).
[1797] G. Tenenbaum. Sur la densité divisorielle d'une suite d'entiers. *J. Number Theory*, 15:331–346, 1982.
[1798] E. Teuffel. Einige asymptotische Eigenschaften der Quadratwurzelschnecke. *Math. Semesterber.*, 28:39–51, 1981.
[1799] S. Tezuka. Walsh-spectral test for GFSR peusdorandom numbers. *Commun. ACM*, 30:731–735, 1987.
[1800] S. Tezuka. The k-dimensional distribution of combined GFSR sequences. *Math. Comput.*, 62:809–817, 1994.
[1801] S. Tezuka. A unified view of long-period random number generators. *J. Oper. Res. Soc. Japan*, 137:211–227, 1994.
[1802] S. Tezuka. *Uniform random numbers. Theory and practice.* Kluwer Academic Publishers, Dordrecht, 1995.
[1803] S. Tezuka and M. Fushimi. A method of designing cellular automata as pseudroandom number generators for built-in-self-test for VLSI. *Contemp. Math.*, 168:363–367, 1994.
[1804] S. Tezuka and P. L'Ecuyer. Efficient and portable combined Tausworthe random number generators. *ACM Trans. Model. Comput. Simulation*, 1:99–112, 1991.
[1805] A. Thomas. Dimension de Hausdorff. *Bull. Soc. Math. Fr.*, 37:161–167, 1974.
[1806] A. Thomas. Discrepance en dimension un. *Ann. Fac. Sci. Toulouse, Math.*, 10:369–399, 1989.
[1807] J. P. Thouvenot. La convergence presque sure des moyennes ergodiques suivant certaines sous-suites d'entries (d'apres J. Bourgain). *Seminaire Bourbaki*, 189-190:133–153, 1990.
[1808] R.F. Tichy. Three examples of triangular arrays with optimal discrepancy and linear recurrences. *preprint.*
[1809] R.F. Tichy. Gleichverteilung von Mehrfachfolgen und Ketten. *Anz. Österr. Akad. Wiss.*, 174–179, 1978.
[1810] R.F. Tichy. Diskrepanz von Ketten. *Monatsh. Math.*, 87:317–324, 1979.
[1811] R.F. Tichy. Ein Analogon zum Satz von Koksma in Quaternionen. *Arch. Math.*, 32:346–348, 1979.
[1812] R.F. Tichy. Ein metrischer Satz in der Theorie der Gleichverteilung. *Österr. Akad. Wiss. SB II*, 188:317–327, 1979.
[1813] R.F. Tichy. Diophantische Approximation und Gleichverteilung in Oktaven. *Österr. Akad. Wiss. SB II*, 189:83–94, 1980.
[1814] R.F. Tichy. Diskrepanz von Ketten.II. *Monatsh. Math.*, 89:131–140, 1980.
[1815] R.F. Tichy. Zur Diskrepanz in Formalen Sprachen. *Österr. Akad. Wiss. SB II*, 189:203–207, 1980.
[1816] R.F. Tichy. Ein metrischer Satz in der Theorie der Gleichverteilung.II. *Anz. Österr. Akad. Wiss.*, 151–153, 1982.
[1817] R.F. Tichy. Einige Beiträge zur Gleichverteilung modulo Eins. II. *Anz. Österr. Akad. Wiss.*, 113–117, 1982.
[1818] R.F. Tichy. Einige Beträge zur Gleichverteilung Modulo Eins. *Anz. Österr. Akad. Wiss.*, 9–13, 1982.
[1819] R.F. Tichy. Zur Diskrepanz bezüglich gewichteter Mittel. *Manuscr. Math.*, 37:393–413, 1982.
[1820] R.F. Tichy. Diskrepanz bezüglich gewichteter Mittel und Konvergenzverhalten von Potenzreihen. *Manuscr. Math.*, 44:265–277, 1983.
[1821] R.F. Tichy. Zur Gleichverteilung bezüglich gewichteter Mittel. *Österr. Akad. Wiss. SB II*, 192:131–141, 1983.
[1822] R.F. Tichy. Beiträge zur Polynomdiskrepanz. *Österr. Akad. Wiss. SB II*, 193:513–519, 1984.
[1823] R.F. Tichy. Einige zahlentheoretische Ungleichungen. *Anz. Österr. Akad. Wiss.*, 89–96, 1984.
[1824] R.F. Tichy. Bemerkung zu einem Lemma aus der Variationsrechnung. *Lect. Notes Math.*, 1114:154–157, 1985.

[1825] R.F. Tichy. Gleichverteilung und zahlentheoretische Ungleichungen II. *Anz. Österr. Akad. Wiss.*, 95–99, 1985.
[1826] R.F. Tichy. Über eine zahlentheoretische Methode zur numerischen Integration und zur Behandlung von Integralgleichungen. *Österr. Akad. Wiss. SB II*, 193:329–358, 1985.
[1827] R.F. Tichy. Uniform distribution and diophantine inequalities. *Monatsh. Math.*, 99:147–152, 1985.
[1828] R.F. Tichy. Zur asymptotischen Verteilung linearer Rekursionen. *Anz. Österr. Akad. Wiss.*, 35–39, 1985.
[1829] R.F. Tichy. Beiträge zur C-Gleichverteilung. *Österr. Akad. Wiss. SB II.*, 195:179–190, 1986.
[1830] R.F. Tichy. On the asymptotic distribution of linear recurrence sequences. *Fibonacci Numbers and Their Applications*, Math. Appl., D. Reidel Publ. Co., 28:273–291, 1986.
[1831] R.F. Tichy. A criterion for the uniform distribution of sequences in compact metric spaces. *Rend. Circ. Mat. Palermo*, 36:332–342, 1987.
[1832] R.F. Tichy. Ein metrischer Satz über vollständig gleichverteilte Folgen. *Acta Arith.*, 48:197–207, 1987.
[1833] R.F. Tichy. Einige Bemerkungen über stetige Funktionen auf topologischen Gruppen. *Lect. Notes Math.*, 1262:139–143, 1987.
[1834] R.F. Tichy. Gleichverteilung zum Summierungsverfahren H_∞. *Math. Nachr.*, 131:119–125, 1987.
[1835] R.F. Tichy. Rekursive Folgen in Dedekind Ringen - einige Resultate und Probleme. *Contributions to General Algebra*, 5:401–406, 1987.
[1836] R.F. Tichy. Statistische Resultate über computergerechte Darstellungen von Zahlen. *Anz. Österr. Akad. Wiss.*, 1–8, 1987.
[1837] R.F. Tichy. Ein Approximationsverfahren zur Lösung spezieller partieller Differentialgleichungen. *Z. Angew. Math. Mech.*, 68:187–188, 1988.
[1838] R.F. Tichy. Random points in the cube and on the sphere with applications to numerical analysis. *J. Comput. Appl. Math.*, 31:191–197, 1990.
[1839] R.F. Tichy. Random points on the sphere with applications to numerical analysis. *Z. Angew. Math. Mech.*, 70:T642–T646, 1990.
[1840] R.F. Tichy. Contributions to uniformly distributed functions I. *Proc. Japan Acad. A*, 67:49–54, 1991.
[1841] R.F. Tichy. A general inequality with applications to the discrepancy of sequences. *Grazer Math. Ber.*, 313:65–72, 1991.
[1842] R.F. Tichy. Two distribution problems for polynomials. *Applications of Fibonacci numbers*, 5:561–568, 1992.
[1843] R.F. Tichy. Stability of a class of non-uniform random number generators. *J. Math. Anal. Appl.*, 181:546–561, 1994.
[1844] R.F. Tichy and K. W. Tomantschger. A Quasi-Monte-Carlo approach for an initial-value-probelm on Riemannian manifolds. *Österr. Akad. Wiss. SB II*, 201:89–95, 1992.
[1845] R.F. Tichy and G. Turnwald. Uniform distribution of recurrences in Dedekind domains. *Acta Arith.*, 46:81–89, 1985.
[1846] R.F. Tichy and G. Turnwald. Gleichmäßige Diskrepanzabschätzung für Ziffernsummen. *Anz. Österr. Akad. Wiss.*, 17–21, 1986.
[1847] R.F. Tichy and G. Turnwald. Logarithmic uniform distribution $(\alpha n + \beta \log n)$. *Tsukuba J. Math.*, 10:351–366, 1986.
[1848] R.F. Tichy and G. Turnwald. On the discrepancy of some special sequences. *J. Number Theory*, 26:68–78, 1987.
[1849] R.F. Tichy and G. Turnwald. Weak uniform distribution of $u_{n+1} = au_n + b$ in Dedekind domains. *Manuscr. Math.*, 61:11–22, 1988.
[1850] R.F. Tichy and R. Winkler. Gleichverteilung in hyperkomplexen Systemen. *Österr. Akad. Wiss. SB II*, 197:417–432, 1988.
[1851] R.F. Tichy and R. Winkler. Uniform distribution preserving mappings. *Acta Arith.*, 60:177–189, 1991.
[1852] R. Tijdeman. Note on Mahler's $\frac{3}{2}$-problem. *K. Norske Vidensk. Selsk. Skr.*, 16:1–4, 1972.
[1853] R. Tijdeman. On a distribution problem in finite and countable sets. *J. Comb. Theory, Ser. A*, 15:129–137, 1973.
[1854] R. Tijdeman. On integers with small prime factors. *Compos. Math.*, 26:319–330, 1973.

[1855] R. Tijdeman. The chairman assignment problem. *Discrete Math.*, 32:323–330, 1980.
[1856] R. Tijdeman. On the discrepancy of a sequence in [0,1]. *Theorie des nombres, Semin. Delange-Pisot-Poitou, Paris 1979-80, Prog. Math.*, 12:317–329,1981.
[1857] R. Tijdeman. A progress report on discrepancy. *Journées arithmétiques, Metz 1981, Asterisque*, 94:175–185,1982.
[1858] R. Tijdeman. On complementary triples of Sturmian bisequences. preprint.
[1859] R. Tijdeman and M. Voorhoeve. Bounded discrepancy sets. *Compos. Math.*, 42:375–389, 1981.
[1860] R. Tijdeman and G. Wagner. A sequence has almost nowhere small discrepancy. *Monatsh. Math.*, 90:315–329, 1980.
[1861] E.C. Titchmarsh. *The theory of the Riemann Zeta-Function.* Clarenden Pr., Oxford, 1988.
[1862] Ph. Toffin. Conditions suffisantes d'équirépartition modulo 1. Probleme de Waring-Goldbach pour $f(x) = x^c$, c non entier. *Sémin. Théor. Nombres Bordeaux*, Exp.21:8 p., 1975.
[1863] Ph. Toffin. Conditions suffisantes pour qu'une suite réelle soit équirépartie modulo 1. *C.R. Acad. Sci., Paris*, 280:693–696, 1975.
[1864] Ph. Toffin. Conditions suffisantes d'équirépartition modulo 1 de suites $(f(n))_{n\in N}$ et $(f(p_n))_{n\in N}$. *Acta Arith.*, 32:365–385, 1977.
[1865] Z. Toleuov and A.S. Fainleib. Eine Verschärfung eines Grenzwertsatzes für additive Funktionen, die auf der Wertemenge eines Polynoms von gewissen Folgen gegeben sind. *Litov. Mat. Sb.*, 11:367–382, 1971.
[1866] D.I. Tolev. On the simultaneous distribution of the fractional parts of different powers of prime numbers. *Number Theory*, 37:298–306, 1991.
[1867] Y. H.. Too. On the uniform distribution modulo one of some log-like sequences. *Proc. Japan Acad. A*, 68:269–272, 1992.
[1868] A. Topuzoglu. On u.d. mod 1 of sequences $(a_n x)$. *Indag. Math.*, 43:231–236, 1981.
[1869] A. Topuzoglu. Repeated rearrangements of uniformly distributed sequences. *Fac. Arts Sci.*, 8:15–24, 1985.
[1870] E. Topuzoglu. Discrepancy in compact Abelian groups. *Karadeniz Univ. Math. J.*, 5:275–284, 1983.
[1871] E. Topuzoglu. Estimates of discrepancy mod 1 of slowly growing sequences $(a_n x)$. *J. Pure Appl. Sci., Ankara*, 16:241–252, 1983.
[1872] J. F. Traub, G.W. Wasilkowski, and H. Woźniakowski. *Information-Based Complexity.* Academic Press, London, 1980.
[1873] J.F. Traub and H. Wozniakowski. *A general theory of optimal algorithms.* Academic Press, New York, 1980.
[1874] N. Trendafilov. Uniformly distributed sequences and vector-valued integration. *C.R. Acad. Bulg. Sci.*, 40:23–25, 1987.
[1875] N. Trendafilov. The pettis integration via uniform distributed sequences and its application. *Österr. Akad. Wiss. SB II*, 198:317–325, 1989.
[1876] A. Tripathi. A comparison of dispersion and Markov constants. *Acta Arith.*, 63:193–203, 1993.
[1877] M. Tsuji. On the uniform distribution of numbers mod 1. *J. Math. Soc. Japan*, 4:313–322, 1952.
[1878] P. Turán. On some phenomena in the theory of partitions. *Astérisque*, 24-25:311–319, 1975.
[1879] G. Turnwald. Gleichverteilung von linearen rekursiven Folgen. *Österr. Akad. Wiss. SB II*, 193:201–245, 1985.
[1880] G. Turnwald. Uniform distribution of second-order linear recurring sequences. *Proc. Am. Math. Soc.*, 96:189–198, 1986.
[1881] G. Turnwald. Einige Bemerkungen zur diskreten Gleichverteilung. *Lect. Notes Math.*, 1262:144–149, 1987.
[1882] G. Turnwald. Weak uniform distribution of second-order linear recurring sequences. *Lect. Notes Math.*, 1380:242–253, 1989.
[1883] S. Uchiyama. A note on the uniform distribution of sequences of integers. *J. Fac. Sci.*, 3:163–169, 1968.
[1884] J. D. Vaaler. A Tauberian theorem related to Weyl's criterion. *Lect. Notes Math.*, 475:253–258, 1975.

[1885] J.D. Vaaler. A Tauberian theorem related to Weyl's criterion. *J. Number Theory*, 9:71–78, 1977.
[1886] J.D. Vaaler. Limit theorems for uniformly distributed p-adic sequences. *Acta Arith.*, 39:83–94, 1981.
[1887] J.D. Vaaler. Some extremal functions in Fourier analysis. *Bull. Am. Math. Soc.*, 12:183–216, 1985.
[1888] J.D. Vaaler. Refinements of the Erdős-Turán inequality. *Number Theory with an Emphasis on the Markoff-Spectrum, W. Moran and A. Pollington (eds.)*, 263–269, 1993.
[1889] T. van Aardenne-Ehrenfest. Proof of the impossibility of a just distribution of an infinite sequence of points over an interval. *Proc. Kon. Ned. Akad. v. Wetensch.*, 48:266–71, 1945.
[1890] T. van Aardenne-Ehrenfest. On the impossibility of a just distribution. *Proc. Kon. Ned. Akad. v. Wetensch.*, 52:734–739, 1949.
[1891] J.G. van der Corput. Verteilungsfunktionen. I. *Proc. Kon. Ned. Akad. v. Wetensch.*, 38:813–821, 1935.
[1892] J.G. van der Corput. Verteilungsfunktionen. II. *Proc. Kon. Ned. Akad. v. Wetensch.*, 38:1058–1066, 1935.
[1893] J.G. Van der Corput. Verteilungsfunktionen. III. *Proc. Akad. Amsterdam*, 39:10–19, 1936.
[1894] J.G. Van der Corput. Verteilungsfunktionen. IV. *Proc. Akad. Amsterdam*, 39:19–26, 1936.
[1895] J.G. Van der Corput. Verteilungsfunktionen. V. *Proc. Akad. Amsterdam*, 39:149–153, 1936.
[1896] J.G. Van der Corput. Verteilungsfunktionen. VI. *Proc. Akad. Amsterdam*, 39:339–344, 1936.
[1897] J.G. Van der Corput. Verteilungsfunktionen. VII. *Proc. Akad. Amsterdam*, 39:489–494, 1936.
[1898] J.G. Van der Corput. Verteilungsfunktionen. VIII. *Proc. Akad. Amsterdam*, 39:579–590, 1936.
[1899] J.G. Van der Corput and C. Pisot. Sur la discrépance mudoulo un. *Indag. Math.*, 1:260–269, 1939.
[1900] V.N. Vapnik and A.Ya. Chervonenkis. On the uniform convergence of relative frequencies of events to their probabilities. *Theory Prob. Appl.*, 16:264–280, 1971.
[1901] I. Vardi. A relation between Dedekind sums and Kloosterman sums. *Duke Math.J.*, 55:189–197, 1987.
[1902] R.C. Vaughan. On the distribution of αp modulo 1. *Mathematika*, 24:135–141, 1978.
[1903] W. A. Veech. Some questions on uniform distribution. *Ann. of Math.*, 94:125–138, 1971.
[1904] W. A. Veech. Well distributed sequences of integers. *Trans. Am. Math. Soc.*, 161:63–70, 1971.
[1905] W.A. Veech. Some questions of uniform distribution. *Trans. Am. Math. Soc.*, 140:1–33, 1969.
[1906] W.A. Veech. Topological dynamics. *Bull. Am. Math. Soc.*, 83:775–830, 1977.
[1907] W.A. Veech. Ergodic theory and uniform distribution. *Astérisque*, 61:223–234, 1979.
[1908] W.Y. Velez. Uniform distribution of two-term recurrence sequences. *Trans. Am. Math. Soc.*, 301:37–45, 1987.
[1909] A. M. Vershik and N.A. Sidorov. Arithmetic expansions associated with the rotation of a circle and continued fractions. *Algebra Anal.*, 5:97–115, 1993.
[1910] N.J. Vilenkin. *Special Funktions and the Theory of Group Representations.* Am. Math. Soc., Providence, 1968.
[1911] I.M. Vinogradov. *The method of trigonometrical sums in the theory of numbers.* New York, 1954.
[1912] I.M. Vinogradov. Estimate of a prime-number trigonometric sum. *Izv. Akad. Nauk SSSR, Ser. Mat.*, 23:157–164, 1959.
[1913] I.M. Vinogradov. On the estimation of trigonometric sums. *Izv. Akad. Nauk SSSR, Ser. Mat.*, 29:493–504, 1965.
[1914] B. Volkmann. Verallgemeinerung eines Satzes von Maxfield. *J. Reine Angew. Math.*, 271:203–213, 1974.
[1915] B. Volkmann. Un problème concernant des nombres normaux. *Sémin. Théor. Nombres 1977-1978*, Exp. 27:3 p., 1978.
[1916] B. Volkmann. On modifying constructed normal numbers. *Ann. Fac. Sci. Toulouse, Math.*, 1:269–285, 1979.
[1917] B. Volkmann and P. Szüsz. On numbers containing each block infinitely often. *J. Reine Angew. Math.*, 339:199–206, 1983.

[1918] B. Volkmann and P. Szüsz. Sur des nombes normaux d'efnis par un polynôme. *Sémin. Théor. Nombres Bordeaux*, Exp. 42:4 p., 1988.
[1919] G. Wagner. On a problem of Erdős in diophantine approximation. *Bull. London Math. Soc.*, 12:81–88, 1980.
[1920] G. Wagner. Note on irregularities of distribution. *Monatsh. Math.*, 92:239–245, 1981.
[1921] G. Wagner. Irregularities of charge distributions. *Topics in classical number theory, Colloq. Budapest 1981, Vol. I, Colloq. Math. Soc. Janos Bolyai*, 34:1603–1616, 1984.
[1922] G. Wagner. Mixing properties of the linear permutation group. *Lect. Notes Math.*, 1068:260–267, 1984.
[1923] G. Wagner. On an imbalance problem in the theory of point distribution. *(Irregularities of partitions). Algorithms Comb.*, 8:153-160, 1989.
[1924] G. Wagner. On the product of distances to a point set on a sphere. *J. Austral. Math. Soc. A*, 47:466–482, 1989.
[1925] G. Wagner. On means of distances on the surface of a sphere. *Pac.J.Math.*, 144:389–398, 1990.
[1926] G. Wagner. On averaging sets. *Monatsh. Math.*, 111:69–78, 1991.
[1927] G. Wagner. On the maximal modulus of polynomials on Cantor sets. *J. Approximation Theory*, 67:1–18, 1991.
[1928] G. Wagner. Erdős-Turán inequalities for distance function on spheres. *Michigan Math. J.*, 39:17–34, 1992.
[1929] G. Wagner. On means of distances on the surface of a sphere ii. *Pac. J. Math.*, 154:381–396, 1992.
[1930] G. Wagner. On a new method for constructing good point sets on spheres. *Discr. Comput. Geom.*, 9:111–129, 1993.
[1931] G. Wagner. On rings of numbers which are normal to one base but non-normal to another. *J. Number Theory*, 54:211–231, 1995.
[1932] S. Wagon. Is π normal? *Math. Intell.*, 7:65–67, 1985.
[1933] S. Wagon. *The Banach-Tarski paradox.* Cambridge University Press, Cambridge, 1986.
[1934] S. S. Wagstaff. The period of the Bell exponential integers modulo a prime. *Math. of Comput. 1943-1993: a half century of comput. math.*, 595–598, 1993.
[1935] P. Walters. *An Intorduction to Ergodic Theory.* Springer, New York, 1975.
[1936] Y. Wang. On diophantine approximation and approximate analysis. *Kexue Tongbao*, 27:468–472, 1982.
[1937] Y. Wang. On diophantine approximation and approximate analysis. I. *Acta Math. Sin.*, 25:248-256, 1982.
[1938] Y. Wang. On diophantine approximation and approximate analysis. II. *Acta Math. Sin.*, 25:323-332, 1982.
[1939] Y. Wang. On number theoretic methods of numerical integration in multi-dimensional space I, II. *Acta Math. Appl. Sin.*, 1:106–114, 1982.
[1940] Y. Wang. Number theoretic method in numerical analysis. *Contemp.Math.*, 77:63–82, 1988.
[1941] Y. Wang and K. Fang. The number-theoretic method in applied statistics. *J. Sichuan Univ., Nat. Sci.Ed.*, 26:86–89, 1989.
[1942] Y. Wang and K. Fang. Number theoretic method in applied statistics. *Chin. Ann. Math.*, B 11:51-65, 1990.
[1943] Y. Wang and K. Yu. A note on some metrical theorems in diophantine approximation. *Bures-sur-Yvette*, page 18, 1979.
[1944] F. Warner. *Foundations of Differentiable Manifolds and Lie Groups.* Springer, Berlin, 1983.
[1945] T. T. Warnock. *Computational investigations of low discrepancy point sets.* Applications of Number Theory to Numerical Analysis. Academic Press, New York, 1972.
[1946] L.C. Washington. Benford's law for Fibonacci and Lucas numbers. *Fibonacci Q.*, 19:175–177, 1981.
[1947] G.W. Wasilkowski and Woźniakowski. The exponent of discrepancy is at most 1.4778 .... *Math. Comput.*, to appear, 1996.
[1948] W. Webb. Uniformly distributed functions in GF$[q,x]$ and GF$\{q,x\}$. *Ann. Mat. Pura Appl.*, 95:285–291, 1973.
[1949] W. Webb. Distribution of the first digits of Fibonacci numbers. *Fibonacci Q.*, 13:334–336, 1975.

[1950] W. Webb and C. T. Long. Distribution modulo p^h of the general linear second order recurrence. *Atti Accad.*, 58:92–100, 1975.

[1951] W.A. Webb. Uniformly distributed sequences of algebraic integers. *Bull. Inst. Math., Acad. Sin.*, 13:349–355, 1985.

[1952] T. Werbinski. Uniform distribution of some sets of integers. *Funct. Approximation, Comment. Math.*, 8:9–11, 1980.

[1953] H. Weyl. Über die Gleichverteilung von Zahlen mod. Eins. *Math. Ann.*, 77:313–352, 1916.

[1954] B.E. White. Mean-square discrepancies of the Hammersley and Zaremba sequences for arbitrary radix. *Monatsh. Math.*, 80:219–229, 1975.

[1955] B.E. White. On optimal extreme-discrepancy point sets in the square. *Numer. Math.*, 27:157–164, 1977.

[1956] B.A. Wichmann and I.D. Hill. An efficient and portable pseudo-random number generator. *Appl. Statist.*, 31:188–190, 1982.

[1957] B.A. Wichmann and I.D. Hill. Building a random-number generator. *Byte*, 12:128–128, 1987.

[1958] N. Wiener. The spectrum of an array and its application to the study of the translation poperties of a simple class of arithmetical functions. *Math. and Phys.*, 6:145–157, 1927.

[1959] N. Wiener. *The Fourier integral and certain of its applications.* Cambridge University Press, Cambridge, 1933.

[1960] R.S. Wikramaratna. ACORN - a new method for generating sequences of uniformly distributed pseudo-random numbers. *J. Comput. Phys.*, 83:16–31, 1989.

[1961] R. Winkler. Distribution preserving sequences of maps and almost constant sequences on finite sets. *Monatsh. Math., to appear.*

[1962] R. Winkler. Two results in uniform distribution theory. *Anz. Österr. Akad. Wiss.*, 57–59, 1987.

[1963] R. Winkler. Some constructive examples in uniform distribution of finite sets and normal numbers. *Anz. Österr. Akad. Wiss.*, 1:1–8, 1989.

[1964] R. Winkler. Strong Weyl property in uniform distribution. *Lect. Notes Math.*, 1452:208–220, 1990.

[1965] R. Winkler. Some remarks on pseudorandom sequences. *Math. Slovaca*, 43:493–512, 1993.

[1966] R. Winkler. Monotonic functions and an inequality of Myerson on point distributions. *Indag. Math.*, 6:247–255, 1995.

[1967] R. Winkler. Distribution preserving sequences of maps and almost constant sequences on finite sets. *Monatsh. Math., to appear*, 1996.

[1968] G. Wirsching. An improved estimate concerning $3n+1$ predecessor sets. *Acta Arith.*, 63:205–210, 1993.

[1969] M.A. Wodzak. Primes in arithmetic progression and uniform distribution. *Proc. Am. Math. Soc.*, 122:313–315, 1994.

[1970] S. Wolfram. Random sequence generation by cellular automata. *Adv. Appl. Math.*, 7:123–169, 1986.

[1971] D. Wolke. Zur Gleichverteilung einiger Zahlenfolgen. *Math. Z.*, 142:181–184, 1975.

[1972] H. Woźniakowski. Average case complexity of multivariate integration. *Bull. Am. Math. Soc.*, 24:185–194, 1991.

[1973] Y.-J. Xiao. Suites équiréparties associées aux automorphismes du tore. *C.R. Acad. Sci., Paris*, 311:579–582, 1990.

[1974] Y.-J. Xiao. Estimates for the volume of points of $(0, s)$-sequences in base $b \geq s \geq 2$. *Lect. Notes Stat.*, 106:362–372, 1995.

[1975] C. Xing and H. Niederreiter. A construction of low-discrepancy sequences using global function fields. *Acta Arith.*, 73:87–102, 1995.

[1976] K.Yu. Yavid. Estimate of the Hausdorff dimension of a set of singular vectors. *Dokl.Akad. Nauk BSSR*, 31:777–780, 1987.

[1977] K. Yosida. *Functional Analyis.* Springer, Berlin, 1974.

[1978] K. Yu. A note on a problem of Baker in metrical number theory. *Math. Proc. Cambridge Phil. Soc.*, 90:215–227, 1981.

[1979] K. Yu. Two results in metrical diophantine approximation. *Sémin. Théor. Nombres 1980-1981*, Exp. 28:8 p., 1981.

[1980] C.-K. Yuen. Testing random number generators by Walsh transform. *IEEE Trans. Comput.*, 26:329–333, 1977.

[1981] V.V. Yurinskii. On inequalities for large deviations for certain statistics. *Theory Prob. Appl.*, 16:385–387, 1971.

[1982] A. Zaharescu. Small values of $n^2\alpha$ (mod 1). *Invent. Math.*, 121:379–388, 1995.

[1983] A. Zame. On normal sets of numbers. *Acta Arith.*, 20:147–154, 1972.

[1984] S. K. Zaremba. Some applications of multidimensional integration by parts. *Ann. Pol. Math.*, 21:95–96, 1968.

[1985] S. K. Zaremba. Gleichverteilung, gute Gitterpunkte und Integration durch Monte-Carlo Methoden. *Grazer Math. Ber.*, 124:11, 1979.

[1986] S.K. Zaremba. Le discrépance isotrope et l'intégration numerique. *Ann. Mat. Pura Appl.*, 87:125–136, 1970.

[1987] S.K. Zaremba. La méthode des "bons treillis" pour le calcul des intégrales multiples. *Appl.Number Theory*, 39–119, 1972.

[1988] S.K. Zaremba. Computing the isotropic discrepancy of point sets in two dimensions. *Discrete Math.*, 11:79–92, 1975.

[1989] K. Zeller and W. Beekmann. *Theorie der Limitierungsverfahren*. Springer, Berlin, 1970.

[1990] Q. Zhang. On convergence of the averages. *Monatsh. Math.*, 122:275–300, 1996.

[1991] S. Zhang and S. Zhou. A pseudo-random number generator. *Math. Pract. Theory*, 1:33–34, 1992.

[1992] Z. Zheng. Dedekind sums and uniform distribution (mod 1). *Acta Math. Sin.*, 11:62–67, 1995.

[1993] Y. Zhu. Construction of normal numbers. *Acta Math. Sin.*, 24:508–515, 1981.

[1994] Y. Zhu. On some multidimensional quadrature formulas with number-theoretic nets. *Acta Math. Appl. Sin.*, 9:335–347, 1993.

[1995] Y. Zhu. On some multidimensional quadrature formulas with number-theoretic nets. II. *Acta Math. Sin.*, 10:86–98, 1994.

[1996] Y. Zhu. Formulas of exact calculation of discrepancy of low-dimensional finite point sets II. *Chin. Sci. Bull.*, 40:610–612, 1995.

[1997] P. Zinterhof. Über einige Abschätzungen bei der Approximation von Funktionen mit Gleichverteilungsmethoden. *Österr. Akad. Wiss. SB II*, 185:121–132, 1976.

[1998] P. Zinterhof. Zahlentheoretische Methoden in der numerischen Mathematik - Theorie und Anwendungen. *Schriftenr. Österr. Comput. Ges.*, 12:127–149, 1981.

[1999] P. Zinterhof. Gratis lattice points for multidimenional integration. *Computing*, 38:347–353, 1987.

[2000] P. Zinterhof and H. Stegbuchner. Trigonometrische Approximation mit Gleichverteilungsmethoden. *Stud. Sci. Math. Hung.*, 13:273–289, 1978.

List of Symbols

$\mathbf{Z}$	integers, 1
$\mathbf{N} = \mathbf{Z}^+$	positive integers (natural numbers), 107
$\mathbf{N}_0 = \mathbf{Z}_0^+$	non-negative integers, 99
$\mathbf{Q}$	rational numbers, 48
$\mathbf{R}$	real numbers, 1
$\mathbf{R}_+$	positive real numbers, 209
$\mathbf{C}$	complex numbers, 14
k	dimension, 1
$\mathbf{R}^k$	k-dimensional real space, 1
$\mathbf{R}^k/\mathbf{Z}^k$	k-dimensional torus, 1
$\mathbf{T}^k$	k-dimensional torus, 1
$\mathbf{S}_k$	k-dimensional unit sphere, 231
$\emptyset$	empty set, 2
I, J	intervals, 1
$I = [0,1]$	(closed) unit interval, 181
$I^k = [0,1]^k$	k-dimensional unit cube, 269
$\mathbf{U} = [0,1)$	(half open) unit interval, 1
$\mathbf{U}^k = [0,1)^k$	k-dimensional unit cube, 1
χ_M	characteristic function of the set M, 1
$\|M\|$	cardinality of the set M, 88
$A(I, N, \mathbf{x}_n)$	number of points $\mathbf{x}_1, \ldots, \mathbf{x}_N$ in I, 1
λ_k	k-dimensional Lebesgue measure, 1
λ_∞	infinite dimensional Lebesgue measure, 266
μ_∞	infinite dimensional product measure, 324
σ_k	surface measure, 231
σ_k^*	normalized surface measure, 231
δ_x	DIRAC measure concentrated on x, 30
$\mathbf{P}$	probability, 96
$\mathbf{E}$	expected value, 197
$x, y, z, \ldots$	real numbers, 1
$\mathbf{x}, \mathbf{y}, \mathbf{z}, \ldots$	k-dimensional real vectors, 1
$\underline{\alpha}, \underline{\beta}, \ldots$	k-dimentional real vectors, 78

$\mathbf{0} = (0, 0, \dots, 0)$	zero vector, 5
$\mathbf{1} = (1, 1, \dots, 1)$	1-vector, 391
$\delta_q(\underline{\alpha})$	approximation norm of $\underline{\alpha}$, 67
$\mathbf{x} \cdot \mathbf{y}$	scalar product, 15
$\lvert\mathbf{x}\rvert$	absolute value, Euclidean norm, 1
$\lVert\mathbf{x}\rVert_p$	L^p-norm, 5
$\lVert\mathbf{x}\rVert_\infty$	supremum norm, 11
$[\mathbf{x}]$	(vector of) integral part(s), 1
$\{\mathbf{x}\}$	(vector of) fractional part(s), 1
$\lVert x\rVert$	$\min\{\lvert x-n\rvert : n \in \mathbf{Z}\}$, 65
$\mathbf{h} = (h_1, \dots, h_k)$	k-dimensional integer point, 15
$r(\mathbf{h})$	product norm of $\mathbf{h}$, 15
N	number of points, 1
$D_N(\mathbf{x}_n)$	discrepancy, 4
$D_N^*(\mathbf{x}_n)$	star-discrepancy, 5
$D_N^{(p)}(\mathbf{x}_n)$	L^p-discrepancy, 5
$J_N(\mathbf{x}_n)$	isotropic discrepancy, 9
$V(f)$	variation of f, 10
$d_N^{(\delta)}(\mathbf{x}_n)$	dispersion with respect to δ, 11
$d_N(\mathbf{x}_n)$	usual dispersion (d_N^∞), 11
$F_N(\mathbf{x}_n)$	diaphony, 24
L^p	L^p space, 5
$\omega_f^{(\delta)}(c)$	modulus of continuity of f, 12
$e(x)$	$e^{2\pi i x}$, 15
supp	support, 307
$H(\mathbf{u})$	support function of a convex body, 220
$R_{\mathbf{r}}(\mathbf{x})$	RADEMACHER function, 45
$\hat{f}(\mathbf{t})$	Fourier transform, 17
$\hat{\mu}(t)$	Fourier transform of the measure μ, 30
$f * g$	convolution of functions, 15
$f * (d\mu)$	convolution with measure μ, 30
$\mathbf{E}_\alpha^k(C)$	specific class of periodic functions, 25
$SO(k)$	group of rotations in $\mathbf{R}^k$, 212
diam	diameter, 73
∂A	boundary of A, 204
det	determinant, 226
tr	trace, 234
$\mathbf{F}_b$	finite field of order b, 384
$\mathbf{Z}_b$	integers modulo b, 426
$R[x]$	ring of polynomials in x over R, 346
gcd	greatest common divisor, 67
O, o	LANDAU notation, 4
$\gg, \ll$	VINOGRADOV notation, 26

Subject Index

Druck: STRAUSS OFFSETDRUCK, MÖRLENBACH
Verarbeitung: GANSERT, WEINHEIM/SULZBACH